59th International Meat Industry Conference (MEATCON 2017)

IOP Conference Series: Earth and Environmental Science
Volume 85

Zlatibor, Serbia
1 - 4 October 2017

Editor:

Sheryl Avery

ISBN: 978-1-5108-4891-7
ISSN: 1755-1307

Printed from e-media with permission by:

Curran Associates, Inc.
57 Morehouse Lane
Red Hook, NY 12571

Printed by Curran Associates, Inc. (2017)

For permission requests, please contact the Institute of Physics
at the address below.

Institute of Physics
Dirac House, Temple Back
Bristol BS1 6BE UK

Phone: 44 1 17 929 7481
Fax: 44 1 17 920 0979

techtracking@iop.org

Additional copies of this publication are available from:

Curran Associates, Inc.
57 Morehouse Lane
Red Hook, NY 12571 USA
Phone: 845-758-0400
Fax: 845-758-2633
Email: curran@proceedings.com
Web: www.proceedings.com

Table of contents

Volume 85

59th International Meat Industry Conference MEATCON2017
1–4 October 2017, Zlatibor, Serbia

Accepted papers received: 14 September 2017
Published online: 26 September 2017

Preface

Papers

IOP Conf. Series: Earth and Environmental Science **85** (2017) 011001 doi:10.1088/1755-1315/85/1/011001

Preface

It is my great pleasure and honour to welcome you to the 59th International Meat Industry Conference (MEATCON2017) which will be held in Mt. Zlatibor, Serbia during October 1-4, 2017. MEATCON2017 is dedicated to issues related to food safety and quality, but also to food processing technology and food-related legislative.

The major goal and feature of the conference is to bring academic scientists, governmental decision makers and food business operators together to exchange and share their experiences, ideas and research results, and discuss the practical challenges encountered and the solutions adopted. Distinguished experts from EFSA, Directorate General for Health and Food Safety of the European Commission, TNO, NRL's and POPs Environmental Consulting and professors from all over Europe were invited to deliver plenary lectures regarding latest information in their respective expertise areas. It will be a golden opportunity for the students, researchers and food business operators to interact with the experts and specialists to get their advice or consultation on different food related topics.

These proceedings present a selection from papers submitted to the conference Programme Committee from universities and research institutes. All of the papers were subjected to peer-review by conference Programme Committee members and international reviewers. The papers selected depended on their quality and their relevancy to the conference. The volume tends to present to the readers the recent advances in all fields of Food Safety, Food Quality, Food Technology, Food related Legislative, Food & Environment. etc.

I would like to thank all the authors who have contributed to this volume and also to the Organizing Committee, reviewers, speakers, chairpersons, sponsors and all the conference participants for their support to MEATCON2017.

Dr Branko Velebit

Institute of Meat Hygiene and Technology, Serbia

September 12, 2017

English language editorial services provided by

Dr Sheryl Avery

Avery Buncic Scientific & English Editorial Services (ABSeeS)

Hamilton

New Zealand

e-mail: absees.editorial@gmail.com

web site: http://absees http://absees-editorial.my-free.website/contact-us

Peer review statement

All papers published in this volume of *IOP Conference Series: Earth and Environmental Science* have been peer reviewed through processes administered by the proceedings Editors. Reviews were conducted by expert referees to the professional and scientific standards expected of a proceedings journal published by IOP Publishing.

The importance of data collection for timely and accurate risk assessment

MB Gilsenan

European Food Safety Authority (EFSA), Via Carlo Magno 1A, Parma, Italy

E-mail: mary.gilsenan@efsa.europa.eu

Abstract. The European Food Safety Authority (EFSA) is responsible for food safety risk assessments at EU level. It provides independent scientific advice on risks associated with the food chain to support EU risk management decisions. Since its establishment, EFSA has amassed a wealth of data to underpin its risk assessments, such as food consumption data, monitoring data and experimental data. Increasing transparency of its risk assessments is a core objective of EFSA. EFSA aims to enhance the quality and transparency of its outputs by giving insofar as possible access to data and methods underpinning its scientific outputs. This paper provides an overview of the role of EFSA, its core data collections and their regulatory framework, as well as data quality and standardisation aspects. Finally, the paper elaborates on EFSA's 2020 strategy in relation to data, and describes EFSA scientific data warehouse and Knowledge Junction in this regard.

1. European Food Safety Authority mandate

The European Food Safety Authority (EFSA) is responsible for EU food safety risk assessments to protect consumers. It provides independent scientific advice on questions relating to food and feed safety, animal health and welfare, plant health, nutrition, and associated environmental issues and emerging risks. In addition, EFSA communicates on its scientific assessments to all interested stakeholders, including risk managers, scientists and the public. The authority was established in 2002 in the aftermath of the BSE (Bovine Spongiform Encephalopathy) crisis in Europe, and constitutes an integral part of the EU food safety system to protect public health.

EFSA's scientific advice informs EU risk managers (the European Commission, Parliament and EU Member States) who make regulatory decisions relating to food and feed safety (e.g. authorisation of a food ingredient or a new health claim, or the establishment or amendment of maximum legal limits for a food contaminant). The European Commission, Parliament and competent authorities in EU Member States are the only bodies entitled to ask scientific advice from EFSA. Once a request for scientific advice is accepted by EFSA, it becomes a mandate and is assigned to one or more Scientific Panels or EFSA's overarching scientific committee depending on the scientific area to which the question pertains. EFSA's scientific advice is provided in the form of scientific opinions or statements by one or more of its scientific panels or its scientific committee. These expert groups are composed of independent experts appointed through an open selection procedure. Experts mainly come from European universities, public research organisations, national authorities and food safety agencies. EFSA's panels and its scientific committee also carry out scientific work on their own initiative, so called self-tasking, to examine emerging issues and new hazards or to update risk assessment approaches for example. EFSA staff also produces scientific outputs such as peer reviews of the risk assessment of active substances in pesticides or dietary exposure to food-borne substances. Since its creation, EFSA's experts and staff have produced assessments for more than 4,400 substances in over 1,650 scientific opinions, statements and conclusions. These assessments are published in the open access EFSA Journal. Openness and transparency are core values of EFSA and are embedded in all activities undertaken by the authority. Adherence to these values helps to ensure a high level of trust in EFSA's work as well as accountability to society.

2. Food safety risk assessments

EFSA applies the overarching principles of the risk assessment paradigm in its food safety assessments: hazard identification and hazard characterisation (hazard assessment) determine safe levels of exposure of substances for human health, animal health or the environment. These safe levels are then combined with exposure estimates (exposure assessment) to determine the likelihood of risk associated with a given exposure (risk characterisation). Hazard assessments typically rely on toxicology data from animal studies. Dietary exposure assessments require data on the occurrence and levels of microbiological or chemical substances in food, as well as food consumption data.

EFSA's scientific committee develops cross-cutting guidance documents to be implemented in the routine work of panels to further improve openness, transparency and robustness in the risk assessment process and to achieve a more harmonised approach to risk assessments across panels and in the wider risk assessment community [1]. Recent examples include guidance on use of the benchmark dose approach, use of the weight of approach in scientific assessments, assessment of biological relevance and uncertainty analysis.

In 2015 EFSA published principles and processes for dealing with data and evidence in risk assessments [2] which describe the process for performing fit for purpose risk assessments. A key component is that data used as evidence in scientific assessments are both relevant and reliable.

3. Data collection regulatory framework

EFSA's founding regulation defines an overarching legal obligation on EFSA to collect, collate and summarise scientific and technical data to inform EU risk assessments and to work in close co-operation with all operators in the field of data collection to achieve this goal [3]. This framework is reinforced by sector specific EU legislation pertaining to chemical and biological hazards.

In the case of pesticide residues, Member States have a legal obligation to monitor pesticide residues in food commodities from national and EU co-ordinated sampling programmes and to submit monitoring results to EFSA and to the European Commission [4]. Some 18 million analytical records from European data providers are sent directly to EFSA annually.

In area of biological hazards, Member States have a legal obligation to monitor trends and sources of zoonoses, zoonotic agents and antimicrobial resistance and to transmit the results of monitoring programmes to the European Commission [5,6]. The European Commission entrusts the task of European data collation directly to EFSA.

To support EFSA's risk assessments of intentionally added food ingredients such as food additives and flavourings, Member States are required to maintain systems to monitor the consumption and use of these intentionally added ingredients using a risk-based approach and to report their findings with appropriate frequency to the European Commission and to EFSA [7,8].

In the domain of chemical contaminants, Commission Regulation (EC) No. 1881/2006 [9] complemented by several pieces of contaminant specific legislation requests Member States to monitor contaminant occurrence in food and feed and to report findings to the Commission or EFSA. The European Commission entrusts EFSA with the task of collecting all contaminant data from Member State data providers on the occurrence of contaminants in foodstuffs on an annual basis.

In the field of veterinary medicinal product residues, Member States must submit to the Commission, on an annual basis, national monitoring plans together with the results of monitoring for the previous year [10]. The European Commission subsequently entrusted the task of data collection (at sample level) to EFSA. EFSA is currently piloting a European data collection at sample based level.

For some of its risk assessments (e.g. in the regulated product area), EFSA relies on published information as well as on scientific studies sponsored and submitted by industry. Data requirements for regulated product areas are outlined in sector specific guidance documents issued by EFSA.

4. EFSA data collections

Since its foundation, EFSA and Member States have made significant progress in the area of data collection for risk assessment and monitoring. EFSA has become a central hub of European data on food consumption and monitoring data of food-borne hazards to support its scientific activities. The Evidence Management unit of EFSA is primarily responsible for collecting monitoring and food consumption data from EU Member States. Data are collected through various channels such as voluntary submissions from the food industry and academia in response to calls for data on EFSA's website, direct submission of data from competent authorities in Member States to EFSA, as well as provision of funding to Member States via grants / procurements for data generation and collection.
EFSA receives support for its data collection activities from EFSA's advisory forum, the EFSA focal point network and EFSA scientific data networks composed of respective representatives from Member States. EFSA also engages with industry and other stakeholders through its stakeholder forum to collect food additive and contaminant data for its dietary exposure assessments.

EFSA's Comprehensive European Food Consumption Database [11] contains individual dietary records from some 100,000 subjects (infants to elderly) from more than 510 million inhabitants in 23 Member States. This database is used for EFSA exposure assessments. One of the limitations of the database is that dietary surveys use different methodologies, e.g. 24h recall vs. 7 day food record. To address this, EFSA launched its EU Menu project which aims to provide more standardised and higher quality data on food consumption using FoodEx2 food classification and description [12] across the EU, following the EU Menu methodology [13]. EFSA provides guidance and financial support to participating countries to collect these food consumption data. To date, some 30 surveys covering different population groups (e.g. children, adults) are ongoing in the EU Menu project. EFSA aims to gradually update its comprehensive database with national surveys conducted according to the EU Menu project methodology.

EFSA's databases on contaminant and pesticide residue occurrence are populated on an annual basis with some one million and 18 million analytical records, respectively. In order to standardise data on contaminants and pesticide residues transmitted to EFSA, EFSA developed the Standard Sample Description (SSD) data model [14]. Many public and private laboratories in Europe are now sending (electronically) laboratory results to EFSA using the SSD format. The inclusion of the SSD specification in pesticides legislation [15] has facilitated the adoption of this specification by laboratories. Veterinary medicinal product residue data are also reported in SSD format to EFSA. Using EFSA's FoodEx2 food classification system, the monitoring results are combined with EFSA's EU wide food consumption databases to estimate consumer exposure.

EFSA collects annually data on zoonoses, food-borne outbreaks, animal populations as well as data on antimicrobial resistance [16], and uses these data to generate annual EU summary reports together with ECDC (the European Centre for Disease Prevention and Control). EFSA has established a joint molecular typing database with ECDC, to share common data in a joint repository so that microbiological data from humans can be linked to similar data from the food chain. The goal is to enable early detection and investigation of cross border food-borne outbreaks and contribute to source attribution studies to better understand the epidemiology of food-borne pathogens [17].

Regarding data for hazard assessments, EFSA developed an open source chemical hazards database called OpenFoodTox. This database is available as a simple viewer through Microstrategy or as a downloadable database from EFSA's Knowledge Junction. OpenFoodTox summarises the outcome of hazard characterisation for human health and, for some compounds, animal health and the environment (feed additives and pesticides). The data cover some 4400 substances including information on substance characterisation and toxicological data (reference points, reference values and genotoxicity) linked to EFSA outputs [18]. EFSA also published a database on botanical species reported to contain naturally occurring substances of possible concern for human health when present in food. The database supports the safety assessment (hazard identification) of botanicals and botanical preparations intended for use in food.

4.1. Data quality

Data quality is defined as the fitness for purpose of the data for a specific risk assessment question. Many aspects of a data collection can affect data quality, e.g. the representativeness of a sample, the temporal relevance of a sample or the sensitivity of an analytical method used to detect a hazard in a sample. Timeliness and completeness are important aspects of data quality relevant to EFSA's work, in particular given that many of its assessments pertaining to a specific time period are bound by legal deadlines. Good metadata allow end users to assess the fitness for purpose of a dataset for a particular assessment. EFSA's standard for monitoring data transmission, the SSD, was developed to standardise and improve the quality of incoming data. The SSD contains some 80 different fields with metadata describing an analytical sample (e.g. year of sampling, analytical method, limit of detection, sampling strategy, etc.).

Poor data management handling can also introduce errors into datasets that can affect data quality. EFSA continually improves its data management processes and tools to detect any anomalies in the data received that may impact data quality. Data are mainly submitted through EFSA's Data Collection Framework (DCF), a web interface to receive data from data providers. The DCF contains in-built business (validation) rules to perform automatic checks on incoming data. Automatic feedback is sent to data providers and files are rejected in case issues are identified. Additional quality checks are performed on the data following standard operating procedures to identify any anomalies in the data that may impact its quality.

Close cooperation with data providers is important to ensure data quality at source. EFSA regularly publishes guidelines for data providers on data reporting and invests in training on data reporting and transmission. Training on data collection and reporting is also extended to pre-accession countries in preparation for their future participation in the EU. In 2017, EFSA started to piloting a project with five Member States with a view to improving data quality at Member State level.

5. EFSA 2020 strategy: data

EFSA has accumulated a wealth of risk assessment data since its foundation. In parallel, the work of EFSA is increasingly subject to demands for more openness and transparency across its spectrum of stakeholders. EFSA's 2020 strategy includes a strategic objective to 'widen EFSA's evidence base and optimise access to its data'. Under this objective, EFSA aims to enhance the quality and transparency of its outputs by giving insofar as possible access to data and evidence underpinning its scientific outputs [19, 20]. Many of EFSA's data collections can now be accessed through its Knowledge Junction or its scientific data warehouse. EFSA's Knowledge Junction is a community on the EU-funded Zenodo research-sharing open data platform. Uploaded items have a unique digital object identifier to make them citable. The contents can then be cited and re-used by the scientific community and other interested stakeholders. EFSA's scientific data warehouse currently contains monitoring, food consumption and hazard data, and is publically accessible from EFSA's website. Access rules define different levels of data access for different stakeholders [21]. In 2017, EFSA became a partner of GODAN (Global Open Data for Agriculture and Nutrition) which advocates open data and open access policies by default while respecting legitimate concerns such as relating to privacy, security and commercial sensitivity. EFSA also plans to migrate insofar as possible from unstructured to structured scientific data to increase efficiency, innovation and enable data re-use. Within this context a transformational project is currently ongoing on electronic submission of industry dossier data.

Acknowledgements
All EFSA data providers as well as staff in EFSA's Evidence Management unit are gratefully acknowledged.

References

[1] European Food Safety Authority 2009 Guidance of the scientific committee on transparency in the scientific aspects of risk assessments carried out by EFSA Part 2: general principles *EFSA J.* **1051** 1
[2] European Food Safety Authority 2015. Scientific report on principles and process for dealing with data and evidence in scientific assessments *EFSA J.* **13** 4121
[3] Regulation (EC) No 178/2002 of the European Parliament and of the Council of 28 January 2002 laying down the general principles and requirements of food law, establishing the European Food Safety Authority and laying down procedures in matters of food safety *OJ* L **31**
[4] Regulation (EC) No 396/2005 of the European Parliament and of the Council of 23 February 2005 on maximum residue levels of pesticides in or on food and feed of plant and animal origin and amending Council Directive 91/414/EEC *OJ* L**70**
[5] Directive 2003/99/EC of the European Parliament and of the Council of 17 November 2003 on the monitoring of zoonoses and zoonotic agents, amending Council Decision 90/424/EEC and repealing Council Directive 92/117/EEC *OJ* L **325**
[6] Commission Implementing Decision 2013/652/EU of 12 November 2013 on the monitoring and reporting of antimicrobial resistance in zoonotic and commensal bacteria *OJ* L **303**
[7] Regulation (EC) No 1333/2008 of the European Parliament and of the Council of 16 December 2008 on food additives *OJ* **L** 354
[8] Regulation (EC) No 1334/2008 of the European Parliament and of the Council of 16 December 2008 on flavourings and certain food ingredients with flavouring properties for use in and on foods and amending Council Regulation (EEC) No 1601/91, Regulations (EC) No 2232/96 and (EC) No 110/2008 and Directive 2000/13/EC *OJ* L **354**
[9] Commission Regulation (EC) No 1881/2006 of 19 December 2006 setting maximum levels for certain contaminants in foodstuffs *OJ* L **364**
[10] Council Directive 96/23/EC on measures to monitor certain substances and residues thereof in live animals and animal products and repealing Directives 85/358/EEC and 86/469/EEC and Decisions 89/187/EEC and 91/664/EEC *OJ* L **125**
[11] Merten C, Ferrari P, Bakker M, Boss A, Hearty A, Leclercq C, Lindtner O, Tlustos C, Verger P, Volatier JL and Arcella D 2011 Methodological characteristics of the national dietary surveys carried out in the European Union as included in the European Food Safety Authority (EFSA) comprehensive European food consumption database *Food Addit Contam* **28** 975
[12] European Food Safety Authority 2015 The food classification and description system FoodEx2 (revision 2) EFSA supporting publications EN-804
[13] European Food Safety Authority 2014 Guidance on the EU menu methodology *EFSA J.* **12** 3944
[14] European Food Safety Authority 2013 Standard sample description version 2.0 *EFSA J.* **11** 3424
[15] Commission Implementing Regulation (EU) No 2016/662 of 1 April 2016 concerning a coordinated multiannual control programme of the Union for 2017, 2018 and 2019 to ensure compliance with maximum residue levels of pesticides and to assess the consumer exposure to pesticide residues in and on food of plant and animal origin *OJ* L **115**

[16] Boelaert F, Amore G, Van der Stede Y and Hugas M 2016 EU wide monitoring of biological hazards along the food chain: achievements, challenges and EFSA vision for the future *Curr Opin Food Sci* **12** 52

[17] Rizzi V, Da Silva Felicio T, Felix B, Gossner CM, Jacobs W, Johansson K, Kotila S' Michelon D, Monguidi M, Mooijman K, Morabito S, Pasinato L, Torgny Björkman J, Torpdahl M, Tozzoli R, Van Walle I 2017 The ECDC-EFSA molecular typing database for EU public health protection *Euroreference* **2** 4

[18] Dorne JL, Richardson J, Kass G, Georgiadis N, Monguidi M, Pasinato L, Cappe S, Verhagen H and Robinson T 2017 Editorial: Open FoodTox: EFSA'S open source toxicological database on chemical hazards in food and feed **15** e15011

[19] Gilsenan MB, Abbinante F, O'Dea E, Canals A and Tritscher A 2016 Open risk assessment: data. *EFSA J.* **14** s0509

[20] Verloo D, Meyvis T and Smith A 2016 Open risk assessment: methods and expertise *EFSA J.* **14** s0505

[21] European Food Safety Authority 2015 The EFSA data warehouse access rules. EFSA supporting publications EN-768

Learning from Dioxin & PCBs in meat – problems ahead?

R Weber
POPs Environmental Consulting, Lindenfirststr. 23, Schwäbisch Gmünd, Germany

E-mail: roland.weber10@gmail.com

Abstract. Persistent organic pollutants (POPs) including polychlorinated dibenzo-p-dioxins and polychlorinated dibenzofurans (PCDD/Fs; "Dioxins"), or polychlorinated biphenyls (PCBs) are widely recognized environmental and food contaminants. More than 90% of PCDD/Fs and PCB exposure of the average population stem from animal based food including meat. While average PCDD/F and PCB levels have decreased compared to levels 1980s, still contamination above regulatory limits are observed and a share of the population is above the tolerable daily intake recommended by the WHO. For PCBs the contamination of feed and food along the life cycle from production, use, recycling, end of life and related contaminated sites has been documented and can be seen as a model. Furthermore, it has been recently discovered that levels of PCBs in feed and soil below regulatory limits can result in meat contamination above EU regulatory limits. In particular, beef meat and chicken meat/eggs have been found very sensitive towards PCB contamination in the environment (soil and feed) but also in stables (paints and sealants). For PCDD/Fs, the major exposure pathways are feed, feed additives and contaminated sites.

Chlorinated paraffins have substituted PCBs the last 40 years in open application and short chain chlorinated paraffins (SCCPs) were recently (05/2017) listed in the Stockholm Convention. Furthermore, brominated and fluorinated POPs have been listed in the Convention. All these POPs groups can accumulate in meat animals. For these new listed POPs no regulatory limits in food including meat has been established yet. Initial information on presence and risk of new listed POPs to food animals is compiled. A more systematic assessment of exposure and risks of POPs to food animals/meat is needed.

1. Introduction

Polychlorinated dibenzo-p-dioxins, polychlorinated dibenzofurans (PCDD/Fs), and polychlorinated biphenyls (PCBs) are widely recognized environmental and food contaminants. In the EU, the Commission Regulation (EC) No 1881/2006 of 19 December 2006 sets maximum levels for PCDD/Fs and the sum of PCDD/Fs and dioxin-like PCBs (dl-PCBs) in certain foodstuffs (European Commission 2006). The regulation was amended by Commission Regulation (EU) No 1259/2011 of 2 December 2011 introducing new EU maximum levels for PCDD/Fs, for the sum of PCDD/Fs and dl-PCBs (based on WHO toxicity equivalency factors established in 2005, TEF_{2005}) and for non-dioxin-like PCBs (ndl-PCBs) (European Commission 2011).

Most of the meat and milk samples on the European market meet the regulatory limits (EFSA 2012). Average PCDD/F and PCB levels have decreased compared to levels 1990s which is also reflected in decreasing PCB and PCDD/F levels in human milk (UNEP 2013). However, in recent 10 years more frequent contamination of meat and eggs with PCBs were detected (Weber et al. 2015). This was trigged by the inclusion of PCBs in the regulation since 2006.

In the past often feed incidents were responsible for exceeding maximum levels of PCDD/Fs and PCBs in food of animal origin (Tlustos et al. 2013; Malisch & Kotz 2014). In recent years also sheep (in particular liver) (EFSA 2011) and beef (Kamphues et al. 2011; Weber et al. 2014) from free-range production exceeded the existing maximum limits. Depending on the source, PCDD/Fs or dioxin-like PCBs can contribute in various ratios to TEQ with dl-PCB often as main contributor.

In addition to these chlorinated persistent organic pollutants in the last 8 years the first brominated and fluorinated POPs have been listed in the Stockholm Convention.

In the current paper experiences with PCBs, PCDD/Fs and selected new Stockholm Convention listed POPs and contamination of food animals/meat is compiled.

2. Samples of food contamination resulting from POPs

2.1. Experiences and lessons learnt from PCBs

PCB is the best assessed industrial POP in food and feed due to regulatory requirements and related monitoring and control. Food animals and related meat and other food products can become PCB contaminated along the life cycle of PCBs (Weber et al. 2015):

Exposure from PCB production sites via food animals

Animals and humans were exposed to PCBs around production sites. For the PCB production sites the contamination and exposure via animals/cattle in Anniston (ATSDR 2015) and Brescia/Italy have been documented (Turrio-Baldassarri et al. 2009). For the former PCB production site in Slovakia elevated human exposure was found up to 70 km from the plant with food animals as likely major human exposure pathway (Wimmerová et al. 2015).

Exposure of food animals from PCB use

PCBs have been used in "closed applications" (e.g. transformers, capacitors) and "open applications" (e.g. sealants and paints). Already in the 1970s/80s it had been discovered that PCB-paints used in fodder silos had contaminated cattle (Willett & Hass 1975; Deutscher Bundestag 1989). The recent cases of meat or egg[1] contamination with PCB-paints on farms in Germany (Weber et al. 2015; Winkler 2016), France (Marchant 2017), Netherlands (Hoogenboom et al. 2014) and Switzerland (Zennegg et al. 2014) demonstrate that still PCB in open applications in particular paints exist and need further assessment and management.

PCB exposure of food animals from recycling of PCB containing oils and wastes

Although the recycling of PCB containing oils is not allowed, it can enters recycling cycles and has in some cases contaminated feed and related meat and other animal products. The largest and most costly (approx. 1 billion US$) PCB/Dioxin food scandal occurred in Belgium in 1999 where PCB oil were mixed with food fat impacting meat and eggs from more than 1500 farms (Fiedler et al. 2000; Larebeke et al. 2001). Another large meat contamination case with associated 100 million $ damage cost resulted from use/recycling of PCB-contaminating oil for drying of animal feed contaminating the Irish pork/meat with related recalls (Heres et al. 2010; Marnane 2012). Today in particular in developing countries still 14 million tonnes of PCB contaminated oil and equipment exist which is partly recycled in an uncontrolled manner. However, no monitoring of PCBs in food is established in developing countries and meat or other food contamination is normally not discovered.

PCB exposure of food animals from waste treatment and from contaminated soils

Free-range cattle are ingesting a considerable amount of soil when grazing. In recent survey in Germany contamination of beef meat from a range of sources mainly related to soil contamination were discovered:

- Flood plains of industrially impacted rivers
- Application of sewage sludge from 1960s and 1970s
- Sediment deposits on agricultural land
- Construction debris scattered and incorporated into soil of a pasture area
- Long term emission/deposition from industrial facilities (metal industry; incinerators)
- Areas where (agricultural) machinery was parked (dripping machinery/hydraulic oil)
- Use of former PCB-contaminated scrap yard as storage area for dung

[1] If an egg is above the EU regulatory limit, also the chicken meat is above the regulatory limit.

- Former use as military area
- Area close to street with PCB road marking

The cases showed that in particular for suckling herds the dl-PCB levels of soils were mostly below 5 ng WHO-PCB-TEQ/kg dm and the feed in average below 0.2 ng WHO-PCB-TEQ/kg dm. As conclusion, meat of free range cattle in particular when calves are fed by milk of grazing cows for a longer period can exceed the EU-regulatory limits at relatively low soil levels (below 5 ng WHO-PCB-TEQ/kg dm) in combination with grass/feed levels around 0.15 ng TEQ/kg dm considerably below the EU-regulatory limits for feed (Weber et al. 2015). When calculating the total intake of the cows (consumption of 10 kg dm of grass/hay containing approx. 3 wt-% soil), a total intake of approx. 2 ng WHO-PCB-TEQ/day from soil and feed might be critical with regard to possible exceedance of the maximum limits for meat from beef in these cases. A systematic screening of PCB contaminated sites is needed for promoting safe feed and food (including meat) production.

2.2. Experience with polychlorinated dibenzo-p-dioxins and dibenzofurans (PCDD/Fs)

PCDD/Fs are not produced intentionally but are emitted from a wide range of sources. The exposure of cattle and chicken and related meat can result from (Fiedler et al. 2000; Huwe et al. 2004; Malisch & Kotz 2014; Tlustos et al. 2013; Weber et al. 2015):

- PCDD/F contamination of feed
- PCDD/F contamination of feed additives (e.g. ZnO or $CuSO_4$)
- PCDD/F contaminated in stables and in animal bedding (e.g. PCP treated wood)
- PCDD/F contamination of soils and exposure from grazing (+ feed contamination)

All four exposure routes have resulted in meat contamination above regulatory limits. Since the largest amount of PCDD/Fs are from historic releases, the contamination of soils and related exposure of feed and cattle is likely the most relevant PCDD/F exposure route (Weber et al 2008, 2015).

2.3. New listed POPs of emerging concern

Perfluorooctanesulfonic acid (PFOS) and other per-/polyfluorinated alkylated substances (PFAS)

Since 2009 the first fluorinated POPs (PFOS and related precursor chemicals) have been listed in the Stockholm Convention and wide environmental pollution is discovered. Also these POPs can bio-accumulate in meat (Lupton et al. 2014). For PFOS transfer factors from environmental contamination to food animals have been established (Brambilla et al. 2014). Meat and other animal food can become major human exposure pathways for PFOS (Brambilla et al. 2014). Assessments of a PFOS/PFAS contaminated site impacting the groundwater of a farm has shown highest PFOS/PFAS levels in cattle of the farms up to ppm level in blood (Bräuning et al. 2017). Currently approx. 3000 PFAS are in use which are highly persistent or have highly persistent degradation products. There are currently no limit values in food including meat but only limit values for PFOS/selected PFAS in drinking water.

Polybrominated flame retardants

Since 2009 also some brominated flame retardants have been listed as POPs in the Stockholm Convention including polybrominated biphenyl ethers (PBDEs), polybrominated biphenyls (PBBs) and hexabromocyclododecane (HBCD). The most known cattle/meat incident with BFRs occurred 1973 when the Michigan Chemical Company producing PBB and magnesium oxide (a cattle feed supplement) by mistake sent 230 to 500 kg of PBB instead of MgO accidentally to Michigan Farm Bureau Services (Michigan Department of Community Health (2011). As a result of this incident, over 500 contaminated Michigan farms were quarantined, and approximately 30,000 cattle, 4,500 swine, 1,500 sheep, and 1.5 million chickens were destroyed, along with over 800 tons of animal feed, 8165 kg of cheese, 1134 kg butter, 5 million eggs, and 15400 kg of dried milk products (Michigan Department of Community Health 2011). 60% of the Michigan population still have high PBB levels 40 years after the incident (Department of Epidemiology - The Michigan PBB Registry 2017).

For PBDEs it has been shown that poultry and red meat consumption is a major contributor for PBDEs burden of the population in the US. Blood levels of five major PBDE congeners were associated with consumption of poultry fat and red meat fat (Fraser et al. 2009). ΣPBDE serum concentrations among vegetarians were approx. 25% lower than among omnivores (Frazer et al. 2009).
In North America where the largest amount of PBDEs have been used, PBDE exposure to cattle/meat resulted from application of sewage sludge/biosolids (Rawn et al. 2017).

Chlorinated paraffins

Chlorinated paraffins are the chlorinated semivolatile organic compounds with highest production volume (>1 million tonne/year) having substituted PCBs in most open applications (Glüge et al. 2016). Most recently short chain chlorinated paraffins (SCCP) have been listed as POPs in the Stockholm Convention. Chlorinated paraffins can bioaccumulate in meat/fat (Ueberschär & Matthes 2004) and might pose a risk for future safe meat production considering the large use volumes in open applications.

2.4 Problematic chemicals in meat production with lower persistence not considered here

There are a wide range of problematic chemicals which are a threat for meat production including pesticides/biocides (e.g. recently fipronil, glyphosate), antibiotics and other veterinary drugs, PAHs or heavy metals. These chemicals are not addressed in this abstract with the focus on POPs but need to be controlled and reduced for healthy and environmental friendly meat production.

3. Governmental guidance documents and support

To support safe food production and to avoid dioxin and PCB contamination of food of animal origin (meat, milk and eggs) different governmental institutions have published guidance documents.

On international level the "Code of Practice for the Prevention and Reduction of Dioxin and Dioxin-like PCB Contamination in Food and Feeds" have been published 2006 (FAO 2006).

Several documents were developed on EU E.g. "Evaluation of the Occurrence of PCDD/PCDF and POPs in Wastes and Their Potential to Enter the Food Chain" (Fiedler et al. 2000) or "Guidelines for the enforcement of provisions on dioxins in the event non-compliance with the maximum levels for dioxins in food (DG Sanco 2004).

Also guidance documents have been developed on national level. E.g. the German Environmental Ministry has published a guidance on "Environmental protection – pillar for food safety to avoiding dioxin and PCB entry" (BMU 2013). The Chamber of Agriculture of the federal state of Lower Saxony developed leaflet for cattle breeder to control Dioxin and PCBs input or on cultivation on contaminated land (Landwirtschaftskammer Niedersachsen 2011; 2014).

In the frame of the Stockholm Convention inventory guidance documents for individual POPs have been developed include a chapter for contaminated sites for PFOS, PBDEs, HBCD, PCNs, PCP and HCBD (Secretariat of the Stockholm Convention 2017). The development of inventories of POPs contaminated sites and the securing and remediation of these sites can improve feed and food/meat safety contributing to several Sustainable Development Goals (SDGs) (Bell et al. 2016).

References

[1] ATSDR (Agency for Toxic Substances & Disease Registry) 2015 Anniston Community Health Survey. https://www.atsdr.cdc.gov/sites/anniston_community_health_survey/overview.html

[2] Bell L, Weber R, De Borst B, Paun MC, Holoubek I, Watson A, Vijgen J (2016) Assessment of POPs contaminated sites and the need for stringent soil standards for food and feed safety. Working document for UNEP Dioxin Toolkit and BAT/BEP group. October 2016.

[3] Brambilla G, D'Hollander W, Oliaei F, Stahl T, Weber R 2015 *Chemosphere* **129**, 192-202

[4] Bräunig J, Baduel C, Heffernan A, Rotander A, Donaldson E, Mueller JF 2017 Fate and redistribution of perfluoroalkyl acids through AFFF-impacted groundwater. *Sci Total Environ.* **596-597**, 360-368

[5] BMU (Bundesministerium für Umwelt, Naturschutz, und Reaktorsicherheit) 2013 Umweltschutz - Standbein der Lebensmittelsicherheit – Dioxin- und PCB-Einträge vermeiden. 5. aktualisierte Auflage, Januar 2013

[6] Commission Regulation (EC) No 1881/2006 of 19 December 2006 setting maximum levels for certain contaminants in foodstuffs, Official Journal of the European Union L 364/5

[7] Commission Regulation (EU) No 1259/2011 of 2 December 2011 amending Regulation (EC) No 1881/2006 as regards maximum levels for dioxins, dioxin-like PCBs and non dioxin-like PCBs in foodstuffs, *Official Journal of the European Union*, **L 320/18**

[8] DG Sanco 2004 Guidelines for the enforcement of provisions on dioxins in the event non-compliance with the maximum levels for dioxins in food. https://www.fsai.ie/uploadedFiles/ECguidelines-july_2004.pdf

[9] Deutscher Bundestag 1989 Drucksache 1114276 —Drucksache 11/4352; 14.04.1989

[10] Department of Epidemiology - The Michigan PBB Registry 2017 http://pbbregistry.emory.edu/

[11] EFSA (European Food Safety Authority) 2011 *EFSA Journal* **9**: 2297, 1-71

[12] EFSA (European Food Safety Authority) 2012 *EFSA Journal* **10**: 2832

[13] FAO 2006 Code of Practice for the Prevention and Reduction of Dioxin and Dioxin-like PCB Contamination in Food and Feeds. 1st edition http://www.fao.org/ag/againfo/home/en/news_archive/2009_IN_dioxin.html

[14] Fiedler H, Hutzinger O, Welsch-Pausch K, Schmiedinger A 2000 Evaluation of the Occurrence of PCDD/F and POPs in Wastes and Their Potential to Enter the Foodchain. Study on behalf of the European Commission

[15] Fraser AJ, Webster TF, McClean MD 2009 *Environ Health Perspect.* **117**, 1520-1525

[16] Glüge J, Wang Z, Bogdal C, Scheringer M, Hungerbühler K 2016 *Sci Total Environ.* **573,** 1132-1146

[17] Heres L, Hoogenboom R, Herbes R, Traag W, Urlings B. 2010 *Food Addit Contam Part A Chem Anal Control Expo Risk Assess.* **27**, 1733-1744

[18] Huwe JK, Davison K, Feil VJ, Larsen G, Lorentzsen M, et al 2004 *Food Additives and Contaminants* **21**, 182-194

[19] Hoogenboom R, ten Dam G, Immerzeel J and Traag W 2014 *Organohalogen Compounds* **76**, 1700-1703

[20] Kamphues J, Schulz AJ, Gude K, Bruns-Weller E, et al. 2011 *Organohalogen Compounds* **73**, 2040-2041

[21] Van Larebeke N, Hens L, Schepens P, Covaci A, Baeyens J, Everaert K, Bernheim J L, Vlietinck R, De Poorter G 2001 *Environ Health Perspect.* **109**, 265–273

[22] Landwirtschaftskammer Niedersachsen 2011 Bewirtschaftung von schadstoffbelasteten Standorten in Niedersachsen. Oldenburg, Stand: August 2011

[23] Landwirtschaftskammer Niedersachsen 2014 Merkblatt PCB und Dioxine in Rindfleisch - Haltungsempfehlungen für Rinderhalter. Stand 22.01.2014

[24] Lupton SJ, Huwe JK, Smith DJ, Dearfield KL, Johnston JJ 2014 *J Agric Food Chem.* **62**, 1167-1173

[25] Malisch R, Kotz A 2014 *Sci Total Environ* **491–492**, 2-10

[26] Marchant P 2017 Decontamination of pigs exposed to an environmental source of PCB. 37th International Symposium on Halogenated Persistent Organic Pollutants (POPs). August 20-25 Vancouver Canada

[27] Marnane I 2012 *J. Environ. Monit.* **14**, 2551-2556

[28] Michigan Department of Community Health 2011 PBBs (Polybrominated Biphenyls) in Michigan Frequently Asked Questions – 2011 update https://www.michigan.gov/documents/mdch_PBB_FAQ_92051_7.pdf

[29] Rawn DFK 2017 PBDE concentrations in cattle fat and meat following exposure to biosolid amended pasture lands. 37th Intern. Symposium on Halogenated Persistent Organic Pollutants. August 20-25 Vancouver Canada.

[30] Secretariat of the Stockholm Convention 2017 Guidance for National Implementation Plans (NIPs). http://www.pops.int/Implementation/NationalImplementationPlans/Guidance/tabid/2882/Default.aspx

[31] Tlustos C, Anderson W and Evans R "Responding to food contamination incidents: principles and examples from cases involving dioxins", chapter 5 in "Persistent organic pollutants and toxic metals in foods", Rose, M., and Fernandes, A. (editors) (2013), Woodhead Publishing Ltd.

[32] Turrio-Baldassarri L, Alivernini S, Carasi S, Casella M, Fuselli S, Iacovella N, Iamiceli AL, La Rocca C, Scarcella C, Battistelli CL. 2009 *Chemosphere* **76**, 278-285

[33] Ueberschär KH, Matthes S 2004 *Food Addit Contam.* **21**, 943-948

[34] Weber R, Gaus C, Tysklind M, et al. 2008 *Env Sci Pollut Res* **15**, 363-393

[35] Weber R, Albrecht M, Ballschmiter K, Berger J et al. 2014 *Organohalogen Compounds* **76**, 815-818

[36] Weber R, Hollert H, Kamphues J, Ballschmiter K, Blepp M, Herold C 2015 Analyse und Trendabschätzung der Belastung der Umwelt und von Lebensmitteln mit ausgewählten POPs und Erweiterung des Datenbestandes der POP-Dioxin-Datenbank des Bundes und der Länder mit dem Ziel pfadbezogener Ursachenaufklärung. R&D Report for German EPA (UBA), FKZ 371265407/01

[37] Willett LB, Hess JF 1975 *Residue Reviews* **55**, 135–147

[38] Wimmerová S, Watson A, Drobná B, Šovčíková E, Weber R, et al. 2015 *Environ Sci Pollut Res Int.* **22**, 14405-15

[39] Winkler J 2015 *Environ Int.* **80**, 72-78

[40] Zennegg M, Schmid P, Kuchen A, Beer M, Tamborini L, Beckmann M, Arpagaus S, Caduff A, Lanfranchi M 2014 *Organohalogen Compounds* **76**, 118-121

Food allergy and risk assessment: Current status and future directions

Benjamin C. Remington

TNO, Zeist, The Netherlands

Abstract. Risk analysis is a three part, interactive process that consists of a scientific risk assessment, a risk management strategy and an exchange of information through risk communication. Quantitative risk assessment methodologies are now available and widely used for assessing risks regarding the unintentional consumption of major, regulated allergens but new or modified proteins can also pose a risk of *de-novo* sensitization. The risks due to *de-novo* sensitization to new food allergies are harder to quantify. There is a need for a systematic, comprehensive battery of tests and assessment strategy to identify and characterise *de-novo* sensitization to new proteins and the risks associated with them. A risk assessment must be attuned to answer the risk management questions and needs. Consequently, the hazard and risk assessment methods applied and the desired information are determined by the requested outcome for risk management purposes and decisions to be made. The COST Action network (ImpARAS, www.imparas.eu) has recently started to discuss these risk management criteria from first principles and will continue with the broader subject of improving strategies for allergen risk assessment throughout 2016-2018/9.

Risk analysis is a three part, interactive process that consists of a scientific risk assessment, a risk management strategy and an exchange of information through risk communication [1]. In its purest sense, risk assessment is the scientific evaluation of known or potential adverse health effects resulting from human exposure to foodborne hazards. Risk assessment consists of four steps: hazard identification, hazard characterization, exposure assessment, and risk characterization [1]. Risk management is the process of weighing policy alternatives to accept, minimize, or reduce assessed risks and to select and implement appropriate options. In risk management, one accounts for the risk assessment results and other legitimate factors (political, technical, economical, societal, etc.) may be considered.

Risk communication is the interactive process exchanging information and opinion on risk among risk assessors, risk managers and other interested parties. As all manner of risks are evaluated with the same process, risk analysis for food allergens does not differ in concept from other risks associated with foods. However, while methodologies and principles of risk assessment in food safety have developed and become harmonized to a large extent worldwide over the past half century, the risks addressed are mainly those posed by chemical, microbiological and physical hazards.

Within risk assessment, hazard identification is the recognition of a particular component in foods associated with potential or known health effects [1]. The hazard in food allergy is a specific protein (or perhaps carbohydrate) that can cause sensitization and allergic reactions on subsequent exposures. Multiple proteins within a single food can be sensitizing agents and any of them can be the cause of an allergic reaction [2]. Additionally, proteins from one allergenic food can cross react with proteins in foods from a number of categories including fresh fruits, vegetables and legumes, as is known with the major allergen in birch pollen, Bet v 1 (3). Sensitivity to a cross-reactive carbohydrate determinant (CCD) can have broad implications and lead to reactions after consumption of multiple foods previously allowed in the diet (i.e. alpha-gal in beef, pork or lamb) [4]. Food allergens and exposure to them are not a risk to the majority of the population; however, individuals within the food-allergic population risk potentially life-threatening reactions upon consumption and must take their avoidance diets seriously.

Hazard characterization is the qualitative and/or quantitative evaluation of the nature of the adverse effects. If data are obtainable, a dose-response assessment should be performed [1]. As commonly known and reported by others, a wide range of symptoms can be experienced by food-allergic individuals upon exposure to the offending food. It is important to note that not all allergic reactions are life-threatening. Some food-allergic individuals will never experience a severe reaction and their symptoms range from very mild, such as itching and flush, while others can experience a severe drop in blood pressure and bronchospasm. The reasons for the differences in reaction severity are not fully understood, but many factors are expected to play a role [5]. The minimal eliciting dose (MED), or threshold, also varies widely across foods and across the entire population of individuals allergic to any specific food [6–9]. Double-blind, placebo-controlled food challenges (DBPCFC) can be used to quantify an individual food allergen thresholds in a clinical setting, preferably with objective symptoms as the endpoint. Dose-response curve assessments can then be conducted by risk assessors using the DBPCFC results to determine the population threshold for a particular allergen. Data now exist to conduct quantitative, dose-response based risk assessments for a number of food allergens.

Exposure assessment is the qualitative and/or quantitative evaluation of the likely intake via food and other sources if relevant [1]. For a food allergen risk assessment, two main variables shape the exposure patterns: likelihood of consumption and the amount eaten. First, exposure will only be relevant if the allergic individual at risk consumes the particular allergen-containing product of interest. Second, the outcome of the risk assessment will be influenced by the amount of food eaten by the individual and thus the amount of allergen. Population-based dietary intake surveys exist for a number of countries and are regularly used by risk assessors in all fields but there is no consumption database available solely for food-allergic consumers. Until food allergy specific surveys are completed, with proper controls, risk assessors must assume that if an allergic individual chooses to consume a product, they consume it in the same amount as non-allergic individuals. Assumptions regarding frequency of consumption are more difficult as it is well known that allergic consumers are brand loyal, share experiences and will avoid perceived "risky" products and product categories. Additionally, some allergic consumers will ignore warning labels and purchase products that have allergen advisory statements [10]. While uncertainty exists regarding the consumption patterns of allergic consumers, use of the overall population consumption patterns is considered a suitable surrogate until dietary surveys are designed specifically for the allergic individual. However, as with all parts of a risk assessment, the assumptions involved during the exposure assessment must be stated and understood.

Risk characterization is the integration of hazard identification, hazard characterization and exposure assessment into a qualitative and/or quantitative estimation of the adverse effects likely to occur in a given population, with the attendant uncertainties [1]. Food allergy was a relatively late arrival to the field of food safety hazards and real progress in method development to ensure consumer protection is fairly recent in comparison to other fields (chemical, microbiological and physical hazards). In the past 30 years, true progress has been made regarding analytical methods for detecting the presence of allergens in foods and a deeper understanding of food allergy has been obtained through clinical research. In the past 15 years, knowledge has begun to accumulate regarding the sensitivity of food-allergic individuals from observed thresholds during large scale, structured clinical food challenges (DBPCFC) [6–9]. This growth of knowledge has enabled the development of methods for assessing the risk to food-allergic consumers from oral exposure to known allergenic food proteins already present in the diet [11–14]. Quantitative risk assessment methodologies are now available and widely used for assessing risks regarding the unintentional consumption of major, regulated (EC 1169/2011) food allergens [15,16]. Similar combinations of DBPCFCs and quantitative risk assessment methods may also be applied to assessing potential allergenic risks due to cross-reactivity between new or modified proteins when the new protein will likely to be a risk for persons with food allergies to similar known allergens, e.g. the novel food mealworm cross-reacting with shrimp [17,18]. However, new or modified proteins can also pose a risk of *de-novo* sensitization, leading to the development of new food allergies. The risks due to these possible new food allergies are harder to quantify.

Adaptive immune responses, such as allergy, consist of two phases: sensitization and elicitation, which must be analysed separately. An expert panel of ILSI Europe recently published a framework allowing for the categorisation and prioritisation of allergenic foods in accordance with their importance to public health. Within that framework, the expert panel also proposed a scheme to also systematically apply the terms and principles to the interconnected risk analysis processes for the distinctly separate sensitization and the elicitation phases of IgE-mediated food allergy [19]. The first phase to be analysed in the risk analysis process is that of sensitisation to the allergenic food. The prevalence of allergy and the sensitivity of the allergic population (i.e. individual thresholds during DBPCFC) are the outcomes from the risk analysis for the sensitisation phase and are also the hazard input for the elicitation phase risk analysis (figure 1). During the sensitization phase, an allergenic food's sensitising potency, in combination with the pattern of exposure, determine the prevalence of sensitisation, subsequent prevalence of allergy and the sensitivity of the allergic individuals. During the elicitation phase, the prevalence and sensitivity of the allergic individuals, in combination with exposure, determine the frequency and severity of allergic reactions. During the elicitation phase (after the risk has been identified), exposure is not an independent determinant, as risk management measures (labelling, education) could influence the level of exposure in the allergic subpopulation. Thus, the frequency and severity of allergic reactions may be a reflection of risk management techniques and not only the potency of the allergen. Quantitative approaches to assess the risk posed by substances in the eliciting phase (reactions in already sensitized individuals) are proving very successful [15,20,21]. However, dose-response relationships in the sensitization phase are harder to study. Sensitization seems to be a non-linear response, as similar levels of exposure can lead to either tolerance or sensitization. To complicate matters further, the relationship between sensitization and elicitation is complex in allergic individuals. Additionally, the form and route of exposure to new proteins also impacts sensitization and allergenicity risks.

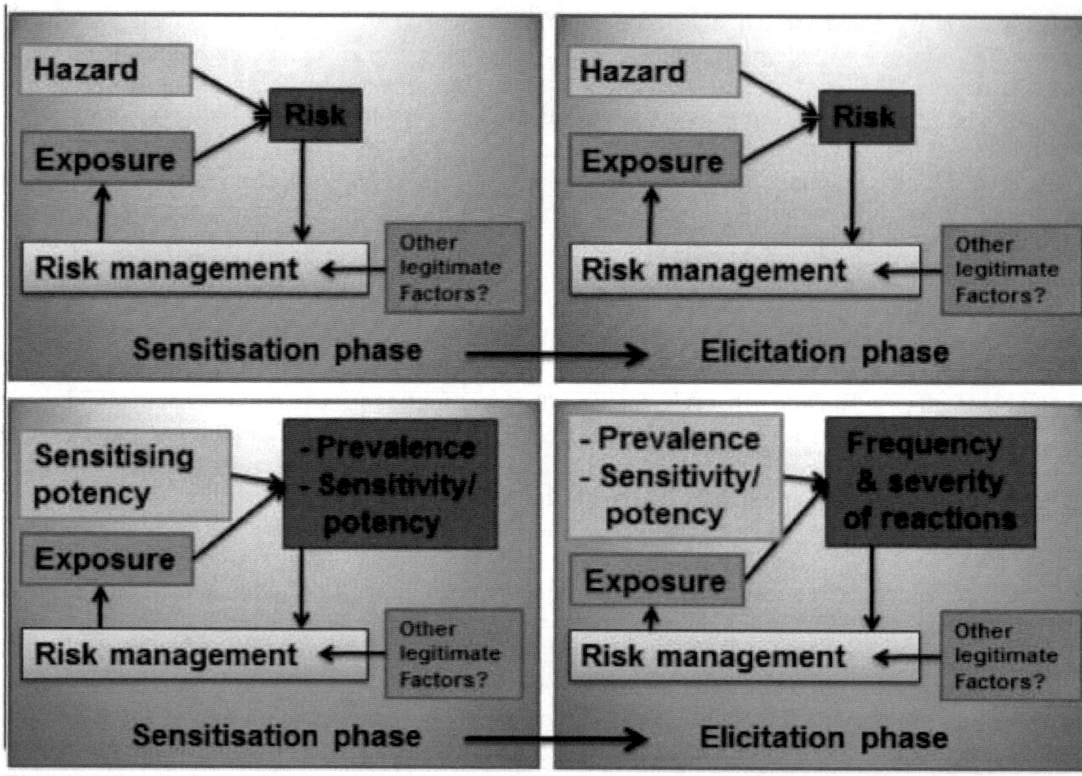

Figure 1. Reproduced from Houben *et al.* [19]. The generic risk analysis cycle (top) applied to food allergy (bottom). The prevalence of allergy and the sensitivity of the allergic population (i.e. individual thresholds during DBPCFC) are the outcomes of the risk analysis for the sensitisation phase and are also the hazard input for the elicitation phase risk analysis.

Previous and current guidelines developed for food allergy focus on hazard analysis in already sensitised individuals and rely mainly on structural characteristics of the protein involved. Similarly, the available tools to predict allergenicity of a new food or protein analyse cross-reactivity due to their reliance on properties of known allergens. Current testing of new proteins assesses risks due to cross-reactivity/co-sensitization, but fewer options are available to identify and characterise the risks from *de-novo* sensitization. No single test is currently available, nor is one expected in the near future, for predicting and characterising *de-novo* sensitization potencies of new proteins. To date, there are no *in-silico* or *in-vitro* approaches that could be used to identify the sensitization potential of a protein or the subsequent potential to elicit a clinical reaction. The only tests currently fit to identify induction of specific IgE from a new protein are *in-vivo* models, either in animals or humans, although many limitations exist both scientifically and ethically with regard to these tests. Additionally, uncertainties in characterising differences between tolerance and allergenicity are present in these *in-vivo* models. Allergic responses in the human body are extremely complex and an overall mechanistic model (or adverse outcome pathway, AOP) for food allergy does not yet exist, despite extensive research. Arguably, the development and use of novel protein sources is restricted due to the limited ability of current testing strategies to predict sensitization. However, for the safety of our future food supply, it is critical to find more sustainable protein sources. Thus, there is a need for a systematic, comprehensive battery of tests and assessment strategy to identify and characterise *de-novo* sensitization to new proteins and the risks associated with them. This overall strategy should incorporate all relevant intrinsic protein properties, aspects of exposure and matrix/processing effects.

New potential allergenic hazards are currently managed through avoiding exposure. Risk managers can avoid exposure of an allergen to the population by not authorizing the introduction of a protein identified as a new allergen into the marketplace or possibly earlier in the process by identifying potential allergenicity early in development and cancelling the project. Labelling can be utilized to alert consumers with existing sensitivities to the presence of a potential hazard (e.g. rapeseed protein isolate and individuals with existing mustard seed allergies potentially at risk for reaction). However, food proteins with a high sensitizing potency cannot have their public risks mitigated through labelling alone. Post-launch monitoring exercises should be considered and are expected to provide an early indication of any unexpected development of allergy after introduction of a novel food to a new population or market. Any indication of allergenicity should then trigger the initiation of risk management measures. While resource intensive, post-launch monitoring is viewed as a necessary tool and others have previously discussed the possibilities and limitations in the context of novel foods and unintended health effects or allergenicity have been critically discussed [22,23].

Development of a comprehensive, coherent risk assessment strategy would benefit greatly from a clear definition of criteria for distinguishing between proteins of high and low allergenicity (i.e. ability to induce IgE, potency to induce IgE, expected prevalence of IgE-sensitization, expected prevalence of allergy, expected exposure, expected eliciting potency, expected frequency of reactions, expected frequency of severe reactions). It is important to establish first what we need to test for in order to define the requirements of new methods, and this is, in turn, dependant on what risk we want to manage. Do we want to prevent only the extremes (i.e. that no people die or that no single individual is sensitized) or could we accept a certain level of risk regarding sensitization and allergic reactions? As indicated above, the risk analysis outcome is based on an interactive process between risk assessment and risk management. These are not independent of each other, as a risk assessment must to be attuned to answer the risk management questions and needs. Consequently, the hazard and risk assessment methods applied and the desired information are determined by the requested outcome for risk management purposes and decisions to be made. Likewise, if new methods for hazard and risk assessment are to be developed, then the methods, the information needed and the requested outcomes all depend on the information that is going to be requested for the risk management goal or decision to be made. It is an important step in the coming period regarding novel food allergen risk assessment and it is critical that parameters and criteria for risk management decision making are clearly outlined. Once risk management criteria are defined, risk assessors could apply or develop the appropriate tests as needed to investigate the relevant protein characteristics. The COST Action network (ImpARAS, www.imparas.eu) has recently started to discuss these criteria from first principles and will continue with the broader subject of improving strategies for allergen risk assessment throughout 2016-2018/9.

It is important to demonstrate a proof-of-principle for any chosen approach, to establish that the methodologies are able to distinguish between, and rank allergens of different potency appropriately. As scientific knowledge progresses, improvements to the methodologies used in allergenicity risk assessment should be possible. New development of biologically relevant *in-vitro* or *in-silico* methods should open new possibilities, and reduce animal and human testing while improving the safety and risk management of introducing novel foods into the diet. It is going to be crucial to identify those approaches, methods and technologies on which future research efforts should be focussed and a better understanding of AOPs could guide the development of better *in vitro/vivo* allergenicity testing methods. As the perfect tests have not yet been identified, it is, therefore, important to leave flexibility within any regulation or guidance to account for improvement of methodologies regarding food allergy and allergen risk assessment.

Acknowledgement

The author is part of the COST Action FA1402 entitled: Improving Allergy Risk Assessment Strategy for New Food Proteins (ImpARAS).

References

[1] FAO/WHO 2008 *Codex Alimentarius Commission procedural manual Food and Agriculture*

Organization of the United Nations Codex Alimentarius Commission Rome
[2] Taylor S L and Hefle S L 2006 *Food allergies and intolerances,* ed Shils M E et al. (Lippincott Williams and Wilkins Baltimore) 1512–30
[3] Geroldinger-Simic M et al. 2011 Birch pollen–related food allergy: Clinical aspects and the role of allergen-specific IgE and IgG_4 antibodies *J Allergy Clin Immunol.* **127** 616–22
[4] Commins S P and Platts-Mills T A E 2009 Anaphylaxis syndromes related to a new mammalian cross-reactive carbohydrate determinant *J Allergy Clin Immunol.* **124** 652–7
[5] Turner P J et al. 2016 Can we identify patients at risk of life-threatening allergic reactions to food? *Allergy Eur J Allergy Clin Immunol* **72** 9-12
[6] Crevel R, Moneret-Vautrin D A, Morisset M, Sheffield D, Taylor S L and Baumert J L A 2010 preliminary analysis of the evolution of peanut thresholds over repeated challenges in a population of consecutive clinic patients *J Allergy Clin Immunol [Internet]* **4** 125
[7] Ballmer-Weber et al. 2015 How much is too much? Threshold dose distributions for 5 food allergens *J Allergy Clin Immunol.* **135** 964–71
[8] Taylor S L et al. 2010 Threshold dose for peanut: Risk characterization based upon diagnostic oral challenge of a series of 286 peanut-allergic individuals. *Food Chem Toxicol* [Internet]. **48** 814–9
[9] Blom W M, Vlieg-Boerstra B J, Kruizinga A G, Heide S Van Der, Houben G F and Dubois A E J 2013 Threshold dose distributions for 5 major allergenic foods in children *J Allergy Clin Immunol* **131** 172–9
[10] Hefle S L, Furlong T J, Niemann L, Lemon-Mule H, Sicherer S and Taylor S L 2007 Consumer attitudes and risks associated with packaged foods having advisory labeling regarding the presence of peanuts *J Allergy Clin Immunol.* **120** 171–6
[11] Kruizinga A G, Briggs D, Crevel R W R, Knulst A C, Bosch L M C Van Den and Houben G F 2008 Probabilistic risk assessment model for allergens in food : sensitivity analysis of the minimum eliciting dose and food consumption *Food Chem. Toxicol* **46** 1437–43
[12] Spanjersberg M Q I, Kruizinga A G, Rennen M A J and Houben G F 2007 Risk assessment and food allergy : the probabilistic model applied to allergens *Food Chem. Toxicol* **45** 49–54
[13] Spanjersberg M Q I, Knulst a C, Kruizinga a G, Van Duijn G and Houben G F2010 Concentrations of undeclared allergens in food products can reach levels that are relevant for public health F*ood Addit Contam Part A Chem Anal Control Expo Risk Assess* [Internet] **27** 169–74
[14] Rimbaud L et al. 2010 Quantitative risk assessment relating to adventitious presence of allergens in food: A probabilistic model applied to peanut in chocolate *Risk Anal.* **30** 7–19
[15] Crevel R W R, et al. 2014 Development and evolution of risk assessment for food allergens *Food Chem Toxicol* [Internet] **67** 262–76
[16] Remington B C, Baumert J L, Blom W M, Houben G F, Taylor S L and Kruizinga G. 2015 Unintended allergens in precautionary labelled and unlabelled products pose significant risks to UK allergic consumers *Allergy Eur J Allergy Clin Immunol.* **70** 813–9
[17] Verhoeckx K, Broekman H, Knulst A and Houben G 2016 Allergenicity assessment strategy for novel food proteins and protein sources *Regul Toxicol Pharmacol.* **79** 118–24
[18] Broekman H et al. 2016 Majority of shrimp-allergic patients are allergic to mealworm *J Allergy Clin Immunol.* **137** 1261–3
[19] Houben G et al. 2016 Prioritisation of allergenic foods with respect to public health relevance: Report from an ILSI Europe Food Allergy Task Force Expert Group *Food Chem Toxicol.* **89** 8–18
[20] Taylor S L et al. Establishment of Reference 2014 Doses for residues of allergenic foods: Report of the VITAL Expert Panel *Food Chem Toxicol.* **63** 9–17
[21] Allen K J et al. 2014 Allergen reference doses for precautionary labeling (VITAL 2.0)Clinical implications. *J Allergy Clin Immunol.* **133** 156–64
[22] Hepburn P et al. 2008 The application of post-market monitoring to novel foods *Food Chem*

Toxicol. **46** 9–33
[23] Wal J-M, Hepburn P A, Lea L J and Crevel R W R 2003 Post-market surveillance of GM foods: applicability and limitations of schemes used with pharmaceuticals and some non-GM novel foods *Regul Toxicol Pharmacol.* **38** 98–104

IOP Conf. Series: Earth and Environmental Science **85** (2017) 012004 doi:10.1088/1755-1315/85/1/012004

Predicting the role of veterinary medicine in future health and food safety challenges

N Fejzić and S Šerić-Haračić

University of Sarajevo Veterinary Faculty, Zmaja od Bosne 90, Sarajevo, Bosnia and Herzegovina

Email: nihad.fejzic@vfs.unsa.ba

Abstract. Animals have always been a source of food, materials, protection and wellbeing for humans; however, animal diseases, including zoonoses, have both direct and indirect negative effects on human health, economy and the society. Since its establishment, the veterinary profession has provided crucial input in eradicating disease, increasing animal production and reducing losses due to diseases. Currently, foodborne diseases and zoonoses have raised awareness in developed countries, which have excellent systems for disease surveillance and reporting both in humans and animal populations. Due to lack of modern, integrated surveillance and reporting, the burden of zoonoses and foodborne diseases in developing European countries is much harder to assess. Differences in countries' animal health status (demonstrated through disease surveillance) have been a main pivot point for international trade of animals and animal products. However, rapid and dramatic evolvement of the health trends in the world changed the principles of animal disease surveillance. Approaches requiring lower cost (i.e. risk-based surveillance) are now proposed, not only due to less available public funding, but also because the costs are harder to justify to policy makers if a disease is exotic and/or rare. Therefore, the veterinary profession has faced insufficient interest of governments and funds for further research into many persistent endemic animal diseases and zoonoses. On the other hand, eradication of selected diseases in some areas while elsewhere they still persist, and the continuous emergence of new diseases, cannot guarantee permanent epidemiological stability. As food safety and security become more important, global trends and events have highlighted the biological, health and economic inseparability of the relationships between humans, animals as pets and/or food sources and wildlife within the social and ecological framework of living space that these species share. Veterinarians are called on and expected to offer strategic and operational solutions for better integration of public health systems (i.e. One Health), animal health, food safety and environmental protection. At the same time, the profession faces challenges in the organisation and implementation of surveillance and disease mitigation measures.

1. Animals as source of food, prosperity, health and disease

Humans have been both oriented and dependent on animals as source of food, materials, protection and wellbeing throughout history. Even though it could be argued whether many of the animal species would be better off without sharing habitats with us, it is certain that without animals, we would not have reached today's anthropological, civilizational and technological development, perhaps even mankind's survival would be at stake [1].

The importance of animals as a source of protein for human consumption has significant impact on development of the veterinary profession. The 1996 World Food Summit defined food security as "a situation that exists when all people, at all times, have physical, social and economic access to sufficient, safe and nutritious foods that meets their dietary needs and food preference for a healthy life" [2]. However, modern times have brought new challenges such as overpopulation, climate change, urbanisation, globalisation and consequent disease (re)emergence in our efforts to ensure sufficient food supply [3].

By integrating animals in our lives and societies, we unknowingly integrated their specific microbiota. Hence, from early times, the occurrence and spread of human diseases was closely related to microorganisms found in animals. Due to the small size, density and few contacts of the early human communities, in order to survive, causative agents of diseases had to adapt either as chronic agents or to find secondary hosts. Humans also became secondary hosts to many animal diseases. Therefore, it is not surprising that over 60% of infectious human diseases today are zoonotic [4].

Without underestimating or understating the positive effects of microorganisms, especially in production of food, the resulting death toll in human populations far surpasses any other cause of human mortality such as war or famine [5]. Besides direct effects on human health and life, animal infectious agents, zoonotic or not, can cause death of animals, decline in animal production and degradation of the quality of animal products, leading in the end to less available food for humans (reduced food security). Animal diseases are also obstacles to international trade of animals and animal products, as well as substantial financial burdens to national veterinary and public health services and producers/industry. In the wider perspective, this has huge influence on the economy of region/countries and, therefore, affects the overall wellbeing of mankind.

These threats are currently addressed in two ways: in developed countries, activities are focused on reducing and controlling animal diseases endangering human health, compromising trade status and animal welfare, while in developing countries, the main issues remain increasing animal production in order to supply increasing demand for animal protein. In both scenarios, the veterinary profession has crucial input with responsibility to provide disease eradication, increasing animal production and reducing losses due to diseases.

2. Zoonoses and foodborne diseases in Europe today

According to the joint report of the European Food Safety Agency (EFSA) and the European Centre for Disease Prevention and Control (ECDC), the most common zoonotic disease in the European Union (EU) is campylobacteriosis, followed by salmonellosis, yersiniosis, vero toxic *Escherichia coli*, listeriosis, echinococcosis, Q fewer, brucellosis, West Nile virus disease, tularaemia, trichinellosis, tuberculosis (*M. bovis*) and rabies [6] (Table 1).

Table 1 Reported cases, hospitalisation and fatal outcomes of zoonotic diseases in humans in the European Union during 2013 [6]

Disease	Number of cases	Number hospitalised	Fatal outcome	
			Number	Case fatality rate (%)
Campylobacteriosis	214,779	11,922	56	0.05
Salmonellosis	82,694	7,841	59	0.14
Yersiniosis	6,471	481	2	0.05
Vero toxic *Escherichia coli*	6,043	922	13	0.36
Listeriosis	1,763	735	191	15.6
Echinococcosis	794	127	2	0.88
Q fever	648	NA	2	0.61
Brucellosis	357	139	1	0.99
Tularaemia	279	39	0	0
West Nile virus disease	250	52	16	3.4
Trichinellosis	217	106	1	0.56
Tuberculosis (*M. bovis*)	134	-	-	-
Rabies	1	1	1	100

The top five diseases from this list are predominantly transmitted to humans through contaminated food of animal origin, while the frequency of finding the causative agents in animal population is also increasing (i.e. in the same period *Campylobacter* has been found in 30.4% of animal samples in Nordic countries) [7].

Foodborne diseases and zoonoses have raised awareness and are very important in developed countries, which have advanced systems of data gathering, disease surveillance and disease reporting both in human and animal populations. Simultaneously in these countries, animal health protection is vastly regulated, and technology of animal origin food production incorporates many systems of safety and quality assurance (i.e. farm biosafety, good production/management/agricultural practices, HACCP and traceability from farm to fork) [8].

Health management at the country level (or even at the EU level, represented by the Directorate General for Health and Food Safety – DG SANTE of the European Commission) for humans and animals is increasingly integrated through the "One Health" approach, which together with oversight of the entire food production chain, enables high protection of consumers, early response in case of incidents (therefore reducing the potential economic losses) and better access to international markets. The achievements of veterinary medicine in laboratory diagnostics, surveillance, food production practices and food quality standards had led to beneficial effects on animal health status and improved food safety. On the down side, there is reduced interest, justification and available funding for further investigation of infectious diseases of food animals [9].

The burden of zoonoses and foodborne diseases in developing European countries is much harder to assess, due to inadequate disease surveillance systems and less regulated and implemented corrective measures to ensure animal health and food safety, with less oversight. Many countries report worsening of the animal health and public health situations. However, these observed trends are not easy to interpret. Firstly, in many of these countries, animal disease reporting systems are based on reporting the absolute frequency (number) of cases or outbreaks (infected households/farms), while the population data (denominator) is generally unavailable or unreliable. Secondly, epidemiological intelligence campaigns are heavily dependent on current general public/political focus, and due to lack of available public funds, are never truly systematic. Monitoring and disease control programs are very expensive and technically demanding activities which exceed the competencies, budget capacities and public fund management of health services (animal and human) in many developing countries.

3. Disease surveillance

Animal disease is an economic problem with veterinary implications, not a veterinary problem with economic implications, because it affects people's wellbeing. More than anything else, coherent disease surveillance and reporting systems have influenced the current difference in health and consumer protection status of many EU and other developed countries compared to the rest of the world [10]. In contrast to the clinical approach of understanding diseases, the population-based approach (epidemiological surveillance) starts with the animal unit, moves to the level of farm, then to the local area, the country or region or even the whole world. Animal disease surveillance, according to the World Organisation for Animal Health (OIE) (Chapter 1.4. of the Terrestrial manual), is aimed at demonstrating the absence of disease or infection, determining the presence or distribution of disease or infection or detecting as early as possible exotic or emerging diseases. This is a science-based tool to monitor disease trends, to facilitate the control of disease or infection, to provide data for use in risk analysis for animal or public health purposes, and to substantiate the rationale for sanitary measures.

Together with rapid and dramatic evolvement of health trends in the world, the principles of animal disease surveillance have also changed. Due to expansion of international trade and globalisation, disease agents are transferred through food, vectors, reservoirs, latent carriers and diseased individuals in short times over great distances, potentially able to arrive in our backyard from exotic or far-flung destinations with different standards of health protection. At the same time, due to the integration and industrialisation of food production, raw materials originating from around the world are used for mass production of food and reach huge numbers of world consumers.

By exclusion of tariff-based trade barriers (adoption of the World Trade Organisation Agreement on the Application of Sanitary and Phytosanitary Measures) and recognition of international standard-setting bodies (i.e. the OIE for animal health, *Codex Alimentarius* for food and feed), differences in countries' animal health status (demonstrated through disease surveillance), has been a main pivot point for international trade of animals and animal products. Hence, countries that have employed scientific solutions and cost-effective decision-making mainly provided by veterinary services/professionals not only have disease surveillance systems that enable them to export, but also can influence the standard-setting processes at the international level. For example, the currently propagated risk-based surveillance applied in demonstrating disease absence enables the garnering of sufficient and suitable scientific evidence, similarly to the population-based studies [11]. However, with risk-based surveillance, much smaller sample sizes are required (because suitable population groups are sampled according to estimation of disease risk), and therefore, costs are much less than with the older system [11]. The need to reduce costs is not only due to reduced availability of public funding, but also because resources allocated to disease surveillance (preventing losses) have opportunity costs, where the benefits can be foregone as a result of allocating them to disease mitigation instead of other productive use [10].

On the other hand, as is evident by the emergence or re-emergence of many diseases even in developed countries, eradication of selected diseases in some areas while elsewhere they still persist, and the continuous emergence of new diseases, cannot guarantee permanent epidemiological stability. Therefore, it remains for the veterinary services of developing countries to make a progressive leap in focusing their disease surveillance strategies on sound epidemiological data and move away from improvisation for the benefit of their own national consumers, producers and economies. In addition, more interdisciplinary and collaborative efforts and funding are needed to ensure worldwide inclusiveness in achieving and upgrading animal health and food safety to international standards. In decision-making regarding disease surveillance, balance should be reached between the appropriate level of protection and cost-effectiveness. Improvisation in corrective measures is not justified by an "anything is better than nothing" approach, because such measures are not only inefficient, but often have adverse effects with dramatic consequences on public health and animal production. This can lead to the situation whereby eradication of disease becomes unattainable for a very long time [12].

4. What is our future role in food related issues?

Public veterinary services have complex structures and competencies, not only because they are already a part of a defined and complex administrative framework, but because of the wide spectrum of veterinary activities and responsibilities which are steadily increasing throughout the world. From the establishment of the first Veterinary Faculty in 1761, in Lyon (France), the prime role of veterinarians for many decades (treating horses) gradually expanded into clinical practice for other types of domestic animals as their breeding and their farm exploitation grew in economic importance [13].

At the beginning of the last century, the global commitment to eradicate the frequent diseases of animals and people such as rinderpest, foot and mouth disease, tuberculosis and brucellosis, to name just a few, initiated development and implementation of national, and partly global, programs for the control of these diseases. At the same time, with the development of animal production, the diseases (including the cost of treatment) were recognised as a significant burden on profitability, and therefore, veterinarians were redirected to the population approach and prevention. Later, commercial farms introduced computer programs and modern technology for monitoring animal health and production data with a focus on the control and reduction of "production" diseases (such as mastitis, reproduction disorders etc.) (14). The highly-contagious animal diseases and zoonoses have been eradicated in most of the developed countries, and disease-free status was then maintained by the "zero" risk principle (demanding the absolute evidence of non-existence of diseases in the exporting country) in international trade. The private sector interest in preventing the emergence of diseases, as well as shifting responsibility and funding for animal health protection and food safety from governments to producers, have directly affected the trend (recorded in recent decades) of veterinarians orienting towards more profitable jobs (companion animal medicine). At the end of the 1900s, the veterinary profession was faced with insufficient interest of governments and funds for further research into many persistent endemic animal diseases and zoonoses [9], neglecting the role of wild animals as reservoirs of disease agents (15), as well as the lack and/or weak development of interdisciplinary approaches. Instead, the primary political interest was in preserving the sanitary status of states/regions/countries (free of disease) strictly in the interests of economics and trade.

Shortly after that, the world was faced with old and new health challenges, now with even more dramatic consequences. Global trends and events have highlighted the biological, health and economic inseparability of the relationships between humans, animals as pets and/or food sources and wildlife within the social and ecological framework of living spaces that these species share. Food safety and security have become more important than ever in human history, and scientists have identified fully some new and extremely influential concerns for which we remain unprepared: overpopulation, climate and ecosystem change and urbanisation. However, some other evolving issues, with more local/national influence must not be underestimated in the global arena, such as: biosecurity, bioterrorism, antimicrobial resistance, animal welfare, increasing volume of international trade and influence of media and politics. Concerns in the sphere of food security, besides the formerly defined components (availability of food, access to food and for the food to be culturally appropriate) might be expanded today to biological and health safety. At first glance, food security appears to be primarily a supply-and-demand-related issue. However, new evidence reveals that it is a hybrid of supply, quality and health issues as well, so compromised food safety and health is also a food security issue (Table 2).

Table 2 Review of current and expected challenges and concerns of food veterinarians in the sphere of food safety, health and food security

Challenges facing veterinary medicine	Possible consequences	Action required and demands
Overpopulation; Climate change and ecosystem; Urbanisation; Bioterrorism; New agents; Emerging diseases; Fast international spread; Antimicrobial resistance; Animal welfare; Biological and chemical hazards; Lack of research funding.	Economical and social insecurity; Disturbance in food supply; International trade; Case fatality rate; Health vulnerability; Lost of public trust.	Changes in education for food veterinarians; One health; Interdisciplinary approach; Leading role of veterinarians in food security; New surveillance and survey systems; Access to and proper distribution of information; New biosecurity and biodefense frame.

Besides prediction of new emerging disease agents, for which control approaches have not yet been identified or tested, many zoonoses have been relatively well-known (although they can never be fully known) and recognised for the last 50 years or more, and they are the subjects of long-term monitoring and control programs in public and animal health and food surveillance systems. However, the observed trends indicate that even these diseases continue to appear and increase, so therefore, the scientific discipline and knowledge expected to support the mitigation and prevention of these zoonoses still requires upgrading and further research. Significant improvements in knowledge and scientific infrastructure have been made in dealing with health and food safety issues, but clearly, health hazards and foodborne illness will not disappear. Obviously, problems related to the occurrence and spread of zoonoses among humans and animals without discrimination occupy both developing and developed countries, with subtle diversity of consequences for each society or community. Even the most developed countries with advanced surveillance systems grossly under-report cases of foodborne illnesses [3].

In conclusion, zoonoses are still, and will remain, a global health issue, common in their nature (animal to human transmission), but not with the same solutions for the poor and the rich, private sector (producers) and government, animal health carers and human health providers. The demand for better knowledge and more efficient solutions and approaches is intensive and dynamic and today is directed by national and global food and health security. Veterinarians are called on and expected to offer strategic and operational solutions for assessing risks, analysing challenges and applying scientific knowledge. The resulting need for better integration of public health systems, animal and plant health protection, food safety and environmental protection, has led to new approaches (One Health) and challenges in the organisation and implementation of mitigation measures and surveillance. Our role, as veterinarians, is development of the new type of competencies and institutions required to deliver the appropriate science and technology, risk-based regulatory management and communication policies and practices related to agriculture, animal and public health, environment and climate change, wildlife and biodiversity, social and economic development and growth.

References

[1] Caras, R.A.: A perfect harmony: The intertwining lives of animals and humans throughout history. Purdue University Press, 1996.
[2] Barrett, C.B.: Measuring food insecurity. Science 327 (5967): 825-828, 2010
[3] Havas K., Salman M.: Food security: its components and challenges. Int. J. Food Safety, Nutrition and Public Health 4 (1): 4-11, 2011
[4] Taylor, L.H., Latham S.M., Mark E.J.: Risk factors for human disease emergence. Philosophical Transactions of the Royal Society of London B: Biological Sciences 356(1411): 983-989, 2001
[5] Wolfe, N.D., Dunavan, C.P., Diamond J.: Origins of major human infectious diseases. Nature 447(7142): 279-283, 2007
[6] EFSA/ECDC. The European Union summary report on trends and sources of zoonoses, zoonotic agents and food-borne outbreaks in 2013. EFSA Journal 13(1), 2015
[7] Osimani, A., Aquilanti, L., Pasquini, M., Clementi, F.: Prevalence and risk factors for thermotolerant species of Campylobacter in poultry meat at retail in Europe. Poultry Science 2017
[8] Unnevehr, L.J., Jensen, H.H.: The economic implications of using HACCP as a food safety regulatory standard. Food policy 24(6): 625-635, 1999
[9] Ducrot C., Bed'hom B., Beringue V., Coulon, J.B., Fourichon, C., Guérin, J.L., Krebs, S., Rainard, P., Schwartz-Cornil, I., Torny, D., Vayssier-Taussat, M.: Issues and special features of animal health research. Veterinary Research 42(1): 96, 2011
[10] Howe, K.S., Häsler, B., Stärk K.D.C.: Economic principles for resource allocation decisions at national level to mitigate the effects of disease in farm animal populations" Epidemiology & Infection 141(1): 91-101, 2013
[11] Oidtmann B., Peeler E., Lyngstad T., Brun E., Jensen B.B., Stärk K.D.C.: Risk-based methods for fish and terrestrial animal disease surveillance. Preventive veterinary medicine112(1): 13-26, 2013
[12] Godfroid, J., Al Dahouk, S., Pappas, G., Roth, F., Matope, G., Muma, J., Marcotty, T., Pfeiffer, D., Skjerve, E.: A "One Health" surveillance and control of brucellosis in developing countries: moving away from improvisation. Comparative immunology, microbiology and infectious diseases 36(3): 241-248. 2013

[13] Zinsstag, J., Schelling, E., Waltner-Toews, D., Tanner, M.: From "one medicine" to "one health" and systemic approaches to health and well-being. Preventive veterinary medicine 101(3):148-156. 2011

[14] Morris, R. S.: The application of economics in animal health programmes: a practical guide." Revue scientifique et technique-Office international des épizooties 18: 305-314, 1999

[15] Travis, D.A., Watson, R.P., Tauer, A.: The spread of pathogens through trade in wildlife. Revue Scientifique et Technique-OIE 30(1): 219, 2011

IOP Conf. Series: Earth and Environmental Science **85** (2017) 012005 doi:10.1088/1755-1315/85/1/012005

Toxoplasmosis as a food-borne infection

O Đurković-Đaković

National Reference Laboratory for Toxoplasmosis, Center of Excellence for Food- and Vector-Borne Zoonoses, Institute for Medical Research, University of Belgrade, Serbia

E-mail: olgicadj@imi.bg.ac.rs

Abstract. *Toxoplasma gondii* is a globally distributed parasite that infects all mammals, including one third of the world population. Long known to cause disease in the developing foetus and in immunosuppressed individuals, a body of data that has emerged in the past decades suggests its role in human pathology may be even more important. The WHO and FAO have recently established toxoplasmosis as a foodborne infection of global concern, with a disease burden the greatest of all parasitic infections. Transmission of toxoplasmosis occurs by ingesting tissue cysts from undercooked meat and meat products, and oocysts from the environment with contaminated fresh produce or water. This review provides an update on the current understanding of toxoplasmosis, focusing on the risk of infection from food of animal origin, with particular reference to the risk in Serbia and the region of South-East Europe.

1. Introduction

Known for over a century, *Toxoplasma gondii* is considered one of the most successful parasites on Earth due to its omnipresence and widest array of hosts, including all mammals.

Domestic cats or other *Felidae* are the definitive host of the *Toxoplasma gondii* parasite, in whose intestines the sexual cycle takes place, and they are the primary source of infection for all animals including humans, by contaminating the environment with oocysts excreted in the faeces. Herbivores are infected by ingestion of food and water contaminated by oocysts, and carnivores by eating tissue cysts present in the flesh of infected animals. Humans (and other omnivores) can be infected both ways, by food contaminated with oocysts (improperly washed vegetables or fruits), contaminated water or hands, and by eating improperly processed or raw meat of infected animals.

Human infection is widespread; it has been estimated that one third of the global population is infected [1]. Infection is generally mild and self-limiting in the immunocompetent host, and according to the classical understanding, of no further concern. Once infected, the parasite is thought to persist for the lifetime of its host, under control by the host immune response. This explains the vulnerability of population categories with an incompetent immune system, including the unborn baby and immunosuppressed individuals, to serious, even life-threatening *Toxoplasma*-induced disease. Congenital toxoplasmosis (CT), for which preventive strategies have long been implemented, is quite rare, but the population of immunosuppressed individuals is increasing, with the AIDS pandemic, and with the advance in transplantation medicine, which depends on immunosuppressive treatment to avoid graft rejection. In this context, it is important that treatment options have not advanced much for decades and there is still no drug able to eliminate encysted parasite.

A body of data has emerged in recent decades, however, which sheds different light on some existing concepts. The concept that *T. gondii* infections are mostly subclinical has been challenged by a current intriguing line of research on the potential role of toxoplasmosis in the pathogenesis of neurologic and psychiatric diseases, including schizophrenia, Parkinson's disease and depression [2]. Others stem from the insight into the molecular epidemiology of *T. gondii* infection.

The WHO and FAO have recently established toxoplasmosis as a foodborne infection of global concern, with a disease burden, similar to that of salmonellosis and campylobacteriosis, the greatest of all parasitic infections [3]. This has naturally renewed the interest in this zoonosis, particularly from the aspect of food safety, and calls for new strategic approaches in both management and prevention.

This paper reviews the novel understanding of *T. gondii* infection, focusing on the risk of infection from food of animal origin, particularly in the region of South East Europe (SEE).

2. *T. gondii* population structure and implications

The *T. gondii* genus comprises a single species infective for all hosts, with limited genetic diversity characterised by clonal lineages designated as types I, II and III predominating in Europe and North America (and a fourth one in the latter), and by a higher frequency of non-clonal, atypical strains in sub-tropical and tropical countries, and a wider genetic diversity characterised by non-clonal, atypical strains in South America and Africa. The diversity encountered in the tropical and subtropical areas has been associated with the presence of diverse *Felidae* in which sexual reproduction, and consequentially, genetic recombinations, occur [4].

Insight into the *T. gondii* population structure may have huge implications on our understanding of this infection. Differences in infecting genotype have been related to the clinical presentation. Atypical strains have been associated with more severe ocular toxoplasmosis, atypical presentations and even life-threatening disease in both immunocompetent and immunosuppressed individuals. The concept of only primary maternal infection during pregnancy being a risk for foetal infection has been challenged by cases of CT in babies born to immunised mothers, re-infected with atypical strains [5].

This, along with the widely varying geographically-dependent prevalence of infection and the geographic differences in the *T. gondii* population structure, determines toxoplasmosis as a travel risk [6]. Moreover, globalisation of food, including importation of meats from areas of a highly divergent population structure, may also present a risk factor for severe infections.

3. Toxoplasmosis as a food-borne infection

3.1. Basic epidemiology of human toxoplasmosis

Transmission of toxoplasmosis occurs by ingesting tissue cysts from undercooked meat and meat products, and oocysts from the environment along with contaminated fresh produce or water.

Notably, there has been a decreasing trend of *T. gondii* infection prevalence over the last 30 years across Europe [7]. The same has been noted in SEE. Continuous monitoring of the prevalence of *T. gondii* infection in women of childbearing age in Slovenia, Serbia and Greece has shown a significant decrease in the infection prevalence since the eighties onwards [rev. In 8]. The largest decrease, from 86% in 1988 to 31% in 2007 and beyond, was noted in Serbia.

Many factors may have contributed to such a change, including increased public awareness as a result of health education, better hygiene on livestock farms, and more frequent use of frozen meat (freezers now present in most households).

Importantly, consumption of undercooked or insufficiently cured meat has repeatedly been shown to be a major risk factor for human infection with *T. gondii* [9].

3.2. Toxoplasmosis in meat animals in SEE

A major reason for the control of *T. gondii* infection in meat animals is the reduction of the reservoir of human infection. Accordingly, the meat route has been intensively explored, in parallel with the development of serological and molecular tools that allowed for the analysis of meat/meat products.

Earlier serological investigations of various animal species in SEE countries were generally carried out in limited samples sizes and using different and non-standardised methodologies. Thus, only studies in the last 20 years are referred to here.

In Serbia, the first nation-wide cross-sectional survey on the seroprevalence of *T. gondii* infection in cattle, sheep and pigs carried out in 2002/2003 showed a seroprevalence of 76.3% in cattle, 84.5% in sheep and 28.9% in pigs [10]. The antibody levels ranged from 1:25 to 1:400 in cattle, and up to 1:25,600 in sheep and to 1:12,800 in pigs. Among the seropositive animals, the proportion of high antibody levels (≥1:1600), suggestive of acute infection, and indicating continuous presence of infection reservoirs in the environment, was 10% in sheep, and 4% in pigs.

Risk factors for cattle were small herd size and farm location in Western Serbia, while housing in stables with access to outside pens was protective. In sheep, an increased risk of infection was found in ewes from state-owned flocks vs. Private flocks, and, interestingly, also in those from Western Serbia. In pigs, the risk of infection was highly increased in adult animals, as well as in those from finishing type farms.

Despite the high prevalence of 85% in sheep of which 10% had high antibody levels suggestive of acute infection, correlation with ovine abortions could not be established, since aetiological laboratory diagnosis of ovine abortions in Serbia does not include diagnosis of *T. gondii*.

An outbreak of toxoplasmosis in sheep has recently been reported; massive abortions (60%) occurred in a flock of 500 dairy sheep in Northern Greece at 110-130 days of pregnancy, diagnosed upon observation of tissue cysts in brain smears of aborted foetuses, and by serological (ELISA) examinations of mother and foetal serum samples. The abortion rate declined immediately upon instituting sulphadimidine therapy [11].

Another study showed a high prevalence of 74.7% in goats in Serbia as well [12], higher than in Greece (62.9%), Bulgaria (59.8%) and very markedly so than in Croatia – 8.4% [13].

A later study on *T. gondii* infection in slaughter pigs in Serbia [14], showed, however, a three-fold lower prevalence of 9.2% in a total of 488 swine from abattoirs in the vicinity of Belgrade. This difference was largely attributed to the difference in the studied samples since the latter research consisted of a large majority (96%) of market-weight pigs, which generally have a much lower prevalence than adult pigs. Similarly to the 2006 study, risk factors for infection in slaughter pigs included age and farm type, with a 41-fold higher likelihood of infection in adult *vs.* Market-weight pigs, and a 15-fold higher likelihood of infection in pigs of all ages from smallholders' finishing type farms *vs.* Those from farrow-to-finish intensive farms. It was proposed that a national strategy to reduce the level of *T. gondii* infection in pigs should include a shift towards the development of more farrow-to-finish farms, as well as vigilance in farm management and implementation of zoo-hygienic measures at finishing farms. Damriyasa et al. [15] stated that *T. gondii* seropositivity is an indicator of the hygienic status of the pig farm. In the 2011 study [14], the demonstration of viable *Toxoplasma* in blood samples of seropositive swine indicated that slaughtering took place during parasitaemia, which suggests frequent swine reinfections. The study concluded that swine meat is a significant potential risk for human *Toxoplasma* infection, but that the risk cannot be predicted on the basis of specific antibody level in the pigs.

On the other hand, the modern approach in farm management to provide for the welfare of the animals as well as organic food for human consumption is to develop animal-friendly (organic) farms. According to experience from the Netherlands [16], development of such farms can result in an increase in *T. gondii* infection. Nevertheless, a report from organic sheep and goat farms in Greece [17] showed similar prevalence rates to those in animals from conventionally managed farms [18].

Cattle are generally thought not to be significant in this context [19]. However, beef is often consumed undercooked (rare beef steaks, roast beef, steak tartar), and at least one outbreak of toxoplasmosis whose source was raw beef has been documented [20]. In addition, one out of four beef samples randomly chosen from UK retail outlets tested positive for *T. gondii* by PCR [21]. These facts, along with the circumstantial evidence provided by the high prevalence of cattle infection of 92% in Italy and 69% in France [22], and in Serbia, countries in which human infection is highly prevalent as well, all suggest a role for cattle as a *T. gondii* reservoir for human infection. In addition, Bobić et al. [9] have demonstrated that among all the meat consumed, undercooked beef presents the highest risk for human infection in Serbia. Similarly, although Opsteegh et al. [23] did not establish a correlation between seropositivity and the detection of parasites in cattle, a study in which the relative contribution of sheep, beef and pork products to human *T. gondii* infection in the Netherlands was quantified (by Quantitative Microbial Risk Assessment), showed that beef is indeed an important source even if the seroprevalence in cattle is low [24].

On the other hand, according to official statistical reports (RZS, 2006–2010), pork comprises approximately 50% of all meat consumed in Serbia. Thus, although pigs were the least infected of the examined species, and given the findings that the prevalence increases with age and reaches 41% in sows [10], pork consumption could significantly contribute to human infection. When used for cooking, pork is generally properly thermally processed, but in most of the SEE countries' traditions, mature pork is also highly valued for making delicatessen meat products. Raw or improperly cured sausages and ham are the source of small (family) epidemics of trichinellosis which, in spite of mandatory meat examination for *Trichinella spiralis*, occasionally occur in Serbia [25], and thus are a quite plausible source of human *T. gondii* infection too.

Horsemeat, on the other hand, is typically consumed rare or undercooked. In horses, which generally have lower seroprevalence values than small ruminants [26], a prevalence of only 1.7% has been determined in Greece [27]. However, a study which included 105 horses from all regions of Serbia slaughtered at two abattoirs between 2013 and 2015 showed a seropositivity of 48.6% at the 1:6 MAT cut-off, but of only 12.4% at the standard 1:25 cut-off. Importantly, viable parasites were isolated from two grade type mares; according to microsatellite genotyping, both isolates were type III, of which one was similar to a strain isolated from a duck in Iran, and the other one was identical in all markers to three strains isolated from a goat from Gabon, a sheep from France and a pig from Portugal. Interestingly, one of the source horses was seronegative, the other weakly positive. The isolation of viable *T. gondii* parasites from slaughter horses points to horsemeat as a potential source of human infection, but the fact that viable parasites were isolated from horses with only a serological trace of *T. gondii* infection presents further evidence that serology may not be adequate to assess the risk of toxoplasmosis from horsemeat consumption.

For most meat animals, there is no visible reduction in *T. gondii* prevalence, as opposed to the decreasing trend in humans (explained, among other reasons, by increased frozen meat use and better farm management). Farming practices and level of environmental contamination have not changed much, and except for the intensive pig farms in which a major reduction in *T. gondii* prevalence has occurred, a decline in *T. gondii* prevalence in meat animals is yet to be achieved. Moreover, for strictly herbivorous species that require outdoor access, this is probably impossible [28].

Ubiquitous contamination of the environment is also shown by the presence of *T. gondii* in both farm and urban rodents, pigeons and dogs [rev. In 13]. As long as there is such widespread environmental contamination in SEE, a decrease of *T. gondii* prevalence in meat animals may hardly be expected in the absence of energetic and systematic prevention measures throughout the region.

3.3. Food-borne is not only meat-borne: other food sources

In contrast to the availability of various serological and molecular tools which have allowed study of the *T. gondii* transmission meat route, it is precisely the lack of such tools that has left the oocyst stage much less studied, and until recently, it was virtually impossible to distinguish if a case of toxoplasmosis was caused by the tissue cyst or the oocyst. Several lines of evidence have linked oocyst infection to cases of acute toxoplasmosis, including hydric epidemics, case-control studies which showed consumption of fresh produce as an infection risk factor, and experimental studies showing the adherence of oocysts to fruits such as berries that produced infection in mice, (although the latter has yet to be confirmed in natural settings). A recently developed sporocyst-specific antibody test enabled clinical studies to be carried out, which showed that 43% of infections in pregnant women in Chile and 78% in mothers giving birth to congenitally infected children in the USA were caused by oocysts [29], but this has not been studied extensively, nor is such an antibody commercially available. Importantly, oocyst-induced infections appear to be clinically more serious than those caused by tissue cysts, and in fact, most often induce clinical infection.

Oocyst sources for human infection include soil and water, directly or indirectly through contamination of produce. Additional sources increasingly gaining importance include marine mammals, as well as filter-feeding invertebrates such as mussels and oysters, that do not get infected but can concentrate viable oocysts and serve as an infection reservoir for marine predators and humans. Continuous climate and man-made environmental changes favour an increase in oocyst-induced infections in both humans and animals, calling for the development of commercial technologies to detect oocysts in produce as well as for strategies for large-scale detection of oocysts in terrestrial and aquatic environments.

4. Conclusions

Strategies for the prevention of toxoplasmosis generally target only CT, and even their implementation, although clearly diminishing the incidence, has not eradicated CT. In addition, in the light of the new understanding of *T. gondii* infection, there may be additional risks calling for prevention of toxoplasmosis in other population categories. This clearly shows that new, comprehensive prevention strategies are needed, based on accurate and validated data on: (1) the routes and risk factors for human infection at the local level, which would allow for more efficient health education; (2) routes and risk factors for meat animal infection to diminish infection reservoirs, and; (3) environmental contamination. Such a complex task may best be accomplished within the One Health concept, with physicians, veterinarians, food technologists and ecologists acting in concert to prevent this important zoonosis.

Acknowledgement
This work has been supported by project grant III 41019 from the Ministry of Education, Science and Technological Development of Serbia.

References

[1] Montoya J G, Liesenfeld O Toxoplasmosis 2004 *Lancet.* **363** 1965
[2] Sutterland A L, Fond G, Kuin A, Koeter M W, Lutter R, van Gool T, Yolken R, Szoke A, Leboyer M and de Haan L Beyond the association. *Toxoplasma gondii* in schizophrenia, bipolar disorder, and addiction: systematic review and meta-analysis 2015 *Acta. Psychiatr. Scand.* **132** 161
[3] FAO/WHO *Multicriteria-based ranking for risk management of food-borne parasites. Microbiological Risk Assessment Series* 2014 www.fao.org
[4] Sibley L D, Khan A, Ajioka J W and Rosenthal B M 2009 Genetic diversity of *Toxoplasma gondii* in animals and humans *Philos. Trans. R. Soc. Lond. B. Biol. Sci.* **364** 2749
Su C et al. 2012 Globally diverse *Toxoplasma gondii* isolates comprise six major clades originating from a small number of distinct ancestral lineages *Proc. Natl. Acad. Sci. USA.* **109** 5844
[5] Elbez-Rubinstein A, Ajzenberg D, Darde M L, Cohen R, Dumètre A, Yera H, Gondon E, Janaud J C and Thulliez P 2009 Congenital toxoplasmosis and reinfection during pregnancy: case report, strain characterization, experimental model of reinfection *J. Infect. Dis.* **199** 280
[6] Sepúlveda-Arias J C, Gómez-Marin J E, Bobić B, Naranjo-Galvis C A and Djurković-Djaković O 2014 Toxoplasmosis as a travel risk *Travel. Med. Infect. Dis.* **12** 592
[7] Aspöck H and Pollak A 1992 Prevention of prenatal toxoplasmosis by serological screening of pregnant women in Austria*Scand. J. Infect. Dis.* **84** 32
Nowakowska D, Stray-Pedersen B, Śpiewak E, Sobala W, Małafiej E and Wilczyński J 2006. Prevalence and estimated incidence of *Toxoplasma* infection among pregnant women in Poland: a decreasing trend in the younger population *Clin. Microbiol. Infect.* **12** 913

Berger F, Goulet V, Le Strat Y and Desenclos J R 2009 Toxoplasmosis among pregnant women in France: risk factors and change of prevalence between 1995 and 2003 *Rev. Epidemiol. Sante. Publique.* **57** 241

[8] Bobić B, Nikolić A, Klun I and Djurković-Djaković O 2011 Kinetics of *Toxoplasma* infection in the Balkans *Wien. Klin. Wochenschr.* **123** 2

[9] Bobić B, Jevremović I, Marinković J, Šibalić D and Djurković-Djaković O 1998 Risk factors for *Toxoplasma* infection in a reproductive age female population in the area of Belgrade, Yugoslavia *Eur. J. Epidemiol.* **14** 605

Cook A J et al. 2000 Sources of *Toxoplasma* infection in pregnant women: a European multicentre case-control study *Br. Med. J.* **15** 142

Bobić B, Nikolić A, Klun I, Vujanić M and Djurković-Djaković O 2007 Undercooked meat consumption remains the major risk factor for *Toxoplasma* infection in Serbia *Parassitologia.* **49** 227

[10] Klun I, Djurković-Djaković O, Katić-Radivojević S and Nikolić A 2006 Cross-sectional survey on *Toxoplasma gondii* infection in cattle, sheep and pigs in Serbia: seroprevalence and risk factors *Vet. Parasitol.* **135** 121

[11] Giadinis N D, Terpsidis K, Diakou A, Siarkou V, Loukopoulos P, Osman R, Karatzias H and Papazahariadou M 2011 Massive toxoplasma abortions in a dairy sheep flock and therapeutic approach with different doses of sulfadimidine *Turk. J. Vet. Anim. Sci.* **35** 207

[12] Djokić V, Klun I, Musella V, Rinaldi L, Cringoli G, Sotiraki S and Djurković-Djaković O 2014 Spatial epidemiology of *Toxoplasma gondii* infection in goats in Serbia *Geospat. Health.* **8** 479

[13] Bobić B, Klun I, Nikolić A and Djurković-Djaković O 2012 *Toxoplasmosis – Recent Advances* (ed O Djurković-Djaković Rijeka, Croatia: Intech) pp 37-54

[14] Klun I, Vujanić M, Yera H, Nikolić A, Ivović V, Bobić B, Bradonjić S, Dupouy-Camet J and Djurković-Djaković O 2011 *Toxoplasma gondii* infection in slaughter pigs in Serbia: seroprevalence and demonstration of parasites in blood *Vet. Res.* **42** 17

[15] Damriyasa I M, Bauer C, Edelhofer R, Failing K, Lind P, Petersen E, Schares G, Tenter AM, Volmer R and Zahner H 2011 Cross-sectional survey in pig breeding farms in Hesse, Germany: seroprevalence and risk factors of infections with *Toxoplasma gondii*, *Sarcocystis spp.* and *Neospora caninum* in sows *Vet. Parasitol.* **126** 271

[16] Kijlstra A, Eissen O A, Cornelissen J, Munniksma K, Eijck I and Kortbeek T 2004 *Toxoplasma gondii* infection in animal-friendly pig production systems *Invest. Ophthalmol. Vis. Sci.* **45** 3165

[17] Ntafis V, Xylouri E, Diakou A, Sotirakoglou K, Kritikos I, Georgakilas E and Menegatos I 2007 Serological survey of antibodies against *Toxoplasma gondii* in organic sheep and goat farms in Greece *J. Hell. Vet. Med. Soc.* **58** 22

[18] Kontos V, Boutsini S, Haralabidis S, Diakou A, Athanasiou L, Magana O and Nomikou K 2001 Ovine Toxoplasmosis. An Epizootiological Research *Proc. 3rd Hellenic Symp. Farm Animals Vet. Med.* (Thessaloniki, Greece, November 9-11) p. 81

Diakou A, Papadopoulos E, Panousis N, Giadinis N and Karatzias C 2005 *Toxoplasma gondii* and *Neospora* spp. infection in sheep and goats mixed stock farming *Proc. 6th Int. Sheep Vet Congress* (Hersonissos, Greece, June 17-21) p. 170

[19] Dubey J P and Thulliez P 1993 Persistence of tissue cysts in edible tissues of cattle fed *Toxoplasma gondii* oocysts*Am. J. Vet. Res.* **54** 270

[20] Smith J L 1993 Documented outbreaks of toxoplasmosis: Transmission of *Toxoplasma gondii* to humans *J. Food. Protect.* **56** 630

[21] Aspinall T V, Marlee D, Hyde J E, Sims P F 2002 Prevalence of *Toxoplasma gondii* in commercial meat products as monitored by polymerase chain reaction – food for thought? *Int. J. Parasitol.* **32** 1193

[22] Tenter A M, Heckeroth A R and Weiss L M 2000 *Toxoplasma gondii*: from animals to humans.

Int. J. Parasitol. **30** 1217
[23] Opsteegh M, Teunis P, Züchner L, Koets A, Langelaar M and van der Giessen J 2011 Low predictive value of seroprevalence of *Toxoplasma gondii* in cattle for detection of parasite DNA*Intl. J. Parasitol.* **41** 343
[24] Opsteegh M, Prickaerts S, Frankena K and Evers E G 2011 A quantitative microbial risk assessment for meatborne *Toxoplasma gondii* infection in The Netherlands *Intl. J. Food. Microbiol.* **150** 103
[25] Djordjević M 1989 Swine trichinellosis in some enzootic areas in Serbia and comparison of some direct diagnostic methods *PhD Thesis* (University of Belgrade School of Veterinary Medicine)
Čuperlović K, Djordjević M and Pavlovic S 2005 Re-emergence of trichinellosis in southeastern Europe due to political and economic changes *Vet. Parasitol.* **132** 159
[26] van Knapen F, Franchimont J H and van der Lugt G 1982 Prevalence of antibodies to toxoplasma in farm animals in the Netherlands and its implication for meat inspection 1982 *Vet. Q.* **4** 101
[27] Kouam M K, Diakou A, Kanzoura V, Papadopoulos E, Gajadhar A A and Theodoropoulos G 2010 A seroepidemiological study of exposure to *Toxoplasma*, *Leishmania*, *Echinococcus* and *Trichinella* in equids in Greece and analysis of risk factors *Vet. Parasitol.* **170** 170
[28] Kijlstra A and Jongert E 2009 Toxoplasma-safe meat: close to reality? *Trends. Parasitol.* **25** 18
[29] Munoz-Zanzi C, Fry P, Lesina B and Hill D2010 *Toxoplasma gondii* oocyst-specific antibodies and source of infection *Emerg. Infect. Dis.* **16** 1591

IOP Conf. Series: Earth and Environmental Science **85** (2017) 012006 doi:10.1088/1755-1315/85/1/012006

Identification of species- and tissue-specific proteins using proteomic strategy

I M Chernukha[1], N L Vostrikova[1], L I Kovalev[2], S S Shishkin[2], M A Kovaleva[2], Y S Manukhin[1]

[1] The V.M. Gorbatov All-Russian Meat Research Institute, 26, Talalikhina str, Moscow, Russian Federation
[2] The Federal Research Centre "Fundamentals of Biotechnology" of the Russian Academy of Sciences, Leninsky prospect, 33, build. 2, Moscow, Russian Federation

E-mail: imcher@inbox.ru

Abstract. Proteomic technologies have proven to be very effective for detecting_biochemical changes in meat products, such as changes in tissue- and species-specific proteins. In the tissues of cattle, pig, horse and camel *M. longissimus dorsi* both tissue- and species specific proteins were detected using two dimensional electrophoresis. Species-specific isoforms of several muscle proteins were also identified. The identified and described proteins of cattle, pig, horse and camel skeletal muscles (including mass spectra of the tryptic peptides) were added to the national free access database "Muscle organ proteomics". This research has enabled the development of new highly sensitive technologies for meat product quality control against food fraud.

1.Introduction

The aim of proteomics is identification of all proteins, their biological activity, post-translation modifications and interactions in a cell, as well as identification of changes in the proteome as a response to changed biological conditions. The typical workflow in proteomics includes protein extraction and separation, protein and peptide identification and data analysis. The most common method used for detection of proteins or peptides in proteomics is mass-spectrometry. This strategy has multiple applications including for meat science research; however, it is limited by huge biochemical heterogeneity of proteins and inability to precisely detect low-abundance proteins.

In recent decades, the scientific community witnessed the rapid development and improvement of "-omic" methods with high throughput. The development of these methods has also changed the experimental approaches in food science [1,2].

In life sciences, including agriculture, food and animal sciences, the use of proteomics is a great step forward both for safe and high quality food production and improvement of animal husbandry and sustainability. The meat composition, sensory characteristics and nutritional value are important characteristicsto determine meat quality and consumer acceptability. Meat quality is closely linked with biological peculiarities of an animal. Obviously meat quality characteristics (like tenderness, water binding capacity, nutrient composition, autolytic changes etc.) are complex and multicomponent systems. Their detailed description would provide the next step toward understanding processes that cause changes in their characteristics and subsequent meat quality management [3,4,5].Proteomics is a prospective approach to studying mechanisms that are the basis for various meat quality traits.

At present, several methods based on identification of species-specific DNA (different variations of polymerase chain reaction) as well as the alternative method of enzyme-linked immunosorbent assay (ELISA) based on the specific reaction antigen-antibody are used in the laboratory analysis of meat and meat-based products. Despite all their advantages, these methods have several significant drawbacks. A large part of Brucellosis, one of the most widespread zoonosis, is a contagious chained disease affecting a great number of animals and, in a smaller proportion, people [10, 22]. In the animals, this disease is usually manifested as either chronic or latent infection. Causal agents of the infection are bacteria from Brucella genus which can have different virulence and host affinities.

The appearance of brucellosis in human population is closely related to its incidence in animal populations, and this correlation is the only way that the disease can be observed, studied and controlled. Humans are most commonly affected by consuming either meat or dairy such as milk or young cheese produced from uncooked goat or sheep milk. Transmission of brucellosis from animal to human occurs through the air or via skin wounds. People working in the higher risk professions, such as the farming or meat industry as well as the veterinary or lab professionals, show higher incidence rate [8, 18]. An estimate of a half of a million people per year seeking medical attention due to brucellosis has been given by World Health Organisation (WHO), although it is believed that the number of affected people is up to 25 times greater.

Brucellosis, although in many countries controlled or completely eradicated, remains great health and economic problem. This is especially true in the regions in which highly contagious *Brucella melitensis*, causing disease in sheep, goats, and humans.
studies in proteomics is performed using two-dimensional electrophoresis (2DE). The Brucellosis, one of the most widespread zoonosis, is a contagious chained disease affecting a great number of animals and, in a smaller proportion, people [10, 22]. In the animals, this disease is usually manifested as either chronic or latent infection. Causal agents of the infection are bacteria from Brucella genus which can have different virulence and host affinities.
The appearance of brucellosis in human population is closely related to its incidence in animal populations, and this correlation is the only way that the disease can be observed, studied and controlled. Humans are most commonly affected by consuming either meat or dairy such as milk or young cheese produced from uncooked goat or sheep milk. Transmission of brucellosis from animal to human occurs through the air or via skin wounds. People working in the higher risk professions, such as the farming or meat industry as well as the veterinary or lab professionals, show higher incidence rate [8, 18].

An estimate of a half of a million people per year seeking medical attention due to brucellosis has been given by World Health Organisation (WHO), although it is believed that the number of affected people is up to 25 times greater.

Brucellosis, although in many countries controlled or completely eradicated, remains great health and economic problem. This is especially true in the regions in which highly contagious *Brucella melitensis*, causing disease in sheep, goats, and humans. Combination of high-performance liquid chromatography (HPLC) and tandem mass-spectrometry is also a very promising approach to animal protein analysis.

The aim of this research was to identify potential biomarkers of meat (beef, pork, horse and camel meat) species-and tissue- specificity based on the proteomic profiles of their skeletal muscles.

2. Materials and methods

The following muscle types and animal specieswere used in the research:

cattle (*Bos taurus*) *m. Longissimus dorsi* (OOO KRROS, Moscow region);
pig (*Sus scrofa*) *m. Longissimus dorsi* (OOO Velcom, Moscow region);
horse (*Equus caballus*) *m. Longissimus dorsi* (MPZSafa, Moscow region);
camel (*Camelus bactrianus*) *m. Longissimus dorsi* (Almaty Technological University).

The 2DE by O'Farrell with isoelectrofocusing in ampholine (IEF-PAGE) or immobiline (IPG-PAGE) pH gradients was used as the main proteomic technology; the following protein detection was carried out by staining with Coomassie R-250.

For 2DE, 100 mg of a mixed musclewas homogenized in the Teflon-glass system in 2 ml of the lysing solution: 9 M urea, 5% mercaptoethanol, 2% triton X-100, 2% ampholines with pH 3.5-10 (IEF-PAGE). Homogenate was then clarified by centrifugation at 800 g for 5 min and the supernatant withsolubilized proteins (theextract) was used for separation. In the immobiline pH gradient (IPG-PAGE), the lysing solution with 9 M urea, 4% CHAPS, 2% ampholines with pH 3.5-10 and 0.6% dithiothreitol was used.

Protein identification was performed after tryptic proteolysis by MALDI-TOF MS and MS/MS mass-spectrometry using a MALDI-TOF mass-spectrometer Ultraflex (Bruker, Germany) with UV-laser (336 nm) in the positive ion mode and a mass range of 500-8000 Da with their calibration according to the known trypsin autolysis peaks.

The mass-spectra of tryptic peptideswere analyzed by theMascot software, Peptide Fingerprint option (Matrix Science, USA) with accuracy of MH+ mass detection of 0.01%,with the useof the database of the National Center for Biotechnology Information (NCBI).

3. Results and discussion

Up to 170 protein fractions were obtained and over 120 proteins were identified (table 1 and figure 1).

Table 1. Number of identified proteins extracted from cattle,pig, horse and camel *m.Longissimus dorsi.*

	Detected proteins	Identified proteins
cattle	115	51
pig	145	108
horse	130	61
camel	170	114

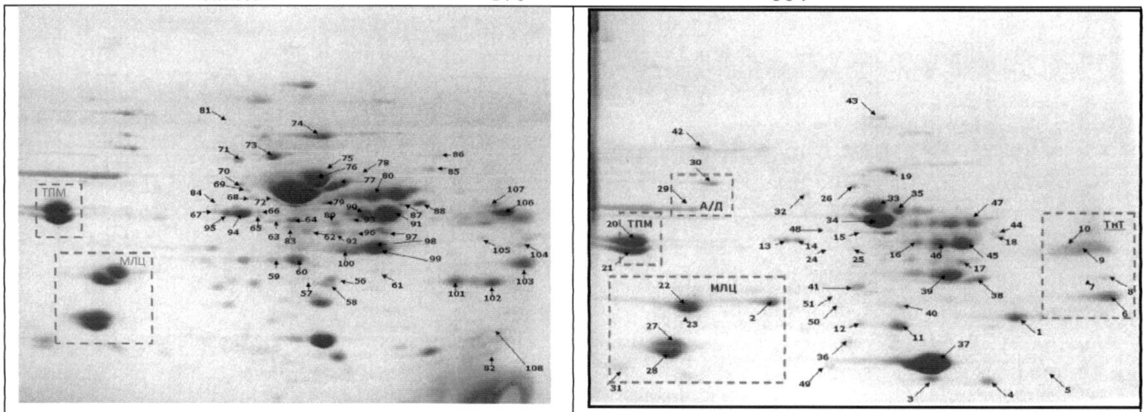

A. Pig *Sus scrofa*	B. Cattle *Bos taurus*

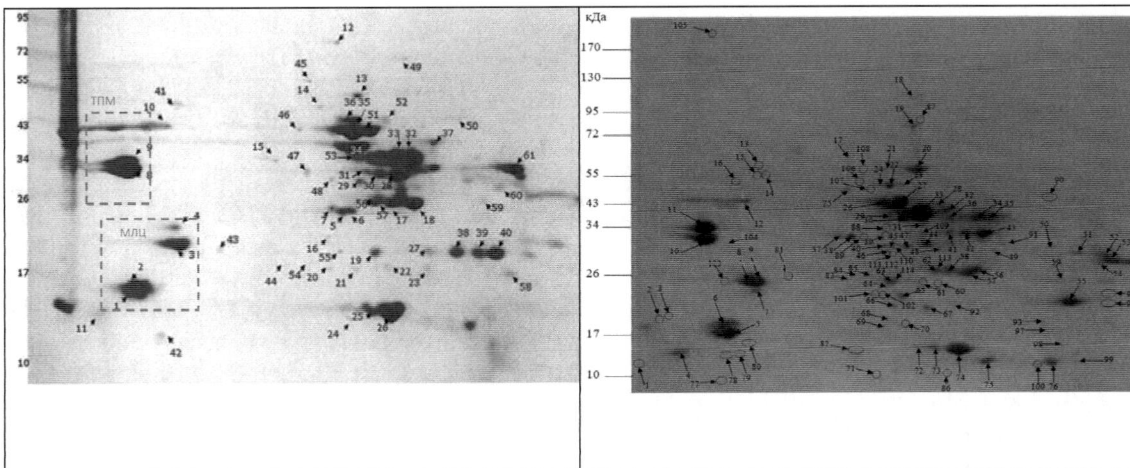

C. Horse *Equus caballus*	D. Camel *Camelus bactrianus*

Figure 1. Proteomic maps of *m. Longissimus dorsi* of pigs (A), cattle (B), horses (C) and camels (D). Staining with Coomassie R-250. Arrows and numbers (1-61; 1-55;1-51) indicate fractions identified by MALDI-TOF MS (see the information module "Skeletal muscle proteins" in the free-access database "Muscle organ proteomics",http://mp.inbi.ras.ru).ТПМ – tropomyosin; МЛЦ – myosin light chains; ТнТ – troponin T; А/Д – actin- desmin.

Among 108 totally identified protein fractions on the two-dimensional electrophoregrams of porcine skeletal muscles, 18 were isoforms of different troponin proteins known as strictly specific for the mammalian muscle tissues [6]. Among these proteins, eight were characterized as products of TNNT1 gene expression, seven as products of TNNT3 gene, and three as products of TNNI2 gene.

Information on proteomic studies of pig troponins is still very limited. In the pig (*Sus scrofa*) genome, eight genes encoding different troponinshave been found to date.

In the analyzed muscles, three protein products of genes encoding troponins I(No. 84, 94 and 95) were identified in addition to troponins T. Two of them (No. 101 and 102) corresponded to transcript 73853890 / Q4JH15 (gene TNNI2), and the third one to transcript 47522664 / B3VFA9 (gene TNNI1).

As a result,this researchestablished that there are three genes encoding troponins T (TNNT1, TNNT2, TNNT3), three genes for troponins I (TNNI1, TNNI2, TNNI3) and two genes for troponins C (TNNC1, TNNC2) in the pig genome.

Two protein fractions (No. 68 and 79) identified as protein products of the porcine TNNT3 gene corresponded to transcript 55741811 / Q75NG9, and no special information in the NCBI and UniProt databases was found. Taking into consideration the experimentally determined pI value, these fractions were named astroponin T fast skeletal muscle acidic (fTnt-a) and troponin T fast skeletal muscle neutral (fTnt-n).Ozgur Ogut et al., while examining chicken (*Gallus domesticus*) muscle, assumed that both acidic and basic TnT isoforms would modulate the Ca $^{2+}$ sensitivity of muscle contraction [7].

Proteins specific to muscle type were also found. For example, pig muscle contained pig Troponin I, fast skeletal muscle, (*TNNI2*) TNNI2 with Mm/pI 21.1/9.00 and pig Troponin I, slow skeletal muscle, isoform TnI-S4 (*TNNI1*) TNNI1 with Mm/pI 23.0/9.50 (positions 101and 102 in figure 1A,respectively).

Protein No 13 is of special interest to us. It wasidentified by Mascot program as a hypothetical protein containing a crystalline domain – a product of a gene from locus LOC494560 (protein sequence coverage 77%, score – 371, taking into account the MS/MS results). Therefore, this is first real evidence of a protein product that corresponds to the predicted hypothetical protein. We regard this significant resultas a certain contribution to the full annotation of the *Sus scrofa* genome.

Among the identified major bovine muscle proteins (figure 1B), nine (No. 6-10, 15, 16, 18 and 48) were the products of the TNNT3 gene. At the same time, the multiplicity of fast Troponin T isoformswas revealed, which was also noted by other authors [8,9].

This research into the horse meat proteome led, in general, to identification of 61 protein fractions (figure 1 C). New direct data was obtained regarding equine proteins, theexistence of which was earlier only predicted [10].Of particular interest among the identified horse meat proteins is protein Dj-1 (Figure 1 C No 54). This protein is known to protect human cells from oxidative stress and prevents the development of apoptosis, while mutations in the gene encoding Dj-1 are a cause of several forms of Parkinsonism (see Q99497 UniProt). The obtained data can be considered an indirect confirmation of the hypoallergenic properties of horse meat.

Although among the identified proteins were, for example, myosin light chains, which are important components of the main muscle engine – the actomyosin complex – direct studies of even major horse skeletal muscle proteins appears to be extremely limited. For instance in the PubMed database only one studywas found that reported the myosin light chain 3 detected in the biopsies of the horse muscle using proteomic technique [11].Thus,the results obtained in our study can be considered novel.

When studying the camel proteome (figure 1D), in 16 identified protein fractions, different types of post-translational modifications were found. Among them were acetylation of N-terminal amino acids [+Acetyl (Protein N-term)], phosphorylation ofserine and threonine residues [+Phospho (ST)] and others not discussed herein. Y. Zahedi et al. identified two full-length slow TnT and one fragmentoffast TnT in camel [12].As a result of our research, 18 camel mitochondrial proteins were identified. This result is of a significant interest, as direct data on these proteins are unavailable.

In addition to protein polymorphism analysis and other basicresearch in the field of muscular biochemistry, the proteomic strategies are challenging new applications in applied science, from protein biomarkers and pharmacological targets to developing food quality and composition control methods [13,14].

As a result of identification in the current study, typicalmammalian muscle tissue proteins were revealed. These proteins are not species-specific, for example, tropomyosins (see table 2) with regard to tropomyosin β-chain (TPM2) with the following sequence of amino acids:
0001MDAIKKKMQM LKLDKENAID RAEQAEADKK QAEDRCKQLE EEQQALQKKL
0051KGTEDEVEKY SESVKDAQEK LEQAEKKATD AEADVASLNR RIQLVEEELD
0101RAQERLATAL QKLEEAEKAA DESERGLKVI ENRAMKDEEK MELQEMQEMQ
0151LKEAKHIAED SDRKYEEVAR KLVILEGELE RSEERAEVAE SRARQLEEEL
0201RTMDQALKSL MASEEEYSTK EDKYEEEIKL LEEKLKEAET RAEFAERSVA
0251KLEKTIDDLE ETLASAKEEN VEIHQTLDQT LLELNNL

Table 2. Comparative data on identification of tissue-specific β-Tropomyosin (TPM2)of cattle, pig, camel and horse skeletal muscles.

Animal species	Number of amino acid residues	Mm	pI	Match, %
Sus scrofa	287	33.5	4.80	29.0
Bos Taurus	284	32.8	4.66	89.0
Equus caballus	284	32.8	4.66	64.0
Camelus bactrianus	284	33.0	4.66	85.0

By analyzing the proteomic profiles, we revealed several proteins that could be suitable as markers of mammalian muscle tissue. Among them are β-enolase, myosin light chain, troponin I, 5-

triosephosphate isomerase. A search for other protein biomarkers is being carried out. A necessary prerequisite for determination of these markers is protein thermal stability. It can be of scientific interest to study fast myosin behavior during thermal and other types of meat processing, since according to Guerrero, M. et al., fast myosin is an exceptional marker for skeletal muscles [15].The high sensitivity of the protein separation by 2DE makes it possible to identify theanimal species origin of these marker proteins (figure 2).

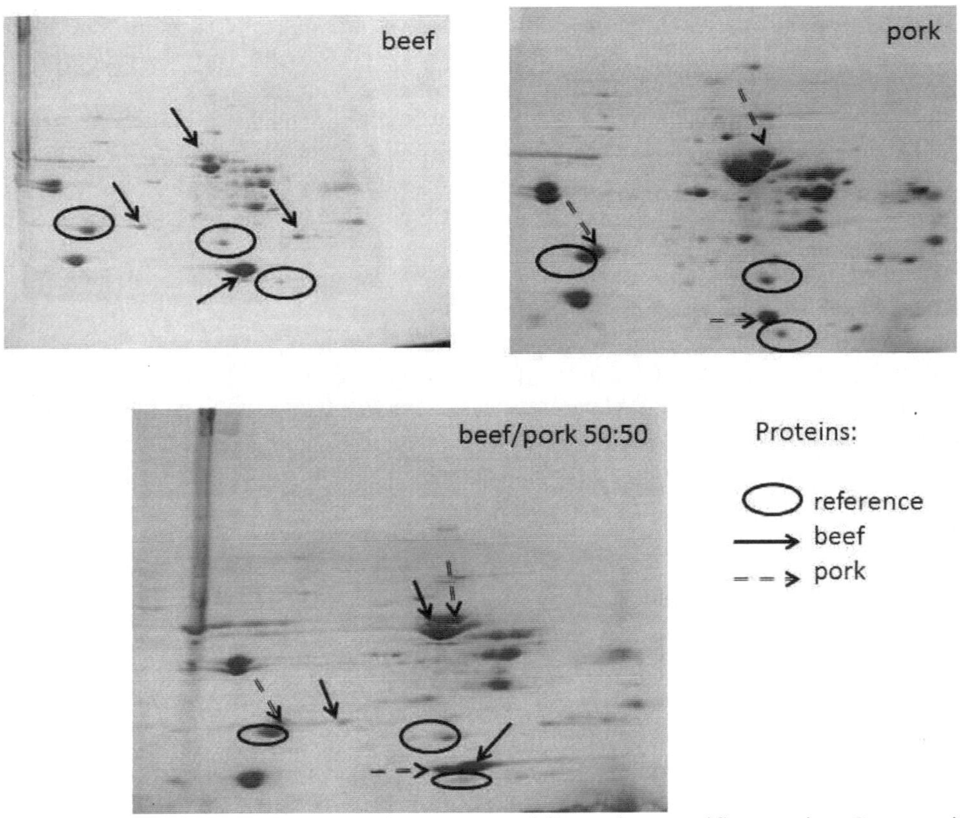

Figure 2. Two-dimensional electrophoregrams with species-specific proteins, Coomassie 250 stain.

Species-specific differences in the electrophoretic mobility of myoglobins as well as troponins, β-enolase isoforms and myosin light chain isoforms was evident when using 2DEto examine beef and pork muscle, as shown in figure 2.

The results of some species-specific muscle protein identification are presented in table 3.

Table 3. Skeletal muscle proteins identified by mass-spectrometry (MALDI-TOF MS and MS/MS) methods.

Protein(name in *NCBI*) / **gene symbol**	Species	Mм/pI
Troponin T fast skeletal muscle type (*TNNT3*)/ **fTnT3/17**	*Bos taurus*	32.0/8.60
	Suss scrofa	34.0/8.65
	Equus caballus	32.0/9.10
	Camelus bactrianus	31.5/10.20
Myosin light chain 1/3, skeletal	*Bos taurus*	21.5/5.10
	Sus scrofa	21.0/4.90

muscle isoform (MLC1)/**MYL1**	*Equus caballus*	20.0/5.15
	Camelus bactrianus	22.0/5.40

4. Conclusion

The results of the presented proteomic analysis have opened the way to the development of new highly sensitive technologies for meat product quality control, based on the analysis of species-specific isoforms of several muscle proteins.

Proteins that can be considered as markers of the presence of mammalian muscle tissue in meat products were identified. The information obtained will aid in quantifying the proportion of muscle tissue in meat products, including emulsified and heat-treated products.

Proteins of the contractile actomyosin complex (myosin light chains, tropomyosins, troponins) and enzymes participating in numerous interdependent carbohydrate metabolic pathways (glyceraldehyde 3-phosphate dehydrogenase, β-enolase) were chosen as the most informative proteins in terms of species specificity. In addition, α- and β-hemoglobins were added to the list as highly species-specific proteins for horse and camel meat.

The identified and described proteinsofcattle, pig, horse and camel skeletal muscles (including mass spectra of the tryptic peptides)were added to the national free access database “Muscle organ proteomics” (http://mp.inbi.ras.ru). Information modules: Proteins of porcine skeletal muscle (*Sus scrofa*), Proteins of bovine skeletal muscle (*Bos taurus*), Proteins of horse skeletal muscle (*Equus caballus*) and Proteins of camel skeletal muscles *Camelus bactrianus*) within the database were constructed.

Acknowledgement
This work was supported by the Russian Science Foundation (project No. 16-16-10073).

References

[1] Vostrikova N L and Chernukha I M 2017 *Theory and practice of meat processing* **2**pp 4-17
[2] D'Alessandro A and Zolla L 2013 *J. Proteomics* **78** 558-77
[3] Chernukha I M 2012 *Vsye o myase* **6** 32
[4] Picard B, Lebret B, Cassar-Malek I, Liaubet L, Berri C, Nihan-Duval le B, Hocquette J F and Renand G 2015 *Meat Sci.* **109** 18
[5] Zhang R et al. 2015 *Meat Sci.* **107** 26
[6] Schiaffino S and Reggiani C. 1996 *Physiol. Rev.***76** pp. 371–423
[7] Ozgur O, Granzier H and Jian-Ping Jin 1999 *American Journal of Physiology - Cell Physiology* **276**№ 5 pp1162-1170
[8] Hartner Karl-Thomas, KirschbaumBernhard J and PetteDirk *Eur. J. Biochem.*1989 **179** pp 32 -8
[9] Wells Scott M and Sleeper Meg *Journal of Veterinary Emergency and Critical Care* 2008 **18** pp pp. 235–245
[10] Lubec G, Afjehi-Sadat L, Yang JW and John J P 2005 *Prog Neurobiol.* **77** pp 90-127
[11] Bouwman FG, van Ginneken MM, van der Kolk JH, van Breda E and Mariman EC 2010 *Comp Biochem Physiol Part D Genomics Proteomics* **5** pp178
[12] Zahedi Y, Varidi Mohammad-Javad,Varidi Mehdi 2016 *Food Technol. Biotech.***54** pp 324–334
[13] Anderson N L et al. 2004 *Mol. Cell Proteomics.* **3** pp 311-326
[14] Shishkin S S, Kovaleva M A, Eryomina L S, Lisitskaya K V and Kovalev L I 2013 *Current Proteomics* **10** 165
[15] Guerrero M, Guiu-Comadevall M, Cadefau J A, Parra J, Balius R, Estruch A, Rodas G, Bedini J L, Cussó R 2008 *British Journal of Sports Medicine***42** pp581-584

Research to lessen the amounts of curing agents in processed meat through use of rock salt and carbon monoxide

R Sakata[1], **S Takeda**[1], **Y Kinoshita**[1, 2] **and M Waga**[1]

[1] Laboratory of Food Science, School of Veterinary Medicine, Azabu University, Sagamihara, Kanagawa 252-5201, Japan
[2] Daisen Ham Co., Ltd., Yonago, Tottori 683-0651, Japan

Abstract. This study was carried out to examine the reddening of meat products due to the addition of natural yellow salt (YS) and carbon monoxide (CO). Following YS or NaCl addition at 2% to pork subsequent to nitrite (0~100 ppm) treatment, color development due to this addition was analyzed visually. Heme pigment content in the meat was also determined spectrophotometrically. YS was found to bring about greater reddening than NaCl, indicating residual nitrite and nitrate content to be significantly higher in meat containing YS, through the amount of either was quite small. The amount of nitrite required for a red color to develop was noted to vary significantly from one meat product to another. CO treatment of pork caused the formation of carboxy myoglobin (COMb) with consequent reddening of the meat. COMb was shown to be heat-stable and form stably at pH 5.0 to ~8.0 and to be extractable with water, but was barely extractable at all with acetone. Nitric oxide was found to have greater affinity toward myoglobin (Mb) than CO. Nitrosyl Mb was noted to be stable in all meat products examined. CO was seen to be capable of controlling the extent of lipid oxidation.

1. Introduction

The purchase of meat products is initially based on color prior to any assessment of parameters such as odor, taste or texture. Meat color is the primary determinant of any decision to purchase a meat product. Color developing agents such as nitrite and nitrate have been found quite useful but food safety considerations limit the extent of their usage.

In recent years, attention has come to be increasingly directed to food safety in response to consumer demands in this regard. Accordingly, techniques for effectively enhancing the red color of meat products through the least possible use of nitrite and nitrate, but with usage of naturally occurring ingredients, have thus become modern day focal points of emphasis [1,2]. As substitutes for curing agents, natural mineral substances are being avidly examined for potential application, both in Japan and foreign countries. The findings should prove valuable for enhancing and maintaining good food appearance as well as being in the best interests of human health.

However, unexpected reddening and discoloration problems have recently been pointed out and effective solutions must be found. The reddening effect of Himalayan rock salt on meat products was previously reported [3]. In uncured meat products, carboxy myoglobin (COMb) can be considered a major reason for unexpected meat reddening. Carbon monoxide (CO) binds strongly to myoglobin (Mb) to form COMb, which shows a stable bright red color. The addition of CO to food is prohibited by law in Japan, though in foreign countries, this compound is applied as a gas in food packing, and its use in meat products is being carefully examined [4]. Should CO be capable of lowering nitrite content in meat, its use should not meet with any consumer objections.

Natural salts, particularly those in Himalayan rock salts, should prove favorable candidates for reddening of meat products, since they contain nitrite along with nitrate in only small amounts, plus minerals. Yellow salt (YS) from Himalayan products was, therefore, examined in this study for its capacity to bring about a red coloration in meat products and the results were compared with those for ordinary cooking salts.

Additionally, the characteristics of CO, such as its strong binding to Mb, leading to an attractive red meat color, were also investigated in this study. A comparison was made of parameters such as heat stability, extractability and formation of COMb at different pHs. Mb derivatives were then investigated spectrophotometrically subsequent to CO flushing into nitrosylmyoglobin (NOMb). The antioxidative effects of CO on meat were also examined.

2. Materials and methods

2.1. Natural salts experiment

Meat taken from pig leg was depleted as much as possible of its fat and connective tissue and then minced. YS or NaCl was added at 2% to each meat sample, along with 0.1% sodium ascorbate and $NaNO_2$ in the concentration range: 0, 10, 30, 50 or 100 ppm. After 4 days storage at 4°C under anaerobic conditions, the mince samples were cooked at 75°C for 30 min. After sample cooling, color was assessed visually and then with a spectral colorimeter.

The color forming ratio (CFR) [5], and heme pigment content were measured by acetone extraction. Residual nitrite and nitrate content were determined according to the method of Mirna and Schütz [6] and by copper-cadmium column reduction, respectively.

2.2. CO experiment

COMb was prepared in a model solution (0.1% Mb-0.2% $Na_2SO_2O_4$, pH 5.5 with CO gas flushing) and cooked at 70°C for 20 min. The absorption spectra of COMb at pHs adjusted to 5.0-9.0 were monitored.

CO-treated meat and cured meat samples were prepared with pig thigh flushed with CO, to which 100ppm $NaNO_2$ had been added, and the meats then kept at 2°C for 4 days. COMb and NOMb were extracted with water or 75% acetone and the absorption spectra of the solutions thus obtained were used to determine the extent of COMb formation. Lipid oxidation in each meat sample was also evaluated by 2-thiobarbituric acid (TBA) analysis [7].

Absorption spectra of NOMb solution (0.1% Mb-1% Na ascorbate-0.1% $NaNO_2$, pH 5.5) previously flushed with CO was measured to determine the extent of COMb formation.

3. Results and Discussion

3.1. Effects of yellow rock salt and low levels of nitrite and nitrate on meat color

In mince with added NaCl and without nitrite, the CFR and a* (redness) were less than those measured in YS mince samples. In YS mince samples with and without nitrite, meat redness and CFR were found to exceed 70%, regardless of the low nitrite range, from 0 to 100 ppm (figure 1). Compared to NaCl mince samples, residual nitrite content was usually higher in mince with added YS (figure 2). The YS itself contained more nitrite than nitrate. YS was, thus, shown to effectively enhance meat reddening. Small amounts of nitrite, nitrate and minerals in the YS could possibly have contributed to this finding.

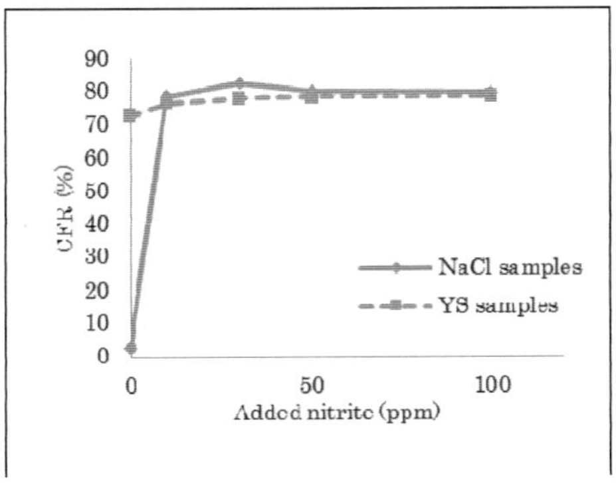

Figure 1. Color forming ratio (CFR).

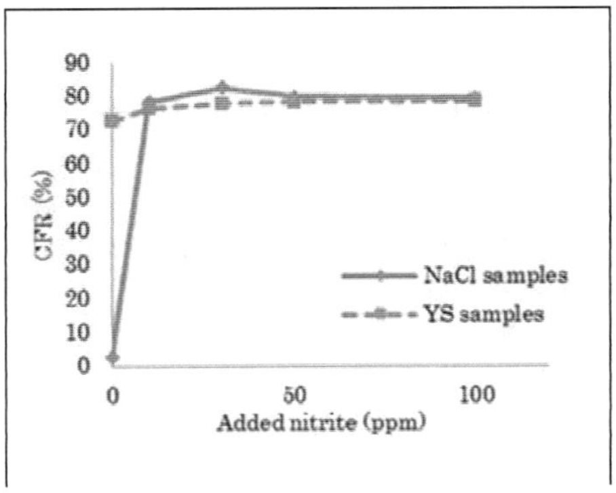

Figure 2 Residual nitrite levels

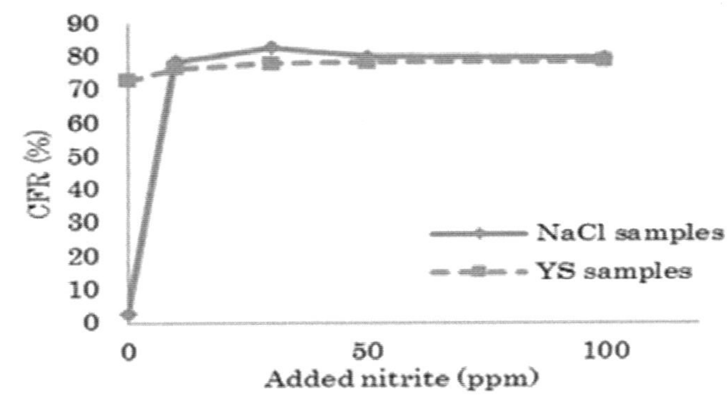

Figure 3 Colour forming ratio as a function of heme pigment content

3.2. Effects of yellow rock salt and low levels of nitrite and nitrate on meat color

In mince with added NaCl and without nitrite, the CFR and a* (redness) were less than those measured in YS mince samples. In YS mince samples with and without nitrite, meat redness and CFR were found to exceed 70%, regardless of the low nitrite range, from 0 to 100 ppm (figure 1). Compared to NaCl mince samples, residual nitrite content was usually higher in mince with added YS (figure 2). The YS itself contained more nitrite than nitrate. YS was, thus, shown to effectively enhance meat reddening. Small amounts of nitrite, nitrate and minerals in the YS could possibly have contributed to this finding.

The CFR was noted to change with increase in heme pigment content in pork mince with added YS. When the content of heme was quite small, the CFR in the mince samples was larger than in mince with more heme (figure 3). This could possibly have been due to the limited capacity of heme nitrosation to bring about meat reddening.

3.3. Effects of CO gas on meat and NOMb

The meat sample extracts showed the specific absorption spectra of COMb. In a model solution, COMb was heat stable and stable throughout the given cooling times (figure 4), with pH maintained within a range of 5.0 to ~8.0. Meat pH is normally approximately 5.5, so this indicates the red color of meat should be maintained in the presence of CO.

The extraction of COMb was possible with water, though virtually impossible with acetone (figure 5), in contrast to NOMb (data not shown). Thus, the extent of COMb formation compared to total Mb could not be determined in the present study.

Even when CO was added to NOMb solution, the absorption spectrum of NOMb showed no change (data not shown), thus indicating the higher affinity of NO than CO toward Mb. The addition of $NaNO_2$ (100 ppm) to COMb resulted in a change in the absorption spectrum of the NOMb formed (data not shown).

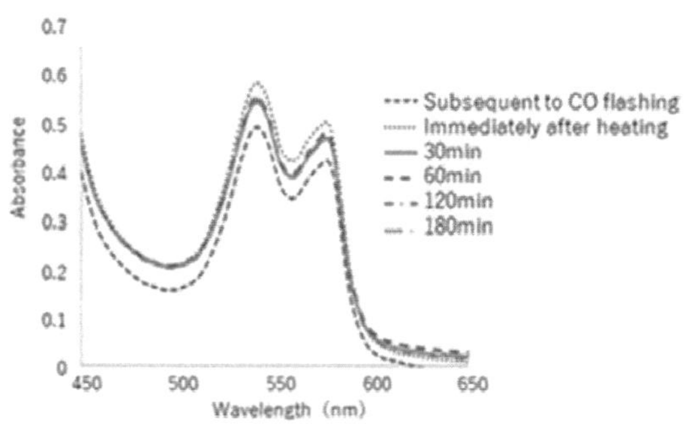

Figure 4. Absorption spectra of COMb solution after heating and cooling for up to 180 mins.

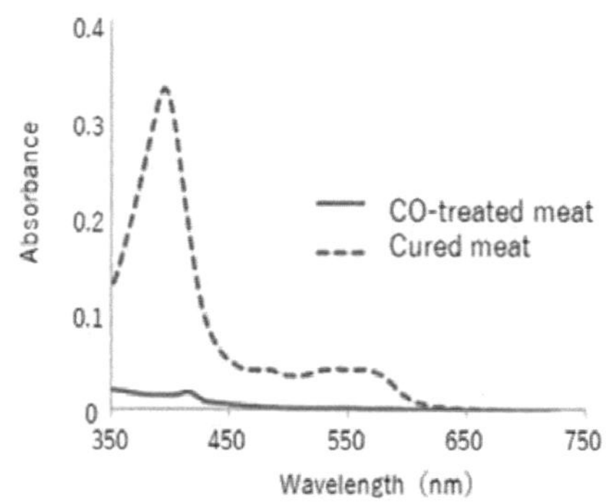

Figure 5. Absorption spectra of CO-treate and cured meat after extraction with.

TBA values of raw meat increased significantly during storage, but TBA values for CO-treated pork decreased to nearly the same as those observed in the NO-group, compared to the control, from day 7 to day 21 of storage (figure 5). Therefore, CO was shown to be as capable as nitrite of controlling lipid oxidation in raw pork.

CO proved inhibitory to microorganism growth, as was also noted for nitrite (data not shown).

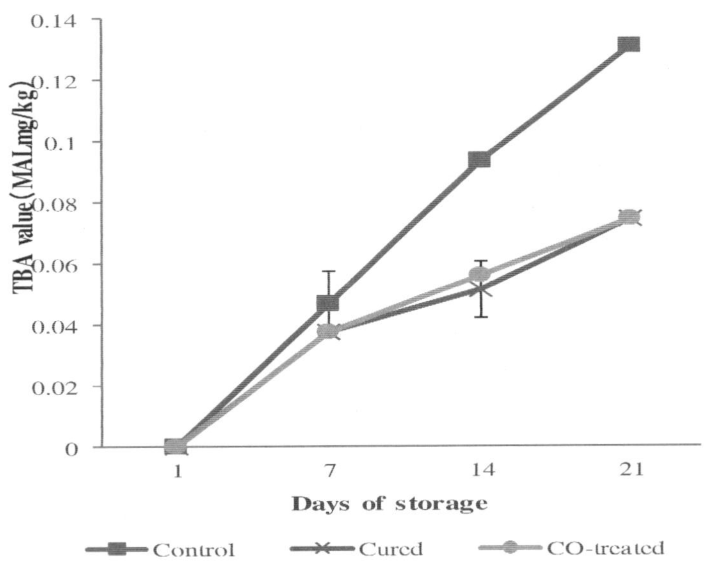

Figure 6. Change in pork TBA value during storage.

4. Conclusion

Himalayan rock salt was found to be effective for enhancing the red color of meat at very small nitrite content or even in the absence of nitrite. The CFR in cooked meat changed with heme pigment content in meat, owing to the low degree of nitrosation in the meat.

COMb was noted to form rapidly in meat when Mb and CO were combined, giving rise to a bright red color in the meat. It follows then, that CO could serve to significantly lessen the amount of nitrite added to meat products, while still supporting formation of an acceptable red color. CO could also function in a very similar fashion as nitrite in meat. CO itself could possibly be a factor for the unexpected reddening of meat as noted at food service facilities, but confirmation of this point will require additional research.

Acknowledgements

The authors are thankful to Dr. M Motoyama (NARO, Japan), Mr. Y Miki and Miss M Kaneko (the latter two are graduates of the Azabu University, Japan) for their assistance in conducting the CO treatment.

References

[1] Mancini R A and Hunt M C 2005 *Meat Sci.* **1** 100-121
[2] Sakata R 2010 *Meat Sci.* **86** 243-248
[3] Kaneko M and Sakata R 2014 *Proc. 60th Inter. Cong. Meat Sci. Technol., Punta del Uste, Uruguay,* CD 068
[4] Suman S P, Hunt M C, Nair M N and Rentfrow G 2014 *Meat Sci.* **98** 490-504
[5] Sakata R, Ohso M and Nagata Y 1981 *Agric. Biol. Chem.* **45** 2077-2081
[6] Mirna A and Schütz G 1972 *Fleischwirt.* **52** 337-1338
[7] Yamauchi K, Murata H, Ohashi T, Katayama H, Pearson A L, Okada T and Yamakura T 1991 *Nippon Shokuhin Kogyo Gakkaishi* **38** 545-552

Can long chain *n*-3 fatty acids from feed be converted into very long chain *n*-3 fatty acids in fillets from farmed rainbow trout (*Oncorhynchus mykiss*)?

M Lušnic Polak, L Demšar, U Luzar and T Polak

Biotechnical Faculty, University of Ljubljana, Jamnikarjeva 101, Ljubljana, Slovenia

E-mail: tomaz.polak@bf.uni-lj.si

Abstract. The link between the basic chemical and fatty acid composition of trout feed on one hand and trout (*Oncorhynchus mykiss*) meat (fillet) was investigated.. The content of 52 fatty acids from feed and trout meat lipids was determined by *in-situ* transesterification and capillary column gas-liquid chromatography. On average, 100 g of trout feed contained 7.4 g of moisture, 47.7 g of proteins, 6.09 g of ash, 21.4 g of fat, and as for fatty acid composition, 47.8 wt. % were monounsaturated, 34.0 wt. % were polyunsaturated and 18.1 wt. % were saturated fatty acids, with the PS ratio 1.88, *n*-6/*n*-3 ratio 1.74, 0.80 wt. % of *trans* and 3.28 wt. % of very long chain *n*-3 fatty acids. On average, 100 g of trout meat contained 76.1 g of moisture, 21.4 g of proteins, 1.34 g of ash, 2.52 g of fat, and in the fatty acid composition 42.1 wt. % were monounsaturated, 38.2 wt. % were polyunsaturated and 18.9 wt. % were saturated fatty acids, with the PS ratio 2.02, *n*-6/*n*-3 ratio 0.98, 0.95 wt. % of *trans* and 13.25 wt. % of very long chain *n*-3 fatty acids.

1. Introduction

Fish farming has been developed into a highly productive and efficient industry for the production of animal protein for human consumption. However, in the modern way of life, care for human health is very important. It is accepted that fish consumption has nutritional and health benefits in humans [1,2]. Fish meat is considered to be a good source of proteins of high biological value, polyunsaturated *n*-3 fatty acids (*n*-3 PUFA), minerals and vitamins. Preventive effects of *n*-3 PUFA on atherosclerosis, thrombosis or hypertension have been reported in numerous studies [3,4]. Additionally, it has been suggested that *n*-3 PUFA may have a favorable influence on diseases, such as asthma, stroke, cancer or diabetes [5].

Besides the fat content, their fatty acid (FA) composition, and above all the proportion and ratio of *n*-3 to *n*-6 FAs, as well as between saturated and unsaturated FA, the most important fish quality parameters are microbiological safety, color, texture and content of essential minerals; all of them contribute to high nutritional value of fish meat and their positive nutritional effects [6]. However, the composition of fish meat is highly variable; some factors of variability are the age, size or part of the fish [7], the sex, the season of the year [8], the quality of water [9], the diet [10,11] and the feeding system [10,12].

Research found that the histological changes observed suggest an effect of dietary lipid sources on the transport and metabolism of fat in the fish, but further studies are required to clarify this [10].

Consequently, the aim of this preliminary study was to investigate the FAs and basic chemical composition of trout feed on one hand and farmed trout meat on the other, suggest a possible link between FA profiles of feed and meat, as well to obtain some information about nutrient content in the muscle of fillets from farmed trout in Slovenia.

2. Materials and methods

Freshwater-reared rainbow trout (n = 10) were randomly selected from stocks of ready-for-sale animals obtained from Slovenian commercial farms, producing for the domestic market. At the same time, feed, with which fish were farmed, was collected. The sampling was carried out between March and April, 2017.

Fish were slaughtered in water and ice, packed in polystyrene boxes, and covered with ice. Boxes with fish were immediately transported to the laboratory where fish were weighed and processed. The peritoneal cavity was opened along a ventral midline incision; the entire visceral mass was discarded. An incision along the dorsal fin up to the caudal fin, and another incision behind the opercula, excluding lateral and ventral fins, were made to separate both fillets from each carcass. Each fillet was cut along the insertion line of the ribs to obtain a dorsal and a ventral fillet (figure 1). After skinning, separately the two dorsal and two ventral fillets from each fish were joined. On the dorsal fillets, FA composition and on the ventral fillets, basic chemical composition was analyzed. Ten representative fish (dorsal and ventral fillets) and ten associated fish feed samples were prepared for analysis in accordance with ISO standard method [13]. The samples were homogenized by using a homogenizer Grindomix GM 200 (Retch, Germany) at 5000-6000 rpm for 20 s and stored vacuum packed in plastic bags at -18 °C until further analysis.

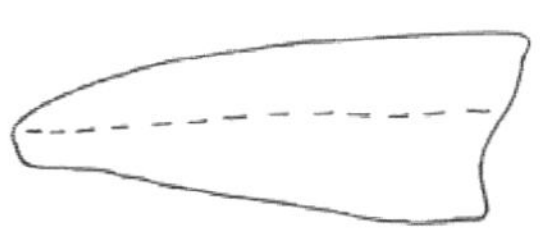

Figure 1. Insertion line of ribs rainbow trout [7].

2.1 Chemical composition

The moisture content of feed and fillet was determined on 5 g of previously homogenized samples dried in an oven at 105 °C (according to Association of Official Analytical Chemists (AOAC) 950.46) [14]. The total protein content (crude protein, N × 6.25) was determined by using the Kjeldahl method (according to AOAC 928.08) [15], and the ash content was determined by mineralization of the samples at 550 °C (according to AOAC 920.153) [16]. The fat content in feed and fillet was determined by the method described as AOAC Official Method 991.36 Fat (crude) in Meat and Meat Products [17]. Lipids were extracted with petroleum ether (boiling point ranged from 40 to 60 °C) after hydrolysis of the sample with hydrochloric acid. Data from the basic chemical analyses were expressed on a wet matter basis.

2.2 Fatty acid analysis

The FA composition of feed and fillet was determined by gas chromatography, where *in situ* transesterification [18] was used, modified by Polak et al. [19]. The FAMEs were determined by capillary gas chromatography on GC Agilent Technologies 6890 with a flame ionization detector and HP-88 capillary column (100 m × 0.25 mm × 0.20μm, Agilent Technologies). Separation and detection were performed under the following temperature conditions: 150° C, hold 10 min, rate 1.5 °C min^{-1} to 180° C, hold 40 min, 3° C min^{-1} to final temperature of 240° C. Total analysis time was 95 min. The injector and detector temperatures were 250° C and 280° C, respectively. The carrier gas was helium at a flow rate of 2.3 mL min^{-1}. Injected volume was 1 μL and injector split ratio was 1:30.

Nitrogen was used as the make-up gas at a flow rate of 45 mL min^{-1}; detector gases were hydrogen and synthetic air (21% O_2) at a rate of 40 mL min^{-1} and 450 mL min^{-1}, respectively.

The FAMEs were determined through their retention times in comparison to the relevant standard mixtures using: 37 Components FAME mix (Supelco, Bellefonte, USA); PUFA No. 1-animal source (Supelco, Bellefonte, USA); linoleic acid methyl ester *cis/trans* isomer Mix (Supelco, Bellefonte, USA); *cis*-7-octadecenoic methyl ester (Supelco, Bellefonte, USA) and *cis*-11-octadecenoic methyl ester (Supelco, Bellefonte, USA); methyl stearidonate (Fluka, Switzerland); Nu-Chek standards GLC-68D, GLC-85, GLC-411g (Nu-Chek, Minnesota, USA). The GLC-68D and GLC-85 standard mixtures were used to determine the response factor for each FA. The weight of each FA in the feed and fillets was determined using the response factor and the transformation factor of the FA content from the FAME content. The samples of feed and fillets were analyzed in duplicate. The FAMEs were expressed as weight percentages of the total FA content.

2.3 Data analysis

The data were analyzed for normal distributions using the UNIVARIATE procedure (SAS/STAT, USA). The differences according to the samples were analyzed through a general linear model procedure and Duncan test (SAS/STAT), with a 0.05 level of significance.

3. Results and discussion

Basic chemical and FA compositions of trout feed are presented in tables 1 and 2, respectively.

On average, 100 g of wet weight of trout feed contained 7.4±2.1 g of moisture, 47.7±5.0 g of proteins, 6.09±1.37 g of ash, 21.4±5.72 g of fat, and as for FA composition, 47.8±3.2 wt. % of total FAs were monounsaturated (MUFA), 34.0±2.4 wt. % were polyunsaturated (PUFA) and 18.1±1.1 wt. % were saturated fatty acid (SFA), with the PS ratio 1.88±0.12, *n*-6/*n*-3 ratio 1.74±0.34, 0.80±0.19 wt. % of *trans* FA and 3.28±0.94 wt. % of very long chain *n*-3 PUFA (data are not presented in tables).

The chemical composition of the ten trout feeds significantly differed ($P < 0{,}001$). The protein content in trout feed ranged from 39.9 to 55.0 g 100 g^{-1}, fat content from 12.9 to 29.1 g 100 g^{-1}, moisture content from 3.7 to 9.8 g 100 g^{-1} and ash content from 5.0 to 8.9 g 100 g^{-1}. These amounts for chemical parameters found are in full agreement with data reported by Rasmussen et al. [20], who stated following data for rainbow trout feed: content of protein 47.0 g 100 g^{-1}, fat 26.0 g 100 g^{-1}, dry matter 94.3 g 100 g^{-1}, and ash 7.5 g 100 g^{-1}.

In all, 53 FAs were detected (> 0.01 g 100 g^{-1} FA) in the trout feeds; only thirteen of them are presented in table 2. For almost all FAs, their proportions and calculated nutrition information significantly differed between the different trout feeds ($P \leq 0.001$) except for C22:4*n*-6 ($P = 0.355$). The primary FA was oleic acid (C18:1*cis*-9), the content of which was 37.27±1.76 g 100 g^{-1} FA, followed by linoleic (C18:2*cc n*-6; 20.01±2.42 g 100 g^{-1} FA), palmitic (C16:0; 11.07±0.61 g 100 g^{-1} FA), α-linolenic (C18:3*n*-3; 5.12±0.44 g 100 g^{-1} FA), and FAs in amount under 4 g 100 g^{-1} FA, such as stearic (C18:0; 3.43±0.58 g 100 g^{-1} FA), eicosapentaenoic (C20:5*n*-3; 2.90±0.58 g 100 g^{-1} FA), vaccenic (C18:1*cis*-11; 2.77±0.08 g 100 g^{-1} FA), docosahexaenoic (C22:6*n*-3; 2.61±0.81 g 100 g^{-1} FA), palmitoleic (C16:1; 2.43±0.30 g 100 g^{-1} FA), gondoic (C20:1*cis*-11; 2.12±0.96 g 100 g^{-1} FA), and myristic acid (C14:0; 1.97±0.36 g 100 g^{-1} FA).

Basic chemical and FA compositions of farmed rainbow trout fillets are presented in tables 1 and 3, respectively.

On average, 100 g of wet weight of trout fillet contained 76.1±1.2 g of moisture, 21.4±0.9 g of proteins, 1.34±0.08 g of ash, 2.52±1.24 g of fat, and in FA composition, 42.1±5.1 wt. % of total FA were MUFA, 38.2±4.6 wt. % were PUFA, and 18.9±1.6 wt. % were SFA, with the PS ratio 2.01±0.27, *n*-6/*n*-3 ratio 0.98±0.28, 0.95±0.19 wt. % of *trans* FA and 13.25±4.72 wt. % of very long chain *n*-3 PUFA (data are not presented in tables).

The basic chemical composition of farmed rainbow fillets trout significantly differed between samples ($P < 0.01$), except for protein content ($P = 0.071$). The protein content in trout fillets ranged from 20.3 to 23.6 g 100 g^{-1}, the fat content from 0.8 to 5.0 g 100 g^{-1}, moisture content from 74.4 to

78.1 g 100 g^{-1} and ash content from 1.24 g 100 g^{-1} to 1.54 g 100 g^{-1}. These amounts for chemical parameters found are not in full agreement with data reported elsewhere. Protein contents were reported to be at the lower limit of those in our study (in the range of 19.3 to 20.3 g 100 g^{-1}), while the fat contents reported were at the upper limit compared to our study (in the range of 4.0 to 6.7 g 100 g^{-1}) for dorsal and ventral fish fillets or fillets from fish reared in standing moisture [7,20].

In all, 52 FAs were detected (> 0.01 g 100 g^{-1} FA) in the fillets from farmed rainbow trout; twelve of them are presented in table 3. For almost all fatty acids, their proportions and calculated nutrition information significantly differed between different trout fillets ($P \leq 0.05$) except for C13:0, C17:1*cis*-10 and C18:1*trans*-11 ($P > 0.05$). The primary FA was oleic acid (C18:1*cis*-9), content of which ranged from 24.33 to 36.68 g 100 g^{-1} FA, followed by linoleic (C18:2*cc n*-6; from 14.57 to 19.99 g 100 g^{-1} FA), palmitic (C16:0; from 11.07 to 15.56 g 100 g^{-1} FA), and docosahexaenoic acid (C22:6*n*-3; from 7.04 to 21.61 g 100 g^{-1} FA). Our results are in line with those reported in literature [2,10,21] for the relative concentrations of individual FAs in the lipid fraction. In contrast, other researchers found that C22:6*n*-3 followed by C16:0 were the predominant FAs in the dorsal muscle of cultivated rainbow trout [22]. This is probably due to the use of different lipid sources in the diet because the FA composition of the muscular tissue in fish reflects, to a large extent, that of the diet [23].

The lower part of (table 3) gives the indicators related to human health for the trout fillets. The nutritional quality of fat has been evaluated in terms of the ratio of PUFA:SFA (PS), the atherogenicity index (AI) [24], and the ratio of *n*-6/*n*-3 FA. In a balanced diet, the recommended ratio for PS is 0.4 or higher [25,26], for AI as low as possible, and for ratio *n*-6:*n*-3 less than 4 [27]. In this regard, our results showed that differences ($P \leq 0.001$) in the PS ratio occurred between the different rainbow trout fillets, but the PS ratios ranged from 1.69 to 2.46, which is within the recommended range. Also determined AI values (0.20 to 0.28), which is considered as the rightful estimation for lipid nutritional quality, is also quite comparable with rabbit meat (0.70), deer meat (0.40 to 0.72), beef (0.51), lamb (1.07) and chicken (0.42) [28,29,30,31,32]. An unfavorable *n*-6/*n*-3 ratio of the PUFAs is considered to be a risk factor for cancer and coronary heart disease, so it is recommended that this *n*-6/*n*-3 ratio is < 4.0 [27]. In the present study, an average *n*-6/*n*-3 ratio of 0.98 was achieved, while in the literature that value was 0.62 for cultured rainbow trout by Stancheva et al. [33].

FA chains differ in length, often categorized from short to very long. Figure 2 shows the percentages of medium- (with aliphatic tails of 6 to 12 carbons), long- (with aliphatic tails of 13 to 21 carbons) and very long-chain (with aliphatic tails of 22 or more carbons) FA in feed (diet) source and in trout fillet lipids. On average, a greater percentage of very long chain FAs (i.e. C22:5*n*-3, C22:6*n*-3) in trout fillet in comparison with feed was detected, reflecting the lower percentage of some long chain FAs (i.e. C18:3*n*-3) in fillet compared to feed (figure 3). Furthermore, the *n*-3/*n*-6 ratios and percentages of very long chain *n*-3 PUFA revealed that these values were significantly higher ($P \leq 0.001$) in the trout fillets than in the feeds (2.62 *vs.* 1.96; 16.8 *vs.* 7.4). According to Aslan et al. [34], FA composition of fish from aquaculture does not always depend on that of feed because of the fish metabolism. However, our data on the percentage of C18:3*n*-3, C20:5*n*-3 and C22:6*n*-3 observed in feed and fillets suggest an effect of feed (diet) source on metabolism of fat in trout fillet (Figure 3). It can be concluded that long chain *n*-3 PUFA from feed can be converted into very long chain *n*-3 PUFA in farmed rainbow trout fillets. This fact was also seen by Rebolé et al. [2], who showed that the level of C18:3*n*-6 was lower, whereas the level of C22:6*n*-3 was higher in the muscle than in the feed. This fact seems to support the documented effectiveness of rainbow trout and other freshwater fish species in elongating and desaturating precursor shorter-chain PUFAs to longer derived homologs [10,35].

There is scientific evidence that *trans* FA intake is associated with cardiovascular diseases in different ways [36]. Therefore, the recommendation for introducing *trans* FA in human body is limited to 1% of energy [37,38]. Naturally-occurring *trans* FAs produced in the gut of some animals and foods made from these animals (e.g., milk and meat products) can contain small quantities of these fats; low contents of *trans* FAs were also detected in our study (feed: 0.80±0.19 g 100 g^{-1} FA; fillet: 0.95±0.19 g 100 g^{-1} FA; data are not presented in tables). Twelve *trans* FA were detected in the feed and farmed rainbow trout fillets, but just eight of them were present in amounts under the limit of detection (> 0.01 g 100 g^{-1} FA) and taken into account in the calculation.

Table 1. Basic chemical composition of feed and farmed rainbow trout fillets.

Parameter (g 100 g^{-1})	Sample ^{x} S1	S1	S3	S4	S5	S6	S7	S8	S9	S10	SE ^{y}	*P* value ^{z}
Feed												
Protein	$54.99^{a\,u}$	47.85^{de}	39.86^{h}	49.59^{dc}	47.99^{de}	51.57^{bc}	53.93^{ba}	41.51^{hg}	43.87^{fg}	45.87^{fe}	1.37	≤ 0.001
Fat	22.74^{c}	28.66^{a}	22.58^{dc}	29.07^{a}	12.88^{f}	25.28^{b}	23.39^{c}	21.20^{d}	13.86^{fe}	14.64^{e}	0.18	≤ 0.001
Moisture	4.97^{e}	5.32^{ced}	5.17^{ed}	5.71^{cb}	8.53^{a}	5.29^{ed}	5.54^{cbd}	8.86^{a}	5.72^{cb}	5.79^{b}	0.63	≤ 0.001
Ash	8.39^{d}	9.26^{b}	8.62^{cd}	6.70^{e}	3.86^{f}	8.94^{cb}	8.22^{d}	3.67^{f}	9.81^{a}	6.96^{e}	0.22	≤ 0.001
Trout fillet												
Protein	20.83^{bc}	20.47^{c}	21.06^{bac}	21.99^{bac}	21.42^{bac}	21.85^{bac}	22.56^{a}	22.49^{ba}	20.68^{c}	20.43^{c}	0.70	0.071
Fat	4.39^{b}	3.24^{c}	0.87^{f}	1.68^{e}	4.75^{a}	1.36^{e}	2.61^{d}	2.29^{d}	1.63^{e}	2.41^{d}	0.15	≤ 0.001
Moisture	75.23^{cb}	75.11^{cb}	77.91^{a}	77.07^{a}	74.60^{c}	77.40^{a}	74.94^{cb}	75.79^{b}	75.84^{b}	77.31^{a}	0.45	≤ 0.001
Ash	1.29^{bc}	1.27^{c}	1.27^{c}	1.38^{ba}	1.41^{a}	1.39^{ba}	1.38^{ba}	1.46^{a}	1.27^{c}	1.24^{c}	0.04	0.004

^{x} S – rainbow trout (*Oncorhynchus mykiss*)
^{y} Standard error of mean.
^{z} Statistical probability of sample effect.
^{u} Means with a different superscript within rows ($^{a-i}$) differ significantly.

Table 2. Fatty acid composition (selected fatty acids) and calculated nutritional information of fish feed.

Fatty acid (FA) (g 100 g^{-1} total FA)	Feed sample ^{x} S1	S2	S3	S4	S5	S6	S7	S8	S9	S10	SE ^{y}	*P* value ^{z}
C14:0	1.22^{h}	1.97^{d}	1.90^{e}	$2.55^{a\,u}$	2.01^{c}	1.98^{d}	1.84^{f}	2.01^{c}	1.75^{g}	2.50^{b}	0.01	≤ 0.001
C14:1*cis*-7	0.04^{g}	0.04^{f}	0.04^{d}	0.08^{b}	0.07^{c}	0.04^{f}	0.04^{f}	0.07^{c}	0.04^{e}	0.10^{a}	0.00	≤ 0.001
C15:1*cis*-5	0.01^{e}	0.01^{dc}	0.01^{dc}	0.03^{a}	0.01^{c}	0.01^{dc}	0.01^{dc}	0.01^{dc}	0.01^{d}	0.02^{b}	0.00	≤ 0.001
C16:0	10.59^{e}	11.29^{d}	11.88^{a}	9.87^{f}	10.62^{e}	11.33^{dc}	11.55^{b}	10.57^{e}	11.40^{c}	11.59^{b}	0.05	≤ 0.001
C18:0	3.35^{f}	3.91^{d}	3.61^{e}	2.89^{g}	2.76^{i}	3.95^{c}	4.03^{b}	2.75^{i}	4.28^{a}	2.80^{h}	0.01	≤ 0.001
C18:1*cis*-9	40.08^{b}	37.67^{c}	35.22^{f}	36.92^{d}	40.51^{a}	37.61^{c}	36.80^{d}	40.47^{a}	36.33^{e}	31.09^{g}	0.06	≤ 0.001
C18:2 *cc n*-6	23.58^{a}	20.18^{e}	22.95^{b}	16.68^{i}	17.04^{g}	20.14^{e}	21.68^{c}	16.96^{h}	21.46^{d}	19.44^{f}	0.02	≤ 0.001
C18:3*n*-3	5.87^{a}	5.33^{c}	5.23^{e}	5.08^{f}	4.59^{g}	5.33^{c}	5.41^{b}	4.60^{g}	5.30^{d}	4.41^{h}	0.01	≤ 0.001
C20:4*n*-6	0.28^{e}	0.69^{c}	0.25^{f}	0.23^{g}	0.22^{h}	0.70^{b}	0.58^{d}	0.22^{h}	0.91^{a}	0.23^{g}	0.00	≤ 0.001
C20:5*n*-3	1.65^{i}	3.46^{b}	3.32^{c}	2.34^{h}	2.66^{g}	3.46^{b}	3.06^{d}	2.71^{f}	2.82^{e}	3.48^{a}	0.01	≤ 0.001
C22:4*n*-6	0.13^{ba}	0.14^{ba}	0.12^{ba}	0.07^{b}	0.12^{ba}	0.15^{a}	0.13^{ba}	0.12^{ba}	0.14^{a}	0.16^{a}	0.03	0.355

C22:5*n*-3	0.45^{f}	0.57^{d}	0.58^{d}	0.61^{c}	0.78^{b}	0.57^{d}	0.54^{e}	0.78^{b}	0.57^{d}	0.88^{a}	0.01	$\leq$0.001
C22:6*n*-3	1.99^{g}	2.05^{f}	2.20^{e}	2.61^{d}	3.47^{c}	2.05^{f}	1.96^{h}	3.55^{b}	2.01^{g}	4.21^{a}	0.01	$\leq$0.001
SFA^{w}	16.71^{ef}	19.11^{ba}	19.16^{ba}	16.68^{f}	16.89^{d}	19.02^{b}	19.22^{a}	16.84^{ed}	19.25^{a}	18.37^{c}	0.07	$\leq$0.001
MUFA†	47.82^{c}	46.89^{d}	44.48^{g}	54.53^{a}	50.93^{b}	46.96^{d}	45.76^{e}	50.93^{b}	45.60^{f}	44.33^{h}	0.07	$\leq$0.001
PUFA§	35.40^{c}	33.91^{e}	36.28^{b}	28.69^{g}	32.08^{f}	33.93^{e}	34.93^{d}	32.14^{f}	35.05^{d}	37.16^{a}	0.06	$\leq$0.001
n-3$^{\#}$	10.67^{f}	12.20^{d}	12.23^{d}	10.76^{f}	13.80^{c}	12.20^{d}	11.75^{e}	13.94^{b}	11.77^{e}	16.32^{a}	0.04	$\leq$0.001
n-6	24.80^{a}	21.80^{e}	24.14^{b}	18.04^{h}	18.38^{g}	21.82^{e}	23.27^{d}	18.30^{g}	20.98^{f}	23.38^{c}	0.05	$\leq$0.001
n-6/*n*-3	2.32^{a}	1.79^{c}	1.97^{b}	1.68^{d}	1.33^{e}	1.79^{c}	1.98^{b}	1.31^{f}	1.29^{g}	1.99^{b}	0.01	$\leq$0.001
PS	2.12^{a}	1.77^{e}	1.89^{c}	1.72^{f}	1.90^{c}	1.78^{e}	1.82^{d}	1.91^{c}	1.82^{d}	2.02^{b}	0.01	$\leq$0.001
AI	0.19^{g}	0.25^{c}	0.25^{b}	0.25^{dc}	0.23^{f}	0.25^{c}	0.25^{d}	0.23^{f}	0.27^{a}	0.24^{e}	0.00	$\leq$0.001
trans$^{\$}$	0.70^{c}	1.01^{a}	1.00^{ba}	0.61^{de}	0.57^{e}	1.01^{a}	0.97^{ba}	0.57^{e}	0.95^{b}	0.66^{dc}	0.02	$\leq$0.001
VLC *n*-3$^{\&}$	2.46^{h}	2.65^{f}	2.80^{e}	3.26^{d}	4.30^{c}	2.65^{f}	2.53^{g}	4.37^{b}	2.60^{f}	5.15^{a}	0.02	$\leq$0.001

x S – rainbow trout (*Oncorhynchus mykiss).*
y Standard error of mean.
z Statistical probability of sample effect.
u Means with a different superscript within rows ($^{a-h}$) differ significantly.
w Saturated fatty acid: C8:0 + C10:0 + C11:0 + C12:0 + C13:0 + C14:0 + C15:0 + C16:0 + C17:0 + C18:0 + C20:0 + C21:0 + C22:0 + C23:0 + C24:0.
† Monounsaturated fatty acid: C12:1*cis*-11 + C14:1*trans*-7 + C14:1*cis*-7 + C15:1*cis*-5 + C15:1*cis*-10 + C16:1*trans*-7 + C16:1*trans*-9 + C16:1*cis*-9 + C17:1*trans*-10 + C17:1*cis*-10 + C18:1*trans*-7 + C18:1*trans*-8 + C18:1*trans*-9 + C18:1*trans*-10 + C18:1*trans*-11 + C18:1*cis*-7 + C18:1*cis*-8 + C18:1*cis*-9 + C18:1*cis*-11 + C20:1*cis*-5 + C20:1*cis*-8 + C20:1*cis*-11 + C21:1*trans*-12 + C21:1*cis*-12 + C22:1*cis*-13 + C24:1*cis*-15.
§ Polyunsaturated fatty acid: C18:2*tt n*-6 + C18:2*tc n*-6 + C18:2*ct n*-6 + C18:2*cc n*-6 + C18:3*n*-6 + C18:4*n*-6 + C18:3*n*-3 + C18:4*n*-3 + C20:2*n*-6 + C20:3*n*-6 + C20:4*n*-6 + C20:3*n*-3 + C22:2*n*-6 + C20:5*n*-3 + C22:3*n*-3 + C22:4*n*-6 + C22:5*n*-6 + C22:5*n*-3 + C22:6*n*-3.
\# C18:3*n*-3 + C18:4*n*-3 + C20:3*n*-3 + C20:5*n*-3 + C22:3*n*-3 + C22:5*n*-3 + C22:6*n*-3.
C18:2*tt n*-6 + C18:2*tc n*-6 + C18:2*ct n*-6 + C18:2*cc n*-6 + C18:3*n*-6 + C18:4*n*-6 + C20:2*n*-6 + C20:3*n*-6 + C20:4*n*-6 + C22:2*n*-6 + C22:4*n*-6 + C22:5*n*-6.
Index of atherogenicity: (C12:0 + 4 C14:0 + C16:0 + Σ(*trans*))/(Σ(*n*-6) + Σ(*n*-3) + C18:1*cis*-9 + other MUFA) [24].
$ C14:1*trans*-7 + C16:1*trans*-7 + C16:1*trans*-9 + C17:1*trans*-10 + C18:1*trans*-7 + C18:1*trans*-8 + C18:1*trans*-9 + C18:1*trans*-10 + C18:1*trans*-11 +C18:2 *trans* + C18:2 *trans*-7 *cis*-9 + C21:1*trans*-12.
& Very long chain *n*-3: C22:3*n*-3 + C22:5*n*-3 + C22:6*n*-3.

Table 3. Fatty acid composition and calculated nutritional information of farmed rainbow trout fillets.

Fatty acid (FA) (g 100 g^{-1} total FA)	Trout sample											
	S1	S2	S3	S4	S5	S6	S7	S8	S9	S10	SE	*P* value
C14:0	1.15^{d}	1.67^{a}	1.19^{d}	1.19^{d}	1.70^{a}	0.99^{e}	1.51^{b}	1.48^{b}	1.31^{c}	1.41^{cb}	0.05	$\leq$0.001
C14:1*cis*-7	0.03^{c}	0.04^{b}	0.03^{d}	0.04^{b}	0.05^{a}	0.03^{d}	0.04^{b}	0.04^{b}	0.04^{b}	0.03^{c}	0.00	$\leq$0.001
C16:0	11.49^{ef}	12.14^{cd}	11.07^{f}	13.15^{b}	11.88^{ed}	11.75^{ed}	12.64^{cb}	11.39^{ef}	15.56^{a}	11.42^{ef}	0.26	$\leq$0.001

C18:0	3.27^{d}	3.29^{d}	3.49^{c}	3.06^{e}	3.31^{d}	3.10^{e}	3.71^{b}	3.05^{e}	3.95^{a}	2.85^{f}	0.06	≤ 0.001
C18:1*cis*-9	36.08^{a}	35.68^{a}	29.40^{c}	24.33^{e}	36.68^{a}	27.86^{dc}	32.60^{b}	36.14^{a}	26.51^{d}	32.55^{b}	0.82	≤ 0.001
C18:2*cc n*-6	19.88^{a}	14.76^{e}	18.06^{b}	14.57^{e}	15.31^{de}	16.24^{dc}	16.88^{c}	14.86^{e}	14.35^{e}	19.23^{a}	0.49	≤ 0.001
C18:3*n*-3	4.21^{a}	3.01^{e}	3.06^{e}	3.18^{dce}	3.27^{dc}	3.28^{c}	3.46^{b}	3.14^{dce}	3.09^{de}	4.34^{a}	0.08	≤ 0.001
C20:4*n*-6	0.14^{a}	0.12^{bac}	0.07^{e}	0.13^{ba}	0.13^{ba}	0.07^{ed}	0.11^{bac}	0.10^{dc}	0.10^{bc}	0.11^{bc}	0.01	0.0013
C20:5*n*-3	2.07^{gf}	2.18^{f}	3.48^{b}	4.10^{a}	1.79^{g}	3.16^{cb}	2.85^{cd}	2.35^{ef}	2.97^{cd}	2.62^{ed}	0.17	≤ 0.001
C22:4*n*-6	0.06^{a}	0.05^{b}	0.00^{d}	0.00^{d}	0.05^{a}	0.00^{d}	0.04^{c}	0.04^{c}	0.00^{d}	0.04^{c}	0.00	≤ 0.001
C22:5*n*-3	0.60^{c}	0.68^{c}	0.95^{b}	0.94^{b}	0.59^{c}	1.22^{a}	0.87^{b}	0.91^{b}	0.85^{b}	0.87^{b}	0.04	≤ 0.001
C22:6*n*-3	7.04^{f}	9.63^{e}	13.10^{dc}	21.61^{a}	8.57^{fe}	17.98^{b}	9.87^{e}	10.94^{de}	15.28^{c}	9.28^{fe}	1.07	≤ 0.001
SFA	17.67^{de}	18.94^{c}	18.63^{c}	19.06^{c}	18.66^{c}	18.33^{dc}	20.02^{b}	17.64^{de}	23.08^{a}	17.44^{e}	0.34	≤ 0.001
MUFA	45.60^{b}	47.37^{ba}	39.15^{d}	33.38^{e}	48.30^{a}	37.15^{d}	42.95^{c}	47.06^{ba}	37.09^{d}	43.32^{c}	1.00	≤ 0.001
PUFA	36.05^{e}	32.97^{f}	40.99^{c}	46.96^{a}	32.36^{f}	43.67^{b}	36.34^{e}	34.77^{e}	38.96^{d}	38.50^{d}	0.70	≤ 0.001
n-3	14.55^{g}	16.63^{feg}	21.07^{dc}	31.25^{a}	15.49^{fg}	26.14^{b}	17.77^{fe}	18.47^{de}	23.26^{c}	17.87^{fe}	1.17	≤ 0.001
n-6	22.19^{a}	17.06^{ed}	21.16^{a}	16.31^{e}	17.54^{cd}	18.37^{cb}	19.26^{b}	16.83^{ed}	16.57^{ed}	21.37^{a}	0.48	≤ 0.001
n-6/*n*-3	1.53^{a}	1.03^{cd}	1.01^{cd}	0.52^{f}	1.13^{cb}	0.71^{e}	1.09^{cb}	0.91^{d}	0.71^{e}	1.20^{b}	0.06	≤ 0.001
PS	2.04^{d}	1.74^{g}	2.20^{c}	2.46^{a}	1.73^{g}	2.38^{b}	1.81^{f}	1.97^{e}	1.69^{g}	2.21^{c}	0.03	≤ 0.001
AI	0.21^{gf}	0.25^{b}	0.21^{ef}	0.23^{c}	0.24^{b}	0.20^{g}	0.25^{b}	0.22^{d}	0.28^{a}	0.22^{ed}	0.00	≤ 0.001
trans	0.91^{cbd}	1.05^{b}	1.41^{a}	0.71^{e}	0.96^{cbd}	0.84^{ced}	0.99^{cb}	0.81^{ed}	0.86^{cd}	0.93^{cbd}	0.07	≤ 0.001
VLC *n*-3	7.70^{f}	10.37^{e}	14.10^{dc}	22.62^{a}	9.24^{fe}	19.28^{b}	10.80^{e}	11.91^{de}	16.22^{c}	10.24^{fe}	1.11	≤ 0.001

Abbreviations are explained in the legend of table 2.

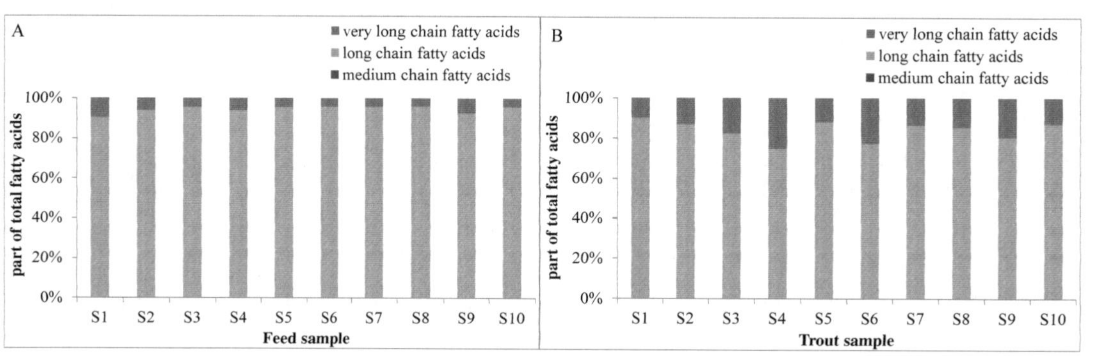

Figure 2. Percentage of medium-, long- and very long-chain fatty acids among total fatty acids in feed (diet) source (A) and in trout fillet lipids (B).

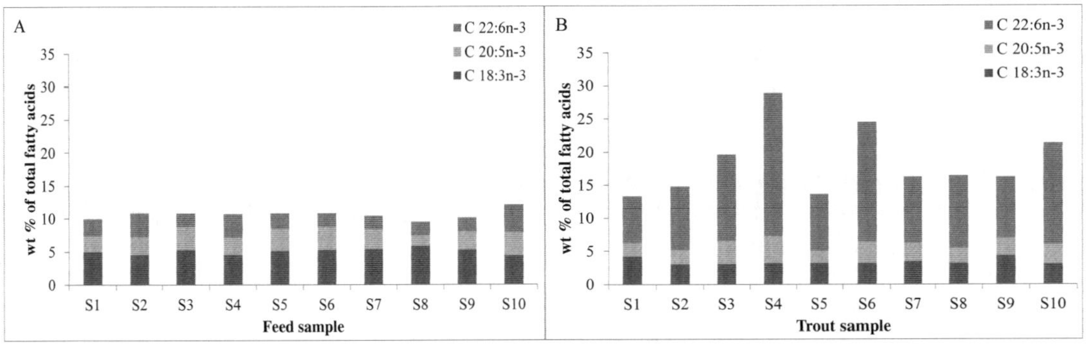

Figure 3. Weight% of C18:3*n*-3 (linolenic FA), C20:5*n*-3 (eicosapentaenoic FA) and C22:6*n*-3 (docosahexaenoic FA) in feed source (A) and in trout fillet lipids (B).

4. Conclusion

Preliminary studies in the field of FA and basic chemical composition of trout feed on one hand, and trout (*Oncorhynchus mykiss*) fillet meat on the other, showed that, despite wide variability between observed parameters, some conclusions can be drawn. In the present study, the FA profile and basic chemical composition of rainbow trout farmed in Slovenia were defined. On the basis of nutritional quality of the fat in the trout fillets, it can be concluded that rainbow trout farmed in Slovenia provides an important source of healthy fats, as it contains favorable *n*-3/*n*-6 ratios and PS and AI indices, all within recommended limits. Our data also suggest that long chain *n*-3 PUFA from feed can be converted into very long chain *n*-3 PUFA in trout fillets. From these points of view, farmed rainbow trout could be a healthy choice in human diet.

References

[1] Burger J and Gochfeld M 2009 *Environ Res* **109** 343
[2] Rebolé A, Velasco S, Rodríguez M L, Treviño J, Alzueta C, Tejedor J L and Ortiz L T 2015 *Food Chem* **174** 614
[3] Kris-Etherton P M, Harris W S and Apple L J 2002 *Circulation* **106** 2747
[4] Mayneris-Perxachs J, Bondia-Pons I, Serra-Majem L and Castellote A I 2010 *Food Chem* **119** 54
[5] Tapiero H, Nguyen B G, Couvreur P and Tew K D 2002 *Biomed Pharmacother* **56** 215
[6] Marin M, Polak T, Demšar L and Žlender B 2010 *Acta Agri Slov* **96** 95
[7] Testi S, Bonaldo A, Gatta P P and Badiani A 2006 *Food Chem* **98** 104
[8] Gökçe M A, Taşbozan O, Çelik M and Tabakoğlu S S 2004 *Food Chem* **88** 419
[9] Haliloğlu H I, Bayır A, Sirkecioğlu A N and Atamanalp M 2004 *Food Chem* **86** 55
[10] Caballero M J, Obach A, Rosenlund G, Montero D, Gisvold M and Izquierdo M S 2002 *Aquaculture* **214** 253
[11] Barrado E, Jiménez F, Prieto F and Nuevo C 2003 *Food Chem* **81** 13
[12] Nielsen N S, Göttsche J R, Holm J, Xu X, Mu H, Jacobsen C 2005 *Aquaculture* **250** 411
[13] ISO 3100-1 1991 Meat and meat products - Sampling and preparation of test samples - Part 1: Sampling
[14] AOAC 950.46 1997 (Washington: AOAC)
[15] AOAC 928.08 1997 (Washington: AOAC)
[16] AOAC 920.153 1997 (Washington: AOAC)
[17] AOAC 991.36 2000 (Arlington: AOAC)
[18] Park P W and Goins R E 1994 *J Food Sci* **59** 1262
[19] Polak T, Rajar A, Gašperlin L and Žlender B 2008 *Meat Sci* **80** 864
[20] Rasmussen R S, Heinrich M T, Hyldig G, Jacobsen C and Jokumsen A 2011 *Aquaculture* **314** 159

[21] Haliloğlu A G 2012 *Pakistan J Zool* **44** 1013
[22] Suzuki K, Mizusawa K, Noble C, and Tabata M 2008 *Fish Sci* **74** 941
[23] Simmons A P, Turk P, Beamer S, Jaczynski J, Semmens K, and Matak K F 2011 *J Food Sci* **76** 192
[24] Ulbricht T L V and Southgate D A T 1991 *Lancet* **338** 985
[25] Santos-Silva J, Bessa R J B and Santos-Silva F 2002 *Livest. Prod. Sci.* 77 187
[26] Wood J D, Richardson R I, Nute G R, Fisher A V, Campo M M, Kasapidou E, Sheard P R and Enser M 2003 *Meat Sci* **66** 21
[27] Enser M, Scollan N, Gulati S, Richardson I, Nute G and Wood J 2001 *Proc. Int. Cong. of Meat Sci. and Tech.* (Kraków) p 186
[28] Polak T, Demšar L, Rajar A and Žlender B 2006 *Food Technol. Biotechnol* **44** 65
[29] Žlender B, Rajar A, Polak T and Gašperlin L 2008 *Meat Sci* **80** 864
[30] Žlender B, Gašperlin L and Polak T 2001 *Proc. Int. Cong. of Meat Sci. and Tech.* (Kraków) p. 190
[31] Žlender B, Rajar A, Polak T, Gašperlin L and Čepin S 2004 *Proc. Int. Cong. of Meat Sci. and Tech.* (Helsinki) p 1198
[32] Žlender B, Holcman A, Stibilj V and Polak T 2000 *Poljoprivreda* **6** 53
[33] Stanchewa M, Dobreva D, Merdzhanava A and Galunska B 2010 *Bulgaria Scientific papers* **37** 117
[34] Aslan S S, Guven K C, Gezgin T, Alpaslan M and Tekinay A 2007 *Fish Sci* **73** 1195
[35] Gregory M K and James M J 2014 *Biochim Biophys Acta* **1841** 1656
[36] Monguchi T, Hara T, Hasokawa M, Nakajima H, Mori K, Toh R, Irino Y, Ishida T, Hirata K-I, and Shinohara M 2017 *J Cardiol* **70** 121
[37] EFSA 2010 *EFSA J* **8** 1461
[38] FAO 2010 [No authors listed] *FAO Food Nutr Pap* **91** 1
[39] Ramos A, Bandarra N M, Rema P, Vaz-Pires P, Nunes M L, Andrade A M, Cordeiro A R, and Valente L M P 2008 *Aquaculture* **274** 366

IOP Conf. Series: Earth and Environmental Science **85** (2017) 012009 doi:10.1088/1755-1315/85/1/012009

Public protection – reliable allergen risk management

V Janković[1], J Popov Raljić[2] and V Đorđević[1]

[1] Institute of Meat Hygiene and Technology, Kaćanskog 13, Belgrade, Serbia
[2] University of Novi Sad, Faculty of Science, Department of Geography, Tourism and Hotel Management, Dositeja Obradovića 3, Novi Sad, Serbia

Email: vesna.jankovic@inmes.rs

Abstract. Consumers with potentially fatal food allergies are dependent on correct product labelling to protect their health. The food industry is responsible for providing every detail consumers need to make informed decisions. Considering public health, food suppliers have to monitor the presence of allergens, prevent cross-contamination and label products accurately. Allergen labelling of food products, drinks and non pre-packed food and drink products is clearly defined with legal regulations. To achieve this, a complete understanding of each product's allergenic ingredients is needed and cross-contamination of food with allergens must be avoided. Raw materials need to be checked, every ingredient must be verified and every single allergen has to be stipulated. A mislabeled product could be recalled at potential cost, financially damaging business and at the same time, negatively impacting brand and reputation.

1. Introduction

Food allergens affect the health and life of people with hypersensitivity caused by some food components, and such allergens are identified as a severe food safety hazard and their management is one of the fundamental areas of food safety management systems [1,2].

The Codex Alimentarius Commission (CAC) of the Food and Agriculture Organization/the World Health Organization (FAO/WHO) [3] recommends that allergen information should be placed on labels of food products. Furthermore, legal regulations in the United States (US) [4], the European Union (EU) [5] and in Serbia [6], define 14 food allergens that can cause allergic reactions or intolerance, and their distribution and that of food derivatives which contain them can trigger allergic reactions impose an obligation on food producers to label food in regard to food allergens. In European countries, the prevalence of food allergy is 1-3% of adults and 4-6% of children, while in the US, 4% of adults and 8% of children under the age of 18 suffer with food allergies [7].

2. The fourteen food allergens

The 14 listed allergens are, according to [5]:

1. Cereals containing gluten, namely wheat (such as spelt and Khorasan wheat), rye, barley, oats and their hybridised strains and products thereof, except: a) wheat based glucose syrups including dextrose b) wheat based maltodextrins c) glucose syrups based on barley d) cereals used for making alcoholic distillates including ethyl alcohol of agricultural origin;
2. Crustaceans and products thereof (for example prawns, lobster, crabs and crayfish);
3. Egg and products thereof;

IOP Conf. Series: Earth and Environmental Science **85** (2017) 012009 doi:10.1088/1755-1315/85/1/012009

4. Fish and products thereof, except: a) fish gelatine used as carrier for vitamin or carotenoid preparations b) fish gelatine or Isinglass used as a fining agent in beer and wine.
5. Peanuts and products thereof;
6. Soybeans and products thereof, except: a) fully refined soybean oil and fat b) natural mixed tocopherols (E306), natural D-alpha tocopherols, natural D-alpha tocopherol acetate and natural D-alpha tocopherol succinate from soybean sources c) vegetable oils derived phytosterols and phytosterol esters from soybean sources d) plant stanol ester produced from vegetable oil sterols from soybean sources;
7. Milk and products thereof (including lactose), except: a) whey used for making alcoholic distillates including ethyl alcohol of agricultural origin b) lactitol;
8. Nuts (namely almond, hazelnut, walnut, cashew, pecan nut, Brazil nut, pistachio nut and Macadamia nut (Queensland nut) and products thereof, except for nuts used for making alcoholic distillates including ethyl alcohol of agricultural origin;
9. Celery and products thereof;
10. Mustard and products thereof;
11. Sesame seeds and products thereof;
12. Sulphur dioxide and/ or sulphites at concentrations of more than 10mg/kg or 10mg/L (litre) in terms of the total SO2 which are to be calculated for products as proposed ready for consumption or as reconstituted according to the instructions of the manufacturers;
13. Lupin and products thereof;
14. 14 Molluscs and products thereof (for example mussels, clams, oysters, scallops, snails and squid).

Eggs, fish, milk, peanuts, soy, shellfish, tree nuts, and wheat are the “big 8” food allergens, which have triggered more than 90% of the food allergy reactions in the US [8].

3. Allergen management in catering

Applying the basic principles of food safety management in the catering industry is based on understanding the nutritional allergies and food production processes leading to the development of systems that can support the production of food which contains or does not contain a particular allergen in circumstances where the risk for the consumer is minimized [9].

A modern approach to food safety management in the catering industry is most often based on the need for a particular risk to be completely preempted and reduced to an acceptable level, before it actually occurs in practice. Allergies and intolerances to foods are one of the security risks which are widely discussed in the food industry. The general view is that in the case of allergenic foods, it is unrealistic to talk about the possibility of zero risk [10,11], which makes it extremely important to set standards in the catering industry that will minimize this risk.

Precisely defined and consistent safety management standards for allergens lead to a consistent and sustainable food safety management in the food industry [12], but also in the production of products for consumption in catering facilities. Allergenic foods can be risky in two cases: when they are directly taken into the body alone or as an integral part of another product, or by cross-contamination of non-allergenic foods with allergenic ones during the food production process [13].

4. Risk management in the catering industry

In the area of risk management of allergenic foods in the food and catering industries, there is an irrational desire for zero risk tolerance that entails a complete avoidance of any food that is potentially a causative allergen [14]. However, the risk of cross-contamination by allergens during food production is present despite the efforts of food producers to comply with all the requirements of good hygiene and good manufacturing practice.

Predictive modeling in risk management of allergenic foods is significantly hampered by a poorly defined method of declaring the foods where the zero risk tolerance for allergens imposes on food producers, while labeling the food products, to use the term “may contain” [15]. Some stakeholders

IOP Conf. Series: Earth and Environmental Science **85** (2017) 012009 doi:10.1088/1755-1315/85/1/012009

are not convinced that the new labelling legislation provides sufficient information to food allergic consumers [16]. Due to fear of cross-contamination, in the absence of accurate precautionary i.e. "may contain" labelling, food allergic consumers are uncertain about product safety, and cannot always clearly understand or interpret the information on the food labels [17]. For food allergic consumers, unintentional exposure to allergens when eating outside in restaurants or catering outlets outside the home is particularly problematic, as unintentional exposures to problematic allergens can occur. The need to establish a reliable system for declaring, labeling and marketing of foods has caused a necessary step ahead in science, such as the determination of eliciting doses (ED) of allergenic food ingredients, which vary depending on individual predispositions and geographical determinants. For the safety of consumers, as the initial EDs of proteins in allergenic foods, the highest ones found by the research group of Allergen Bureau VITAL scientists in Australia were deemed suitable [14].

The VITAL 2.0 program has established reference doses of total allergenic protein intake and defined an action network of risk levels for allergenic foods, calculated using reference doses and reference quantities of food intake/portion size. By determining the reference doses, an effective basis of communication within the risk management of allergenic foods has been set up, which has enabled detailed identification, characterization and significantly easier risk management, weighting and selection, i.e. a detailed risk analysis.

Communication researchers have found that risk communication plays an important role in controlling and preventing negative consequences such as food allergy reactions in restaurants. Establishing proper communication between and among customers and food service employees could be one of the first and most important steps in preventing food allergy reactions in restaurants [18,19]. Proper communication among stakeholders would initiate increased attention to food preparation by service staff when serving customers with food allergies. Although there are other food allergy-related publications available, no research has been published regarding food allergy risk communication.

Researchers found that restaurant staff lacked knowledge regarding food allergens in the menu, ways to prevent cross contact, and the severity of food allergy reactions [20]. One study from the United Kingdom revealed that about 21% of the peanut-free meals that were prepared immediately after peanut-containing meals were contaminated with peanut or peanut protein. Researchers also found that restaurant employees' confidence levels were high even though their knowledge about serving customers with food allergies was not adequate [21]. Specifically, 70% of the respondents in this study felt that they could guarantee a safe meal, while 35% thought that fryer heat could destroy allergens and 25% thought it was safe to remove allergens from a finished meal [21].

Strict avoidance of food allergens and early recognition and response to allergic reactions are extremely important for individuals with food allergies to prevent fatal food allergy reactions [22]. To prevent potential food allergy reactions, customers with food allergies have used various strategies prior to and while dining out. For example, customers chose restaurants with which they were familiar and where they were known by the staff; avoided establishments and cuisines that are considered high-risk such as buffets or ethnic restaurants, and; checked online menus, ingredients, and allergen information before dining out [23].

Despite these prevention strategies, customers with food allergies have experienced communication challenges when dining out because some restaurant staff had insufficient knowledge about food allergies, did not understand special requests, or were not aware of the severity of food allergy reactions [23]. There is a lack of legislation or training guidelines focusing on the risk management of food allergies [11] and risk communication-related issues in restaurants. Yet most food handlers perceive the foodservice industry as a low-risk business, and this preception negatively affects their safe food-handling behaviors [24]. Therefore, food allergy risk communication can be used as a tool to reduce the chance of food allergy reactions caused by the mistakes of restaurant staff when serving customers with food allergies [25].

5. Conclusion

The main features of food allergens (a large number of foods that contain allergens, the minimum amount needed to trigger a reaction, the various spectra of symptoms and the small number of people suffering from food allergies) pose a great challenge to food manufacturers in developing a safety management plan. Safety management of food allergens must be focused on careful risk analysis throughout each segment of the food chain in the catering industry. Guidelines that point to potential risks must be ensured so that food allergens are either clearly and precisely defined or are not present in quantities that will endanger the health of consumers. All segments of the food chain in the catering facility must be monitored, from the design, through the source of foodstuffs, declarations and labeling within the supply, up to the safe food consumption and established accountability.

Foods that contain allergens have specific, characteristics health and safety risks. However, risks caused by allergens can be controlled and minimized using the developed methodologies within the context of other risks. The key basis for security management of allergens in a catering facility is communication – good consumer-to-customer and employee-supplier communication and interpersonal communication. However, the risks which are not under the control of allergen management are: 1) undeclared or wrongly declared allergens, and; 2) unverified allergies. Allergens in foods that are not properly labeled or are not recognizably highlighted can cause significant failures in the safety management system. Another risk that is almost impossible to avoid is that of the first occurrence of an allergic reaction. Because of this risk, there has to be a person in the catering facility at all times who has been trained to recognize such symptoms and respond to them properly and timely.

The factor that is most difficult to control and which can significantly affect food safety system is the human factor. Therefore, one of the primary management tasks in catering facilities is providing appropriate education and training to raise awareness of employees about the risks that could arise from allergenic foods and culinary products.

6. References

[1] Crevel R W R and Cochrane S A 2014 Food Safety Assurance Systems: Management of Allergens in Food Industry *Encyclopedia of Food Safety* Yamine Motarjemi (Elsevier: Netherlands) 254-61

[2] Stein K 2015 Effective allergen management practices to reduce allergens in food *Handbook of Food Allergen Detection and Control* Flanagan S (Woodhead Publishing UK) 103-31

[3] Codex Alimentarius Commission 1999 General Standard for the Labelling of Pre-Packaged Foods *CODEX STAN 1-1985*

[4] USFDA 2004 Food allergen labelling and consumer protection act of 2004 *Public law 108-282* title II

[5] Regulation (EU) No 1169/2011 of the European Parliament and of the Council, *OJ L 304*

[6] Regulations on food declarations, labeling, and advertising 2017 *RS Official Gazzete 19/17*

[7] De Blok M, Vlieg-Boerstra J, Oude Elberink G, Duiverman J, Galvin A, Hourihane B, Cornelisse-Vermaat R, Frewer L, Mills C, Dubois E 2007 A framework for measuring thesocial impact of food allergy across Europe: a Euro Prevall state of the art paper *Allergy* 62 733–37

[8] Sicherer S H, Muñoz-Furlong A, Godbold J H and Sampson H A 2010 US prevalence of self-reported peanut, tree nut, and sesame allergy: 11-year follow-up *J Allergy Clin Immunol.* 125 1322-26

[9] Alldrick J 2006 Managing allergy issues *The World of food ingredients* 64-65

[10] Kroes R, Galli C, Munro I, Schilter B, Tran L and Walker Würtzen G 2000 Threshold of toxicological concern for chemical substances present in the diet:a practical tool for assessing the need for toxicity testing *Food Chemical Toxicology* 38 255–312

[11] Madsen B et al. 2012 Can we define a tolerable level of risk in food allergy? *Clinical & Experimental Allergy* 42 30–37

[12] Hattersley S, Ward R, Baka A, Crevel R 2014 Advances in the risk management of unintended presence of allergenic foods in manufactured food products – An overview *Food and Chemical Toxicology* 67 255-261

[13] Madsen B et al. 2009 Approaches to risk assessment in food allergy *Food and Chemical Toxicology* 47 480–489

[14] Allen K J, Remington B C, Baumert J L, Crevel R W R, Houben GF, Brooke-Taylor S, Kruizinga, AG and Taylor SL 2014 Allergen reference doses for precautionary labeling (VITAL 2.0): clinical implications *J Allergy Clin Immunol.*1331 156-164

[15] Jankovic V, Lakicevic B, Petronijevic R, Spiric D, Đorđević V, Popov- Ralic J 2016 Soybean and gluten in meat products- Consumer Protection Strategy *Agro FOOD Indusrty Hi Tech* 27 **3** 30-32

[16] Mills et al. 2004 Information provision for allergic consumers-where are we going with food allergen labelling? *Allergy* 59 **12** 1262-68

[17] Mils C 2007 Improved quality of life to food allergic consumers *Molecular nutrition food research* 1 **51** 148-149

[18] Leftwich J, Barnett J, Muncer K, Shepherd R, Raats M M, Hazel Gowland M and Lucas J S 2011 The challenges for nut-allergic consumers of eating out *Clin Exp Allergy* 41 **2** 243-49

[19] Han W 2012 *Risk communication when serving customers with food allergies in restaurants in the USA* PhD thesis (University of Houston)

[20] Abbot M, Byrd-Bredbenner C, Grasso D 2007 Know before you serve: developing a food-allergy fact sheet *Cornell Hotel Restaurant Administration* 48 274–283

[21] Ahuja R and Sicherer H 2007 Food allergy management from the perspective of restaurant and food establishment personnel *Ann. Allergy Asthma Immunol.* 98 344–48

[22] Sicherer S 2004 Current approach to the diagnosis and management of adverse reactions to foods *J Allergy Clin Immunol.* 5 **114** 1146-50

[23] Kwon J and Lee Y M 2012 Exploration of past experiences, attitudes and preventive behaviors of consumers with food allergies about dining out: a focus group study *Food Protection Trends* 32 736-46

[24] Clayton D A, Griffith C J Price P and Peters A C 2002 Food handlers' beliefs and selfreported practices *International Journal of Environmental Health Research* 12 25-39

[25] Kronenberg S A 2012 Food allergy risk management: More customers, less liability *Journal of Foodservice Business Research* 15 1117-121

IOP Conf. Series: Earth and Environmental Science **85** (2017) 012010 doi:10.1088/1755-1315/85/1/012010

Rapid and reliable QuEChERS-based LC-MS/MS method for determination of acrylamide in potato chips and roasted coffee

S Stefanović[1], V Đorđevic[1] and V Jelušić[2]

[1] Institute of Meat Hygiene and Technology, Kacanskog 13, Belgrade, Serbia
[2] Veterinary Office of Bosnia and Herzegovina, Maršala Tita 9a/II, Sarajevo, Bosnia and Herzegovina

E-mail: srdjan.stefanovic@inmes.rs

Abstract. The aim of this paper is to verify the performance characteristics and fitness for purpose of rapid and simple QuEChERS-based LC-MS/MS method for determination of acrylamide in potato chips and coffee. LC-MS/MS is by far the most suitable analytical technique for acrylamide measurements given its inherent sensitivity and selectivity, as well as capability of analyzing underivatized molecule. Acrylamide in roasted coffee and potato chips wasextracted with water:acetonitrile mixture using NaCl and $MgSO_4$. Cleanup was carried out with $MgSO_4$ and PSA. Obtained results were satisfactory. Recoveries were in the range of 85-112%, interlaboratory reproducibility (Cv) was 5.8-7.6% and linearity (R^2) was in the range of 0.995-0.999. LoQ was 35 μg $kg^{-1}$for coffee and 20 μg kg^{-1} for potato chips. Performance characteristic of the method are compliant with criteria for analytical methods validation. Presented method for quantitative determination of acrylamide in roasted coffee and potato chips is fit for purposes of self-control in food industry as well as regulatory controls carried out by the governmental agencies.

1. Introduction

Acrylamide (prop-2-enamide, 2-propene amide, CAS No. 79-06-1) isa compound that attracts much attention of the scientific, regulatory and food business operators' communities for the past decade, due to its almost ubiquitous presence in some thermally processed foods, as well as its harmful properties to humans. International Agency for Research of Cancer (IARC) classified acrylamide as "probably carcinogenic to humans", belonging to Group 2A [1]. Besides carcinogenicity, studies have attributed some other harmful properties to acrylamide, such is neurotoxicity [2]. Although the studies from available literature have not yet fully explained all possible formation mechanisms of acrylamide, it is well known that one way of its formation is based on Maillard reaction between amino acid asparagine and carbonyl-group compounds, especially reducing sugars e.g. glucose or fructose, at the temperatures higher than 120°C [3, 4]. Therefore, occurrence of acrylamide is high in carbohydrates-rich,thermally treated foods (potato chips, French fries biscuits, snacks of large varieties, roasted coffee, cereals, etc). On the other hand, Maillard reaction is responsible for desirable sensory properties of processed foods (dark yellow to brown color, crusty surface and rich taste).

However, at the same time, these fried or baked surfaces contain the highest quantities of acrylamide [5].

Since acrylamide presence in processed foods clearly poses a health risk to consumers, due to harmful properties of the molecule itself, as well as considering the abundance of acrylamide-containing product in our nutrition, European Commission initiated monitoring of acrylamide levels in processed foods in 2007 [6]. This survey ended in 2009 followed by another monitoring cycle [7]. Based on its data, in 2012 European Food Safety Agency (EFSA) concluded that no significant reduction of acrylamide levels is observed, except in food for infants and some other categories [8]. The European Commission published in 2013 its second Recommendation on "indicative acrylamide values based on monitoring data from 2007-12" [9] setting the maximum recommended levels e.g. for potato chips to 1000 μg kg^{-1}and for roasted coffee to 450 μg kg^{-1}. Along with recommended levels, EC has promoted "toolbox" compiled by the organization "FoodDrinkEurope" with the aim to help food business operators to reduce acrylamide levels in their products [9].

From the analytical perspective, acrylamide detection and measurement is not without hindrances, regardless of relatively long period of availability of analytical methods. Historically, one of the first protocols describes analysis of acrylamide in drinking water after derivatization (bromination with 2,3-dibromopropionamide), and GC separation using ECD or MS detection [10]. Although this method can be applied to complex matrices such as various foodstuffs [11], procedure is time-consuming and with limited success in respect to reproducibility. HPLC methods with UV detection arealso available, however, lack of chromophore and general limited selectivity and sensitivity of UV detectors make detection of low levels of acrylamide difficult and with questionable reproducibility. Liquid chromatography coupled with tandem-mass spectrometry (LC-MS/MS) is by far the most suitable analytical technique for acrylamide measurements given its inherent sensitivity and selectivity, as well as capability of analyzing underivatized molecule. Since molecular ion of acrylamide is of low weight (m/z = 72 for protonated molecule), it is not specific enough for an unambiguous spectral determination using full scan. Therefore, multiple reaction monitoring (MRM) mode of operation with triple-quadrupole instruments is the technique of choice for accurate and precise measurements of acrylamide content.

Another difficulty encountered in acrylamide analysis is practical impossibility to obtain true blank sample. The need for matrix-matched calibration, especially in mass spectrometry (due to often severe matrix-related effects), implies use of fortified blanks and blank samples in order to properly quantify the compound of interest. However, virtually every thermally processed matrix contains some levels of acrylamide. Furthermore, application of high temperatures during analytical process (e.g. Soxlet extraction or GC injection) can also contribute to the formation of additional acrylamide levels, if precursors for Maillard reaction are present in the sample [2].

The aim of this paper is to verify the performance characteristics and fitness for purpose of rapid and simple QuEChERS-based LC-MS/MS method for determination of acrylamide in two food commodities that are likely to contain significant quantities of this molecule and are, at the same time, rather abundant in everyday nutrition - potato chips and coffee. The applied analytical method is slightly modified LC-MS/MS protocol presented by Mastovska and Lehotay [2]. QuEChERS (**Qu**ick **E**asy **Ch**eap **E**ffective **R**ugged **S**afe) is an extraction and cleanup method which gained much popularity in analytical community in the past decade, due to significant benefits in solvents reduction, costs, turnarround time and efficiency of extraction. Chemically, it is a form of dispersive solid phase extraction and cleanup protocol that provides satisfactory to excellent recoveries and adequate interferences removal. It is developed by the US Department of Agriculture for the purpose of multiresidual pesticide analysis [12]. However, it can be successfully applied to other analytes, including acrylamide. Although it is not suitable for ultra-trace (sub-ppb) levels, recommended maximum levels for acrylamide are quite within reach of its capabilities.

2. Materials and Methods

Samples of potato chips (n=6) and roasted grinded coffee (n=6) were purchased from local supermarkets. Analytical standard of acrylamide (99%) was obtained from Sigma-Aldrich (St. Louis, MO, USA). Acetonitrile, n-hexane and water of HPLC purity were acquired from the same manufacturer. QuEChERS extraction kit (1g NaCl + 4g $MgSO_4$ in 50 mL extraction tube; 50mg PSA + 150mg anhydrous $MgSO_4$ in dispersive 2mL SPE tube) were obtained from Agilent (Agilent Technologies CA, USA).

2.1. Sample preparation - potato chips

One and a half gram of previously grinded potato chips was weighted into 50 mL QuEChERS tube (1g NaCl + 4g $MgSO_4$). Standard of acrylamide was added at this point in QC samples (matrix-matched standards and control spike). HPLC-grade n-hexane (7mL) was added and the tube was vortexed for 1 min. HPLC-grade water and acetonirile (10 mL each) were added to the tube and shaken vigorously for 1 min. This mixture was centrifuged at 4500g for 5 min. Hexane layer (upper) was discarded using Pasteur pipette and 1 mL of water:acetonitrile layer (lower) was transferred into 2mL dispersive SPE tube containing 50mg PSA and 150mg of anhydrous $MgSO_4$. Tubes were vortexed for 1 min and centrifuged at 9000g for 5 min. Supernatant (1 mL) was carefully withdrawn using automatic pipette in order not to disturb precipitate and transferred into HPLC vial.

2.2. Sample preparation - roasted coffee

Coffee samples (2g) were transferred into an empty tube and 10 mL of boiling water was added. Tube was closed and shaken vigorously for one minute. Entire content was transferred to QuEChERS extraction tube. The other steps were identical as in potato chips preparation except for the defattening step with n-hexane which was omitted. During the method development, it was concluded that interferences from fat were much less intense, comparing to potato chips while the peak shape and baseline noise were satisfactory. Therefore, in the interest of simplicity and time-saving, defattening step was skipped.

2.3. LC-MS/MS analysis

LC-MS/MS system consisted of Shimadzu (Shimadzu Corporation, Kyoto, Japan) components: two LC-30AD UPLC pumps connected in binary gradient mode, DGU-20A degassing unit, SIL-30AC autosampler, CTO-20AC column oven, CBM-20A system controller and LCMS 8040 triple-quadrupole mass spectrometer. Merck (Darmstadt, Germany) Purospher® STAR RP-18 encapped column (50x2.1mm, 3μm) was used for acrylamide separation from co-eluting compounds and was maintained at 50°C during analysis. Mobile phase consisted of 2% methanol and 1% acetonitrile in 0.1% formic acid(isocratic mode), flow rate was 300 μL/min. Mass spectrometric determination of acrylamide was performed in electrospray (ESI) positive mode. Interface voltage was set to 4kV. Heat block, interface and dessolvation line temperatures were 400°C, 350°C and 250°C respectively. Nebulizing gas (N_2) flow was 3 L/min while drying gas flow was maintained at 15 L/min. Detector was operating in multiple reaction monitoring (MRM) mode. Collision gas was Ar_2. Two product ions of acrylamide were monitored (m/z 72>55 and 72>27). Dwell time was set to 100 ms. Adequate collision energy, Q1 and Q3 pre bias values were determined using automatic method optimization procedure. LabSolutions software was responsible for data acquisition and processing.

Five-point calibration curve was constructed from pure standards corresponding to final sample concentrations of 0, 50, 100, 500 and 1000 $\mu g\ kg^{-1}$. Samples of potato chips and coffee previously purchased from the supermarkets were analyzed prior to analysis, and specimens with lowest content of acrylamide were selected for blanks. Selected "blank" samples were fortified with acrylamide standard solution (c=1 ng μL^{-1} in acetonitrile) up to the final concentrations in a sample of 100, 200, 500 1000 and 1500 $\mu g\ kg^{-1}$. Spiked samples prepared in such a way were used for construction of matrix-matched calibration curve. Blank substraction and external standardization methods were used for quantification of acrylamide content.

3. Results

Described method showed good performance characteristics for both matrices. Degree of interferences removal during extraction and cleanup stages of sample preparation was sufficient for reliable analyte identification and subsequent quantification. Figure 1 represents MRM chromatogram of total ion current and both transition products for acrylamide pure standard (0.2 ng on column). Retention time of acrylamide was 0.5 min due to short column (50 mm) and addition of 1% acetonitrile to the mobile phase. Initial run time was set to 4 minutes, however subsequent coelution with other compounds was not observed in both sample types. Separation of acrylamide was clear with narrow peak with slight tailing. This could be reduced with alteration of chromatographic parameters but runtime of the analysis would be longer.

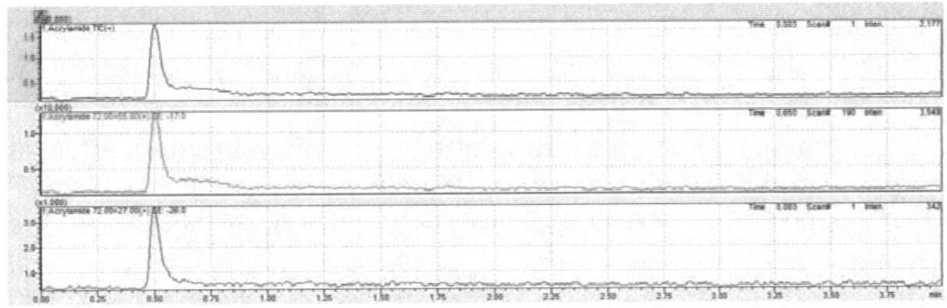

Figure 1. Chromatogram of acrylamide standard solution (0.2 ng on column)

Baseline noise was prominent in coffee samples at low levels (under 50 μg kg^{-1}), while this effect was not observed in potato chips samples. This could be explained by the fact that coffee is rather difficult matrix compared to potato chips with much more coeluting compounds. Omission of defattening step in coffee analysis was excluded as a reason for increased baseline noise, since the same noise was observed in samples subjected to defattening with n-hexane during method development. Even with increased noise, peak was sufficiently resolved and reproducible integration and quantitation could be achieved. Figures 2 and 3 show MRM chromatograms of total ion current and both transition products for acrylamide in coffee and potato chips respectively at concentrations of 100 μg kg^{-1}.

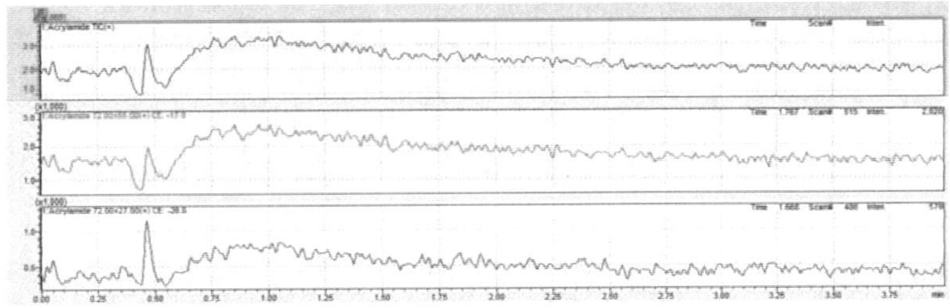

Figure 2. Chromatogram of 100 μg kg^{-1} of acrylamide in roasted coffee sample

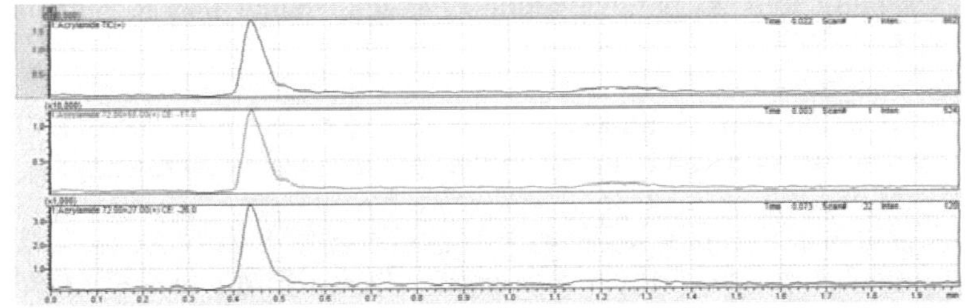

Figure 3. Chromatogram of 100 μg kg^{-1} of acrylamide in potato chips sample

Coffee samples, as stated previously, were harder matrix to analyze comparing to potato chips. One of the main drawbacks was pronounced difficulty in obtaining credible blank samples. Six samples of commercially acquired roasted coffee were analyzed in preliminary investigations. All samples contained acrylamide in various concentrations. Unlike potato chips samples where it was possible to find a sample with low enough concentrations of acrylamide which could be considered as true blank, in the case of coffee samples, blank substraction method and standard addition were employed for reliable quantification.

With every batch of samples, control spike was injected after blank with the aim to assess recoveries of the method. Blank samples were fortified at 50 μg $kg^{-1}$for coffee and 100 μg kg^{-1} for potato chips. Obtained recoveries were in the range of 85-112% for coffee samples and 91-107% for potato chips samples.

Linearity of the method was satisfactory. Five-point calibration curve consisting of matrix-matched standard was constructed at the beginning of each batch. Coefficients of determination (R^2) were in the range of 0.995-0.998 for coffee and 0.997-0.999 for potato chips.

Experimentally determined limit of quantification was 35μg $kg^{-1}$for coffee and 20 μg kg^{-1} for potato chips. Interlaboratory reproducibility (Cv%) at recommended indicative values from the European Commission (450 μg $kg^{-1}$for roasted coffee and 1000 μg kg^{-1} for potato chips) were 7.6% and 5.8% respectively.

Accuracy of the method was initially difficult to assess due to the lack of adequate certified reference material (CRM). Recovery could be a measure of accuracy when CRMs are not available, but practical impossibility to obtain true blank matrix further complicates this process. However, subsequent proficiency test for acrylamide in roasted coffee and z-score of -1 confirmed that applied method was adequate for accurate measurings.

One aspect of quality control was not applied in this method - internal standardization since d_3-acrylamide was not available in laboratory at the time of method development and validation. Even with this, presented method showed adequate performances in analysing roasted coffee and potato chips.

4. Conclusion

Presented method for quantitative determination of acrylamide in roasted coffee and potato chips is fit for purposes of self-control in food industry as well as regulatory controls carried out by the governmental agencies. Performance characteristics of the method are compliant with criteria for analytical methods validation. Low solvent consumption, lack of derivatization, overall low cost of analysis and very high sample throughput in both preparation and analysis steps are key benefits of applying such method in order to protect consumers from high exposure to acrylamide in food.

Acknowledgments
This work was supported by the Ministry of Science and Technological Development of the Republic of Serbia, grant III46009.

References

[1] IARC 1994 Monographs on the evaluation of carcinogenic risks to humans: Some industrial chemicals **60** 389-433
[2] Mastovska K and Lehotay S J 2006 *J. Agric. Food Chem.***54** 7001-8
[3] Claeus W L, Vieeschouwer K de and Hendrick M E 2005 *Trends Food Sci. Technol.* **16** 181-93
[4] Tareke E, Rydberg P, Karlsson P, Eriksson S and Tornquist M 2002 *J. Agric. Food Chem.* **35**(6) 909-12
[5] EPA (Environmental protection Agency) 2010*EPA/635/R-07009F* URL https://cfpub.epa.gov/ncea/iris/iris_documents/documents/toxreviews/0286tr.pdf
[6] EU (European Union) 2007 Commission Recommendation of 3 May 2007 on the monitoring of acrylamide levels in food 2007/331/EC OJ L **123** 33-40

[7] EU (European Union) 2010 Commission Recommendation of 2 June 2010 on the monitoring of acrylamide levels in food 2010/307/EU OJ L **137** 4-10
[8] EFSA (European Food Safety Agency) 2012 Scientific report of EFSA. Update on acrylamide levels in food from monitoring years 2007-2010*EFSA Journal***10** 10 2938
[9] EU (European Union) 2013 Commission Recommendation of 8 November 2013 on investigations into the levels of acrylamide in food 2013/647/EU OJ L **301** 15-17
[10] Andrawes F, Greenhouse S and Draney D 1987 *J. Chromatogr.* **399** 269-75
[11] Castle L, Campos M and Gilbert J 1991 *J. Sci. Food Agric.* **54** 549-55
[12] Anastassiades M, Lehotay S J, Štajnbaher D and Sheck F J 2003 *J. AOAC Int.* **86**(22) 412-31

IOP Conf. Series: Earth and Environmental Science **85** (2017) 012011 doi:10.1088/1755-1315/85/1/012011

The role of food quality assurance and product certification systems on marketing aspects

Z Petrović, D Milićević, I Nastasijević, V Đorđević, D Trbović and B Velebit

Institute of Meat Hygiene and Technology, Kaćanskog 13, Belgrade, Serbia

E-mail: zoran.petrovic@inmes.rs

Abstract. The level of quality that a product offers to consumers is a fundamental aspect of competition in many markets. Consumers' confidence in the safety and quality of foods they buy and consume is a significant support to the economic development of production organizations of this type, and therefore the overall economic development. Consumer concerns about food safety as well as the globalization of food production have also led to the existence of a global internationally linked food production and distribution system. The necessity demanded by the consumer population to provide safe food with consistent quality at an attractive price imposes a choice of an appropriate quality assurance model in accordance with the specific properties of the product and the production processes. Modern trends, especially for the last ten years in quality assurance within specific production, such as the food industry, have marked the trend of hyperproduction and a number of production and safety standards, as well as a change of approach in the certification process of organizations according to one or more standards. This can be an additional source of costs for organizations, and can burden the food business operator`s budget in order to ensure their consistent application and maintenance. Quality assurance (QA) standards are considered to be a proven mechanism for delivering quality of product.

1. Introduction

Generally speaking, food quality is a rather heterogeneous term because it is directly related to the individual perception of the consumer (end-user). Quality of goods is defined as a three-dimensional category consisting of: search, experience and credence [1].

The dimension of search is reflected by the buyer's individual perception of quality at the place of purchase using prior knowledge. The dimension of experience is characteristic in the period after the product has been purchased (after consuming the product and registering its taste and sensory properties). The third dimension (credence) of food quality can be regarded as in the case of an average consumer who rarely or never subjectively perceives the quality of the product, but they buy it because they gathered information from others – for example that organic food is healthy. The person's judgment on the quality of the product is, in fact, made by others [2,3].

As a matter of fact, food safety is an important issue that affects anyone who consumes food. Food safety could be classified as a new component of overall product quality and can make it into the third dimension of food quality. Attributes of the third dimension are characterized by a greater need for information about the product [4].

Some consumers are willing to pay marginally higher prices for quality assurance and hence reduced risk in food, especially during periods of safety concern [5].

Food Assurance Schemes are generally run as product certification schemes and use regular independent inspections to check that participants are meeting specific scheme standards. The scope of assured food schemes covers both primary production and processes covering the rest of the food chain as far as retail sale.

2. Consumers perception of food quality and public interest in safe food

Whereas in the first seven decades of the 20th century, consumers were interested in quantity and mainly sensory aspects of meat quality, the focus of their interest was later shifted to a comprehensive view of quality and especially to safety. Safe food has to be nutritious, and low in chemical contaminants and microbial counts. In order to enforce this among all the participants in the food chain, authenticity and traceability systems from farm to fork, including feed manufacturers, are required nowadays by directives and laws. Labelling systems or even further advanced nutrition and health claims have been introduced [6].

Consuming food, certainly, has great implications to human health. However, the importance between mutual relationships of certain groups of food-related risks is less clear. According to Bunčić [7], the confusion regarding this issue, often presented not only in the media, is greatly contributed to by the disagreement between the opinions of experts (based on research data) and perceptions of consumers (laymen). In a study conducted in the United Kingdom, food experts believed that wrong nutrition causes foodborne illness that can have a lethal outcome, whereas, contrary to that, consumer surveys show that approximately 50% of the population consider pesticides and additives as the greatest food hazards, and about 40% think that those are genetically modified ingredients.

A number of authors have dealt with the question of what really represents the quality of food when considered and applied to food supply chains. There are different interpretations about the general definition of quality [8,9,10]. From the large quantity of literature in regard to this issue, some commonly used food quality parameters have been identified (table 1).

Table 1. Commonly used food quality parameters [11].

Quality Aspect	Description
production method	traditional, ecologically acceptable
production place	regional or locally recognized product
traceability	fully traceable origin and production steps
characteristics of raw materials and authenticity	consumer recognition and standard quality they are used to
food safety	confidence in the safety of processes and technology of production, packaging, labelling, storage and distribution
nutritional properties	provides favourable nutritional properties and proper nutrition
sensory properties	appearance, freshness, texture, colour, smell, taste, consistency, appearance of the package, way of presentation
functional properties	food fulfils the intended purpose
biological value	food has properties that have a beneficial effect on the healthy life style

2.1. Definition of meat quality

The annual surveys conducted to collect data about the expectations and needs of consumers in supermarkets show that the main driving force for the purchase of fresh beef is the "taste" of meat. The taste ranks in first place, as a factor that is defined as "very important" when people make buying decisions. Holding second-, third- and fourth-ranked places as other important factors are: nutritional properties, safety and price, respectively [12,13]. This quality parameter (taste) is often called "tenderness" of meat, i.e. gauging how easily it is chewed or cut. Tenderness is a desirable property, as tender meat is softer, easier to chew, and generally more palatable.

The basic message is that the consumer wants to pay more for desirable taste. The taste is a measure of consumer satisfaction in food. Meat producers who want to be guided by their customer's demands must look at all the factors affecting the taste of their products [14].

3. Accredited product certification – the way to confirm and communicate quality of the product

Quality assurance (QA) standards are considered to be a proven mechanism for delivering quality of product and service [15].

The emerging effort to provide higher food safety and food quality has led to stricter safety specifications and a considerably increased number of quality assurance schemes both in an international and European Union (EU) level [16]. Beyond the usefulness of these quality assurance schemes to the consumers, the aspect of "quality" has also been accepted as an important ingredient of marketing, which offers producers a great opportunity to differentiate themselves in the market and add value to their products [17].

Production standards are set by the assurance scheme and vary across different schemes, generally covering food safety and traceability, animal welfare and environmental protection. Members of a particular scheme can use the scheme's logo on their produce, and/or use a specific claim, to advertise to consumers that the product has been produced to these standards [18].

The "Q Mark" for food products is owned by the scheme owner. The producer who wishes to use this mark on his product is required to obtain formal approval from the scheme owner for the use of the mark only after they have been assessed to be compliant to all the scheme requirements by the certification body (CB).

The International Accreditation Forum (IAF) and International Laboratory Accreditation Cooperation (ILAC) work constantly to maintain a high level in credible certification which is increasingly essential to world trade. This cannot be achieved without the added credibility of accredited certifications. Many mainstream product certification schemes (in the food sector) now have standards that are significantly above the legal minimum [18].

Nowadays, conventional-plus food products are increasingly available on the food market. These are conventional food products that communicate a specific attribute that also applies to corresponding organic products. Thus, conventional-plus products can be considered as products placed between organic and conventional products. Given this overlap with respect to specific attributes, conventional-plus products could compete with organic products. In particular, consumers who occasionally buy organic food might be interested in conventional-plus products [19].

Unlike existing certification schemes in different production sectors which, as a rule, only confirm that the product complies with regulations, there is an idea to develop specific certification schemes and certification procedures of foodstuffs in accordance with EN ISO/IEC 17065:2012. The scheme must be able to present the ways, if any, the scheme standards exceed the legal minimum.

The prerequisite for approval of the certification bodies by the scheme owner are accreditation by the National Accreditation Board for Certification Bodies (NBCB) providing accreditation to Certification Bodies based on assessment of their competence as per the Board's criteria and in accordance with International Standards and Guidelines.

These schemes are conceived as a voluntary product certification and are intended to enable the confirmation and communication of the qualitative characteristics of a particular product that in some way represent a comparative advantage on the market. In short, do you think your product has something that separates it from others on the shelf, and you want it, independently confirmed, to be communicated to customers. The steps for such a certification procedure should be: defining specifications, validation, certification, communication and surveillance. The certification mark (on the basis of the certified characteristics) is applied to the product and directly communicates the quality to the buyer.

4. Conclusion

Accredited product certification is a logical supplement to the existing certification and food assurance systems. The success depends on a lot, but mostly on how much quality surrounds us. If quality management and assurance systems produce what they need, i.e. quality, then there is no remaining work to be done.

Acknowledgments
This study was supported by project No. TR 46009 III funded by the Ministry of Education, Science and Technological Development of the Republic of Serbia.

References

[1] Darby M R and Karni E 1973 Free competition and the optional amount of fraud *J. Law. Econ.* **16** 67–88

[2] Becker T 1999 The Economics of Food Quality Standards *Proc. Int. Workshop on Stand. Res.* University of the Federal Armed Forces May Hamburg Germany pp 24–7

[3] Bech-Larsen T and Grunert K G 2001 Konsumentenentscheidungen bei Vertrauenseigenschaften: eine Untersuchung am Beispiel des Kaufes von ökologischen Lebensmitteln in Deutschland und Dänemark. *Marketing-ZFP* **23** 188–97

[4] Nelson P 1970 Information and Consumer Behavior *J. Polit. Econ.* **78** 311-29

[5] Latouche K, Rainelli P and Vermersch D 1998 Food Safety Issues and the BSE Scare: Some Lessons from the French Case *Food Policy* **23** 347-56

[6] Honikel K O 2005 Food Safety in Meat Production – A Challenge and a Chance for the Future *Tehnologija mesa* **46** 22-8

[7] Bunčić S 2005 Brief Overview of Main Existing and Some Potential Microbial Hazards in Meat *Tehnologija mesa* **46** 29-38

[8] Marsden T and Arce A 1995 Constructing Quality: Emerging Food Networks in the Rural Transition. *Environ. Plann.* A **27** 1261-79

[9] Ilbery B and Kneafsey M 1998 Product and Place: Promoting Quality Products and services in the Lagging Rural Regions of the EU *Eur. Urban Reg. Stud.* **5** 329-41

[10] Goodman D and Watts M 1997 Global appetite, local metabolism: Nature, culture and industry in fin-de-siècle agro-food systems Ch 1 ed D Goodman and M Watts *Globalizing Food: Agrarian Questions and Global Restructuring* (London and New York: Routledge) pp 1-32

[11] Morris C and Young C 2000 "Seed to Shelf", "Teat to Table", "Barley to Beer" and "Womb to Tomb": Discourses of Food Quality and Quality Assurance Schemes in the UK *J. Rural Stud.* **16** 103-15

[12] Food Marketing Institute 1988 Trends 88: Trends in the United States – Consumer Attitudes and the Supermarket 1988 *FMI* Washington D C

[13] Food Marketing Institute 1988 Trends 98: Trends in the United States – Consumer Attitudes and the Supermarket 1988 *FMI* Washington D C

[14] Quinn F 1999 Tracing the future of meat *Proc. 12th World Meat Cong.* Dublin Ireland. Session 2 pp 1-10

[15] Manning L, Baines R N and Chadd S A 2006 Quality Assurance Models in the Food Supply Chain *Brit. Food. J.* **108** 91-104

[16] Botonaki A, Polymeros K, Tsakiridou E and Mattas K 2006 The Role of Food Quality Certification on Consumers' Food Choices *Brit. Food. J.* **108** 77-90

[17] Jervell A M and Borgen S O 2004 New marketing channels for food quality products in *Norway Acta Agric. Scand.* Sect. C Food Economics **1** 108-18

[18] Kirk-Wilson R 2008 Review of Uptake of FSA Food Assurance Scheme Guidance by UK Scheme Operators (Available at: http://collections.europarchive.org/tna/20101209122142/; http://www.food.gov.uk/multimedia/pdfs/foodassureguidance.pdf)

[19] Stolz H. Stolze M, Hamm U, Janssen M and Rutoc E 2011 Consumer Attitudes Towards Organic Versus Conventional Food with Specific Quality Attributes NJAS Wagen. *J. Life Sc.* **58** 67– 72
[20] ISO 2012 EN ISO/IEC 17065: Conformity Assessment-Requirements for Bodies Certifying Products, Processes and Services International Organization for Standardization (Geneva, Switzerland

IOP Conf. Series: Earth and Environmental Science **85** (2017) 012012 doi:10.1088/1755-1315/85/1/012012

Dietary habits of Serbian preschool and schoolchildren with regard to food of animal origin

V Đorđević[1], D Šarčević[1] and M Glišić[2]

[1] Institute of Meat Hygiene and Technology, Kaćanskog 13, 11 000 Belgrade, Serbia;
[2] Department of Food Hygiene and Technology, Faculty of Veterinary Medicine, Bulevar oslobođenja 18, 11 000 Belgrade, Serbia

E-mail: danijela.sarcevic@inmes.rs

Abstract. The goal of this study was to explore attitudes and habits of Serbian preschool and school children in consumption of meat products, milk and milk products, eggs and egg products and honey and bee products. The survey was conducted on a sample of 227 children, divided into three different age groups: preschool (ages 4-6), primary school I-IV grade (ages 7-11) and primary school V-VIII grade (ages 12-15). The results showed that all examined groups of children consumed meat products, milk and milk products, eggs and egg products, and honey and bee products. In all groups of children, the most frequently consumed food (among our food category choices) was dried ham (consumed by 19.64% of preschool children; 23.75% of schoolchildren from I-IV grade; 19.74% of schoolchildren from V-VIII grade). Fewer preschool children consumed sterilized milk compared to children of school age. The results showed that in all three groups of children, the most commonly consumed milk products were yoghurt (from 12.20 to 15.29% of children consumed these) and sour cream (from 11.57 to 12.74% of children consumed this), while kefir was the least-consumed product. In addition, there was no difference in consumption of boiled or fried eggs in the examined groups of children, while the consumption of egg products (mayonnaise) was higher in the group of preschool children than in the group of schoolchildren from V-VIII grade. Preschool children consumed honey 14.99% more often than schoolchildren from I-IV grade, and 14.49% more often than did schoolchildren from grade V-VIII.

1. Introduction

In the process of healthy physiological and psychological development of a person, the period of youth is very important. Children's eating habits are influenced by numerous social and individual factors. Improper eating behaviour in children and youths can cause numerous health complications that manifest themselves in adulthood [1]. The choice of food, its classification, and the methods of obtaining, storing and serving food is a cultural phenomenon [2]. Food preferences play a central role in food choices and consumption, and can be described as general predisposition for a particular food, expressed by degree of liking [3,11]. Food promotion has an effect on children's preferences, purchase behaviour and consumption of food categories, as well as consumption of specific brands within food categories [4].

Due to this complexity of eating behaviour, it is difficult to predict a child's eating behaviour patterns and preferences, or to precisely define the causes of a certain form of behaviour. Understanding the various factors that influence children's eating behaviour is the first step in forming effective interventions intended to improve the eating habits of children and young people (adolescents) [5].

The purpose of this paper was to provide information about Serbian preschool and school children's most consumed foods among meat products, milk and milk products, eggs and egg products and honey and bee products.

2. Materials and Methods

The sample frame for this research consisted of 227 preschool children and primary schoolchildren, from one Belgrade municipality, who participated in the survey. Children were of three different age groups: preschool children (ages 4-6), primary schoolchildren I-IV grade (ages 7-11) and primary schoolchildren V-VIII grade (ages 12-15). Convenience sampling was used, and the questionnaire was distributed to all children who agreed to participate after they had been informed about the goals of the research. The questionnaire consisted of the following groups of questions on children's consumption habits for meat products, milk and milk products, egg and egg products and honey and bee products. The questionnaire was distributed among respondents, self-administrated, and was collected after three days.

Statistical analysis

Statistical analysis of the results was performed using GraphPad Prism software version 6.00 for Windows (GraphPad Software, San Diego, CA, USA, www.graphpad.com). All data were expressed in percentages and the differences were tested by Chi-square test. Values of $P< 0.05$ were considered statistically significant.

3. Results

Answers to the question, "Do you eat meat products, milk and milk products, eggs and egg products or honey and bee products?" (Table 1), showed that there was no difference between the different age groups of surveyed children, and that most of the children reported they ate meat products, milk, and milk products, eggs and egg products, and honey and honey products (P>0.05). All children of preschool age consumed milk and milk products (Table 1). Results related to consumption of honey showed that fewer older schoolchildren from V-VIII grade consumed honey than the younger groups (10.51% and 9.92% fewer, respectively, than preschool children and schoolchildren from I-IV grade).

Table 1. Results (% of yes answers/% of no answers) of different age groups of children in response to the survey question, "Do you eat meat products, milk and milk products, eggs and egg products or honey and bee products?"

Group	Meat Products	Milk and dairy products	Eggs and egg products	Honey and bee products
Preschool children	92.31/7.69	100.00/0.00	94.59/5.41	90.00/10.00
Schoolchildren grade I-IV	95.92/4.08	94.51/5.49	92.54/7.46	89.41/10.59
Schoolchildren grade IV-VIII	93.67/6.33	97.40/2.60	93.06/6.94	79.49/20.51
P-value	0.746	0.278	0.922	0.149

The results in Table 2 showed that there was no difference between the tested groups of children in frequency of yes/no responses related to the consumption of different meat products, with the exception of other products (bacon, chicken breast and pancetta), where there was a significant difference between preschool children and schoolchildren (P<0.05). Among meat and meat products, the group of preschool children most frequently consumed dried ham (19.64% of all preschool children ate this), and then in descending order, the following meats: cooked sausages-frankfurters, cured ham, Čajna fermented sausages, pate, smoked pork neck, Kulen, grilled sausages, other products and mortadella (1.19%). Altogether, 23.75% of schoolchildren from I-IV grade consumed dried ham, but fewer children consumed mortadella and other products (2.01% and 0.67%, respectively). Children of school age from V-VIII grade most frequently consumed dried ham and Čajna fermented sausages (19.74% and 15.46% of children consumed these products, respectively), and the least frequently, mortadella and other products (they were consumed by only 2.96% and 0.33% of these children, respectively).

Table 2. Consumption patterns (% of children reporting they consumed a food) among different age groups of children, distributed by the predominantly consumed meat products

Meat products	Preschool children	School children I-IV grade	School children V-VIII grade	*P*-value
Cured ham	14.29	8.70	8.55	0.093
Dried ham	19.64	23.75	19.74	0.409
Fermented sausages-Čajna	13.69	13.38	15.46	0.743
Fermented sausages-Kulen	5.36	7.36	8.55	0.446
Smoked pork neck	9.52	9.70	9.21	0.979
Mortadella	1.19	2.01	2.96	0.436
Cooked sausages/frankfurters	15.48	14.38	12.17	0.559
Pate	11.31	13.71	13.82	0.708
Grilled sausage	4.76	6.35	9.21	0.159
Other (bacon, prosciutto, chicken breast)	4.76^{a}	0.67^{b}	0.33^{b}	0.0002

[a,b] Within rows, numbers with different lower-case letters are significantly different ($P<0.05$).

The results in Table 3 show that fewer preschool children consumed sterilized milk compared to children of school age ($P<0.05$), while schoolchildren from V-VIII grade liked to consume kefir more than preschool and school children from I-IV grade. The results of all three surveyed groups of children showed that they most commonly consumed yogurt (from 12.20% to 15.29% of all children consumed this) and sour cream (from 11.57% to 12.74%) and least commonly drank kefir (1.27% to 3.36%).

Table 3. Consumption patterns (% of children reporting they consumed a food) among different age groups of children, distributed by predominantly consumed milk and dairy products

Milk and dairy products	Preschool	School children I-IV grade	School children V-VIII grade	*P* value
Milk pasteurized	8.28	8.54	8.02	0.954
Milk sterilized	1.59^{a}	4.36^{b}	5.78^{b}	0.014
Yogurt	15.29	14.80	12.50	0.425
Fruit flavoured yogurt	11.78	7.97	9.89	0.183
Fermented milk	10.19	9.49	7.65	0.388
Kefir	1.27^{a}	1.33^{a}	3.36^{b}	0.036
Sour cream	12.74	12.71	11.57	0.816
Raw cheese	8.92	9.68	8.96	0.900
Caciocavallo	7.32	9.49	10.45	0.318

Cream cheese	5.73	5.69	5.78	0.998
Kaymak	6.05	5.69	5.41	0.926
Butter	6.05	4.36	4.29	0.448
Margarine	4.78	5.88	6.34	0.639

[a,b] Within rows, numbers with different lower-case letters are significantly different ($P<0.05$).

There was no difference in consumption of boiled and fried eggs in the examined groups of children, while the consumption of egg products (mayonnaise) was significantly more common in the group of preschool children compared to the schoolchildren group of V-VIII grade (Table 4).

Table 4. Consumption patterns (% of children reporting they consumed a food) among different age groups of children, distributed by predominantly consumed eggs and egg products

Eggs and egg products	Preschool	School children I-IV grade	School children V-VIII grade	*P*-value
Boiled eggs	46.58	43.84	51.26	0.489
Fried eggs	43.84	52.38	47.90	0.495
Egg products	9.59^{a}	$3.97^{a,b}$	0.84^{b}	0.012

[a,b] Within rows, numbers with different lower-case letters are significantly different ($P<0.05$).

There was no difference between the tested groups of children in their consumption of honey, propolis and royal jelly. Altogether, 14.99% more preschool children consumed honey than did schoolchildren I-IV grade, and 14.49% more preschoolers consumed honey compared to children of grade V-VIII. In addition, the results showed that school-aged children were more likely consumers of propolis than preschool children ($P>0.05$).

Table 5. Consumption patterns (% of children reporting they consumed a food) among different age groups of children, distributed by predominantly consumed honey and bee products

Honey and bee products	Preschool	School children I-IV grade	School children V-VIII grade	*P*-value
Honey	63.93	48.94	49.44	0.135
Propolis	19.67	36.17	31.46	0.089
Royal jelly	16.39	14.89	19.10	0.745

Table 6. Eating frequency of food groups (%) reported by different age groups of children

Food groups	Possible answer	Preschool	School children I-IV grade	School children IV-VIII grade	*P*-value
Meat products	Several times per day	10.00	10.59	10.00	0.990
	Once per day	36.00	42.35	55.00	0.079
	Several times per week	18.00	9.41	6.25	0.094
	Once per week	32.00	36.47	26.25	0.369
	Once per month	4.00	1.18	2.50	0.570
Milk and dairy products	Several times per day	$4.17^{a,b}$	11.76^{a}	1.35^{b}	0.021

	Once per day	$81.25^{a,b}$	69.41^{a}	91.89^{b}	0.002
	Several times per week	14.58^{b}	1.18^{a}	1.35^{a}	0.0004
	Once per week	0.00^{a}	14.12^{b}	5.41^{a}	0.009
	Once per month	0.00	3.53	0.00	-
Eggs	Several times per day	2.33	5.71	7.25	0.536
	Once per day	18.60	32.86	34.78	0.160
	Several times per week	34.88^{a}	4.29^{b}	7.25^{b}	<0.0001
	Once per week	44.19	47.14	43.48	0.902
	Once per month	0.00	10.00	7.25	0.111
Honey	Several times per day	2.22	5.33	0.00	0.138
	Once per day	26.67	36.00	35.29	0.535
	Several times per week	22.22^{a}	6.67^{b}	0.00^{b}	<0.0001
	Once per week	40.00	30.67	32.35	0.559
	Once per month	8.89^{a}	$21.33^{a,b}$	32.35^{b}	0.013

^{a,b} Within rows, numbers with different lower-case letters are significantly different ($P<0.05$).

The survey results show that meat and meat products and milk and dairy products are commonly consumed once a day (from 36% to 55% and from 91.89% to 69.41%, respectively) in all three examined age groups of children.

A greater percentage of preschool children compared with older children consumed eggs and honey several times per week (P<0.05), while schoolchildren from V-VIII grade were more likely than younger children to consume honey once per month.

4. Discussion

This survey of dietary attitudes and habits of Serbian children aged 4-15 indicates that meat products featured once per day in their diets. According to Weichselbaum and Buttriss [6], the contribution of food group intakes in the diets of children aged 4-18 were: meat products 29% (aged 4-10) and 38% (aged 11-18); milk and milk products 21% (aged 4-10) and 14% (aged 11-18). In our research, the most frequently consumed meat, among the entire meat and meat products group, was dried ham. Djordjević et al. found that meat products such as dried ham, Čajna fermented sausages etc. have an acceptable taste and appearance for most children [10], as does mayonnaise which they used in preparing sandwiches. According to Baltic and Boskovic [8], meat has a significant role for maintenance of proper growth, development and health of children. Children and teenagers need to eat well and be active to get the nutrients they need for growth and to reach their full developmental potential without the problem of excessive weight grain [7].

Among the products in our research, milk is one of the most complete foods, providing the body with most of the nutrients growing children need. We found that all children of preschool age consumed milk and milk products. Foods in this group are also a good source of B vitamins, such as riboflavin, B12, vitamin A and protein. During growth in childhood and into teenage years, a good supply of calcium is needed to build healthy bones and teeth. At puberty, the onset of which can begin in children as young as 9 years of age, or can occur in children as old as 18, five servings of milk and yoghurt daily is necessary to meet calcium needs [7]. Among Serbian children, yogurt is the most commonly consumed milk product because it is the most common milk product on the market that is sold in small, ready-to-use packages. Older children often consume kefir, because of the characteristic flavour, to which younger children are not accustomed. In addition, the higher consumption of pasteurized milk compared to sterilized milk among consumers in Serbia can be explained by its lower prices.

Our results related to consumption of honey showed that 10.51% fewer schoolchildren from V-VIII grade consumed honey than did preschool children and 9.92% fewer of the oldest children ate honey than did schoolchildren from I-IV grade. According to Karabasil et al. [9], nutritional and health aspects of the consumption of honey and bee products by children aged 4-15 is significant, because of its important influence of preservation of the immune system. The percentage of children who consumed eggs and egg products, and honey and bee products, was significantly different between preschool and school aged children from V-VIII grade (school aged children consume these foods less frequently). Eggs are a good source of energy and vital fats and minerals for breakfast, but should be limited to no more than seven per week [7].

In addition, we speculate that among children in Serbia, most food products are likely preferred consumer choices because of their sensory characteristics, such as mild taste and soft consistency (pancetta, Čajna sausages, frankfurter sausages, yoghurt and sour cream). An additional factor in favour of these products is their availability on the local market at mostly moderate prices.

5. Conclusion

In conclusion, children aged 4-15 years in Serbia are frequent consumers of foods of animal origin, which is necessary for their proper growth and development. It is very important that this trend continue, of course, as an integral part of a balanced diet, in spite of the negative image and context of meat and meat products in recent years. When it comes to preschool children, particular attention should be paid to parents' education, especially about the frequency with which their children consume foods of animal origin.

Acknowledgments

This paper was supported by the Ministry of Education, Science and Technological Development, Republic of Serbia, through the funding of the Project No III 46009 and Project No 31034.

References

[1] Moreno L A, Sarría A, Fleta J, Rodriguey G, Pérez González J M, and Bueno M 2001. Sociodemographic factors and trends in overweight prevalence in children and adolescents in Aragon (Spain) from 1985 to 1995 *J Clin Epidemiol* **54**(9) pp 921-927;

[2] Sarcevic D, Djordjevic V, Petronijevic R, Matekalo-Sverak V, Karabasil N, Popovic Lj, Jankovic V 2013. The attitudes and habits of Serbian schoolchildren in consumption of meat *Tehnologija mesa* **54** (2) pp 160-167;

[3] Nicklaus S, Boggio V, Chabanet C, Issanchou S 2004. A prospective study of food preferences in childhood *Food Qual Prefer* **15** (7) pp 805-818;

[4] Djordjevic V, Sarcevic D, Petronijevic R 2015. The attitudes and habits of Serbian schoolchildren to consumption of fish *Procedia Food Sci* **5** pp 73-76;

[5] Cullen K W, Rittenberry L, Olivera N, Baranawski T 2000. Environmental influences on children's diets: results from focus with African, Euro and Mexican-American children and their parents *Health Educ Res* **16** (2) pp 187-200;

[6] Weichselbaum E, Buttriss J L 2014. Diet, nutrition and schoolchildren: An update. *Nutrition Bulletin* **39** pp 9-73;

[7] Food Safety Authority of Ireland for children over age of 5 years and teenage 2011, pp 1-92

[8] Baltic M Z and Boskovic M 2015. When Men Meet Meat *Procedia Food Sci* **5** pp 6-9;

[9] Karabasil N, Baltić Ž M, Kilibarda N, Teodorović V, Dimitrijević M 2006. Med i mikrobiološki rizici, XIV Naučno savetovanje sa medjunarodnim učešćem „Zaštita i proizvodnja domaće pčele i meda", Poljoprivredni fakultet, Zemun, *Zbornik plenarnih i naučnih radova* pp 42-49;

[10] Djordjević V, Glišić M, Teodorović V, Mirilović M, Đurić S, Bošković M, Milan Ž B 2017. Quality assessment of Srpska sausage from nine different manufacturers in Serbia *Meat Technology* **58** 2 in press;

[11] Baltić Ž. M., Glišić M., Janjić Jelena, Marković Radmila, Bošković Marija, Dokmanović Marija, Đorđević Jasna. (2015). Značaj informisanja potrošača pri izboru namirnica. *Veterinarski Žurnal Republike Srpske*, 15, 1, 22-40.

Evaluation of nutritive quality of common carp, *Cyprinus carpio* L.

D Ljubojević[1], V Đorđević[2] and M Ćirković[1]

[1] Scientific Veterinary Institute "Novi Sad", Rumenački put 20, 21000 Novi Sad, Republic of Serbia
[2] Institute of Meat Hygiene and Technology, Kaćanskog 13, 11000 Belgrade, Republic of Serbia

E-mail: dragana@niv.ns.ac.rs

Abstract. Common carp is the most important commercial fish species in Serbia. This fish is a valuable source of nutritive components and plays a role in healthy human nutrition. This review evaluates the nutritive quality of common carp including proximate and fatty acid compositions as well as their effects on human health. The fat content and fatty acid composition of carp have been shown to vary due to different environmental factors and particularly due to nutrition. Technology of production and composition of planktonic and benthic organisms in fish ponds have been recognised as significant factors affecting carp meat quality and desirable chemical and fatty acid composition. Carp meat quality but also production parameters and fish health are positively influenced by a balanced feed mixture. Due to the low content of saturated fatty acids and cholesterol plus high levels of unsaturated fatty acids, common carp meat consumption could be linked with reduced risk of different heart diseases in humans. Also, fish proteins can have many beneficial roles in the preservation of human health. This paper emphasises the importance of consumption of common carp in order to prevent many diseases and preserve human health.

1. Introduction

Aquaculture has been one of the fastest growing branches of animal protein production during recent decades, and common carp is one of the most frequently farmed fish worldwide and also in Central and Eastern Europe. It is a highly esteemed fish species due to numerous desirable traits such as fast growth rate, good feed conversion ratio of both natural and supplementary feeds and relative resistance to poor environmental conditions and diseases. Common carp is the most widespread fish species on the market in Serbia and undoubtedly is a valuable nutritional source of proteins, lipids and other nutritive components which play many important roles in human health. Besides that, meat of common carp possesses a specific flavour and is easily digested. Having that in mind, information regarding factors affecting the quality of common carp meat is very necessary. The proximate composition and fatty acid composition of common carp are influenced by age, season, different environmental factors, culture systems and diet. The aim of this review is to evaluate the nutritive quality of common carp. We also highlight the importance of consuming common carp in order to promote human health.

2. Factors which affect nutritional composition of common carp

The fat content and fatty acid composition of different fish species vary significantly [1-4], even among fish belonging to the same family. The other factors which affect nutritional composition of common carp meat are environmental factors, structure of natural food present in fish ponds, cultural system including production technology and particularly, diet [1,5-7]. The use of formulated feed which contained ingredients of both animal and vegetable origin showed many positive effects on fish health, conditions, production parameters and meat quality [3,5,8]. The development of better feeding practice is important prerequisite in sustainable common carp production and enables the improvement of growth performance and chemical and fatty acid composition of carp.

3. The diet effect on meat quality of common carp

A variety of feeds are provided for carp raised in aquaculture, and the feed type mainly depends on the culture system. Traditional culture systems for carp are the extensive culture system, in which only the natural food available in the fish ponds is used, and the semi-intensive culture system, in which, to supplement natural food, cereals such as corn are used as additional feed. Also, simple mixtures of agricultural products which are easily accessible and cheap are often used. These are rice, wheat, barley, peas and defatted meals of oil-producing plants such as soya bean, sunflower, cotton or rapeseed meal. Undoubtedly, the quantity of feed given as well as the percentage of each component in the mixture varies considerably.

The main problem in traditional culture systems is that addition of corn frequently leads to increased accumulation of fat in the meat of common carp and especially around the internal organs [5,9]. That fact further leads to the prejudice that common carp is a very fatty fish. Examination of common carp muscle tissues from fish obtained from different culture systems showed that this really is prejudice, and that the accumulation of lipids in common carp meat is linked with the culture system and, consequently, with nutrition [5,7] (table 1).

During recent decades, due to the expansion and intensification of carp rearing, traditional feed consisting of locally accessible components has been replaced by industrially-produced feed which contains protein components of animal or plant origin. Such feed, used together with the exploitation of natural food, has become more common in the carp facilities in Serbia. The main result has been improvement of meat quality, particularly in terms of lowering the fat content and improving the fatty acid composition in comparison with the traditional semi-intensive system [4,5,9]. The importance of adequate preparation of fish ponds in these culture systems, which achieves a desirable structure of plankton and benthic organisms was described earlier [1]; these organisms are a significant source of nutritive components for carp [8]. Plankton and benthos contain high concentrations of n-3 polyunsaturated fatty acids (n-3 PUFA), including eicosapentaenoic acid (EPA) and docosahexaenoic acid (DHA) [8,10].

The lipid content of common carp fillets can differ greatly, from 6.3 to 15% [3] and from 1 to 13% [11,12] in commercially-sized fish, mainly depending on diet. Furthermore, the high average lipid content in carp fillets was probably the result of the fact that the energy-protein ratio in the fish diet was not well balanced [3]. It is known that feed rich in saccharides leads to an increased percentage of lipids in fish and a simultaneous decrease in the percentage of protein. Unsuitably high levels of carbohydrates and fats in common carp feed negatively affects the meat quality of this fish. Common carp deposits fats mainly around the organs, but also stores it in muscular tissue [13]. When carp were fed only natural food from their fish pond, their fillets contained only 1.76% fat, while fillets of carp fed supplementary maize contained 13.26% fat; fillets of carp fed supplementary wheat contained 11.22% fat and fillets of carp fed supplementary triticale contained 9.72% fat [14].

The fatty acid composition of fish meat corresponds to that of the feed which is ingested. The relatively undesirable composition of fatty acid profile in the lipids of common carp reported earlier [3] is connected with the diet. Also, the ratio of n-3/n-6 fatty acids in common carp varied by feed [5]. Carp grown on natural food had a high content of both n-6 and n-3 fatty acids [1,15]. Supplementary feeding with grains leads to reduced amounts of essential fatty acids in fish meat, and this is due to the lower proportion of natural food in the diet of the carp which received additional grains. Moreover, the PUFA/SFA ratio was the most favourable in the carp fed complete feed mixtures and the least in those fish fed with maize and wheat (table 1) [5]. The recommended ratio of PUFA/SFA, which is an important indicator of the quality of fish fat, should be above 0.4 [16]. Since some meats of terrestrially-farmed animals naturally have a PUFA/SFA ratio of around 0.1 [16], meat has been implicated in causing the imbalanced fatty acid intake of today's consumers. The ratio of unsaturated fatty acids and SFA (USFA/SFA) is also of great importance for assessing the quality of fats. It is assumed that the favourable ratio is above 0.35 [17]. Generally, fish fats contain relatively low percentages of SFA, less than 30% (table 1), with the exception of a few species [18]. Oleic acid was reported as the dominant MUFA in common carp (table 1) [3-6,18]. This fatty acid also plays an important role in the prevention of cardiovascular diseases [19]. High levels of oleic acid, arachidonic acid and palmitoleic acid are characteristic for fat of freshwater fish [3-6,18] (table 1).

Table 1. Content of protein (%), fat (%) and selected fatty acids (%) of total fatty acids) of the muscle of common carp fed different diets, adapted from Ćirković et al. [1] and Ljubojević et al. [4,5,6].

	Wild carp caught in Danube	Carp reared in polyculture on natural food	Carp reared in monoculture on natural food	Carp fed supplementary grains (80% corn and 20% wheat)	Carp fed feed mixtures in earthen ponds	Carp fed feed mixtures in cages
Protein content	16.69	16.21	15.4	15.59	17.1	16.23
Fat content	7.13	2.42	2.07	6.85	3.19	9.79
C16:0	19.4	18.35	20.99	17.31	16.89	12.52
C18:1, cis-9	30.2	19.39	32.58	51.35	34.45	33.55
C18:2, n-6	8.79	10.29	13.49	8.7	22.57	38.43
C18:3, n-3	2.71	5.96	4.59	0.61	2.12	3.16
C20:4, n-6	2.42	6.21	2.79	0.73	1.44	1.13
C20:5, n-3	1.36	4.05	1.17	0.2	0.93	0.20
C22:6, n-3	0.87	5.75	2.22	0.25	1.86	0.43
SFA[a]	27.59	27.15	28.82	24.19	22.4	17.18
MUFA[b]	52.94	28.79	43.49	64.31	45.12	37.25
PUFA[c]	19.7	44.08	27.69	11.53	32.48	45.46
n-6	13.73	22.96	17.93	10.24	26.01	41.56
n-3	5.97	21.12	9.75	1.29	6.48	4.00
n-3/n-6	0.44	0.92	0.54	0.13	0.25	0.1
n-6/n-3	2.30	1.09	1.84	7.99	4.02	10.79
PUFA/SFA	0.71	1.62	0.64	0.18	0.72	2.65
USFA[d]/SFA	2.63	2.68	0.96	0.48	1.45	4.82

[a] SFA – saturated fatty acids.
[b] MUFA – monounsaturated fatty acids.
[c] PUFA – polyunsaturated fatty acids.
[d] USFA – unsaturated fatty acids.

4. Consumer preference regarding meat quality of common carp

In some countries (Asia, Israel, Central and Eastern Europe), common carp is a highly esteemed fish species, and carp meat is highly regarded due to its specific savoury flavour and its high digestibility. In contrast, in other parts of the world, especially in North America and Australia, this species is considered as a weed-inhabiting fish that is not desirable for human consumption. However, many dishes can be prepared using this fish, undoubtedly confirming the gastronomic quality of carp [2].

Common carp meat contains highly valuable proteins, fats and other nutritive substances. It is a medium fatty fish and stores most of its fats as adipose tissue in the abdominal wall [11]. The amount of fat in muscle tissue contributes to its sensory properties, including organoleptic properties, texture and flavour. Meat which is rich in fat is juicy, while lean tissue is dry and often perceived as thickly fibrous [20]. The fat content in fish can sometimes exceed the protein content [21]. Such excessive fat content (>10%) has a negative effect on the sensory properties of common carp meat, which becomes soapy. This can occur when fish are cultured in earthen ponds where the amount of natural food is insufficient and a lot of grain is given as a supplementary feed [21]. On the other hand, an excessively low fat content in carp muscle has a negative impact on the sensory properties, and also consumption of so lean a fish reduces the intake of fatty acids in the human diet.

5. Cholesterol content in common carp and other fish

The amount of total cholesterol was 48.9 mg/100 g in one-year old carp in April and 54.3 mg/100 g in the same aged fish harvested in June [22]. The cholesterol content in female and male carp fillets was 69.4 mg/100g to 77.6 mg/100g [23], and was 55.8 mg/100g in two-year-old carp [1]. The total cholesterol content of common carp was 47mg/100g [24], and cholesterol in carp muscle varied considerably, from 38 to 120 mg/100g, depending on fish breed and age, husbandry system, and harvest season [25]. In other literature on cholesterol in fish, the cholesterol content of many freshwater fish species ranged between 40.99 and 52.79 mg/100g [26], while the total cholesterol in freshwater fish is lower than in marine fish [27]. In humans, daily intake of cholesterol is currently recommended not to exceed 300 mg [4].

6. The beneficial effects of common carp in human diet

Common carp should be included in human diets for at least three reasons: as a general source of nutritional components; as low fat, high protein food; and as source of PUFA. In carp meat, all nutrients and especially essential fatty acids are present in optimum quantities for human needs [1,3,5]. Consumption of n-3 PUFA from fish meat has positive effects on human health, especially in the prevention of heart attack, stroke, atherosclerosis and high blood pressure. Furthermore, there are positive effects on the circulatory system, the process of remembering and learning, reproductive system and photoreceptors [28].

It is well known that fish are the most important source of n-3 long-chain fatty acids and highly unsaturated fatty acids (n-3 HUFA) in human diets. A favourable n-3/n-6 ratio has a positive impact on human health [28]. There are various recommendations from world organizations related to fish consumption and intake of n-3 fatty acids, as well as of appropriate ratios of different groups of fatty acids. The optimal range of n-3/n-6 ratio for human health recommended by WHO/FAO is 0.5 to 0.25 [28]. There are various data regarding the n-3/n-6 ratio of common carp which varies between 0.8 and 2.4 [29]. Other studies reported this ratio is about 0.5 [5,25], or about 0.2 [5,22] (table 1).

The importance of fats and fatty acids from common carp for human health is highlighted in this review, but it is impossible to explain the beneficial effects of fish meat on the human health only in terms of fats, because the edible parts of fish include also muscle tissue, which provides a many other nutritional ingredients, such as proteins. Fish protein, in relation to casein, lowered the level of blood cholesterol in laboratory animals, showed anti-hypertensive properties and other beneficial effects related to cardiovascular diseases as well as showed favourable effects against obesity [30,31]. The percentage of essential amino acids in fish meat is very high, especially, for example, tryptophan, the precursor of serotonin that likely contributes to feelings of well-being in humans [31].

7. Conclusion

Common carp meat is an important source of nutrients in human nutrition. It provides not only n-3 fatty acids but a variety of other nutrients that are important for health. The chemical and fatty acid composition in carp meat varies significantly, which is due to different nutrition and environmental factors. Overall, however, data on the nutritional composition of common carp meat highlight the relative value of this food in balanced healthy human nutrition.

Acknowledgments
This work was supported by grants from the Ministry of Education, Science, and Technological Development of the Republic of Serbia (project no. TR31011)

References

[1] Ćirković M, Ljubojević D, Đorđević V, Novakov N, Petronijević R, Matekalo-Sverak V and Trbović D 2012a *Kafkas Univ Vet Fak Derg* **18** 775
[2] Ćirković M, Ljubojević D, Đorđević V, Novakov N and Petronijević R 2012b *Archiva Zootechnica* **15** 37
[3] Ljubojević D, Ćirković M, Đorđević V, Puvača N, Trbović D, Vukadinov J and Plavša N 2013a *Czech J. Food Sci* **31** 445
[4] Ljubojević D, Trbović D, Lujić J, Bjelić-Čabrilo O, Kostić D, Novakov N and Ćirković M 2013b *Bulgarian Journal of Agricultural Science* **19** 62
[5] Ljubojević D, Ćirković M, Novakov N, Jovanović R, Janković S, Đorđević V and Mašić Z 2013c *Kafkas Univ Vet Fak Derg* **19** 43
[6] Ljubojević D, Ćirković M, Đorđević V, Trbović D, Vranić D, Novakov N and Mašić Z 2013d *Tehnologija mesa* **54** 48
[7] Ljubojević D, Radosavljević V, Puvača N, Živkov Baloš M, Đorđević V, Jovanović R and Ćirković M 2015 *J. Food Comp. Anal.* **37** 44
[8] Ljubojević D, Ćirković M, Novakov N, Puvača N, Aleksić N, Lujić J and Jovanović R 2014 J Appl Ichthyol **30** 50
[9] Trbović D, Marković Z, Milojković-Opsenica D, Petronijević R, Spirić D, Đjinović-Stojanović J and Spirić A 2013 *J. Food Comp. Anal.* **31** 75
[10] Domaizon I, Desvilettes C, Debroas D and Bourdier G 2000 *J. Fish Biol* **57** 417
[11] Fauconneau B, Alami-Durante H, Laroche-Marcel J and Vallot D 1995 *Aquaculture* **129** 265
[12] Romvári R, Hancz C S, Petrási Z S, Molnár T and Horn P 2002 *Aquac. Int.* **10** 231
[13] Vacha F, Vejsada P, Huda J and Hartvich P 2007 *Aquac. Int.* **15** 321
[14] Buchtová H, Svobodová Z, Kocour M and Velíšek J 2010 *Acta Vet Brno* **79** 551
[15] Wood J D, Enser M, Fisher A V, Nute G R, Sheard P R, Richardson R I, Huges S I and Whittington F M 2008 *Meat Sci* **78** 343
[16] Kmínková M, Winterová R and Kučera J 2001 *Czech J. Food Sci* **19** 177
[17] Guler G O, Kiztanir B, Aktumsek A, Citil O B and Ozparlak H 2008 *Food Chem.* **108** 689
[18] Peterson D B, Fisher K, Carter R D and Mann J 1994 *Lancet* **343** 1528
[19] Żmijewski T, Kujawa R, Jankowska B, Kwiatkowska A and Mamcarz A 2006 *J. Food Comp. Anal.* **19** 176
[20] Ćirković M, Ljubojević D, Jovanović R, Janković S, Đorđević V, Novakov N, Trbović D and Lujić J 2012c *Proceedings, The International Conference Biological Food Safety and Quality BFSQ* 177
[21] Trbović D, Vranić D, Đinović J, Borović B, Spirić D, Babić J and Spirić A 2009 *Tehnologija mesa* **50** 276
[22] Komprda T, Zelenka J, Fajmonová E, Bakaj P and Pechová P 2003 *J. Agric. Food Chem.* **51** 7692
[23] Bieniarz K, Koldras M, Kaminski J and Mejza T 2001 *Archives of Polish Fisheries* **9** 5
[24] Kopicova Z and Vavreinova S 2007 *Czech J. Food Sci.* **25** 195
[25] Moreira A B, Visentainer J V, de Souza N E and Matsushita M 2001 *J. Food Comp. Anal.* **14** 565
[26] Luzia A L, Sampaio G R, Castellucci C M N and Torres E A F S 2003 *Food Chem.* **83** 93

[27] Simopoulos A P 2002 *Biomedicine & Pharmacotherapy* **56** 365
[28] Steffens W, Wirth M and Füllner G 2005 *Archives of Polish Fisheries* **13** 15
[29] Boukortt F O, Girard A, Prost J L, Ait-Yahia D, Bouchenak M and Belleville J 2004 *Med Sci Monit.* **10** BR397
[30] Oishi Y and Dohmoto N 2009 *J. Nutr. Sci. Vitaminol.* **55** 156
[31] Gadoth N 2008 *Brain Dev.* **30** 309

IOP Conf. Series: Earth and Environmental Science **85** (2017) 012014 doi:10.1088/1755-1315/85/1/012014

Sensory properties and fatty acids profiles of fermented dry sausages made of pork meat from various breeds

N Parunović[1], Č Radović[2] and R Savić[3]

[1]Institute of Meat Hygiene and Technology, 11000 Belgrade, Serbia
[2]The Institute for Animal Husbandry, 11080 Zemun. Serbia
[3]The Institute of Animal Sciences, Faculty of Agriculture, University of Belgrade, 11081 Zemun.

E-mail: nenad.parunovic@inmes.rs

Abstract. In this study, the parallel examinations on fatty acid profiles and sensory properties of fermented dry sausages are presented. Three types of kulen and sremska sausages were made, which varied depending on the percentage of meat and fat derived from different pig breeds: autochthonous (Swallow-Belly Mangalitsa and Moravka) and commercial (Swedish Landrace). In sausages made from meat of commercial pig breed, the highest cholesterol content was detected. However, sausage made from the Mangalitsa pork meat contained higher levels of monounsaturated fatty acid (MUFA) and unsaturated fatty acid (USFA), and lower saturated fatty acid levels (SFA).The level of polyunsaturated (PUFA) in sausage made of Landrace pork meat was significantly higher than levels in other types. These differences between fatty acids were mostly deriving by higher total *n*-6 PUFA content. In a sausage made of Mangulitsa meat, the values of atherogenic (IA) and thrombogenic (IT) health lipid indexes are lower. Kulen and sremska sausages made from the Mangalitsa pork meat was superior in terms of colour, odour, taste, after taste and overall acceptability. This study demonstrate that pig breed have an effect on the chemical content, fatty acids profiles and sensory properties of dry fermented sausages.

1. Introduction

In Europe, particularly in Mediterranean region, there is a special interest in autochthonous meat products produced from traditional pig breeds extensive way of breeding. Meat and meat products from local pig breeds generally have a good public acceptance and positive media image, and they are often considered to be better quality, than the meat and meat products from conventional, noble pig breeds and crossbreeds.

Mangalitsa is a typical representative fat breed of pig, for example, carcass sides are 65-70% fat and about 30-35% meat [1]. Unlike the colour of other pig meat, the Mangalitsa pig meat had a darker colour. The Mangalitsa fat was whiter and the intramuscular fat content of meat and thickness of back fat was considerably greater than fat of noble pig breeds. The lower saturated fatty acid (SFA) content and higher proportion of unsaturated fatty acid (USFA) compared with that in meat from other fat pig breeds is favorable from a human-nutrition point of view [2, 3].

Cholesterol levels in blood depend not only on dietary cholesterol, but also on the amount of fat and the fatty acid composition of the diet [4]. The influence of fat on cholesterol content can also be observed through the index of atherogenicity (IA) and thrombogenicity (IT), which includes those fatty acids that influence the change of cholesterol. Nutritionists advise a reduction in total fat intake, primarily of SFA and trans fatty acids, which are associated with an increased risk of cardiovascular disease and some kinds of cancers. Moreover, supporting that they reduce their fat intake, nutritionists force consumers to increase their intake of polyunsaturated fatty acids (PUFA), specifically *n*-3 PUFA, on account of *n*-6 PUFA. The PUFA/SFA and *n*-6/*n*-3 PUFA ratios have, accordingly, become some of the most significant parameters in evaluating the nutritional value and healthfulness of foods [5, 6, 7].

However, inline with modern trends directed at revitalizing and advancing traditional food production processes, autochthonous meat products made from local breeds are obtaining importance. Moreover, numerous studies have been conducted to determine the characteristics of traditional and naturally fermented sausages throughout the world [8,9,10].

2. Materials and methods

2.1. Animals and Diets

In this study, thirty mature male pigs, ten of each breed, were included: Swallow-Belly Mangalitsa, Moravka and Swedish Landrace. Pigs were housed in an experimental animal house, equipped appropriately to meet the requirements of the study, at the Institute for Animal Husbandry, Zemun, Serbia. All pigs had access to green forages (pasture, clover) ad libitum, with the addition of a feed mixture based on corn and wheat. Animals were stunned, slaughtered and exsanguinated at a local slaughterhouse at 105.2 ± 5.0 kg live weight. Meat was processed 24 h after slaughter and cooling.

2.2. Sausage Formulation and Processing

The examined variants of kulen and sremska sausages (Table 1) were manufactured in a processing plant of the Institute for Animal Husbandry. Three sausages were taken from each variant for all analyses and each analysis was done in duplicate.

Table 1. The percentage of meat originating from specified pig breeds in different types of fermented dry sausages

Pig meat different breeds	Fermented dry sausage types					
	Kulen			Sremska sausage		
	KM	KL	KMM	SM	SL	SMM
Mangalitsa %	100	-	50	100	-	50
Moravka %	-	-	50	-	-	50
Swedish Landrace %	-	100	-	-	100	-

Meat used in the production of kulen sausage contained little fat or connective tissue, being primarily meat from the leg, shoulder and some parts of neck and a firm backfat tissue, was also used. Muscle and adipose tissue (75:25) was chopped in a cutter (Seydelman K60, Germany) to achieve 10 mm granulation. After chopping, the chopped meat was transferred to a mixer and the other ingredients of the filling were added: 2.3% table salt, 0.4% saccharose, 0.3% garlic (powder), 0.3% pepper and 0.8% ground sweet and hot red paprika. The filling was then firmly stuffed into the natural pig colon casings. Smoking and maturation of sausages was carried out in the winter period. The smokehouse temperature ranged from 10 to 15° C, and humidity from 75 to 90%. Kulen sausages were smoked during the first 4 weeks. Subsequently, sausages were transferred to the ripening room at a temperature of 10 to 12° C. The ripening was carried out in a drying chamber under controlled conditions (Maurer, Germany).

The examined variants of sremska sausage were produced on the same day and in an identical manner. Meat and fat (85:15) were ground in a cutter (Seydelman K60, Germany) to 8 mm. The same amounts of ingredients were added to sausage variants: 2.3% salt, 0.011% NaNO2, 0.3% dextrose, 0.20% garlic and 0.5% sweet red paprika. The mixture was filled in pig small intestines of around 32 mm diameter. After stuffing, the sausages were hung on sticks and the ripening was carried out in a drying chamber under controlled conditions (Maurer, Germany).

2.3. Analyses

2.3.1. Chemical Analysis.

The following measurements were taken to analyse the chemical composition of the Kulen sausages after 90 days' maturation: protein, water, total fat, ash, total fatty acid and cholesterol concentrations. Chemical composition was determined by the methods defined by the Association of Official Analytical Chemists [11].

2.3.2. Extraction of Total Lipids to Determine Fatty Acids.

To determine the concentration of fatty acids, total lipids were extracted by the rapid extraction method, using solvents on the Dionex ASE 200. A homogenized sample, mixed with diatomaceous earth, was extracted with a mixture of hexane and isopropanol (60 : 40 v/v) in a 33 mL extraction cell at a temperature of 100 °C and under nitrogen pressure of 10.3 MPa. The extract thus obtained was steamed in a nitrogen flow at a temperature of 50 C until dry fat remains were obtained [12].

2.3.3. Determination of Fatty Acids.

Fatty acids as methyl esters were detected by capillary gas chromatography with a flame ionization detector. A predetermined quantity of lipid extracts, obtained by the rapid extraction method, was dissolved in tert-butyl methyl ether. Fatty acids were converted to fatty acids methyl esters (FAME) with trimethylsulphonium hydroxide, according to the SRPS EN ISO 5509:2007 method. FAMEs were analysed with the GC-FID Shimadzu 2010 device (Kyoto, Japan) on a cyanopropyl-aryl column HP-88 (column length 100, internal diameter 0.25 mm, film thickness 0.20 μm). The injected volume was 1 μL. Temperatures of the injector and detector were 250 C and 280 C, respectively. Nitrogen was used as a carrier gas, 1.33 mL/min, with a split ratio of 1 : 50, while hydrogen and air were used as detector gases. The temperature of the column furnace was programmed to range between 120 °C and 230 °C. The total duration of analysis was 50.5 min. Methyl esters of acids were identified according to their retention times, which were compared with those of the mixture of methyl esters of fatty acids in the standard Supelco 37 Component FAME Mix [12].

From the data on the fatty acid composition, the following were calculated:

1) Index of atherogenicity (IA): indicating the relationship between the sum of the main saturated fatty acids and that of the main classes of unsaturated [35, 36].
The following equation was applied:

$$IA = \frac{[(4 \times C14{:}0) + C16{:}0 + C18{:}0]}{[\Sigma MUFA + \Sigma PUFA\text{-}n6 + \Sigma PUFA\text{-}n3]}$$

2) Index of thrombogenicity (IT): showing the tendency to form clots in the blood vessels. This is defined as the relationship between the pro-thrombogenetic (saturated) and the anti-thrombogenetic fatty acids (MUFAs, PUFAs-*n*6 and PUFAs-*n*3), [35, 36].
The following equation was applied:

$$IT = \frac{[C14{:}0 + C16{:}0 + C18{:}0]}{[0.5 \times MUFA + 0.5 \times PUFA\text{-}n6 + 3 \times PUFA\text{-}n3 + PUFA\text{-}n3/PUFA\text{-}n6]}$$

2.3.4. Determination of Cholesterol Content.

Cholesterol content was measured with a HPLC/PDA on the HPLC Waters 2695 Separations Module, with a Waters 2996 Photo Diode Array Detector, as defined by Maraschiello et al.[13].

2.3.5. Sensory Analysis.
For sensory evaluation of kulen and sremska sausages, quantitative descriptive analysis (QDA) was used. The evaluation of sensory properties of sausages were conducted by 10 selected and trained professional assessors (three males and seven females with an average age of 34.5 years) with previous experience in the evaluation of fermented dry sausages. During testing, one sample of sausage was presented at a time and the assessors were asked to rate the following nine attributes of the sausages on a numeric-descriptive scale from 1 (extremely unacceptable) to 7 (extremely acceptable): appearance, cross-section, colour intensity, odour intensity, taste, consistency, acidity, aftertaste and overall acceptability. Samples were taken from the middle of the sausages after 90 days maturation.The sausages were cut into pieces of approximately of 5 mm. These samples, individually labeled with three-digit random numbers, were left for 20 min to reach room temperature. Finally, the samples were served at room temperature on white plastic dishes. Evaluations were performed in individual sensory test booths, prepared as described in accordance with ISO 6658:2005. The assessors were given unsalted crackers and room temperature water to clean the palates and remove residual flavors at the beginning of the session and in between samples. White fluorescent light was used during tests.

2.3.6. Statistical Analysis.
Statistical analysis of experimental data was carried out by analysis of variance (ANOVA) (SAS 9.1.3 software package, 2002–2003). The differences between the different types of sausage were tested using Tukey's test. The effect of sausage type on fatty acids, chemical content and sensory scores was determined.

3. Results and discussion

3.1. Basic Chemical Composition.
The chemical composition of the different types of kulen and sremska sausages are shown in Table 2. The lowest protein content was found in sremska sausage type SLM and the highest in the kulen type KL. Kulen and sremska sausages made from the meat of Mangalitsa and Moravka pig breeds (type KM and SM) had the lowest moisture content, plus the highest fat content, and these were significantly different in comparison with other types of sausages ($P<0.001$). Some studies have indicated the occurrence of lower protein content in Mangalitsa meat (*m. longissimus*), compared to the meat of commercial pig breeds [1, 14,15], thus partly explaining the slightly lower protein content in the kulen types KM and all types of sremska sausages in the current study. Low moisture content is typical of similar products from Greece, Hungary and Croatia [16], and is a consequence of not only drying, but also of a relatively high fat content. In our study, the ash content of the kulen and sremska sausages varied among the all types of sausages ($P<0.001$). The lowest ash content was found in sremska sausage type SM, and the highest in the sremska type SL. The established pH values ranged from 4.72 (sremska type SL) to 5.25 (sremska type SM) ($P<0.001$). At the results of studies Vuković *et al.*[17], beginning of the ripening process of traditional kulen, pH values range from 5.6 to 5.8, which correspond to the pH value of cooled pork meat, and subsequently they start to decline. Certainly, the sugars (fructose, glucose, sucrose) which are natural ingredients of the paprika spice, have a major impact on the decline of pH value in ripening kulen. According to literature data [18], ground paprika contains approximately 15% of sugars, whereas the total sugar content in locally-produced ground paprika is higher and amounts to approximately 25%.The lowest ash content was found in sremska sausage type SM, and the highest in the sremska type SL.

The naturally fermented dry sausages from Mediterranean region are generally characterised by low acidity with a final pH ranging from 5.2 to 6.4, in Italy [19,20], Greece [21], Spain [22; 23] and France [24].

Table 2. Chemical composition of different type fermented dry sausages (means ± standard error)

Traits	Fermented dry sausages[1]						
	KM	KL	KMM	SM	SL	SLM	P[2]
Protein (%)	27.20±0.17^{a}	35.79±0.17^{b}	34.24±0.17^{c}	29.16±0.17^{d}	28.04±0.17^{e}	23.20±0.17^{f}	***
Water (%)	23.62±0.20^{a}	37.92±0.20^{b}	35.70±0.20^{c}	21.67±0.20^{d}	39.41±0.20^{e}	33.30±0.20^{f}	***
Fat (%)	39.30±0.22^{a}	15.02±0.22^{b}	21.00±0.22^{c}	39.45±0.22^{da}	22.00±0.22^{e}	34.92±0.22^{f}	***
Ash (%)	4.61±0.09^{a}	5.25±0.09^{b}	5.67±0.09^{c}	4.48±0.09^{da}	5.83±0.09^{ec}	4.86±0.09^{fad}	***
pH	5.24±0.10^{a}	5.05±0.10^{b}	5.03±0.10^{c}	5.25±0.10^{da}	4.72±0.10^{eb}	4.73±0.10^{fb}	***
F/Pratio[3]	1.45±0.01^{a}	0.42±0.01^{b}	0.61±0.01^{c}	1.35±0.01^{d}	0.78±0.01^{e}	1.51±0.01^{f}	***

[1]Fermented sausages type dependent on the percentage of meat from differing pig breeds
[2]NS- not significant ($P \geq 0.05$);*: Statistical significance at the level of P<0.05;**: Statistical significance at the level of P<0.01;***: Statistical significance at the level of P<0.001;
[3]Fat/Protein ratio; $^{a-e}$ Means in the same row with different letters are significantly different (P< 0.05).

3.2. Fatty Acid Composition.

The fatty acid profiles of the six different fermented dry sausages are introduced in Table 3. Palmitic acid (C16:0) was the most represented SFA, oleic acid (C18:1 n-9) and linoleic acid (C18:2 n-6) were the most represented MUFA and PUFA for the all types of sausages.

The levels of PUFA in fermented dry sausages made from the meat of Mangalitsa pig breed was significantly lower (P<0.001) than levels in other types of fermented dry sausages. These differences were mainly produced by lower total n-3 PUFA content (P<0.01), and also by lower levels of total n-6 PUFA (P<0.001). These led to lower n-6/n-3 ratios in fermented dry sausages type KM (16.96) and SM (14.38). In spite of that though, the n-6/n-3 ratio of unsaturated fatty acids in other types of sausages were between 25 and 37. In separate trials, Hoz [25] and Valencia*et al.* [26], both found, in their control groups of fermented dry sausages, lower ratios of n-6/n-3 fatty acids (12.05 and 13.86, respectively), compared to our findings. MUFA values ranged from 43.49 to 52.80 percent. The sausage types SM and KM, made from the meat of Mangalitsa pigs, contained higher levels of MUFA (P<0.001) than the other types of sausages. Additionally, oleic acid (C18:1 cis-9), cis-vaccenic acid, (C18:1 cis-11) and palmitic acid (C16:1) levels in these types of sausages were considerably higher than in the other types. Kulen and sremska sausages type KM and SM had higher USFA (P<0.01) and lower SFA (P<0.001) levels. Overall, the USFA was definitely significantly higher in types of sausages made from the meat of Mangalitsa pig breeds.

Table 3. Fatty acid composition (%), cholesterol content (mg/100g), Index of atherogenicity (IA) and Index of thrombogenicity (IT) of different fermented dry sremska sausages.

Traits	Fermented dry sausages[1]						
	KM	KL	KMM	SM	SL	SLM	P[2]
C14:0	1.21±0.05	1.18±0.05	1.02±0.05	1.18±0.05	1.02±0.05	1.09±0.05	NS
C16:0	26.27±0.09^{a}	24.77±0.09^{b}	25.55±0.09^{c}	25.88±0.09^{dc}	23.99±0.09^{e}	25.26±0.09^{fc}	***
C16:1	3.87±0.07^{a}	1.86±0.07^{b}	2.67±0.07^{c}	3.87±0.07^{da}	1.76±0.07^{eb}	2.11±0.07^{fb}	***
C17:0	0.31±0.03	0.35±0.03	0.24±0.03	0.29±0.03	0.30±0.03	0.31±0.03	NS
C18:0	11.25±0.08^{a}	14.12±0.08^{b}	13.22±0.08^{c}	10.88±0.08^{da}	14.19±0.08^{eb}	14.09±0.08^{fb}	***
C18:1c-9	42.73±0.14^{a}	39.47±0.14^{b}	39.01±0.14^{cb}	43.41±0.14^{d}	37.74±0.14^{e}	38.77±0.14^{fc}	***
C18:1c-11	4.38±0.07^{a}	3.26±0.07^{b}	3.42±0.07^{cb}	4.55±0.07^{da}	2.91±0.07^{ef}	3.17±0.07^{fb}	***
C18:2n-6	6.37±0.10^{a}	11.66±0.10^{b}	11.46±0.10^{cb}	6.58±0.10^{da}	14.40±0.10^{e}	11.91±0.10^{fb}	***
C18:3n-6	ND	ND	ND	ND	ND	ND	
C18:3n-3	0.39±0.03	0.35±0.03^{b}	0.36±0.03	0.47±0.03	0.44±0.03	0.35±0.03^{f}	*
C20:0	0.17±0.02	0.18±0.02	0.18±0.02	0.17±0.02	0.21±0.02	0.19±0.02	NS
C20:1	0.85±0.03	0.79±0.03	0.69±0.03	0.84±0.03	0.72±0.03	0.73±0.03	NS
C20:2	0.63±0.07	0.70±0.07	0.72±0.07	0.54±0.07^{d}	0.91±0.07^{e}	0.83±0.07	**
C20:3n-6	1.33±0.06^{a}	0.67±0.06^{b}	1.09±0.06^{ca}	1.10±0.06^{dac}	1.03±0.06^{ec}	0.91±0.06^{fbce}	***
C20:3n-3	0.08±0.02	0.15±0.04	ND	0.09±0.02	ND	ND	**
C22:1/C20:4	0.14±0.02^{a}	0.48±0.02^{b}	0.37±0.02^{c}	0.13±0.02^{da}	0.37±0.02^{ec}	0.26±0.02^{f}	***
SFA	39.22±0.17^{a}	40.60±0.17^{b}	40.21±0.17^{cb}	38.40±0.17^{d}	39.71±0.17^{ea}	40.94±0.17^{fb}	***
MUFA	51.97±0.20^{a}	45.86±0.20^{b}	46.16±0.20^{cb}	52.80±0.20^{da}	43.49±0.20^{e}	45.04±0.20^{fb}	***
PUFA	8.80±0.16^{a}	13.53±0.16^{b}	13.63±0.16^{cb}	8.78±0.16^{da}	16.78±0.16^{e}	14.00±0.16^{fb}	***
USFA	60.78±0.31^{a}	59.39±0.31^{c}	59.79±0.31^{b}	61.58±0.31^{a}	60.27±0.31^{c}	59.05±0.31^{b}	**
MUFA/PUFA	5.94±0.09^{a}	3.39±0.09^{b}	3.36±0.09^{c}	6.02±0.09^{a}	2.59±0.09^{b}	3.22±0.09^{c}	***
MUFA/SFA	1.33±0.01^{a}	1.13±0.01^{b}	1.15±0.01^{b}	1.38±0.01^{a}	1.10±0.01^{b}	1.10±0.01^{b}	***
PUFA/SFA	0.22±0.00^{c}	0.33±0.00^{b}	0.34±0.00^{a}	0.23±0.00^{c}	0.42±0.00^{b}	0.34±0.00^{a}	***
Total n-3	0.47±0.03	0.50±0.03^{b}	0.36±0.03^{c}	0.55±0.03^{db}	0.43±0.03^{ebcd}	0.35±0.03^{fc}	**
Total n-6	7.70±0.12^{a}	12.33±0.12^{b}	12.55±0.12^{cb}	7.69±0.12^{da}	15.43±0.12^{e}	12.82±0.12^{fb}	***
n-6/n-3	16.96±1.98^{a}	25.21±1.98^{ba}	36.07±1.98^{c}	14.38±1.98^{da}	35.86±1.98^{ec}	37.36±1.98^{fc}	***
Cholesterol	50.16±0.20^{a}	61.48±0.20^{b}	66.00±0.20^{c}	59.65±0.20^{d}	64.92±0.20^{e}	53.47±0.20^{f}	***
IA	0.70±0.00^{a}	0.74±0.00^{b}	0.73±0.01^{cb}	0.68±0.00^{d}	0.71±0.00^{eac}	0.75±0.01^{fb}	***
IT	1.24±0.01^{a}	1.31±0.01^{b}	1.31±0.01^{cbe}	1.19±0.01^{d}	1.27±0.01^{eab}	1.35±0.01^{f}	***

Values are means and standard error (mean ± SE).
[1]Fermented sausage type dependent on the percentage of meat from differing pig breeds (Table 1)
[2]NS- not significant ($P \geq 0.05$);[*]: Statistical significance at the level of $P < 0.05$;[**]: Statistical significance at the level of $P < 0.01$;[***]: Statistical significance at the level of $P < 0.001$;
[a- e] Means in the same row with different letters are significantly different ($P < 0.05$).

The PUFA/SFA ratio is, nowadays, recommended to be above 0.4 to 0.5 in order to prevent both an excess of saturated fatty acids with a negative effect on low density lipoprotein (LDL) cholesterol plasmatic level, and an excess of PUFA, some of them being precursors of powerful clotting agents and also being involved in the etiology of some cancers [28]. In our study, the PUFA/SFA ratio was determined to be the lowest in sausages type KM and SM (0.22 and 0.23), with none of the levels in the sausages exceeding 0.40. Baggio and Bragagnolo [29] ascertained that PUFA/SFA for Italian type salami was 0.4 at the end of production.

3.3. Cholesterol Content.
The cholesterol content in kulen and sremska sausages at the conclusion of the production process ranged from 50.16 mg/100 g (KM) to 66.00 mg/100 g (KMM), with significant differences between the samples ($P<0.001$). Based on the obtained results (Table 3), we can conclude that sausages made from the mixed meat and fat of Mangalitsa and Moravka indigenous pig breeds contained the highest percentage of cholesterol, while the lowest cholesterol level was seen in sausages produced from the Mangalitsa pig breed.

For salami Milano, Zanardi *et al.*[30] established cholesterol contents ranging from 94.8 to 110.5 mg/100g. Baggio and Bragagnolo [29] for Italian type salami found the cholesterol content ranged from 48 to 57 mg/100g. Pleadin et al.[31] concluded that the average cholesterol content of industrially fermented sausages was 58.48 to 105.24 mg/100g, until that of homemade fermented sausages was up to 75.07 mg/100g.

3.4. Sensory Properties.
The results of sensory analyses by professional trained assessors are presented in figures 1 and 2. Kulen type KMM and sremska type SLM were awarded the lowest marks and so had the least acceptable appearance, while the kulen type KL and sremska type SL was rated the most acceptable. The cross-section of kulen sausage made from the meat of Mangalitsa pigs was assessed as the poorest. Sremska sausage type SL was the most consistent sausage produced. Product colour was correlated with the colour of the meat used in production. The meat of Mangalitsa pigs was darker than the meat of Swedish Landrace and Moravka; therefore, sausages made from the meat of Mangalitsa breed were assessed as too dark, and received a somewhat lower grade than the other sausage types. Odour was the sensory indicator most affected by the pig breed. The most typical and the best sausages were made from the meat of Mangalitsa breed. The odour of this sausage type was rich and very pronounced, and received a much higher grade than the other kind of sausages. The taste of all sausage types were similar and graded from 5.08 to 6.08, while after taste was graded from 5.17 to 5.33. Professional evaluators gave sremska sausage type SM the highest marks for both taste and after taste. The overall sensory acceptability scores of the examined products showed some differences. Sremska sausage type SM was of the highest stable quality. On the other hand, the quality properties of sausage made from Moravka pork meat were significantly poorer. The current study has shown that the Sremska sausages type SM was good sensory quality, which was in line with Ortiz-Somovilla *et al.* [32] and Moretti *et al.* [33]. Overall, the professional assessors found sremska sausage type SM, was the most acceptable. Sausages made only from the meat of Mangalitsa breed (type KM and SM) had better sensory characteristics, thus confirming the work of Radman *et al.* [34], who found that some pig breeds are suitable for the production of dry fermented pork sausages.

Fig 1. Sensory properties of different types of kulen fermented dry sausage rated by professional assessors (scale test rating)

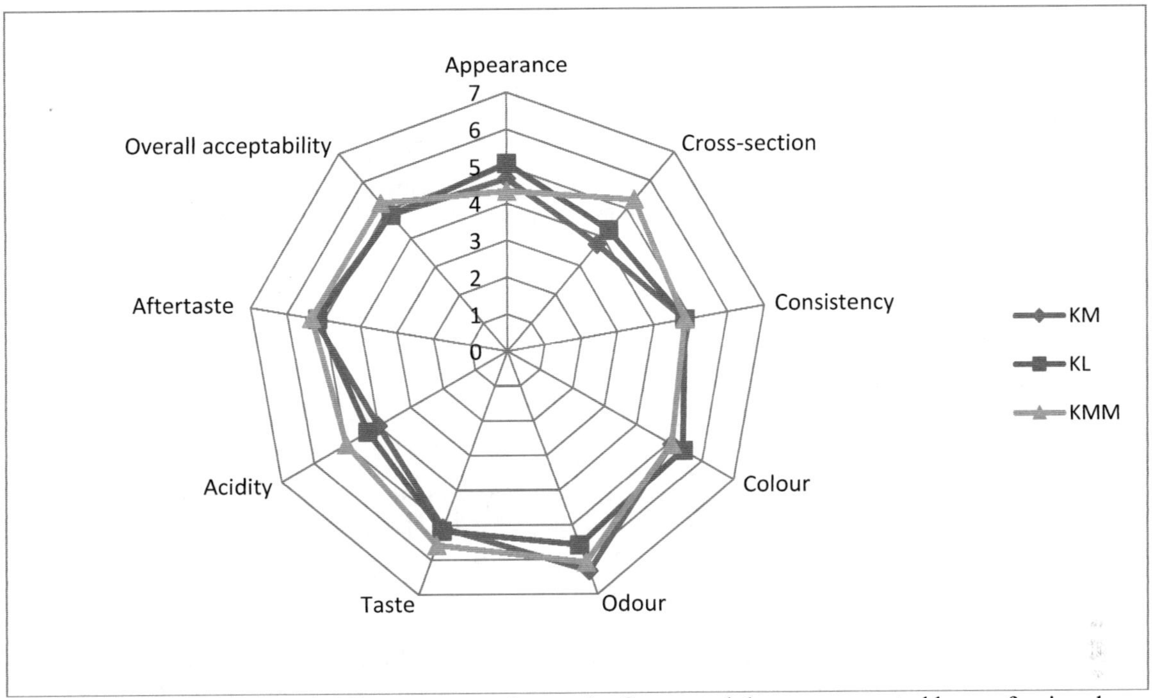

Fig 2. Sensory properties of different types of sremska fermented dry sausage rated by professional assessors (scale test rating)

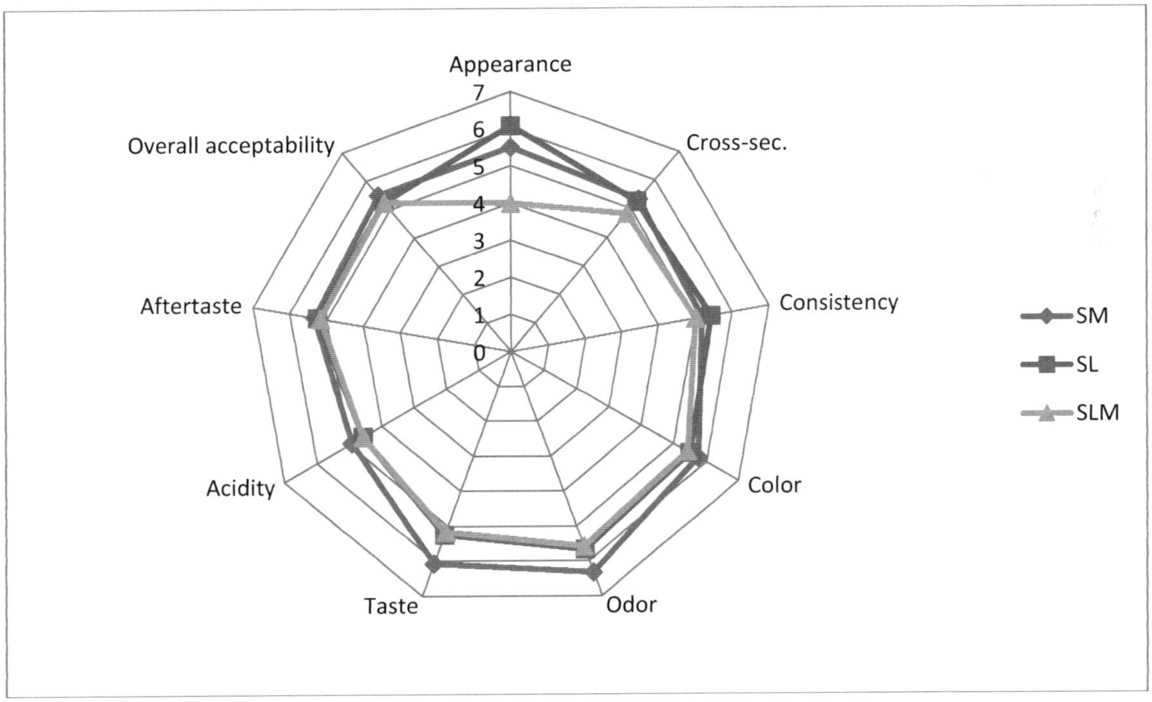

4. Conclusion

The results of this study demonstrated that pig breed have an effect on the sensory and chemical characteristics of fermented dry sausages. According to the results of the present research, it is possible to produce, with the appropriate combination of meat and fat from autochthonous pig breeds alone, kulen and sremska dry fermented sausages, with a respectable chemical content, a favourable and reasonably healthful fatty acid composition and with sensory qualities acceptable to discerning consumers. Provided market possibility exist for kulen and sremska sausages, these results should contribute to encouraging the sustainable breeding of the Moravka and Mangalitsa pigs.

Acknowledgments
This research was part of the project "Application of different breeding, selection and biotechnological methods for refinement of pig", funded by Ministry of Science Republic of Serbia, record number 31081, project Manager: Professor Dr. Milica Petrovic.

References

[1] Egerszegi I, Ratky J, Solti L,and Brussow K P 2003Mangalica – an indigenous swine breed from Hungary (Review). *Arch Tierzucht*, Dummerstorf, **46** pp 245-256.
[2] Holló G, Seregi J, Ender K, Nürnberg K, Wegner J, Seenger J, et al. (2003). Examination of meat quality and fatty acid composition of Mangalitsa. *Acta Agraria Kaposvariensis*, **7** (2) pp 19-32.
[3] Parunović N, Petrović M, Matekalo-Sverak V, Trbović D, Mijatović M andRadović Č 2012b Fatty acid profile and cholesterol content of *m. longissimus* of free-range and conventionally reared Mangalitsa pigs. *S Afr J Anim Sci*, **42** pp 101-113.
[4] Bragagnolo N and Rodriguez-Amaya D B 2002Simultaneous determination of total lipid, cholesterol and fatty acids in meat and backfat of suckling and adult pigs. *Food Chemistry*,**79**(2) pp255-260.
[5] Burlingame B, Nishida C, Uauy R and Weisell R 2009 Fats and fatty acids in human nutrition: introduction. *Ann Nutr Metab*, **55** pp 5-7.
[6] Simopoulos A P 2004 Omega-6/Omega-3 essential fatty acid ratio and chronic diseases. *Food Reviews International*, **20** pp 77-90.
[7] Griffin B A 2008 How relevant is the ratio of dietary n-6 to n-3 polyunsaturated fatty acids to cardiovascular disease risk? Evidence from the OPTILIP study. *Curr Opin Lipidol*, **19** pp 57-62.
[8] Di Cagno R, Lopez CC, Tofalo R, Gallo G, De Angelis M, Paparella A, Hammes WP and Gobbetti M 2008 Comparison of the compositional, microbiological, biochemical and volatile profile characteristics of three Italian PDO fermented sausages. *Meat Sci.*,**79** pp 224–235.
[9] El Malti Jand Amarouch H 2009 Microbiological and physicochemical characterization of the natural fermented camel meat sausage. *African Journal of Biotechnology*,**8** pp 4199–4206.
[10] Parunović N, Petrović M, Matekalo-Sverak V, Radojković D andRadović Č 2014 Fatty acid profiles, chemical content and sensory properties of traditional fermented dry kulen sausages. *Journal of Food Processingand Preservation*, **38** pp 2061-2068.
[11] AOAC 1990. Official methods of analysis of the Association of Official Analytical Chemists. 15th ed. Washington, DC, USA.
[12] Spirić A, Trbovic D, Vranic D, Djinovic J, Petronijevic R and Matekalo-Sverak V 2010 Statistical evaluation of fatty acid profile and cholesterol content in fish (common carp) lipids obtained by different sample preparation procedures. *Anal Chim Acta*, **672** pp 66-71.
[13] Maraschiello C, Diaz I and Regueiro J A G 1996Determination of cholesterol in fat and muscle of pig by HPLC and capillary gas chromatography with solvent venting injection. *J High Resolut Chromatogr*, **19** pp 165-168.
[14] Rahelić S 1984 Growing pigs and meat. Zagreb, School books.
[15] Parunović N, Petrović M, Matekalo-Sverak V, Radojković D, Vranić DandRadović Č2012a.

Cholesterol and total fatty acid content in *m. longissimus dorsi* of mangalitsa and swedish landrace. *Acta Alimentaria***41**(2) pp 161–171.

[16] Kozačinski L, Drosinos E, Čaklovica F, Cocolin L, Gasparik-Reichardt J and Vesković S 2008 Investigation of microbial association of traditionally fermented sausages. *Food Tech. Biotechnol.***46**(1) pp 93–106.

[17] Vuković I, Bunčić O, Babić Lj, Radetić P and Bunčić S 1988 Investigation of major changes during ripening of traditional fermented sausage Lemeški kulen. *Meat Technology***34**(2) pp 34–39.

[18] Oberdick R 1988 Paprika. *Fleischwirtschaft* **68** (9) pp 1086–1096.

[19] Mauriello G, Casaburi A, Blaiotta G and Villani F 2004 Isolation and technological properties of coagulase negative staphylococci from fermented sausages of Southern Italy. *Meat Sci.***67**(1) pp 149–158.

[20] Comi G, Urso R, Iacumin L, Rantsiou K, Cattaneo P, Cantoni C and Cocolin L 2005 Characterisation of naturally fermented sausages produced in the North East of Italy. *Meat Sci.***69**(3) pp 381–392.

[21] Fista GA, Bloukas JG and Siomos AS 2004 Effect of leek and onion on processing and quality characteristics of Greek traditional sausages. *Meat Sci.*68(2) pp 163–172.

[22] Hierro E, De la Hoz Ll and Ordoñez JA 1999 Contribution of the microbial and meat endogenous enzymes to the free amino acid and amine contents of dry fermented sausages. *J. Agr. Food Chem.***47**(3) pp 1156–1161.

[23] Aymerich T, Martín B, Garriga M and Hugas M 2003 Microbial quality and direct PCR identification of lactic acid bacteria and nonpathogenic staphylococci from artisanal low-acid sausages. *Appl. Environ. Microbiol.***69**(8) pp 4583–4594.

[24] Chevallier I, AmmorS, Laguet A, Labayle S, Castanet V, Dufour E and Talon R 2006 Microbial ecology of a small-scale facility producing traditional dry sausage. *Food Contr.*, **17**(6) pp 446–453.

[25] Hoz L 2004 Development of an n-3 fatty acid and tocopherol enriched dry fermented sausage. *Meat Sci.*67, pp 485–495.

[26] Valencia I, Ansorena D and Astiasaran I 2006 Nutritional and sensory properties of dry fermented sausages enriched with n-3 PUFAs. *Meat Sci.***72**, pp 727–733.

[27] Mahan LK and Escott-StumpS 2000 Krause's Food, Nutrition and Diet Therapy. (10th ed.). Pennsylvania: W.B. Saunders Company, USA.

[28] Ansorena D and Astiasaran I 2004 The use of linseed oil improves nutritional quality of the lipid fraction of dry-fermented sausages. *Food Chem.***87**(1) pp 69–74.

[29] Baggio SRand Bragagnolo N 2006 Cholesterol oxide, cholesterol, total lipid and fatty acid contents in processed meat products during storage. *Food Science and Technology (LWT)***39**(5) pp 513–520.

[30] Zanardi E, Novelli E, Ghiretti GP and Chizzolini R 2000 Oxidative stability of lipids and cholesterol in salame Milano, coppa and Parma ham: dietary supplementation with vitamin E and oleic acid. *Meat Sci.***55**(2) pp169–175.

[31] Pleadin J, Vahčić N, Perši N, Vulić A, Volarić M and Vraneš I 2010Sadržaj kolesterola u domaćim i industrijskim kobasicama. *Meso*, **12**(3)pp 156-161.

[32] Ortiz-Somovilla V, Espana-Espana F, De Pedro-Sanz EJ and Gaitan-Jurado AJ2005 Meat mixture detection in Iberian pork sausages. *Meat Sci.***71**(3) pp 490–497.

[33] Moretti V M, Madonia G, Diaferia C, Mentasti T, Paleari M A, Panseri S, Pirone G and Gandini G 2004 Chemical and microbiological parameters and sensory attributes of a typical Sicilian salami ripened in different conditions. *Meat Sci.***66** (4) pp 845–854.

[34] Radman M, Karolyi D and Kovacic D 2005 Consumer satisfaction with Slavonian Kulen from Black Slavonian or modern pigs. *Ital. J. Anim. Sci.* **4** (3) pp 181–183.

[35] Ulbritch T L V and Southgate D A T 1991 Coronary Heart Disease: Seven Dietary Factors, *Lancet*, **338**, pp 985-992.

[36] SensoL, Suarez M D, Ruiz-Cara T and Garcia- Gallego M 2007 On the Possible Effects of Harvesting Season and Chilled Storage on the Fatty Acid Profile of the Fillet of Farmed Gilthead Sea Bream (*Sparus aurata*), Food Chem, **101**(1) pp 298-307.

IOP Conf. Series: Earth and Environmental Science **85** (2017) 012015 doi:10.1088/1755-1315/85/1/012015

Environmental footprints in the meat chain

I Đekić and I Tomašević

Faculty of Agriculture, University of Belgrade, Nemanjina 6, 11080 Beograd, Serbia

E-mail: idjekic@agrif.bg.ac.rs

Abstract. The objective of this paper was to present environmental performance of the meat chain and highlight main environmental footprints. The meat sector is recognized as one of the leading polluting sectors in the food industry. The meat chain was analyzed from a five-link perspective introducing the following actors: farm(er)s, slaughterhouses, meat processors, customers and consumers. Meat production needs natural resources (water and energy) resulting in waste and waste water discharge. As an outcome it has a high influence on climate change in respect to global warming, acidification and eutrophication potentials and ozone depletion substances.

1. Introduction

Meat production is increasing as a result of world's population growth and consumption of meat per capita [1]. However, meat is considered as a food product with the greatest environmental impact [2]. Regardless of the role in the meat chain, this type of production seeks for natural resources (land, water and energy) and emits various pollutants into the environment [air, water, land] [3, 4]. Steinfeld et al. assume that all livestock systems occupy up to 30 per cent of the planet's ice-free terrestrial surface area [5]. The future will bring us differences in livestock production between countries [developed vs. developing] and production systems (highly intensive production systems vs. smallholder systems) [6]. Livestock production is characterized with the inefficiency of animals in converting feed to meat since over 75% of the energy consumed is lost in body maintenance, manure and by-products such as skin and bones [2]. At the farm level, manure management is mostly responsible for polluting the environment. Main environmental impacts in slaughterhouses and meat processing plants are usage of energy, usage of water, waste handling and wastewater discharge [7]. The availability of environmental indicators allows comparing the environmental performance over time and against other food companies, highlighting optimization potentials [8].

The meat chain has five main links – farm(er)s, slaughterhouses, meat processors, customers [supermarkets, butcheries, retailers] and consumers [9, 10]. 'Farming' includes all livestock activities which take place in a farm, covering also contribution of feed production and waste/manure management [11]. 'Slaughterhouse' covers reception of live animals, livestock handling, animal welfare, slaughtering and chilling [12]. 'Meat processing plant' starts with reception of carcasses and ends up with the storage of final meat products [12]. 'Customers' are points of sale and cover supermarkets and grocery shops or specialized shops selling meat [10]. 'Consumers' cover refrigeration of food [13], meat preparation and cooking [14]. Meat is consumed for a number of reasons such as nutritional needs and dietary patterns [15] sensory attributes and cultural habits, religion beliefs and wealth [16, 17]. This clarifies why the discussion on the nutritional benefits versus the environmental effects of meat consumption is opposed [18].

2. Materials and methods

A literature review was performed by analyzing scientific manuscripts in the domains of environmental impacts in the meat chain published in databases such as Web of Science, EBSCO, ScienceDirect and GoogleScholar. No geographical restrictions were applied.

2.1. Environmental footprints

Depending on the approach, there are different methods on how to evaluate environmental impacts [19]. The basic approach is in calculating environmental performance indicators [EPIs]. As referred to the latest ISO 14001 standard, EPI is a measurable representation of the status of operations, management or conditions related to environmental aspects [20]. Reasons for calculating reliable numeric indicators are for organization's legal responsibility on environmental issues and for ensuring achievement of certain environmental objectives [21]. Guidance on the design and use of EPIs within an organization on both continual improvement and prevention of pollution is outlined in ISO 14031 [22]. Environmental practices in meat companies show two performance dimensions – environmental and economic [7]. Financial indicators are perceived as backward looking, lacking predictive ability to explain future [environmental] performance, being too summarized to guide managerial action and providing no guidance to evaluate intangible assets [21]. Rule of the thumb for all EPIs are that they should be (i) measurable; (ii) objective; (iii) verifiable; (iv) repeatable and (v) technically feasible [23]. In general there are three levels of EPIs that are related to the maturity of implemented environmental practices, Figure 1.

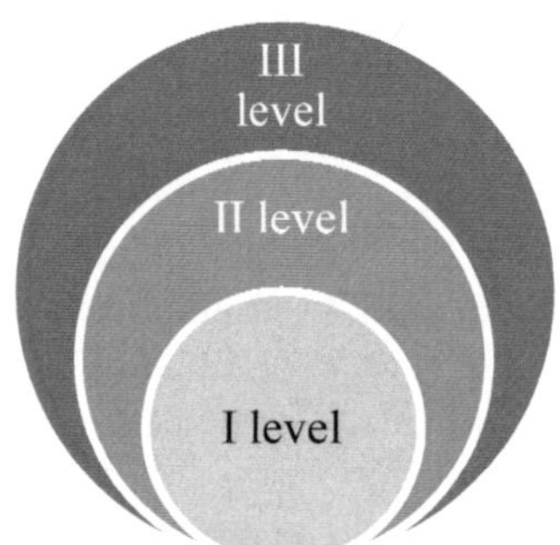

Figure 1. Levels o EPIs

First level of EPIs are elementary indicators showing only figures such as energy and water consumption, waste water discharge or annual food production with no connection between them. Second level introduces functional units such as one kg of livestock [24, 25]; one kg of carcass [26, 27] and one kg of meat [28] and enables correlation between first level EPIs. Most common used second level EPIs are water consumption per functional unit (m^3 of water/kg of food), energy consumption per functional unit (MJ of energy/kg of food), etc. This approach to environmental performance shows the relationship between the organization and the environment, including environmental effects of resources consumed and the environmental impacts of the organizational processes [29]. Third level of EPIs provide information on different footprints on the environment such as ecological footprint, water footprint and carbon footprint as the three most recognized members of the so-called footprint family [30]. The ecological footprint refers to the number of individuals who can be supported in a given area within natural resource limits, and without degrading the natural, social, cultural and economic environment for present and future generations [31]. This footprint is not commonly used in the meat sector. The water footprint is built on the concept of 'virtual' water at a [meat] company level, and the indicator can be estimated for a business or a product by calculating the total water used during the production of goods and services in the entire supply chain [30]. Carbon footprint measures the total set of greenhouse gas emissions caused directly and indirectly by an individual, event, organization or product and is expressed in CO_2 equivalent since the largest single contributor to climate change is CO_2 [30]. This footprint is very often used in presenting the environmental impact of meat production.

2.2. *Meat chain framework*

There are three main environmental research perspectives recognized in the meat chain. The first analyzes the meat product perspective through life cycle assessment. This technique calculates environmental indicators in relation to the meat product to assess the potential environmental impacts and consumption of resources [4]. The second focuses on manufacturing processes recognized in the meat industry. This perspective analyzes specific environmental impacts connected with recognized processes that occur on site during meat production / processing [32]. Finally, the third explores the environmental systems in which the meat companies operate [32].

Most common second level of EPIs in the meat production are meat yield (share of lean meat in live animals and/or in carcass), solid output [in farming mostly manure, in slaughtering/deboning percentage of by-product such as offal, bones, fat and skin] and energy consumption [electric and thermal]. Besides these EPIs, meat companies calculate various consumptions and discharges per functional unit such as energy-to-meat ratio, water consumption, waste water discharge and waste water load [chemical oxygen demand] and chemical usage [12, 33, 34]. Simplified generic model of the environmental impact of the meat chain is presented in Figure 2.

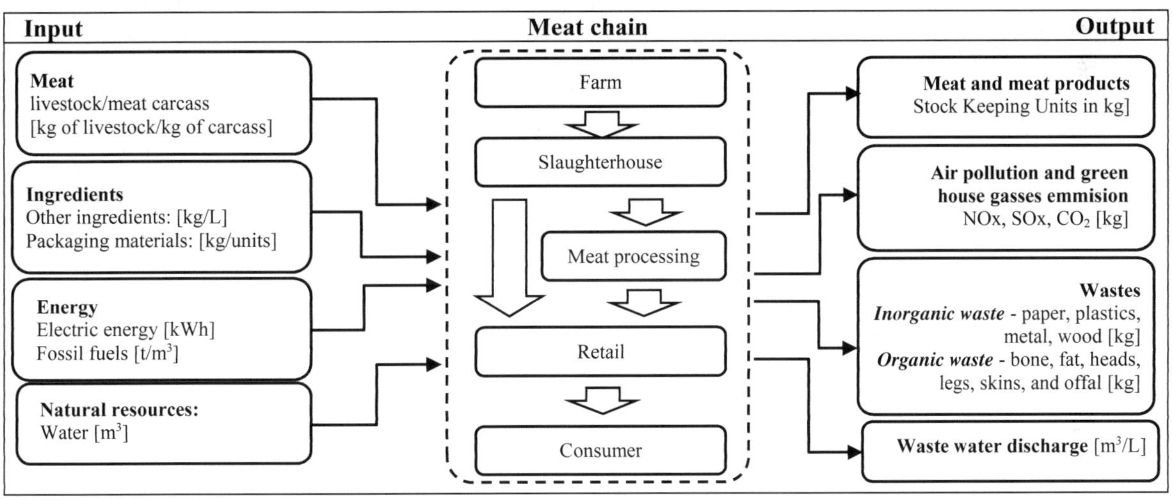

Figure 2: Simplified generic model of the environmental impact of the meat chain

3. Environmental footprints in the meat chain

Although literature review confirms that the greatest environmental impacts arise at farms as a result of livestock production, the entire meat chain has a significant contribution to pollution. The main environmental impacts related to pork production are global warming potential, acidification and eutrophication potential as well as consumption of water and energy [35, 36]. Livestock contributes to global warming potential directly coming from enteric fermentation and manure management and indirectly as a result of feed production [2, 37]. Ammonia is the dominant source of acidifying emissions during animal production [12]. It is released from manure in farms and during manure handling and is dependent on several factor such as physical state, temperature and pH [2]. Liquid manure handling systems emit less ammonia than solid manure handling but liquid/slurry storage stimulates CH_4 production, due to anaerobe conditions [38]. Nitrate leaching from fields during feed production and ammonia release from manure handling dominate the emissions of eutrophying substances as the main contributors to eutrophication in meat production [2]. Two main improvement streams regarding global warming potential and acidification in meat production are [i] manure management and [ii] feeding strategy [12].

Energy is used throughout the meat chain by the machines and equipment, for controlling temperature regimes (heating / refrigerating) and for transportation purposes [32, 33]. Water is used in all activities in the meat chain. It starts with live animals entering the slaughterhouse, through hygiene and sanitation in slaughterhouses, meat processing plants and retail and finishes at the final – consumption stage [10, 33]. Waste water is a result of various cleaning and sanitation activities such as washing of livestock, carcasses and offal, cleaning and sanitation of equipment and work environment and workers' personal hygiene [39]. At slaughterhouses, water becomes an effluent with high levels of organic load from manure, blood and fat and undigested stomach contents [34].

Speaking about waste in the meat industry, literature recognized two main types - inedible products, mostly bones, heads, legs, hair and offal and various packaging materials [7, 39]. Since consumers prefer lean meat, this causes production of waste in slaughterhouses/meat processing plants [40]. Handling this type of animal by-products is regulated by the law in developed markets, like the EU.

It is known that keeping products at low temperatures inhibits growth of potentially harmful microorganisms [41]. However, the cold chain requirements with their impact on ozone layer depletion due to the use of refrigerants in the processes of chilling / freezing affect the entire meat chain [12]. Development of new refrigerants with low GWP and promotion of natural refrigerants throughout the cold chain is expected [14]. Generic figure of deployed levels of EPIs in meat industry are presented in Figure 3.

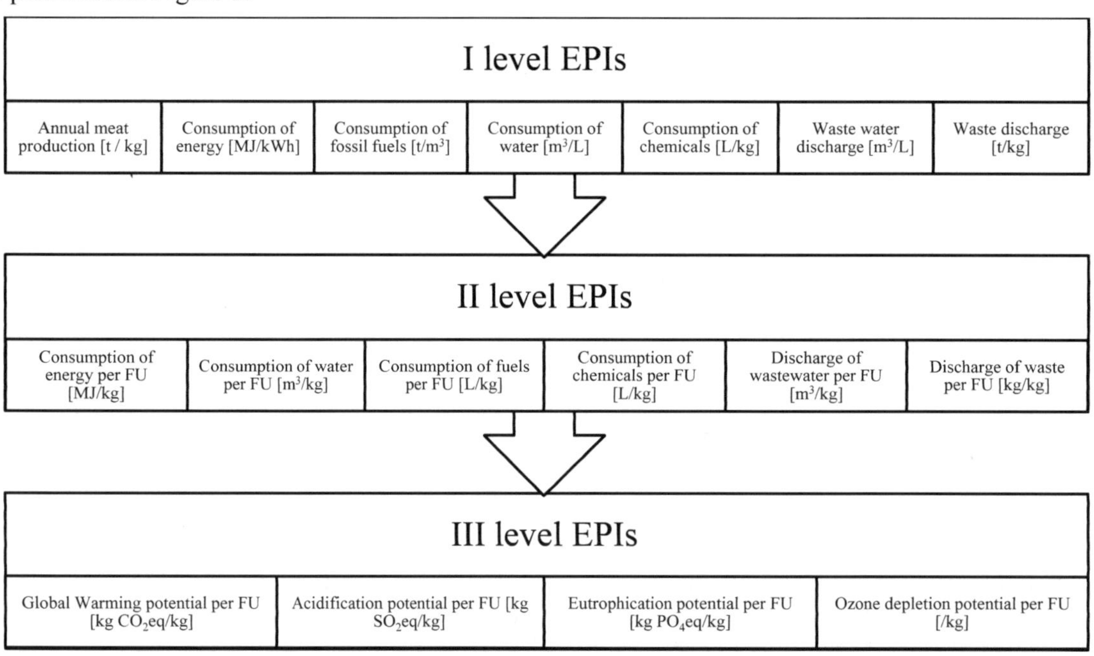

Figure 3: Deployment of three levels of EPIs in the meat chain

Legend: FU – functional unit. In meat industry it is 1 kg of livestock or 1 kg of carcass or 1 kg of meat product [depending on the role in the meat chain].

4. Conclusion

The meat sector is one of the food sectors with global environmental impacts. Regardless of the type of meat produced and technology applied, similar actors in the food chain exist and similar environmental impacts occur. This type of production influences climate change in respect to global warming, acidification and eutrophication potentials and ozone depletion substances and has a high ratio of consumption of water and energy resulting in waste and waste water discharge. Regardless of differences in meat technology, eating habits and cultural diversity, environmentally sound production is one of the greatest meat chain challenges in 21st century.

References

[1] Henchion M, McCarthy M, Resconi VC, Troy D. Meat consumption: Trends and quality matters. *Meat Sci.* 2014;98(3):561-8.
[2] Röös E, Sundberg C, Tidåker P, Strid I, Hansson P-A. Can carbon footprint serve as an indicator of the environmental impact of meat production? *Ecol Indic.* 2013;24(0):573-81.
[3] de Vries M, de Boer IJM. Comparing environmental impacts for livestock products: A review of life cycle assessments. *Livest Sci.* 2010;128(1-3):1-11.
[4] Lopez-Ridaura S, Werf H, Paillat JM, Le Bris B. Environmental evaluation of transfer and treatment of excess pig slurry by life cycle assessment. *J Environ Manage.* 2009;90(2):1296-304.
[5] Steinfeld H, Gerber P, Wassenaar T, Castel V, Rosales M, de Haan C. *Livestock's Long Shadow. Environmental Issues and Options.* Rome, Italy: FAO - Food and Agriculture Organization of the United Nations, 2006.
[6] Thornton PK. Livestock production: recent trends, future prospects. Philosophical Transactions of the Royal Society of London B: *Biological Sciences.* 2010;365(1554):2853-67.
[7] Djekic I, Blagojevic B, Antic D, Cegar S, Tomasevic I, Smigic N. Assessment of environmental practices in Serbian meat companies. *J Clean Prod.* 2016;112, Part 4:2495-504.
[8] Jasch C. Environmental performance evaluation and indicators. *J Clean Prod.* 2000;8(1):79-88.
[9] Borrisser-Pairó F, Kallas Z, Panella-Riera N, Avena M, Ibáñez M, Olivares A, et al. Towards entire male pigs in Europe: A perspective from the Spanish supply chain. *Res Vet Sci.* 2016;107:20-9.
[10] Djekic I, Tomasevic I. Environmental impacts of the meat chain – Current status and future perspectives. *Trends Food Sci Tech.* 2016;54:94-102.
[11] McAuliffe GA, Chapman DV, Sage CL. A thematic review of life cycle assessment (LCA) applied to pig production. *Environ Impact Asses.* 2016;56:12-22.
[12] Djekic I, Radović Č, Lukić M, Stanišić N, Lilić S. Environmental life-cycle assessment in production of pork products. *Meso.* 2015;XVII(5):345-51.
[13] Coulomb D. Refrigeration and cold chain serving the global food industry and creating a better future: two key IIR challenges for improved health and environment. *Trends Food Sci Tech.* 2008;19(8):413-7.
[14] Xu Z, Sun D-W, Zhang Z, Zhu Z. Research developments in methods to reduce carbon footprint of cooking operations: A review. *Trends Food Sci Tech.* 2015;44(1):49-57.
[15] Hawkesworth S, Dangour AD, Johnston D, Lock K, Poole N, Rushton J, et al. Feeding the world healthily: the challenge of measuring the effects of agriculture on health. Philosophical Transactions of the Royal Society of London B: *Biological Sciences.* 2010;365(1554):3083-97.
[16] Font-i-Furnols M, Guerrero L. Consumer preference, behavior and perception about meat and meat products: An overview. *Meat Sci.* 2014;98(3):361-71.
[17] Richardson NJ, Shepherd R, Elliman NA. Current attitudes and future influences on meat consumption in the U.K. *Appetite.* 1993;21(1):41-51.
[18] De Smet S, Vossen E. Meat: The balance between nutrition and health. A review. *Meat Sci.* 2016;120:145-156.
[19] Carvalho A, Mimoso AF, Mendes AN, Matos HA. From a literature review to a framework for environmental process impact assessment index. *J Clean Prod.* 2014;64:36-62.
[20] ISO 14001:2004 *Environmental management systems - Requirements with guidance for use.* Geneva, Switzerland: International Organization for Standardization; 2015.

[21] Henri J-F, Journeault M. Environmental performance indicators: An empirical study of Canadian manufacturing firms. *J Environ Manage*. 2008;87(1):165-76.
[22] ISO 14031:2016 *Environmental management - Environmental performance evaluation - Guidelines*. Geneva, Switzerland: International Organization for Standardization; 2016.
[23] Đekić I. *Upravljanje zaštitom životne sredine u proizvodnji hrane*. 1 izdanje. Beograd, Srbija: Poljoprivredni fakultet Univerziteta u Beogradu; 2009. 280 p.
[24] Dalgaard R, Halberg N, Hermansen JE. *Danish pork production - An environmental assessment*. University of Aarhus - Faculty of Agricultural Sciences, 2007.
[25] Basset-Mens C, van der Werf HMG. Scenario-based environmental assessment of farming systems: the case of pig production in France. *Agr Ecosyst Environ*. 2005;105(1–2):127-44.
[26] Williams AG, Audsley E, Sandars DL. *Determining the environmental burdens and resource use in the production of agricultural and horticultural commodities*. Bedford: Cranfield University and Defra, 2006.
[27] Nguyen TLT, Hermansen JE, Mogensen L. *Environmental Assessment of Danish Pork*. Aarhus University, Aarhus, Denmark, 2011.
[28] Cederberg C, Flysjö A. *Environmental Assessment of Future Pig Farming Systems –* Quantifications of Three Scenarios from the FOOD 21 Synthesis Work The Swedish Institute for food and agriculture, 2004.
[29] Dubey R, Gunasekaran A, Samar Ali S. Exploring the relationship between leadership, operational practices, institutional pressures and environmental performance: A framework for green supply chain. *Int J Prod Econ*. 2015;160(0):120-32.
[30] Herva M, Franco A, Carrasco EF, Roca E. Review of corporate environmental indicators. *J Clean Prod*. 2011;19(15):1687-99.
[31] Kratena K. From ecological footprint to ecological rent: An economic indicator for resource constraints. *Ecol Econ*. 2008;64(3):507-16.
[32] Djekic I. Environmental Impact of Meat Industry – Current Status and Future Perspectives. *Procedia Food Science*. 2015;5:61-4.
[33] *Integrated Pollution Prevention and Control. Reference Document on Best Available Techniques in the Food, Drink and Milk Industries*. Seville, Spain: European Commission; 2006.
[34] *Cleaner Production Assessment in Meat Processing*. United Nations Environment Programme Division of Technology, Industry and Economics; 2000. Paris. France
[35] Reckmann K, Traulsen I, Krieter J. Environmental Impact Assessment – methodology with special emphasis on European pork production. *J Environ Manage*. 2012;107(0):102-9.
[36] Nguyen TLT, Hermansen JE, Mogensen L. Environmental costs of meat production: the case of typical EU pork production. *J Clean Prod*. 2012;28(0):168-76.
[37] Gerber PJ, Mottet A, Opio CI, Falcucci A, Teillard F. Environmental impacts of beef production: Review of challenges and perspectives for durability. *Meat Sci*. 2015;109:2-12.
[38] *IPCC Guidelines for National Greenhouse Gas Inventories*. Intergovernmental Panel on Climate Change; 2006.
[39] Kupusovic T, Midzic S, Silajdzic I, Bjelavac J. Cleaner production measures in small-scale slaughterhouse industry – case study in Bosnia and Herzegovina. *J Clean Prod*. 2007;15(4):378-83.
[40] Rahman Uu, Sahar A, Khan MA. Recovery and utilization of effluents from meat processing industries. *Food Res Int*. 2014;65, Part C(0):322-8.
[41] Sofos JN. Chapter 6 - *Meat and Meat Products*. Food Safety Management. San Diego: Academic Press; 2014. p. 119-62.

IOP Conf. Series: Earth and Environmental Science **85** (2017) 012016 doi:10.1088/1755-1315/85/1/012016

Inactivation of pathogenic bacteria in food matrices: high pressure processing, photodynamic inactivation and pressure-assisted photodynamic inactivation

A Cunha[1], J Couceiro[1,2], D Bonifácio[1,2], C Martins[1,2], A Almeida[1], M G P M S Neves[2], M A F Faustino[2] and J A Saraiva[2]

[1] Biology Department and CESAM, University of Aveiro, 3810-193 Aveiro, Portugal

[2] Chemistry Department and QOPNA, University of Aveiro, 3810-193 Aveiro, Portugal

E-mail: acunha@ua.pt

Abstract. Traditional food processing methods frequently depend on the application of high temperature. However, heat may cause undesirable changes in food properties and often has a negative impact on nutritional value and organoleptic characteristics. Therefore, reducing the microbial load without compromising the desirable properties of food products is still a technological challenge. High-pressure processing (HPP) can be classified as a cold pasteurization technique, since it is a non-thermal food preservation method that uses hydrostatic pressure to inactivate spoilage microorganisms. At the same time, it increases shelf life and retains the original features of food. Photodynamic inactivation (PDI) is also regarded as promising approach for the decontamination of food matrices. In this case, the inactivation of bacterial cells is achieved by the cytotoxic effects of reactive oxygens species (ROS) produced from the combined interaction of a photosensitizer molecule, light and oxygen. This short review examines some recent developments on the application of HPP and PDI with food-grade photosensitizers for the inactivation of listeriae, taken as a food pathogen model. The results of a proof-of-concept trial of the use of high-pressure as a coadjutant to increase the efficiency of photodynamic inactivation of bacterial endospores is also addressed.

1. Introduction

Over time and starting from the moment of harvest, fresh foodstuff loses quality by physical, chemical, and biological processes. Microorganisms and enzymes are key agents of deterioration and consequently priority targets of preservation techniques.

Traditional food processing methods frequently depend on the application of high temperature. However, heat may cause undesirable changes in food properties and often has a negative impact on nutritional value and organoleptic characteristics. Therefore, reducing the microbial load without compromising the desirable properties of food products is still a technological challenge.

An ideal method of food preservation should be inexpensive and convenient to apply to ensure the inactivation of microorganisms and enzymes and/or inhibition of microbial activity and growth [1] while extending shelf life, preserving organoleptic and nutritional attributes, without leaving chemical residues and not raising objections from consumers and legislators [1, 2]. By using temperatures lower than those typically used in thermal processing, non-thermal food processing technologies are expected to exert a minimal impact on the nutritional, physical, chemical and sensory properties of food [3]. High pressure processing (HPP) and photodynamic inactivation (PDI) are examples of non-thermal techniques that inactivate/eliminate microorganisms at ambient temperatures [3, 4].

HPP uses hydrostatic pressure - force per unit area applied in a direction perpendicular to the surface - to inactivate pathogens and vegetative forms of spoilage microorganisms [5]. Additionally, it increases shelf life and retains the original features of food. This technique can be used in different types of solid or liquid food matrices, at pressure values between 100 and 1000 MPa (1 MPa = 0.101 atm = 0.1 bar = 6.89×10^{-3} psi) in a range of temperature between -20 and 80 °C, during periods that can range from seconds to minutes [6, 7]. During the pressurization, a decrease in food volume proportional to the pressure applied occurs but food material returns to its initial volume during decompression [8]. Pressure is applied in an isostatic mode, *i.e.* the transmission of pressure occurs uniformly and almost instantly through the food material regardless of its shape and size, making this technique suitable for the inactivation of pathogens present at the surface or imbed in the food matrix [9]. The level of bacterial inactivation depends on the type of microorganism, but also on the composition and pH of the food matrix and therefore, it is necessary to carefully choose the appropriate processing protocol [2, 10].

PDI is also regarded as a promising approach for the decontamination of food matrices. In this case, the inactivation of bacterial cells is achieved by the cytotoxic effects of reactive oxygen species (ROS) produced from the combined interaction of a photosensitizer (PS) molecule, light and oxygen. The PS is excited by light and changes to a long-lived triplet state. Molecular oxygen is regarded as a key factor in PDI. In the type I photochemical mechanism, the PS in the excited state interacts with molecular oxygen by electron or hydrogen transfer, generating radical species including the superoxide anion, which can further originate other ROS like hydroxyl radicals. Type II mechanism is associated to the interaction of the excited state of the PS with molecular oxygen, in this case by energy transfer, which leads to the production of singlet oxygen [11]. PDI has been demonstrated as an efficient alternative for the inactivation of virus, bacteria and fungi [12]. However, the application of this technique to the inactivation of microbes in food products is dependent on the availability of food-compatible PSs that do not compromise food composition, appearance, taste and flavour. Some natural compounds with photosensitizing potential, such as hypericin, chlorophyll, riboflavin and curcumin, are common additives in foods and drinks that have been proposed to overcome this problem [13].

2. Non-thermal inactivation of listeriae

The genus *Listeria* is composed of short Gram positive rods with 0.4 – 0.5 μm of diameter and 1.2 μm of length, appearing singly or in short chains. This genus is closely related to other food pathogens like *Bacillus*, *Clostridium*, *Enterococcus*, *Streptococcus* and *Staphylococcus*. *Listeria* spp. are facultative anaerobes, do not produce spores and grow at temperatures between 0 and 45 °C, although their optimum temperature for growth is between 30 and 37 °C. Listeriae easily form biofilms and grow in high concentrations of salt (10% NaCl) and at pH values between 4.5 and 9. This genus entails six species: *L. monocytogenes*, *L. ivanovii*, *L. innocua*, *L. grayi*, *L. welshimeri* and *L. seeligeri*. Only the first two species are pathogenic [14, 16]. *L. monocytogenes* is a food pathogen responsible for an opportunistic infection called listeriosis, most often transmitted by raw food like soft cheese, fruits and vegetables but also by cooked meat [17, 18]. Although listeriosis is not very frequent, it can be lethal among adults and neonates [19]. *L. innocua*, is a non-pathogenic species that has been often used as a surrogate for the pathogenic *L. monocytogenes* in biological studies, since it presents similar responses to chemical or thermal treatments [20].

2.1. Inactivation of listeriae by HPP

The exposure of bacterial cells to high pressure causes damage to the cell membrane and denaturation of proteins, affects the function of enzymes and ribosomes, deregulates homeostasis and may ultimately cause cell disruption [6, 21]. With low pressure values (<50 MPa), processes like gene expression and protein synthesis are affected but at pressure values above 400 MPa significant structural damage occurs [22]. Although not all the previously mentioned effects are demonstrated for listeriae, morphological, structural, physiological and genetic changes have been reported [23].

The inactivation efficiency is strongly affected by intrinsic (related with the bacterial cell) and/or extrinsic (related with the extracellular medium) factors. *L. innocua* cells in exponential growth phase are more susceptible that stationary phase cells probably because of the stress response proteins expressed by the latter [24, 25]. The composition of the food matrix, particularly in terms of pH and a_w has also a significant effect on the efficiency of inactivation of listeriae. In general, lower pH like in fruit juices, enhances inactivation since the survival of pressure-damaged cells is reduced in acid media [22]. Lower a_w decreases susceptibility to HPP, although the magnitude of the effect seems to be dependent on the type of solute [26, 27]. As an example, literature data indicate that milk significantly protects listeriae from HPP but this effect may not be observed in milk derivatives such as cheese, because of the particular properties of this product in terms of pH and a_w [23]. Cells that have been previously exposed to low temperatures (10 - 25 °C) are more susceptible to HPP than cells grown at temperatures near 43 °C [28]. This may implicate that contamination originating from warm-blood animals may be more difficult to control than contaminants from other environmental sources.

Pressurization parameters are of major importance in the inactivation of listeriae by HPP. In general, *L. innocua* is completely inactivated in food matrices at 400-600 MPa [29]. There is a direct relation between holding time (the time during which pressure is applied) and the efficiency of inactivation. At moderate pressure values, extending the holding time from 5 to 15 minutes may increase the factor of inactivation of *L. innocua* by as much as 5 log but at very high pressure values (>500 MPa), holding time losses relative importance as an operational parameter [30]. A study conducted with stationary phase cells of *L. innocua* demonstrated that lower compression/decompression rates increased inactivation efficiency [29]. A large set of data obtained in different food matrices and with different combinations of pressure and temperature indicates that inactivation efficiency seems to be the highest at refrigeration temperatures (4-10 °C) or when high pressure is combined with heat (50 °C) [23, 29]. All this further stresses the importance of a careful design of HPP protocols.

2.2. Photodynamic inactivation of listeriae

Literature reports of successful photodynamic inactivation of listeriae are strikingly scarce. Tests conducted with *Escherichia coli O157:H7* and *L monocytogenes* revealed that the latter was more easily inactivated than the former, which was attributed to the differences in the composition of the cell wall. Also, photosensitization was more efficient with the dyes toluidine blue O (TBO) and methylene blue trihydrate (MB), than with a porphyrin, the tetratosylate salt of 5,10,15,20-tetrakis(1-methylpyridinium-4-yl)porphyrin (**Tetra-Py$^+$-Me**) [31]. However, TBO and MB cannot be used in foodstuff. The results of experiments in which a colourless and odourless solution of 5-aminolevulinic acid (ALA) was used to induce the synthesis of endogenous porphyrins demonstrated that plankton cells and biofilms of *L. monocytogenes* could be efficiently inactivated [32]. But again, the use of natural compounds, already approved as food additives, is preferable and more easily accepted by consumers.

Curcumin (food additive E100) is a yellow pigment and the active constituent of turmeric, which is obtained from the dried rhizome of *Curcuma longa*. Turmeric powder is an essential ingredient in curry, and it is often used in ready-to-eat meat dishes. Curcumin is a natural PS, able to generate cytotoxic ROS when activated with blue light [33] and photosensitization with curcumin has been successfully tested on yeasts and bacteria [34-36].

In recent experiments conducted by our group, irradiation of biofilms of *L. innocua* with blue light in presence of 10 μM of commercial curcumin caused an approximate 5 log reduction in the concentration of viable cells, whereas with an equivalent concentration of the reference cationic porphyrin, the tetraiodide salt of **Tetra-Py$^+$-Me**, the inactivation factor was only of ~1 log. The planktonic form was much more susceptible to PDI and in planktonic cells, photosensitization was more efficient with the porphyrin, which caused complete inactivation (unpublished). Bacterial biofilms represent a considerable challenge in terms of PDI because the extracellular matrix limits the penetration of the photosensitizer and the diffusion of oxygen to inner layers and cytotoxic ROS are efficiently captured by the extracellular polymeric substances

[37]. Our results indicate that curcumin may be a particularly advantageous PS for the control of biofilm bacteria in packaged ready-to-eat meat meals.

3. Pressure-assisted photodynamic inactivation of bacterial endospores

Sporulation is a biological mechanism of resistance that enhances bacterial survival in harsh environmental conditions for considerably long periods of time. They represent a challenge in terms of inactivation since endospores resist pasteurization temperatures, desiccation, and chemical biocides like organic acids, alcohols and phenols, which easily destroy vegetative cells [38]. *Bacillus cereus*, *Clostridium perfringens* and *C. botulinum* are examples of foodborne endospore producing bacteria that due to their ubiquity in the environment and resistance to thermal processing represent a major concern in terms of the safety and stability of foods [39].

HPP inactivation of endospores is a two-step process in which an initial treatment with moderate to low temperature and pressure (100-200 MPa, T<50 °C) induces spore germination and higher pressure and temperature (400-600 MPa, T>60 °C) cause significant endospore inactivation [40]. In general, the process requires the application of very high pressure values and quite long holding times which makes it costly in terms equipment and energy. PDI of bacterial endospores has also received attention in the perspective of an alternative non-thermal inactivation approach. Although much less susceptible than vegetative cells, endospores of *Bacillus* spp. have been successfully inactivated with phenothiazine dyes and cationic porphyrins [41, 42]. However, complete inactivation is not easily achieved and the binding of the PS to the essentially inert spore coating has been identified as a limiting step in the photosensitization process [42].

Using endospores of *B. cereus* as biological models, we tested the hypothesis that high pressure could be used as a physical coadjutant to enhance PS binding and therefore to improve the PDI of bacterial endospores [pressure-assisted photodynamic inactivation (HP-PDI)]. In repeated independent trials, pressurized endospore suspensions (300 MPa for 30 minutes in the presence of 20 μM **Tetra-Py$^+$-Me**) revealed up to 76-fold increases in PS binding, in relation to the controls in which the dark exposure to PS was conducted at atmospheric pressure, and a significant increase in photosensitization efficiency upon irradiation (unpublished). Considering that the proof-of-concept of HP-PDI was successful, the challenge is now to optimize the approach using food-grade photosensitizers for endospore inactivation in food products.

4. Future perspectives

HPP and PDI represent promising non-thermal alternatives for the control of bacterial pathogens and food-spoilage microorganisms, such as inferred from experiments with *Listeria* spp. Both approaches still face technical challenges. Future developments of HPP protocols must address the problem of protein denaturation, so that it can be used in a wider variety of food products, and also the cost effectiveness of the process, in terms of equipment and energy requirements. A major step forward in PDI of food pathogens will be the development of efficient food-grade photosensitizers and the combined application of physical co-adjutants, like high-pressure or heat, to overcome the problem of more resistant forms like microbial biofilms, protozoan cysts, fungal spores and bacterial endospores.

Acknowledgments

The authors are thankful to the University of Aveiro, to FCT/MEC for the financial support to the CESAM unit (project Pest-C/MAR/LA0017/2013) and QOPNA research Unit (FCT UID/QUI/00062/2013), through national founds and, where applicable, co-financed by the FEDER, within the PT2020 Partnership Agreement, and also to the Portuguese NMR Network.

References

[1] Van Schothorst M, Zwietering M, Ross T, Buchanan R and Cole M 2009 *Food Control* **20** 967-79
[2] Raso J and Barbosa-Cánovas GV 2003 *CRC Cr Rev Food Sci* **43** 265-85
[3] Morris C, Brody AL and Wicker L 2007 *Packag Technol Sci* **20** 275-86
[4] Kreitner M, Wagner K-H, Alth G, Ebermann R, Foißy H and Elmadfa I 2001 *Food Control* **12** 529-33
[5] Rivalain N, Roquain J and Demazeau G 2010 *Biotechnol Adv* **28** 659-72
[6] San Martin M, Barbosa-Cánovas G and Swanson B 2002 *CRC Cr Rev Food Sci* **42** 627-45
[7] Yamamoto K 2017 *Biosci Biotechnol Biochem* **81** 672-9
[8] Patterson M, Ledward D and Rogers N 2006 High Pressure Processing In: *Food Processing Handbook* ed J Brennan (New York, Wiley & Sons) pp 173-200
[9] Rastogi N, Raghavarao K, Balasubramaniam V, Niranjan K and Knorr D 2007 *CRC Cr Rev Food Sci* **47** 69-112
[10] Zhang HQ, Balasubramaniam VB, Dunne CP, Farkas DF and Yuan JT 2011 *Nonthermal Processing Technologies for Food (*New York, Wiley & Sons) 672p.
[11] Craig RA, McCoy CP, Gorman SP and Jones DS 2015 *Expert Opin Drug Del* **12** 85-101
[12] Alves E, Faustino MA, Neves MG, Cunha Â, Nadais H and Almeida A 2015 *Journal of Photochemistry and Photobiology C: Photochemistry Reviews* **22** 34-57
[13] Luksiene Z and Brovko L 2013 *Food Eng Rev* **5** 185-99
[14] Vázquez-Boland JA, Domínguez-Bernal G, González-Zorn B, Kreft J and Goebel W 2001 *Microb Infect* **3** 571-84
[15] Hain T, Steinweg C and Chakraborty T 2006 *J Biotechnol* **126** 37-51
[16] Rocourt J and Buchrieser C 2007 The Genus *Listeria* and *Listeria Monocytogenes*: Phylogenetic Position, Taxonomy, and Identification In: *Listeria, Listeriosis, and Food Safety* ed Ryser E and Marth E (Boca Raton, CRC Press,) 1-20
[17] Todd E, Notermans S 2011 *Food Control* **22** 1484-90
[18] Evans EW and Redmond EC 2014 *J Food Prot* **77** 510-21
[19] Cartwright EJ, Jackson KA, Johnson SD, Graves LM, Silk BJ and Mahon BE 2013 *Emerg Infect Dis* **19** 1-9
[20] Milillo SR, Friedly EC, Saldivar JC, Muthaiyan A, O'Bryan C, Crandall PG, Johnson MG and Ricke SC 2012 *CRC Cr Rev Food Sci* **52** 712-25
[21] Huang H-W and Lung H-M, Yang BB and Wang C-Y 2014 *Food Control* **40** 250-9
[22] Smelt J 1998 *Trends Food Sci Technol* **9** 152-8
[23] Ferreira M, Almeida A, Delgadillo I, Saraiva J and Cunha Â 2016 *Food Rev Int* **32** 377-99
[24] Wemekamp-Kamphuis HH, Wouters JA, de Leeuw PP, Hain T, Chakraborty T and Abee T 2004 *Appl Environ Microb* **70** 3457-66
[25] Saucedo-Reyes D, Marco-Celdrán A, Pina-Pérez MC, Rodrigo D and Martínez-López A 2009 *Innov Food Sci Emerg* **10** 135-41
[26] Molina-Höppner A, Doster W, Vogel RF and Gänzle MG 2004 *Appl Environ Microb* **70** 2013-20
[27] Koseki S and Yamamoto K 2007 *Int J Food Microbiol* **115** 43-7
[28] Shearer AE, Neetoo HS and Chen H 2010 *Int J Food Microbiol* **136** 359-63
[29] Ferreira M, Pereira S, Almeida A, Queirós R, Delgadillo I, Saraiva J and Cunha A 2016 *Acta Aliment Hung* **45** 61-8
[30] Yuste J, Mor-Mur M, Capellas M and Pla R 1999 *Meat Sci* **53** 251-7
[31] Romanova NA, Brovko LY, Moore L, Pometun E, Savitsky AP, Ugarova NN and Griffiths MW 2003 *Appl Environ Microb* **69** 6393-8
[32] Buchovec I, Paskeviciute E and Luksiene Z 2010 *Food Technol Biotechnol* **48** 207-13
[33] Chignell CF, Bilskj P, Reszka KJ, Motten AG, Sik RH and Dahl TA 1994 *Photochem Photobiol* **59** 295-302

[34] Dovigo LN, Pavarina AC, Ribeiro APD, Brunetti IL, Costa CAdS, Jacomassi DP, Bagnato VS and Kurachi C 2011 *Photochem Photobiol* **87** 895-903
[35] Ribeiro APD, Pavarina AC, Dovigo LN, Brunetti IL, Bagnato VS, Vergani CE and de Souza-Costa CA 2013 *Laser Med Sci* **28** 391-8
[36] Tortik N, Steinbacher P, Maisch T, Spaeth A andPlaetzer K 2016 *Photoch Photobio Sci* **15** 187-95
[37] Biel MA 2010 Photodynamic Therapy of Bacterial and Fungal Biofilm Infections In: *Photodynamic Therapy: Methods and Protocols* ed Gomer C J (Humana Press, New York) 175-94
[39] Leggett MJ, McDonnell G, Denyer SP, Setlow P and Maillard JY 2012 *J Appl Microbiol* **113** 485-98
[40] Heyndrickx M 2011 *Appl Environ Soil Sci* **2011** Article ID 561975
[41] Reineke K, Mathys A, Heinz V and Knorr D 2012 *Trends Microbiol* **20** 1-9
[42] Demidova TN and Hamblin MR 2005 *Appl Environ Microb* **71** 6918-25
[43] Oliveira A, Almeida A, Carvalho CMB, Tomé JPC, Faustino MAF, Neves MGPMS, Tomé AC, Cavaleiro JAS and Cunha A 2009 *J Appl Microbiol* **106** 1986-95

Microbial consortia in meat processing environments

V Alessandria[1], K Rantsiou[1], M C Cavallero[2], S Riva[3] and L Cocolin[1]

[1] Disafa, University of Turin, Largo Braccini 2, Grugliasco (To), Italy
[2] Tecnogranda, Via G.B.Conte 19, Dronero (Cn), Italy
[3] Veterinary Food Safety, Via Monviso 7, Salmour (Cn), Italy

E-mail: valessan@unito.it

Abstract. Microbial contamination in food processing plants can play a fundamental role in food quality and safety. The description of the microbial consortia in the meat processing environment is important since it is a first step in understanding possible routes of product contamination. Furthermore, it may contribute in the development of sanitation programs for effective pathogen removal. The purpose of this study was to characterize the type of microbiota in the environment of meat processing plants: the microbiota of three different meat plants was studied by both traditional and molecular methods (PCR-DGGE) in two different periods. Different levels of contamination emerged between the three plants as well as between the two sampling periods. Conventional methods of killing free-living bacteria through antimicrobial agents and disinfection are often ineffective against bacteria within a biofilm. The use of gas-discharge plasmas potentially can offer a good alternative to conventional sterilization methods. The purpose of this study was to measure the effectiveness of Atmospheric Pressure Plasma (APP) surface treatments against bacteria in biofilms. Biofilms produced by three different *L. monocytogenes* strains on stainless steel surface were subjected to three different conditions (power, exposure time) of APP. Our results showed how most of the culturable cells are inactivated after the Plasma exposure but the RNA analysis by qPCR highlighted the entrance of the cells in the viable-but non culturable (VBNC) state, confirming the hypothesis that cells are damaged after plasma treatment, but in a first step, still remain alive. The understanding of the effects of APP on the *L. monocytogenes* biofilm can improve the development of sanitation programs with the use of APP for effective pathogen removal.

1. Introduction

Spoilage organisms can be transferred from the environment to intermediates of production and may negatively affect the production process and the quality of the final product. The first purpose of this study was to characterize the type of microbiota in the environment of meat processing plants situate in north west of Italy and gain insights regarding potential microbial contamination risks for the final products.

The second step of the wok focused on the evaluation of an alternative sanitization technique that have become an important aspect in the food environment. The development of antimicrobial measures that are not subject to evolving microbial resistance represent a new challenge in the food control. In this context, the cold atmospheric pressure plasma (APP) is a relatively new antimicrobial technique that has been recently adopted also for applications in the food industry. The reactive free radicals and H_2O_2 produced during APP generation play the main role in bacterial inactivation [1] together with the oxidation of amino acids and nucleic acids that are involved in microbial death or injury [2]. Evaluating any sanitization process, particular attention has to be focused in the understanding of the state of the cells after the treatments: cells can be stressed and non culturable but may not necessarily be dead. Variation in the surrounding conditions can influence bacterial counts and because of the environmental instability, the bacteria can enter in a viable but non-culturable state (VBNC) [3].

IOP Conf. Series: Earth and Environmental Science **85** (2017) 012017 doi:10.1088/1755-1315/85/1/012017

The purpose of this second part of the study was to measure the effectiveness of APP treatments against bacteria organized in biofilm on surfaces, evaluating also the individual susceptibility of different *L. monocytogenes* strains. The attention focused in particular on the state of the cells after the treatment and on their possible entry in the VBNC state by traditional and molecular methods.

2. Material and methods

2.1. Traditional and molecular methods in the analysis of food processing plants

The microbiota of three different meat plants was studied by both traditional and molecular methods (PCR-DGGE) in two different periods (winter and summer). Environmental samples from surfaces and tools (swabs from knifes, blades, walls, conveyor belts, hoppers, bagging, saw bones, table surfaces, cutter, meat mixers) were analyzed and the occurrence of pathogens (*Listeria monocytogenes* and *Salmonella* spp.) was also investigated.

Plate Count Agar was used for total mesophilic aerobic count. *L. monocytogenes* and *Salmonella* spp. were investigated following the ISO methods.

DNA was extracted directly from swabs and culture-independent analyses were carried out by PCR-DGGE of V3 region of 16S rRNA gene [4].

2.2. Plasma treatment

Three *L. monocytogenes* strains and in particular a collection strain (EDGe) and two from the Unito culture collection (3 and 36) were selected for the evaluation of their ability to attach to abiotic surface. Different conditions of APP were tested on the biofilm produced on stainless steel (SS) coupons by these three strains after 144 hours. The plasma was generated at an input power of 1154, 760 or 430W for a time period of 10 min or 2 min each side of the SS coupon (Table 1). Detachment of attached cells from the SS coupons and plate counts were performed by using the bead vortexing method [5]. with some modifications. The cell suspension obtained was also conserved at -80°C in the presence of RNA*later* (Ambion, Italy) for future use in RNA extraction.

Table 1. The APP conditions on SS coupons: time and power

	conditions
CONTROL	No treatment
A	10 min each side-431W- 159 KHz
B	10 min each side-724W-151KHz
C	2 min each side- 1154W-142KHz
D	2min each side-741W-151KHz
E	2min each side-431W-159KHz

2.3 Resuscitation of the VBNC cells and enumeration of L. monocytogenes

After an APP exposure for 10 min, coupons were aseptically inserted in BHI broth in order to evaluate the vitality of the cells. This resuscitation step was performed leaving the coupons in the medium at 37°C for 24 hours. In order to count the viable cells, the same protocol described above with bead vortexing was used.

RNA Purification Kit (Epicentre, Madison,WI, USA) following the manufacturer's instructions was adopted for the RNA extraction. Reverse transcription (RT) reactions were performed and one μL of the obtained complementary DNA was used as template for the qPCR amplification of the bacterial V3 region of the 16S rRNA, using primers 338f and 518r [4]. Amplifications were performed with the use of SSo Advanced Sybr Green Supermix (Biorad, Italy). Samples were amplified in triplicate using the following conditions: initial denaturation at 95 °C for 5 min and 40 cycles of 95 °C for 15 s, 60 °C for 30 s, according to [6].

In order to obtain a culture-independent enumeration of the active *L. monocytogenes* cells, a standard curve was constructed. Ten-fold serial dilutions of an overnight culture of *L. monocytogenes* strains were performed in Ringer's solution (Oxoid, Milan, Italy). One ml of each dilution was subjected to

RNA extraction as described above and the resulting cDNA sample was submitted to qPCR. Standard curves were constructed by plotting the threshold cycle (Ct) values obtained against CFU/ml, as determined on BHI agar from each dilution. Correlation coefficients (R^2) and efficiency of amplification were calculated as previously described [7].

3. Results and Discussion

3.1. Microbiota in food processing plants

Differences in the microbial populations (composition, load) between the three plants among the two periods were detected by traditional methods. More than 50% of samples analyzed in Plant 1 showed the presence of *L. monocytogenes* during the first sampling. Also *Salmonella* spp. was found with a high occurrence in Plant 1.

Plant 1 and 2 showed differences in the mesophilic aerobic counts between the two seasons, characterized by a high incidence of samples (more than 50%) with values > 10^4 cfu/cm^2 in the summer. Plant 3 did not report relevant differences among the two seasons.

By PCR-DGGE, tools and environmental samples were found to be characterized by the presence of several contaminants such as *Pseudomonas, Acinetobacter, Brochotrix.* Genera *Staphyloccoccus* and *Lactobacillus* were detected mainly in Plant 1. Swabs analyzed in winter and summer, for the same plant, showed a degree of microbial diversity (Table 2).

Table 2. Results of the identification of selected PCR-DGGE band sequencing

	Plant 1		Plant 2		Plant 3	
closest relative	winter	summer	winter	summer	winter	summer
Pseudomonas sp.	■		■	■	■	
Pseudomonas migulae			■			
Pseudomonas graminis			■		■	
Acinetobacter xiamenensis	■					
Acinetobacter johnsonni	■		■	■		
Acinetobacter haemolyticus				■	■	
Acinetobacter calcoaceticus						■
Acinetobacter sp.		■			■	■
Staphylococcus saprophyticus	■					
Staphylococcus xylosus	■					
Staphylococcus lentus		■				
Staphylococus sciuri		■				
Staphylococcus sp.						■
Leuconostoc citreum		■				
Lactobacillus plantarum		■				
Lactococcus piscium			■			
Micrococcus sp.				■		
Bacillus sp.	■				■	
Psychrobacter sp.	■					
Klebsiella sp.				■		
Moraxella sp.	■					
Corynebacterium sp.	■	■			■	
Blastocloris sp.	■					
Flavobacterium sp.			■			
Micrococcus sp.				■		
Uncultured *Brochotrix* sp.				■	■	
Exiguobacterium sp.				■		
Arthobacter sp.					■	
Spingomonas sp.					■	
Pantoea vagans				■		

The presence of the black box indicates the presence of the bands in the DGGE profiles. The sequences obtained were aligned with those in GenBank with Blast program.

This study underlined the high incidence of spoilage microorganisms in the environment of meat processing plants and tools.

Traditional methods showed the presence of pathogenic bacteria such as *L. monocytogenes* and S*almonella* spp. Differences emerged between the two sampling periods. The corrective actions taken after the first sampling, allowed a % decrease of contaminated samples. The mesophilic count was higher in the samples analyzed in the summer.

DGGE analysis underlined the occurrence of members of *Enterobacteriaceae*, lactic acid bacteria, *Pseudomonas* spp. and *Brochothrix* spp. that are recognized as the principal players in meat spoilage.

3.2. Plasma treatment

Due to the microbiota contamination found in the cleaned environment, the second step of the wok focused on the evaluation of new sanitization techniques. In this context, the APP is a relatively new antimicrobial technique that has been recently adopted also for applications in the food industry.

This study investigated the effect of different combinations of time/intensity APP treatments on *L. monocytogenes* cells attached on stainless steel surface. After 10 minutes of AAP treatment, for all 3 intensities tested, *L. monocytogenes* was not detected by plate counts. In the case of shorter APP exposure time (2 min), the lowest power (431W-159Hz) reduced the count more than 2 Log CFU/cm^2 (data not shown).

After the APP exposure for 10 min, coupons were inserted in BHI broth in order to evaluate if cells could be resuscitated. Growth in BHI was observed in the coupons treated with the lowest plasma power underlining the capability of cells to survive for as much as 10 minutes at the APP treatment (data not shown). This resuscitation step confirmed the entrance of these cells in the VBNC state after APP treatment since no *L. monocytogenes* growth was observed without incubation in BHI at 37 °C for 24 hours.

In order to enumerate *L. monocytogenes* in a culture-independent way, qPCR was applied on RNA extracted from cell suspensions recover from the SS coupons. The results obtained analysing the RNA by the amplification of the 16S showed the presence of viable cells also in the coupon treated with the highest power for 10 minutes. Non treated SS coupons showed the highest count (6 Log CFU/cm^2 for strain 36). Differences between strains were observed. Regarding strain 3, the conditions D and E (treatment for 2 minutes) were no different compared to the control. In the case of strains 36 and EGDe, all the Plasma conditions reduced the count significantly compared to the control (Table 3).

Table 3. Inactivation kinetics after the APP treatments as resulted by molecular methods expressed as CFU/cm^2

		Plasma conditions						sig.
		Control	**A**	**B**	**C**	**D**	**E**	
strains	**3**	5,48 b	3,82 a	3,91 a	3,65 a	5,13 b	4,99 ab	p<0,05
	36	6,04 b	4,5 a	3,825 a	4,125 a	4,095 a	4,79 a	p<0,05
	EGDe	5,44 b	4,57 a	4,22 ab	4,34 a	4,88 a	4,68 a	p<0,05

The Plasma condition (A-E) are reported in Table 1. Values with different letters for are significantly different, P < 0,05.

Traditional methods showed how most of the culturable cells are inactivated after the Plasma exposure but the RNA analysis obtained by q(PCR) highlighted the entrance of the cells in the viable-but non culturable (VBNC) state, confirming the hypothesis that cells are damaged after plasma treatment, but still remain alive.

The results showed that bacterial biofilms can be reduced by using gas-discharge plasma thus confirming the potential of plasma as an alternative sterilization method. However, discrepancies were observed between the two microbial enumeration methods employed: the plate count highlighted the suitability of the APP in eliminating *L. monocytogenes* cells organized in biofilms while by targeting the 16S rRNA by qPCR, the presence of VBNC populations was revealed and no significant differences emerged between the different conditions of the treatments. These results were at least partly confirmed by the resuscitation experiment: incubating the cell suspensions, after APP treatment (at low intensity) in BHI broth, plate count results showed that part of the *L. monocytogenes* population survived the treatment. Therefore, plasma treatment damaged but did not eliminate the *L. monocytogenes* cells attached to SS coupons.

Vitality of bacteria is an important aspect, especially in the food safety sector. Cells that appear unculturable in laboratory conditions may still possess several functions and activities typical of living cells [8]. The resuscitation in medium was not obtained for coupons treated at high APP intensity but the results of qPCR counts with about 4 Log CFU/cm^2 showed the vitality of cells. We can state that the conventional cultivation methods overestimate the decontamination efficiency of the APP, and must therefore be complemented by alternative techniques capable of detecting viable but non-culturable bacteria. Notwithstanding the discrepancies observed between culture dependent and independent approach the APP resulted to be effective in decreasing the load of attached cells. The untreated sample has shown higher counts by both traditional and molecular methods, compared with those that were treated, confirming that APP activity may challenge the physiology of microorganisms.

4. Conclusion

The results of this study highlight how it is vital for food producing companies to have control of the contaminants in the plants. This can be achieved with the implementation of adequate cleaning and disinfection procedures ant it may contribute in the development of programs for effective pathogen removal.

The description of the microbial consortia in the meat processing environment is important since it is a first step in understanding possible routes of product contamination. Sanitizing procedures and VBNC state of the cells cover a fundamental role in the food safety: considering that VBNC populations can subsequently recover and grow [9]. As also affirmed by [10], it is important to use methods independent of cell culturability to monitor pathogens in food processing plants since cultivation may underestimate the microbial load.

Reference

[1] Yong H I, Kim H J, Park S, Kim K, Choe W, Yoo S J and Jo C 2015. Pathogen inactivation and quality changes in sliced cheddar cheese treated using flexible thin-layer dielectric barrier discharge plasma. *Food Res. Int.* **69** 57-63

[2] Yun H, Kim B, Jung S, Kruk Z A, Kim K, Choe W and Jo C 2010. Inactivation of Listeria monocytogenes inoculated on disposable plastic tray, aluminum foil, and paper cup by atmospheric pressure plasma. *Food Control* **21** 1182-86

[3] Oliver J D 2005. The viable but nonculturable state in bacteria. *J Microbiol* **43** 93-100

[4] Cocolin, L., Manzano, M., Cantoni, C., Comi, G., 2001. Denaturing gradient gel electrophoresis analysis of the 16S rRNA gene V1 region to monitor dynamic changes in the bacterial population during fermentation of Italian sausages. Applied and Environmental Microbiology **67** 5113-21

[5] Giaouris E, Chorianopoulos N and Nychas GJ 2014. Impact of acid adaptation on attachment of Listeria monocytogenes to stainless steel during long-term incubation under low or moderate temperature conditions and on subsequent recalcitrance of attached cells to lethal acid treatments. *Int J Food Microbiol* **171** 1-7

[6] Alessandria V, Ferrocino I, De Filippis F, Fontana M, Rantsiou K, Ercolini D and Cocolin L 2016. Microbiota of an Italian Grana-Like cheese during manufacture and ripening, unraveled by 16S rRNA-Based approaches. *Appl Environ Microbiol* **82** 3988-95

[7] Higuchi R, Fockler C, Dollinger G and Watson R 1993. Kinetic pcr analysis - real-time monitoring of dna amplification reactions. *Biotechnology* **11** 1026-30

[8] Lahtinen S J, Ahokoski H, Reinikainen JP, Gueimonde M, Nurmi J, Ouwehand AC and Salminen SJ 2008. Degradation of 16S rRNA and attributes of viability of viable but nonculturable probiotic bacteria. *Lett Appl Microbiol* **46** 693-98

[9] Rowan NJ 2004. Viable but nonculturable forms of food and waterborne bacteria: Quo Vadis? Trends Food Sci Technol **15** 462-67

[10] Overney A, Jacques-Andre-Coquin J, Ng P, Carpentier B, Guillier L and Firmesse O 2017. Impact of environmental factors on the culturability and viability of Listeria monocytogenes under conditions encountered in food processing plants. *Int J Food Microbiol* **244** 74-81.

Hepatitis E – a "new" foodborne disease

A Kirbiš and P Raspor Lainšček

Institute for food safety, feed and environment, Veterinary faculty, University of Ljubljana, Gerbičeva 60, 1000 Ljubljana, Slovenia

E-mail: andrej.kirbis@vf.uni-lj.si

Abstract. Hepatitis E (HE) is a zoonosis caused by hepatitis E virus (HEV). The disease that used to be problematic only in developing regions with inadequate water supplies and poor sanitary conditions is now considered one of the foodborne diseases in industrialized countries as well. According to current knowledge, the main reservoir of the virus is linked to domestic swine and wild boar. Consumption of raw or undercooked pork meat and liver is considered as a risk factor for HE human infection, together with some other sources of infection like blood transfusion or organ transplantation. Although the number of cases has been rising in the last decade, HEV is still a generally unknown virus among the general public. Consumers need to be warned and educated about HEV and its potential sources of contamination within the food supply chain.

1. Introduction

Hepatitis E virus (HEV) is a non-enveloped virus with a single-stranded, positive sense RNA genome. The genome is 7.2 kb long and codes for three open reading frames [1]. It belongs to the *Hepeviridae* family, which is divided into the genera *Ortohepevirus* (species A-D) and *Piscihepevirus*. *Ortohepevirus* A has 7 genotypes [2]. Genotypes 1 and 2 infect only humans and are distributed mainly in Asia, Africa and North America. They can cause large waterborne epidemics due to insufficient hygiene. Genotypes 3 and 4 are zoonotic, their main reservoirs are swine and wild boar, and they can cause sporadic cases of hepatitis E (HE) in human. Genotype 3 is distributed worldwide [1, 3]. Genotype 4 is found mainly in Asia [1], but it has already been detected in swine populations in Italy, Belgium and Denmark [4]. HEV from genotypes 5 and 6 is found in wild boars and genotype 7 in camels [5].

2. Hepatitis E infection in human

HEV in humans causes acute icteric disease with a very variable clinical picture [6]. In most cases the infection is asymptomatic [3], while the symptomatic patient shows the following clinical signs: icterus with nausea, fever, abdominal pain, vomiting, hepatomegaly, pale stools and darkened urine [5]. The disease is mostly self-limiting [3]. In low levels, with immunosuppressed patients (organ transplant recipients, lymphoma/leukemia patients, patients with human immunodeficiency virus infection), it can develop into chronic infection and progress to severe fulminant liver failure with fatal outcomes [6]. The mortality is approximately 2% [3,5]. Besides typical symptoms, HEV infection has been associated with neurological disorders, such as Guillain-Barre and Parsonage Turner syndromes, neuralgic amyotrophy, bilateral brachial neuritis, peripheral neuropathy and encephalitis [7].

3. Animal reservoirs

Although domestic swine, wild boar and deer are the only animal species that have been directly linked to zoonotic hepatitis, viral RNA has been found in many other animal species such as rabbit, mongoose, moose, fox, chicken, rat, ferrets, mink and bats. Anti-HEV antibodies have been found in dogs, cats, cows, buffalo, goat, sheep and horses [8].

As domestic swine are considered the main reservoir of HEV, many facts about the dynamics of infection with HEV in pigs have already been discovered. After the level of maternal antibodies drop, young piglets become susceptible to infections, HE being one of them. When they come in contact with the virus, the infection occurs, which usually happens at around 8 weeks of age. Pigs start to shed the virus in feces between one and two weeks after the infection and the shedding lasts for up to 7 weeks [4,6,9]. The virus can be found in muscles and liver until 4 weeks after the onset of fecal HEV excretion, meaning that the risk of HEV-contaminated meat and entrails entering stores is greater when younger animals are slaughtered [4,10]. The age at which pigs are slaughtered varies slightly from country to country, but overall there are three age groups – around the age of 3 months, 6 months and sows/boars. Many countries have conducted an overview of the state of HE infections either on farms or in slaughterhouses and all of them came to the same conclusions: the percentage of viremic pigs in the acute stage of infection (meaning that the viral load in meat and organs is high and can cause an infection in human through consumption of such food) is significantly higher at the age of 3 months than at the age of 6 months [11,12,13,14,15,16,17,18].

A very important fact that strongly contributes to the foodborne nature of HE infection is that swine infected with HEV show no clinical symptoms at all. Only some microscopic lesions have been found in experimentally infected pigs – mild multifocal lymphoplasmatic hepatitis was observed in liver tissue [5].

4. Routes of transmission

HEV has many routes of transmission (Figure 1). In developing countries, the main source of infection is water contaminated by sewage. Poor sanitary conditions in connection with several environment settings (heavy monsoon rains and floods) lead to frequent waterborne epidemics [19]. In industrialized countries, the situation is different – two decades ago it was believed that HE infections occurred after trips to developing countries. When more information was gathered from these patients, it was clear that some people never travelled outside the country, which led to new findings about the possible sources of infection [20,21]. Now we know that autochthonous cases can be acquired through blood transfusion or organ transplantation, and even vertical transmission from mother to child has been observed, but the most important route is ingestion of infected raw or undercooked pork meat/liver. Besides meat products, HEV RNA has also been detected on leafy green vegetables and strawberries (probably caused by the use of contaminated irrigation water or contaminated manure) [22,23].

5. Foodborne disease

With increasing interest in HEV, and also more human cases in Europe revealed, it now appears that foodborne transmission is a major route of infection in Europe. The European Food Safety Authority (EFSA) reported that more than 21000 clinical cases with 28 fatalities have been reported in the last 10 years and a 10-fold increase in reported HE cases has been observed [7]. France, Germany and the UK are countries with the majority of reported cases. Austria, Czech Republic, Hungary, Italy and Spain have also reported outbreaks and/or sporadic foodborne cases [7]. However, it should be noted that the monitoring of HEV differs between EU countries and the overall number of cases is most likely even higher. Sometimes the source of infection is directly identified by detection of viral RNA, while other times it remains unknown or the suspected source is only epidemiologically determined. In reported outbreaks/sporadic cases, tripe sausages, undercooked pork meat, liver sausage, raw figatelli, pig liver, wild boar meat and shellfish were the sources of infection. The most frequently reported food products causing HE infection are raw or undercooked pork meat or pork liver [7].

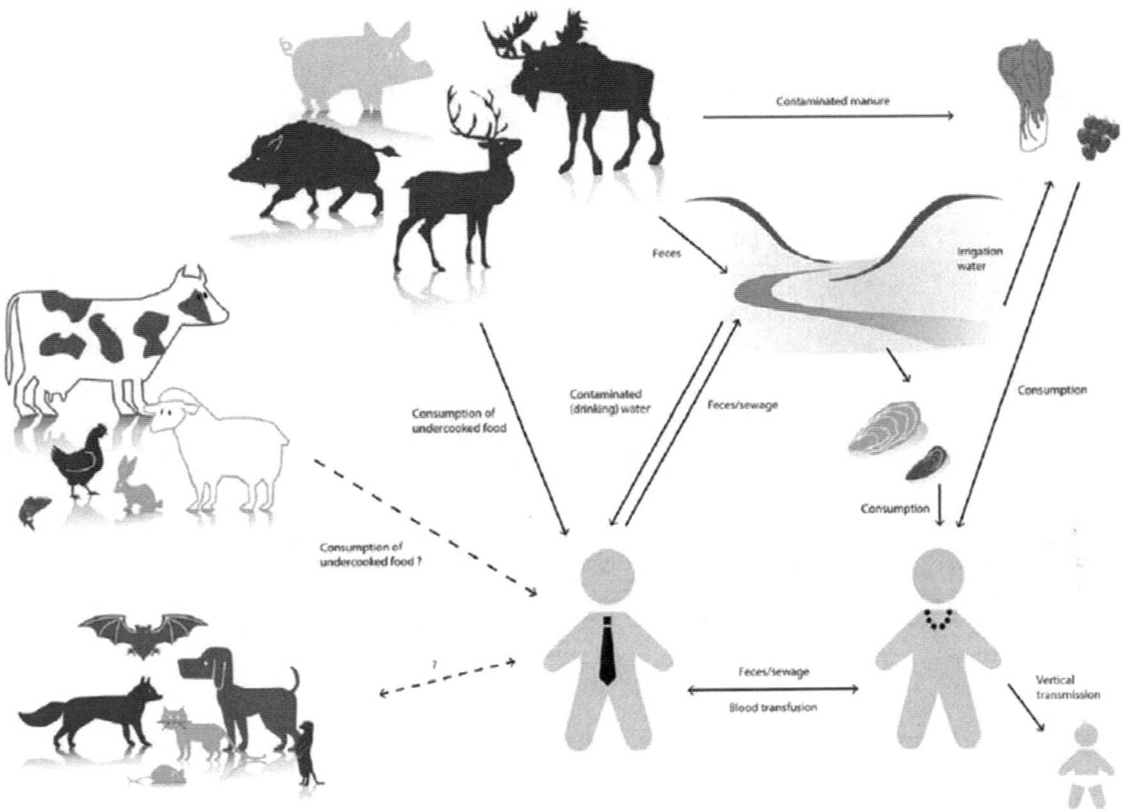

Figure 1. Different routes of hepatitis E virus transmission. Solid arrows indicate proven models of transmission, dashed arrows indicate the potential routes of spread [23].

6. Preventive measures – from stable to table

Hepatitis E, as many other foodborne diseases, should be treated comprehensively. Handling only one element of the whole issue will not make a drastic improvement. Responsible professionals involved in the pig chain should be aware of the critical control points that contribute to efficient management of this foodborne zoonosis (Table 1).

6.1. Hepatitis E virus on farms

Clear and focused measures for reducing HE infections should start at the farm level. Pigs shed the virus with feces and urine, so hygiene is of crucial importance. First, housing and equipment should always be disinfected before new animals are brought to the farm, otherwise the viral load in the environment can be high enough to cause infection in the new production batch of animals [10,24]. Secondly, farm workers have to understand biosecurity and practice it. If they follow the simple rules of changing clothes and shoes when entering the farm or when crossing from one part of the farm to another, the chances of spreading the infection between different age groups of animals will be reduced [25].

6.2. Hepatitis E virus in slaughterhouses

Even though *ante* and *post mortem* inspection in slaughterhouse is mandatory and samples are taken to monitor the zoonoses, this is not enough to detect pigs infected with HEV. As mentioned before, animals show no symptoms at all and we have no way of knowing when HEV-positive animals come to slaughter. To avoid possible cross contamination, good practices should always be followed, not only at the slaughterhouse, but also during transport and lairage of the animals and processing and storage of pork meat. Cross contamination with feces has to be avoided at all times.

After stunning and bleeding, pigs are either scalded and dehaired or skinned. If the pig's skin is removed without hot steaming or burning, viruses on the skin (if the pigs are contaminated with feces) can be transferred to the meat surface. If a scald tank (with water temperature of 64°C) is used, the chance of the virus on the skin surface surviving is probably lower, although it is still not clear what temperature and time of exposure to a particular temperature eliminates the virus in the technological environment.

Another critical control point on the slaughtering line is the removal of the intestines. If they are perforated, the feces can contaminate meat, organs and also the equipment used by workers. The same care should be taken when removing a gall bladder – bile has an even higher viral load than feces and perforation of the gall bladder can lead to cross contamination of liver and meat with HEV-laden bile [7].

6.3. Hepatitis E virus at home

In the end, the only efficient control for HE infection is the right heat treatment of food before eating. Pork meat, liver or meat products have to be thoroughly thermally processed (cooked, baked). The exact temperature and time period that should be used to efficiently eliminate the virus is still not clearly determined. Many experiments have been conducted and the results suggest that the heat resistance of HEV is variable – it depends on the strain or genotype and matrix as well. However, no matter how many experiments have been performed, there is a gap that still needs to be filled – a robust cell culture system for HEV that will allow the determination of HEV infectivity in laboratory circumstances with all practical parameters in real practice is urgently needed [26].

Table 1. Critical control points and preventive measures in pork chain production regarding hepatitis E virus (HEV)

	Pig farm	Pig slaughter house	Pork consumption
HEV Problem	Circulation of HEV on the farm: 1. high viral load in the environment (feces and urine) 2. transmission through direct contact between the animals (mingling of animals from different pens) 3. cross contamination caused by workers on the farm	Cross contamination of individual pigs and meat during slaughter and co-infection of meat/entrails via faces, bile or blood	Consumption of undercooked or raw pork meat, liver or pork meat products
Preventive measure	Biosecurity: 1. hygiene measures (washing of the transport vehicles, housing disinfection) 2. protection of the herd from wild animals 3. showering and changing clothes before entering the farm	Education of slaughterhouse workers about good practices to help avoid cross contamination	Education of consumers about the pork production chain HEV infection risk (importance of correct thermal treatment) and cross contamination risks during food preparation

7. Conclusion

A lot of new information has been discovered about HEV in the last 20 years and this "new" foodborne disease is getting more attention from day to day. The implementation of these discoveries in the practical pork supply chain still needs a professional push. Also of crucial importance is the fact that consumers are still not familiar with HE; most of them have never heard of it. This issue needs to be addressed and people need to be educated, so they will have the knowledge to protect themselves from HEV when it occurs in foods coming from the pork production chain.

References

[1] Doceul V, Bagdassarian E, Demange A and Pavio N 2016 Zoonotic hepatitis E virus: classification, animal reservoirs and transmission routes *Viruses* **8** (10)

[2] Smith D B, Simmonds P, Jameel S, Emerson S U, Harrison T J, Meng X-J, Okamoto H, Van der Poel W H M and Purdy M A 2014 Consensus proposals for classification of the family Hepeviridae *J. Gen. Virol.* **95** 2223-32

[3] Sridhar S, Lau S K P and Woo P C Y 2015 Hepatitis E: a disease of reemerging importance *J. Formos. Med. Assoc.* **114** 681–90

[4] Widen F 2016 Hepatitis E as a zoonosis *Adv. Exp. Med. Biol.* **948** 61–71

[5] Park W-J, Park B-J, Ahn H-S, Lee J-B, Park S-Y, Song C-S, Lee S-W, Yoo H-S and Choi I-S 2016 Hepatitis E virus as an emerging zoonotic pathogen *J. Vet. Sci.* **17** 1-11

[6] Yugo D M and Meng X-J 2013 Hepatitis E virus: foodborne, waterborne and zoonotic transmission *Int. J. Environ. Res. Public Health* **10** 4507-33

[7] Ricci A *et al.* 2017 Public health risks associated with hepatitis E virus (HEV) as a food-borne pathogen *EFSA Journal* **15** (7)

[8] Pavio N, Meng X-J and Doceul V 2015 Zoonotic origin of hepatitis E *Curr. Opin. Virol.* **10** 34-41

[9] Meng X-J 2011 From barnyard to food table: the omnipresence of hepatitis E virus and risk for zoonotic infection and food safety *Virus Res.* **161** 23-30

[10] Bouwknegt M, Rutjes S A, Reusken C B E M, Stockhofe-Zurwieden N, Frankena K, de Jong M C M, de Roda Husman A M and van der Poel W H M 2009 The course of hepatitis E virus infection in pigs after contact-infection and intravenous inoculation *BMC Vet. Res.* **5** (7)

[11] Casas M, Cortés R, Pina S, Peralta B, Allepuz A, Cortey M, Casal J and Martín M 2011 Longitudinal study of hepatitis E virus infection in Spanish farrow-to-finish swine herds *Vet. Microbiol.* **148** 27–34

[12] Breum S, Hjulsager C K , de Deus N, Segalés J and Larsen L E 2010 Hepatitis E virus is highly prevalent in the Danish pig population *Vet. Microbiol.* **146** 144–149

[13] Berto A, Backer J A, Mesquita J R, Nascimento M S, Banks M, Martelli F, Ostanello F, Angeloni G, Di Bartolo I, Ruggeri F M, Vasickova P, Diez-Valcarce M, Hernandez M, Rodriguez-Lazaro D, van der Poel W H 2012 Prevalence and transmission of hepatitis E virus in domestic swine populations in different European countries *BMC Res. Notes* **5** (190)

[14] Berto A, Martelli F, Grierson S and Banks M 2012 Hepatitis E virus in pork food chain, United Kingdom 2009-2010 *Emerg. Infect. Dis.* **18** 1358–1360

[15] Berto A, Mesquita J R, Hakze-van der Honing R, Nascimento M S J and van der Poel W H M 2012 Detection and characterization of hepatitis E virus in domestic pigs of different ages in Portugal *Zoonoses Public Health* **59** 447–481

[16] Rutjes S A, Lodder W J, Lodder-Verschoor F, Van Den Berg H H J L, Vennema H, Duizer E, Koopmans M and De Husman A M R 2009 Sources of hepatitis E virus genotype 3 in the Netherlands *Emerg. Infect. Dis.* **15** 381–387

[17] Di Bartolo I *et al.* 2012 Hepatitis E virus in pork production chain in Czech Republic, Italy and Spain, 2010 *Emerg. Infect. Dis.* **18** (8)

[18] Machnowska P, Ellerbroek L and Johne R 2014 Detection and characterization of potentially zoonotic viruses in faeces of pigs at slaughter in Germany *Vet. Microbiol.* **168** 60–68
[19] Khuroo M S, Khuroo M S and Khuroo N S 2016. Hepatitis E: Discovery, global impact, control and cure *World J. Gastroenterol.* **22** 7030–45
[20] Dalton H R, Bendall R , Ijaz S and Banks M 2008 Hepatitis E: an emerging infection in developed countries *Lancet Infect. Dis.* **8** 698-709
[21] Meng X-J 2009 Recent advances in hepatitis E virus *J. Viral. Hepat.* **17** 153-161
[22] Donnelly M C, Scobie L, Crossan C L, Dalton H, Hayes P C and Simpson K J 2017 *Aliment. Pharmacol. Ther.* **46** 126-41
[23] Sayed I M, Vercouter A-S, Abdelwahab S A, Vercauteren K and Meuleman P 2015 Is hepatitis E virus an emerging problem in Industrialized countries? *Hepatology* **62** 1883-92
[24] Andraud M, Dumarest M, Cariolet R, Aylaj B, Barnaud E, Eono F, Pavio N and Rose N 2013 Direct contact and environmental contaminations are responsible for HEV transmission in pigs *Vet. Res.* **44** 102
[25] The pig site: pig health biosecurity http://www.thepigsite.com/pighealth/article/462/biosecurity/ 9.8.2017
[26] Cook N and van der Poel W H M 2015 Survival and elimination of hepatitis E virus: a review Food Environ. *Virol.* **7** 189-94

Brucellosis - the past, the present, the future

M Petrović[1] and Ž Cvetnić[2]

[1] Veterinarski specijalistički Institut Niš, Dimitrija Tucovića 175, Niš, Serbia
[2] Hrvatski veterinarski Institut, Savska cesta 143, Zagreb, Croatia

Brucellosis, one of the most widespread zoonosis, is a contagious chained disease affecting a great number of animals and, in a smaller proportion, people [1, 2]. In the animals, this disease is usually manifested as either chronic or latent infection. Causal agents of the infection are bacteria from Brucella genus which can have different virulence and host affinities.

The appearance of brucellosis in human population is closely related to its incidence in animal populations, and this correlation is the only way that the disease can be observed, studied and controlled. Humans are most commonly affected by consuming either meat or dairy such as milk or young cheese produced from uncooked goat or sheep milk. Transmission of brucellosis from animal to human occurs through the air or via skin wounds. People working in the higher risk professions, such as the farming or meat industry as well as the veterinary or lab professionals, show higher incidence rate [3, 4].

An estimate of a half of a million people per year seeking medical attention due to brucellosis has been given by World Health Organisation (WHO), although it is believed that the number of affected people is up to 25 times greater.

Brucellosis, although in many countries controlled or completely eradicated, remains great health and economic problem. This is especially true in the regions in which highly contagious *Brucella melitensis*, causing disease in sheep, goats, and humans.

It is believed that the first description of brucellosis was by Hippocrates (450-370BC) in his „Epidemic". The disease was found on the island of Thasos, Greece, around 430BC, causing long-term or occasional fluctuating fever, which could fit the clinical expression of brucellosis. Most specialists agree that the first correct description of this illness expression was given in 1861. by Jeffery Allen Marston, assistant surgeon at the British Army Medical Department. The disease has had concerning proportions and caused a death of a great number of British soldiers situated at Malta [5, 6]. Up until 1887. the real cause of their death remained unknown. This changed when British military general Sir David Bruce isolated bacterium, which he later called *Micrococcus melitensis*, from the spleen of the soldier that died from the Maltese fever. [3, 7]. In 1905. The mistocles Zammit, a Maltese member of the Mediterranean Fever Commission, reveals the zoonotic potential of the disease.

By isolating *Brucella melitensis* from goat's milk, Zammit tags goats as the brucellosis reservoir on Malta and stresses their role in the prevalence/spread of the Mediterranean Fever [8] The vet from Copenhagen, Bernhard Lauritz Frederik Bang isolates *Bacillus abortus* from foetus, placenta and uterine secretions in cows that have had miscarriages, and marks this bacteria in 1897. as the cause of infectious abortion in cattle in Denmark [9, 10, 11]. It has later been shown that this pathogen can affect people as well. Alice Evans, an American microbiologist, has described the relationship between this, Bangs disease and Maltese Fever in 1918. She stated that these two pathogens, *Micrococcus melitensis*, and *Bacillus abortus*, are related closely to their morphological and cultural aspects [9, 7]. Two years after this statement, Mayer and Shaw suggest Brucella as the name of the newly classified bacteria, in honour to David Bruce [9, 12] In 1914, Traum isolates bacterium he called *Bacillus abortussuis*, from the liver and the aborted material in the pigs [8, 13]. *B. ovis* in sheep was discovered by McFarlane with co-operators in 1952. at New Zealand and Simmons G.C. and Hall W.T. in 1953. in Australia [14, 10, 7]. Stoenner and Lackman then followed with the discovery of the *Brucella neotomae* found in a species of desert rat in the south of USA in 1957 [15]. Almost a decade after, in 1968., *Brucella canis* was isolated by Carmichael and Bruner from the aborted dog foetuses [16].

In the last decade, four new Brucella types have been described, *Brucella ceti* and *Brucella pinnipedialis* which affect sea mammals [17, 18], *Brucella microti* which was isolated from rodents, *Microtus alvaris* found in the central Europe [18, 19, 20] and *Brucella inopinata* isolated from infected breast implants from the female patients with clinical presentation of brucellosis [21, 22].
According to the FAO/OIE/WHO [6], infection of sheep *B. melitensis* is found in the enzootic area of Mediterranean region, especially along the northern and east coast, spreading through Central Asia up to the Arabic peninsula and east to Mongolia. The disease is also found in the countries of Latin America, Africa and in India.

It is also still found today in the small and large ruminants of Northern Ireland, Portugal, Spain, Italy and Greece, and in sheep and goats of Portugal, Spain, and Italy.
In the territory of the Kingdom of Yugoslavia, brucellosis was first found in animal and then in human population in the years of 1912. and 1924. respectively in the surrounding of Kicevo. Reports of the occasional cases of brucellosis at Hvar and in the surroundings of Subotica and Maribor were submitted shortly after. In 1947., according to the data from the Federal Institute for Health Protection, the disease started having epidemic proportions in Istra and Slovenian seaside where it became of endemic character. In the same year, brucellosis was diagnosed in the Macedonian beef cattle.
For a long period, brucellosis was not diagnosed on the SFRY territory. It was eradicated by planned control of animal health. Unfortunately, in 1968, brucellosis was reported again in sheep in Macedonia and in 1971. the sheep farm in the surroundings of Bitolj which imported infected sheep from Israel. Without appropriate restrictions, and taking into consideration that the sheep were used for breeding, brucellosis started spreading into the other parts of SFRY. Sampling sheep and goats for brucellosis in 1992. in Macedonia lead to diagnosing 6890 infected animals. In the same year, 907 people got infected, which was the largest number found in the region until that point. Between 1980. and 1996., 5447 human individuals were found infected by brucellosis in Macedonia [18].
In the 80's, brucellosis was widespread in the neighbouring territories of Greece, Albania, as well as in Kosovo and southern municipalities of Serbia. The first cases of brucellosis in human populations of Kosovo were identified in 1980. During the 1981, brucellosis was identified on the beef cattle farm in Dusanovo near Prizren. In that occasion, the prevalence of the disease on the farm was found to be greater than 70%. From 1985. to 1994., 789 people were affected and one has died as a consequence of the brucellosis in the Kosovo.

In the southern Serbia, first cases were documented in 1988., where the family members were affected in Konculj village of Bujanovac municipality. Serological testing has shown that the families buffalo and goat were Brucella positive.

Traditional farming methods, involving keeping sheep on the shared pastures, with constant mixing of the flocks and breeding material, poor keeping conditions etc., have all contributed to the uncontrolled spread of brucellosis in Balkan area. Epizooty has been recorded in Istra once again in 1990. followed by the one in the northern Croatia in 1991. In the years between 2004. and 2007. Bosnia and Herzegovina are affected by epizooty and epidemics of the large-scale. In Slovenia, brucellosis caused by *B. melitensis* and *B. abortus* has been eradicated in the earlier years which gave her status of a country free of brucellosis. Occasional episiotomies, caused by *B. suis* occur in many European countries, and are usually caused by *B.suis* biovar 2 specifically. B. suisbiovar 3 was also detected in swine and equine populations in Croatia, which was the first evidence of this strain.Great threat to the sheep farming is posed by *B. ovis* which can bring great losses to the industry

In order to improve surveillance and control of brucellosis, European Union (EU) has implemented various laws and restrictions regarding import and export of cattle and pig (EC 64/432), sheep and goats (EC 91/68), as well as regulations regarding products of animal origin, animal identification, tagging and etc.

Various different agencies, as ECDC: European Centre for Disease Prevention & Control (2005) and EFSA: European Food Safety Authority (2002), are formed in the recent years in order to aid the control of zoonosis. Apart from these, much other organization has formed that are financed from the programs that support the eradication of brucellosis (Task Force for Monitoring Diseases Eradication (2000), Eradication programs, expert groups (2012).

In 2006., EU referral lab was established with the task to coordinate the work of all the national brucellosis labs in the EU (organising proficiency tests, workshops, staff training and help in the case of appearance of brucellosis in the member countries). There is an established strategy for control and eradication of brucellosis in the EU. Where low prevalence of the disease is found, testing and culling are done. In the case of medium prevalence vaccination of the breeding females (S19, RB51 and REV 1 vaccines for small and large ruminants) is also done, as well as already mentioned strategies, while in the case of high prevalence mass vaccination strategy is implemented (Greece and certain regions of Portugal). Conductions of common control and eradication politics in the case of brucellosis have led to a decrease in the incidence of the disease in the human population living in the areas. In the 1999., around 4000 people were affected by the disease, while in 2011. this number reduced almost ten times, with around 400 people affected in this year. The last ten years have shown great success in terms of the development of the strategies of livestock movements (identification, regular health checks, import and export controls) and reintroduction of vaccination as an important tool in disease control. Results of these actions are best seen in France and Cyprus where brucellosis has been eradicated, but also in NorthernIreland (eradication of *B. abortus* close to being accomplished) and Spain. A bit slower development is observed in Portugal and Italy, while Greece did not manage to get satisfying results. The appropriate strategies have been efficient, and constant surveillance and control are the priority in the areas that are disease-free.

The threat of the disease remains constant in the countries that are still not affected or have had brucellosis extradited. Less strict border control, as in the case of Northern and Republic of Ireland and UK, or Belgium and France, BiH and Croatia (border areas), can lead to the spread of the brucellosis from the affected to the unaffected areas which was the scenario seen between the countries mentioned above. Wild reservoirs of the disease (*B. abortus*/*B. melitensis*) in the EU have not been identified until recently. Wild ruminants, in which few sporadic cases of *B. abortus* and *B. melitensis*have been discovered, were considered the dead-end hosts In the December of 2011. *B. melitensis* was found in the people living in the alpine areas of France. Following the discovery, control of the wild ruminants was enforced. In 2012. an*Alpine Ibex* was identified as the primary reservoir of the disease. (22 positive results from 77 tested animals).

Although each country has its own regulations regarding the disease, livestock movement, tagging etc., all those regulations are closely related to the ones issued by the EU. However, strategies for disease control are differentiated in different countries, as it is dependent on the status of the disease in the country, it's economic, social and political circumstances and other factors important for implementation of advanced measures involved in the control of brucellosis.

"TASK FORCE" for the region? Formation of the group of individuals with various professional backgrounds, to meet occasionally in order to share information (professional workshops on different topics), exchange of the staff (lab training) etc.

"TOP DIAGNOSTICS- molecular epidemiology" –Insufficient research of the regional epidemiology enables efficient disease control.

Classic serological and microbiological techniques are insufficient. Molecular research using PCR (BRU-UP, BRU-LOW, Bruce ladder, Multiplex suis, RFLP) is a better option, as the genetic homogeneity of the Brucella genus (>90%) poses great obstacles with subtyping using microbiological or serological methods. Identifying differences between the strains, finding the source of the infection, differentiation between the vaccinated and wild strains is essential. Investigation of the locus number 16 which shows tandem repeats (MLVA-multiple locus number tandem repeat analysis) has come out as the most suitable method of genotyping as it also contains an international genotype base. We have proven the methods diagnostic ability ourselves. Strains are grouped together according to their phylogenetical and geographical connection, revealing their regional specificity which shows us the possible ways of the disease transmission between the animals. We have added more than 50 new genotypes in the international base and have introduced a new method that is now used routinely. These steps should contribute to the global efforts of the control and eradication of brucellosis, especially if they are followed by the extensive regional research.

References

[1] Joint Food and Agriculture Organization of the United Nations/World Health Organization 1986 Expert Committee on Brucellosis *Technical Report Series 740* Sixth Report (WHO, Geneva, Switzerland)
[2] Young E J 1983 Human brucellosis *Rev Infect Dis.* **5** 821-842
[3] Hughes M L 1893 The natural history of certain fevers occurring in the Mediterranean *Mediterranean Nat.* **2** 325-327
[4] Sokolovski B, Nikolovski B and Angelovski A 1994 Bruceloza i njezinite karakteristiki vo Republika Makedonija (Prličev, Skopje)
[5] Marston J A 1861 Report on fever (Malta) *Royal Army Med Dept Rep.* **3** 486-521
[6] Purcell B K, Hoover D L and Friedlander AM 2007 *Brucellosis Medical aspects of biological warfare* (Washington, DC: Borden Institute, Walter Reed Army Medical Center) pp 167-184
[7] Nielsen K and Duncan J R 1990 Animal Brucellosis *Ontario: CRC Press* 107-20
[8] Wyatt H V 2010Sir Themistocles Zammit: his medical and scientific career *Malta Medical Journal* 22
[9] Alice C and Evans 1947 Brucellosis in the United States *American Journal of Public Health and The Nation's Health*
[10] Bang B 1897 The etiology of epizootic abortion *J Comp Path.* **10** 125-149
[11] Burgess G W 1982 Ovine contagious epididymitis: areview *Vet. Microbiol.* **7** 551- 575
[12] Meyer K F and Shaw E B 1920 A comparison of the morphologic, cultural and biochemical characteristics of *B. abortus* and *B. melitensis J Infect Dis.* **27** 173-184
[13] Traum J 1914 Immature and hairless pigs *Annual Report of the Chief Bureau Animal industry* (U.S. Department of Agriculture, Washington D.C) 30
[14] Buddle M B 1956 Studies on *Brucella ovis* (n. sp.), a Cause of Genital Disease of Sheep in New Zealand and Australia *J Hyg.* **54** 351-364
[15] Stoenner H G and Lackman D B 1957 A new species of *Brucella* isolated from the desert wood rat, Neotoma lepida Thomas *Am. J. Vet. Research* **18** 947-951
[16] Carmichael LE and Bruner DW 1968 Characteristics of a newly recognized species of *Brucella* responsible for infectious canine abortions *Cornell Veterinarian* **58** 579-592
[17] Foster G, Osterman B S, Godfroid J, Jacques I and Cloeckaert A 2007 *Brucella ceti* sp. nov. and *Brucella pinnipedialis* sp. nov. for *Brucella* strains with cetaceans and seals as their preferred hosts *Int. J. Syst. Evol. Microbiol.* **57** 2688–2693
[18] International Committee on Systematics of Prokaryotes Subcommittee on the Taxonomy of *Brucella* http://www.the-icsp.org/subcoms/Brucella
[19] Scholz H C et al. 2008 Brucella microti sp. nov., isolated from the common vole Microtus arvalis *Int. J. Syst. Evol. Microbiol.* **58** 375–382
[20] Scholz H C et al. 2008 Isolation of *Brucella microti* from soil. *Emerg. Infect. Dis.* **14** 8 1316–7
[21] Scholz HC et al. 2010 Brucella inopinata sp. nov., isolated from a breast implant infection *Int J Syst Evol Microbiol.* **60** 801-8
[22] FAO/OIE/WHO Animal Health Yearbook, FAO Animal Production and Health Series, FAO, Rome, Italy, 1997

Safety in Serbian animal source food industry and the impact of hazard analysis and critical control points: A review

I Tomašević and I Đekić

Faculty of Agriculture, University of Belgrade, Nemanjina 6, 11080 Beograd, Serbia

E-mail: tbigor@agrif.bg.ac.rs

Abstract. There is a significant lack of HACCP-educated and/or HACCP-highly trained personnel within the Serbian animal source food workforces and veterinary inspectors, and this can present problems, particularly in hazard identification and assessment activities. However, despite obvious difficulties, HACCP benefits to the Serbian dairy industry are widespread and significant. Improving prerequisite programmes on the farms, mainly through infrastructural investments in milk collectors and transportation vehicles on one hand, and increasing hygiene awareness of farmers through training on the other hand has improved the safety of milk. The decline in bacterial numbers on meat contact surfaces, meat handlers' hands and cooling facilities presents strong evidence of improved process hygiene and justifies the adoption of HACCP in Serbian meat establishments. Apart from the absence of national food poisoning statistics or national foodborne disease databases, the main obstacle to fully recognising the impact of HACCP on the safety of animal source food in Serbia is the lack of research regarding the occurrence of chemical and/or physical hazards interrelated with its production.

1. Introduction

Hazard Analysis and Critical Control Points (HACCP) seven principles today are read like the "ten commandments" among food producers whenever they discuss food safety. HACCP, which started as an idea between Howard Bauman from Pillsbury Company and NASA, has attained significant authority and will remain as a legacy that will not fade away. However, the way it has been implemented in different parts of a world, especially its inspection and control, although seven principles are widely understood and accepted, varies greatly [1]. HACCP is capable of accommodating changes, such as advances in equipment design and processing procedures, as well as technological developments. It can be tailored to address individual product or process. It is the best system currently available for maximising the safety of meat and meat products, as well as food in general, which is why it has been recommended for use in the food industry and promoted by governments and scientific groups for decades [2].

Once a visionary approach and today almost 60 years old, HACCP is an integral part of the American food industry but above all, the food safety inspection system. Richard Rominger – the Acting Secretary of Agriculture in its time, said: "We are proposing to reinvent the meat and poultry inspection system. HACCP will fundamentally reform our inspection system into a science-based system – a system which will ensure an even safer food supply." [3]. For this purpose, the Food Safety and Inspection Service (FSIS) created a new brigade of several thousand newly-employed Consumer Safety Officers (CSOs) who were charged with keeping meat safe from contamination.

In 1995, the US Food and Drug Administration (FDA) issued regulations that made HACCP mandatory for fish and seafood products, and in 2001, they issued regulations for mandatory HACCP in juice processing and packaging plants. HACCP has also been implemented by the United States Department of Agriculture (USDA). In 1998, USDA's FSIS mandated HACCP for the nation's meat and poultry processing plants. In addition, a voluntary HACCP program was implemented in 2001 for Grade A fluid milk and milk products under the cooperative federal/state National Conference on Interstate Milk Shipments (NCIMS) program [4].

The Republic of Serbia, within its process of legal harmonization with the EU, recently reorganized its food safety system to comply with EU regulations according to *acquis communautaire*. The main goal of the on-going harmonization is to allow subjects in the food chain to perform their activities according to the EU regulatory requirements. Among the legislative changes, a new Food Safety Law was introduced in 2009 [5]. This law mandatory requires implementation of a food safety system based on HACCP principles for all subjects in the food chain, except primary production. Full implementation of this law was set at June 1 2011 when the inspection service started on-site verification [6].

The introduction of this new Food Safety Law and mandatory HACCP implementation for all food producers, regardless of their type and size, achieved one goal with certainty and that was to fundamentally shift the burden of food safety regulation from the government onto the Serbian food industry. Rather than having the government assess the risk inherent in different food production difficulties (and bear the responsibility if it was wrong), Serbian food producers became responsible for the risk assessment. However, as in most developing countries, Serbian food processing industries lack the necessary basic scientific information (i.e., national food poisoning statistics or national foodborne disease databases) required to develop reliable hazard assessments [7] or to effectively assess the benefits of mandatory HACCP in terms of its ability to reduce the occurrence of foodborne illness.

In 1995, US Government officials said that the system shaped by the new red tape – known as HACCP – will mean a safer food supply, if not immediately, then after the 3.5 year phase-in period [3]. Nowadays, we are celebrating the 6th birthday of HACCP in the Serbian food industry and the only thing we can be sure about is that it has not revolutionized our food safety inspection. Since its phase-in period was exceeded years ago, we believe that it is the time to evaluate the impact of HACCP on safety in the Serbian animal source food industry, at least.

2. Materials and methods

2.1. Survey on HACCP implementation in animal source food industries

For both meat and dairy industries, a questionnaire was developed to identify the effects of HACCP implementation in Serbian meat industry, which included the issues regarding general information about the food business operators (such as the number of employees, status of HACCP system, information about the educational level, age and work experience of HACCP team members). Questions related to prerequisite programs (PRPs) that are implemented in the company were also included. Further questions were related to the incentives for implementing the HACCP system, costs, benefits and difficulties in implementation/operation of HACCP. All producers were visited and on-site interviews were performed. Respondents were mainly HACCP team leaders, production managers or owners. The respondents ranked these factors in order of importance according to their own conditions and experiences.

2.2. Meat industry sampling and methods

A total of 48,246 swab samples were analysed from two types of meat establishments: 130 meat plants and 220 meat retailers. The period covered was seven years long (from 2008 to 2014) divided into two terms:

first from January 1 2008 until May 31 2011 (a period of 41 months before mandatory HACCP implementation) and second from June 1 2011 until December 31 2014 (a period of 43 months after HACCP became obligatory). Samples were taken from three types of surfaces: food (meat) contact surfaces (cutting boards, machines, knives and slicers, tables and containers), cooling facilities (refrigerators, freezers and other meat and meat products cooling devices) and meat handlers' hands.
During the first term (January 1 2008 – May 31 2011) samples were analysed according to the Regulation (8) in force, for Aerobic Colony Count (ACC) using methods coherent with ISO 4833:2003. The new Regulation (9), effective from June 1 2011, legally prescribed methods that were used in the second term of our investigation. Samples were analysed for Aerobic Colony Count (ACC) according to ISO 4833:2003, *Enterobacteriaceae* (ISO 21528-2:2004), coagulase positive *Staphylococcus* (ISO 6888-1:1999), *Salmonella* (ISO 6579:2002) and *Listeria monocytogenes* (ISO 11290-1:1998). Samples were examined in an ISO/IEC 17025:2005 accredited laboratory.
For the purpose of our investigation all the results and the respective number of microorganisms detected (n) were divided into four classes as follows: Class I ($n \leq 1$ log10 CFU/cm2); Class II (1 $\log_{10}$ CFU/$cm^2 < n \leq 2$ $\log_{10}$ CFU/cm^2); Class III (2 $\log_{10}$ CFU/$cm^2 < n \leq 2.7$ $\log_{10}$ CFU/cm_2); Class IV ($n \geq 2.7$ $\log_{10}$ CFU/cm^2).

2.3. *Dairy industry sampling and methods*

Samples of raw milk were analysed at the reception of dairy plants and total of 45,600 samples from different locations were collected on a daily basis during four years (from 2006 to 2009). Samples of pasteurized milk were analysed after packaging in the dairy plants over three different periods. A total of 558 samples were collected in the period before HACCP system implementation, 260 samples during first eight months following HACCP implementation and the final 677 samples in the period after additional infrastructural investments in the dairy plants.
Samples of raw and pasteurized milk were analysed for total plate count (TPC), while pasteurized milk samples were also analysed for the presence of coagulase positive *Staphylococcus*, sulphite-reducing *Clostridia*, *Proteus* spp. and *Escherichia coli*. Methods for the laboratory determination of TPC, *E. coli*, coagulase positive *Staphylococcus*, sulphite-reducing *Clostridia*, *Proteus spp.* were adjusted to the National Regulation on microbiological methods for analysis of food (Regulation on methods for microbiological analysis and super analysis of food. Official Gazette SFRY No 25/1980. All raw milk samples were also investigated for the presence of antibiotics using the commercial SNAP tests (IDEXX Laboratories, USA). Tests were performed according to the manufacturer's protocol. The acidity of milk was analysed by titratable method and expressed in Soxlet-Henkel degrees (°SH). Somatic cell count was determined using Fossomatic Minor (Foss, Denmark). Additionally, temperature of raw milk at the reception was also determined.

Based on the TPC and somatic cell count, raw milk was classified into four different quality categories being extra class with TPC not exceeding 100,000 CFU/ml, I class with TPC between 100,001 and 500,000 CFU/ml, II class with TPC between 500,001 and 1,000,000 CFU/ml, and III class with TPC greater than 1,000,000 CFU/ ml. Requirements for the somatic cell count were always the same, less than 400,000 cells/ml. The percentage of raw milk which belongs to the specific category was calculated as the amount in the total quantity of received raw milk for each year during the examined period. The percentage of raw milk that did not comply with the given specifications regarding presence of antibiotics, acidity > 6.8°SH or temperature of raw milk >10°C was calculated as the quantity of non-conforming raw milk in the total quantity of received raw milk.

2.4. *Analysis of results*

The results obtained for TPC in pasteurized milk during the examined period were expressed as $\log_{10}$ CFU/ml, and used for the calculation of mean values, standard deviation and significance of difference between means using one-way ANOVA. Categorical variables (classes of surface hygiene in meat producing facilities) were expressed as percentages. Chi-Square test for association was used to discover possible relationships between results of meat process hygiene and the period (time) they were sampled. Yate's correction was calculated when the expected frequency was less than 5. The level of statistical significance for all was set at 0.05. Statistical processing was performed using Microsoft Excel 2010 and SPSS Statistics 17.0.

3. Results and discussion

3.1. Impact of HACCP in the Serbian Dairy Industry

From 27 companies that responded to the survey, 19 companies (70.4% of the respondents) claimed that they had a fully operational and certified HACCP system in place, while 8 companies (29.6%) implemented HACCP, but they had no third party certification at the time the survey was conducted. Almost 60% of the Serbian dairy industry respondents estimated that it had taken 12 months or less to implement HACCP system, while only 14.8% estimated that it took more than that. The most important motive for implementing HACCP system for Serbian dairy producers was to increase and improve safety of their products. The second most important incentive was the quality increase of their products, which was unexpected since HACCP was designed primarily with the food safety in mind. The cost of product investigation/analysis was on the top of their financial concerns, although it was logical to assume that HACCP is a system that minimizes testing by focusing on critical control points. The cost of hiring external consultants was rated as the second most important expense, in spite of the relatively high education level of their respective HACCP team members. A major difficulty encountered during HACCP implementation and operation was associated with the attitude/motivation of production staff and the need to retrain production and managerial staff. Cleaning and sanitation control of health and hygiene of employees, equipment maintenance and calibration, pest, water and temperature control together with traceability was used by almost 100% of the Serbian dairy plants covered by our survey. The most important identified benefits were increased safety of dairy products followed by increased customer confidence [10].

Our microbiological test results for TPC in raw milk indicated that in 2006, more than 55% of accepted milk belonged to the III class, which did not satisfy even official Serbian requirements. In the same period, only 36% of milk was of high quality (extra and I class). When the microbiological results of raw milk were examined for the following three years (from 2007 till 2009), the percentage of class III milk in total raw milk decreased, being only 20.3% in 2009. At the same time, the percentage of raw milk of extra and I class increased from 36.5% in 2006 to 53.7% in 2009. In addition, the percentage of milk that was classified as non-conforming (related to the specification set up for antibiotic residues, SH-acidity or temperature of raw milk), decreased or fell to completely non-existent levels [11].

Microbiological analysis of pasteurized milk samples was also performed in the periods before and after HACCP implementation. The results obtained indicated a significant decrease ($p < 0.05$) in TPC from 3.32±0.48 to 3.11±0.30 log CFU/ml for results obtained before HACCP implementation and during the first eight months after implementation. The other investigated bacteria were not present in all investigated periods. The reason the results indicated only 0.21 log CFU/ml decrease in TPC after HACCP was implemented can be explained by the fact that the time/temperature regime during pasteurization process had been already followed before HACCP was put in place. As a result of HACCP implementation, pasteurization of milk has been more carefully followed and monitored, resulting in the obtained decrease. Eight months after HACCP implementation, the results showed even more pronounced decrease, from 3.11±0.30 to 2.18±0.54 log CFU/ml ($p < 0.05$). This improvement in microbiological quality of pasteurized milk was possibly related to the infrastructural investments in the factory, mainly concerning new pasteurization units and automated cleaning and disinfection systems [11].

3.2. Impact of HACCP in the Serbian Meat Industry

From 77 companies that responded to the survey, 72 companies (93.5% of the respondents) claimed that they had a fully operational and certified HACCP system in place, while 5 companies (6.5% of the respondents) implemented HACCP, but they had no third party certification at the time the survey was conducted Almost 50% of the Serbian meat industry respondents estimated that it had taken 12 months or less to implement HACCP system, while only 11.7% estimated that it took more than that. It was apparent that regulation is a very important incentive for HACCP implementation and the degree of enforcement can cause even the smallest of enterprises to comply without question, as was the case with 100% of the meat producers in our investigation. The other major incentive for the implementation of HACCP was an increase in product safety and quality. Major costs of implementing HACCP were associated with investment in new equipment while major difficulties experienced while implementing it were associated with recouping its costs. It was obvious that most of the costs involved with HACCP could not be recouped in the short term [12].

The most widely followed PRPs in the Serbian meat industry were cleaning and sanitation, temperature control, pest control and the control of the health and hygiene of the employees, all above 90% among the surveyed meat producers. Only 60% of respondents used waste and wastewater management and less than 30% of respondents regularly controlled air in their processing area or had allergen declaration [13].

The most evident impact of HACCP on meat process hygiene was observed in meat contact surfaces, where 90.45% of the ACC samples taken in meat plants and 98.3% of the ACC samples taken in meat retail were above 2 $\log_{10}$ CFU/cm^2 before HACCP. After mandatory HACCP implementation, these values dropped below 2 $\log_{10}$ CFU/cm^2 in 96.38% of the cases for meat plants and 85.8% of the cases for meat retail. Class IV ($n \geq 2.7 \log_{10}$ CFU/cm^2) of the hand swab ACC results was the prevailing category in Serbian meat plants (52.71%) and meat retailers (51.10%) until May 31 2011. In the period of 43 months that followed, Class III (2 $\log_{10}$ CFU/cm$^2 < n \leq 2.7 \log_{10}$ CFU/cm^2) took precedence, accounting for 97% and 98.96% of the hand swab ACC samples for meat plants and meat retailers, respectively. In the same period, the number of surfaces positive for *Enterobacteriaceae* steadily declined in both meat plants and meat retailers for all types of surfaces examined. The improvement in process hygiene was even more obvious in meat retail facilities, especially regarding food contact surfaces where the percentage of positives dropped from 31.4% in the second half of 2011 to 14.4% in 2014. The same period for meat plants indicated a 10% decrease of *Enterobacteriaceae*-positive food contact surfaces [14].

Because meat handlers have a very important role in the prevention of meat poisoning, as they can introduce pathogens within the entire food chain, we also investigated the level of food safety knowledge among meat handlers in the Serbian meat industry, in different stages of the meat chain, i.e. in slaughterhouses, meat processing plants and retail stores. The average food safety knowledge score for all 352 Serbian meat handlers surveyed was 64%, whereas handlers from slaughterhouses and meat processing plants obtained significantly better scores (65% and 66%, respectively) than handlers from retail (60%, $p<0.05$). The knowledge score among all meat handlers was significantly associated with the age, education and previous food safety training. Meat handlers with the lowest education (only primary school) scored the lowest values (59%), and the highest educated participants (holding university degrees) scored the highest values (76%). Results indicated that 57.9% of meat handlers could identify that bacteria will readily multiply at 25°C, but they did not understand the manifestation of bacterial growth and incidence in food, as only 5.5% of all meat handlers knew that food contaminated with food poisoning bacteria cannot be recognized by visual, olfactory or taste checks [15].

4. Conclusions

There is a significant lack of HACCP-educated and/or HACCP-highly trained personnel within the Serbian animal source food workforces and veterinary inspectors, and this can present problems, particularly in hazard identification and assessment activities. The supply of Serbian language HACCP guidelines is also very limited and therefore a heavy reliance upon foreign language (predominantly English) documentation is evident.

However, despite obvious difficulties, HACCP benefits to the Serbian dairy industry are widespread and significant. In particular, Serbian dairy producers reported increased safety and quality of the products, increased customer confidence and better discipline of the employees. Increased product sales and the ability to use HACCP as a legal instrument against complaints were equally important. Improving PRPs on the farms, mainly through infrastructural investments in milk collectors and transportation vehicles on one hand, and increasing hygiene awareness of farmers through training on the other hand improved the safety of milk.

The main reasons for successful implementation of HACCP among the Serbian meat producers were well placed and closely followed PRPs. The decline in bacterial numbers on meat contact surfaces, meat handlers' hands and cooling facilities presents strong evidence of improved process hygiene and justifies the adoption of HACCP in Serbian meat establishments.

Apart from the absence of national food poisoning statistics or national foodborne disease databases, the main obstacle to fully understanding the impact of HACCP on safety of animal source food in Serbia is the lack of research regarding the occurrence of chemical and/or physical hazards interrelated with its production.

Acknowledgements
This work was performed within the National Project number TR31034, supported by the Ministry of the Education and Science, Republic of Serbia.

References

[1] Tomašević I 2007 HACCP worldwide and in Serbia. Proc. I Int. Cong. "Food technology, quality and safety" Novi Sad, Serbia pp 165-176
[2] Tomašević I, Adamović J, Šakota T, Radovanović R 2003 HACCP plan for the "Serbian Sausage" traditional semi-dry product. Proc. 49th Int. Cong. of Meat Science and Technology (ICoMST) São Paulo, Brazil pp 443-444
[3] Dunn EC 2004 Travelling facts, negotiated meaning and the social construction of safe meat Travelling Facts: The Social Construction, Distribution, and Accumulation of Knowledge New York, USA.
[4] Tomasevic I and Djekic I 2016 HACCP in Fermented Meat Production. In: Zdolec N, editor. Fermented Meat Products: Health Aspects CRC Press Taylor&Francis Group, UK
[5] Food Safety Law No. 41/09 Serbia
[6] Djekic I, Kuzmanović J, Anđelković A, Saračević M, Stojanović MM, Tomašević I. 2016 Effects of HACCP on process hygiene in different types of Serbian food establishments. Food Control 60 pp 131-138
[7] Ropkins K, Beck AJ 2000 Evaluation of worldwide approaches to the use of HACCP to control food safety Trends in Food Science & Technology 11(1) pp 10-21
[8] Regulation on methods for microbiological analysis and superanalysis of food, No 25/1980 Serbia
[9] Regulation on general and specific food hygiene requirements at any stage of production, processing and trade, No. 72/2010 Serbia
[10] Tomašević I, Šmigić N, Đekić I, Zarić V, Tomić N, Miocinovic J 2016 Evaluation of food safety management systems in Serbian dairy industry Mljekarstvo 66 (1) pp 48-58
[11] Smigic N, Djekic I, Tomasevic I, Miocinovic J, Gvozdenovic R 2012 Implication of food safety

measures on microbiological quality of raw and pasteurized milk Food Control 25 (2) pp 728-31
[12] Tomašević I, Šmigić N, Dekić I, Zarić V, Tomić N, Rajković A 2013 Serbian meat industry: A survey on food safety management systems implementation Food Control 32 (1) pp 25-30
[13] Tomasevic I, Smigic N, Rajkovic A, Djekic I, Tomic N, Radovanovic R 2012 Serbian meat industry: A survey on prerequisite programmes. Proc. Int. Conf. of Biological food safety and quality Novi Sad, Serbia
[14] Tomasevic I, Kuzmanović J, Anđelković A, Saračević M, Stojanović MM, Djekic I. 2016 The effects of mandatory HACCP implementation on microbiological indicators of process hygiene in meat processing and retail establishments in Serbia Meat Science 114 pp 54-7
[15] Smigic N, Antic D, Blagojevic B, Tomasevic I, Djekic I. 2016 The level of food safety knowledge among meat handlers British Food Journal 118 (1) pp 9-25

Non-thermal inactivation of Noroviruses in food

B Velebit[1], R Petronijević[1] and T Bošković[2]

[1] Institute of Meat Hygiene and Technology, Kaćanskog 13, Belgrade, Serbia
[2] Ministry of Agriculture, Forestry and Water Management, Veterinary Directorate, Food Safety Unit, Nemanjina 22-26, Belgrade, Serbia

E-mail: branko.velebit@inmes.rs

Abstract. An increased incidence of foodborne illnesses caused by Norovirus and consumer demand for fresh, convenient, and safe foods have prompted research into alternative antiviral processing technologies. Chlorine dioxide, UV treatment and thermal processing are standard antinoroviral technologies that have been employed for a while; however, they tend to be non-effective in modern processing due to residue concerns (ClO_2), shadowing effects (UV) and low-energy efficiency (heat treatment). Alternative technologies have been validated such as ozone treatment, high pressure processing and pulse electric fields. Although these techniques are promising, none of them individually can deem food free of Norovirus. Further research on the effects on Norovirus in various food matrices is required. Good manufacturing practices and proper sanitation procedures remain the "gold" safety tools in food business.

1. Introduction

Noroviruses (NoV) are small, non-enveloped viruses with a single-stranded RNA genome that make up the genus norovirus of the family *Caliciviridae*. These viruses cause acute diarrhoea with an incubation period of 24 to 48 hours [1]. NoV usually have a low infectious dose of as few as 10 virus copies [2], they are highly infectious, and can be spread by water, aerosol vomitus droplets, person-to-person contact, fomites and food. Outbreak control is hindered by the low infectious dose and environmental persistence of the virus.

When it comes to NoV shedding in human infection, it was determined that the amount of norovirus shed was approximately 9.5×10^9 genomic copies/g faeces as determined using quantitative reverse transcriptase-PCR (qRT-PCR). Viral shedding usually lasts up to 10 days after infection [3]. In contrast to some of the other enteric viruses, the genetic type of NoV may impact its resistance or sensitivity to inactivation treatments. They are divided into six major genogroups designated GI through GVI. GI and GII contain the majority of NoV strains associated with human illness and are further divided into about 30 genotypes. Genogroup V includes murine norovirus (MNV), which is the most genetically similar laboratory surrogate for human noroviruses.

Since NoV have an RNA genome and replicate rapidly using a polymerase that lacks 3'-5' proofreading activity, they would be expected to mutate quickly. This was demonstrated by Nilsson et al. [4], who found 32 amino acid changes in the capsid protein of a norovirus shed chronically over 1 year by an immunocompromised patient. A single genotype, GII.4, has been associated with the majority of worldwide outbreaks since the mid-1990's, when active surveillance with reverse transcriptase-PCR (RT-PCR) was initiated. Thanks to implementation of these molecular diagnostic techniques during 1990s, an increase in the percentage of all documented laboratory-confirmed outbreaks due to noroviruses increased from more than 10-fold in 2000s [5]. However, regardless of this increase, gastrointestinal illnesses from noroviruses are still grossly underreported. As predicted by Widdowson [5], the CDC [6] confirmed that noroviruses were the cause of >50% of foodborne disease outbreaks. Greater than 56% of norovirus outbreaks are associated with eating salads, sandwiches, or fresh produce indicating that contamination of foods usually requires handling without an intervening heating step [5].

Many industrial techniques for controlling bacterial levels in food, including pH, temperature, and water activity are mostly ineffective barriers against NoV viral transmission to human hosts [7]. Up-to-date food hygiene criteria, which are optimized for the prevention of bacterial growth, may not be effective against NoV [1]. Since enteric viruses must survive the enzymatic and extreme pH conditions of the upper gastrointestinal tract to infect a host, they are resistant to a wide range of commonly used food processing, preservation, and storage treatments. Additionally, ingredients of the food itself may provide protection to virus particles against processing methods and host ingestion.

There has been increasing consumer's demand for food that is minimally processed, additive free, and that has an extended shelf life. Therefore, there also has been greater interest in the commercial development of non-thermal processing technologies. This interest is motivated by the advantages of a preservative process that effectively inactivates problematic microorganisms and proteinaceous metabolites, increases shelf life while retaining the sensory attributes and nutrient content similar to the raw or fresh product. There are many non-thermal treatments that are in various stages of development that have the potential to destroy pathogens and retain food quality [8]. Key role in NoV inactivation plays either changes in protein capsid layer or damage of NoV RNA. Structural changes in capsid cause failure to recognize receptor on host cell thus disabling virus to enter host, while treatments damaging NoV RNA render virus incapable to enter lysogenic/lytic cycle.

In this paper, we will discuss on three currently employed non-thermal noroviral inactivation treatments in food.

2. Ozone

Ozone (O3), a strong oxidizing agent which has been widely used as a disinfectant in drinking water treatment plants in many European countries, receiving more attention after the discovery of potentially harmful chlorine by-products [9,10]. Owing to its high effectiveness and lack of residue after disinfection (unlike chlorine dioxide), ozone can be used for both surface and groundwater disinfection.

Ozone is a naturally occurring form of oxygen that exists as a bluish gas with a strong odour. It is moderately soluble in water and is effective at killing a range of microorganisms through the oxidation of cellular membranes [11]. Due to its oxidizing potential is 2.07 mV, it is considered as one of the most powerful common oxidizing agents available [12]. The antimicrobial effectiveness of ozone can be up to 52% stronger than chlorine [13], and it is active over a much wider spectrum of microorganisms than chlorine and other disinfectants.

The potential targets of viral inactivation by ozone treatment are the viral capsid proteins, antigenic sites for host cell receptor attachment and nucleic acid of non-enveloped viruses such as NoV. The interaction of ozone with the viral capsid proteins and antigenic sites may affect the adsorption and penetration of the virus into host cells. However, ozone also does penetrate through the capsid to RNA core inside virus, thus rendering RNA defective and prevent the replication of the virus after infection [14,15]. Typical ozone treatment (0.51 mg/L) at 20°C and pH 7.2 with a 2-min contact time resulted in approximately 75% of viruses still capable of cell attachment but >90% of these were non-infective.

Ozone decomposes in the water phase of foods very fast that which results in expression of its antiviral effect mainly at the surface of solid foods. Main product of O_3 disintegration is plain oxygen, so there are no safety issues related to residual hazardous ozone residues in the treated food products or surfaces [16]. What also a positive effect of ozone is, the rapid breakdown prevents the accumulation of waste products from the process in the environment [17]. Thanks to its high oxidation potential, ozone quickly kills microorganisms; yet, it also reacts rapidly with complex organic compounds as found in foods. The existence of organic substances can neutralize the ozone required to inactivate microorganisms once being mixed with food.

Ozone should not be used directly on high ozone-demand foods, such as meat products, because the application of ozone may change sensory qualities of the product [18]. Aqueous ozone is applicable on products that have smooth and intact surfaces with low ozone demand such as fruits and vegetables [17,19]. Ozone is applicable in the industry to treat process water, as a fruit and vegetable wash, in fruit and vegetable storage, and in recycled water where fresh fruits and vegetables are first washed by ozonated water [20].

3. High pressure processing

High pressure processing (HPP) has emerged as a novel technology for food processing where foods can maintain their raw character and flavour. Applications of HPP include its use as a "cold pasteurization" method for fruit juices, a means of sanitizing packaged ready-to-eat meats, and inactivation of spoilage enzymes to enhance refrigeration shelf-life of avocados and guacamole. High pressure can also separate raw shellfish meat from its shell [21].

Successful inactivation of NoV varies depending on time, temperature, pressure level, and characteristics of the food matrix. The main mechanisms of noroviral inactivation by HPP is altering the virus capsid or protein coat surrounding the positive-stranded RNA. Enteric viruses are non-enveloped and, by definition, do not contain lipid envelopes. Therefore, HPP inactivation of foodborne virus, unlike foodborne bacteria, has no lipid-specific component. High pressure usually does not disrupt covalent bonds and it is hypothesised that high pressure does not damage the primary structure of nucleic acids, such as the RNA encoded within these viruses. It makes sense, therefore, that HPP inactivation is a function of pressure's effect on virus protein conformations. Pressure-denaturation of proteins is a complex event and is dependent upon the structure of the protein, pressure range, temperature, pH, and solvent composition (including the presence of sugars, salts, and other additives) Pressure-induced changes to the viral coat can be discreet modifications to capsid proteins or receptor recognition proteins that result in loss of NoV infectivity.

Viewed from a capsid function perspective, the virus must attach to its host cell receptor, penetrate the cell membrane, and then release the RNA into the cytosol of the cell. Once inside the cytosol, the virus RNA genomes of picornaviruses and caliciviruses are functional mRNAs that are sufficient to initiate transcription and subsequent virus replication [22]. Thus, high pressure must cause a protein-mediated effect that prevents virus attachment, penetration of the host cell, or uncoating once the virus has entered the cell.

An increase in pressure level has a greater effect on viral inactivation compared to increases in treatment time [23]. Typical pressures used for commercial food processing machines are as high as 600 MPa (1 MPa equals 9.87 bar). Commercial HPP units are quite large with capacities exceeding several hundred litres. Processing is by the batch with machines filled, treated for short intervals (usually less than 5-min), and then emptied. Commercial units are almost exclusively water based, but research units can use water, oil, or alcohol as the pressure application medium. Although HPP is classified as a non-thermal process, an adiabatic heating effect occurs under pressure that can be substantial with increasingly greater adiabatic heating effects observed for water, oil, and alcohol pressure medium [24-26]. Therefore, while the temperature before pressure application of 600-MPa may be at 25°C, the expected temperature achieved under pressure assuming a 3.5°C adiabatic heating per 100-MPa for a water-based unit, would increase to approximately 46°C [21].

Some viruses are extremely pressure-resistant, while others are sensitive; FCV is completely inactivated by pressures as low as 275 Megapascals (MPa) for 5 min [22].

A human volunteer study involving HPP treatment of NoV-contaminated oysters was performed [27]. This was accomplished by injecting 10^4 RT-PCR units of GI.1 NoV into pressure-shucked oysters. A 5-min, 400-MPa treatment at 25°C was not sufficient to inactivate the virus. Testing a second volunteer group with 5-min-, 600-MPa-treated virus at 6°C indicated that the virus was completely inactivated. A third group was fed NoV-contaminated oysters after a 5-min, 400-MPa treatment at 6°C. This treatment reduced the numbers of volunteers who became sick, but did not completely protect all volunteers. Therefore, pressures of at least 400-MPa or higher would be required to make human NoV-contaminated shellfish safe for consumption. Based on the reduction of human volunteers, it was postulated that the 400-MPa, 6°C treatment probably inactivates between 3- and 4 log_{10} of human NoVs [27]. This conclusion is also supported by subsequent research which has shown a dramatic drop in NoV's ability to bind to virus receptor-like swine mucin glycoproteins after a 5-min, 400-MPa treatment at 5°C [28]. This drop was not observed for a 5-min, 300-MPa treatment at 5°C [28] which is not sufficient to inactivate NoV. The volunteer study also confirmed that colder temperatures did enhance the inactivation of human norovirus as was observed for the NoV surrogates FCV and MNV since complete inactivation of NoV was observed when pressure was applied at 6°C and not 25°C for 400-MPa treatments.

In conclusion, NoV can be inactivated by HPP. However, given the complexities of food matrices (water activity, carbohydrates, ionic strength) and variable response of different viruses, direct validation of HPP conditions within the food or food matrix being produced will be required in future.

4. Pulsed electric fields

Pulsed electric fields (PEFs) are another non-thermal process that can be used to inactivate foodborne pathogens in foods and beverages. Pulsed electric field processing is conducted by passing short high voltage electrical pulses through a fluid. These short electrical pulses may range anywhere between microseconds up to milliseconds, and the food may be processed at either ambient or refrigerated temperatures. The pulse process typically occurs in milliseconds, but the localized heating of the food from this short pulse is significant and should be measured or controlled.

In order to be successful, food preservation requires the inactivation of not only spoilage microorganisms and pathogenic microorganisms, but also enzymes that catalyse unwanted chemical degradation in foods limiting their shelf life. PEF has the advantage of being able to inactivate both enzymes and microorganisms. Although the antibacterial effects of pulsed electric field treatment are well documented, little information is available on its antiviral effects. The few studies carried out on enteric viruses show that pulsed electric fields of strengths up to 29 kV/cm do not produce the minimal 3 log_{10} cycle reduction of infectious particles required to warrant further attention [18,29].

Mechanism of activity relies upon the "dielectric rupture theory," proposed by Zimmerman [30]. The external electric field will induce a change in transmembrane potential (TMP) across the cell membrane of a target organism. When this TMP reaches a threshold value, pore formation or electroporation of the cell membrane may occur. This will in turn induce an increase in cell membrane permeability, which can be lethal to the cell. It is hypothesized that the mechanism is somewhat similar in viruses, where PEF affects the outermost protein capsid of the virus, allowing its genetic material to escape the cell making it non-infectious.

5. Conclusion

Traditional thermal processing is highly efficient in inactivation of noroviruses although it may not be the most effective way for optimisation of food safety. Having in mind recent outbreaks in fruit and other RTE food, new technologies should be taken in account in NoV combat. These emerging technologies will help food business operators firmly to hold balance between safe NoV inactivation and the preservation of sensory properties.

For sure, no technology itself can deem complete NoV inactivation. This is due to specificity of non-envelope enteric virus, influence of food matrix, mechanism of inactivation, and virus strain. Further research on the effects on viruses in various food matrices is required. While all these techniques improve antiviral safety of food, good manufacturing practices and proper sanitation procedures remain the "gold" safety tools in food business.

References

[1] Koopmans M, Duizer E. 2004 Foodborne viruses: an emerging problem. *Int J Food Microbiol* **90**:23–41.

[2] Caul EO. 1996 Viral gastroenteritis: small round structured viruses, caliciviruses and astroviruses. Part I. The clinical and diagnostic perspective. *J Clin Pathol* **49**:874–80.

[3] Atmar RL, Opekun AR, Gilger MA, Estes MK, Crawford SE, Neill FH, Graham DY. 2008 Norwalk virus shedding after experimental human infection. *Emerg Infect Dis* **14**:1553–7

[4] Nilsson, M., K. O. Hedlund, M. Thorhagen, G. Larson, K. Johansen, A. Ekspong, and L. Svensson. 2003 Evolution of human calicivirus RNA in vivo: accumulation of mutations in the protruding P2 domain of the capsid leads to structural changes and possibly a new phenotype. *J Virol* **77:**13117-13124.

[5] Widdowson MA, Sulka A, Bulens SN, Beard RS, Chaves SS, Hammond R, Salehi EDP, Swanson E, Totaro J, Woron R, Mead PS, Bresee JS, Monroe SS, Glass RI. 2005 Noroviruses and foodborne disease, United States, 1991–2000. *Emerg Infect Dis* **11**:95–102.

[6] CDC Centers for Disease Control and Prevention. 2009 Surveillance for foodborne disease outbreaks—United States, 2006. *Morb Mort Wkly Rep* **58**:609–15.

[7] Grove SF, Lee A, Lewis T, Stewart CM, Chen H, Hoover DG. 2006 Inactivation of food- borne viruses of significance by high pressure and other processes. *J Food Prot* **69**:957– 68.

[8] Murchie LW, Cruz-Romero M, Kerry JP, Linton M, Patterson MF, Smiddy M & Kelly AL 2005 High pressure processing of shellfish: a review of microbiological and other quality aspects. *Inn Food Sci Emerg Technol* **6**: 257–270.

[9] Von Gunten, U. 2003 Ozonation of drinking water: Part I. Oxidation kinetics and product formation. *Water research*, **37(7)**, 1443-1467.

[10] Von Gunten, U. 2003 Ozonation of drinking water: Part II. Disinfection and by-product formation in presence of bromide, iodide or chlorine. *Water Research*, **37(7)**, 1469-1487.

[11] Langlais B, Reckhow DA, Brink DR. 1991. Practical application of ozone: principle and case study. Chelsea, Mich.: Lewis Publishers.

[12] Manley TC, Niegowski SJ. 1967. Ozone. In: Encyclopedia of chemical technology, Vol 14, 2nd ed. New York: Wiley. p 410–32.

[13] Castell-Perez ME, Moreira RG. 2004 Decontamination systems. In: Beier RC, Pillai SD, Phillips TD, editors. Preharvest and postharvest food safety. 1st ed. Oxford: Blackwell Publishing and the Inst of Food Technologists. p 337–64.

[14] Roy D, Wong PKY, Engelbrecht RS, Chian ESK. 1981. Mechanisms of enteroviral inactivation by ozone. *Appl Environ Microbiol* **41**:718–23.

[15] Vaughn JM, Chen YS, Novotny JF, Strout D. 1990. Effects of ozone treatment on the infectivity of hepatitis A virus. *Can J Microbiol* **36**:557–60.

[16] Graham DM. 1997. Use of ozone for food processing. *Food Technol* **51**(6):72–5.

[17] Kim JG, Yousef AE, Dave SA. 1999a. Application of ozone for enhancing the microbiological safety and quality of foods: a review. J Food Prot 62:1071–87.

[18] Khadre MA, Yousef AE, Kim JG. 2001 Microbiological aspects of ozone applications in food: a review. *J Food Sci* **65**:1242–52.

[19] Achen M, Yousef AE. 2001 Efficiency of ozone against *Escherichia coli* O157:H7 on apples. *J Food Sci* **66**:1380–4.

[20] Xu L. 1999. Use of ozone to improve the safety of fresh fruits and vegetables. *Food Technol* **53**(10): 58–62.

[21] Kingsley, D H 2013 High Pressure Processing and its Application to the Challenge of Virus-Contaminated Foods. *Food and Environmental Virology*, **5**(1), 1–12.

[22] Racaniello VR. 2001 Picornaviridae: The viruses and their replication. In: Knipe DM, Howley PM, editors. Fields Virology. Philadelphia: Lippincott, Williams, and Wilkins Inc 685–722.

[23] Chen H, Joerger RD, Kingsley DH, Hoover DG. 2004 Pressure inactivation kinetics of phage Lambda cI 857. *J Food Prot* **67**:505–11.

[24] Silva JL, Luan P, Glaser M, Voss EW, Weber G 1992 Effects of hydrostatic pressure on a membrane-enveloped virus: high immunogenicity of the pressure-inactivated virus. *J Virol* **66**:2111–7.

[25] Kingsley DH, Chen H, Hoover DG. 2004 Inactivation of selected picornaviruses by high hydrostatic pressure. *Virus Res* **102**:221–4.

[26] Balasubramanian S, Balasubramaniam VM. 2003 Compression heating influence of pressure transmitting fluids on bacteria inactivation during high pressure processing. *Food Research International* **36**:661–668.

[27] Leon JS, Kingsley DH, Montes JS, Richards GP, Lyon GM, Abdulhafid GM, Seitz SR, Fernandez ML, Teunis PF, Flick GJ, Moe CL. 2011 Randomized, double-blinded clinical trial for human norovirus inactivation in oysters by high hydrostatic pressure processing. *Applied and Environmental Microbiology* **77**:5476–5482.

[28] Dancho BA, Chen H, Kingsley DH. 2012 A method to discriminate between infectious and inactive human noroviruses. *International Journal of Food Microbiology* **155**:222–226.

[29] Drees KP, Abbaszadegan M, and Maier RM (2003) Comparative electrochemical inactivation of bacteria and bacteriophage. Water Research. 37(10) 2291-2300

[30] Zimmermann U. 1986 Electric breakdown, electropermeabilization and electrofusion, *Rev. Phys Biochem Pharmacol* **105**: 196-256.

IOP Conf. Series: Earth and Environmental Science **85** (2017) 012022 doi:10.1088/1755-1315/85/1/012022

Cold chain management in meat storage, distribution and retail: A review

I Nastasijević, B Lakićević and Z Petrović

Institute of Meat Hygiene and Technology, Kacanskog 13, Belgrade, Serbia

E-mail: ivan.nastasijevic@inmes.rs

Abstract. Meat is a perishable product with a short shelf life and therefore short selling times. Therefore, cold chain management in meat supply is of utmost importance for the maintenance of quality and safety of meat/meat products. Raw meat/meat products are likely to support the growth of pathogenic microorganisms and/or spoilage bacteria, and should be kept at temperatures that do not result in a risk to health. The cold chain should not be interrupted at all times along the meat distribution chain. The complexity of global meat supply chain, with frequently long distribution chains associated with transportation of the product within one country, from one to another country and from one to another continent, makes the solutions for the chilling and freezing regimes, as well as monitoring of time-temperature profiles, very important for the overall success in delivery of product which will be accepted by consumer for its freshness and safety levels. From recently, there are several available options for control and management of the cold chain, such as chilled and frozen storage combinations, superchilling, ionizing radiation, biopreservation, high hydrostatic pressure (HHP), active packaging, wireless sensors, supported with the software-based cold chain database (CCD).

1. Introduction

Meat is a perishable product with a short shelf life and therefore short selling times. In contrast to fresh fruit and vegetables, packaged meat has to be declared with a `use by` date [1]. The maintenance of the cold chain is also one of the main principles and basic requirements of European Union (EU) legislation on food hygiene [2]. Raw materials, ingredients, intermediate products and finished products that are likely to support the growth of pathogenic microorganisms and/or spoilage bacteria, are to be kept at temperatures that do not result in a risk to health. The cold chain should not be interrupted at all times along the meat distribution chain [3].

It is known that shelf life of chilled fresh meat can be extended by various packaging solutions, such as vacuum or modified atmosphere packaging (MAP) [4, 5, 6]. However, a freshness of chilled meat is strongly influenced by temperature. Inadequate storage, distribution and retail temperatures can lead to a significant reduction in shelf life and early spoilage of meat and meat products [6].

Healthy animals that are hygienically slaughtered after proper resting and fasting provide a practically aseptic meat. However, during slaughter, evisceration and dressing operations the microbial cross-contamination usually occurs, especially on the surface of meat, via contact with equipment, tools, hands, clothes, objects, etc. Meat is a particularly favorable substrate for the growth of microorganisms due to its chemical composition, e.g. rich in proteins, lipids and water. The lipid content of meat also makes it very sensitive to oxidation (the reaction of oxygen with fatty acids) and subsequent production of peroxides. The breakdown products of the peroxides produce the characteristic objectionable odor and flavor of rancid meat.

Several weak points exist in meat cold chain, such as the chilling of products during storage - before shipping, temperature abuse during transport and transferring products from one actor to another and waiting times during consolidation and deconsolidation at retail [7, 8, 9]. Temperature abuses result in variations of product quality during distribution and at the end of shelf life and may cause spoilage before the use by date is reached, leading to food waste and economical losses. The vulnerability of meat cold chain became very important in modern, global meat trade where distribution chain is sometimes very complex and long (slaughterhouse-transportation/distribution-retail-consumer continuum), where meat has to be shipped from one to another country or from one to another continent (Figure 1). Therefore, the management of cold chain is of paramount importance and presents a permanent challenge to maintain the safety and freshness of chilled fresh meat until it reaches the final consumer.

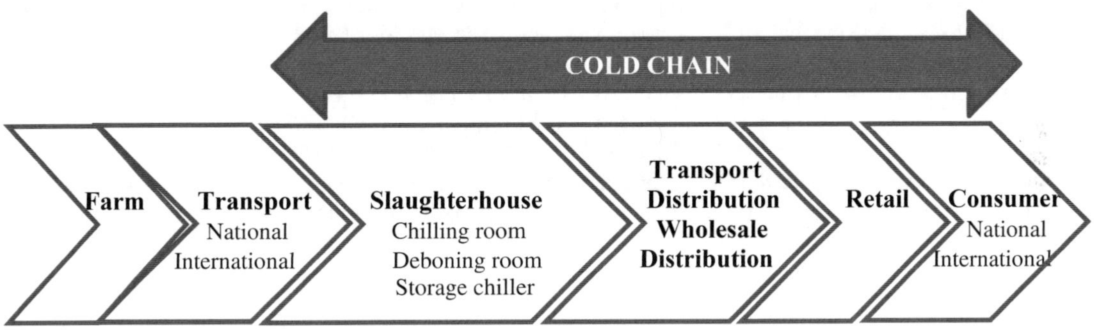

Figure 1. The schematic layout of meat supply cold chain

2. Meat spoilage

Pork, beef, lamb and poultry carcasses are chilled immediately after *post-mortem* inspection at slaughter line and/or in slaughterhouse chilling rooms. The chilling process is aerobic as the carcasses are exposed to air circulation. Most often, after 24-96h of chilling, the carcasses are usually moved to a boning/cutting room where they are further cut into primary cuts (primals). The primals can be typically stored up to 6 weeks in vacuum packs under anaerobic conditions [3]. Ground meat products may be prepared from trimmings from deboning and/or trimmings from primals after 6 weeks of anaerobic storage; these may be stored aerobically or anaerobically (Figure 2).

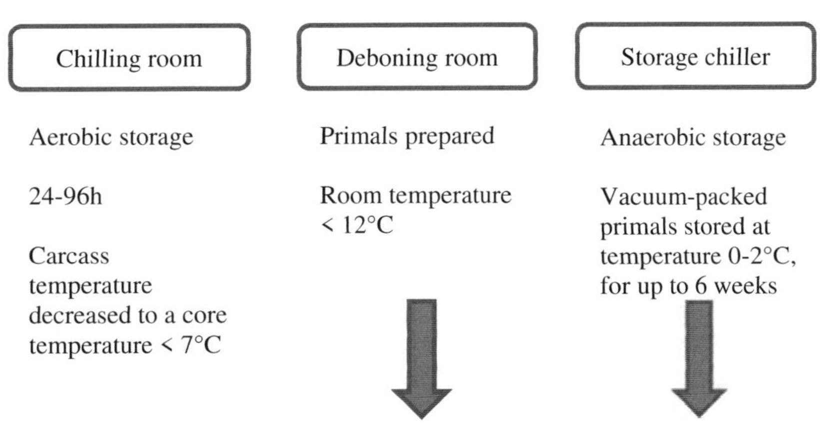

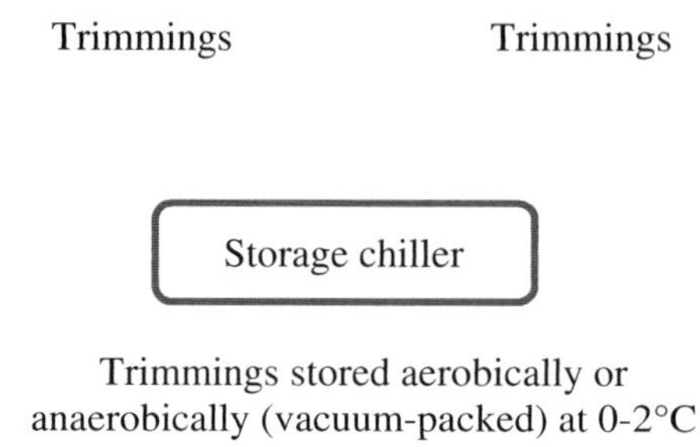

Figure 2. The chilling and chilled storage conditions for beef, pork and lamb carcasses and their primals and trimmings (adapted from [3])

It is well known that chilling of red meat and poultry carcasses is essential to retard bacterial growth. Chilling is also required for appearance and eating quality of meat. Most frequently bovine, pork and lamb carcasses are chilled using a forced convection of chilled air [10], although spray chilling (application of a fine spray) may also be used since it is faster than air chilling and it is primarily used in poultry, but may be also used in beef, pork and lamb processing plants.

According to Regulation (EC) 853/2004/EC the carcasses should be immediately chilled after post-mortem inspection to ensure that the temperature throughout the meat is < 7°C and < 3°C for offal. Interestingly, the provision on time limit by when this temperature must be achieved is not defined. For example, beef and lamb carcasses are usually not chilled to < 10°C (core temperature) within the first 10h to avoid cold shortening and toughening of the meat. Therefore, such conditions are favorable for the bacterial growth on the surface of the carcass, until the temperature is sufficiently reduced to retard bacterial activity [3].

Meat is considered to be spoiled when certain sensory changes occur, e.g. discoloration, off-odor and/or slime development and is usually primarily triggered by spoilage bacteria (Table 1) although indigenous enzymes may also be involved [11]. For instance *Pseudomonads*, *Lactobacillus* and *Enterococcus* produce slime on meat, while *Enterococcus* produce hydrogen peroxide greening spots, similar to greening caused by *Clostridium* spp. The growth of bacteria on meat surface is influenced by temperature, pH, water activity, nutrient availability, storage atmosphere (aerobic or anaerobic) and competition with other microbiota present on meat [12].

Table 1. The overview of the main meat/meat products spoilage defects and causal bacteria (adapted from [8]).

Defect	Meat/Meat product	Causal bacteria
Slime	Fresh meat	*Pseudomonads, Lactobacillus, Enterococcus, Weissella, Brochothrix*
Hydrogen peroxide greening	Fresh meat	*Weissella, Leuconostoc, Enterococcus, Lactobacillus*
Hydrogen sulfide production	Cured meats	*Vibrio, Enterobacteriaceae*
Sulfide odour	Vacuum-packed fresh meat	*Clostridium, Hafnia*
Cabbage odour	Bacon	*Providencia*
Cheesy or dairy odour	Vacuum-packed fresh meat	*Brochothrix thermosphacta*
Putrefaction	Ham	*Enterobacteriaceae, Proteus*
Bone taint	Whole meats	*Clostridium, Enterococcus*
Souring	Vacuum-packed meats	*Lactic acid bacteria, Enterococcus, Micrococcus,*

	Bacillus, Clostridium

3. Meat supply cold chain management

Preservation of red and poultry meat for trade and export purposes as anaerobic vacuum-packed product in either chilled or frozen form at sufficiently cold temperatures is essential for successful national and international trade [13]. The impact of long-term chilled storage, as well as frozen storage for export purposes was thoroughly investigated [12, 14]. It was proved that frozen storage durations were extended for more than one year [15]. In spite of these advances, improvement of preservation technologies within the current export cold chain remains an issue, in particular at industrial level, with potential for advancements encompassing superchilling, ionizing radiation, biopreservation and high hydrostatic pressure or high pressure processing (HPP) [16].

3.1. Cold chain in slaughterhouse

In slaughterhouse the meat chill chain starts with two main steps: (a) the primary chilling (rapid cooling of meat carcasses after slaughter) so that the warmest point of the carcass (center of the hind leg) has to reach a temperature of about < 7°C / 3°C for edible offal and < 4°C for poultry carcasses, to prevent microbial growth and extend the shelf life; with current technology these temperatures can be arrived at in 16–24 hours in small carcasses (lamb), in less than 48 hours in large carcasses (beef, pork) and less than 2h for poultry carcasses (internal deep breast); average and surface temperatures are obviously much lower, reaching 0°C on the surface within four hours; this is very important to slow microbial proliferation; and (b) secondary chilling (maintenance of the meat temperature below 7°C (red meat) and below 4°C (poultry meat) at all times during chill storage, cutting, deboning, mincing), before transportation.

The most common technologies to chill meat before transportation are: (i) air chilling, (ii) immersion chilling, (iii) spray chilling, and (iv) vacuum chilling [17].

3.1.1. Air chilling. The effectiveness of air chilling applications depends on a number of factors including air temperature and velocity, relative humidity, weight, fat cover of carcasses and loading of chilling chamber. Air temperature must be within the region of 0°C, with no decrease below -1°C, which could freeze the meat surface and impair its appearance. Air speed can range from 0.25 to 3.0 m/s. However, for economic reasons the most common speeds in use are from 0.75 to 1.5 m/s in the empty section of the cold chamber. Relative humidity during the chilling operation should be kept high to prevent excessive weight loss. The recommended rate is between 90 and 95 %, though this is the most difficult factor to control. It is worth of note that sometimes quick chilling has its problems because of the `cold shortening`. Cold shortening can often be seen in beef and mutton, when the meat, still in its pre-rigor phase, reaches temperatures of 10°C or lower. These conditions cause irreversible contractions of the muscle tissue which toughen the meat even after prolonged ripening. Quick primary chilling also signifies an increase in investment and higher operational costs. The chilling period can be reduced by lowering the air temperature (surface freezing risks) or increasing air speed (higher operational costs) or both. Occasionally cold chambers are refrigerated in advance to reach lower temperatures than those in operation (-5°C/-6°C for beef; -10°C/-12°C for pork), taking advantage of thermal inertia to offset the effect of warm meat loads [18].

3.1.2. Immersion chilling. This is the oldest chilling method and also the least expensive method and provides very rapid cooling with no risk of freezing. Immersion chilling is an off-line system that does not allow automatic weighing before and after chilling and is commonly used for the chilling of poultry carcasses, predominantly in USA. Immersion chillers use little floor space and are very energy efficient. The process can be accomplished using just tap water with or without flaked ice or pre-chilled water to create lower temperatures in the immersion bath. Such system is capable of lowering the temperature of the tap water to 1°C. This process can result in product weight increase by means of a controllable water pick-up. An increase of up to 12% can be achieved.

3.1.3. Spray chilling. This is an alternative method to immersion chilling which has been increasingly used especially in the USA for the chilling of poultry carcasses. Spray chilling can be applied in the processing and production of `frozen products` or `hard scalded birds`, as well as 'fresh products' or `soft scalded birds`. This chilling method is based on combination of sprays and air during the initial stage of the chilling cycle and the use of air for the rest of the chilling period; it uses a cold airflow across the surface of the product and intermittent spraying with cold water. This system prevents the skin from drying out and speeds up the chilling process. With this technique, the discolouration of the skin, which can occur with hard scalded birds, is prevented. Usually, spray chilling tunnels are designed to take one layer of birds in order to prevent cross contamination. Two layers are possible, but the bottom layer must be positioned in between the lines of the top layer to prevent drip contamination and additional drip troughs are required.

3.1.4. Vacuum chilling. This is a rapid batch process whereby moist products containing free water are cooled by evaporation of moisture under vacuum. The advantage of this technology is that it significantly reduces bacterial counts of phychrophiles and mesophiles after the meat was stored for several days. Vacuum packed chilled meat has significantly longer shelf life compared with conventionally chilled wrapped meat. For instance, vacuum packed beef cuts can be stored for up to 12 weeks, while lamb and pork cuts can have a shelf life for up to 5 and 8 weeks, stored at 0°C, respectively. The disadvantage is the large weight loss of meats.

3.2. Cold chain in meat distribution
During meat distribution (transportation) route to the final user - wholesale cold storage and/or display at retail, the cold chain must be maintained vigorously. Industrial and/or truck chambers have different characteristics and performances. Its size, initial temperature of incoming meat, targeted temperature during transportation, mechanical characteristics (e.g. power of compressors, ventilation and insulation), as well as energy/cost matters are issues of first priority when considering the meat distribution/transportation [17]. In general, the vehicle must be provided with a good refrigerated system capable to maintain the required temperature of meat/offal at all times during distribution (see 3.1.).

3.3. Cold chain in meat retail
The maintenance of cold chain during display at retail is of crucial importance for prevention of microbial spoilage, as well as maintenance of meat freshness and safety. The size and capacity of chilling chambers at retail establishment, the size of cabinets, initial temperature of incoming meat, meat handling procedures (cutting, mincing), temperatures of surroundings, location of refrigeration machinery, ventilation and light are the possible weak points to be addressed. The special focus should be put on temperature consolidation/deconsolidation during meat handling (e.g. daily transfer of meat cuts from the chilling chamber to the retail cabinet and vice versa, where internal temperature of meat cuts should be at all times < 7°C). The management approach that dominates in the meat market is related to the principle "First In – First Out". However, such approach should also adhere to all stages of the cold chain (see Figure 1) and has to be achieved through properly designed handling procedures in the chill storage rooms and retail cabinets. In all, different points of transport, from cold storage in slaughterhouse to retail outlet, and then to consumer refrigerator, are critical points for the overall meat quality and safety.

4. Available options in meat supply cold chain management
As said, if the cold supply chain is breached, meat and poultry products will suffer from a range of quality problems such as shrinkage, rotting, trim loss, unpleasant odor, color and texture changes, as

well as exacerbated health risks from pathogens such as *Salmonella*, *Campylobacter*, Shiga toxin-producing *E. coli* (STEC) and *Listeria monocytogenes*. Despite a numerous studies targeted to better understanding of meat microbial ecology (spoilage microorganisms and food borne pathogens) and its relation to temperature/time within the complex meat supply chain, e.g. meat storage, transportation, distribution and retail, the maintenance of cold chain, as well as meat quality and safety remains a challenge. From recently, there are several options/control measures which can be effectively and synergistically applied to improve the control over the `classic` meat supply cold chain.

4.1. Chilled and frozen storage combinations

Over the previous two decades a numerous studies had been carried out to assess the enhancement of shelf life, as well as quality and safety of meat achieved by chilling and freezing regimes. However, a limited number of studies were conducted to evaluate combined chilled and frozen storage practices, particularly regarding long-term chilled-than-frozen beyond 3-4 months of total duration.

It appeared that prolonged vacuum-packed chilled storage (two weeks for beef and lamb, at 4°C and even up to 30 weeks for beef, at -0.5°C and 7 weeks for lamb, at -1.5°C) combined with subsequent frozen storage (up to one year, at -18°C) had the beneficial effect to quality parameters (shear force, tenderness, fluid levels, flavor and color) of meat; the improved tenderness was confirmed as a major achievement [15, 19, 20]. Obviously, the prolonged chilled-frozen storage combinations can have a beneficial effect for export of fresh meat. Further and deeper studies are needed to investigate instrumental color and microbial profile of chilled-then-frozen meat, as these studies are highly relevant for consumer approval and health issues [21].

4.2. Superchilling

Superchilling is used for preserving foods by process in which the temperature of a food product is lowered to 1–2°C below the initial freezing point (for most foods between 0.5°C and 2.8°C). The superchilling technology combines the favorable effect of low temperatures with the conversion of some water into ice, which makes it less available for deteriorative processes. Superchilling gives the food product an internal ice reservoir so that there is no need for external ice around the product during transportation or storage for shorter periods. Generally, superchilling is positioned between freezing and refrigeration (conventional chilling), where the surrounding temperature is set below the initial freezing point. Storing food at superchilling temperature has three major advantages: maintaining food freshness, retaining high food quality and suppressing growth of harmful microbes. It can also reduce the use of freezing/thawing for production and thereby increase yield, reduce energy, labor and transport costs [22].

4.3. Ionizing radiation

Processing of food by ionizing radiation offers arrange of beneficial effects, which can`t be achieved by other and, in particular, traditional techniques such as chilling alone. Food radiation, if applied at low dose (up to1kGy) can be very effective against parasites presented in red meat, as well as for inactivation of spoilage and pathogen bacteria in chilled and/or frozen meat and poultry.

On the other hand, several consumer organizations raised concerns from the beginning whether it would be 'safe' to consume irradiated food? However, after more than100 years of research, the questions raised are resolved and the consumers accept irradiated food where it has become available together with an understandable explanation of the new technology. For example, in China very special items, as pickled chicken feet (irradiated), appear to have a certain market share. In the European Union the amount of irradiated food on the market place is marginal; however, varying drastically between Member States. More recent consumer studies are not available for the EU. The use of ionizing radiation in food processing remains still to be an under-estimated and under-exploited technology, in spite of its great potential [23].

4.4. Biopreservation

Natural compounds, such as essential oils, nisin, lysozyme, as well as natural or controlled microbiota, e.g. lactic acid bacteria (LAB) and their antimicrobial products such as lactic acid and bacteriocins, were investigated to extend the shelf life of meat/meat products and to obtain `green label` products. For example, nisin is only commercially available bacteriocin and showed good antibacterial action in artificially contaminated pork and in combination with 2% sodium chloride an excellent anti-listerial effect in minced beef; pentocin 31-1 (produced by Lactobacillus pentosus 31-1 isolated from the traditional Chinese fermented Xuanwei ham) effectively inhibited volatile basic nitrogen (VBN) and suppressed the growth of indigenous microbiota, especially *Listeria* and *Pseudomonas*, in chilled pork storage [16].

4.5. High hydrostatic pressure (HHP)
The meat preservation process using HHP, a non-thermal technology, can effectively inactivate product-spoiling microorganisms and enzymes at low temperatures without changing dramatically the sensory or nutritional characteristics of the product; the tenderness is even improved, while the fresh meat color is slightly changed, after the HHP treatment. HHP is also a powerful tool to control risks associated with *Salmonella* and *Listeria monocytogenes* in raw or marinated meats [24]. Pressure processing is usually carried out in a steel cylinder containing a liquid pressure-transmitting medium, e.g. water, while the product is protected from direct contact by using sealed flexible packaging [16]. The recommended pressure/time combination for meat/meat products of 600–700 MPa for 2–5 minutes showed quick inactivation of *L. monocytogenes*, as well as other indigenous microbiota [25].

4.6. Active packaging
Active packaging is an innovative technology that allows the product and its environment to interact to extend the product shelf life and to ensure its microbial safety, while maintaining the quality of packed food. According to EU legislation, active packaging is a type of food packaging with an extra function, in addition to that of providing a protective barrier against external influence. It means that the packaging can absorb food- and environment-derived chemicals within the packaging surrounding the food or its releases substances into the food or the environment surrounding the food, e.g. preservatives, antioxidants, and flavorings [26]. There are three main types of active packaging which were developed so far: antimicrobial active packaging, antioxidant active packaging, carbon-dioxide emitting/generating packaging.

4.6.1. Antimicrobial active packaging. This is one of the most important concepts of active packaging of meat. There are four basic categories of antimicrobial packaging: (i) incorporation of antimicrobial substances into a pad inside the package, with aim to provide a slow release of antimicrobials, (ii) direct incorporation of the antimicrobial agents into the packaging film, by heat treatment (co-extrusion of packaging films with antimicrobials), so that antimicrobials can be gradually released from the packaging films to the packaging head space or food surface, (iii) coating of packaging with a matrix that acts as a carrier for antimicrobial agents, so to allow the release of antimicrobials onto the food surface through evaporation into the headspace (volatile substances) or migration into the food through diffusion (non-volatile substances), and (iv) use of polymers that have antimicrobial activity (e.g. chitosan, poly-l-lysine). In addition, a large number of antimicrobials were tested for inhibiting the growth of microorganisms in food, e.g. ethanol, carbon dioxide, silver ions, chlorine dioxide, antibiotics, bacteriocins, organic acids, essential oils, spices, plant extracts (rosemary), peptides, etc [27].

4.6.2. Antioxidant active packaging. High levels of oxygen in meat packaging can facilitate microbial growth, lipid oxidation, development of off-flavors and off-odors, color changes and nutritional losses. Therefore, control of oxygen level in meat packaging is important to prevent/retard the deterioration and spoilage of meat. Antioxidant active packaging systems can be classified into 2 groups: (i) independent antioxidant devices (sachets, pads or labels with oxygen scavengers – fine iron powder and ferrous oxide), and (ii) antioxidant packaging materials (antioxidant active agent is incorporated into the walls of the packaging film (e.g. terpenoids from the propolis).

4.6.3. Carbon-dioxide emitting/generating packaging. Carbon dioxide has proven inhibitory effect for a range of aerobic bacteria and fungi via reduced oxygen level and through direct antimicrobial effect (by prolonging a lag phase and generation time during the logarithmic phase of microbial growth). Therefore, a CO_2 generating packaging system is a technique complementary to oxygen scavenging. The levels of CO_2 applied for meat and poultry preservation are relatively high (10-80%). For example 10-20% of CO_2 is needed for the inhibition of *Pseudomonas*, while 50% of CO_2 is necessary for control of proliferation of *C. perfringens*, *C. botulinum* and *Listeria monocytogenes*.

4.7. Wireless sensors

The manufacturers in the food industry frequently face the dilemma regarding the type of cold chain management in delivering products to retailers or end consumers, having to choose between frozen storage and cool storage (cooling, freezing, delivering and storage). Temperature is the main post-processing parameter in the determination of shelf-life in a cold chain of chilled and frozen food products. Frozen storage includes high-energy consumption for the preservation of food products, whereas cool storage involves the constant threat of bacterial-induced spoilage. Contemporary cold chain management encompasses temperature control and is focused on single logistic chain rather than serving multiple channels. In order to overcome the aforementioned deficiency, a time-temperature indicator (TTI) based cold-chain system is developed, which uses wireless sensors for collecting temperature data along the meat supply chain (from cold storage in slaughterhouse to retail) and implements the formulation of Critical Control Point (CCP) criteria throughout the entire delivery process. Under strict temperature monitoring, switching a number of products from frozen storage to cool storage seems to be feasible to improve the shortcomings associated with frozen storage (high-energy consumption, deterioration of taste, limited number of sale channels). Control charts are formulated for monitoring each point in the process. This approach is based on Internet of Things (IoT) architecture and international food standard (ISO 22000). IoT is a growing trend with a powerful influence in shaping the development of the information and communication technology (ICT) sector, e.g. radio frequency identification (RFID) tags, sensors, actuators, and even smart devices like mobile phones [28]. In practice, the IoT is expected to develop in areas such as wireless sensor networks with the aim of collecting contextual data. Further, a software approach to expanding web-based services using the capabilities of IoT (Web of Things,WoT) were recently developed. Lastly, a new business models in the food industry have been also developed: (i) cold chain home delivery service; (ii) convenience store (CVS) indirect delivery; (iii) CVS direct delivery; (iv) flight kitchen service [29].

4.8. Cold chain database (CCD)

The effective cold chain management can optimize freshness and safety of the product from farm/slaughterhouse up to the final consumer. The targeted data acquisition of time-temperature (t - T) profiles along the meat supply chain, as well as within the specific module of the cold chain stage (e.g. cold storage in slaughterhouse, transportation/delivery, retail) can serve as a valuable input for prediction of a product shelf-life status. From recently, a web-based platform was developed, within FRISBEE European project (http://frisbee-project.eu), for temperature conditions data collection throughout the chilled and frozen food supply chain. Data including all cold chain modules (industry, distributors, retailers and consumers), were collected to create the extensive database comprising more than 14,000 time-Temperature (t - T) profiles. Such platform can serve as a valuable Cold Chain Management tool. The Cold Chain Predictor (CCP) software based on the Cold Chain Database (CCD) allows calculation of product shelf-life status at different cold chain stages based on existing or user defined kinetic data. The developed tools offer the potential to run simulation scenarios based on real cold chain data and contribute to effective cold chain improvement and management [30].

5. Conclusion

The cold chain management in meat supply is of utmost importance for the maintenance of quality and safety of meat/meat products. The complexity of global meat supply chain, with frequently long distribution chains associated with transportation of the product within one country, from one to another country and from one to another continent, makes the solutions for the chilling and freezing regimes, as well as monitoring of time-temperature profiles, very important for the overall success in delivery of product which will be accepted by consumer for its freshness and safety levels. Although the importance of chilling and freezing regime effectiveness and temperature monitoring along the meat supply chain are well know, it remains a permanent challenge for the industry, distributors, retailers and consumers. From recently, there are several available options for control and management of the cold chain, such as chilled and frozen storage combinations, superchilling, ionizing radiation, biopreservation, high hydrostatic pressure (HHP), active packaging, wireless sensors, supported with the software-based cold chain database (CCD).

References

[1] The European Parliament and the Council. 2011 Regulation (EU) No 1169/2011 of 25 October 2011 on the provision of food information to consumers. *Off. J. Eur. Union* OJ **(L 304)** 18–63.

[2] Regulation (EC) 853/2004 of the European Parliament and of the Council of 29 April 2004, laying down specific hygiene rules for on the hygiene of foodstuffs. *Off. J. Eur. Union* OJ **(L 304)** L139/55.

[3] EFSA 2016 Growth of spoilage bacteria during storage and transport of meat. *EFSA Journal* **14** (6) 4523.

[4] Lambert AD, Smith JP and Dodds KL 1991 Shelf life extension and microbiological safety of fresh meat—a review. *Food Microbiol.* **8** 267–97.

[5] Kiermeier A, Tamplin M, May D, Holds G, Williams M and Dann A 2013 Microbial growth, communities and sensory characteristics of vacuum andmodified atmosphere packaged lamb shoulders. *Food Microbiol.* **36** 305–15.

[6] Berruga MI, Vergara H and Gallego L 2005 Influence of packaging conditions on microbial and lipid oxidation in lamb meat. *Small Ruminant Res.* **57** 257–64.

[7] Raab V, Petersen B and Kreyenschmidt J 2011 Temperature monitoring in meat supply chains. *Br. Food J.* **113** 1267–89.

[8] Nychas GE, Skandamis PN, Tassou CC and Koutsoumanis KP 2008 Meat spoilage during distribution. *Meat Sci.* **78** 77–89.

[9] Olsson A 2004 Temperature controlled supply chains call for improved knowledge and shared responsibilities. In *Conf. Proc. NOFOMA 2004* (ed. H Aronsson), 569–82. Sweden: Linköping.

[10] James SJ and James C, 2004 Meat marketing (d) transport of meat and meat products. In: Jensen WK, Devine C and Dikeman M (eds.). Encyclopedia of Meat Sciences. Academic Press, Elsevier Science, Ltd. 696–702.

[11] Tsigarida E and Nychas GJE 2001 Ecophysiological attributes of a Lactobacillus spp. and a Pseudomonas spp. on sterile beef fillets in relation to storage temperature and film permeability. *Journal of Appl. Microbiol.* **90** 696–705.

[12] Sumner J and Jenson I 2011 The effect of storage temperature on shelf life of vacuum-packed lamb shoulders. *Food Australia* **63** 249–51.

[13] Deards B, Leith R, Mifsud C, Murray C, Martin P and Gleeson T 2014 Live export trade assessment.Canberra:DepartmentofAgriculture.http://www.herefordsaustralia.com.au/Portals/0/PDF/ABARES%20Live%20Export%20Trade%20Assessment.pdf (accessed on 30 June 2 017)

[14] Kiermeier A, Tamplin M, May D, Holds G, Williams M and Dann A 2013 Microbial growth, communities, and sensory characteristics of vacuum and modified atmosphere packaged lamb shoulders. *Food Microbiol.* **36** 305-15.

[15] Muela E, Monge P, Safiudo C, Campo M M and Beltran J A 2015 Meat quality of lamb frozen stored up to 21 months: instrumental analysis on thawed meat during display. *Meat Sci.* **102** 35-40.

[16] Zhou GH, Xu XL, Liu Y 2010 Preservation technologies for fresh meat – A review. *Meat Sci.* **86**(1), 119-28.

[17] Nychas G J E, Skandamis P N, Tassou C C and Koutsoumanis K P 2008 Meat spoilage during distribution. *Meat Sci.* **78** 77–89.

[18] UN FAO, 1991. Manual on meat cold storage operation and management. http://www.fao.org/docrep/004/T0098E/T0098E02.htm (accessed on 29 June 2017).

[19] Hagyard C J, Keiller A H, Cummings T L and Chrystall B B 1993 Frozen storage conditions and rancid flavor development in lamb. *Meat Sci.* **35** 305-12.

[20] Muela E, Monge P, Sanudo C, Campo M M, Medel I and Beltran J A 2016 Sensory quality of lamb following long-term frozen storage. *Meat Sci.* **114** 32-7.

[21] Coombs C E O, Holman B W B, Friend M A and Hopkins D L 2017 Long-term red meat preservation using chilled and frozen storage combinations: A review. *Meat Sci.* **125** 84-94.

[22] Kaale L D, Eikevik T M, Rustad T and Kolsaker K 2011 Superchilling of food: A review. *J. Food Eng.* **107** 141–46.

[23] Ehlermann D A E 2016 Particular applications of food irradiation: Meat, fish and others. *Radiat. Phys. Chem.* **129** 53–7.

[24] Garriga M, Grebol N, Aymerich M T, Monfort J M and Hugas M 2004 Microbial inactivation after high-pressure processing at 600 MPa in commercial meat products over its shelf life. *Innov. Food Sci. Emerg.* **5** 451-57.

[25] Lucore L A, Shellhammer T H and Yousef, A E 2000 Inactivation of *Listeria monocytogenes* on artificially contaminated frankfurters by high-pressure processing. *J. Food Prot.* **63** (5), 662–64.

[26] EU, 2009. Guidance to the Commission Regulation (EC) 450/2009 on active and intelligent materials and articles intended to come into contact with food. Version 10. European Commission Health and Consumers Directorate, General Directorate E – Safety of the Food Chain, E6 – Innovation and sustainability.

[27] Fang, Z, Zhao Y, Warner R D and Johnson S K 2017 Active and intelligent packaging in meat industry. *Trends Food Sci. Tech.* **61** 60-71.

[28] Kang Y S & Lee Y H 2013 Development of generic RFID traceability services, *Comput Ind.* **64** (5), 609–23.

[29] Shih C W, Wang C H 2016 Integrating wireless sensor networks with statistical quality control to develop a cold chain system in food industries. *Comp. Stand. Inter.* **45** 62–78.

[30] Gogou E, Katsaros G, Derens E, Alvarez G, Taoukis P S 2015 Cold chain database development and application as a tool for the cold chain management and food quality evaluation. *Int. J. Refrig.* **52** 109-21.

Food safety – the roles and responsibilities of different sectors

N Karabasil[1], T Bošković[2], M Dimitrijević[1], D Vasilev[1], V Đorđević[3], B Lakićević[3], V Teodorović[1]

[1] University of Belgrade, Faculty of Veterinary Medicine, Department of Food Hygiene and Technology, Bulevar oslobodjenja 18, Belgrade, Serbia
[2] Ministry of Agriculture and Environment, Veterinary Directorate, Food Safety Unit, Nemanjina 22-26, Belgrade, Serbia
[3] Institute of Meat Hygiene and Technology, Kaćanskog 13, Belgrade, Serbia

E-mail: nedja@vet.bg.ac.rs

Abstract. Serbia is a relatively small country but with a long tradition in food production, especially meat and meat products. Serbia, as part of its open negotiation process as a candidate country with the European Union (EU), started to harmonise its legislation with the EU, and has published a set of laws and regulations relating to the hygiene of food production and food safety, the official control of production and the welfare of animals. Therefore, the food safety system in Serbia is based on principles established in the EU. There is a need for cooperation of different sectors (government, food business operators and consumers) in the management of food safety, and every sector has its role and responsibility. This paper aims to provide analytical support for the process of upgrading safety and quality in Serbia's food sector and explains the roles and responsibilities of different sectors in the food chain.

1. Introduction

Serbia is a relatively small country but with a long tradition in food production, especially meat and meat products [1,2]. In the period of pre-accession to the European Union (EU), Serbia, as a candidate country, is required to adopt a strategy for implementation of EU food hygiene standards in establishments for the production of food of animal origin. The National Programme for the upgrading of establishments should enable the food business operators (FBOs) covered by the National Programme to fulfil new legal requirements and to continue their operations in an equal and competitive manner on internal and on EU markets.

The principal objective of the new food hygiene rules is to ensure a high level of consumer protection with regard to food safety [1,3]. Every FBO along the food chain should ensure that food safety is not compromised. With regard to public health, these rules and procedures contain common principles, in particular in relation to the manufacturers and competent authorities' responsibilities, structural, operational and hygiene requirements for establishments, requirements for storage and transport and procedures for registration and approval of establishments.

This paper aims to provide analytical support for the process of upgrading safety and quality in Serbia's food processing sector, briefly describes the current level of compliance of the Serbian regulation with the EU, and the means of implementation and enforcement, and as well as, particularly, roles and responsibilities of different sectors in the food chain.

2. Analysis of the legal framework

Serbia, as part of its open negotiation process with the EU, started to harmonise its legislation with the EU, and has published a set of laws and regulations relating to the hygiene of food production and food safety, the official control of production and the welfare of animals. Requirements in the areas of food safety are set out in Chapter 12 of the EU *acquis* (Food Safety – Veterinary – Phytosanitary Policy), which is an integrated approach "from farm to fork" to ensure a high level of public health, animal health, animal welfare and plant health. The applied system of binding measures in the entire chain of primary production, processing and distribution of food to the consumer, as well as the appropriate methods of controls, provide a high level of protection of human life and health as the main objectives.

The food safety system in Serbia is based on principles established in the EU. The primary laws that form the backbone of the food safety system in Serbia are the Law on Food Safety (Official Gazette of the Republic of Serbia, No. 41/09), Law on Veterinary Matters (Official Gazette of the Republic of Serbia, No.91/05, 30/10, 93/12) and Animal Welfare Law (Official Gazette of the Republic of Serbia, No.41/09) [4-6].

Serbia has adopted general and specific rules on food hygiene which is partially (almost completely) in line with the EU's Hygiene Package.

2.1. The Law on Food Safety

The Law on Food Safety [5] defines the general conditions for the safety of food and feed, duties and responsibilities of food and feed business operators, rapid alert system, emergency measures and crisis management, as well as food and feed hygiene and quality. The purpose of this Law is to provide a high level of protection of human life and health and protection of the consumers' interests, including the principle of honesty and scrupulousness in the food circulation. It should also take into account whenever possible the protection of the health and well-being of animals, as well as the health of plants and environmental protection.

The provisions of this Law do not apply to primary food production and preparation, handling and storage of food for private domestic use, nor to feed for non food-producing animals. This Law also restricts the production and circulation of food and feed to legal persons or entrepreneurs that are recorded with the Register of Economic Operators and the Register of Facilities/Establishments kept by the Ministry of Agriculture.

2.2. The Law on Veterinary Matters

The Law on Veterinary Matters [4] defines the protection and enhancement of animal health and welfare; it identifies animal infectious diseases and measures for prevention, detection, containment, control and eradication of infectious diseases and diseases that can be transmitted to humans, veterinary-sanitary/official control and conditions under which animals and animal products, food of animal origin and animal feed may be produced and distributed.

The Law on Veterinary Matters is particularly focused on general and specific hygiene conditions in establishments for production of food of animal origin, covering the following areas: slaughtering the animals; production and processing of food of animal origin (e.g. cutting meat, processing); and distribution of food of animal origin on wholesale and retail level, including storage.

Veterinary Law defines the procedure for approval of establishments' producing/processing/storage of food of animal origin, including the obligation that all establishments handling food of animal products have to be registered in the Register of Establishments or in the Register of Approved Establishments.

In addition, there is a legal basis for the adoption of detailed rules prescribing the types of facilities that are registered or approved, methods and procedures of registration and approval of establishments, as well as the form of application for registration of registered or approved facilities. There is also a legal basis for the adoption of regulations that speak to the approval process for establishments that export products, according to the national veterinary and food hygiene requirements and specific requirements of the importing country. These two regulations should be drafted and adopted soon.

2.3. The Animal Welfare Law

The Animal Welfare Law [6] regulates animal welfare, rights, obligations and liability of legal entities and individuals, and entrepreneurs for the benefit of animals, treatment of animals and the protection of animals from abuse, protection of animal welfare during slaughter and killing, keeping, breeding, transport, slaughter and conducting experiments on animals, as well as other issues important for the protection of animal welfare.

3. Food sectors – roles and responsibilities

According to the Codex Alimentarius Commission, the definition of safe food is that "food will not cause harm to the consumer when it is prepared and/or eaten according to its intended use" [7]. Food is considered safe when it is evidenced that all measures have been taken to prevent contamination. In the 21st century, there is a need for cooperation between different sectors in the management of food safety. There are roles and responsibilities for each sector: government, FBOs and consumers [8].

3.1. Government and Regulatory Bodies

Government and regulatory bodies have the key role in food safety management from the primary production until the consumption [8-9]. This sector has to adopt and enforce legislation, with the food safety and public health issue as a priority. There should be transparent approach and provision of advice to producers.

The Ministry of Agriculture and Environmental Protection and the Veterinary Directorate are prepared to promote the improvement of hygiene standards in the establishments in the area of Veterinary Public Health. This strategic approach and consistent enforcement of decisions should prevent the closure of a high number of establishments, which would have a negative impact on the national economy and social status of the employees in the area of production, processing and trade of food of animal origin, especially in the meat sector. The harmonisation process in these areas with the EU legislation requires that the Serbian competent authority adapts its legal system to ensure the full implementation of EU regulations, which includes official inspection authorities, provision of relevant sources of information and data for risk assessment, as well as competent laboratory capacities and accredited test methods. The obligation of transposition and harmonisation of legislation also includes that all FBOs and inspection authorities to have implemented and maintained continuous improvement on the relevant knowledge and skills necessary to perform tasks from their activities and competence.

Communication with the consumers is very important and education of this sector is part of the government responsibility. One of the good examples is the Food Safety Council, which has to provide advice to the government sector in the area of risk assessment, but as well as to provide consumers with information about the possible risks. Risk assessment includes information such as scope, food sector, type of food, type of products, target population, scientific opinion, preparation of reports and the method of presenting data.

3.2. Food business operators

FBOs have a responsibility to put on the market product which will not have a negative effect on consumer [8,10-14]. They have to operate in hygienic environments, including processing, storage, transport/distribution. In ensuring food safety of the product, they should comply with appropriate criteria, like microbiological criteria for biological hazards, adequate labelling, have under control specific processing step in the flow diagram of production to be sure that the hazard as the potential risk is eliminated and as well recall a product in the case of an incident. FBOs have to implement the regulatory requirements, code of good manufacturing/hygiene practice and HACCP as a good methodology in ensuring food safety issue.

For a small country, Serbia has a large number of facilities for the production of food of animal origin, especially meat and meat products. Unfortunately, only a few meet all regulatory requirements and have been approved for export to the EU. Most of the establishments predominantly sell on local markets, are an important resource for supplying food of animal origin and employ a significant workforce for their processes. It speaks to the fact that food made by small producers can have an impact on a large number of consumers. Therefore, the processes of hygiene management and HACCP principles are critical in ensuring safe products. Serbia has started the development of national legislation and/or procedures and guidelines for FBOs with direct sales to final consumers and establishments covered within the definition of marginal, localised or restricted businesses, and likewise in the area of flexibilities and derogations for low capacity establishments and traditional methods of production.

3.3. Consumers

Consumers have an equally important role as the previous two sectors [8]. Sometimes, consumers in Serbia do not have enough information and/or adequate education about the food products of animal origin or their stability. There should be an improvement in consumer education about their role in food safety. Consumers have to apply good hygiene practice in food preparation and carefully read and understand the information given by the FBOs on food labels.

References

[1] Karabasil N, Boskovic T, Dimitrijevic M, Vasilev D, Teodorovic V, Ilic N. and Djordjevic, V 2015. Food hygiene – Flexibility in traditional and small meat establishments. *Procedia Food Science* **5** 140-43.

[2] Serbia 2015 Statistical Yearbook of the Republic of Serbia (Chapter Agriculture) *Statistical Office of the Republic of Serbia* Belgrade.

[3] Moy G G and Motarjemi Y 2014 Public Health Measures: International Standards and Harmonization of Food Safety Legislation. Encyclopedia of Food Safety Waltham: Academic Press 39-45.

[4] Serbia 2005 Law on Veterinary Matters Official Gazette of the Republic of Serbia No. 91/05, 30/10 and 93/12 Belgrade.

[5] Serbia 2009a Food Safety Law Official gazette Republic of Serbia No. 41/09 Belgrade.

[6] Serbia. 2009b Animal Welfare Law Official Gazette of the Republic of Serbia No. 41/09 Belgrade.

[7] CAC/RCP 1969 General Principles of Food Hygiene Codex Alimentarius Commission 1-1969, rev. 2003.

[8] Motarjemi Y and Lelieveld H. 2014 Chapter 1 - Fundamentals in Management of Food Safety in the Industrial Setting: Challenges and Outlook of the 21st Century Food Safety Management. San Diego: Academic Press 1-20.

[9] Boström M, Jönsson A M, Lockie S, Mol A P J and Oosterveer P 2015 Sustainable and responsible supply chain governance: challenges and opportunities. *Journal of Cleaner Production* **107** 1-7.

[10] Smigic N, Antic D, Blagojevic B, Tomasevic I and Djekic I 2016 The level of food safety knowledge among meat handlers. *Brit. Food J.* ***118***(1) 9-25.

[11] Tomasevic I and Djekic I 2016 HACCP in Fermented Meat Production. In N. Zdolec (Ed.), *Fermented Meat Products: Health Aspects* 512-34 CRC Press.

[12] Tomašević I, Šmigić N, Đekić I, Zarić V, Tomić N and Rajković A. 2013. Serbian meat industry: a survey on food safety management systems implementation. *Food Control,* ***32***(1), 25-30.

[13] Henderikx F. 2017 Labelling of food: a challenge for many *Veterinarski Glasnik* **71**(1) 16-23.

[14] MAEP 2015 Registry of Approved Establishments in the Veterinary Directorate. *Ministry of Agriculture and Environmental Protection*, Republic of Serbia.

Shedding light on food fraud: spectrophotometric and spectroscopic methods as a tool against economically motivated adulteration of food

R B Petronijević, B Velebit and T Baltić

Institute of Meat Hygiene and Technology, Kaćanskog 13, 11000 Belgrade, Serbia

E-mail: radivoj.petronijevic@inmes.rs

Abstract. Intentional modification of food or substitution of food ingredients with the aim of gaining profit is food fraud or economically motivated adulteration (EMA). EMA appeared in the food supply chain, and following the global expansion of the food market, has become a world-scale problem for the global economy. Food frauds have involved oils, milk and meat products, infant formula, honey, juices, spices, etc. New legislation was enacted in the last decade in order to fight EMA. Effective analytical methods for food fraud detection are few and still in development. The majority of the methods in common use today for EMA detection are time consuming and inappropriate for use on the production line or out of the laboratory. The next step in the evolution of analytical techniques to combat food fraud is development of fast, accurate methods applicable using portable or handheld devices. Spectrophotometric and spectroscopic methods combined with chemometric analysis, and perhaps in combination with other rapid physico-chemical techniques, could be the answer. This review discusses some analytical techniques based on spectrophotometry and spectroscopy, which are used to reveal food fraud and EMA.

1. Introduction

Cases of food adulteration in order to increase profit margins are dated back to the earlier ages of civilization. Intentional modification of food or substitution of food ingredients with the aim of gaining profit is food fraud or economically motivated adulteration (EMA) [1]. Unfortunately, EMA occurs today and, following advances in food technology, has become more sophisticated. EMA appears in food supply chains and is conducted by food producers, processors, distributors and/or retailers [2]. Following the global expansion of the food market, the consequences of EMA have emerged into a world-scale problem and have great impact on the global economy.

EMA inevitably leads to production of food with lowered nutritional value. This type of fraud not only has financial impacts on consumers, and society in general, but, more importantly, has adverse effects on health and food safety. Prolonged diet with such food could have devastating consequences to public health [3]. Recent reports exposed that food frauds and adulterations include a vast variety of food, such as oils, milk and meat products, infant formula, honey, juices and spices. In order to protect consumers and eliminate EMA, or to lessen its possibility, new legislation relating to the adulteration of food and food labeling was enacted in last decade [4,5]. Legislation for containing food fraud is still in development but, in order to be effective, needs strong logistic support. In addition to government agencies and inspectorates, essential support in revealing EMA includes development of analytical

IOP Conf. Series: Earth and Environmental Science **85** (2017) 012024 doi:10.1088/1755-1315/85/1/012024

methods to detect food frauds. As it is difficult to formulate any effective legal basis for suppressing EMA, it is also a burdensome task to develop and apply appropriate analytical methods for its detection [6].

Knowledge of the specificity and characteristics of analytical techniques is essential for detecting EMA of food. Such knowledge should be helpful in choosing appropriate analytical techniques for detection of adulterants of interest in a specified food matrix by controlling laboratories. Also, this could be a guideline for research laboratories to focus their exploration into development of new or improvement of existing methods. In a reciprocal way, it could help regulators to establish legal frameworks for revealing food frauds and EMA.

Various methods and techniques of chemical analysis are used to reveal adulterants in food [7]. They can be divided into several groups based on: absorption of electromagnetic radiation – colorimetric and spectroscopic methods and nuclear magnetic resonance (NMR); chromatographic; molecular and immunological methods; elemental, isotopic and mass spectrometry techniques; and chemometric methods based on statistical analysis of data obtained by some of these previously mentioned techniques. Especially in recent years, the need for rapid methods for food authentication and food fraud has arisen, as well as the need for methods that can be performed on-site. Therefore, the aim of this review is not to list all the methods for the detection of EMA, but to indicate approaches that can lead us to, or at least closer to, achieving that goal. The most promising methods are spectroscopic methods. They are quick, simple, require small quantities of the food/sample, and are usually non-destructive and non-invasive. Miniaturization of spectrophotometers and the manufacture of portable or hand-held instruments means *in-situ* measurements have become a reality. There are many portable spectrophotometers on market today. They can measure reflectance or transmittance of food/samples and can be used for direct measurement without sample pre-treatment. This makes them suitable for control on production lines, in storage areas and shipping departments, or indeed, everywhere where it is necessary to perform fast measurements. Furthermore, an overview of recent methods based on the absorption of electromagnetic radiation in the infrared (IR), visible or ultraviolet ranges used to detect EMA of food is presented.

2. Spectrophotometric methods

Methods such as sensitive spectrophotometry, differential spectrophotometry, derivative spectrophotometry, and spectrophotometry using chemometrics [8] can be used to reveal EMA in beverages, drinks, confectionery, spices, milk, olive oil, jams and sweets. Spectrophotometric measurement of specific extinction increase at 270 nm can be used to detect adulteration of extra-virgin olive oil by the addition of refined olive oil or other vegetable oils. Stigmasta-3,5-diene, a characteristic compound of refined oils, has an absorption peak at 270 nm [9]. The method is simple, rapid, low-cost, and eminently suitable for *in-situ* measurement.

Combining spectrophotometric methods with other analytical techniques enhances their possible application in detecting food EMA. Spectrophotometric methods and thin layer chromatography (TLC) [10] were applied to detect sweets and jams adulteration with synthetic colorants. Eight synthetic food dyes were used in the reported research [10]. The Kjeldahl method and spectrophotometric methods were employed to evaluate detection of milk adulteration by nitrogen-rich compounds such as melamine, ammonium sulfate or urea [11]. Standard (Kjeldahl) methods fail to identify adulteration by nitrogen compounds, and classical spectrophotometric methods for protein are not sensitive to these compounds. In combination, the proposed methods were able to recognize nitrogen added to milk and/or whey. It is suggested that this combination of Kjeldahl and spectrophotometric methods should be used to screen for milk adulteration by these compounds [11].

However, spectrophotometric techniques have limited application and often, they must necessarily be combined with other analytical techniques.

3. Color measurement

Measurement of color in HunterLab and CIE L*a*b* systems were used as supplementary techniques, mainly to describe visual properties of food [12]. Color of food can sometimes be improved by adding prohibited dyes [13]. Adulteration by adding coloring stuffs in food is maybe the oldest and most common means of food fraud. Colorings are added (fradulently and legally) to various foods from beverages to meat products and from confectionery to spices and oils. Reasons for the widespread usage of dyes in EMA is twofold: first, synthetic dyes are cheap and easily available, and, second, staining of lower quality food or food manufactured from low-grade ingredients provides higher profit. Therefore, the measurement of the surface color of food in HunterLab and CIE L*a*b* systems is usually used in assessment of food quality and safety [14]. However, the potential of such methods for detecting EMA is not yet fully recognized. Some efforts were made to employ food color measurements for detecting EMA and compliance with regulatory requirements [15]. The rapid method based on CIE L*a*b* measurements combined with chemometrics was used to estimate food coloring contents in meat products [15]. The method was developed to determine the presence and quantification of one permitted and two prohibited food dyes in frankfurters and fine-chopped boiled sausages, and it might also be useful during sensory evaluation of meat products for the assessment of the added food colorants.

4. Fluorescence spectroscopy

Fluorescence spectroscopy has the lowest limits of detection of all spectroscopic analytical techniques, but it is limited to polyaromatic hydrocarbons (PAHs) and heterocyclic compounds with rigid skeletons. It is a sensitive, selective and nondestructive technique. Development of new generation fluorescence spectrometers with powerful software for acquisition and interpretation of spectra resulted in application of this technique for EMA detection. Therefore, total synchronous fluorescence (TsyF) with the partial least square (PLS) model was used for the differentiation of virgin olive oil from olive-pomace, corn, sunflower, soybean, rapeseed and walnut oils [16]. Other applications of fluorescence spectroscopy combined with extensive use of chemometric tools were in food quality assessment of animal (dairy, meat, fish and egg) and vegetable (oils, cereal, sugar, fruit and vegetable) products [17-19].

The combination of fluorescence spectroscopy with other techniques such as Fourier transformed infrared spectroscopy (FTIR) was reported for authentication of extra-virgin olive oil. PLS method was used for both fluorescence and FTIR spectra analysis. The results suggest that FTIR and fluorescence could be a useful tool for analysis and detecting adulteration of extra-virgin olive oil with pomace oil.

5. Infrared spectroscopy

Infrared spectroscopy (IR), based on absorption of heat rays by molecular bonds with dipole moment, has been used for food quality and food authenticity assessments. In accordance with the applied spectral range, IR can be divided into two techniques – mid infrared (MIR) and near infrared (NIR). In MIR (fingerprint range), IR radiation is absorbed as a consequence of fundamental vibrations of atoms in the molecules. NIR spectra are the result of superposition of composite, overtone and high frequency absorption bands at shorter wavelengths.

Methods based on IR spectroscopy are among the most widely used methods for revealing EMA of food. Both MIR and NIR were used for detection of food frauds and authenticity of honey, meat, meat products, cheeses, milk, fish, spices, olive oil, and so forth.

The application of attenuated total reflectance (ATR) MIR microspectroscopy was evaluated as a rapid method for detection and quantification of milk adulteration [20]. Milk samples were spiked with different concentrations of whey, hydrogen peroxide, synthetic urine, urea and synthetic milk. Soft Independent Modeling of Class Analogy (SIMCA) pattern recognition analysis with Partial Least Squares Regression (PLSR) were used for estimation of adulterant levels. Results showed that MIR-microspectroscopy can provide an alternative methodology to the dairy industry for screening potential fraudulent practice for economic adulteration of cow's milk [20].

NIR methods with latent variable models applied to the spectral data were developed and used to estimate proximate composition, fatty acid profile, fillet yield and cooking loss, and to classify the available dataset by the rearing farm and genetic strain of raw and cooked freeze-dried rainbow trout (*Oncorhynchus mykiss*) fillets [21]. No major differences were observed between the results obtained from raw freeze-dried fillets and those obtained from cooked freeze-dried fillets, except for the estimation of some chemical constituents of interest, such as C22:6n-3 and polyunsaturated fatty acid content, both of which were better estimated in cooked freeze-dried fillets [21].

In a recent study, methodology based on the combination of MIR and NIR measurements and chemometric data processing was reported [22]. Such approach could overcome the disadvantages of both techniques and offer more accurate and reliable results in detecting the EMA.

6. Raman spectroscopy

Raman spectroscopy (RS), similar to IR, gives spectral information on fundamental vibrations of functional groups in a molecule. It is based on the inelastic scattering of incident radiation through its interaction with vibrating molecules [23]. RS and MIR are complementary techniques, but RS has some advantages over MIR. Water has weak absorption in RS and does not cause interference in Raman spectra. Taking into account the fact that most foods contain more or less water, the lack of interference due to water is essential for the reliability of methods for determining food authenticity. Absorption bands in Raman spectra are narrower than IR spectral bands. This is of significant importance for RS application, because narrow bands mean less overlapping, especially in complex matrices such as foods. RS also can be applied to analysis of foods directly through plastic or glass bottles and wrappers, bags and other transparent packaging materials [24].

RS scattering is weak, and new, improved techniques like Surface Enhanced Raman Spectroscopy (SERS), in contrast to classic Raman spectroscopy, provide lower detection limits, allowing applications in food adulterant determination [7]. For example, three Raman modalities, normal Raman, FT-Raman and SERS, combined with principal component analysis (PCA), were evaluated as a tool for detecting Sudan I dye in culinary spices. The results show that SERS is the most appropriate modality capable of providing a proper Raman signal when a complex matrix is analyzed [25]. RS combined with chemometrics was also applied to examine adulterations of honey by syrup and sugar solutions, olive oil by vegetable oils or pomace olive oils, and meat and fish [26].

7. Chemometrics

Almost all recent spectroscopic methods for food authentication use the chemometric approach for development and validation processes. This brief review would not be complete without a section dedicated to the chemometric application for food authenticity analysis. The reason for employing chemometrics in spectrophotometric and spectroscopic techniques is that spectra obtained by these aforementioned techniques are large datasets which cannot be resolved by conventional analysis. Chemometric tools, such as data mining, data fusion etc., are crucial for handling datasets generated by analytical methods for EMA detection.

A sample spectrum is, in fact, an "image" of the sample and can be treated in the same way as any digital image analysis. Data pre-processing, enhancement, segmentation, feature extraction and data classification for generation of applicable methods to spectroscopic data are based on similar mathematical models to those used in digital image processing for facial recognition or satellite image analysis [27]. Chemometrics resembles digital extraction and clean-up of the sample instead of using physical manipulation for clean-up; this is a major obstacle for common application of chemometric methods, because their calibration and validation necessitates a large number of samples. They are matrix-dependent too, and every change in matrix composition means a new calibration and validation process. Despite these disadvantages, the development of analytical methods based on spectroscopy combined with chemometrics is becoming more important, and the further efforts are being directed toward improvement of chemometric tools to become more robust and less sensitive to the influence of the matrix.

8. Conclusion

General features of spectrophotometric and spectroscopic methods are their simplicity, rapidity, and ability to be directly applied to foods/ingredients in different aggregate states, with fast, minimal or without any preparation. Production of new, cheaper and miniature, portable or handheld instruments for spectroscopy made them applicable on production lines, outdoors, in storage areas, etc. Their potential dual measurement modality, transmittance and reflectance, make spectroscopic methods useful in the resolution of various analytical tasks.

The combination of spectroscopic techniques with chemometrics has made them applicable for revelation of EMA and authentication of food. The majority of new spectroscopic methods for detecting food frauds are based on chemometric analysis. Although they are still under development, some of these methods have already been incorporated into benchtop and portable instruments. Further improvement of spectroscopy combined with chemometrics will require development of methods which are more robust and less sensitive to the influence of the matrix.

The next aspect to be addressed is the combination of spectroscopy with other analytical techniques, such as measurement of thermal, electrical or electrochemical properties of the food. This combination could improve reliability of results, and accuracy of determination and classification without loss of rapidity, simplicity or other advantages. Any additional data on the characteristic of the food, or by extension, other sample types, will contribute to a better and more accurate answer as to whether the food/sample is authentic or not. As the coupling of chromatographic techniques with mass spectrometry has made a significant breakthrough in analytical practice in the past, the integration of spectroscopic methods, either with one another or with other physico-chemical techniques, could make significant progress in detecting EMA of food in the future. That future requires rapid and reliable methodology for revealing EMA, which is fundamental for the production of safe and healthy food.

References

[1] Johnson R 2014 *Congressional Research Service Report* R43358
[2] Moore J C, Spink J and Lipp M 2012 *J. Food Sci.* **77** R118-R126
[3] Spink J and Moyer D C 2011 *J. Food Sci.* **76** R157-R163
[4] US Food and Drug Administration 2010 Adulterated food *Federal Food, Drug, and Cosmetic Act*
[5] REGULATION (EU) No 1169/2011 *O. J.* L 304/18
[6] Everstine K, Spink J and Kennedy S 2013 *J. Food Prot.* **76** 723-735
[7] Danezis G P, Tsagkaris A S, Camin F, Brusic V and Georgiou C A 2016 *TRAC-Trend. Anal. Chem.* **85** A 123-132
[8] Sorouraddin M H, Saadati M, Mirabi F 2015 *J. Food Drug. Anal.* **23** 447-452
[9] Jabeur H, Zribi A and Bouaziz M 2016 *Food Anal. Methods* **9** 712–723
[10] Bachalla N 2016 *IAIM* **3** 182-191
[11] De Lourdes Mendes Finete V, Gouvêa M M, De Carvalho Marques F F, Netto A D P 2013 *Food Chem.* **141**(4) 3649-3655
[12] Tomasevic I 2015 *Meat Technology* **56** 1-7
[13] Abhishek R U, Thippeswamy S, Manjunath K and Mohana D C 2014 *Proc. Indian Natn. Sci. Acad.* **80** 931-935
[14] Jha S N 2010 Colour Measurements and Modeling *Nondestructive Evaluation of Food Quality Theory and Practice* (Berlin: Springer) chapter 2 17-40
[15] Petronijević R B, Matekalo-Sverak V F, Spirić A T, Babić J A, Milijašević M P, Trbović D K and Vuković I K 2014 *Hem. Ind.* **68** 781-791
[16] Poulli K I, Mousdis G A and Georgiou C A 2007 *Food Chem.* **105** 369–375.
[17] Karoui R and Blecker C 2011 *Food Bioprocess Technol.* **4** 364-386
[18] Dankowska A, Małecka M and Kowalewski W 2015 *Dairy Sci. Technol.* **95** 413-424
[19] El-Bialee N M, Rania K I, El-Bialee A M, Harith M A and Darwish N E M 2013 *Australian Journal of Basic and Applied Sciences* **7** 132-138

[20] Santos P M, Pereira-Filho E R and Rodrigez-Saona 2013 *Food Chem.* **138** 19-24
[21] Dalle Zotte A, Ottavian M, Concollato A, Serva L, Martelli R and Parisi G 2014 *Food Res. Int.* **60** 180-188
[22] Lin C A, Ayvaz H and Rodriguez-Saona L E 2014 *Food Anal. Meth.* **7** 1407–1414
[23] Bumbrah G S, Sharma R M 2016 *Egyptian Journal of Forensic Sciences* **6** 209-215
[24] Ellis D I, Muhamadali H, Haughey S A, Elliott C T and Goodacre R 2015 *Anal. Methods* **7** 9401-9414
[25] Di Anibal C V, Marsal L F, Callao M P and Ruisánchez I 2012 *Spectrochimica Acta Part A: Molecular and Biomolecular Spectroscopy* **87** 135-141
[26] Herrero A M 2008 *Food Chem.* **107** 1642-1651
[27] Chitradevi B and Srimathi P 2014 *International Journal of Innovative Research in Computer and Communication Engineering* **2** 6466-6472

IOP Conf. Series: Earth and Environmental Science **85** (2017) 012025 doi:10.1088/1755-1315/85/1/012025

Chromatography methods and chemometrics for determination of milk fat adulterants

D Trbović, R Petronijević and V Đorđević

Institute of Meat Hygiene and Technology, Kaćanskog 13, Belgrade, Serbia

E mail: dejana.trbovic@inmes.rs

Abstract. Milk and milk-based products are among the leading food categories according to reported cases of food adulteration. Although many authentication problems exist in all areas of the food industry, adequate control methods are required to evaluate the authenticity of milk and milk products in the dairy industry. Moreover, gas chromatography (GC) analysis of triacylglycerols (TAGs) or fatty acid (FA) profiles of milk fat (MF) in combination with multivariate statistical data processing have been used to detect adulterations of milk and dairy products with foreign fats. The adulteration of milk and butter is a major issue for the dairy industry. The major adulterants of MF are vegetable oils (soybean, sunflower, groundnut, coconut, palm and peanut oil) and animal fat (cow tallow and pork lard). Multivariate analysis enables adulterated MF to be distinguished from authentic MF, while taking into account many analytical factors. Various multivariate analysis methods have been proposed to quantitatively detect levels of adulterant non-MFs, with multiple linear regression (MLR) seemingly the most suitable. There is a need for increased use of chemometric data analyses to detect adulterated MF in foods and for their expanded use in routine quality assurance testing.

1. Introduction

Milk fat (MF) is largely composed of triacylglycerols (TAGs) (these comprise about 98% of the total fat), while phospholipids account for only 0.8% of milk lipids. Sterols are also a minor component, comprising about 0.3% of the fat, cholesterol being the principal sterol [1]. The fat content of milk can vary from about 3.0% to 6.0%, but typically is in the range 3.5 to 4.7% [2]. In addition to the well-characterized differences among and within breeds of dairy cattle in MF content and fatty acid (FA) composition i.e. genetic factors, differences can also occur due to stage of lactation and diet [3-7].

MF is a good source of fat-soluble vitamins and essential FAs [8]. Fat is an essential component of the human diet and inclusion of MF as part of a balanced diet should be advantageous rather than detrimental. To date, no scientific study has produced evidence of any increased risk of disease associated with milk consumption [9]. Cow's milk is an important component of the human diet because of its high nutritional value. Its nutritionally balanced composition makes it one of the most complete foods available [10]. Ruminant MF contains butyric acid (C4:0), which is an important anti-cancer agent. However, longer-chain FAs can be problematic; e.g., myristic (C14:0) and palmitic acids (C16:0) are considered harmful, while stearic acid (C18:0) and short-to-medium-chain (C4–C10) FAs are deemed neutral [9]. Up to one-third of the FAs in MF have a chain-length of 14 carbons or less. These FAs are oxidized rapidly in the liver, have a lower energy value and are oxidized more readily than long-chain FAs. It follows that MF should contribute less to obesity than an equivalent amount of other dietary fats [11]. This reinforces the need for the dietetic community to reconsider current recommendations on dairy fat and human health on the basis of scientific evidence [2]. In conclusion, whereas future studies will help to elucidate the role of milk and dairy products in human health, their use within a balanced diet should be considered in the absence of clear recommendations [9, 11].

MF contains several compounds that have demonstrated anticancer activity in animal models [12]. The more important ones are rumenic acid (RA, *cis*-9, *trans*-11 conjugated linoleic acid, CLA) [13-14], a potent inhibitor of mammary tumorigenesis, sphingomyelin and other sphingolipids that prevent the development of intestinal tumors and butyric acid (C4:0), which prevents colon and mammary tumor development [4, 14-17]. Cows' diets have a major influence on the CLA content of MF, and these effects have been recently summarized [4, 14, 17-18]. Emerging evidence suggests that MF can prevent intestinal infections, particularly in children, prevent allergic disorders, such as asthma and improve the level of long-chain n-3 polyunsaturated fatty acids in blood [19].

This paper discusses analytical methods for detecting MF adulteration, with an emphasis on GC analysis of TAG and FA in combination with multivariate statistical data processing. Such data could be potentially useful in detecting foreign fats in the milk and dairy industries. Based on this preliminary investigation, the usefulness of this approach could be examined in the future for other foreign fats and oils, including vegetable and animal fat.

2. Adulteration of milk and dairy products

Milk adulteration is a current fraudulent practice to mask the quality parameters (e.g. protein and fat content) and increase the product shelf life. Milk and milk-based products are among the leading food categories according to reported cases of food adulteration [20]. Perhaps the most high-profile case involved the addition of melamine to high-protein feed and milk-based products to artificially inflate protein values in products that may have been diluted [21]. Melamine, an organic base, is widely used in plastics, adhesives, and other consumer products, and is known to pose a public health threat [22]. Adulterated milk could also be added into infant formula and other milk-based products. Baby formula is a common target for retail fraud, often by tampering with the sell-by codes to move expired product. The safety and integrity of dairy products is of particular interest, because these foods play an important role in feeding the population and are essential for certain groups of consumers, such as women, children and the elderly. Milk is a fairly expensive raw material, and from an economic point of view it could, therefore, be attractive to fraudulently modify its composition, replacing part of it with other dairy or non-dairy ingredients [23, 24].

The major adulterants of MF are vegetable oils (soybean, sunflower, groundnut, coconut, palm and peanut oil) and animal fat (cow tallow and pork lard). Butter is made from milk, whereas butter substitutes, also called imitation butters, are generally manufactured from non-dairy fats or other suitable components to make butter-like products [25-27]. Dairy products have been traded for hundreds of years and make up a large proportion of the food industry trade. However, the adulteration of milk and butter is a major issue for the dairy industry. Adulteration of milk used to manufacture butter can result in an inferior final product that fails to meet consumer expectations. Adulterated milk and butter contain added substances such as water, neutralizers to mask acidity, salt or sugar to mask extra water or high solid contents, whey and hydrogen peroxide, among others [28]. Although many authentication problems exist in all areas of the food industry, adequate control methods are required to evaluate the authenticity of milk and milk products in the dairy industry. One method to detect adulteration of milk with water is measurement of osmolality [29]. Rezende *et al.* [30] stated that the refractive index method for water adulteration could adequate such as density and freezing point determinations. Mid infrared (MIR) spectroscopy combined with pattern recognition analysis were used to classify and quantify milk adulteration with whey, synthetic urea or hydrogen peroxide [28]. The addition of NaOH to milk, which aims to mask acid formation, is very easy to determine using principal component analysis (PCA) to separate control and adulterated milks [31]. However, it is important to mention that the goal is not to identify a specific adulterant and its concentration, but the presence of a group of adulterants.

3. Methods for detecting the authenticity of milk and dairy products

Quality assurance (QA) and the methods used to authenticate foodstuffs are of great interest both from commercial and legal points of view [32]. In Europe, origin is one of the main authenticity issues concerning food. Determination of food authenticity is an important issue for both QA and food safety. Interest in Europe concerning food authentication is also shown by continuous funding of this topic, from FP 5 to the Horizon 2020 initiative. Authenticity testing is a quality criterion for food and food ingredients and is increasingly a result of legislative protection of regional foods. Thus, there is a pressing need for accurate, standardized food authentication techniques [32].

Over the last decade, several analytical procedures have been proposed for rapid screening or selective confirmation of the quality and authenticity of milk such as liquid chromatography (LC) and GC, especially coupled with mass spectrometry (MS). The studies are often supported by a chemometric approach allowing reliable qualitative (classification) and quantitative (multivariate calibration) procedures. GC flame ionization detection is exploited for analysis of MF because the milk FAs and TAGs can be monitored and compared with reference standards. This procedure can be used to discriminate the source of adulteration. GC separation of the TAG classes in MF according to their carbon number (CN) (C24–C54) has been used to determine milk origins and the potential adulteration of dairy products with foreign fats [25, 33]. Adulteration of expensive oils and fats, such as MF, has always been a serious problem because of the economic advantages of replacing high-priced fats and oils with low-priced oils, including soybean oil or corn oil, without labeling the product accordingly [25, 27]. Consequently, the European Union made the GC methodology official, converting it to a reference method for the detection of foreign fat in MF by TAG analysis using short capillary column GC (more polar polysiloxane phases containing a higher proportion of phenyl groups (50–65%) and low blending at temperatures as high as 370-400°C) [34]. Because of the wide variety of FAs contained in MF, the characterization of TAGs in MF is a complex and difficult task [35]. Before quantitative analysis, TAGs must be grouped on the basis of some of their common characteristics (molecular weight, degree of unsaturation, etc.). Moreover, the GC analysis of TAG or FA profiles of MF in combination with multivariate statistical data processing has been used to detect adulterant fats in milk and dairy products [25, 27, 36, 37, 39]. GC analysis of TAG was also used to detect goat cheese adulterated with cheaper cow`s milk [40].

4. Chemometrics as a tool to identify milk fat adulteration

Chemometrics is an interdisciplinary research field that involves multivariate statistics, mathematical modelling and computing especially applied to chemical data, and is required for food authentication or identity confirmation. It must be combined with suitable database infrastructure and uses appropriate mathematical tools [41]. Chemometrics has been useful in evaluating the quality and identity control of processing parameters for dairy products [41-44]. For this approach, a large set of analyses and many pattern classification procedures, such as PCA, linear discriminant analysis (LDA), hierarchical cluster analysis (HCA), soft independent modelling of class analogies (SIMCA), partial least squares regression (PLS), canonical variate analysis (CVA), and artificial neural network (ANN) can be utilized. Pattern analyses are applied to a dataset to compare similarities or differences of sample data with original data. GC analysis of the FA profile is widely used to detect adulteration of MF with foreign fats. For example, Rebechi *et al.* [37] artificially adulterated MF with 0, 2, 5, 10 and 15% tallow or lard. Their multiple linear regression (MLR) detected adulterations of MF at levels greater than 10% for tallow and 5% for lard. Gutiérrez *et al.* [25] adulterated raw MF with 0, 5, 10, 15, and 20% non-MFs. When LDA was used, the global percentage of satisfactory classification was 94.4%; consequently, LDA was effective in detecting adulterations at levels <10%. Kim *et al.* [27] used specific bio-markers (FAs, TAG and cholesterol) which enabled the detection of adulteration as low as 10% of non-MFs in MF. The validity of the classification rule was also tested by 206 gravimetrically prepared fat mixtures. These data can be potentially useful in detecting foreign fats in butter products. In the work of Lipp [36], GC analysis of TAG was further analyzed by PLS and ANN to identify mixtures of butter fat with foreign fat. While ANN was most suitable for classification, quantitative results were obtained by PLS. Souza *et al.* [42] identified groups of adulterants, including formaldehyde, starch, urine, hydrogen peroxide and chlorine, by physicochemical analysis and application of PCA and HCA.

The results obtained in this research should contribute to a proposal for a national standard to verify MF authenticity in milk and dairy products. Theoretically, after multivariate analysis, and taking into account many analytical factors, adulterated MF should form singular groups that can be easily distinguished from authentic MF.

5. Conclusion

Cow's milk is an important component of the human diet because its nutritionally balanced composition makes it one of the most complete foods available. Milk and milk-based products are among the leading food categories according to reported cases of food adulteration. Although many authentication problems exist in all areas of the food industry, adequate control methods are required to evaluate the authenticity of milk and milk products in the dairy industry. This paper discusses analytical methods for detecting MF adulteration with an emphasis on using GC to analyze TAG and FA in combination with multivariate statistical data processing. Using multivariate statistical methods such as MLR, adulterations of MF at levels as low as 10% of non-MF can be detected. With respect to aduteration, chemometrics is a powerful data reduction tool used to qualitatively group or classify unknown MF samples with similar characteristics and to quantitatively determine levels of adulterant analytes in MF. The results obtained in this research should contribute to a proposal for a national standard to verify MF authenticity in milk and dairy products.

References

[1] Mac Gibbon A K H and Taylor M W 2006 *Advanced Dairy Chemistry, Volume 2, Lipids,* ed P F Fox and P L H McSweeney (New York, Springer, USA) pp 1-42
[2] Lock A L and Bauman D E 2011 *Proc. Cornell Nutr. Conf. for Feed Manuf.* (New York, Cornell University Ithaca) pp 126-35
[3] White S L, Bertrand J A, Wade M R, Washburn S P, Green Jr J T and Jenkins T C 2001 *J. Dairy Sci.* **84** 2295-301
[4] Kelsey J A, Corl B A, Collier R J and Bauman D E 2003 *J. Dairy Sci.* **86** 2588-97
[5] Auldist M J, Johnston K A, White N J, Fitzsimons W P and Boland M J 2004 *J. Dairy Res.* **71** 51-7
[6] Kay J K, Weber W J, Moore C E, Bauman D E, Hansen L B, Chester-Jones H, Crooker B A and Baumgard L H 2005 *J. Dairy Sci.* **88** 3886- 93
[7] Soyeurt H, Dardenne P, Gillon A, Croquet C, Vanderick S, Mayeres P, Bertozzi C. and Gengler N 2006 *J. Dairy Sci.* **89** 4858-65.
[8] Jensen R G, Ferris A M and Lammi-Keefe C J 1991 *J. Dairy Sci.* **74** 3228–43
[9] Visioli F and Strata A 2014 *Adv. Nutr.* **5** 131-43
[10] Palmquist D L 2006 *Advanced Dairy Chemistry, Volume. 2, Lipids, ed* P F Fox and P L H McSweeney (New York, Springer, USA) pp 43-80
[11] Parodi P W 2004 *Aust. J. Dairy Technol.* **59** 3-59
[12] Parodi P W 1997 *J. Nutr.* **127** 1055–60
[13] Ip C, Banni S, Angioni E, Carta G, McGinley J, Thompson H J, Barbano D and Bauman D 1999 *J. Nutr.* **129** 2135-42
[14] Griinari J M and Bauman D E 1999 *Advances in Conjugated Linoleic Acid Research, Vol. 1, ed* M P Yurawecz, M M Mossoba *et al.* (Champaign, IL, AOCS Press, USA) pp 180-200

[15] Parodi P W 1999 *J. Dairy Sci.* **82** 1339–49

[16] Bauman D E, Baumgard L H, Corl B A and Griinari J M 1999 *Proc. Amer. Soc. Anim. Sci.* (Champaign, IL USA) pp1-15

[17] Bauman D E, Corl B A, Baumgard L H and Griinari J M 2001 *Recent Advances in Animal Nutrition*, ed P C Garnsworthy and J Wiseman (Nottingham, Nottingham University Press, UK) pp 221-50

[18] Bauman D E and Lock A L 2006 *Advanced Dairy Chemistry, Volume. 2, Lipids, ed* P F Fox and P L H McSweeney (New York, Springer, USA) pp 93-136

[19] Parodi P W 2006. *Advanced Dairy Chemistry, Volume 2 Lipids,* ed P F Fox and P L H McSweeney (New York, Springer, USA) pp 601-39

[20] Johnson R 2014. *Food Fraud and "Economically Motivated Adulteration" of Food and Food Ingredients* (Washington D.C, Congressional Research Service, Library of Congress, USA) pp 1-40

[21] Sharma K and Paradakar M 2010 *Food Secur.* **2** 97–107

[22] Bottemiller H 2011 Food Safety News, April 29, (Available at: http://www.foodsafetynews.com/2011/04/chinese-authorities-seize-more-melamine-tainted-dairy/#.WMfu5G8rLcs)

[23] De La Fuente M A and Juárez M 2005 *Crit. Rev. Food Sci.* **45** 563–85

[24] Karouri R and Baedemaeker J 2007 *Food Chem.* **102** 621–40

[25] Gutiérrez R, Vega S, Díaz G, Sánchez J, Coronado M, Ramírez A, Pérez J, González M and Schettino B 2009 *J. Dairy Sci.* **92** 1846–55

[26] Molkentin J 2007 *Eur. J. Lipid Sci. Technol.* **109** 505-10

[27] Kim J M, Kim H J and Park J M 2015 *J. Food Sci.* **80** C1945- 51

[28] Santos P M, Pereira-Filho E R and Rodriguez-Saona L E 2013 *Food Chem.* **138** 19–24

[29] Büttel B, Fuchs M and Holz B 2008 *Chem. Cent. J.* **2** 6-12

[30] Rezende P S, Carmo G P and Esteves E G 2015 *Talanta* **138** 196–202

[31] Santos P M, Wentzell P D and Pereira-Filho E R 2012 Scanner digital images combined with color parameters: A case study to detect adulterations in liquid cow's milk. *Food Anal. Method* **5** 89–95

[32] Danezis G P, Tsagkaris A S, Camin F, Brusic V. and Georgiou C A 2016 *TrAC Trends Anal. Chem.* **85** Part A 123–32

[33] Fontecha J, Ríos J J, Lozada L, Fraga M J and Juárez M 2000 *Int Dairy J.* **10** 119-28

[34] Commission Regulation (EC) No 273/2008 *Official Journal of the European Union* 2008, *Annex XX, Article 16,* L88/97-L88/106

[35] Mottram H R and Evershed R P 2001 *J. Chromatogr.* A **926** 239–253

[36] Lipp M 1996 *Food Chem.* **55** 389-395

[37] Rebechi S R, Vélez A, Vaira S and Perotti M C 2016 *Food Chem.* **192** 1025–32

[38] Fraga M J, Fontecha J, Lozada L and Juárez M 1998 *J. Agr. Food Chem.* **46** 1836–43

[39] Vieitez I, Irigaray B, Callejas N, González V, Gimenez S, Arechavaleta A, Grompone M and Gámbaro A 2016 *J. Food Compos Anal.* **48** 95–101

[40] Bertacchini L, Cocchi M, Vigni M L, Marchetti A, Salvatore E, Sighinolfi S, Silvestri M and Durante C 2013 *Chemometrics in Food Chemistry,* ed F Marini (Oxford, Elsevier, UK) pp 371-410

[41] Faye B, Konuspayeva G, Messad S and Loiseau G 2008 *Dairy Sci. Technol.* **88** 607–17

[42] Souza S S, Cruz A G, Walter E H M, Faria J A F, Celeghini R M S, Ferreira M M C, Granato D and de S. Sant'Ana A 2011 *Food Chem.* **124** 692–5

[43] Sacco D, Brescia M A, Sgaramella S, Casiello G, Buccolieri A, Ogrinc N and Sacco A 2009 *Food Chem.* **114** 1559–63

[44] Sola-Lanarranga C and Navarro-Blasco I 2009 *Food Chem.* **112** 189–96

Effects of Inulin and Sodium Carbonate in Phosphate-Free Restructured Poultry Steaks

B Öztürk and M Serdaroğlu

Ege University Engineering Faculty Food Engineering 35100 Bornova-Izmir/Turkey

E-mail: burcu.ozturk@ege.edu.tr

Abstract. Recently inorganic phosphates used in meat product formulations have caused negative impact on consumers due to their potential health risks. Therefore, utilization of natural ingredients as phosphate replacers has come into prominence as a novel research topic to meet consumer demands for clean-label trends. In this study, we objected to investigate the effects of inulin utilization either in the powder or gelled form, alone or in combination with sodium carbonate on quality of phosphate-free restructured chicken steaks. Total moisture, protein, lipid and ash values of the trial groups were in the range of 71.54-75.46%, 22.60-24.31%, 0.94-1.70% and 1.45-2.13%, respectively. pH of the samples was between 6.18-6.39, significant increments were recorded in samples containing inulin with sodium carbonate. L*, a* and b* values were recorded as 78.92-81.05, 1.76-3.05 and 10.80-11.94, respectively, where use of gelled inulin resulted in changes of L* and a* values. Utilization of inulin in combination with sodium carbonate decreased cook loss and enhanced product yield. Sensory scores in control group with phosphate showed a similar pattern to sensory scores in groups with inulin and sodium carbonate. During storage, purge loss and lipid oxidation rate were similar in control and inulin + sodium carbonate samples. The results showed that use of inulin in combination with sodium carbonate provided equivalent physical, chemical and sensory quality to phosphates in restructured chicken steaks.

1. Introduction

In the last years, there has been a considerable increase in consumption of ready-to-eat poultry products in line with technological improvements, modernization of everyday life and consumer habits. However, some of the food additives used in formulation of further processed products cause negative impact on consumers due to potential health concerns. Hence, recently consumption of clean-label products in which synthetic additives are minimized has been a growing trend. Accordingly, a novel reformulation strategy is the reduction of widely used additives in poultry product formulations such as salt, nitrite and phosphate or replacement of them with alternative natural ingredients [1, 2].

Utilization of phosphates lead to change pH value that is highly related to water-holding capacity, breakdown the calcium bridges of actomyosin complex and improve the functionality of meat proteins, such as water-holding, solubility, emulsification and gelling. Thus, in various meat products phosphates could enhance product yield, texture, flavor and colour, decrease cook loss, thaw loss and purge loss, as well as they act as antioxidants extending shelf life [3, 4]. However, recently inorganic phosphates used in meat product formulations have caused negative impact on consumers due to their potential health risks [4, 5]. Therefore, utilization of natural ingredients as phosphate replacers has come into prominence as a novel research topic to meet consumer demands for clean-label trends.

Inulin is a soluble and health-promoting prebiotic dietary fiber industrially extracted by a washing process mainly from chicory (*Cichorium intybus*) roots and consisted of oligo and polysaccharides [6, 7]. Inulin was reported to enhance various quality parameters like water-holding capacity, emulsion stability, textural and sensory attributes in different types of meat products [6, 8-10]. In this study, we aimed to research the effects of inulin (powder/gelled) utilization with or without sodium carbonate on quality characteristics of restructured chicken steaks.

2. Material and Methods

Fresh boneless post-rigor chicken breast cubes (2x3x3 cm) were supplied from the production line of Lezita Integrated Meat Processing Plant (Abalıoğlu Co., İzmir) and stored at 4°C prior to production. Food-grade sodium tripolyphosphate (STPP) and sodium caseinate was donated by Pacovis Food Co. (Izmir, Turkey), Orafti-HP inulin was purchased from Artisan Food Co. (Istanbul, Turkey), rosemary extract was supplied from Frutarom-Etol Co. (Kocaeli, Turkey) and microbial transglutaminase was purchased from Stern Ingredients (Izmir, Turkey). Sodium carbonate (SC) and other ingredients were purchased from local market. For production of control treatment (P), breast cubes were tumbled with NaCl (1%), STPP (0.5%), rosemary extract (0.015%) and ice (added to have 72% initial moisture) using a tumbler (Vakona GmbH, Germany) operated at 50 rpm under vacuum for 45 min. Other treatments were formulated with: 0.2% SC (C), 4.5% powder inulin (I), 4.5% powder inulin and 0.2% SC (IC), 4.5% gelled inulin (GI), 4.5% gelled inulin and 0.2% SC (GIC) as phosphate replacers. Gelled inulin was prepared as 30% (w/v) aqueous solution. All treatments were also contained sodium caseinate (1.5%) and microbial transglutaminase (0.7%) to enhance meat binding. The tumbled mixture was stuffed into synthetic ham casings, cold-setting was applied for 18h at 4°C and then the products were steam-cooked for 4-6 h at 85°C. After that, the samples were cooled to 4°C, casings were removed and samples were sliced into 1 cm thickness. The steaks were finally packaged under modified atmosphere consisted of 70% N_2 and 30% CO_2 (Multivac 240, Germany) and stored for 15 days at 2±2°C.

In restructured poultry steaks, total moisture, lipid and ash analysis were carried out according to AOAC [11] and protein content was determined according to AOCS [12]. pH was measured by using a pH-meter (WTW, Germany) equipped with a penetration probe. Hunter colour (L*, a*, b*) parameters of the emulsions was measured with a portable colorimeter (Konica Minolta, Japan). Cook yield was determined as a percentage from the weight loss between un-cooked and dry-cooked samples. A 9-point hedonic scale was used to evaluate sensory characteristics of the samples. Thiobarbituric Acid Reactive Substances (TBARS) was measured according to Witte *et al.* [13] to determine lipid oxidation secondary products during storage. Data was statistically analysed by Analysis of Variance (ANOVA) and Duncan Post-Hoc tests using the SPSS software.

3. Results and Discussion

Chemical composition of the treatments is presented in Table 1. Total moisture, protein, fat and ash values ranged between 71.54-75.46%, 22.60-24.31%, 0.94-1.70% and 1.45-2.13%, respectively. Samples containing STPP (P group) and sodium carbonate (C group) had higher moisture content compared to others ($P<0.05$). Protein content of the samples generally did not differ, except GI samples having higher protein content compared to P ($P<0.05$). The highest fat content was obtained in I, IC and GI samples, while the lowest fat content was in P and GIC samples ($P<0.05$). However, it was found that fat content of the samples was generally pretty low since no added fat was included in the formulations. P samples had the highest ash content among treatments ($P<0.05$), probably due to higher inorganic content.

Table 1. Chemical composition of restructured poultry steaks.

Treatments	Moisture (%)	Protein (%)	Fat (%)	Ash (%)
P	75.46±0.33^{a}	22.60±0.37^{b}	0.94±0.06^{c}	2.13±0.09^{a}
C	72.99±0.81^{b}	23.25±1.26^{ab}	1.48±0.09^{b}	1.71±0.04^{b}
I	70.52±0.29^{d}	23.68±1.17^{ab}	1.63±0.09^{a}	1.45±0.05^{d}
IC	72.23±0.74^{bc}	22.64±0.69^{b}	1.70±0.06^{a}	1.51±0.03^{cd}
GI	71.55±0.37^{cd}	24.31±0.28^{a}	1.68±0.11^{a}	1.59±0.03^{c}
GIC	71.54±1.00^{cd}	23.36±0.47^{ab}	1.04±0.05^{c}	1.70±0.02^{b}

a, b, c: Different letters indicate significant difference (P<0.05). Data is presented as mean values ± standard deviation.

pH values of restructured steaks are presented in Figure 1. The values were between 6.18-6.39 and significant differences were obtained depending on formulations (P<0.05). The lowest pH values were belonged to I and GI samples (P<0.05). However, pH of the treatments formulated with inulin and SC (IC and GIC samples) were significantly increased (P<0.05). Control samples with STPP (P group) had significantly lower pH compared to C, IC and GIC (P<0.05). Therefore, it was found that utilization of inulin and SC combination in restructured poultry could meet the effect of phosphates on pH value, regardless of the form of inulin (powder/gelled).

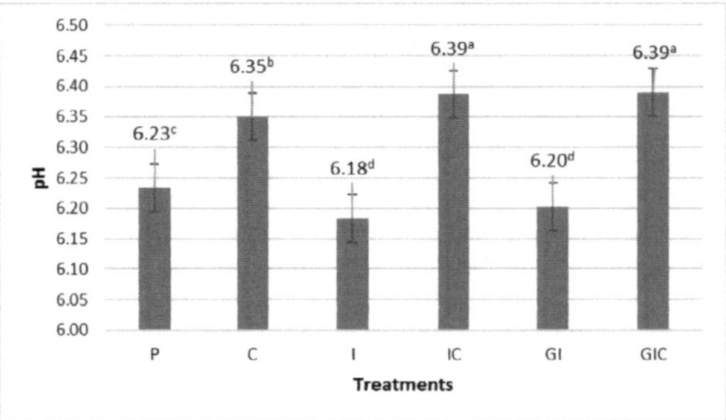

Figure 1. pH value of restructured chicken steaks. a, b, c, d: Different letters indicate significant difference (P<0.05).

The surface colour parameters of restructured steaks are shown in Table 2. L*, a* and b* values were between 78.92-81.05, 1.76-3.05 and 10.80-11.94, respectively. GI samples with gelled inulin had higher L* values compared to P samples with STPP (P<0.05), while the other treatments had similar values. This result showed that utilization of gelled inulin alone could favour lighter colour probably due to the opaque gel colour. a* values of GIC samples were higher than P, C and GI samples (P<0.05). No significant differences were obtained in b* values of the samples. A study reported that inulin utilization in fermented chicken sausages had an increment effect on L* and a* values [14].

Table 2. Surface colour of restructured poultry steaks.

Treatments	L*	a*	b*
P	78.92±0.61^{b}	1.92±0.72^{b}	10.80±0.49
C	80.98±0.87^{ab}	1.76±0.61^{b}	11.92±0.85
I	80.07±1.00^{ab}	2.49±0.58^{ab}	11.94±0.27
IC	80.95±0.93^{ab}	2.34±0.81^{ab}	11.76±0.65
GI	81.05±1.75^{a}	2.03±0.23^{b}	11.79±1.07
GIC	79.28±1.94^{ab}	3.05±0.64^{a}	11.64±0.73

a, b, c: Different letters indicate significant difference (*P*<0.05). Data is presented as mean values ± standard deviation.

Figure 2 presents the results of cook yield of restructured steaks, which were between 71.72-85.83%. Compared to P groups; C, IC, GI and GIC samples had higher cook yields (*P*<0.05). Therefore, the combined addition of inulin and SC presented an advantage to increase cook yields, which is probably due to the synergistic effect of pH increment by SC and binding ability of inulin. The form of inulin was also effective on cook loss, when used solely and in powder form, cook yield was significantly decreased (*P*<0.05). Thus, when used in gelled form, since inulin was homogenously mixed with meat it could favour meat binding and thus product yield. On the other hand, powder form of inulin likely was not able to provide homogenous distribution and thereby decreased the yield.

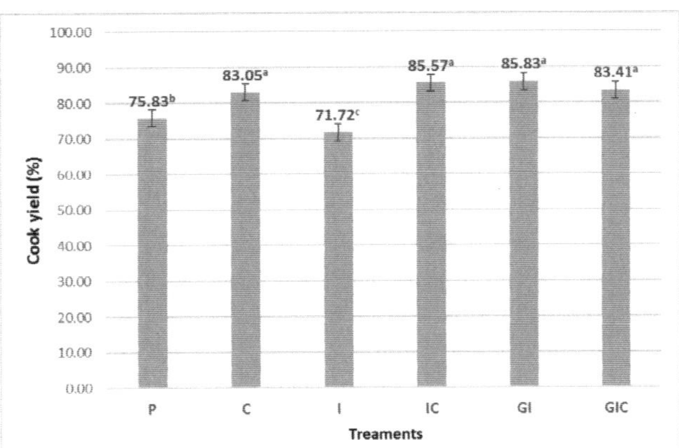

Figure 2. Cook yield (%) of restructured chicken steaks. a, b: Different letters indicate significant difference (*P*<0.05).

The results of sensory analysis are presented in Figure 3. Appearance, colour, texture, juiciness, saltiness, sweetness, flavour and general acceptability of the samples were scored between 6.44-7.33, 6.56-6.89, 6.22-7.33, 5.44-7.67, 6.22-7.44, 6.78-7.56, 5.89-7.67 and 5.89-7.67, respectively. The scores were generally found to be in acceptable ranges. The highest acceptability was obtained in samples formulated with STPP (P groups) and formulated with inulin and SC (IC and GIC groups) (*P*<0.05). Therefore, use of inulin and SC together should provide equivalent sensory characteristics to phosphates. In samples containing inulin alone (I and GI), juiciness, flavour and general acceptability was significantly lower than other samples (*P*<0.05). Previously, inulin was reported to enhance sensory quality in different kind of meat product formulations [9, 14]. However, in phosphate-free formulations it could be suggested to use inulin in combination with SC to obtain similar sensory quality to phosphate containing products.

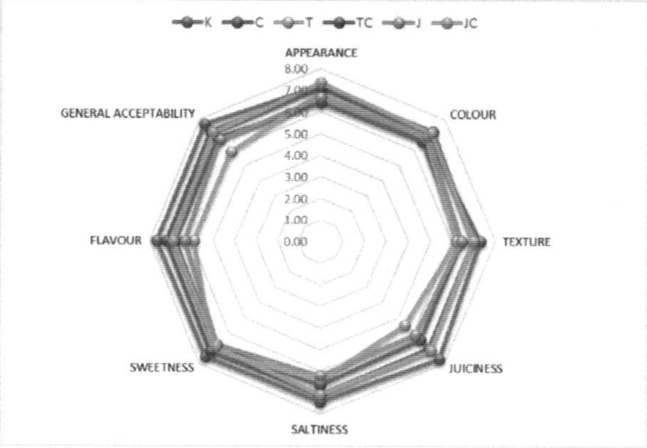

Figure 3. Sensory scores of restructured chicken steaks.

TBARS values of the samples analysed during storage are presented in Figure 4. The initial TBARS values were measured between 0.24-0.56 mg malonaldehyde/kg, while the values were between 0.73-0.90 mg malonaldehyde/kg in the end of the storage period. In final products, it was found that P and C samples had higher TBARS values compared to IC, GI and GIC ($P<0.05$). During the storage period, significant increments were detected in samples depending on the propagation of oxidation ($P<0.05$). Final values were generally similar in treatments. Therefore, it was indicated that inulin could compensate for the antioxidant effect of phosphates and have the potential to meet oxidative quality.

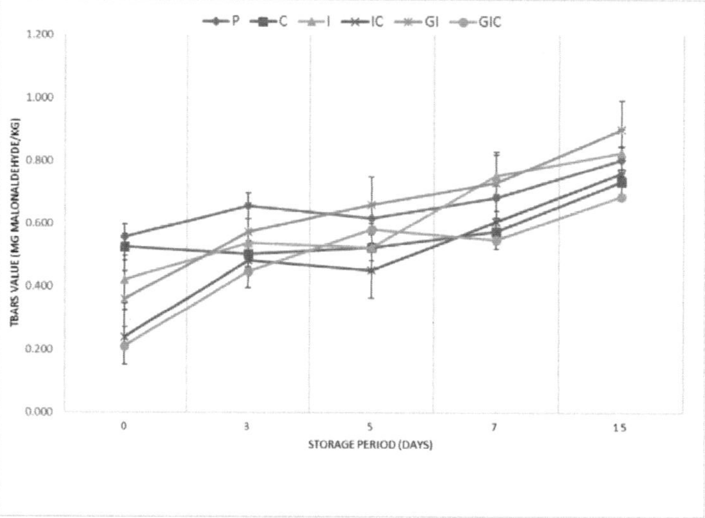

Figure 4. TBARS values (mg malonaldehyde/kg) of restructured chicken steaks during storage period.

4. Conclusion

Our study indicated that inulin and SC as clean-label ingredients presented favourable results as phosphate replacers in restructured chicken steaks. Although proximate composition of the products generally did not seem to change intensely, pH of the products was significantly increased with combination of inulin and SC. Colour values were similar to control samples in most of the phosphate-free formulations. The combined usage of inulin and SC had an increment effect in cook yield. It was also observed that inulin should favour product yield when used in gelled form and presented advantages over powder form. Sensory analysis and lipid oxidation rate showed that usage of inulin and SC could enhance sensory parameters and oxidative quality and showed similar acceptability to phosphate containing products. The results indicated that utilization of inulin and SC in restructured chicken steaks have good potential to enhance physical, chemical, technological and sensory quality and offers a novel possibility for phosphate replacement in formulation of healthier poultry products.

Acknowledgement
Authors would like to thank to Republic of Turkey, Ministry of Science, Industry and Technology with Project No: 0764.STZ.2014 (SAN-TEZ Program).

References

[1] Olmedilla-Alonso B, Jiménez-Colmenero F and Sanchez-Muniz FJ 2013 *Meat Sci.* **95** 919
[2] Young JF, Therkildsen M, Ekstrand B, Che BN, Larsen MK, Oksbjerg N and Stagsted J 2013 *Meat Sci.* **95** 904
[3] Sebranek JG 2009 *Ingredients in Meat Products: Properties, Functionality and Applications*, ed Tarté R (USA: Springer Publishing)
[4] Petracci M, Bianchi M, Mudalal S, Cavani C 2013 *Trends Food Sci Tech.* **33** 27
[5] Dimitrakopoulou MA, Ambrosiadis JA, Zetou FK, Bloukas JG 2005 *Meat Sci.* **70** 743
[6] Bodner JM, Sieg J 2009 *Ingredients in Meat Products: Properties, Functionality and Applications*, ed Tarté R (USA: Springer Publishing)
[7] Öztürk B, Serdaroğlu M 2017 *J Food Health Sci.* **3** 12
[8] Álvarez D, Barbut S 2013 *Meat Sci.* **94** 320
[9] Angiolillo L, Conte A, Del Nobile MA 2015 *LWT- Food Sci Technol.* **62** 697
[10] Keenan DF, Resconi VC, Kerry JP, Hamill RM 2014 *Meat Sci.* **96** 1384
[11] AOAC 2012 *Official Methods of Analysis* (Gaithersburg, MD, USA)
[12] AOCS 2009 *Official Methods and Recommended Practices* (American Oil Chemists' Society)
[13] Witte VC, Krause GF, Bailey ME 1970 *J Food Sci.* **35** 582
[14] Menegas LZ, Pimentel TC, Garcia S, Prudencio SH 2013 *Meat Sci.* **93** 501

Yersinia enterocolitica in fermented sausages

R Mitrović[1], V Janković[1], B Baltić[1] and J Ivanović[2]

[1]Institute of Meat Hygiene and Technology, Kaćanskog 13, Belgrade 11000, Serbia
[2]Department of Food Hygiene and Technology, Faculty of Veterinary Medicine, University of Belgrade, Bulevar Oslobođenja 18, Belgrade 11000, Serbia

E-mail: radmila.mitrovic@inmes.rs

Abstract. Different types of food, among them meat, can be the cause of food-borne diseases, and infections are commonly caused by *Campylobacter, Salmonella, Yersinia enterocolitica,* verotoxic *Escherichia coli* and *Listeria monocytogenes.* All these bacteria, depending on a number of factors, including animal species, geographical origin, climatic factors, methods of animal breeding and meat production, could cause disease. Here, we summarise results on production of different groups of sausages produced with or without added starter culture, and contaminated with *Y.enterocolitica* (control sausages were not contaminated). During the ripening, changes in the microbiological status of the fermented sausages and their physical and chemical properties were monitored. For all tests, standard methods were used. In these fermented sausages, the number of *Y. enterocolitica* decreased during ripening. The number of *Y. enterocolitica* was statistically significantly lower in sausages with added starter culture on all days of the study Zoonotic pathogens in meat should be controlled through the complete production chain, from the farms to consumers, in order to reduce the probability of disease in humans. However, the necessary controls in the production chain are not the same for all bacteria.

1. Introduction

Meat processing came about from the need to conserve meat. The ancient Romans conquered the territory of the then known "Old World", by, among other things, knowing how to conserve meat (they produced dried ham and fermented sausages), so that these products were a source of food of animal origin without the danger of spoiling. Lowering the water activity (a_w), combined with lowering pH, can be considered the oldest meat conservation technology. Dried products were produced by drying in the air, and the a_w was also reduced by salting, whether it was wet or dry salting. Also, in the case of fermented products, salting was used mainly to prevent the growth of bacteria. Little was known about the process of fermentation, the penetration of salt into the meat, or drying. Meat processing was seen as an art, a skill. In those times, smoke was an additional conservation method, especially significant for surface spoilage (mould). Therefore, some importance was given to smoking of meat, which contributes to a more acceptable smell and taste.

2. Fermented sausages

Fermented sausages are, in fact, raw sausages, which are prepared from minced meat, solid fat, nitrite salt, sugar and spices. The spontaneous fermentation in this type of sausage is conditioned by the presence of lactic acid bacteria and coagulase negative cocci (including *Kocuria* spp.) in sausage filling (raw materials).

Fermentation is more successfully controlled using starter cultures. *Lactobacillus sakei, L. curvatis, L. plantarum, L. pentosus, L. casei, Pediococcus pentosaceus* and *Pediococcus acidilactici* [1-4] are most commonly used as the starter cultures. After filling into casings, sausages are subjected to ripening, during which there are physical, chemical and enzymatic changes that ensure the shelf life of the finished product, as well as the characteristic sensory properties. Usually, sausage filling intended for the production of sausages of a bigger diameter is coarsely minced, while the filling for smaller diameter sausages is prepared from the meat that is finely minced. In the case of bigger diameter sausages, fermentation processes are more noticeable, while in smaller diameter sausages, the external physical processes, i.e. drying of the product, is more prevailing.

Differences in microbiota, physical and chemical changes occur during ripening, which also affect the quality of the finished product. It is understandable that given the place of production, the production conditions, especially those related to the ripening processes, can be controlled (air-conditioned chambers), i.e. the temperature conditions during ripening, circulation and humidity of air, as well as the time and temperature of smoking can be monitored. Such controlled production conditions apply to both bigger and smaller industrial facilities.

The introduction of glucono-delta-lactone (GdL), or starter cultures, as well as the use of other additives, allowed development of the industrial production of sausages with significantly different properties compared to traditional products [2,3,5,6].

The traditional production of raw fermented sausages, which is quite common in households in Serbia, takes place under uncontrolled conditions (temperature, humidity, fermentation) during the cooler seasons. This process does not rely on starter cultures, but on the activity of fermentative bacteria that are naturally present in the initial microbiota of the meat and environment of the production area; it is these bacteria which start and perform fermentation, i.e. decomposition of carbohydrates present in the filling, mainly to lactic acid [7-9].

In recent years efforts have been made to develop autochthonous starter cultures, i.e. starter culture from microorganisms isolated from these traditional sausages. Such starter cultures should positively affect the safety, while the typical sensory characteristics of traditional sausages remain preserved [10-12].

In many European countries, increased demand for traditional products has been observed. These products are foods with strong regional characteristics originating from non-industrial processes and which are produced in small batches with limited equipment. Preservation of traditional production and authentic products can help small producers and local economies, and it is also important to preserve traditional knowledge, cultural heritage and the regional identity of small and often underdeveloped communities [5,8,10,13,14].

3. *Yersinia* in fermented sausages

The genus *Yersinia* belongs to the family *Enterobacteriacae*. In this genus, 3 of the 12 species are pathogens, among them *Y. enterocolitica*, and 6 biotypes are recognised (5 pathogenic and 1 non-pathogenic). This Gram negative, facultative anaerobe grows in a wide range of temperatures: optimal 28-29°C, minimum -2°C [15]. There have been reported data that *Yersinia* grows at temperatures from -2 to 42°C [16,17] and even at temperatures of below 0°C [18]. It grows in a wide pH range, 4.2-10 and at a minimum a_w of 0.96 [15].

It is widespread in nature and can be found in various types of foods, as evidenced by the fact that yersiniosis in humans has been caused by the consumption of milk and milk products, soy products, various types of salads, pork meat products and various ready-to-eat products [19]. After *Salmonella* and *Campylobacter*, *Y. enterocolitica* occupies third place among the biological pathogens which cause foodborne disease in humans. The number of cases in the European Union in 2011 was 7,017. Probably there were more cases but some were likely not registered (not reported, the cause was not proven). Due to the importance of *Yersinia* in meat, the European Food Safety Authority (EFSA) recommended that in addition to controlling the presence of *Salmonella*, it is mandatory to examine for the presence of *Y. enterocolitica* in pig carcasses [20]. As *Y. enterocolitica* can frequently be found in pig meat, there is also an understandable interest in its presence in fermented sausages, since they are not thermally treated, and are ready-to-eat.

The number of *Y. enterocolitica* during fermentation of sausages was studied earlier [21,22]. The sausage filling was contaminated with *Y. enterocolitica*, so the number of these bacteria in the filling was 6.17 log CFU/g. The filling was filled into smaller (36 mm) diameter or bigger (50 mm) diameter casings, with subgroups of sausages with or without added starter cultures for both groups. On day 7, the number of *Y. enterocolitica* in smaller diameter sausages, without a starter culture, decreased by 0.87 log CFU/g, and in smaller diameter sausages with added starter culture, decreased by 1.18 log CFU/g from the starting number. In sausages with the bigger diameter, without added starter culture, the number *Y. enterocolitica* on day 18 of ripening decreased by 1.87 log CFU/g, and in sausages with added starter cultures, decreased by 2.10 log CFU/g, from the starting number. In sausages of a smaller diameter, both with and without added starter cultures, *Y. enterocolitica* was not detected on day 12 of ripening, and in sausages with a bigger diameter, on day 25 of ripening.

The authors [21,22] explained the negative finding of *Y. enterocolitica* in sausage by the fact that the number of lactic acid bacteria in the sausages of both diameters was close to or above 9 log CFU/g, that sausage pHs were below 5.5, and the a_w was 0.93 and 0.95 in smaller and bigger diameter sausages, respectively. In several studies, the possibility of inactivating *Y. enterocolitica* in fermented sausages was examined, whereby the reason for the inactivation was considered the presence of nitrites, starter cultures, pH values, and spice mixtures.

In Turkish fermented sausage, *L. sakei* was used as a starter culture, and *Y. enterocolitica* (about 5 log CFU/g) was inoculated into the sausages [23]. After only three days, the pH was reduced from 6.3 to 4.7 and *Y. enterocolitica* was not found in sausage filling. On the contrary, in the case of sausages without added starter culture (final pH 5.6), *Y. enterocolitica* was confirmed on days 4 and 12 of ripening.

An even more efficient reduction in the number of *Y. enterocolitica* was determined by using *P. acidilactici* starter culture [24]. The number of *Y. enterocolitica* was reduced from 5.1 log CFU/g (day 0) to 1.4 log CFU/g (day 1), so that on days 2 and 3, the number of these bacteria was below 1.0 log CFU/g. After day 4, the presence of *Y. enterocolitica* was not detected in the sujuk sausages. From these results, it can be concluded that the number *L. sakei* or *P. acidilactici* in sujuk sausages completely eliminated *Y. enterocolitica* after just four days of fermentation. The rapid decrease in the number of *Y. enterocolitica* is mostly due to the low pH, i.e. the presence of lactic acid. The presence of salt, nitrite and spice also contributes to the reduction in the number of *Y. enterocolitis* in sujuk sausage [25].

Similar results were obtained by other authors [26]. However, a short period of ripening followed by storage at 8°C could not inactivate *Y. enterocolitica* in sausages. Reduction of *Y. enterocolitica* numbers in sausages is made possible by increasing the temperature and length of ripening [27]. Therefore, *Y. enterocolitica* presents a real danger to human health in fermented sausage spread, where the fermentation time is short, and an extension of sustainability requires a lower storage temperature. In the literature, there are data on the influence of different levels of nitrites and different starter cultures on *Y. enterocolitica* [28], and the activity of *L. sakei*, pH and lactic acid on *Y. enterocolitica* [29].

In traditional dry sausage production, fermentation based on the initial microbiota [9] showed that the number of these microorganisms in the filling of French fermented sausages was, on average, 4.0 log CFU/g, and among them, in addition to the microorganisms useful for fermentation and development of organoleptic properties of sausages (lactic acid bacteria, coagulase negative cocci, yeast and mould), there were also spoilage bacteria (*Pseudomonas* spp., *Enterobacteriaceae*) and enterococci.

The type of microbiota that will develop, dominate or survive in traditional sausages depends on the production recipe, as well as the conditions of fermentation and ripening, and these conditions can be very different in terms of temperature, duration and relative humidity. *Enterobacteriaceae* are common meat contaminants and, therefore, can be found in filling and sausages in numbers that depend on the initial loading of raw materials, sausage type and phase of ripening. These microorganisms usually are eliminated from fermented sausages during ripening as a result of the combined effects of low pH, low temperatures and low a_w values.

The presence of nitrite and lactic acid, as well as other metabolites produced under the influence of starter cultures, further emphasise this phenomenon. Lactic acid bacteria constitute the dominant group of microorganisms after 28 days of ripening, which could be explained by the good adaptation of these microorganisms to the given environment [30,31]. These authors, like others, stated that during the ripening of fermented sausages, the number of all bacterial species decreased except for lactic acid bacteria; these, by producing organic acids and antibacterial substances, had a suppressive effect on Gram negative microorganisms and *Y. enterocolitica* [6,21,32,33,34,35].

In sausages with the addition of an autochthonous starter culture, the number of bacteria was 3.1 log CFU/g on day 0, 4.6 log CFU/g after the fermentation process and 2.0 log CFU/g on day 50 of production. In control sausages, the number of lactic acid bacteria was 3.0 log CFU/g on day 0 day, 5.1 log CFU/g after the fermentation process and 2.8 log CFU/g on day 50 [10]. According to our results for *Enterobacteriaceae* [21], these organisms were not detected in control or experimental sausages during later stages of fermentation (in smaller diameter sausages after day 12, and in bigger diameter sausages after day 25).

4. Conclusion

The aim of the fermentation of raw sausages is to obtain safer products, that is, products which do not contain foodborne pathogens which could cause disease, among which is *Y. enterocolitica*. In fermented sausages that are produced without thermal treatment, during ripening and drying, different parameters are used which could stop/slow the growth of pathogenic bacteria are used, due to the simultaneous action of several factors such as pH decrease, presence of lactic acid, decreases in a_w, inhibitory effect of smoke (if the sausages are smoked), presence of starter culture and action of metabolites (bacteriocins), etc.

Inactivation of pathogenic bacteria during the ripening of sausages involves control of their growth and it is a key step in the production of fermented sausages. Reduction of their number ends or is insufficient if the ripening process does not last long enough or is not optimal (incorrect starter culture, temperature, humidity, circulation), which is why this process should be controlled.

References

[1] Anon EFSA 2006b The Community Summary Report on Trends and Sources of Zoonoses, Zoonotic Agents, Antimicrobial Resistance and Foodborne Outbreaks in the European Union in 2005 *The EFSA Journal* **94** 3–288
[2] Radetić Petar 1997 *Raw sausages* monographs (author)
[3] Vuković I 2012 *Meat technology* (VKS) Belgrade
[4] Barbuti S and G. Parolari 2002 Validation of manufacturing process to control pathogenic bacteria in typical dry fermented products *Meat Science* **62** 323–329
[5] Baltic Z M, Pecanac B, Saric M, Mandic S, Filipovic I, Djuric J and Dojcinovic S. 2011ermented sausage – products with tradition *Veterinarski zurnal Republike Srpske* **15**-11.
[6] Comi G, Urso R, Iacumin L, Rantsio K, Cattaneo P, Cantoni C and Cocolin L 2005 Characterisation of naturally fermented sausages produced in the North East of Italy *Meat Science* **69** 381–392
[7] Heinz G and Hautzinger P 2007 *Meat processing technology for small- to medium-scale producers* (Food and Agriculture Organization of the United Nations Regional office for Asia and Pacific RAP Publication) – 2007/20 Bangkok
[8] Talon R, I Lebert, A Lebert, S Leroy, M Garriga, T Aymerich, EDrosinos, E Zanardi, A. Ianieri, MFraqueza, L Patarata and A Lauková 2007/a Traditional dry fermented sausages

produced in small-scale processing units in Mediterranean countries and Slovakia 1: Microbial ecosystems of processing environments *Meat Science* **77** 570–579

[9] Talon R, S Leroy, I Lebert 2007/b Microbial ecosystems of traditional fermented meat products The importance of indigenous starters *Meat Science* **77** 55-62

[10] Talon R, S Leroy, I Lebert, P Giammarinaro, J Chacornac, M Latorre-Moratalla, C Vidal-Carou, E Zanardi, M Conter, A Lebecque 2008 Safety improvement and preservation of typical sensory qualities of traditional dry fermented sausages using autochthonous starter cultures *International Journal of Food Microbiology* **126** 227–234

[11] Lebert I, S Leroy, P Giammarinaro, A Lebert, J Chacornac, S. Bover-Cid, M Vidal-carou, R. Talon 2007 Diversity of microorganisms in the environment and dry fermented sausages of small traditional French processing unit *Meat Science* **76** 112–122

[12] Leroy F, Verluyten J, De Vuyst L 2006 Functional meat starter cultures for improved sausage fermentation *Int J Food Microbiol* 10 **63** 270-285

[13] Tregear A, SKuznesof, A Moxey 1998 Policy initiatives for regional foods: some insights from consumer research *Food Policy* **23** 383–394

[14] Baltić Ž M, Baltić T, Mitrović R, Mitrović-Stanivuk M, Popović Lj. 2009 Banijska kobasica – proizvod sa tradicijom. 55th International Meat Industry Conference 15th – 17th of June Tara 66-68

[15] Anon EFSA Panel on Biological Hazards 2014 Scientific Opinion on the Public Health Risks Related to the Maintenance of the Cold Chain during Storage and Transport of Meat: Part 1, Meat of Domestic Ungulates *EFSA Journal* 12 **3** p 3601

[16] Bercovier H and Mollaret HH 1984 Genus XIV *Yersinia* in Krieg N R, Bergeys manual of systematic bacteriology vol. 1 Baltimore (Md Williams &Wilkins) 498-506

[17] Bottone E J 1997 *Yersinia enterocolitica*: the charisma continues *Clinical Microbiology Reviews* 10 **2** 257–76

[18] Lee W H M E Harris and D McClain 1980 Two modified selenite media for the recovery of *Yersinia enterocolitica* from meats *Applied and Environmental Microbiology* 39 **1**205–209

[19] Zadernowska A, Chajęcka-Wierzchowska W Łaniewska-Trokenheim 2014 *Yersinia enterocolitica*: a dangerous, but often ignored, foodborne pathogen *Food Reviews International* 30 **1** 53-70

[20] EFSA 2012 Scientific Opinion on animal health risk mitigation treatments as regards imports of animal casings *EFSA Journal* 107 p 2820

[21] Mitrović R 2016 *Testing the possibilities of inactivation of Yersinia enterocolitica in fermented sausages* PhD (Faculty of Veterinary Medicine, University of Belgrade)

[22] Ivanovic J, Janjic J, Boskovic M, Baltic, M, Dokmanovic M, Djordjevic V and Glamoclija N. 2014 Survival *Yersinia enterocolitica* in ground pork meat in different package *J Pure Appl Microbiol* 8 4317-23

[23] Ceylan E and Fung D 2000 Destruction of *Yersinia enterocolitica* by *Lactobacillus sake* and *Pediococcus acidilactici* During Low-temperature Fermentation of Turkish Dry Sausage (sucuk) *J. Food Sci.* 65**5** 876-79

[24] Enver BB, Gurhan C, Funda YE, Hasret Y,Osman Y, Gulay MB, Gulcan D 2014 Effect of starter cultures combinations on lipolytic activity and ripening of dry fermented sausages *Ital J Anim Sci* 13 776-781

[25] Ionescu A, Zara M, Aprodu I, Vasile A, Istrate R 2007 The effect of starter cultures on the physical and biochemical characteristics of dried sausages Scientific Study & Research **VIII** (3) 329-342

[26] Kumar R and Munstedt H 2005 Silver ion release from antimicrobial polyamide/silver composites *Biomaterials* 26 2081-88

[27] Lindqvist R and Lindblad M 2009 Inactivation of *Escherichia coli, Listeria monocytogenes* and *Yersinia enterocolitica* in fermented sausages during maturation/storage *Int J Food Microbiol.* 129 1 59-67

[28] Gomolka-Pawlicka M and Uradzinski J 2003 Antagonistic effect of chosen lactic acid bacteria strains on *Yersinia enterocolitica* species in model set-up, meat and fermented sausages *Pol J Vet Sci* 6 99-108

[29] Janssen M, Geeraerd A H, Logist F, DE Visscher Y, Vereecken K M, Debevere J and Van Impe J F 2006 Modelling *Yersinia enterocolitica* in activation in coculture experiments with *Lactobacillus sakei* as based on pH and lactic acid profiles *Int J Food Microbiol.* 11**11** 59-72

[30] Incze K 1998 Dry Fermented Sausages *Meat Science* 49 169-177

[31] Rubio B, B Martínez, M Sanchez M D Garcia-Cachan, J Rovira and I Jaime 2007 Study of the shelf life of a dry fermented sausage "salchichon" made from raw material enriched in monounsaturated and polyunsaturated fatty acids and stored under modified atmospheres *Meat Science* 76 128–37

[32] Ivanovic J, Mitrovic R, Janjic J, Boskovic M, Djordjevic V, Djordjevic J, Baltic T and Baltic M 2015 Inactivation of *Yersinia enterocolitica* in Fermented Sausages during Fermentation *Journal of Agricultural Science and Technology* 5 626-34

[33] Coppola S, G Mauriello M Aponte, G Moschetti and F Villani 2000 Microbial succession during ripening of Naples-type salami, a southern Italian fermented sausage *Meat Science* 56 321–29

[34] Hwang C A, ACS Porto-Fett, V K Juneja, SC Ingham, B Ingham and J B Luchansky 2009 Modeling the survival of *Escherichia coli* O157:H7, *Listeria monocytogenes* and *Salmonella* Typhimurium during fermentation, drying, and storage of soudjouk-style fermented sausage *International Journal of Food Microbiology* 129 244–52

[35] Ivanović J, Baltić M Ž, Karabasil N, Dimitrijević M, Antić N, Janjić J and Đorđević J 2013 Investigation of the microbiological contamination of contact surfaces in meat processing facilities *Tehnologija mesa* 54 **2** 110-16

IOP Conf. Series: Earth and Environmental Science **85** (2017) 012028 doi:10.1088/1755-1315/85/1/012028

Prevalence and antibiotic resistance of *Salmonella* spp. in meat products, meat preparations and minced meat

M Rašeta[1], Mrdović B[1], V Janković[1], Z Bečkei[2], B Lakićević[1], D Vidanović[3] and V Polaček[4]

[1] Institute of Meat Hygiene and Technology, 13 Kaćanskog, Belgrade 11000, Serbia
[2] Faculty of Veterinary Medicine, Bulevar Oslobođenja 18, Belgrade 11000, Serbia
[3] Veterinary Specialist Institute "Kraljevo", Žička 34, 36000 Kraljevo, Serbia
[4] Scientific Veterinary Institute "Novi Sad", Rumenački put 20, 21000 Novi Sad, Serbia

E-mail: mladen.raseta@inmes.rs

Abstract. This study aimed to determine *Salmonella* spp. prevalence in meat products, meat preparations and minced meat. Over a period of three years, a total of 300 samples were taken (100 RTE meat products, 100 meat preparations and 100 minced meat) and examined for the presence of *Salmonella* spp. Sampling was carried out at the warehouses of the food manufacturers. *Salmonella* spp. were not detected in RTE meat products, while 7% of semi-finished meat products (fresh sausages, grill meat formed and unformed) contained *Salmonella*, as did 18% of minced meats (minced pork II category, minced beef II category, mixed minced meat). The 25 *Salmonella* isolates obtained were examined for antibiotic resistance by the disk diffusion test, according to the NCCLS and CLSI guidelines. Isolates showed resistance to ampicillin and nalidixic acid (80%), tetracycline (72%), cefotaxime/clavulanic acid (48%), but not to gentamicin (8%) or trimethoprim/sulfamethoxazole (0%).

1. Introduction

Salmonella is an important cause of foodborne disease in humans throughout the world and is a significant cause of morbidity, mortality and economic losses. *Salmonella* problems can occur in all segments of the food chain [1]. According to the European Food Safety Authority (EFSA) report on trends and sources of zoonoses, zoonotic agents and food-borne outbreaks in 2015, a total of 94,625 confirmed human salmonellosis cases were reported by 28 European Union (EU) member states, resulting in an EU notification rate of 21.2 cases per 100,000 population [2]. This was a 1.9% increase in the EU notification rate compared with 2014. There was a statistically significant decreasing trend of salmonellosis in the 8-year period between 2008 and 2015 [2]. Meat during production, preparation and in retail comes into direct contact with microorganisms (different types and strains) which affect its shelf life and safety [3]. Foodborne pathogens are major causes of human illnesses in developing countries, causing high medical and hospitalization expenses [4]. A large number of foods, particularly meat and broiler meat products, are the most important sources of human *Salmonella* contamination [5].

In recent years, an additional health problem is the emergence of multi-resistant strains of pathogenic bacteria, including *Salmonella* and especially in foods [6]. Extensive and intensive use of antibiotics for preventive and therapeutic purposes in veterinary medicine, as well as growth promoters in animal feedstuffs, contributed to the emergence of resistant bacteria, including zoonotic pathogens, in animals, and that can be transmitted in the food chain to humans [7]. *Salmonella* spp. possess highly efficient mobile genome parts (plasmids, genomic islands, transposons) with ability to exchange and keep different genes responsible for antimicrobial resistance [8,9]. Almost every day, new mutations in genes responsible for antimicrobial resistance are happening, induced by antibiotics used to combat *Salmonella* spp. [10]. EFSA and the European Centre for Disease Prevention and Control recommended antibiotics of interest for determining *Salmonella* spp. resistance: ampicillin, cefotaxime, chloramphenicol, gentamicin, nalidixic acid, sulphonamides and tetracycline [11].

The aim of this study was to determine the prevalence and antibiotic resistance of *Salmonella* spp. in retail-ready RTE meat products, and in meat preparations and minced meat (these latter two categories were not RTE).

2. Materials and Methods

In order to determine the presence of Salmonella spp. in RTE meat products, and in non-RTE meat preparations and minced meat, during a three-year period, 300 samples were taken in producers' warehouses. The meats were originally-packed products, taken from the batches intended for transport to retail. The meats were transported in a cold chain and delivered to the laboratory within a day. Determination of the Salmonella prevalence in the meats was performed in accordance with ISO standard [25].

Samples were suspended with 250 ml of buffered peptone water, homogenized for 30 seconds, and after that, were incubated at 37°C for 24 h. After pre-enrichment and incubation, 0.1 ml of the slurry was transferred into 10 ml Rappaport-Vassiliadis medium (bioMérieux, France), which was then incubated at 42°C, for a period of 24-48 h. After incubation, enrichment cultures were seeded onto differential Rambach and XLD agars, which were incubated overnight at 37°C. Colonies with typical growth and clearly differentiated were transferred into cryogenic vials for further testing.

Antibiotic resistance testing was performed by disk diffusion according to NCCLS recommendations using commercial discs and Mueller-Hinton agar (Bio-Rad, USA). The investigated isolates were first subcultured on trypticase soy agar (TSA) and incubated for 24 h at 37°C. Typical *Salmonella* colonies were suspended in physiological saline to 0.5 McFarland standard density. *Salmonella* suspensions were transferred by sterile swabs onto Mueller-Hinton agar, followed by antibiotic disk application (automatic applicator, Oxoid, UK). The following antibiotic disks (Oxoid Ltd., Basingstoke, UK) were used: nalidixic acid (quinolones) 30 μg, ampicillin (penicillin) 10 μg, cefotaxime/clavulanic acid (cephalosporins) 30 μg, gentamicin (aminoglycoside antibiotics) 10 μg, tetracycline (tetracycline) 30 μg, trimethoprim/sulfamethoxazole (inhibitors of folic acid) 30 μg. After 18 h of incubation, inhibition of *Salmonella* growth was measured, and the results were interpreted according to NCCLS (National Committee for Clinical Laboratory Standards) or CLSI (Clinical Laboratory Standard Institute) 2006 recommendations as sensitive, intermediate sensitive and resistant (Table 1).

Table 1. Limits of *Salmonella* growth inhibition for determining antimicrobial resistance in accordance with CLSI recommendations

Antibiotics	Inhibition growth zone (mm)		
	Resistant	Intermediate sensitivity	Sensitive
Ampicillin 10 μg	≤ 13	14 – 16	≥ 17

Nalidixic acid 30 µg	≤ 13	14 – 18	≥ 19
Cefotaxime 30 µg	≤ 22	23 – 25	≥ 26
Gentamicin 10 µg	≤ 12	13 – 14	≥ 15
Tetracycline 30 µg	≤ 11	12 – 14	≥ 15
Trimethoprim/Sulfamethoxazole 30 µg	≤ 10	11 – 15	≥ 16

3. Results

Results of *Salmonella* spp. presence in the meats examined are presented in Table 2.

Table 2. *Salmonella* spp. presence in RTE meat products, meat preparations and minced meat

Meat type	No. of samples	*Salmonella* spp. positive samples	
		Number	%
RTE meat products	100	0	0
Meat preparations	100	7	7
Minced meat	100	18	18
Total	*300*	*25*	*8.33*

Antibiotic resistance results of *Salmonella* spp. isolates from the meat examined are presented in Table 3.

Table 3. Antibiotic resistance of *Salmonella* isolates from meat preparations and minced meat

Antibiotic	No. of Isolates	Sensitive		Intermediate		Resistant	
		Number	%	Number	%	Numbe r	%
Nalidixic acid	25	0	0	5	20	20	80
Ampicillin	25	2	8	3	12	20	80
Tetracycline	25	2	8	5	20	18	72
Cefotaxime	25	7	28	6	24	12	48
Gentamicin	25	18	72	5	20	2	8
Trimethoprim/Sulfamethoxaz ole	25	20	80	5	20	0	0

4. Discussion

RTE foods pose a direct risk to consumers, and according to an EFSA annual report[2], in 2015, 1.1% and 0.7% positive samples were found for RTE food from broilers and pig meat, respectively, whereas one positive sample and no positive samples were found for RTE food from turkey and cattle meat, respectively. In our study, there were no positive samples for Salmonella spp. in RTE meat products (boiled sausages, cooked sausages, pâtés) which were thermally processed and originally packed. The results obtained indicate that any Salmonella spp. presence in RTE meats could be more related to inappropriate conditions or use of RTE food in retail (secondary contamination, disruption of cold chain during storage and contamination after package is opened). Similar results to those in our study were found in Latvia [12]. There, a total of 3,152 samples of raw and RTE meats were collected during the official control and in-house control procedures in 2015. The prevalence of Salmonella was

0.8% (25/3152). The highest prevalence (1.5%) of Salmonella was found in minced meat and meat preparations (7/481), while the lowest (0%) was in frozen meat, meat preparations (0/349) and RTE meats (0/364) [12]. In a ten year study in the United States (US), the Food Safety and Inspection Service conducted microbiological testing programs for RTE meat and poultry products produced at approximately 1,800 federally inspected establishments [28]. The cumulative 10-year Salmonella prevalences were as follows: jerky, 0.31%; cooked, uncured poultry products, 0.10%; large-diameter cooked sausages, 0.07%; small-diameter cooked sausages, 0.20%; cooked beef, roast beef, and cooked corned beef, 0.22%; salads, spreads, and pâtés, 0.05%; and sliced ham and luncheon meat, 0.22%. The cumulative 3-year Salmonella prevalence for dry and semidry fermented sausages was 1.43%. [28]. The prevalence data have certain limitations that restrict statistical interpretations, because these RTE product-testing programs are strictly regulatory in nature and not statistically designed.

In the EU, the highest occurrence of samples not-compliant with *Salmonella* criteria was found in foods of meat origin which are intended to be cooked before consumption [2]. Among these foods, 'minced meat and meat preparations from poultry' had a notable level of noncompliance (6.8% of single samples and 5.1% of batches) [2]. A study in the US in the period 2005-2007 determined an overall *Salmonella* prevalence of 4.2% in minced beef meat [14]. Enumeration showed that 94.2% of these contained *Salmonella* levels below 2 CFU/g. Regional monthly prevalences of *Salmonella* varied from 1.8% to 6.5% but were not statistically different [14]. The results obtained in our study show a similar *Salmonella* prevalence in meat preparations (7% in semi-finished meat preparations), but in minced meat, our *Salmonella* prevalence was significantly higher (18%) compared to other studies [12,14,26]. In a similar study in Belgium, *Salmonella* prevalences in minced meat at retail level ranged from 0.3% to 4.3% [26]. In an investigation in Poland [29], significantly lower prevalences of *Salmonella* spp. in meat preparations were determined (0.4-0.7% in porcine meat preparations and fresh sausages) than in our study.

The presence of *Salmonella* spp. in meat preparations and minced meat poses a risk to human health. The in-laboratory testing for *Salmonella* using reference method EN ISO 6579:2008 [25], lasts 4-5 days, so there is not enough time to prevent exposure of consumers to contaminated meat preparations or minced meat, if these are sold soon after production. Although meat preparations and minced meat are intended to use after thermal processing, and *Salmonella* is thermally sensitive, the presence of *Salmonella* spp. is considered a food safety problem, as is stated in Serbian legislation [27].

The presence of *Salmonella* spp. in minced meat, meat preparations and meat products is related to the origin of meat used in production (epizootic situation, primary production, slaughterline, cutting, cold storage, hygiene practice of employees). *Salmonella* contamination in the food chain was examined in Brazil using a meta-analysis model, and Monte Carlo simulation estimated the *Salmonella* prevalence in beef cuts from processing plants was ~6.1% (95% probability) [13]. This was in reasonable agreement with a pool (n = 105) of survey data for *Salmonella* prevalence in Brazilian beef cuts (~4.9%; 95% probability) carried out in commercial establishments. The results not only underscored the significant increase in *Salmonella* prevalence that can occur during evisceration/splitting and boning but also reinforced that, when hygienic slaughter procedures are properly implemented, the load of *Salmonella* can be reduced at dehiding, rinsing and chilling [13]. A slightly better situation was determined in Belgium [15]. A constant and significant decrease in *Salmonella* prevalence was observed for pork carcasses, trimmings, and minced meat and for beef minced meat. Less than 3% of beef carcasses and trimming samples were positive for *Salmonella* spp. From 1997 to 1999, the prevalence of *Salmonella* spp. was assessed at different stages through the pork, poultry, and beef meat production chains, and initial prevalences were 20 to 26% [15]. Based on this introductory study, a new sampling plan was used from 2000 to 2003. This new plan was suitable for monitoring zoonoses, because it was representative of nationwide production processes, covered all periods of the year, was executed by trained samplers and the analyses were carried out by recognized laboratories using an identical analytical method [15].

In our study, *Salmonella* isolates were sensitive (at intermediate or sensitive levels) to trimethoprim/sulfamethoxazole (100%) or gentamicin (92%), were sensitive to a lesser degree to cefotaxime (52% of isolates were sensitive) but very low percentages of our isolates were sensitive to tetracycline (28%), ampicillin (20%) or nalidixic acid (20%). Similar results were gained by Wang et. al [16] when they determined their *Salmonella* isolates had low ampicillin sensitivity (21.7%). In contrast, Nogrady et. al [17] determined *Salmonella* isolates originating from primary production in Hungary were largely resistant to sulphonamides [17]. In Italy during 2005-2006, *Salmonella* isolates from primary production and retail showed multiresistance to ampicillin, chloramphenicol, streptomycin, sulphonamides, tetracycline, kanamycin and trimethoprim/sulfamethoxazole [18]. In Poland during 2008-2012, 106 *Salmonella* isolates were sensitive to nalidixic acid (47%), tetracycline (68%), ampicillin (72%), sulphonamides (74%) and cefotaxime (100%) [19]. Finally, in Thailand *Salmonella* isolates were sensitive to tetracycline (27%), nalidixic acid (46%), gentamicin (64%), ampicillin (73%) and trimethoprim/sulfamethoxazole (73%) [20].

Salmonella presence in primary production is a significant public health problem, particularly in countries without adequate control measures or in the areas where climate favours *Salmonella* survival and growth [21]. Biosecurity measures in primary production are a line of well-designed obstacles in order to prevent *Salmonella* contamination and spread [22,23]. Continual education of employees is the basis for implementing biosecurity measures, as is stated in primary production biosecurity protocols [24].

5. Conclusion

This study shows that the presence of Salmonella spp. in meat preparations and in minced meat, together with the high prevalence of antibiotic-resistant strains, is a significant public health issue in Serbia (18% of minced meat and 7% of meat preparations contained the organism). Salmonella spp. were not detected in RTE meat products.

The presence of *Salmonella* spp. in minced meat and meat preparations is a safety issue, although these types of foods are intended to be used after thermal processing (as stated on the food declarations).

"*Salmonella* free" status, under current production conditions for Serbian fresh meat, meat preparations and minced meat, is most likely unachievable. Therefore, *Salmonella* prevalence in meat production at different production stages in the food chain must be determined and monitored under the *Salmonella* National Control Program.

The testing time using the reference method for *Salmonella* determination [25] lasts longer (4-5 days) than the shelf life of minced meat or meat preparations (usually 48-72 h), so there is not enough time to take any corrective measures, in order to prevent exposure of consumers to potentially harmful food. Therefore, other analytical methods to determine *Salmonella*, which provide results in a shorter time and with similar levels of reliability as the reference method, must be applied. This would enable appropriate corrective measures to be taken, and should result in less risk to consumers' health.

Acknowledgment

This study was supported by project No. TR 46009 III funded by the Ministry of Education, Science and Technological Development of the Republic of Serbia.

References

[1] Conchello L 2011 *Salmonella* control throughout the Poultry Feed Chain http://www.poultryworld.net/Special-Focus/*Salmonella*-special/*Salmonella*-control-throughout-the-Poultry-Feed-Chain/ (Accessed on 11th July 2017)

[2] EFSA 2016 The European Union Summary Report on Trends and Sources of Zoonoses, Zoonotic Agents and Food-borne Outbreaks in 2015, EFSA Journal 2016 **14** (12) : 4634 231

[3] Dimitrijević D 2000 Sanitation procedures in meat production and processing *Tehnologija mesa* **41** 1–3 39– 48

[4] Fratamico P M, Bhunia A K, Smith J L 2005 Foodborne Pathogens in *Microbiology and Molecular Biology* Caister Academic Press Wymondham Norfolk UK 273

[5] Chittick P, Sulka A, Tauxe RV, Fry A M, 2006 A summary of national reports of foodborne outbreaks of Salmonella Heidelberg infections in the United States: clues for disease prevention *J. Food Protect.* **69** 1150–53

[6] Busani L, Graziani C, Battisti A, Franco A, Ricci A, Vio D 2004 Antibiotic resistance in *Salmonella* enterica serotypes typhimurium, enteritidis and infantis from human infections, foodstuffs and farm animals in Italy *Epidemiol. Infect.* **132** 245–51

[7] Su L H, Chiu C H, Chu C, Ou J T, 2004 Antimicrobial resistance in nontyphoidal *Salmonella* serotypes: A global challenge *Clin. Infect. Dis.* **39** 546–51

[8] Guerra B, Junker E, Miko A, Helmuth R, Mendoza M C 2004 Characterization and localization of drug resistance determinants in multidrug resistant integron-carrying *Salmonella* enterica serotype typhimurium strains *Microb. Drug Resist.* **10** 1 83-91

[9] Velebit B, Lilić S, Borović B, 2010 Antimicrobial resistance of *Salmonella* spp isolates collected from cattle carcasses *Tehnologija mesa* **51** 2 154 -58

[10] Velhner M, Potkonjak D, Stojanović D, Mitevski D, Stojanov I, Petrović J 2013 Resistance to antimicrobial drugs and control measures of *Salmonella* spp. in the poultry industry *Veterinarski Glasnik* **67** 1-2 87-95

[11] EFSA 2010 The European Union Summary Report on Trends and Sources of Zoonoses, Zoonotic Agents and Food-borne Outbreaks in 2010 *The EFSA Journal* 2597 442

[12] Terentjeva M, Avsejenko J, Streikiša M, Utināne A, Kovaļenko K, Bērziņš A 2017 Prevalence and antimicrobial resistance of *Salmonella* in meat and meat products in Latvia *Ann. Agric. Environ. Med.* **24** 2 317-21

[13] Gonzales-Barrón U, Piza L, Xavier C, Costa E, Cadavez V 2016 An exposure assessment model of the prevalence of *Salmonella* spp. along the processing stages of Brazilian beef *Food Sci. Technol. Int.* **22** 1 10-20

[14] Bosilevac J M, Guerini M N, Kalchayanand N, Koohmaraie M 2009 Prevalence and characterization of salmonellae in commercial ground beef in the United States *Appl. Environ. Microb.* **75** 7 1892-900

[15] Ghafir Y, China B, Korsak N, Dierick K, Collard J M, Godard C, De Zutter L, Daube G 2005 Belgian surveillance plans to assess changes in *Salmonella* prevalence in meat at different production stages *J. Food Protect.* **68** 11 2269-77

[16] Wang H, Ye K, Wei X, Cao J, Xu X, Zhou G 2013 Occurrence, antimicrobial resistance and biofilm formation of *Salmonella* isolates from a chicken slaughter plant in China *Food Control* **33** 378–84

[17] Nógrády N, Kardos G, Bistyák A, Turcsányi I, Mészáros J, Galántai Z, Juhász A, Samu P, Kaszanyitzky J E, Pászti J, Kiss I 2008 Prevalence and characterization of *Salmonella* infantis isolates originating from different points of the broiler chicken-human food chain in Hungary *Int. J. Food Microbiol.* **127** 1-2 162–67

[18] Dionisi A M, Lucarelli C, Benedetti I, Owczarek S, Luzzi I 2011 Molecular characterisation of multidrug-resistant *Salmonella enteric*a serotype Infantis from humans, animals and the environment in Italy *Int. J. Antimicrob. Ag.* **38** 5 384–89

[19] Łukasz M Maćkiwa E, Ścieżyńskaa H, Pawłowskaa K, Popowska M 2013 Antimicrobial susceptibility of *Salmonella* strains isolated from retail meat products in Poland between 2008 and 2012 *Food Control* **36** 199–204

[20] Akbar A, Kumar Anal A 2013 Prevalence and antibiogram study of *Salmonella* and *Staphylococcus aureus* in poultry meat *Asian Pac .J. Trop. Biomed.* **3** 2 163–68

[21] Barrow P A, Freitas Neto O C 2011 Pullorum disease and fowl typhoid new thoughts on old diseases: a review *Avian Pathol.* **40** 1–13

[22] Maslić-Strižak D, Spalević Lj, Resanović R 2012 Biosecurity measures in poultry industry *Živinarstvo* **46** 9/10 2–15

[23] Maslić-Strižak D, Spalević Lj, Rašeta M, Branković-Lazić I 2012 Breeding broiler chickens in industrial poultry production *Tehnologija mesa* **53** 1 1–7
[24] Racicot M, Venne D, Durivage A, Vaillancourt J P 2012 Evaluation of strategies to enhance biosecurity compliance on poultry farms in Quebec: Effect of audits and cameras *Prev. Vet. Med.* **103** 208–18
[25] EN ISO 6579: 2008 Microbiology of food and animal feeding stuffs – Horizontal method for the detection of *Salmonella* spp, Processes and Services International Organization for Standardization Geneva Switzerland
[26] Delhalle L, Saegerman C, Farnir F, Korsak N, Maes D, Messens W, De Sadeleer L, De Zutter L, Daube G 2009 *Salmonella* surveillance and control at post-harvest in the Belgian pork meat chain *Food Microbiol.* **26** 3 265–71
[27] Pravilnik o opštim i posebnim uslovima higijene hrane u bilo kojoj fazi prozivodnje, prerade i prometa *Official Gazette of RS* br 72/10)
[28] Levine P, Rose B, Green S, Ransom G, Hill W 2001 Pathogen testing of ready-to-eat meat and poultry products collected at federally inspected establishments in the United States, 1990 to 1999 *J. Food Protect.* **64** 8 1188–93
[29] Modzelewska-Kapituła M, Maj-Sobotka W 2014 *Salmonella* spp occurrence in Minced Meat, Meat Preparations and Mechanically Separated Meat in Poland *J. Food Safety* **34** 2 126–31

The most important parasites in Serbia involving the foodborne route of transmission

J M Petrović[1,3], J Z Prodanov-Radulović[1] and S D Vasilev[2]

[1]Scientific Veterinary Institute "Novi Sad", Rumenački put 20, Novi Sad, Serbia
[2]Institute for the Application of Nuclear Energy – INEP, University of Belgrade, Banatska 31b, Zemun, Serbia

E-mail: jelena@niv.ns.ac.rs

Abstract. Food can be an important route for transmission of parasites to humans. Compared to other foodborne pathogens in Serbia, foodborne (or potentially foodborne) parasites do not get the attention they undoubtedly deserve. The aim of this article is to give an overview of the most important parasitic pathogens that can be transmitted by food, and that cause disease in humans: *Echinococcus*, *Trichinella*, *Taenia solium* and *Toxoplasma gondii.* For each of these pathogens, the severity of human diseases they cause, incidence, mortality and case fatality rate among humans in Serbia as well as their prevalence in animal species in Serbia are described. Some of the described foodborne parasites can induce severe disease symptoms in humans associated with high case fatality rates, while others can cause massive outbreaks. All of the aforementioned parasites occur throughout Serbia and cause both severe public health problems and substantial economic losses in livestock production. In conclusion, the control measures of foodborne parasites certainly need to include education of farmers and improvement of veterinary sanitary measures in animal farming and animal waste control.

1. Introduction

Within the system of public health control, foodborne parasites do not attract the same level of attention as the bacterial and viral hazards. However, parasites can cause severe clinical symptoms, can have a prolonged disease course and sometimes have fatal outcomes. The primary sources of food contamination are animals and/or humans. Parasites do not replicate in food; however, during food processing, accidental contamination of large amounts of safe, parasite-free food can occur. This situation is commonly seen with *Trichinella*. Mixing of meat from one *Trichinella*-infected pig with meat that originates from healthy animals can result in several tens of kilograms of contaminated products, which can cause massive epidemics affecting hundreds of diseased individuals [1]. The transmission of parasites by food is commonly associated with faecal contamination and poor, inadequate hygiene. Although the prevalence of foodborne parasitic diseases remains unreported, it is considered that the global annual prevalence in humans of parasitic trematodes is 56.2 million disease cases of which 7.8 million manifest severe clinical sequelae, with a total of 7,158 fatal outcomes [2].

The complex epidemiology and life cycles of parasites play central roles in the identification, prevention and control of risks associated with foodborne parasitic diseases. Monitoring parasitic diseases is further complicated by their (often) prolonged incubation periods, subclinical course of infections and diagnostic failures in chronic sequelae [3].

The Food and Agriculture Organization of the United Nations (FAO) and World Health Organization (WHO) established a joint international expert group that reported on worldwide foodborne parasite hazards in 2014 [4]. In this report, 24 parasites important for food safety were highlighted, including (ordered by their importance): 1. *Taenia solium,* 2. *Echinococcus granulosus,* 3. *Echinococcus multilocularis,* 4. *Toxoplasma gondii,* 5. *Cryptosporidium spp.* Currently, within the framework of COST Action FA1408, "A European Network for Foodborne Parasites (Euro-FBP)", 25 parasites are ranked at pan-European and regional level.

The aim of this study is to give an overview of the most important foodborne parasitic pathogens which can cause human diseases in Serbia. So far, several tens of foodborne parasites which belong to the groups of protozoa, nematodes, trematodes and cestodes were identified in Serbia. The identified parasites can be transmitted by the meat of pigs, cattle, sheep and fish and/or by food of plant origin. In this article, four foodborne parasites were selected as the most important. Criteria for selection were: severity of diseases and potential fatal outcomes in humans, the mandatory reporting of human and/or animal diseases and availability of relevant data on their prevalence and incidence in animal and human populations in Serbia. The data were obtained from the official web-sites of authorized institutions such as the Institute of Public Health of Serbia, Dr Milan Jovanović Batut (hereafter called Institute Batut) and the Ministry of Agriculture and Environmental Protection, Veterinary Directorate for the period spanning the beginning of 2005 to the end of 2015, if such data were not available from relevant scientific literature.

2. Foodborne parasites

2.1. Echinococcus

The genus *Echinococcus* includes six species, and two of them are the most important from the human health point of view – *E. granulosus* which causes cystic echinococcosis and *E. multilocularis* which causes alveolar echinococcosis. To complete their life cycle, *Echinococcus* requires intermediate and definitive hosts. The definitive hosts – carnivores (most commonly dogs) – excrete, *via* their faeces, either gravid proglottids containing eggs or eggs. These are deposited directly into the environment. Eggs enter the intermediate host by ingestion. The predominant intermediate host species in Serbia are sheep, cattle and pigs as well as humans [5]. Eggs enter the intermediate host by ingestion. Larvae (metacestodes) and infectious elements (proscolex) develop in the bodies of intermediate hosts.

2.1.1. Human echinococcosis in Serbia. Basic criteria for categorising *Echinococcus granulosus* as an important foodborne parasite in Serbia are severity of human disease and case fatality rate.

Human infection can develop after egg ingestion *via* contaminated water, unwashed fruits and vegetables or *via* direct contact with dog hair. After ingestion, eggs release oncospheres that pass from the intestines mostly to the liver, where oncospheres develop into cysts. Besides liver (65%), the cysts can also develop in lungs (25%) and, less frequently, in the spleen, heart or kidneys. Infection is associated with development of single or multiple cysts, which range in size from the size of a walnut to the size of a child's head [6].

The incubation period commonly lasts from 2 to 15 years, and the clinical manifestation of the disease is determined by the size and location of the cyst. The cyst growth rate is usually 1 to 5 cm per year [6]. In Serbia, the incidence of the disease is higher in females (68.9% in childhood) than in males [7]. The infection can remain asymptomatic for a long period of time, which can result in the formation of cysts containing several litres of fluid full of infectious protoscoleces [6].

The disease often takes a chronic course, wherein clinical manifestations include liver enlargement, abdominal pain, and bile duct obstruction associated with jaundice and vomiting. A pulmonary location of cysts is characterized by severe pain in the chest associated with bloody sputum. Cyst rupture results in anaphylactic shock and death of the diseased person [8]. Cysts in the brain or eye, which are common cyst locations in children, produce clinical manifestations quite rapidly [6].

According to official records from the Institute Batut, in the period 2005 to 2016, three persons in Serbia died from echinococcosis. The mortality rate (number of deaths per 100,000 population) was not high, ranging from 0.01 to 0.05; however, the case fatality rate (number of deaths related to number of sick persons with the disease) was extremely high, ranging from 3.13 to 5.88% in hepatic echinococcosis, over 25.00% in *Echinococcosis alia non specificata* and as much as 33.33% in pulmonary echinococcosis. According to these data, pulmonary echinococcosis in Serbia has a fatal outcome in one third of the diseased patients.

Throughout the 11-year monitoring period (2005 to 2016), the average incidence of echinococcosis in Serbia was 0.96 cases per 100,000 population (Table 1). The incidence in Vojvodina was higher than in Central Serbia. The highest incidence was recorded in Vojvodina in 2014, reaching 1.26 cases per 100,000 population.

The first diagnosis of *Echinococcosis multilocularis* in beaver [9] in Serbia was established in 2015, which is of significant epidemiological importance. This species causes human alveolar echinococcosis, an extremely severe disease with poor prognosis. Intensive lateral asexual multiplication in the human host results in a significant risk for human health – the spread of foreign tissue is aggressive, like a malignant tumour and can progress from the liver to lungs, spleen and kidneys [10]. The presence of alveolar echinococcosis is confirmed in neighbouring countries; thus, the future occurrence of the disease in humans in Serbia is likely.

2.1.2. Animal echinococcosis in Serbia. Serbia, as well as other parts of the Balkans and the Mediterranean basin, is considered an endemic region for echinococcosis. According to research conducted in Serbia, the domestic dog is the predominant definitive host for *E. granulosus*, with an infection prevalence in dogs ranging from 15% in Valjevo to 48% in Požarevac [5].

Echinococcosis is a mandatory reportable disease for both humans and all animal species in Serbia. According to data from the Veterinary Directorate, during 2005 to 2016, examination on livestock slaughter lines revealed only 11 echinococcosis foci, with a total of 62 infected pigs. However, these official data do not reflect the true picture of the infection prevalence in animals. Several scientific publications reported the prevalence of echinococcosis in animals in Serbia. According to Debeljak [5], the prevalence of echinococcosis in Serbia ranges from 70.64 to 95.0% in older sheep categories, 5.76 to 7.27% in beef cattle, 56.95 to 94.61% in cattle and 2.41 to 34.8% in pigs.

The high prevalence of echinococcosis in animals is largely due to the lack of knowledge among Serbian livestock farmers and dog owners. Namely, slaughter of sheep and pigs in households for their own needs is mostly performed without veterinary-sanitary examination. The disposal of offals from intermediate host animals (especially liver that contains *Echinococcosis* cysts) is inadequate; they often end up as dog feed. Moreover, dog owners often do not practice dehelminthisation and the population of stray dogs is still not under control [5].

Table 1. Incidence in Serbia of human parasitic illness (per 100,000 population).

Year	*Toxoplasmosis (non specific)*			*Trichinellosis*			*Echinococcosis*		
	S[a]	CS[b]	Vojv[c]	S	CS	Vojv	S	CS	Vojv
2005	0.75	0.57	1.23	4.52	1.13	13.63	0.63	0.51	0.93
2006	0.92	0.90	0.99	2.53	1.66	4.87	0.81	0.45	0.7
2007	1.18	1.18	1.11	2.38	2.21	2.83	0.34	0.33	0.35
2008	1.06	1.04	1.10	1.23	0.67	2.76	0.49	0.43	0.7
2009	0.99	1.01	0.96	1.52	1.89	0.51	0.44	0.48	0.3
2010	1.04	1.08	0.91	1.52	1.89	0.51	0.38	0.24	0.76
2011	0.74	0.77	0.66	1.74	1.09	3.52	0.45	0.38	0.66
2012	0.86	0.89	0.78	0.64	0.70	0.47	0.54	0.61	0.47
2013	0.77	0.74	0.83	1.32	0.49	3.57	0.57	0.46	0.83

2014	1.06	0.99	1.26	1.20	0.95	1.88	0.63	0.58	0.78
2015	1.15	1.17	1.10	1.15	0.55	2.79	0.73	0.62	1.05
Average	0.96	0.94	0.99	1.80	1.20	3.39	0.55	0.46	0.68
SD[d]	0.16	0.19	0.19	1.05	0.61	3.68	0.15	0.12	0.23
Min	0.74	0.57	0.66	0.64	0.49	0.47	0.34	0.24	0.3
Max	1.18	1.18	1.26	4.52	2.21	13.63	0.81	0.62	1.05

a – Republic of Serbia
b – Central Serbia region
c – Vojvodina region
d – Standard deviation

2.2. *Trichinella*

To date, eight species belonging to the genus *Trichinella* and three still unclassified genotypes (T6, T8 and T9) have been identified [11]. The most important species in Serbia are *Trichinella spiralis* and *Trichinella britovi*. *Trichinella* are transmitted by consumption of meat containing infectious larvae, and susceptible animals include all mammal species, reptiles and birds [12].

2.2.1. Human trichinosis in Serbia. Basic criteria for categorising *Trichinella* as an important foodborne parasite in Serbia are the number of affected individuals, parasite distribution over the entire country in both humans and animals and the case fatality rate. Throughout the 11-year monitoring period (2005 to 2016), the average incidence in Serbia was 1.80 cases per 100,000 population (Table 1). The incidence in Vojvodina region was somewhat higher than in Central Serbia (Table 1). The highest incidence was recorded in 2005 in Vojvodina, being 13.63 cases per 100,000 population.

When comparing the incidence of rates for the three parasitoses monitored during 2005 to 2016, the incidence of trichinellosis was higher than that of echinococcosis and toxoplasmosis (Table 1). The average annual incidence of trichinellosis was 1.8 cases per 100,000 population. The highest incidence of trichinellosis (13.63 cases per 100,000 population; 2005) was far higher than that of the other two monitored parasitoses (echinococcosis – 1.26 cases per 100,000 population and toxoplasmosis – 0.81 cases per 100,000 population).

Human trichinellosis develops after ingestion of live *Trichinella* larvae, most commonly *via* the meat or meat products. Traditional smoked pork products are the main source of infection (76%). The occurrence of human trichinellosis in Serbia shows strong seasonality, and most of the cases occur during winter [13].

The disease in humans takes an acute course. The incubation period ranges between one and 33 days. The mean time between onset of symptoms and admission is nine days. Family outbreaks are the most frequent, rather than individual cases or large-scale outbreaks. Fever is the most frequent clinical manifestation (90%), followed by myalgia (80%) and periorbital oedema (76%) [13]. The clinical manifestations of disease are determined by the number of ingested larvae, but also by the immune status of the individual. Infrequently, the disease progresses to a chronic course and manifests as myalgia and fatigue. Complications from trichinellosis include myocarditis, endocarditis, encephalitis, and meningitis, all with an accompanying poor prognosis [14]. Fatal outcomes of trichinellosis occur in third to fifth week of infection, though not very frequently. According to official data, from 2005 to 2016, three fatal outcomes were recorded in Serbia in 2005. The mortality and case fatality rates were not high, being 0.02 to 0.10 and 0.72 to 1.61, respectively. However, the very existence of fatal outcomes in humans emphasizes the importance of this foodborne parasite.

2.2.2. Animal trichinellosis in Serbia. Serbia is considered an endemic area for trichinellosis. In Serbia, *Trichinella* has been reported in domestic pigs, horses, wild boars, foxes, jackals, raccoons, wolves, and bears [15, 16]. Domestic pigs are the predominant reservoir of human infection.

Trichinellosis is a mandatory reportable disease for both humans and all animal species in Serbia. According to official data from the Veterinary Directorate, during the monitored period (2005 to 2016), 2,307 trichinellosis foci were reported, and a total of 3,084 infected pigs. According to the literature data, the prevalence of trichinellosis in domestic pigs in Serbia is around 0.02%, but it can reach approximately 0.5% in endemic regions and about 1% in wild boars [17,18].

Endemic persistence of trichinellosis in Serbia is determined by a range of factors, including poor socioeconomic conditions, inadequate education of livestock farmers and food business operators, insufficient veterinary control and improper disposal of dead animals [16].

2.3. Taenia solium

Tapeworms *Taenia saginata* and *Taenia solium* are zoonotic parasites and both are present in Serbia. Humans are the definitive host for both *Taenia* species, and cattle and pigs are intermediate hosts for *T. saginata* and *T. solium*, respectively. Tapeworms can cause two diseases – taeniasis and cysticercosis.

Taeniasis is a parasitic disease of the definitive host, humans. The disease is induced by adult tapeworm (either *T. saginata* or *T. solium*) inhabiting the intestines, and is typically associated with mild clinical manifestations. Taeniasis develops in humans after consumption of undercooked meat containing cysts enclosing viable larvae [4]. *T. saginata* poses comparably lower severity of foodborne disease than *T. solium*, as it causes only taeniasis.

Cysticercosis is a parasitic disease of the intermediate hosts, cattle and pigs. In cattle, cysticercosis is caused by ingestion of *T. saginata* eggs and in pigs by ingestion of *T. solium* eggs [4]. Rarely, *T. solium* cysticercosis also occurs in humans, after autoinfection by ingesting eggs excreted from *Taenia* living in their own or somebody else's intestines. Therefore, t

he severity of human disease caused by *T. solium* is comparably much higher than that caused by *T. saginata*, as it can cause not only taeniasis but also human cysticercosis. Cysticercosis in humans potentially affects vital organs and has associated severe complications.

2.3.1. Human cysticercosis in Serbia. Basic criteria for categorising *Taenia* as an important foodborne parasite in Serbia are the severity of the disease in humans, distribution over the entire country and case fatality rate.

According to the FAO, *T. solium* is ranked as the most important foodborne parasite in the world. *T. solium* infections are endemic in a number of underdeveloped countries in Africa, Asia, and South America and are associated with poverty, poor sanitary and living conditions as well as inadequate health protection [19]. Cysticercosis is considered an endemic disease in Europe. The regions of the Mediterranean basin and Southeastern Europe (including Serbia) are areas with sporadic cases of autochthonous infection. In the European Union, a substantial number of imported cases of the disease was recorded in humans with history of travel or stay in underdeveloped countries [19].

Humans become infected by ingesting eggs of *T. solium* either directly *via* the faecal-oral route, *via* contaminated food and/or water, or *via* autoinfection in individuals with taeniasis (see above). Once eggs are in the human intestines, they release embryos that spread over the entire body to form cysts. In humans, the central nervous system is the primary site of cyst formation; however, cysts can also occur in striated muscles [20].

The symptoms of human cysticercosis are not pathognomonic and are determined by the location, number and size of lesions as well as the immune status of the diseased individual. The most severe form of cysticercosis is neurocysticercosis, which can lead to epileptic seizures and fatal outcomes. In underdeveloped countries, cysticercosis causes severe human diseases, and it is considered the leading cause of acquired epilepsy in developing countries (30%). It is believed that neurocysticercosis is associated with around 6.3% of reported epilepsy cases in developed countries [19]. The case fatality rate is not high, being <1% [4].

Human cysticercosis is not a mandatory reportable disease in Serbia, and therefore, there are no official data on its incidence among the human population in Serbia. Available data on human cysticercosis in Serbia were obtained from scientific literature. According to Bobić et al. [21], the annual incidence of cysticercosis varies between 0 and 0.29 cases per 100,000 population. During the 1990s, a statistically significant increase in the number of cases was observed. The situation has changed after 2000, and ever since, a statistically significant decrease cases has been recorded. In Serbia, the majority of hospitalized cases (67.2%) were aged 30-59, and the incidence was the same among genders [21]. Less than 9% of cases belonged to professionally endangered categories (e.g. farmers) [21]. According to data from http://global-disease-burden.healthgrove.com the mortality rate in Serbia has decreased, with the following values: 0.0005 deaths per 100,000 population in 2005; 0.0003 deaths per 100,000 population in 2010, and; 0.0002 deaths per 100,000 population in 2013. The majority of deaths associated with cysticercosis were recorded in individuals more than 80 years old.

2.3.2. Animal cysticercosis in Serbia. Animals acquire cysticercosis either from the environment or by direct contact with faeces from infected humans. In undeveloped countries, cysticercosis causes substantial economic losses in pig production. Neurocysticercosis in pigs can lead to seizures similar to epileptic seizure in humans with neurocysticercosis. Pigs manifest tonic muscle contractions followed by sudden diminution in all muscle tone leading to collapse, and they typically walk in circles [22].

Bovine and porcine cysticercosis are mandatory reportable diseases in Serbia and meat inspection is performed to determine the disease on livestock slaughter lines. During 2005 to 2016, according to official data, seven cases of bovine cysticercosis were reported in Serbia. According to the literature data, a low prevalence of bovine cysticercosis is continuously present in the country [23]. According to data from the World Organisation for Animal Health (www.oie.int) from 2005, Serbia reported 65 cases of porcine cysticercosis in 2009 and 4 cases in 2006.

Risk factors for spreading cysticercosis include free access of animals to pastures and risky waters as well as uncontrolled human defecation in the proximity of farms [24]. In developed countries, application of proper sanitation measures, adoption of intensive pig production systems and adequate control on the livestock slaughter line resulted in either complete eradication or control of cysticercosis [25].

Even if the incidence of cysticercosis in Serbia is decreasing, the disease has not yet been fully eradicated. Similarly to the other foodborne parasites, adequate education of livestock farmers and upgrades to veterinary-sanitary measures are the prerequisites to improve the epidemiological situation.

2.4. Toxoplasma gondii

Toxoplasma gondii is a protozoan parasite infecting virtually all warm blooded species, including man and food producing animals. The genus *Toxoplasma* encompasses only one species, *Toxoplasma gondii*; however, there are several genotypes (I, II, III). The definitive hosts of *Toxoplasma* are Felidae family members. Definitive hosts shed oocysts into the environment in their faeces, thus infecting intermediate hosts [26].

2.4.1. Human toxoplasmosis in Serbia. Basic criteria for categorising *Toxoplasma* as an important foodborne parasite in Serbia are severity of the disease in humans, distribution over the entire country and number of affected individuals.

2. Throughout the 11-year monitoring period (2005 to 2016), the average incidence in Serbia was 0.96 cases per 100,000 population (Table 1). The incidence rate in Vojvodina is somewhat higher than

IOP Conf. Series: Earth and Environmental Science **85** (2017) 012029 doi:10.1088/1755-1315/85/1/012029

that in Central Serbia. The highest incidence was recorded in Vojvodina in 2014, being 1.26 cases per 100,000 population.

Human toxoplasmosis has a worldwide distribution with highly variable seroprevalence ranging from 10 to 70%. The three infectious forms of *Toxoplasma* are tachyzoites, bradyzoites and oocysts. Bradyzoites are transmitted by tissues of intermediate hosts, and oocysts by faecally contaminated food and water. An outbreak of toxoplasmosis which was caused by unpasteurized goat milk with confirmed tachyzoite finding was reported by Guy et al. [26]. According to research conducted in Serbia, consumption of undercooked meat (mainly beef) is the main risk factor for the development of toxoplasmosis [27].

The clinical picture of alimentary toxoplasmosis caused by bradyzoites in meat varies according to the degree of meat invasion, cooking method and meal preparation technique as well as immune status of the infected individual and the genotype of *Toxoplasma* [26]; this makes the clinical course of the disease highly variable.

In immunocompetent persons, clinical manifestations are mild, encompassing moderately elevated body temperature, fatigue, throat and muscle pain and headache.

In immunocompromised individuals (patients with AIDS or those undergoing immunosuppressive therapy or treatment), clinical manifestations of toxoplasmosis can be extremely severe, including encephalitis, retinochoroiditis and pneumonitis [26,27]. Toxoplasmic encephalitis is one of the leading causes of deaths of patients with AIDS.

Ocular toxoplasmosis can produce a complicated clinical picture, and it sometimes results in partial or total loss of vision. Even though available literature mostly reports ocular toxoplasmosis occurring in Africa and South America, cases of this disease were also recorded in Serbia.

Toxoplasmic encephalitis is one of the leading causes of deaths of patients with AIDS. Congenital toxoplasmosis is a particularly severe form of toxoplasmosis. The disease is not foodborne, but can occur as the consequence of underlying foodborne infection of the mother. Transplacental infection can lead to abortion or, if the foetus survives, occurrence of hydrocephalus, seizures, retinochoroiditis, spasticity, deafness, and hepatosplenomegaly [26,27]. In some cases, newborns do not manifest disease symptoms at birth; however, later in life, they can manifest symptoms of mental retardation and retinochoroidal lesions. If the infection is acquired at some later stage of pregnancy, the child can remain without symptoms or have only mild complications. Clinical picture is determined by the genotype of *Toxoplasma* [26].

2.4.2. Animal toxoplasmosis in Serbia. Toxoplasmosis in animals occurs worldwide, including in Serbia. Animals can acquire toxoplasmosis either from the environment or by direct exposure to infected cat faeces. Clinical manifestations of the disease in animals have not been recorded. In some countries, sheep and goats are the main source of human infections, whereas predominant reservoirs in other countries are pigs and cattle. In theory, poultry meat could be considered a source of infection; however, there are no records of toxoplasmosis epidemics associated with poultry meat so far [28].

Animal toxoplasmosis is not a mandatory reportable disease in Serbia, but according to the Veterinary Directorate, there were two reported cases of sheep toxoplasmosis in Serbia (one each in Zlatibor and Srem Districts). Seroprevalent animals were reported in 2007 and 2015 including five sheep. Scientific literature reported seroprevalences in domestic animals in Serbia as: 84.5% of sheep, 76.3% of cattle, and 28.9% of pigs being seropositive [27].

Toxoplasmosis shows the worldwide distribution including our country as well. Decrease of toxoplasmosis prevalence in domestic animals can be accomplished primarily by improvement of biosecurity measures on farms. This requires excluding cats as a means of rodent population control on farms.

3. Conclusions

The described foodborne parasites, *Echinococcus*, *Trichinella*, *Taenia solium* and *Toxoplasma gondii*, are endemically present in Serbia. Their distribution and persistence within the domestic animal population is directly associated with inadequate zoohygienic conditions including contamination of irrigation water, usage of manure for fertilization purposes, poor personal hygiene or inadequate food preparation practices. Moreover, the level of education of farmers and pet owners regarding personal hygiene and keeping livestock and/or pets in a healthy and hygienically responsible manner is not adequate.

Improvement of veterinary-sanitary controls for animal farming systems and safer handling and disposal of animal waste from farms and slaughterhouses are prerequisites for enhancing the epidemiological situation in Serbia with respect to parasites.

Acknowledgments

This work was supported by the Ministry of Science and Technological Development of the Republic of Serbia, grant TR 31084.

References

[1] Petrović J, Grgić Ž, Prodanov-Radulović J, Pušić I, Ratajac R, Bošnjak D and Medić S 2017 Neke od karakteristika epidemije trihineloze u Čajetini. *Zbornik radova 15. Simpozijuma Zdravstvena zaštita, selekcija i reprodukcija svinja, Srebrno jezero* (Požarevac: Veterinarski specijalistički institut Požarevac) 102

[2] Furst T, Keiser J and Utzinger J 2012 Global burden of human food-borne trematodiasis: a systematic review and meta-analysis *Lancet Infect Dis* **12** (3) 210–21

[3] Broglia A and Kapel C 2011 Changing dietary habits in a changing world: Emerging drivers for the transmission of food-borne parasitic zoonoses *Vet Parasit* **182** 2–13

[4] FAO/WHO (Food and Agriculture Organisation of the United Nations/World Health Organization) 2014 Multicriteria-based ranking for risk management of food-borne parasites *Microbial Risk Assessment Series* No 23 (Rome: WHO press) 302

[5] Debeljak Z 2016 Parazitološka i molekularna ispitivanja genotipova i halotipova metacestoda Echinococcus granulosus sensu lato i epizootiološke karakteristike hidatidoze kod različitih vrsta životinja *doctoral dissertation* (Belgrade: University of Belgrade) 143

[6] Moro P and Schantz P M 2009 Echinococcosis: a review *Int J Infect Dis* **13** (2) 125–33.

[7] Djuričić S, Grebeldinger S, Kafka D I, Djan I, Vukadin M and Vasiljević Z V 2010 Cystic echinococcosis in children – The seventeen-year experience of two large medical centers in Serbia *Parasit Int* **59** (2) 257-61

[8] Bouraoui H, Trimeche B and Mahdhaoui A 2005 Echinococcosis of the heart: clinical and echocardiographic features in 12 patients *Acta Cardiol* **60** (1) 39–41

[9] Ćirović D, Pavlović I, Kulišić Z, Ivetić V, Penezić A and Ćosić N 2012 *Echinococcus multilocularis* in the European beaver (*Castor fibre* L.) from Serbia: first report *Vet Rec* http://dx.doi.org/10.1136/vr.100879

[10] Kern P 2010 Clinical features and treatment of alveolar echinococcosis *Curr Opin Infect Dis* **23** 505–12

[11] Murrell K, Lichtenfels R, Zarelenga D and Pozio E 2000 The systematics of the genus *Trichinella* with key to species *Vet Parasitol* **93** (3-4) 293-307

[12] Pozio E 2013 The opportunistic nature of *Trichinella* – Exploitation of new geographies and habitats *Vet Parasitol* **194** (2–4) 128-32

[13] Ofori-Belić I, Korać M, Milosević B, Djurković-Djaković O, Dulović O, Dakić Z, Poluga J and Brmbolić B 2010 Seasonality of trichinellosis in patients hospitalized in Belgrade, Serbia *Parasite* **17** (3) 199-204

[14] Capo V and Despommier DD 1996 Clinical aspects of infection with *Trichinella* spp *Clin*

Microbiol Rev **9** (1) 47-54

[15] Cvetković J, Teodorović V, Gianluca M, Vasilev D, Vasilev S, Ćirović D and Sofronić-Milosavljević Lj 2011 First report of *Trichinella britovi* in Serbia *Acta Parasitol* **56** (2) 232-5

[16] Petrović J, Grgić Ž, Pušić I and Urošević M 2014 Sylvatic trichinellosis in the Vojvodina region (Serbia) *Trends in game meat hygiene: From forest to fork* ed Paulsen P et al (Wageningen: Wageningen Academic Publishers doi: 10.3920/978-90-8686-790-5 175-82

[17] Petrović J, Pušić I, Apić J, Milanov D, Grgić Ž, Đorđević V and Matekalo-Šverak V 2012 Silvatična trihineloza – uloga divljih životinja u ciklusu širenja trihienloze u Srbiji *Veterinarski glasnik* **66** (3-4) 175-83

[18] Sofronic-Milosavljevic Lj, Djordjevic M, Plavsic B and Grgic B. 2013 *Trichinella* infection in Serbia in the first decade of the twenty-first century *Vet Parasitol* **194** (2-4) 145-9

[19] Zammarchi L, Strohmeyer M, Bartalesi F, Bruno E, Munoh J, Buonfrate D, Nicoletti A, Garcıa H H, Pozio E and Bartolon A 2013 Epidemiology and Management of Cysticercosis and *Taenia solium* Taeniasis in Europe, Systematic Review 1990–2011 *PLoS One* **8** (7): e69537

[20] Garcia H H and Del Brutto O H 2000 *Taenia solium* cysticercosis *Infect Dis Clin North Am* **14** (97) 119

[21] Bobić B, Dakić Z, Klun I, Nikolić A, Dulović O, Turkulov V and Djurković-Djaković O 2015 Human cysticercosis in Serbia - a 25 year national study *Proc 1ST Cystinet Int Conf Taeniosis and cysticercosis: a one health challenge* (Belgrade: Institute for medical research) 27

[22] Trevisan C, Mkupasi E M, Ngowi H, Forkman B and Johansen M V 2015 Severe seizures in pigs naturally infected with *Taenia solium Proc 1ST Cystinet Int Conf Taeniosis and cysticercosis: a one health challenge* (Belgrade: Institute for medical research) 30

[23] Živojinović M, Stokić Nikolić S, Rogožarski D, Lazić M, Dobrosavljević I, Rajković G and Paunović S 2015 Cysticercosis at the slaughterhouses of the Braničevo district, Serbia *Proc 1ST Cystinet Int Conf Taeniosis and cysticercosis: a one health challenge* (Belgrade: Institute for medical research) 40

[24] Laranjo-González M, Devleesschauwer B, Gabriël S, Dorny P and Allepuz A 2016 Epidemiology, impact and control of bovine cysticercosis in Europe: a systematic review *Parasit Vectors* **9** (81) doi:10.1186/s13071-016-1362-3

[25] Vilhena M, Fonseca A G, Marques da Silva J R, Dias S S and Torgal J 2015 Characterizing human cysticercosis in Portugal 2006 *Proc 1ST Cystinet Int Conf Taeniosis and cysticercosis: a one health challenge* (Belgrade: Institute for medical research) 25

[26] Guy E, Dubey J P and Hill D E 2012 *Toxoplasma gondii Food-borne Protozoan Parasites* ed Robertson L J and Smith H V (New York: Nova Biomedical) chapter 6 167-88

[27] Djurković-Djaković O 2010 Toxoplasmosis as a public health issue in Serbia *Scientia Medica* **20** (1) 108-12

[28] Kijlstra A and Jongert E 2008 Control of the risk of human toxoplasmosis transmitted by meat *Int J Parasit* **38** (12) 1359–70

IOP Conf. Series: Earth and Environmental Science **85** (2017) 012030 doi:10.1088/1755-1315/85/1/012030

Plants as natural antioxidants for meat products

V Tomović, M Jokanović, B Šojić, S Škaljac and M Ivić

Faculty of Technology, University of Novi Sad, Bulevar cara Lazara 1, Novi Sad, Serbia

Email: tomovic@uns.ac.rs

Abstract. The meat industry is demanding antioxidants from natural sources to replace synthetic antioxidants because of the negative health consequences or beliefs regarding some synthetic ones. Plants materials provide good alternatives. Spices and herbs, generally used for their flavouring characteristics, can be added to meat products in various forms: whole, ground, or as isolates from their extracts. These natural antioxidants contain some active compounds, which exert antioxidative potential in meat products. This antioxidant activity is most often due to phenolic acids, phenolic diterpenes, flavonoids and volatile oils. Each of these compounds often has strong H-donating activity, thus making them extremely effective antioxidants; some compounds can chelate metals and donate H to oxygen radicals, thus slowing oxidation via two mechanisms.

Numerous studies have demonstrated the efficacy of natural antioxidants when used in meat products. Based on this literature review, it can be concluded that natural antioxidants are added to fresh and processed meat and meat products to delay, retard, or prevent lipid oxidation, retard development of off-flavours (rancidity), improve colour stability, improve microbiological quality and extend shelf-life, without any damage to the sensory or nutritional properties.

1. Muscle lipids and lipids oxidation

Fresh and processed meats offer numerous nutritional and health benefits and provide unique eating satisfaction in the lifestyle of the modern society. However, consumption of red meat, including processed products, is subjected to increasing scrutiny due to the health risks associated with toxins that potentially could be generated during meat preparation and storage [1].

Lipids and their derivative fatty acids are present in muscles as structural components of muscle membranes, as storage droplets of triacylglycerol between muscle fibres and as adipose tissue [2]. The fatty acids in the lipids of muscle tissues can be saturated or unsaturated. Free fatty acids are electron-deficient at the oxygen atom of the carbonyl group (C=O); unsaturated fatty acids are also electron-deficient at points of carbon–carbon unsaturation (C=C). These electron deficient regions make fatty acids susceptible to attack by a variety of oxidizing and high-energy agents generating free radicals [3]. Triglycerides contain straight chains of primarily 16- to 18-carbon fatty acids and minimal amounts of unsaturated fatty acids. Phospholipids in tissue membranes contain up to 15 times the amount of unsaturated fatty acids (C18:4, C20:4, C20:5, C22:5 and C22:6) found in triglycerides. They are much more susceptible to oxidation because of the number of C=C they contain [4].

IOP Conf. Series: Earth and Environmental Science **85** (2017) 012030 doi:10.1088/1755-1315/85/1/012030

Lipid oxidation is described as an oxygen-dependent, oxidative deterioration of saturated and unsaturated fatty acids. This modification of fatty acid is principally carried out by an autocatalytic mechanism of free radicals, called auto-oxidation and consisting of three phases: initiation, propagation and termination. In the first reaction, the presence of prooxidants, or reactive oxygen species, or any other oxidation-favourable condition, results in the loss of a hydrogen radical from unsaturated fatty acids. In the absence of such oxidation-favourable conditions, the reaction between fatty acids and oxygen molecules cannot occur because of the unequal electronic state and spin barrier posed by these ground states. Thus, the reactive oxygen species or other prooxidants, after thermal, redox or light reaction, can produce free radicals and thus start the primary reaction of lipid oxidation. In the second stage, molecular oxygen reacts with the alkyl radical of an unsaturated fatty acid and results in peroxide radical formation. In a subsequent reaction, the formation of hydroperoxides occurs. These are primary products of lipid oxidation and are relatively stable at moderate reaction conditions (low temperature/absence of prooxidative metal ions). However, because of the adverse conditions present in the muscle foods, the hydroperoxides become susceptible to further free radical chain reactions, such as isomerization and decomposition. This produces secondary products, including pentanal, hexanal, 4-hydroxynonenal and malondialdehyde. The last stage is known as termination reaction, during which the free radicals react in various combinations to form stable products. Other unstable compounds are also formed during the termination reaction, which also affect the quality of meat products and give rise to an unpleasant flavour (taste and odour) [2].

Many compounds formed during lipid oxidation (aldehydes, ketones) contribute off-odours that are perceptible at very low concentrations. Odour detection thresholds for pentanal, hexanal and heptanal, compounds typically generated from the breakdown of oxidized linoleic acid have been reported to be <34, <38 and 62 ppb, respectively, in a gelatine model system [5].

The lipid oxidation rate is directly proportional to the unsaturation of fatty acids. Lipid oxidation also depends upon: the level of antioxidants (internal or external) and the presence of prooxidants, such as free iron [6].

When meat ages it turns brown as the myoglobin is converted to metmyoglobin (oxidized form). Lipid oxidation increases the rate of metmyoglobin formation; metmyoglobin acts as a catalyst for lipid oxidation, which further increases the rate of lipid oxidation and deterioration of product colour and flavour occurs [2].

Thus, lipid oxidation is responsible for development of primary and secondary oxidation products, reduction in nutritional quality, as well as changes in flavour and colour, which can precipitate health hazards and economic losses in terms of inferior product quality.

2. Natural antioxidants

Antioxidants are compounds or systems that delay autoxidation by inhibiting formation of free radicals or by interrupting propagation of the free radical by one (or more) of several mechanisms: (1) scavenging species that initiate peroxidation, (2) chelating metal ions such that they are unable to generate reactive species or decompose lipid peroxides, (3) quenching $^{\cdot}O_2^-$, so preventing formation of peroxides, (4) breaking the autoxidative chain reaction and/or (5) reducing localized O_2 concentrations [3]. Chain-breaking antioxidants differ in their antioxidative effectiveness depending on their chemical characteristics and physical location within a food (proximity to membrane phospholipids, emulsion interfaces, or in the aqueous phase). The chemical potency of an antioxidant and its solubility in oil influence its accessibility to peroxy radicals, especially in membrane, micellar and emulsion systems and its amphiphilic character required for effectiveness in these systems [7].

Based on their mode of action, antioxidants inhibit or prevent oxidation; they are again classified into two groups. The first group is primary antioxidants, which react directly with lipid radicals and convert them into relatively stable products; these are also called chain-breaking antioxidative compounds. The second group is secondary antioxidants, which can reduce the rate of oxidation by different mechanisms of action. Most primary antioxidants act by donating a hydrogen atom ($H^{\cdot}$). Secondary antioxidants can act by binding metal ions (Fe^{2+}, Fe^{3+} and Cu^{2+}) able to catalyse oxidative processes, by scavenging oxygen, by absorbing UV radiation, by inhibiting enzymes, or by decomposing hydroperoxides [8].

Antioxidant effectiveness is related to activation energy, rate constants, oxidation–reduction potential, ease with which the antioxidant is lost or destroyed (volatility and heat susceptibility) and antioxidant solubility. The most effective antioxidants are those that interrupt the free radical chain reaction. Usually containing aromatic or phenolic rings, these antioxidants donate $H^{\cdot}$ to the free radicals formed during oxidation, so becoming a radical themselves. These radical intermediates are stabilised by the resonance delocalisation of the electron within the aromatic ring and formation of quinone structures [3]. Thus, antioxidants are compounds that are capable of donating hydrogen ($H^{\cdot}$) radicals for pairing with other available free radicals to prevent the propagation reaction during the oxidation process.

In human bodies, the antioxidant defence system includes enzymes (e.g., superoxide dismutase, glutathione peroxidases and catalase), iron and copper-binding extracellular proteins (e.g., albumin, transferrin, lactoferrin, haptoglobin and ceruloplasmin), antioxidant vitamins (e.g., vitamin C, vitamin E and β-carotene) and other cellular compounds (e.g., quinones, glutathione, uric acid and bilirubin) [9].

In the food industry, antioxidants can be divided into natural and synthetic antioxidants. BHA (butylated hydroxyanisole), BHT (butylated hydroxytoluene), PG (propyl gallate) and TBHQ (tert-butylhydroquinone) are examples of synthetic antioxidants. Ingredients obtained from natural sources which exhibit antioxidative potential in a food model system are considered as natural antioxidants. These antioxidants play a very important role in the food industry. However, synthetic antioxidants have been identified as toxicological and carcinogenic agents in some studies [10-13]. For these reasons, some synthetic antioxidants (BHA, BHT, PG and EDTA) are regulated by the law as direct food additives. Thus, the food industry now chooses natural products over synthetic ones. Consequently, the food market is demanding natural antioxidants, free of synthetic additives and still capable of diminishing oxidation processes in high-fat meat and meat products [2].

Natural antioxidants are produced in living cells to maintain a delicate oxidation–reduction balance in the process of nutrient metabolism and immune function. Upon oxidative stress, antioxidants will react with radical and non-radical species to initiate defence mechanisms for the protection of both intracellular and extracellular components. The plant kingdom is the most abundant source of natural antioxidants, which are abundantly present in spices (seeds), herbs and essential oils used in meat products for sensory purposes. Certain fruits and vegetables are also good sources of antioxidants and other phytochemicals. Many tree leaves, although not used for flavouring, are also good sources of phenolic compounds, and tea is an excellent example of this plant antioxidant family. Some minerals (Se, Zn) and vitamins (vitamin A – β-carotene as precursor, vitamin C – ascorbic acid and vitamin E – α-tocopherol) function as co-factors for antioxidant enzymes and, therefore, are also considered natural antioxidants. Nature has also produced a number of multi-functional short peptides that are capable of neutralising free radicals and chelating prooxidative metal ions. The latter has led to the preparation of 'natural' antioxidant peptides through enzymatic hydrolysis of proteins [1].

Antioxidant activity is widely observed in plants (herbs and spices), for example, oregano, rosemary, thyme, cinnamon, pepper, nutmeg, liquorice, aniseed, cassia bark, fennel, prickly ash, round cardamom, basil, garlic, coriander and ginger. The activity is attributed to various phenolic compounds, which are structurally related but differ in quantity and type, depending on the specific source. The major phenolic constituents in herbs and spices are phenolic acids (gallic, protochatechuic, caffeic and rosmarinic acids), phenolic diterpenes (carnosol and carnosic acid), flavonoids – flavones, flavonols, flavanols and flavanones (quercetin, catechin, apigenin, kaempferol, naringenin and hesperetin) and volatile oils (eugenol, carvacrol, thymol and menthol) [14].

The phenolics present in the natural antioxidants have strong $H^{\cdot}$ donating activity [15] or have high radical-absorbance capacity. Some phenolics prevent the formation of free radicals and propagation of reactive oxygen species, whereas others scavenge free radicals and chelate prooxidants (transition metals) [16]. Phenolic acids trap free radicals; flavonoids scavenge free radicals and chelate metals

(Fe^{2+}, Fe^{3+} and Cu^{2+}) as well. The antioxidant potential of these natural compounds (phenolics) depends on their skeleton structure and pattern of functional groups on this skeleton [17]. For instance, the number and location of free hydroxyl (–OH) groups on flavonoid skeleton decide the free radical-scavenging potential [18]. The presence of multiple – OH groups and ortho-3,4-dihydroxy structures enhances the antioxidant potential of plant-based phenolics [19,20]. Polymeric structures (containing more –OH groups) possess more antioxidant potential [21], whereas glycosylation of functional groups (reduction of –OH groups) usually decreases antioxidant effectiveness. Plant-derived pigments (anthocyanins and their hydrolysed products, anthocyanidins) also contain –OH groups, which can donate H˙, and thus, possess antioxidant properties. Some phenolics also contain vicinal –OH groups attached to aromatic ring. These phenolics donate H˙ as well as vicinal –OH groups that can chelate metals, thus preventing oxidation *via* more than one method. This type of natural antioxidant (for example, carnosol and carnosic acid) has several times the antioxidant activity as BHA and BHT because the latter do not have vicinal –OH groups, thus do not chelate metals and their antioxidant properties depend only on their H˙ donation mechanism [2]. The natural antioxidants ascorbic acid and α-tocopherol each have one aromatic ring as well.

Antioxidant (and flavour) components of herbs and spices can be removed/concentrated as extracts, essential oils or resins. Extracts are soluble fractions that can be removed from plant materials by solubilizing the component(s) of interest in an aqueous, lipid, alcohol, solvent or supercritical CO_2 phase, then removing it. Essential oils are the volatile oils and often contain isoprenoid compounds. Chemically, essential oils are extremely complex mixtures containing compounds of every major functional group class. Essential oils are isolated by steam distillation, extraction (solvent or CO_2), or mechanical expression from the plant material. Plants also contain resins that are nonvolatile, high molecular weight, amorphous solids, or semisolids that flow when subjected to heat or stress. Most resins are bicyclic terpenes (alpha and beta-pinene, delta-3 carene and sabinene), monocyclic terpenes (limonene and terpinolene) and tricyclic sesquiterpenes (longifolene, caryophyllene and delta-cadinene). They are soluble in most organic solvents but not in water. Resins can contain small amounts of volatile phenolic compounds [14].

It should be noted that antioxidant activity of food extracts can be determined using a variety of tests (stable free radical scavengers: galvinoxyl, diphenyl-b-picrylhydrazyl [DPPH]; lipid oxidation: peroxide oxygen, conjugated dienes, rancimat [measurements of oxygen consumption of a linoleic acid emulsion and oxidation induction period in lard at 100°C], oxygen radical absorbance capacity [ORAC] values), active oxygen method, iodine value (measure of the change in number of double bonds that bind I), anisidine value (reaction of acetic acid *p*-anisidine and aldehydes to produce a yellow colour that absorbs at 350 nm), measurement of absorbance at 234 nm (conjugated dienes) and 268 nm (conjugated trienes) to assess oxidation in the early stages and chromatographic methods; however, extraction procedures strongly influence the composition of the extracts and, therefore, also influence the antioxidant activity results [22-24]. In addition, the effect of the antioxidant compound in a food matrix can be significantly different than the activity of a purified extract [14].

3. Use of natural antioxidants for meat and meat products

The demonstrated efficacy of natural antioxidants, in the form of either a pure extract, a blend of active components, or a powder of the original seeds, leaves, etc., to retard lipid oxidation, colour and flavour deterioration in meat products has stimulated a broad interest within the meat industry to explore non-traditional food ingredient strategies [1]. Effects of natural antioxidants on lipid oxidation, development of rancidity and off-flavours and colour stability have been demonstrated in numerous studies (table 1). However, depending on the product type and added amount, it should be noted that various natural antioxidants could also have negative effect on the colour and sensory properties of the meat products.

Several studies have also shown the efficacy of natural antioxidants for preventing protein oxidation under meat processing conditions [27,78-80]. Besides antioxidant activities, natural antioxidants have potential to reduce numbers of bacteria and improve shelf-life of meat products [26,28,43,45,49,52,57,60,70,74-76,81-83].

Table 1. Natural antioxidants used to inhibit oxidation in processed meat products.

Antioxidant category	Meat/Meat products	Reference
Ascorbic acid, α-tocopherol and sesamol	Irradiated ground beef rounds	[25]
Basil essential oil	Cooked sausage	[26]
Black currant extract	Raw pork patties	[27]
Brassica nigra, *Cinnamomum cassia*, *Origanum vulgare* and *Syzygium aromaticum*	Raw chicken meat	[28]
Butterbur and broccoli extracts	Ground beef patties	[29]
Caraway essential oil	Cooked sausage	[30,31]
Caraway essential oil	Dry fermented sausage	[32]
Caraway essential oil	Dry fermented sausage	[33]
Cinnamon extract	Chicken meatballs	[34]
Cinnamon stick, oregano, clove, pomegranate peel and grape seed extracts	Raw pork	[35]
Citrus extract	Cooked turkey meat	[36]
Combination of sage oregano and honey	Cooked chicken meat	[37]
Du-zhong extracts (leaf, roasted cortex and seed)	Raw pork patties	[38]
Echinacea angustifolia extracts	Cooked chicken meat	[39]
Grape seed and green tea extract	Cooked pork meatballs	[40]
Grape seed extract	Ground chicken thigh meat	[41]
Grape seed extract and pine bark extract	Cooked ground beef	[42]
Grape seed extract, oleoresin rosemary and oregano extract	Cooked beef and pork patties	[43]
Grape seed extract, oleoresin rosemary and oregano extract	Raw beef and pork patties	[44]
Grape seed extract, pine bark extract and oleoresin rosemary	Cooked ground beef	[45]
Green tea or commercial grape seed extract	Ground fresh goat meat	[46]
Hypericum perforatum L. extract	Heated pork meat batters	[47]
Isabel and Niagara grape seed and peel extracts	Raw and cooked processed chicken meat	[48]
Juniperus communis L.	Cooked sausages	[49]
Mushroom extract	Beef and fish meat	[50]
Mushroom extract	Bigeye tuna	[51]
Nutmeg	Cooked sausage	[52]
Oregano and sage essential oils	Porcine and bovine ground meat	[53]
Oregano essential oil	Dry fermented sausage	[54,55]
Oregano extract	Fresh beef steaks	[56]
Peel, pulp and seed from two avocado	Pork patties	[57]
Pomegranate rind powder extract, pomegranate juice and pomegranate seed powder extract	Raw ground pork	[58]
Pomegranate seed extract	Beef and chicken meatballs	[59]
Purple rice bran extract	Restructured patties formulated with minced channel catfish belly flap meat	[60]
Red grape pomace extracts	Pork burgers	[61]

Rooibos forms (dried leaves, water extract and freeze-dried extract)	Ostrich meat patties	[62]
Rooibos tea extract	Ostrich droewors	[63]

Table 1. Natural antioxidants used to inhibit oxidation in processed meat products (continued).

Antioxidant category	Meat/Meat products	Reference
Rosemary	Mechanically deboned turkey meat	[64]
Rosemary and lemon balm	Cooked pork meat patties	[65]
Rosemary and oregano extracts	Irradiated beefburger	[66]
Rosemary extract	Frozen and precooked-frozen pork sausage	[67]
Rosemary, carnosine and taurine, together with ascorbic acid	Fresh beef patties	[68]
Rosemary, green tea, coffee and grape skin extracts	Precooked pork patties	[69]
Rosemary powder, rosemary extract and α-tocopherol and their combinations	Turkish sucuk	[70]
Sage and garlic	Chicken meat	[71]
Spice extracts (n = 13)	Liposome system and cooked pork patties	[72]
Tamarillo	Cooked beef meat	[73]
Thyme and balm essential oils	Fresh chicken breast meat	[74]
Vegetables extracts (n = 10)	Raw beef patties	[75]
Vegetable powders	Patties	[76]
Winter savory essential oil	Mortadella-type sausages	[77]

Finally, when used as antioxidants for product quality preservation, these natural compounds can also be regarded as nutraceutical ingredients or supplements for health promotion. Indeed, plant-derived antioxidants provide meat processors with the flexibility to develop novel products with enhanced nutritional value and health benefits and an attractive overall quality profile.

4. Conclusion

Animal lipids are especially susceptible to oxidation because of their electron-deficient double bonds. The breakdown products of primary and secondary oxidation can produce off-odours, new flavours, loss of nutrient content and colour deterioration. Synthetic phenolic antioxidants effectively inhibit oxidation. Spices and herbs, used in meat products for their flavour, often contain high concentrations of phenolic compounds (phenolic acids, phenolic diterpenes, flavonoids and volatile oils). The fact that they are natural and have antioxidative activity that is as good as or better than the synthetic antioxidants makes them particularly attractive for meat processors because of consumer demand for natural ingredients.

Acknowledgement

The research in this paper was financed by the Provincial Secretariat for Higher Education and Scientific Research, Autonomous Province of Vojvodina, Republic of Serbia No 142-451-3626/2016-01.

References

[1] Jiang J and Xiong Y L 2016 *Meat Sci.* **120** 107–17
[2] Kumar Y, Yadav D N, Ahmad T and Narsaiah K 2015 *Compr. Rev. Food Sci. Food Saf.* **14** 796–812

[3] Nawar W W 1996 Lipids *Food chemistry* ed OR Fennema (New York, Basel, Honk Kong: Marcel Dekker, Inc) chapter 5 pp 225–320
[4] Elmore S, Mottram D S, Enser M and Wood J D 1999 *J. Agric. Food Chem.* **47** 1619–25
[5] Vega J D and Brewer M S 1995 *J. Food Sci.* **60** 592–5
[6] Morrissey P A, Sheehy P J A, Galvin K, Kerry J P and Buckley D J 1998 *Meat Sci.* **49** S73–86
[7] Wanatabe Y, Nakanashi H, Goto N, Otsuka K, Kimura T and Adachi S 2010 *J. Am. Oil Chem. Soc.* **85** 1475–80
[8] Rice-Evans C, Miller N and Paganga G 1997 *Trends Plant Sci.* **2** 152–9
[9] Krinsky N I 1992 *Exp. Biol. Med.* **200** 248–54
[10] Abraham R, Benitz K F, Patil G and Lyon R 1986 *Exp. Mol. Pathol.* **44** 14–20
[11] Ahmad I, Krishnamurthi K, Arif J M, Ashquin M, Mahmood N, Athar M and Rahman Q 1995 *Food Chem. Toxicol.* **33** 209–15
[12] Sarafian T A, Kouyoumjian S, Tashkin D and Roth M D 2002 *Toxicol. Lett.* **133** 171–9
[13] Faine L A, Rodrigues H G, Galhardi C M, Ebaid G M X, Diniz Y S, Fernandes A A H and Novelli E L B 2006 *Exp. Toxicol. Pathol.* **57** 221–6
[14] Brewer M S 2011 *Compr. Rev. Food Sci. Food Saf.* **10** 221–47
[15] Muchuweti M, Kativu E, Mupure C H, Chidewe C, Ndhlala A R and Benhura M A N 2007 *Am. J. Food Technol.* **2** 414–20
[16] Ozsoy N, Candoken E and Akev N 2009 *Oxid. Med. Cell Long.* **2** 99–106
[17] Wojdyło A, Oszmiánski J and Czemerys R 2007 *Food Chem.* **105** 940–9
[18] Lupea A X, Pop M and Cacig S 2008 *Rev. Chim.* **59** 309–13
[19] Geldof N and Engeseth N J 2002 *J. Agric. Food Chem.* **50** 3050–5
[20] Brown J E and Kelly M F 2007 *Eur. J. Lipid Sci. Techno.* **109** 66–71
[21] Ursini F, Rapuzzi I, Toniolo R, Tubaro F and Bontempelli G 2001 *Meth. Enzymol.* **335** 338–50
[22] Halliwell B 1997 *Annu. Rev. Nutr.* **16** 33–50
[23] Schwarz K, Bertelsen G, Nissen L R, Gardner P T, Heinonen M I, Huynh-Ba A H, Lambelet P, McPhail D, Skibsted L H and Tijburg L 2001 *Eur. Food Res. Technol.* **212** 319–28
[24] Trojakova L, Reblova Z, Nguyen H T T and Pokorny J 2001 *J. Food Lipids* **8** 1–13
[25] Ismail H A, Lee E J, Ko K Y and Ahn D U 2008 *Meat Sci.* **80** 582–91
[26] Tomović V et al. 2017 *Proc. V Int. Congress "Engineering, Environment and Materials in Processing Industry"*, (*Jahorina, Republic of Srpska, Bosnia and Hercegovina*) (Zvornik: Faculty of Technology Zvornik) pp 60–5
[27] Jia N, Kong B, Liu Q, Diao X and Xia X 2012 *Meat Sci.* **91** 533–9
[28] Radha Krishnan K, Babuskin S, Azhagu Saravana Babu P, Sasikala M, Sabina K, Archana G, Sivarajan M and Sukumar M 2014 *Int. J. Food Microbiol.* **171** 32–40
[29] Kim S-J, Min SC, Shin H-J, Lee Y-J, Cho A R, Kim S Y and Han J 2013 *Meat Sci.* **93** 715–22
[30] Šojić B, Tomović V, Džinić N, Savanović J and Savanović D 2016 *Proc. XI Conf.* XI *of Chemists, Technologists and Environmentalists of Republic of Srpska*, (*Teslić, Bosnia and Herzegovina*) (Banja Luka: Faculty of Technology) pp 295–9
[31] Šojić B, Džinić N, Tomović V, Kocić-Tanackov S, Ikonić P, Jokanović M, Ivić M, Škaljac S, Okanović Đ and Tasić T 2017 *Proc. 58th Oil Ind. Conf.* (*Herceg Novi, Montenegro*) (Novi Sad: University of Novi Sad, Faculty of Technology Novi Sad/Institute of field and vegetable crops/Business association ''Industrial plants'' Novi Sad) pp 309–14
[32] Hromiš N M, Šojić B V, Škaljac S B, Lazić V L, Džinić N R, Šuput D Z and Popović S Z 2013 *Acta Period. Technol.* **44** 57–66
[33] Krkić N, Šojić B, Lazić V, Petrović Lj, Mandić A, Sedej I, Tomović V and Džinić N 2013a *Food Control* **32** 719–23
[34] Chan K, Khong NH, Iqbal S, Ch'ng S, Younas U and Babji A 2012 *J. Food Sci. Technol.* **51** 3269–76
[35] Shan B, Cai Y-Z, Brooks JD and Corke H 2009 *J. Sci. Food Agric.* **89** 1879–85

[36] Contini C, Katsikogianni MG, O'Neill FT, O'Sullivan M, Dowling DP and Monahan FJ 2012 *LWT Food Sci. Technol.* **47** 471–7
[37] Sampaio G R, Saldanha T, Soares R A M and Torres E A F S 2012 *Food Chem.* **135** 1383–90
[38] Xu Z, Tang M, Li Y, Liu F, Li X and Dai R 2010 *J. Agric. Food Chem.* **58** 7289–96
[39] Gallo M, Ferracane R and Naviglio D 2012 J. *Supercrit. Fluids* **72** 198–204
[40] Price A, Díaz P, Bañón S and Garrido M D 2013 *Food Sci. Technol. Int.* **19** 427–38
[41] Brannan R G 2008 *J. Food Sci.* **73** C36–40
[42] Ahn J, Grün I U and Fernando L N 2002 *J. Food Sci.* **67** 1364–9
[43] Rojas M C and Brewer M S. 2007 *J. Food Sci.* **72** S282–8
[44] Rojas M C and Brewer M S 2008 *J. Food Qual.* **31** 173–88
[45] Ahn J, Grün I U and Mustapha A 2007 *Food Microbiol.* **24** 7–14
[46] Rababah T M, Ereifej K I, Alhamad M N, Al-Qudah K M, Rousan L M, Al-Mahasneh M A, Al-u'datt M H and Yang W 2011 *Int. J. Food Prop.* **14** 1208–16
[47] Sánchez-Muniz F J, Olivero-David R, Triki M, Salcedo L, González-Muñoz M J, Cofrades S, Ruiz-Capillas C, Jiménez-Colmenero F and Benedi J 2012 *Food Res. Int.* **48** 909–15
[48] Selani M M, Contreras-Castillo C J, Shirahigue L D, Gallo C R, Plata-Oviedo M and Montes-Villanueva N D 2011 *Meat Sci.* **88** 397–403
[49] Šojić B, Tomović V, Jokanović M, Ikonić P, Džinić N, Kocić-Tanackov S, Popović Lj, Tasić T, Savanović J and Živković Šojić N 2017 *Czech J. Food Sci.* in press
[50] Bao H N D, Ushio H and Ohshima T 2008 *J. Agric. Food Chem.* **56** 10032–40
[51] Bao H N D, Ushio H and Ohshima T 2009 *J. Food Sci.* **74** C162–9
[52] Šojić B, Tomović V, Kocić-Tanackov S, Škaljac S, Ikonić P, Džinić N, Živković N, Jokanović M, Tasić T and Kravić S 2015 *Food Control* **54** 282–6
[53] Fasseas M K, Mountzouris K C, Tarantilis P A, Polissiou M and Zervas G 2008 *Food Chem.* **106** 1188–94
[54] Krkić N, Lazić V, Savatić S, Šojić B, Petrović Lj and Šuput D 2012 *J. Food Nutr. Res.* 51 60–8
[55] Krkić, Šojić B, Lazić V, Petrović Lj, Mandić A, Sedej I and Tomović V 2013b *Meat Sci.* **93** 767–70
[56] Camo J, Lorés A, Djenane D, Beltrán J A and Roncalés P 2011 *Meat Sci.* **88** 174–8
[57] Rodríguez-Carpena J-Gn, Morcuende D, Andrade Ma-Js, Kylli P and Estévez M 2011 *J. Agric. Food Chem.* **59** 5625–35
[58] Qin Y-Y, Zhang Z-H, Li L, Xiong W, Shi J-Y, Zhao T-R and Fan J 2013 *Food Sci. Biotechnol.* **22** 1063–9
[59] Keşkekoğlu H and Üren A 2014 *Meat Sci.* **96** 1446–51
[60] Min B, Chen M H and Green B W 2009 *J. Food Sci.* **74** C268–C77
[61] Garrido M D, Auqui M, Martí N and Linares M B 2011 *LWT-Food Sci. Technol.* **44** 2238–43
[62] Cullere M, Hoffman L C and Dalle Zotte A 2013 *Meat Sci.* **95** 72–7
[63] Hoffman LC, Jones M, Muller N, Joubert E and Sadie A 2014 *Meat Sci.* **96** 1289–96
[64] Mielnik MB, Aaby K and Skrede G 2003 *Meat Sci.* **65** 1147–55
[65] Lara M , Gutierrez J I, Timón M and Andrés A I 2011 *Meat Sci.* **88** 481–8
[66] Trindade R A, Lima A, Andrade-Wartha E R, Oliveira e Silva A M, Mancini-Filho J and Villavicencio ALCH 2009 *Radiat. Phys. Chem.* **78** 293–300
[67] Sebranek J G, Sewalt V J H, Robbins K L and Houser T A 2005 *Meat Sci.* **69** 289–96
[68] Sánchez-Escalante A, Torrescano G, Djenane D, Beltrán J A, Giménez B and Roncalés P 2011 *CyTA – J. Food* **9** 49–57
[69] Nissen L R, Byrne D V, Bertelsen G and Skibsted L H 2004 *Meat Sci.* **68** 485–95
[70] Gök V, Obuz E, Şahin M E and Serteser A 2011 *J. Food Process. Preserv.* **35** 677–90
[71] Mariutti L R B, Nogueira G C and Bragagnolo N 2011 *J. Food Sci.* **76** C909–15
[72] Kong B, Zhang H and Xiong Y L 2010 *Meat Sci.* **85** 772–8
[73] Castro-Vargas H I, Benelli P, Ferreira S R S and Parada-Alfonso F 2013 *J. Supercrit. Fluids* **76** 17–23

[74] Fratianni F, DeMartino L, Melone A, DeFeo V, Coppola R and Nazzaro F 2010 *J. Food Sci.* **75** M528–35
[75] Kim S-J, Cho A R and Han J 2013 *Food Control* **29** 112–20
[76] Duthie G, Campbell F, Bestwick C, Stephen S and Russell W 2013 *Nutrients* **5** 1241–52
[77] Coutinho de Oliveira T L, Malfitano de Carvalho S, deAraújo Soares R, Andrade M A, Cardoso MdG, Ramos E M and Piccoli R H 2012 *LWT-Food Sci. Technol.* **45** 204–12
[78] Ganhão R, Morcuende D and Estévez M 2010 *Meat Sci.* **85** 402–9
[79] Jongberg S, Lund M N,Waterhouse A L and Skibsted L H 2011 *J. Agric. Food Chem.* **59** 10329–35
[80] Tang C B, Zhang W G, Dai C, Li H X, Xu X L and Zhou G H 2015 *J. Agric. Food Chem.* **63** 902–11
[81] Hromiš N, Lazić V, Popović S, Markov S, Vaštag Ž, Šuput D, Bulut S and Tomović V 2016 *J. Food Nutr. Res.* **55** 78–88
[82] Hromiš N, Lazić V, Bulut S, Popović S, Šuput D, Markov S, Džinić N and Tomović V 2017 *Proc. Fifth Int. Conf. Sustainable Postharvest and Food Technologies INOPTEP 2017 and XXIX National Conf. Processing and Energy in Agriculture PTEP 2017*, (*Vršac, Serbia*) (Novi Sad: National Society of Processing and Energy in Agriculture/Faculty of Agriculture) pp 138-43
[83] Vujadinović D, Golić B, Tomović V, Gojković V, Vukić M and Grujić R 2017 *Matica Srpska J. Nat. Sci.* in press

Milk: Past and Present

S Bulajić[1*], J Đorđević[1], T Ledina[1], D Šarčević[2] and M Ž Baltić[1]

[1] Department of Food Hygiene and Technology, Faculty of Veterinary Medicine, Bulevar oslobođenja 18, 11 000 Belgrade, Serbia

[2] Institute of Meat Hygiene and Technology, Kaćanskog 13, 11 000 Belgrade, Serbia

Email: snezab@vet.bg.ac.rs

Abstract. Although milk/dairy consumption is part of many cultures and is recommended in most dietary guidelines around the world, its contribution to overall diet quality remains a matter of controversy, leading to a highly polarized debate within the scientific community, media and public sector. The present article, at first, describes the evolutionary roots of milk consumption, then reviews the milk-derived bioactive peptides as health-promoting components. The third part of the article, in general, presents the associations between milk nutrients, disease prevention, and health promotion.

1. Milk revolution

Humans have used animal milk as a food resource for at least 8500 years. Despite the significant effort of the scientific community, the clear picture of how, where, and when humans initially consumed milk has not been obtained. However, scientists have revealed the very interesting story of the so-called milk revolution [1]. The main premise of this story is that milk consumption is a classic example of how culture has shaped the human genome, or scientifically speaking, created a gene-culture evolution.

As farming started to replace hunting and gathering in the Middle East around 11000 years ago, adoption of animal milk consumption by humans required behavioral adaptations, such as culturing and curdling techniques in order to remove or reduce the lactose content and thereby make dairy products digestible. Otherwise, milk was essentially a toxin to adults because they could not synthesize the lactase enzyme required for lactose degradation. Archaeological findings undoubtedly have proved this first step in the milk revolution. In the 1970s, archaeologist Petar Bogucki, while he was excavating a Stone Age site in central Poland, found a fragment of pottery dotted with tiny holes. He presumed that the pottery might be connected with cheese-making, but at that time, he had no idea how to test his hypothesis. Fortunately, the mysterious pottery was properly stored until 2011, which gave Melanie Roffet-Salque the chance to analyze fatty residues preserved in the clay. The analysis showed the signatures of abundant milk fats, which served as evidence that the early Polish farmers had used the pottery as sieves, trying to separate milk solids from liquid whey [2].

In terms of evolutionary biology, persistent milk consumption is regarded as a novel dietary behavior established by genetic mutation. Lactose tolerance is a classic example of selection-driven evolutionary change in humans from milk-drinking cultures [3]. The trait of lactase persistence (LP) in humans seems to be linked to a single nucleotide in which the DNA base cytosine changed to thymine in a genomic region not far from the lactase gene. According to Itan and his colleagues [4], this genetic mutation spread through Europe about 7500 years ago from the broad, fertile plains of Hungary.

Once the LP allele appeared, it offered a major selective advantage, as it opened up a valuable new source of nutrition. However, the effects were not confined just to a more varied diet. Interestingly, researchers estimated that people with the mutation were more successful in reproduction and would have produced up to 19% more fertile offspring in comparison to those who lacked the new allele [4]. So, taken altogether, gene-culture coevolution began to happen or as Thomas said figuratively, "the practice of dairying and LP allele feed off of each other" [4]. A novel selection pressure was created, favoring genes that extended lactase production into adulthood, especially in populations with long pastoralist traditions that independently evolved lactase persistence – in Europe India, East Africa, and the Arabian Peninsula. According to research results [5], the degree of selection was among the strongest seen to date for any gene in the human genome. However, given that dairying in the Middle East started thousands of years before the LP allele emerged in Europe, the ancient European herders had to find a way to reduce the lactose content of the milk from their animals. Therefore, in this region, the cultural transmission of dairying practice was followed relatively rapidly by the evolution of technological know-how for processing milk into cheese and yogurt. Thus, the selective advantage was not seen, as any adult could obtain nutritional benefits from milk by consuming dairy products. In these populations, adaptive cultural evolution overcame the natural selection acting in the genes. By the late Neolithic and early Bronze Ages, around 5000 years ago, the LP allele was established in most of the northern and central European human populations.

Beja-Pereira and co-workers [6] also argued the coevolutionary hypothesis. The authors found a substantial geographic coincidence between cattle milk protein gene diversity, present-day lactose tolerance in Europeans and the distribution of the Neolithic cattle farming sites. The study results supported the hypothesis that selection driven by the advantages of milk consumption concurrently influenced the frequencies of milk protein genes in cattle and lactase gene in humans [6]. Furthermore, genetic data from another study confirmed archaeological evidence suggesting that the early cattle herders in North-Central Europe were dependent on milk [7]. The presumption that Neolithic cattle herds were managed for early weaning of calves is supported by analysis of intra-tooth changes in nitrogen isotope ratios from archaeological cattle teeth remains.

Although LP is hailed as one of clearest examples of gene-culture coevolution in humans [8], it offers only indirect lines of evidence. Warinner et al. [9] reported the first direct evidence of milk consumption, by identifying the whey protein, β-lactoglobulin (BLG), preserved in human dental calculus from the Bronze Age (ca. 3000 BCE). Therefore, BLG, as a specific milk biomarker, enabled scientists in this field the opportunity to detect complex patterns of milk consumption.

2. Milk-derived bioactive peptides

Bioactive peptides are described as "food components (genuine or generated) that, in addition to their nutritional value, retain many biological properties and exert a physiological effect in the body" [10]. Mellander first reported that ingestion of casein-derived phosphorylated peptides led to enhanced vitamin D-independent calcification in rachitic infants [11]. Since then, a number of food-derived components isolated from various sources (e.g. eggs, bovine blood, collagen, gelatin, various fish species) [12] have been proposed as being bioactive. Nevertheless, bovine milk, especially the milk proteins, is currently the main source of bioactive peptides. Milk-derived bioactive peptides are usually kept inactive within the primary structure of milk protein, and proteolysis is required for their release and activation to exert a physiological response. Generation of the bioactive peptides can occur in the following ways:

a) enzymatic hydrolysis *in vivo* during digestion by digestive enzymes like pepsin, trypsin, chymotrypsin etc., or gut microbial enzymes

b) fermentation of milk with proteolytic lactic acid bacteria (LAB) during milk processing or ripening

c) *in vitro* hydrolysis using isolated enzymes

Numerous beneficial health effects have been attributed to known milk peptide sequences, including antihypertensive, antithrombotic, anti-inflammatory, antioxidative, antimicrobial, opioid, mineral binding (casein phosphopeptides – CPPs), cytomodulatory, immunomodulatory and anti-obesity properties [13-16]. In particular, the cardiovascular system is the main target of milk-derived bioactive peptides, and the blood pressure-reducing peptides that inhibit the angiotensin-converting enzyme I (ACE) are the most widely studied. Antihypertensive properties have been attributed, in particular, to two potent inhibitory tripeptides from bovine casein (casokinins), Val-Pro-Pro (VPP) and Ile-Pro-Pro (IPP), primarily isolated from sour milk fermented with *Lactobacillus helveticus* and *Saccharomyces cerevisiae* [17]). ACE-inhibitory peptides such as α-lactorphin and β-lactorphin are generated from the whey proteins α-lactalbumin and lactoglobulin, respectively [18]. The ACE-inhibitory peptides, IPP and VPP, most likely resist digestion *in vivo* due to the presence of a proline residue at the carboxyl terminal end, which is resistant to the action of Pro-specific peptidases. Antihypertensive peptides influence blood pressure by preventing ACE from synthesizing the vasoconstrictor, angiotensin II, as well as preventing the enzymatic degradation of bradykinin, a vasodilator, but also through mechanisms that are independent of ACE inhibition such as vascular release of endogenous vasodilators (e.g. prostaglandin I2, nitric oxide, and carbon monoxide).

Likewise, yogurt bacteria, cheese starter bacteria, commercial probiotic bacteria as well as non-starter LAB have been demonstrated to produce potent bioactive peptides, through their proteolytic activity. It is noteworthy that the specific peptidase activity of LAB affects the bioactivity of the peptides produced [19]. Interestingly, increased ACE-inhibitory activity has been demonstrated in milk fermented by the mutants of *L. helveticus* CNRZ32 lacking both general aminopeptidase and X-prolyl dipeptidyl aminopeptidase [20]. In that study, it was suggested that both peptidases are involved in the release or degradation of ACE-inhibitory peptides during the fermentation. Undoubtedly, identification of the links between the proteolytic pattern of LAB cultures used in dairy fermentation and resulting milk bioactivity is of the utmost importance.

The physiological effect is not solely linked to the bioactive peptides released by proteolysis. The protein fraction naturally present in milk fat globule membrane (in particular, fatty acid binding protein and glycoprotein), have been demonstrated to exert anti-cancer and antimicrobial properties [21].

As we are witnessing the post-genomics era, molecular studies are needed to assess the underlying mechanisms by which bioactive peptides exert their physiological effects. In this context, nutrigenomics is a promising tool, and by its definition underscores the basic fact that food components can interact with the genome and consequently can influence human health. The ability to moderate the gene expression should be considered one of the major hallmarks of bioactive peptides.

3. Milk and health

The main concern about possible negative effects of milk consumption on cardiovascular health is related to milk's saturated fat content (70% of total milk fat is saturated), resulting in increased blood lipids, especially cholesterol and low-density lipoproteins [22]. However, investigations on the link between dairy consumption and the risk of cardiovascular diseases (CVD) found that milk, cheese, and yogurt intake was inversely associated with CVD risk [23,24]. Several meta-analyses conducted on the relationship between milk intake and risk of CVD showed, in one study, a non-linear dose response relationship between milk intake and risk of stroke [25], while two studies [25,26] showed an inverse connection between cheese intake and stroke. However, the mechanism of the beneficial association of fermented dairy products and reduced CVD risk is uncertain. The reason for this uncertainty, at least in part, could be an effect of the food matrix reducing lipid absorption and short chain fatty acids produced by the bacteria in the large intestine [27]. Moreover, the beneficial effects of cheese can be accounted for by microbial fermentation producing short chain fatty acids such as butyrate [28]. Furthermore, the background, diet and lifestyle characteristics of study participants should be taken into account in the statistical analyses as major confounders [29]. Meta-analysis based on six

IOP Conf. Series: Earth and Environmental Science **85** (2017) 012031 doi:10.1088/1755-1315/85/1/012031

observational studies showed that low-fat, calcium-rich products were generally considered to decrease blood pressure, but there was no association of decreased blood pressure with intake of high-fat dairy products [30]. In accordance with newer meta-analyses, Nordic Nutrition Recommendations have concluded that high consumption of low-fat milk products is associated with reduced risk of hypertension and stroke [31].

Epidemiological and experimental studies have demonstrated that milk and dairy products can have preventive roles in the pathogenesis of colorectal, bladder, gastric, and breast cancer, but have no connections with pancreatic, ovarian or lung cancer [32]. On the contrary, a number of other epidemiological studies have reported an unfavorable effect of milk on prostate cancer risk [33,34,35]. However, experimental results are diverse and are further complicated by several factors, mostly because of the numerous bioactive components of milk, which can possibly interact with other components, including hormones and growth factors, but which remain unproven. Last, but not least, milk digestion and its metabolites must also be considered in order to understand the effect of dietary compounds on specific cells in the human body. Taken altogether, the complex composition of dairy food matrices and the heterogeneity of cancer as multiple diseases make this a challenging area of study.

Milk consumption is also related to lactose intolerance or allergies [36]. Although the most common treatment for lactose intolerance was milk elimination diets, recent studies showed that most individuals with lactose intolerance can tolerate up to 12 g of lactose (250 ml of milk), which would provide 30% of recommended daily calcium intake. In addition, yogurts prepared with *Lactobacillus delbrueckii* subsp. *bulgaricus* and/or *Streptococcus thermophilus* and hard, slow-ripened cheeses contain more predigested lactose and can be more easily tolerated than milk [37]. This is because bacterial lactase survives acidic conditions in the stomach, physically protected within bacterial cells and by the buffering capacity of yogurt. Furthermore, the slower gastrointestinal transit of these products than of milk allows the surviving bacterial lactase to be active and digest lactose [37].

Milk is not "just a food". The growing body of evidence demonstrates that milk is a sophisticated materno-neonatal species-specific signaling system, activating one of the central and key nutrient-sensing pathways – the mammalian target of rapamycin complex 1 (mTORC1) [38]. The activation of the mTOR pathway promotes cell growth, cell division, lipid and nucleotide biosynthesis and gene expression; thus, it positively regulates anabolic processes. Branched-chain amino acids, especially leucine, are known as positive regulators of mTORC1 signaling. Notably, of all animal proteins, whey proteins contain the highest amount of leucine (14%) as compared to meat (8% leucine). As *Bos taurus* duplicates birth weight four times faster than *Homo sapiens*, it is understandable that bovine milk, in comparison to human milk, triggers a much higher magnitude of mTORC1 signaling. Recent evidence from molecular medicine supports the view that persistently increased mTORC1 signaling is regarded as the driving force of non-communicable diseases like hypertension, osteoporosis, obesity, type 2-diabetes, cancer and neurodegenerative disorders [39]. A deeper understanding of milk signaling functions are needed in order to get more insight into the overall life history consequences of milk in the human diet.

4. Conclusion

Given that milk and dairy products have been important components of human diets for over 8000 years, it is conceivable that dairy farming practice could have created the selective pressure under which the LP allele was favored and persisted. Although the milk-derived bioactive peptides, with their physiological versatility, seem useful to target at diet-related chronic diseases, as yet, there has been limited scientific information in this field due to a lack of molecular studies. Overall, the proven health benefits of consuming milk and dairy products greatly outweigh the possible harmful effects. Long-term randomized controlled intervention studies are, however, required to give more conclusive answers on the health aspects of dairy products.

Acknowledgments
This paper was supported by Ministry of Education, Science and Technological Development, Republic of Serbia, through the funding of the Project No III 46009.

References

[1] Curry A 2013 Archaeology: The milk revolution *Nature* **500**(7460) 20–22. doi: 10.1038/500020a
[2] Roffet-Salque M, Bogucki PI, Pyzel J, Sobkowiak-Tabaka I, Grygiel R, Szmyt M and Evershed RP Earliest evidence for cheese making in the sixth millennium BC in northern Europe 2013 *Nature* **493**(4733) 522–5
[3] Feldman MW and Cavalli-Sforza LL1989 *Mathematical evolutionary theory* ed MW Feldman (Princeton University Press, Princeton, New Jersey) p 145–173
[4] Itan Y, Powell A, Beaumont MA, Burger J and Thomas MG 2009 The Origins of Lactase Persistence in Europe *PloS Comp. Biol.* **5**(8) e1000491
[5] Bersaglieri T, Sabeti PC, Patterson N, Vanderploeg T, Schaffner SF, Drake JA, Rhodes M, Reich DE and Hirschhorn JN 2004 *Am. J. Hum. Genet.* **74**(6) 1111–20
[6] Beja-Pereira A, Luikart G, England PR, Bradley DG, Jann OC, Bertorelle G, Chamberlain AT, Nunes TP, Metodiev S, Ferrand N and Erhardt G 2003 Gene-culture coevolution between cattle milk protein genes and human lactase genes *Nature Genetics* **35**(4) 311
[7] Dudd S and Evershed P 1998 Direct demonstration of milk as an element of archaeological economies *Science* **282**(5393) 1478–80
[8] Nielsen R, Hellmann I, Hubisz M, Busramante C and Clark AG 2007 Recent and ongoing selection in the human genome *Nat. Rev. Genet.* **8**(11) 857–68
[9] Wariner C, Hendy J, Speller C, Cappellini E, Fischer R, Trachsel C, Arneborg J, Lynneerup N, Craig OE, Swallow DM, Fotakis A, Christensen RJ, Olsen JV, Liebert A, Montalva N, Fiddyment S, Charlton S, Mackie M, Canci A, Bouwman A, Ruhli F, Gilbert MTP and Collins MJ 2014 Direct evidence of milk consumption from ancient human dental calculus *Sci. Rep.***4** srep07104doi: 10.1038/srep07104
[10] Vermeirssen V, Van Camp J and Verstraete W 2004 Bioavailability of angiotensin I converting enzyme inhibitory peptides *Br. J. Nutr.* **92**(3) 357–66
[11] Mellander O 1950 The physiological importance of the casein phosphopeptide calcium salts. II. Peroral calcium dosage of infants *Acta Soc. Bot. Pol.* **55** 247–55
[12] Baltić ŽM, Bošković M, Ivanović J, Janjić J, Dokmanović M, Marković R and Baltić T 2014 Bioactive peptides from meat and their influence on human health *Tehnologija mesa* **55**(1) 8–21
[13] Lahov E and Regelson W 1996 Antibacterial and immunostimulating casein-derived substances from milk: casecidin, isracidinpeptides *Food Chem. Toxicol.* **34**(1) 131–45
[14] Meisel H 1997 Biochemical properties of regulatory peptides derived from milk proteins *J. Pept. Sci.* **43**(2) 119–28
[15] Dziuba J, Minkiewicz P, Nalecz D and Iwaniak A 1999 Database of biologically active peptide sequences *Mol. Nutr. Food Res.* **43**(3)190–5
[16] Marcone S, Belton O and Fitzgerald D 2017 Milk-derived bioactive peptides and their health promoting effects: a potential role in atherosclerosis *Br. J. Clin. Pharmacol.***83**(1) 152–162
[17] Nakamura Y, Yamamoto N, Sakai K and Takano T 1995 Antihypertensive effect of sour milk and peptides isolated from it that are inhibitors to angiotensin I-converting enzyme *J. Dairy Sci.* **78**(6) 1253–7
[18] Fitzgerald RJ and Meisel H1999 Lactokinins: Whey protein-derived ACE inhibitory peptides *Mol. Nutr. Food Res.* **43**(3)165–7
[19] Algaron F, Miranda G, Le Bars and D Monnet V 2004 Milk fermentation by *Lactococcus lactis*

with modified proteolytic systems to accumulate potentially bio-active peptides *La Lait* **84**(1-2) 115–23

[20] Kilpi EER, Kahala MM, Steele JL, Pihlanto AM and Joutsjoki VV 2007 Angiotensin i-converting enzyme inhibitory activity in milk fermented by wild-type and peptidase-deletion derivatives of *Lactobacillus helveticus* CNRZ32 *Int. Dairy J.* **17**(8) 976–984

[21] Spitsberg VL 2005 Invited review: Bovine milk fat globule membrane as a potential nutraceutical *J. Dairy Sci.* **88**(7) 2289–94

[22] Pereira PC 2014 Milk nutritional composition and its role in human health *Nutrition* **30** 619–27

[23] Huth PJ and Park KM 2012 Influence of dairy product and milk fat consumption on cardiovascular disease risk: a review of the evidence *Adv. Nutr.* **3**(3) 266–85

[24] Rice BH 2014 Dairy and cardiovascular disease: a review of recent observational research *Curr. Nutr. Rep.* **3** 130–38

[25] Hu D, Huang J, Wang Y, Zhang D and Qu Y 2014 Dairy foods and risk of stroke: a meta-analyses of prospective cohort studies *Nutr. Metab. Cardiovasc. Dis.* **24**(5) 460–9

[26] de Goede J, Soedamah-Muthu SS, Pan A, Gijsbers L and Geleijnse JM 2016 Dairy consumption and risk of stroke: a systematic review and updated dose-response meta-analyses of prospective cohort studies *J. Am. Heart Assoc.* **5**(5) e002787 doi: http://dx.doi.org/10.1161/JAHA.115.002787

[27] Veiga P, Pons N, Agrawal A, Oozeer R, Faurie JM, van Hylckama Vlieg JE, Houghton LA, Whorwell PJ, Ehrlich SD and Kennedy SP Changes of the human gut microbiome induced by a fermented milk product *Sci. Rep.* **4** 6328 doi: 10.1038/srep06328

[28] van Hylckama Vlieg JE, Veiga P, Zhang C, Derrien M and Zhao L Impact of microbial transformation of food on health – from fermented foods to fermentation in the gastro-intestinal tract *Curr. Opin. Biotechnol.* **22**(2) 211–9

[29] Guo J, Astrup A, Lovegrove JA, Gijsbers L, Givens DI and Soedamah-Muthu SS 2017 Milk and dairy consumption and risk of cardiovascular diseases and all-cause mortality: dose-response meta-analysis of prospective cohort studies *Eur. J. Epidemiol.* **32** 269–87

[30] Soedamah-Muthu SS, Verberne LD, Ding EL, Engberink MF and Geleijnse JM 2012 Dairy consumption and incidence of hypertension: a dose-response meta analyses of perspective cohort studies *Hypertension* **60**(5) 1131–37

[31] Nordic Council of Ministers 2014 Nordic nutrition recommendations 2012 – integrating nutrition and physical activity, 5th edition. Copenhagen, DK: Nordic Council of Ministers.

[32] Kongerslev Thorning T, Raben A, Tholstrup T, Soedemah-Muthu SS, Givens I and Astrup A 2016 Milk and diary products: good or bad for human health? An assessment of the totality of scientific evidence *Food Nut. Res.* **60** 32527. http://dx.doi.org/10.3402/fnr.v60.32527

[33] Raimondi S, Mabrouk JB, Shatenstein B, Maisonneuve P and Ghadirian P 2010 Diet and prostate cancer risk with specific focus on dairy products and dietary calcium: a case-control study *Prostate* **70** 1054–65

[34] Park SY, Murphy SP, Wilkens LR, Stram DO, Henderson BE and Kolonel LN 2007 Calcium, vitamin D, and dairy product intake and prostate cancer risk: the multiethnic cohort study *Am. J. Epidemiol.* **166**(11) 1259–69

[35] Pettersson A, Kasperzyk JL, Kenfield SA, Richman EL, Chan JM, Willett WC, Stampfer MJ, Mucci LA and Giovannucci EL 2012 Milk and dairy consumption among men with prostate cancer and risk of metastases and prostate cancer death *Cancer Epidemiol. Biomarkers Prev.* **21** 428–436

[36] Lamarche B, Givens DI, Soedamah-Muthu S, Krauss RM, Jakobsen MU, Bischoff-Ferrari HA, Pan A and Després JP2016 Does milk consumption contribute to cardiometabolic health and overall diet quality? *Can. J. Cardiol.* **32**(8) 1026-32

[37] Savaiano DA 2014 Lactose digestion from yogurt: mechanism and relevance *Am. J. Clin. Nutr.* **99**(5 Suppl) 1251S-5S.

[38] Melnik BC 2015 Milk – A nutrient system of mammalian evolution promoting mTORC1-

dependent translation *Int. J. Mol. Sci.* **16**(8) 17048–87
[39] Melnik BC, John SM and Schmitz G 2013 Milk is not just food but most likely a genetic transfection system activating mTORC1 signaling for postnatal growth *Nutr. J.* **12**(1) 103 doi: 10.1186/1475-2891-12-103

New scientific challenges – the possibilities of using selenium in poultry nutrition and impact on meat quality

R Marković, M Glišić, M Bošković and M Ž. Baltić

University of Belgrade, Faculty of Veterinary Medicine, Bulevar oslobođenja 18, Belgrade, Republic of Serbia

E-mail: glisic.mica@gmail.com

Abstract. Physiological stress is one of many concerns facing modern broiler production. In conditions when birds are exposed to stress, supplementation of selenium, which is a crucial glutathione peroxidase enzymatic cofactor, increases the antioxidant capacity of the animals and decreases the harmful effects of free radicals. Dietary selenium improves production performance and health of animals, and positively affects the immune system, the quality, selenium content and fatty acid composition of meat and eggs. There are several different forms of selenium, the most common dietary supplements being an inorganic form (sodium selenite) and anorganic form (selenomethionine). However, in recent years, new forms of selenium, such as a 2-hydroxy-4-methylselenobutanoic acid (HMSeBA) and nanoselenium, which have more bioavailability, bioefficacy, and low toxicity have been designed. In this short comparative overview discusses the effects of inorganic, organic and nanoforms of selenium on production results, glutathione peroxidase activity, meat quality and level of toxicity in poultry.

1. Introduction

It is estimated that the world population in 2025 will number 8 billion inhabitants, and in 2050, up to 9 billion inhabitants [1]. The Food and Agriculture Organisation of the United Nations predicts that annual meat production of 200 million tons will be required by 2050 to respond to this increase in human population [2]. Consequently, there is a forecast that global broiler meat production is going to rise by1% to a record 89.5 million tons in 2017 and that this trend will continue in the coming years [3]. The constant need for more foods of animal origin, especially chicken meat, has led to the development of much research directions to improve production results and health of animals. In recent years, feed formulations were manipulated using different additives (various sources, forms, formulations, etc.) with the aim of improving poultry growth and conversion ratios, and obtaining better quality and value-added products, and therefore, achieving cost-effectiveness of production [1]. Also, given the increasing world population, it will be increasingly necessary to use modern technologies such as nanotechnology and nanobiotechnology in agricultural and food sciences [4].

The efficiency of contemporary poultry production is based on well-balanced nutrition and highly productive lines of birds. This production involves a number of different stress factors to which broilers and hens are exposed. Particularly, in such conditions, the positive effect of selenium on the animals must not be neglected [5,6].

2. Selenium and antioxidant properties

Selenium was discovered by Jacob Berzelius in 1817 and initially considered toxic to humans and animals [7]. The word "*selenium*" originates from the Greek word Selene, which means moon goddess. Later on, selenium was recognized as a nutritionally essential trace element that is important in many biological processes in mammals and birds [8]. Selenium is an integral part of a large number of seleno proteins (more than 30), which participate in the regulation of various functions of the body, including redox balance maintenance and antioxidant defenses, of which glutathione peroxidase (GSH-Px) was the first identified in 1973 [9].Vitamin E is considered to be the most important antioxidant in biological systems, but selenium-dependent GSH-Pxis considered as the second most important factor in antioxidant defense. Hence, dietary selenium, as GSH-Px and as thioredoxin reductase enzymatic cofactor, participates in two levels of antioxidant cell defense (first: detoxification of H_2O_2 resulting from SOD activity; second: detoxification of hydroperoxide) [5].

In poultry, a delicate antioxidant/prooxidant balance in the body is an important determinant of chicken health, embryonic and postnatal development, immunity, muscle function, sperm quality and probably productive and reproductive characteristics of poultry [8]. Poultry diets deficient in selenium result in poor growth and development, increased mortality, reduced egg production, decreased hatchability, nutritional encephalomalacia, nutritional pancreatic atrophy, exudative diathesis, and muscle myopathies [10].

2.1. Selenium in soil and plants

Metallic selenium, selenite (Se^{4+}), selenate (Se^{6+}), and selenide (Se^{2-}) are the inorganic forms of selenium found in soils. Plants absorb inorganic forms of selenium from soil to create organic forms, the seleno-aminoacids, in which selenium is bonded to different amino acids: selenomethionine, Se-methyl-selenomethionine, selenocysteine(the 21st amino acid), and Se-methyl-selenocysteine. Among these, selenomethionine is the most prevalent (it comprises up to 50% of the selenium in cereal grains) [5]. The selenium concentration in soil varies significantly and its availability to plants depends on many factors [11]. In acidic soils with low levels of aeration (low oxidation reduction potential), selenium can form insoluble complexes with iron hydroxide, and in this form, it cannot be absorbed by plants, so therefore, the content of selenium in animal feed also varies. Feed rich in selenium (mainly selenomethionine) includes grasslands, legumes hay, silage, and grain feeds (soybeans), as well as feed of animal origin [5].

The soil in many regions of the world has low levels of selenium, and its low availability for plants has created a problem with selenium levels in humans [12]. It is considered that the selenium deficiency areas contain less than 0.5 mg/kg of selenium in the soil. In the Balkans, areas that are highly deficient in selenium are: Vojvodina (0.024-0.450 mg/kg), Sjenica-Pester plateau (0.046 mg/kg), some parts of Macedonia, the Pozega Valley in Croatia (0.038 mg/kg) and Zeta (Montenegro) (0.280 mg/kg). In Serbia, soil, wheat, and forages are more or less poor in selenium (from 20 to 70 mg Se/kg) [13].

3. Selenium sources and dietary needs

In poultry as well as in other animal species, selenium can be added to diets indifferent forms, on which depends its metabolic fate [14]. Daily selenium requirement during intensive broiler growth is 0.15 mg/kg [15]. Regardless of selenium source, the maximum amount of supplemental selenium that can be added to animal diets is limited to 0.3 mg/kg of diet in the United States [16], while in the European Union, the maximum amount of selenium allowed in animal diets is 0.5 mg/kg of diet[17], and in Serbia, the minimum amount of selenium allowed in animal diets is 0.15 mg/kg of diet [18].

Selenomethionine is highly bioavailable and currently the most suitable form of selenium for nutritional supplementation [19]. In addition to organic selenium compounds, some selenium-enriched feedstuffs have been created: selenium-enriched yeast – an organic form of selenium produced by *Saccharomyces cerevisiae*, selenium-enriched algae *Scenedesmus quadricauda*, selenium-enriched unicellular alga *Chlorella* and selenium chelate [20]. Beside these organic forms of selenium, a new form of organic selenium has been developed based on 2-hydroxy-4-methylselenobutanoic acid

(HMSeBA), which is a hydroxyl-analog of selenomethionine;HMSeBA has already demonstrated high dietary efficacy in poultry nutrition. Namely, HMSeBA, a probable precursor of selenomethionine, enabled more efficient selenium incorporation into proteins in egg and muscle than selenium-enriched yeast [21].

In recent years, the possibility of using nanoparticles as supplements in poultry feed has developed. The particle size of minerals as feed additives in nanoparticle form is typically between 1-100 nm (or more appropriately, 0.2-100.0 nm) and this property distinguishes them with respect to their physical, chemical, and biological properties (great specific surface area, high surface activity, many surface active centers, high catalytic efficiency and strong adsorbing ability) from non-nano, larger particle sizes [20,22]. Nano additives can also be incorporated in micelles or capsules of protein or another natural food/feed ingredient [23]. Nanoparticles have been used in poultry feed to decrease numbers of harmful bacteria in broilers' digestive systems (silver, gold, zinc, copper, metal oxides – Al_2O_3, Fe_3O_4, CeO_2, ZrO_2, MgO), while other nanoparticles, such as nanoselenium, can be used to improve growth and performance [1]. Numerous studies have shown that a new source of elemental selenium, nano selenium, possessed a higher efficiency than selenite, selenomethionine, and methylselenocysteine in upregulating selenoenzymes, higher bioavailability, and exhibited a lower toxicity [7].

3.1. Different forms of selenium – bioavailability and application in poultry nutrition

The absorption mode of various selenium forms is different, leading to different digestibility and bioavailability. Organic selenium is absorbed in the small intestine via the transport mechanism for amino acids, inorganic selenium is absorbed by passive transport, while the nanoparticles have a high specific surface area, small particles, and form nano emulsion droplets that are well absorbed in the intestines [24].

Inorganic forms of selenium can lead to a production of selenocysteine, which is incorporated specifically into selenoproteins, and not to *de novo* synthesis of selenomethionine, whereas organic selenium sources can lead to the production of selenomethionine as well as selenocysteine [8,21]. The cell can nonspecifically incorporate selenomethionine into the structural proteins and, thus, increase the selenium deposit in all tissues [5,8].

3.1.1. Growth performances and glutathione peroxidase activity. Effects of various sources and levels of selenium in poultry diet on production performances have been a subject of a number of studies and results are not uniform, with both negative and positive responses being reported [25,26,27]. Boostani et al. [9] demonstrated that supplementation with selenium (0.3 mg/kg of diet), of different sources increased the antioxidative capacity of broiler chicken under oxidative stress, where the nanoselenium effect was higher than an organic or inorganic source. Other authors determined that organic selenium-enriched yeast (0.2 mg/kg of diet), was more beneficial than inorganic selenium in increasing GSH and GSH-Px activityin blood and liver [28,29].

3.1.2. Selenium and meat quality. The supplementation of selenium, through GSH-Px activity and overall antioxidant defense of muscle against lipid peroxidation that causes excessive cellular damage and drip loss, improves the meat quality (water holding capacity, fatty acid composition, better color stability of heme pigments) and shelf life of poultry meat and also improves selenium retention in muscle [30]. In addition, differences in the antioxidant defense system between animals and muscles would affect calpain activity, proteolysis, and thus quality characteristics influenced by proteolysis such as tenderness and water holding capacity [31]. Visha et al. [32] demonstrated that nanoselenium supplementation (0.3-0.6 mg /kg) caused a significant reduction in breast muscle drip loss and lipid peroxidation as compared with the control and other selenium forms (sodium selenite and selenomethionine). De Medeiros et al. [27] showed that selenium also positively affects organoleptic properties of broiler chicken meat and that selenium-rich meat is juicier, crispy, and better looking.

With the use of special diet enriched with selenium, foods of animal origin with higher nutritive value can be produced, so-called functional foods. Thus, for example, selenium-enriched eggs have

32.6 μg Se, which is about three times the content of selenium in standard eggs [20], while with supplementation of broiler diets with 0.3, 0.6 and 0.9 ppm of selenium-enriched yeast, selenium enriched breast and drumstick meats, containing0.29 to 0.86 mg/kg Se,were obtained, which was up to 8 times more than in standard meat [33,34]. Haug et al. [35] reported that dietary treatment with rapeseed oil, linseed oil and two levels of selenium-enriched yeast (50 mg/kg and 84 mg/kg) resulted in increased concentration of selenium and very long chain fatty acids (eicosapentaenoic, docosapentaenoic and docosahexaenoic acids) in broiler thigh muscle. Also, Jlali et al. [8] showed that HMSeBA has greater ability to increase the selenium concentration in egg and breast muscle of laying hens than selenium-enriched yeast and sodium selenite given at the equivalent doses.

3.1.3. Selenium forms and toxicity. Selenium doses lower than 3-5 mg/kg feed are usually not associated with toxicity, but selenium is toxic to poultry when used in high doses (when dose exceeds the physiological requirement by at least 10 fold), especially inorganic compounds, which are more toxic than organic ones. The molecular mechanisms of selenium toxicity can be explained by substitution of selenium for sulfur, which could result in weakened protein structure, and by reaction between selenite and glutathione with the production of free radicals (i.e. a prooxidant effect)[36].In birds with selenosis, hepatic degeneration, diffuse tubulo-nephrosis, myocardial and skeletal myodegeneration, damage to the bursa Fabricius and cerebellar edema can be observed [5].The most toxic forms of selenium are selenite and selenate, then selenocysteine, while methylated selenium compounds and nanoselenium show the lowest levels of toxicity [37]. Supplementing 0.3 to 0.5 mg/kg of nanoselenium seemed to be effective and advantageous in improving oxidation resistance, and the maximum supplementation of nanoselenium should not be more than 1.0 mg/kg [31].

4. Conclusion

In conclusion, selenium has a very important role in poultry nutrition due to itsantioxidant effects, and its supplementation in animal feed is necessary especially in areas with soils that are deficient in selenium. Also, because of the possible toxic effects and poorer properties of inorganic forms of selenium (sodium selenite), there is a constant need to design new formulations of organic and especially nanoselenium. However, a precautionary approach should be adopted, so further research is needed to determine the safety of their application. In spite of this, in the future, nanotechnology will likely have potential to play a major role in animal nutrition.

Acknowledgment

This paper was supported by Ministry of Education, Science and Technological development, Republic of Serbia, through the funding of Project No 31034.

References

[1] Gangadoo S, Stanley D, Hughes R J, Moore R J and Chapman J 2016 Nanoparticles in feed: Progress and prospects in poultry research *Trends Food Sci. Technol.* **58** pp 115-26

[2] Ghasemzadeh A 2012 Global issues of food production Agrotechnology **1** (102) http://dx.doi.org/10.4172/2168-9881.1000e102

[3] Anon 2017 Foreign Agricultural Service/USDA, Office of Global Analysis, Despite HPAI Outbreaks, Global Broiler Meat Production and Trade Forecast to Rise in 2017

[4] Sekhon B S 2014 Nanotechnology in agri-food production: an overview *Nanotechnol. Sci. Appl.* **7** 31-53

[5] Surai P F 2002a Selenium in poultry nutrition 1. Antioxidant properties, deficiency and toxicity *Worlds Poult. Sci. J.* **58** (3) 333-47

[6] Surai P F 2002b Selenium in poultry nutrition 2. Reproduction, egg and meat quality and practical applications *Worlds Poult. Sci. J.* **58**(4) 431-50

[7] Shi L, Xun W, Yue W, Zhang C, Ren Y, Shi L, Wang Q, Yang R and Lei F 2011 Effect of sodium selenite, Se-yeast and nano-elemental selenium on growth performance, Se

concentration and antioxidant status in growing male goats *Small Rumin. Res.* **96**(1) pp 49-52

[8] Jlali M, Briens M, Rouffineau F, Mercerand F, Geraert P A and Mercier Y 2013 Effect of 2-hydroxy-4-methylselenobutanoic acid as a dietary selenium supplement to improve the selenium concentration of table eggs *J. Anim. Sci.* **91**(4) 1745-52

[9] Boostani A, Sadeghi A A, Mousavi S N, Chamani M and Kashan N 2015 Effects of organic, inorganic, and nano-Se on growth performance, antioxidant capacity, cellular and humoral immune responses in broiler chickens exposed to oxidative stress *Livest. Sci.* **178** 330-6

[10] Tufarelli V, Ceci E and Laudadio V 2016 2-Hydroxy-4-methylselenobutanoic acid as new organic selenium dietary supplement to produce selenium-enriched eggs *Biol. Trace. Elem. Res.* **171**(2) 453-58

[11] Reilly C 1996 Selenium in food and health. Blackie Academic & Professional, an imprint of Chapman & Hall, London.

[12] Rayman M P 2004 The use of high Se yeast to raise Se status: how does it measure up? *Br. J. Nutr.* **92**(04) 557–73

[13] Manojlović M and Singh B R 2012 Trace elements in soils and food chains of the Balkan region *Acta Agric. Scand. B. Soil Plant Sci.* **62**(8) 673-95

[14] Suzuki K T 2005 Metabolomics of selenium: Se metabolites based on speciation studies *J. Health Sci.* **51**(2) 107-14

[15] National Research Council 1994 Nutrient Requirements of Poultry 9th rev ed Washington DC National Academy Press.

[16] Anon 1987 FDA Food additives permitted in feed and drinking water of animals *Federal Register* **52** 10887

[17] Anon 2012 European Food Safety Authority (EFSA) Scientific opinion on safety and efficacy of selenium in the form of organic compounds produced by the selenium-enriched yeast *Saccharomyces cerevisiae* NCYC R646 (Selmax 1000/2000) as feed additive for all species *EFSA J.* **10** 2778

[18] Anon 2010 2012 2014 Regulation on quality for animal feed *Official Gazette of the Republic of Serbia* 4 113-27

[19] Sarkar B, Bhattacharjee S, Daware A, Tribedi P, Krishnani K K and Minhas P S 2015 Selenium nanoparticles for stress-resilient fish and livestock *Nanoscale Re. Lett.* **10** (371) 1-14

[20] Suchý P, Strakova E and Herzig I 2014 Selenium in poultry nutrition: a review *Czech J. Anim. Sci.* **59** 495-503

[21] Briens M, Mercier Y, Rouffineau F, Vacchina V and Geraert P A 2013 Comparative study of a new organic selenium source v. seleno-yeast and mineral selenium sources on muscle selenium enrichment and selenium digestibility in broiler chickens *Br. J. Nutr.* **110**(4) 617-24

[22] Baltić M Ž, Bošković M, Ivanović J, Dokmanović M, Janjić J, Lončina J and Baltić T 2013 Nanotechnology and its potential applications in meat industry *Tehnologija mesa* **54**(2) 168-75

[23] Bunglavan S J, Garg A K, Dass R S and Sameer S 2014 Use of nanoparticles as feed additives to improve digestion and absorption in livestock *Int. J. Livest. Res.* **2**(3) 36-47

[24] Hu C H, Li Y L, Xiong L, Zhang H M, Song J and Xia M S 2012 Comparative effects of nano elemental selenium and sodium selenite on selenium retention in broiler chickens *Anim Feed Sci Technol* **177**(3) 204-210

[25] Baltić M Ž, Dokmanović Starčević M, Bašić M, Zenunović A, Ivanović J, Marković R, Janjić J, Mahmutović H and Glamočlija N 2016 Effects of dietary selenium-yeast concentrations on growth performance and carcass composition of ducks *Anim. Prod. Sci.* doi.org/10.1071/AN16126

[26] Salah-Eldin T A, Hamady G A A, Abdel-Moneim M A, Farroh K Y and El-Reffaei W H M 2015 Nutritional evaluation of Selenium-methionine nanocomposite as a novel dietary

supplement for laying hens *J. Anim. Health. Prod.* **3**(3) 64-72

[27] De Medeiros L G, Oba A, Shimokomaki M, Pinheiro J W, Da Silva C A, Soares A L, Pissinati A and De Almeida M 2012 Performance, broiler carcass and meat quality characteristics, supplemented with organic selenium *Semina: Ciências Agrárias* **33**(6Supl2) 3361-70

[28] Mahmoud K Z and Edens F W 2003 Influence of selenium sources on age-related and mild heat stress-related changes of blood and liver glutathione redox cycle in broiler chickens (*Gallus domesticus*) *Comp. Biochem. Physiol. B. Biochem. Mol. Biol.* **136**(4) 921-34

[29] Baltić M Ž, Dokmanović Starčević M, Bašić M, Zenunović A, Ivanović J, Marković R, Janjić J and Mahmutović H 2015 Effects of selenium yeast level in diet on carcass and meat quality, tissue selenium distribution and glutathione peroxidase activity in ducks *Anim. Feed. Sci. Technol.* **210** 225–33

[30] Smet K, Raes K, Huyghebaert G, Haak L, Arnouts S and De Smet S 2008 Lipid and protein oxidation of broiler meat as influenced by dietary natural antioxidant supplementation *Poultry Sci.* **87**(8) 1682-8

[31] Cai S J, Wu C X, Gong L M, Song T, Wu H and Zhang L Y 2012 Effects of nano-selenium on performance, meat quality, immune function, oxidation resistance, and tissue selenium content in broilers *Poultry Sci.* **91**(10) 2532-9

[32] Visha P, Nanjappan K, Selvaraj P, Jayachandran S and Thavasiappan V 2017 Influence of Dietary Nanoselenium Supplementation on the Meat Characteristics of Broiler Chickens. *Int. J. Curr. Microbiol. App. Sci.* **6**(5) 340-7

[33] Marković R, Ristić M, Drljačić A, Šefer D, Šević K, Pantić S, Đurić J and Baltić M Ž 2014 Effect of different amounts of organic selenium in the diet on broiler carcass parameters *Eur. Poultry Sci.* **78** 1–9. 10.1399/eps.2014.66

[34] Marković R, Baltić ŽM, Šefer D, Radulović S, Drljačić A, Đorđević V and Ristić M 2010 Einfluss der Futterung auf die Qualitat von Broilern *Fleischwirtschaft* **10** 132–6

[35] Haug A, Eich-Greatorex S, Bernhoft A, Wold J P, Hetland H, Christophersen O A and Sogn T 2007 Effect of dietary selenium and omega-3 fatty acids on muscle composition and quality in broilers *Lipids Health Dis.* **6** (29) 1-9

[36] Raisbeck M F 2000 Selenosis *Vet. Clin. North. Am. Food. Anim. Pract.* **16** 465-80

[37] Mohapatra P, Swain R H, Mishra S K, Behera T, Swain P, Mishra S S, Behura Sabat S C, Sethy K, Dhama K, Jayasankar P 2014 Effects of dietary nano-selenium on tissue selenium deposition, antioxidant status and immune functions in layer chicks *Int. J. Pharm.* **10**(3) 160-7

Perspectives in production of functional meat products

D Vasilev[1], M Glišić[1], V Janković[2], M Dimitrijević[1], N Karabasil[1], B Suvajdžić[1] and V Teodorović[1]

[1]Department for Food Hygiene and Technology, Faculty of Veterinary Medicine, University of Belgrade, Bulevar oslobodjenja 18, Belgrade, Serbia
[2]Institute of Meat Hygiene and Technology, Kaćanskog 13, Belgrade, Serbia

E-mail: vasilevd@vet.bg.ac.rs

Abstract. The meat industry has met new challenges since the World Health Organization classified processed meat in carcinogenic Group 1. In relation to this, the functional food concept in meat processing has gained importance, especially in reducing carcinogenic N-nitroso compounds and polycyclic aromatic hydrocarbons (PAHs) as an additional imperative, apart from the usual fat and salt reduction and product enrichment with functional ingredients. PAH reduction relies on control of the smoking process, but there is also a possibility they could be degraded by means of probiotic microorganisms or spices. The reduction of N-nitroso compounds could be provided by lowering the amount of added nitrite/nitrate, using substitutes for these chemicals, and/or by preventing conditions for the creation of N-nitroso compounds. Nevertheless, fat and salt reductions still remain topical, and rely mostly on the use of functional ingredients as their substitutes.

1. Introduction

The functional food concept in meat industry recently met new challenges, especially since the International Agency for Research on Cancer of the World Health Organization classified processed meat as a Group 1 carcinogen for humans in 2015 [1]. This was mainly because of the presence of N-nitroso compounds and polycyclic aromatic hydrocarbons (PAHs) in meat products [1]. As the design of functional meat products includes two main strategies, the addition of functional ingredients and the reduction of potentially harmful components [2], the importance of the latter strategy has especially grown. Previously, research mainly concerned fat or salt reduction [3], but nowadays, special attention is also paid to nitrite [4] and PAH reduction [5] in meat products. Nevertheless, addition of functional ingredients remains the basis for designing healthier meat products [2]. Bearing in mind the aforementioned, the aim of the paper is to discuss recent studies on perspectives for producing functional meat products.

2. Polycyclic aromatic hydrocarbon reduction

The most frequently mentioned source of PAHs in meat products is smoking, but the importance of heat treatment should also be taken into account [6]. Although there are about 660 compounds belonging to the PAH group, only 16 are of priority according to European Union regulations, with special attention paid to benzo[a]pyrene (BaP) as a marker. The content of PAHs in meat products depends on the type of product. In Europe, the content of BaP in smoked meat products is mostly below 0.5 µg/kg, and in Germany, it ranged from 0.01 µg/kg in cooked ham up to 0.4 µg/kg in frankfurter-type sausages [7]. In Serbia, it ranged from 0.54 µg/kg in Čajna sausage (a local, dry fermented sausage) up to 2.94 µg/kg in dried beef [8].

The strategies for PAH reduction in meat products are often based on lowering the PAH content in smoke itself, by controlling the pyrolysis temperature (temperature should be lower than 400°C), smoke density (light smoke contains 3.01-4.31 µg PAH/kg; intensive smoke 4.88-7.42 µg PAH /kg), providing suitable distance or a hurdle between the furnace and the products, smoking duration, as well as cooking method and intensity [6,9]. There is also a significant influence of the casing type on the BaP content in frankfurters (0.08 µg/kg in peeled cellulose casing, compared to 0.81 µg/kg in sheep intestine casing) as well as the fat content in sausages (0.28 µg/kg in low fat sausage compared to 1.37 µg/kg in high fat sausage) [10]. Also, PAH reductions in meat products could be achieved through treatment of fermented sausage surface with some lactic acid bacteria strains (*Lactobacillus sakei* KTU05-6, *Pediococcus acidilactici* KTU05-7 or *Pediococcus pentosaceus* KTU05-9), which resulted in a significant reduction of the concentration of PAHs such as benzo[a]pyrene and chrysene [5]. Also, some spices such as onion (30g/100g) and garlic (15g/100g) provided PAH reductions of 45% and 60% in meat, respectively [6].

3. Reduction of N-nitroso compounds

N-nitroso compounds in meat products develop through reactions between nitrite and amines derived from decarboxylation of amino acids from meat, which occurs during meat ripening and heat treatment. Nitrite and indirectly nitrate, which is a source of nitrite after being reduced by microorganisms, are used as preservatives in meat products and play antimicrobial and antioxidative roles, as well as take part in color and aroma formation in cured meat products [11]. Although the use of nitrite/nitrate is characteristic of industrial production, they can also occur in some traditional sausages that are produced without the addition of these preservatives, but originate from spices like pepper and garlic [12]. N-nitrosamines (NA) are mostly formed during heat treatment of cured meat products, especially baking and frying. For example, frankfurters which are pasteurized contain 0.64 µg/kg total NA, cooked (boiled) sausages contain 1.86 µg/kg, and canned pork 5.09 µg/kg, but in grilled and fried meats, the NA content is up to 18.87 and 32.46 µg/kg, respectively. It is important to mention that fried cured lean pork contains only 5.43 µg/kg, but fried pork fat contains 19.31µg/kg NA [13], which is the result of high content of the amino acids, proline and hydroxyproline, precursors for N-nitrosopyrrolidine formation, in collagen rich fatty tissue [14]. Some products which are not heat-treated also contain NA, such as fermented sausages (salami 3.92 µg/kg) and dry cured ham (7.33 µg/kg) [13]. Such findings are explained by the long ripening period for these products, during which precursors for NA, such as amines derived from free amino acids, are generated [11].

Strategies for reducing N-nitroso compounds in functional meat products include exclusion or at least reduction of the amount of added nitrite/nitrate, as well as prevention of conditions for NA creation. Prevention strategies are alteration of product composition (fatty tissue reduction), appropriate heat treatment and suitable ripening conditions. Total exclusion of nitrite/nitrate is not easy because of their importance for the safety, as well as sensory properties (especially color) of the products [11], so for this, an adequate substitute for nitrite/nitrate is needed. Nitrite-free products lack the desired color of meat products, and this is especially expressed in heat-treated products as a result of metmyoglobin (grey-brown color) formation. Promising results were obtained in nitrite-free cooked sausages produced with 1% acid whey and autoclaved mustard seed, where the antioxidant activity of these ingredients stabilized the porphyrin ring of heme in myoglobin, providing an acceptable color of the product [15]. Nitroso-compounds from the mustard seed also took part in color formation [15]. Low-nitrite cooked sausages with acceptable sensory properties were produced with 1% red grape pomace in combination with 30 mg/kg nitrite, where the grape pomace played antioxidative role, as well as took part in browning of existing tannins (Maillard reaction) during heat treatment [16]. Fermented sausages are produced more easily without added nitrite/nitrate, having in mind that traditional production uses only table salt, and the stable red color is mainly a result of the reduction and stabilization of deoxymyoglobin [12]. It is known that the microbiota in fermented sausages plays a role in the process of reduction. *Lactobacillus fermentum* inoculated (10^8 CFU/g) into nitrite-free fermented sausage provided a similar red color to products produced with 60 mg/kg nitrite [17]. Interestingly, some lactic acid bacteria strains (*Lactobacillus sakei* KTU05-6, *Pediococcus acidilactici* KTU05-7 and *Pediococcus pentosaceus* KTU05-9) are capable of reducing amine formation during ripening; since amines are precursors for N-nitrosamine, this produced low N-nitroso products [5]. Essential oils with antioxidative and antimicrobial activity could play a promising role as nitrite substitutes. Nutmeg essential oil, when added at 20 mg/kg, extended the shelf life of cooked sausages [18].

4. Fat reduction and fatty acid profile improvement

Fat reduction in meat products is widely investigated, since animal fat is recognized as a source of undesirable saturated fatty acids responsible for cardiovascular diseases. Fat reduction has two directions, the first being to use non-fat, energy-low substitutes, and the second being to use polyunsaturated fatty acid (PUFA)-rich marine or plant oils to improve the fatty acid profile of the products [19].

The prebiotic, inulin, is the most appropriate non-fat substitute to date, because it has neutral aroma, white color, and when dissolved in water forms a stable gel similar to fat. Inulin gel (inulin-water ratio 1:1) was frozen and ground in a bowl chopper providing particles that imitated the fatty tissue in fermented sausages, and could replace 1/3 of the fatty tissue without negative influence on sensory properties [20,21,23]. In cooked sausages, inulin can be added in powder form up to 5% [22] or as a gel (inulin-water ratio 1:3) up to 8% [21]; inulin gel is more easily incorporated in sausage stuffing than powder because the gel contains more easily dispersed, dissolved inulin. Apart from inulin, carrageenans and gums have potential as fat replacers because of their good gelling properties [24]

. However, carrageenans [25] and some gums [26] had harmful effects on experimental animals, so there is reason for concern about their influence on consumer health.

For fatty acid profile improvement, plant oils are most commonly used. Initially, tropical plant oils, which have hard consistency at room temperature and are easy to use, were adopted, but the negative side was their high content of saturated fatty acids. The use of hydrogenated plant fats is also not appropriate because of the presence of harmful *trans* fatty acids [27]. Instead of hydrogenated oils, a study about the use of interesterified palm oil as a substitute for beef fat, and which improved the fatty acid profile in beef fermented sausages [28] indicates that the use of interesterified oils could be a better choice. Liquid plant oils rich in omega-3 fatty acids (α linoleic acid – ALA) can be added to fermented sausages only through adequate emulsions (pre-emulsified) with proteins [29] or inulin [30], and good results were obtained with linseed, flaxseed, corn, canola, cottonseed, soybean and olive oil, where the oil content ranged mostly between 3-6% along with 25-35% fatty tissue in the sausage stuffing. As for fish oil, which is the main dietary source of eicosapentaenoic (EPA) and docosahexaenoic (DHA) fatty acids, the greatest problem is the fishy odor, which can be overcome by the use of either deodorized or encapsulated fish oil. In cooked sausages, the same oils were used in a liquid as well as in pre-emulsified form, mostly in amounts of 5-10 % along with 10-20% fatty tissue in the sausage stuffing [29]. The main problem in oil-enriched sausages is their proneness to lipid oxidation [24], but there are a variety of natural antioxidants which could help to overcome this. For example, spices and herbs contain antioxidative phenolic acid and terpenoids, fruits and leaves contain flavonoids and soluble vitamins, nuts and seeds contain tocopherols and tocotrienols, essential oils contain polyphenols, terpenoids, bioactive peptides and protein hydrolysates with antioxidative properties like carnosine and Tyr-Phe-Glu or Tyr-Ser-Thr-Ala [31].

5. Salt reduction

Meat products, especially dried meats and dry fermented sausages, are recognized as high salt foods. Meat products have salt (NaCl) concentrations ranging from 1.6-2.4% in cooked sausages, 3.5-5.0% in fermented sausages to and 4-7% (in some cases even more) in dried meat [32]. The main health issue concerning NaCl is the sodium content, a high intake of which could be responsible for hypertension and other cardiovascular problems in consumers [33]. Meat processing techniques include the regular use of NaCl, on the one hand in order to provide adequate sensory and technological properties (aroma, texture, water holding capacity, protein solubility, stuffing stability), and on the other hand to provide safety of the products by lowering the water activity. Thus, just simply reducing the NaCl content in meat products has limits (up to 5-10% reductions can be possible), while further reduction would lead to unacceptable product properties [34]. Because of that, other salts that do not contain sodium should be used. The main limitation of such salts is that they cannot replace table salt completely but only a certain percentage of it. For example, potassium chloride or potassium lactate can replace no more than 40%, and glycine no more than 30% of the NaCl; these amounts do not cause changes in sensory properties of the products. Furthermore, salt mixtures containing combinations of potassium, calcium or magnesium chloride or lactate could replace about 30-53% of the NaCl [35,36,37]. Concerning the antimicrobial effect of salt replacers, potassium and calcium chloride as well as lactates produced similar effects to NaCl, which confirmed that these substances could not only provide similar sensory properties, but also microbial safety similar to that of conventional meat products [35]. Furthermore, the type of salt substitute influences the oxidative changes in fermented sausages, so products with KCl as a partial NaCl substitute underwent less oxidative change, but on the contrary, sausages containing $CaCl_2$ showed more intensive oxidative change then control sausages [38]. Fermented sausages containing $CaCl_2$ and lactates had lower pHs than conventional sausages, which was explained by the contribution of divalent calcium ions to the pH decrease, but the pH of sausages was not influenced by KCl [36].

6. Functional ingredients

Functional foods contain functional ingredients, which are added to provide health benefits to consumers. Those ingredients are widely investigated and include probiotic bacteria, prebiotics, dietary fiber, synbiotics, antioxidative substances, polyunsaturated (omega-3) fatty acids, plant sterols, bioactive peptides, minerals and vitamins. As for meat products, it is important that addition of functional ingredients does not alter the properties of the product, and that the new ingredients are present in such quantities to provide beneficial effect on consumers [19]. Many functional ingredients could substitute for unwanted compounds in meat products; e.g. prebiotics, dietary fiber or plant oil emulsions act as fatty tissue substitutes [20,21,22,23,29,30]. Similarly, partial NaCl replacement with KCl, $MgCl_2$ or $CaCl_2$ leads to richment of the products enriched with K, Mg or Ca respectively [35,36,37]. The use of probiotic bacteria is possible only in non-heated products such as fermented sausages, which gives this type of meat product great potential to be produced as functional food [17,19,36]. Additionally, some probiotic lactic acid bacteria, as already mentioned, can reduce amine formation in fermented sausages, which indirectly could slow down the creation of harmful N-nitrosamine; these bacteria can also decompose PAHs [5]. The use of functional ingredients carries a legal obligation that these be properly listed on product labels in order not to mislead consumers [39].

7. Conclusions

Functional foods as a concept have gained importance in the meat industry, especially since the World Health Organization classified processed meat in carcinogenic Group 1. This engendered special efforts to reduce amounts of carcinogenic polycyclic aromatic hydrocarbons (PAHs) and N-nitroso compounds in meat products. PAH reduction relies on closer control of the smoking process, but there is also a possibility they could be degraded in meat products by means of probiotic bacteria and/or spices. The reduction of N-nitroso compounds could be provided for by the lowering the amount of added nitrite/nitrate, using adequate substitutes, and preventing conditions for creation of NA. To this end, the product composition (fatty tissue reduction), appropriate heat treatment as well as ripening conditions should be addressed. Nevertheless, fat and salt reductions still remain topical, and rely on the use of functional ingredients as their substitutes.

Acknowledgment
The paper is a result of the work on the research Project No III46009 financed by the Ministry of Education, Science and Technological Development of the Republic of Serbia.

References
[1] Bouvard V, Loomis D, Guyton K Z, Grosse Y, Ghissassi F E, Benbrahim-Tallaa L, Guha N, Mattock H and Straif K 2015 Carcinogenicity of consumption of red and processed meat *Lancet. Oncol.* **16** 1599–600
[2] Olmedilla-Alonso B, Jiménez-Colmenero F, and Sánchez-Muniz F J 2013 Development and assessment of healthy properties of meat and meat products designed as functional foods *Meat Sci..* **95** 919-30
[3] Mora-Gallego H, Dolors Guàrdia M, Serra X, Gou P and Arnau J 2016 Sensory characterisation and consumer acceptability of potassium chloride and sunflower oil addition in small-caliber non-acid fermented sausages with a reduced content of sodium chloride and fat *Meat Sci..* **112** 9–15
[4] Alahakoona AU, Jayasena DD, Ramachandra S, and Cheorun Jo C 2015 Alternatives to nitrite in processed meat: Up to date *Trends Food Sci.. Tech.* **45** 37-49
[5] Bartkiene E, Bartkevics V, Mozuriene E, Krungleviciute V, Novoslavskij A, Santini A, Rozentale I, Juodeikiene G and Cizeikiene D 2017 The impact of lactic acid bacteria with antimicrobial properties on biodegradation of polycyclic aromatic hydrocarbons and biogenic amines in cold smoked pork sausages *Food Control* **71** 285-92
[6] Singh L, Varshney JG and Agarwal T 2016 Polycyclic aromatic hydrocarbons' formation and occurrence in processed food *Food Chem.* **199** 768–81
[7] Andrée S, Jira W, Schwind KH, Wagner H and Schwägele F 2010 Chemical safety of meat and meat products *Meat Sci..* **86** 38–48
[8] Djinovic J, Popovic A and Jira W 2008 Polycyclic aromatic hydrocarbons (PAHs) in different types of smoked meat products from Serbia *Meat Sci..* 80 449–56
[9] Pöhlmann M, Hitzel A, Schwägele F, Speer K and Jira W 2012 Contents of polycyclic aromatic hydrocarbons (PAH) and phenolic substances in Frankfurter-type sausages depending on smoking conditions using glow smoke *Meat Sci..* **90** 176-84.
[10] Pöhlmann M, Hitzel A, Schwägele F, Speer K and Jira W 2013 Polycyclic aromatic hydrocarbons (PAH) and phenolic substances in smoked Frankfurter-type sausages depending on type of casing and fat content *Food Control* **31** 136-44
[11] De Mey E, De Klerck K, De Maere H, Dewulf L, Derdelinckx G, Peeters MC, Fraeye I, Heydenb YV and Paelinck H 2014 The occurrence of N-nitrosamines, residual nitrite and biogenic amines in commercial dry fermented sausages and evaluation of their occasional relation *Meat Sci..* **96** 821–28
[12] Vukovic I, Petrovic LJ, Vasilev D and Saicic S, 2011, Microflora and quality of raw sausages from Northern Serbia produced according to traditional process *Fleischwirtschaft* **91** 118-22
[13] Yurchenko S and Mölder U 2007 The occurrence of volatile N-nitrosamines in Estonian meat products *Food Chem.* **100** 1713–21
[14] Drabik-Markiewicz G, Van den Maagdenberg K, De Meya E, Deprez S, Kowalska T and Paelinck H, 2009 Role of proline and hydroxyproline in N-nitrosamine formation during heating in cured meat *Meat Sci..* **81** 479–86
[15] Wójciak KM, Karwowska M and Dolatowski ZJ 2014 Use of acid whey and mustard seed to replace nitrites during cooked sausage production *Meat Sci.* **96** 750–56
[16] Riazi F, Zeynali F, Hoseini E, Behmadi H and Savadkoohi S 2016 Oxidation phenomena and color properties of grape pomace on nitrite-reduced meat emulsion systems *Meat Sci..* **121** 350–358
[17] Zhang X, Kong B and Xiong YL 2007 Production of cured meat color in nitrite-free Harbin red

sausage by Lactobacillus fermentum fermentation *Meat Sci.*. **77** 593–98

[18] Sojic B, Tomovic V, Kocic-Tanackov S, Skaljac S, Ikonic P, Dzinic N, Zivkovic N, Jokanovic M, Tasic T and Kravic S 2015 Effect of nutmeg (*Myristica fragrans*) essential oil on the oxidative and microbial stability of cooked sausage during refrigerated storage *Food Control* **54** 282-86

[19] Jimenez-Colmenero F, Carballo J and Cofrades S 2001 Healthier meat and meat products: their role as functional foods *Meat Sci.*. **59** 5-13.

[20] Vasilev D, Saičić S and Vasiljević N 2013 Qualität und Nährwert von mit Inulin und Erbsenfasern als Fettgewebe-Ersatzstoffe hergestellten Rohwürsten *Fleischwirtschaft* **93** 123-7.

[21] Vasilev D, Vuković I and Saičić S 2011 Some quality parameters of functional fermented, cooked and liver sausages *Meat Technology* **52** 141–53.

[22] Šojić B. Petrović L. Pešović B, Tomović V, Jokanović M, Džinić N and Salitrežić P 2011 The influence of inulin addition on the physico-chemical and sensory characteristics of reduced-fat cooked sausages *Acta Periodica Technologica* **42** 157- 64

[23] Vasilev D, Djordjević V, Karabasil N, Dimitrijević M, Petrović Z, Velebit B and Teodorović V 2017 Inulin as a prebiotic and fat replacer in meat products, *Theory and Practice of Meat Processing*, **2**, 2, 4-13

[24] Grasso S, Brunton NP, Lyng JG, Lalor F and Monahan FJ 2014 Healthy processed meat product- Regulatory, reformulation and consumer challenges *Trends Food Sci.. Tech.* **39** 4-17

[25] Tobacman JK 2001 Review of harmful gastrointestinal effects of carrageenan in animal experiments *Environ. Health Perspect.* **109** 983–94.

[26] Tetsuguchi M, Nomura S, Katayama M and Sugawa-Katayama Y 1997 Effects of curdlan and gellan gum on the surface structure of intestinal mucosa in rats *J. Nutr. Sci.. Vitaminol.* **43** 515–27

[27] Vasilev D, Vuković I, Tomović V, Jokanović M, Vasiljević N, Milanović-Stevanović M and Tubić M 2009 Some important physical, physico-chemical and sensory properties of functional fermented sausages *Meat Technology* **50** 342-50.

[28] Kılıça B and Özer C 2017 Effects of replacement of beef fat with interesterified palm kernel oil on the quality characteristics of Turkish dry-fermented sausage *Meat Sci.* **131** 18-24.

[29] Jimenez-Colmenero F 2007 Healthier lipid formulation approaches in meat based functional foods. Technological options for replacement of meat fats by non-meat fats *Trends Food Sci. Tech.* **18** 567-78

[30] Menegas LM, Pimentel TC Garcia S and Prudencio SH 2013 Dry-fermented chicken sausage produced with inulin and corn oil: Physicochemical, microbiological, and textural characteristics and acceptability during storage *Meat Sci.* **93** 501–6

[31] Jiang J, Xiong YL 2006 Natural antioxidants as food and feed additives to promote health benefits and quality of meat products: A review *Meat Sci.* **120** 107–17

[32] Lilić S, Borović B and Vranić D 2014 Possibilities for substitution of sodium chloride with some salts in the production of dried pork *Meat Technology* **55** 176–83

[33] Dickinson, BD and Havas S 2007 Reducing the population burden of cardiovascular disease by reducing sodium intake: A report of the Council on Sci.ence and Public Health *Arch. Intern. Med.* **167** 1460–8

[34] Tunieva EK and Gorbunova NA 2017 Alternative methods of technological processing to reduce salt in meat products *Theory and Practice of Meat Processing* **2**, 1, 47-56

[35] Inguglia ES, Zhang Z, Tiwari BK, Kerry JP and Burgess CM 2017 Salt reduction strategies in processed meat products - A review *Trends Food Sci. Tech.* **59** 70-8

[36] Vasilev D, Jovetić M, Vranić D, Tomović V, Jokanović M, Dimitrijevic M, Karabasil N and Vasiljević N 2016 Qualität und Mikroflora von funktionellen Rohwürsten, Untersuchung von Würsten, die mit KCl und $CaCl_2$ als Kochsalz-Ersatzstoffe hergestellt und mit dem

Probiotikum L. Casei LC01 sowie einem Präbiotikum angereichert worden sind. *Fleiscwirtschaft* **96** 96-102
[37] Zanardi E, Ghidini S, Conter M and Ianieri A 2010 Mineral composition of Italian salami and effect of NaCl partial replacement on compositional, physico-chemical and sensory parameters *Meat Sci.* **86** 742–7
[38] Dos Santos BA, Campagnol PCB, Fagundes MB, Wagner R and Pollonio MAR 2015 Generation of volatile compounds in Brazilian low-sodium dry fermented sausages containing blends of NaC1, KC1, and $CaC1_2$ during processing and storage *Food Res. Int.* **74** 306-14
[39] Henderikx F 2017 Labeling of food: a challenge for many *Veterinarski Glasnik* **71** 16-23

Pre-slaughter stress and pork quality

S Stajković, V Teodorović, M Baltić and N Karabasil

Department of Food Hygiene and Technology, Faculty of Veterinary Medicine, University of Belgrade, Bulevar Oslobođenja 18, Belgrade, Serbia

E-mail: silvana@vet.bg.ac.rs

Abstract. Stress is an inevitable consequence of handling of animals for slaughter. Stress conditions during transport, lairage and at slaughter induce undesirable effects on the end quality of meat such as pale, soft, exudative meat and dark firm dry meat. Hence, it is very important to define appropriate parameters for objective assessment of level of stress. Attempts to define measures of stress have been difficult and no physiological parameter has been successfully used to evaluate stress situations. One physiological change in swine associated with animal handling stress and with pork quality is an increase in blood lactate concentration. Plasma cortisol was thought to be an appropriate indicator of stress, but the concentration was not consistently changed by different stressors. Therefore, finding alternative parameters reacting to stressors, such as acute phase proteins, would be of great value for the objective evaluation of level of stress and meat quality. As the stress during pre-slaughter handling is unavoidable, the final goal is to improve transport and slaughter conditions for the animal and, as a consequence, meat quality and animal welfare.

1. Introduction

The handling of animals for slaughter consists of a series of procedures that are unusual for them and, therefore, stressful. Stress conditions during transport, lairage and at slaughter negatively influence meat quality. Bleeding interrupts blood circulation and oxygen supply to the muscle. Under these anaerobic conditions, the breakdown of glycogen/glucose results in an accumulation of lactic acid and induces the progressive acidification of the muscle, denaturation of muscle protein and the conversion of the muscle into meat [1]. If stressful conditions occur immediately prior to slaughter, the presence of high lactic acid concentration reduces muscle pH within the first hour after slaughter, while carcass temperature is still high. The combination of low pH and high temperature in the meat causes the denaturation of some muscle proteins leading to reduction in their water holding capacity and to changes of the colour [2]. Meat becomes pale, soft, and exudative (PSE) [3]. On the other hand, when animals are exposed to chronic or long term stress before slaughter, glycogen is depleted and less lactic acid will be formed post-mortem. At high pH value, relatively few proteins are denatured, so the water is firmly bound, and little or no exudates are formed [1] leading to the occurrence of dry, firm and dark meat (DFD).

The objective of this study was to evaluate the effect of stress induced with pre-slaughter handling on pork quality as well as to evaluate parameters for assessing stress level.

2. Assessment of meat quality defects

To identify meat quality defects, different parameters are used: pH, temperature, drip loss, colour and electrical conductivity. There are no unique standards for assessment of PSE or DFD meat. PSE meat is commonly defined as having a pH at 45 min after slaughter (pH45) lower than 6 [4]. If the drip loss is greater than 5%, that meat can be classified as PSE [5,6,7]. The pH at 24 h (pH24) alone can be used to assess DFD meat. A pH24 greater than 6.0 is related to DFD meat [1]. pH24 greater than 6.2 is related to a serious DFD problem [8]. One of the most important components in physical appearance is colour, which the consumer uses as an indicator for the quality and freshness of the meat [9]. An optimal range of visual colour, measured subjectively according to a reference colour scale, for meat would be around 3-4, but values lower than 3 or higher than 4 would be considered PSE and DFD meat, respectively [4]. For assessing PSE, some authors have proposed a combination of parameters [5,7].

3. Pre-slaughter handling

Meat quality is influenced by multiple interacting factors which include breed, genotype, feeding, pre-slaughter handling, stunning, and slaughter method, chilling and storage conditions. Pre-slaughter handling consists of handling the animals both on the farm and during the transport, at lairage, and finally on their way to be stunned and slaughtered. These handling practices can all induce stress either psychologically or physically and are known to be responsible for the development of aberrant pork quality.

3.1. Transport

The impact of transport on animal welfare must be seen as a multiple challenge, for which a combination of stress factors is responsible for the welfare of animals. Stress caused by transport can result in pig fatigue, injury, poor meat quality and ultimately death [10]. The factors during transport that may compromise pig welfare are loading and unloading, journey duration and ambient temperature, placement on the transporter, stocking density, vibrations, floor type and bedding, mixing animals from different groups and food and water deprivation [11]. The interaction of these factors, plus the time spent in lairage and handling of pigs, makes it difficult to assess the impact of transport on pork quality. The relationship between journey length and transport stress does not appear to be linear [10,12]. However, short transport (<2 h) can cause acute stress when the level of glycogen is still high, and therefore the occurrence of PSE meat, while longer transport (>2 h) can exhaust glycogen depots in the muscles, causing the occurrence of DFD meat [10].

3.2. Lairage

Besides creating a reservoir of animals aimed at maintaining the constant speed of the slaughter line, the function of lairage is to allow the animals to recover from the stress of transport and unloading [13]. When pigs are subjected to highly stressful lairage conditions, lairage can have an additive effect to transport stress, and pigs can still be stressed at slaughter and produce poor pork quality [14]. Inadequate treatment of slaughter pigs in this stage, mixing unfamiliar pigs, pen size, stocking density and floor type and lairage temperatures and humidities can result in additional stress leading to skin damage and poor meat quality. Therefore, proper resting time is very important to relieve stress and improve meat quality. The effects of different lairage times on both animal welfare and meat quality are not well defined. Shorter lairage is associated with more PSE meat, because of insufficient time to relieve stress. Longer lairage can increase the amount of DFD meat and reduce carcass yield [15].

3.3. Race to the stunning chamber and stunning methods

Pre-stunning handling facilities are of primary importance, given the need to handle pigs faster, so as to follow the speed of the slaughter-line. The combination of higher speeds of slaughter lines, poorly designed animal handling systems and the size of the group in the depot affect the welfare of animals and the pork quality, as it increases the use of electric prods, which does not always lead to speeding up pigs coming to the stunner. Indeed, the use of electric prods increases mounting behaviour between pigs in the group, resulting in more fatigued pigs and a higher proportion of bruised carcasses and PSE meat [13].

In order to reduce pain and promote the welfare of animals during slaughter, a series of stunning methods have been designed and described, including electric stunning and stunning with carbon dioxide (CO_2). Electric stunning requires the animal to be restrained, which is a potential stressor. Additionally, increased physical stress just prior to electric stunning and tonic and clonic convulsions accelerate post-mortem glycolysis, leading to the occurrence of PSE meat. During CO_2 stunning, the pigs are moved into the stunning chamber in groups using their natural group behaviour [11]. The development of these systems reduced the stress before slaughter and, consequently, the appearance of PSE and DFD meat.

4. Assessment pre-slaughter handling

During pre-slaughter handling of pigs, they react to different stressors which can be classified as physical or psychological [16]. The psychological state of animals can only be indirectly assessed, by monitoring behaviour and by measuring physiological parameters, such as level of cortisol and catecholamines from plasma (adrenalin and noradrenalin), given that stressful situations increase the concentration of these hormones [17]. Behavioural measurements usually represent adaptive responses to the environment (exploration, flight, immobilization, aggression, etc.) [18]. Behavioural, physiological and metabolic responses to aversive situations, besides the type, duration and intensity of the individual pre-slaughter stressor, depend on genetic background and prior experience of the animals [18]. Therefore, the assessment of the situation and the resulting stress is subjective, that is, dependent on the individual [18]. Differences in the sensory quality of pork can be, at least partly, explained by differences in an animal's reaction to stress as well as in the effect of these reactions on muscle glycolysis [19,20].

4.1. Physiological parameters of stress

4.1.1. Cortisol. Numerous experimental results indicate an increased level of cortisol in pigs caused by stress on day of slaughter, stress just before slaughter and physical activity [20,21,22]. Secretion of cortisol is highly variable and different factors must be taken into account, such as the time elapsed from stress to sampling, variation of concentration due to diurnal secretion, genetics and effects of chronic stress [23]. In addition, the concentrations of cortisol do not correspond to the stress intensity, so only the exposure of the pigs to the new environment is sufficient to increase its concentration to the maximum level [20]. Although several studies have investigated the association between cortisol concentration and meat quality, this is still a topic of debate. Some studies indicate that concentration of cortisol had no effect on the pork quality [22], while others suggest that increased concentrations of cortisol lead to the decrease in pork quality [20]. Moreover, measurement of cortisol levels is not very informative for the detection of chronic stress situations, and in this regard, may not be a good indicator of meat quality and sensory quality in pigs [20].

4.1.2. Lactate. Level of lactate can be a good indicator of physical and psychological stress in pigs. Blood lactate concentration was used for assessing pre-slaughter handling [21], stunning pigs with different concentrations of CO_2 [24], transport [25] and time spent in lairage and different plants [26]. Exsanguination blood lactate changes with physical activity, frequency of the use of electric prods and vocalization [27]. Increase in blood lactate concentration, which is associated with pre-slaughter stress, has been shown to have a negative effect on pork quality. Stress was associated with high concentrations of exsanguination blood lactate and lower meat quality such as decreased water holding capacity, lighter colour [14] and lower pH45 value [28].

4.1.3. Acute phase proteins. Finding alternative stress biomarkers is of great importance for the objective assessment animal welfare and optimization of production systems. Acute phase proteins (APP) are plasma proteins considered to be markers of inflammation, primarily synthesized as part of the acute phase response (APR) [29]. They also have been proposed as indicators for farm animal stress monitoring [30,31]. Stress caused to the animals during transport, the new accommodation and the pre-slaughter handling affects the change in concentration of PAF [32]. Some APPs react to a lesser extent than others to the same stimulus and can also react differently to different types of stimuli [33]. Therefore, the use of more than one APP is proposed in assessment of stress levels. Given that moderate PAF increases its concentration only two to three times during the response, the major PAFs, which increase their concentration ten to one hundred times, are more interesting for assessing the health and welfare of pigs. In swine, among others, major APP are pig-MAP (Major Acute-phase Protein), C-reactive protein (CRP), haptoglobin (Hp) and serum amyloid A (SAA) [29,34]. Hp is the most widely studied PAF in pigs, mainly due to the availability of methods for determining its concentration. Increased levels of pig-MAP and Hp are linked with stress situations such as transport [35], crowding, mixing unfamiliar pigs [36, 37], or an inadequate handling of feed [34]. Pig-MAP was the most sensitive protein in the detection of the stress caused by changes in the feeding pattern [34], in distinguishing healthy from diseased states [33] and was the only APP which showed concentrations changed in pigs housed at different stocking densities [38]. Pig-MAP has advantages over other PAFs, such as Hp, due to a lower degree of variation its basal concentration, which facilitates the establishment of limits for distinguishing normal from pathological states and stress situations [32]. The concentration of SAA increases within four hours and the maximum is reached within 24 to 48 hours after a triggering event [33]. Therefore, this PAF can be used to evaluate novel situations or to evaluate pigs' reaction to environmental change, if the appropriate period between the stressful situation and the sampling is adopted [39]. Elevated levels of saliva SAA are a good marker of short transport stress (physiological-psychological stress) and social isolation (psychological stress) in pigs [39]. SAA is more susceptible to acute rather than chronic inflammation [33], which raises issues related to the sensitivity of this APP to high stress situations involving the use of electric prods and consequently, the appearance of skin damage.

5. Conclusions

Pre-slaughter stress has negative effects on pork quality. Minimizing this stress is important, not only for animal welfare, but also for improving pork quality. Stress reactivity is an individual characteristic and each pig is characterized by an individual specific range of values, within which the stress parameters can vary. There is no consistent association between the stress parameters and meat quality measurements. The relationship between cortisol levels and stress, as well as between cortisol levels and meat quality parameters is not linear. Therefore, it is important to determine the appropriate parameters for assessing the level of stress. Lactate concentrations showed good correlation both with pre-slaughter stress and meat quality. Studies suggest that the APP assay may have great potential for the assessment of level of stress and welfare. However, it is necessary first to establish reference ranges for the concentration of these proteins in the normal state, taking into account factors such as sex, age, herd and farm conditions.

References

[1] Warris PD 2000 *Meat Sci* An Introductory Text, CABI Publishing CAB International
[2] Offer G 1991 *Meat Sci* Modelling of the formation of pale, soft and exudative meat: effects of chilling regime and rate and extent of glycolysis **30** 157–84
[3] Adzitey F and Nurul H 2011 Pale soft exudative (PSE) and dark firm dry (DFD) meats: causes and measures to reduce these incidences - a mini review *Int. Food Res. J.* **18** 11-20
[4] Dalmau A, Velarde A and Gispert M 2009 Standardisation of the measure "meat quality" to assess the welfare of pigs at slaughter. In Forkman B, and Keeling L, Assessment of Animal Welfare Measures for Sows, Piglets and Fattening Pigs, Welfare Quality Reports No. **10** 117-23
[5] Warner RD, Kauffman RG and Greaser ML 1997 Muscle protein changes post mortem in relation to pork quality traits *Meat Sci.* **45** 339–52
[6] Channon H A, Payne A M and Warner R D 2003 Effect of stun duration and current level applied during head to back and head only electrical stunning of pigs on pork quality compared with pigs stunned with CO_2 *Meat Sci.* **65** 1325–33

[7] Simek J, Grolichova M, Steinhauserova I and Steinhauser L 2004 Carcass and meat quality of selected final hybrids of pigs in the Czech Republic *Meat Sci.* **66** 383-6
[8] Guàrdia MD, Estany J, Balash Ș, Oliver MA, Gispert M and Diestre A 2005 Risk assessment of DFD meat due to pre-slaughter conditions in pigs *Meat Sci.* **70** 709–16
[9] Rosenvold K and Andersen J 2003 The significance of pre-slaughter stress and diet on colour and colour stability of pork *Meat Sci.* **63** 199–209
[10] Schwartzkopf-Genswein S, Faucitano L, Dadgar S, Shand P, González A and Crowe G 2012 Road transport of cattle, swine and poultry in North America and its impact on animal welfare, carcass and meat quality *Meat Sci.* **92** 227–43
[11] Brandt P and Aaslyng MD 2015 Welfare measurements of finishing pigs on the day of slaughter *Meat Sci.* **103** 13–23
[12] Haley C, Dewey E, Widowski T and Friendship R 2008 Association between in-transit losses, internal trailer temperature, and distance travelled by Ontario market hogs *Can. J. Vet. Res.* **72** 385–9
[13] Faucitano L 2010 Invited review: Effects of lairage and slaughter conditions on animal welfare and pork quality *Can. J. Anim. Sci.* **90** 461–9
[14] Warriss PD 1994 Ante-mortem handling of pigs. In Cole D J A, Wiseman J and Varley M A (Eds.), Principles of pig science 425–432 Loughborough, UK: Nottingham University Press
[15] Zhen S, Liu Y, Li X, Ge K, Chen H, Li C and Ren F 2013 Effects of lairage time on welfare indicators, energy metabolism and meat quality of pigs in Beijing *Meat Sci.* 93 287–91
[16] Grandin T 1997 Assessment of stress during handling and transport *J. Anim. Sci.* **75** 249-57
[17] Nanni C 2009 Short-term stress: the case of transport and slaughter *Ital. J. Anim. Sci* **8** 241-52
[18] Terlouw C 2005 Stress reactions at slaughter and meat quality in pigs: genetic background and prior experience, A brief review of recent findings *Livest. Sci.* **94** 125-35
[19] Terlouw E M C and Rybarczyk P 2008 Explaining and predicting differences in meat quality through stress reactions at slaughter: The case of large white and duroc pigs *Meat Sci.* **79** 795–805
[20] Choi YM, Jung KC, Choe JH and Kim BC 2012 Effects of muscle cortisol concentration on muscle fiber characteristics, pork quality, and sensory quality of cooked pork *Meat Sci.* **91** 490–8
[21] Hambrecht E, Eissen J, Nooijen J, Ducro J, Smits M, den Hartog A *et al.* 2004 Preslaughter stress and muscle energy largely determine pork quality at two commercial processing plants *J. Anim. Sci.* **82** 1401–09
[22] Foury A, Devillers N, Sanchez MP, Griffon H, Le Roy P and Mormède P 2005 Stress hormones, carcass composition and meat quality in large whitexduroc pigs *Meat Sci.* **69** 703-7
[23] Mormède P, Andanson S, Aupérin B, Beerda B, Guémené D, Malmkvist J, Manteca X, Manteuffel G, Prunet P, van Reenen CG, Richard S and Veissier I 2007 Exploration of the hypothalamic–pituitary–adrenal function as a tool to evaluate animal welfare *Physiol. Behav.* **9** 317–39
[24] Nowak B, Mueffling TV and Hartung J 2007 Effect of different carbon dioxide concentrations and exposure times in stunning of slaughter pigs: Impact on animal welfare and meat quality *Meat Sci.* **75** 290–8
[25] Mota-Rojas D, Becerril-Herrera M, Roldan-Santiago P, Alonso-Spilsbury M, Flores-Peinado S, Ramírez-Necoechea R, Ramírez-Telles JA, Mora-Medina P, Pérez M, Molina E, Soní E and Trujillo–Ortega ME 2012 Effects of long distance transportation and CO_2 stunning on critical blood values in pigs *Meat Sci.* **90** 893–8
[26] Edwards N, Grandin T, Engle E, Porter P, Ritter J, Sosnicki A, Anderson B, Carlson A and Anderson B 2010a The effects of pre-slaughter pig management from the farm to the processing plant on pork quality *J. Animal Sci.* **86**(4) 938-44

[27] Edwards N, Grandin T, Engle E, Porter P, Ritter J, Sosnicki A and Anderson B 2010c Use of exsanguination blood lactate to assess the quality of pre-slaughter pig handling. *Meat Sci.* **86** (2) 384-90

[28] Edwards M, Engle E, Correa A, Paradis MA, Grandin T and Anderson B 2010b The relationship between exsanguination blood lactate concentration and carcass quality in slaughter *Meat Sci.* **85**(3) 435-40

[29] Cray C, Zaias J and Altman NH 2009 Acute Phase Response in Animals: A Review *J Am Assoc. Lab. Anim. Sci.* **59** (6) 517–26

[30] Petersen H H, Nielsen J P and Heegaard P M H 2004 Application of acute phase protein measurements in veterinary clinical chemistry *Vet. Res.* **35** 163-87

[31] Murata H, Shimada N and Yoshioka M 2004 Current research on acute phase proteins in veterinary diagnosis: an overview *Vet. J.* **168**(1) 28-40

[32] Piñeiro M, Morales J, Vizcaíno E, Murillo J A, Klauke T, Petersen B and Piñeiro C 2013 The use of acute phase proteins for monitoring animal health and welfare in the pig production chain: The validation of an immunochromatographic method for the detection of elevated levels of pig-MAP *Meat Sci.* **95** 712–18

[33] Heegaard P M H, Stockmarr A, Piñeiro M, Carpintero R, Lampreave F, Campbell M F, Eckersall P D, Toussaint J M M, Gruys E, Skall N and Sorensen S 2011 Optimal combinations of acute phase proteins for detecting infectious disease in pigs *Vet Res* **42**(50) 1-13

[34] Piñeiro C, Piñeiro M, Morales J, Carpintero R, Campbell FM, Eckersall PD, Toussaint MJ, Alava MA and Lampreave F 2007a Pig Acute-Phase Protein levels after stress induced by changes in the pattern of food administration *Animal* **1** 133–9

[35] Piñeiro M, Piñeiro C, Carpintero R, Morales J, Campbell F M, Eckersall P D, Toussaint M J and Lampreave F 2007b Characterisation of the pig acute phase protein response to road transport *Vet J.* **173** 669–74

[36] Piñeiro C, Morales J, Piñeiro M, Lampreave F and Mateos G G 2004 Effects of induced mixing and pen size on performance and serum concentration of acute phase proteins in growing pigs *J Anim Sci* (Suppl. 1) 505 (Abstr.)

[37] Piñeiro C, Piñeiro M, Morales J, Andres M, Lorenzo E, del Pozo M, Alava M A and Lampreave F 2009 Pig-MAP and haptoglobin concentration reference values in swine from commercial farms *Vet. J.* **179** 78–84

[38] Marco-Ramell A, Pato R, Peña R, Saco Y, Manteca X, Ruiz de la Torre J and Bassols A 2011 Identification of serum stress biomarkers in pigs housed at different stocking densities *Vet. J.* **190** 66–71

[39] Soler L, Gutiérrez A, Escribano D, Fuentes M and Cerón J J 2013 Response of salivary haptoglobin and serum amyloid A to social isolation and short road transport stress in pigs *Res. Vet. Sci.* **95** 298–302

IOP Conf. Series: Earth and Environmental Science **85** (2017) 012035 doi:10.1088/1755-1315/85/1/012035

The role of veterinarian in the monitoring programs of mastitis control

M Maletić[1], V Magaš, J Maletić[2]

[1] Faculty of veterinary medicine, Department of obstetrics, sterility and A.I., Bld. Oslobođenja 18, 11000 Belgrade, Serbia
[2] Scientific Veterinary Institute "Novi Sad", Rumenački put 20, 21000 Novi Sad, Republic of Serbia

Abstract. Mastitis is the most common and the most expensive disease of dairy cows. It is followed by a large number of direct and indirect costs that burden the farm's budget and lead to major economic and health losses. The veterinarian at the farm plays a key role in implementing a protocol of biosecurity measures, a protocol of control, therapy, and the suppression of clinical and subclinical mastitis. In order to successfully implement these measures, a good communication between a veterinarian and a farm staff who performs milking procedures is necessary in order to detect and treat all cases of mastitis in time.

E-mail: maletic@vet.bg.ac.rs

1. Introduction

It is well known that mammary gland diseases can be controlled by preventing the emergence of new infections and by eliminating existing sources of infection through a "five-point plan":

- The udder disinfection before and after milking,
- All-encompassing dry off with antibiotics,
- Adequate treatment of clinical mastitis,
- Exclusion of cows with chronical disease,
- Correct milking equipment.

This approach proved to be very effective in the control of contagious mastitis [1].

Programs with the aim at improving the quality of milk are focused on the prevention of mastitis caused by agents from the environment of the animal, reduced use of antibiotics, etc. Veterinarian as a part of the chain of control of the health of herd, implements and controls the implementation of procedures for detection and identification of animals suspected on subclinical and clinical forms of mastitis in the herd. [2] Successful control of mastitis depends on the speed of detection, accurate diagnosis, selection of the best option for the treatment and introduction of preventive measures in order to reduce the exposure of cows to a specific pathogen [3]. An important role in the early detection of disorders of secretion of the mammary gland has the staff involved in the milking as they carry out on-the-spot checks on a dark media. Also, in monitoring the health of the udder, an important role has the good keeping of records of each cow in lactation.

At the beginning of the introduction of the system of records and identification of cows on the farm, the veterinarian must be sure that he can answer the following questions:

What is the incidence of clinical mastitis? How clearly are the symptoms of inflammation? Which is the most common cause of clinical and subclinical mastitis on a given farm? What are the most common protocols for treating these conditions? How many days is milk discarded due to ab treatment? How many percent of cows in lactation have less than 4 active mammary complexes? Which percentage of cows with problems of the mammary disease are excluded from production or died during one year? [4]

1.1 Monitoring of subclinical mastitis

It is known that for control any subclinical disease it isimportant to have a clear understanding of prevalence and a mechanism to monitor incidence. Prevalence of mastitis depends on incidence, development of new subclinical cases, and duration. In some herds, the prevalence of mastitis can be small due to the existence of a chronic infection by infectious pathogens, while in other prevalence mastitis are high due to the existence of certain environmental factors that lead to the occurrence of acute mastitis. Very important moment in monitoring subclinical mastitis is to ensure that SCC values are routinely obtained from all cows on a regular basis. So that a cow with SCC values>200 000 cells/ml are considering to have subclinical mastitis [4].

Expiriences say that for assessments of subclinical mastitis, the veterinarians should start with the following questions: What is the prevalence of subclinical mastitis based on SCC? What is the incidence of subclinical mastitis based on SCC? What are the most common bacteria recovered from cows with SCC values >200 000 cells/ml? How many subclinical cases persist more than 2 months? Prevalence of subclinical mastitis by days in milk and parity? How many cows have subclinical mastitis at the first test and the last test? [4]

It is very important to monitor the SCC monthly at the herd level and the cow level. At the herd level, monthly monitoring of SCC could be very useful diagnostic tool for determining cows with subclinical mastitis. High prevalence of subclinical mastitis has the herd that has problems with enviromental mastitis pathogens, and veterinarian should investigate housing conditions, udder hygiene and management of dry and periparturient cows. In the cases where contagiuos mastitis is a problem, prevalence increase as lactation progresses and as cow age because of more opportunities for exposure to infected milk. If we suspect on contagious mastitis, we should investigate the transmission of mastitis pathogens during milking. The existence of a large number of cows with chronically increased SCC, indicates that cows are infected with host adapted pathogens transmitted in a contagious manner. At the cow level, it is helful to look at the list of individual cows sorted by SCC to identify cows that require individual therapy. Also, it is helful to use of a rapid cowside quarter level SCC test [5].

2. The treatment of mastitis

The success of the treatment of mastitis depends on a number of factors, such as accurate diagnosis, proper antibiotic selection [6], early onset therapy, udder physiology and pathology, supportive therapies and elimination of predisposing factors [7], as well as factors of pathogenicity of the microorganism itself.

The goal of antibiotic therapy is to destroy pathogenic microorganisms without damaging tissue of the mammary gland. Therapy of subclinical mastitis is not indicated except in cases where the causative agents of mastitis *Staphylococcus aureus* or *Streptococcus agalactie*. Therapy is then administered immediately after the diagnosis is made and should not wait the dry moment. The general principle of subacute mastitis therapy involves intramammary application of the preparation four times during 12 hours for fast-absorbed medicines or three times during 24 h for drugs that are slowly absorbed. Treatment should be continued at least 24 hours after clinical symptoms stop [9]. In peracute and acute mastitis, intramammary therapy is administered 3-5 days, twice a day, and depending on the severity of the clinical symptoms of the disease, antibiotics can be administered and parenterally with compulsory supportive therapy [9].

Four large groups of factors can lead to the failure of the treatment of mastitis:

- Management and iatrogenic factors
- Factors related to the microorganisms themselves
- Factors related to the application of the drug and damaging the mammary gland
- Factors related to the choice and method of application of the drug

Many management failures as well as iatrogenic factors can cause failure in therapy. The literature states the following factors: incorrect diagnosis [incorrectly take samples for microbiological analysis, some microorganisms require special conditions for cultivation and cannot be isolated in standard cultivation], delayed start of therapy, inadequate supportive therapy, duration of treatment, inadequate dose of the drug, improper use of the drug, introduction of new intramammary infections of the over-ester injectors, super infections, reinfection, as well as completely clinical but not microbiological cure [7,11,12].

A large number of different drug-related factors can lead to failure of therapy [7, 8, 12], such as: inadequate drug choice, short drug half-life, inadequate concentration of the drug lócally in the mammary tissue, adverse effects of the drug, low level of pharmacokinetic utilization, poor drug resorption through blood-milk barrier, high degree of binding of milk and serum proteins and combination of mzcocide and mycostatic antimicrobial agents [7].

3. Preventing the emergence of antibiotic residues in milk

Wrong and inadequate antibiotic therapy in the treatment of mastitis entails a large number of potential risks. Antibiotic residues adversely affect fermentation processes that are unavoidable in the production of cheese and yogurt. Residue of antibiotics can cause allergy in susceptible individuals, as well as the occurrence of resistance in some pathogenic microorganisms. The main causes of the persistence of antibiotic residues in milk are the administration of not recommended drugs whose pharmacokinetics is not completely established, excess dose administration, lack of compliance with the milk discard period in the case of treated animals, identification errors of treated animals, and confusion of milk after milking treated animals [13,14]. Also, the big problem on smaller farms is the self-inducing and inefficient application of the injector by irresponsible farmers. Special attention should be paid to veterinarians to monitor the antibiotic withdrawal period due to other therapeutic indications. Such cows must be separately labeled and separated from each healthy mature cow. Milk treated cows must be destroyed in a special way, and in any case, such calves are fed calves or used for other purposes.

In July 2017, Codex Alimentarius Commission prescribed new legislation “Maximum residue limits [MRLs] and risk management recommendations [RMRs] for residues of veterinary drugs in foods” [15]. But in Serbia, according to legislation on the quality of raw milk, the minimum allowed concentrations of residues in milk are not clearly defined [16].

Acknowledgement
This work is supported by project funds III 46002 financed by the Ministry of Education, Science and Technological Development of the Republic of Serbia

4. References

[1] Makovec J A and P L Ruegg 2003 *J Dairy Sci* 86 3466
[2] Lam T J G M, Olde Rekerink R G M, Sampimon O C and Smith H 2009 *Irish veteterinary journal* **62** p 34
[3] Bačić G 2009 *Dijagnostika i liječenje mastitisa u goveda* (Medicinska naklada Zagreb)
[4] Rueeg P 2011 *ANEMBE Congress procedings*, Avilla, Spain
[5] Down P M, Bradley A J, Breen J E, Hudson C D and Green M J 2016 *Veterinary record* **178** 449
[6] Royster E and Wagner S 2015 *Food anim Pract* **31** 17
[7] Du Preez J H 2002 *J South Afr Vet Assoc* **71** 20
[8] Erskine R J, Wagner S and DeGraves F J 2000 *Food Anim Pract* **19** 109
[9] Moore G A and Heider L E 1984 *Large Anim Pract* **6** 323
[10] Prescott F and Baggot J D 1988 *Antimcrobial therapy in veterinary medicine* (Blackwell Scientific Publication Boston, USA)
[11] Pyorala S 2002 *Rep Dom Anim* **37** p 211

[12] Serieys F, Raquet Y, Goby L, Schmidt H and Friton G 2005 *J Dairy Sci* **88** 93
[13] McEwen S A., Black W D and Meek A H 1991 *J Dairy Sci* **74** 2128
[14] Seymour E H, Jones G M and McGilliard M L 1988 *J. Dairy Sci* **71** p 2292.
[15] *Codex alimentarius CAC*/MRL 2-2017
[16] *Službeni glasnik Republike Srbije* 21/2009

IOP Conf. Series: Earth and Environmental Science **85** (2017) 012036 doi:10.1088/1755-1315/85/1/012036

Effect of modified atmosphere and vacuum packaging on TVB-N production of rainbow trout (*Oncorhynchus mykiss*) and carp (*Cyprinus carpio*) cuts

J Babić Milijašević, M Milijašević, J Đinović-Stojanović and D Vranić

Institute of Meat Hygiene and Technology, Kaćanskog 13, Belgrade, Serbia

E-mail: jelena.babic@inmes.rs

Abstract. The aim of our research was to examine the influence of packaging in modified atmosphere and vacuum on the total volatile basic nitrogen (TVB-N) content in muscle of rainbow trout (*Oncorhynchus mykiss*) and common carp (*Cyprinus carpio*), as well as to determine the most suitable gas mixtures for packing of these freshwater species. Three sample groups of trout and carp cuts were investigated. The two groups were packaged in modified atmosphere with different gas ratios: 90%CO_2+10%N_2 (MAP 1) and 60%CO_2+40%N_2 (MAP 2), whereas the third group of fish cuts were vacuum packaged. During trials, the trout and carp cuts were stored in refrigerator at 3°C±0.5°C. Determination of TVB-N was performed on 1, 4, 7, 9, 12 and 14 days of storage. The obtained results indicate that the investigated mixtures of gases and vacuum had a significant influence on the values of TVB-N in trout and carp cuts. The lowest increase in TVB-N was established in trout and carp cuts packaged in MAP 1, whereas the highest increase was established in vacuum packaged cuts. Based on the obtained results, it can be concluded that the gas mixture consisting of 90% CO_2 and 10% N_2 was the most suitable for packaging of fresh trout and carp cuts in terms of TVB-N value.

1. Introduction

Fish are highly perishable and prone to vast variations in quality due to differences in species, environmental habitats, feeding habits and action of autolysis enzymes as well as hydrolytic enzymes of microorganisms on the fish muscle. Deterioration of fish mainly occurs as a result of bacteriological activity, leading to loss of quality and subsequent spoilage [1].

The shelf life of fresh chilled fish can be extended by vacuum packaging or modified atmosphere packaging (MAP). The most frequently used packaging gases include oxygen (O_2), carbon dioxide (CO_2), nitrogen (N_2) and their combinations in different ratios. Gas mixtures with higher levels of CO_2 and N_2 as inert gases have attracted the most attention from researchers. The most frequently used CO_2 concentrations in fish packaging fall between 40 and 60%. CO_2 acts as an antimicrobial agent. It inhibits the growth of spoilage organisms, particularly *Shewanella putrefaciens*, *Pseudomonas*, *Vibrio* and *Aeromonas spp.* that produce H_2S [2]. The bacteriostatic effect of CO_2 depends on its concentration and storage temperature, and the mechanisms of action are based on changing the permeability of the bacterial membrane, inhibition of bacterial enzymes, alteration in the physico-chemical properties of the protein, and changes in the intra-cell pH. N_2 is used for displacing O_2 from packaging, decreasing oxidative rancidness and inhibiting the growth of aerobic microorganisms [3].

The combined total amount of ammonia (NH_3), dimethylamine (DMA) and trimethylamine (TMA) in fish is called the Total Volatile Base Nitrogen (TVB-N); its content in the fish meat is commonly used as parameter for spoilage estimation, and as index for freshness of fish. It is produced during degradation of protein and non-protein nitrogen components, caused mainly by metabolic activity of fish spoilage bacteria and endogenous enzymes action [4]. NH_3 is formed by the bacterial degradation/deamination of proteins, peptides and amino-acids. It is also produced in the autolytic breakdown of adenosine monophosphate. The presence of DMA and TMA in spoiling fish is due to the bacterial reduction of TMA oxide (TMAO) which is naturally present in the living tissue of all marine and many freshwater fish species [5]. In anaerobic conditions, spoilage bacteria are able to utilize TMAO as the terminal electron acceptor which results in off-odours and -flavours due to formation of TMA. TMA and TVB-N are considered responsible for unpleasant 'fishy' odour of spoiled fish [6].

The aim of this research was to examine the influence modified atmosphere and vacuum packaging on changes in TVB-N values of rainbow trout (*Oncorhynchus mykiss*) meat and carp (*Cyprinus carpio*) meat and to determine the most suitable gas mixtures for the packaging of these two freshwater fish species.

2. Materials and Methods

2.1. Sample collection

Rainbow trout (*Oncorhynchus mykiss*) used in the study were all farmed in the same conditions and came from a trout pool located on the slopes of Zlatibor Mountain. Marketable carp (*Cyprinus carpio*) originated from a fish farm pond located in the lowland region of Serbia, where semi-intensive farming was used. In this study, two-year-old carp of average body weight of 2.5 kg were used. Three sample groups of cleaned trout and carp cuts were formed. Group 1 cuts were vacuum packaged and were used as the control. The other two sample groups were packaged in modified atmospheres with different gas ratios: MAP 1: 90% CO_2+10% N_2 and MAP 2: 60% CO_2+40% N_2. The machine used for packaging was Variovac (Variovac Primus, Zarrentin, Germany), and the material used for packaging was foil OPA/EVOH/PE (oriented polyamide/ethylene vinyl alcohol/polyethylene, Dynopack, Polimoon, Kristiansand, Norway) with low gas permeability (degree of permeability for O_2 - 3.2 cm^3/m^2/day at 23°C, for N_2 - 1 cm^3/m^2/day at 23°C, for CO_2 - 14 cm^3/m^2/day at 23°C and for steam 15 g/m^2/day at 38°C). The ratio of gas:fish cuts in the packages was 2:1. All fish cuts were stored in the same conditions at 3°C±0.5°C and on 1, 4, 7, 9, 12 and 14 days of storage, chemical testing was performed.

2.2. Chemical analysis

The total volatile basic nitrogen (TVB-N) was determined by using the official steam distillation method according to Commission Regulation (EC) 2074/2005 and was expressed as mg TVB-N/100 g.

2.3. Statistical analysis

The mean values and standard deviations were calculated by using column statistics with the processing of six values for each analysed group. Significant differences between groups were calculated by using one-way ANOVA analysis by Tukey's comparative test in the program Microsoft Office Excel (2010). Differences were considered as significant when p value was < 0.05.

3. Results and Discussion

At the beginning of the study, concentrations of TVB-N in carp steaks as well as in trout cuts were practically identical ($P > 0.05$). During the storage period, TVB-N value increases were observed in all experimental groups. As shown in figures 1 and 2, TVB-N values were strongly affected by the packaging atmosphere used. Increases in TVB-N values of carp and trout cuts in the different packaging atmospheres followed the order: MAP 1 < MAP 2 < vacuum.

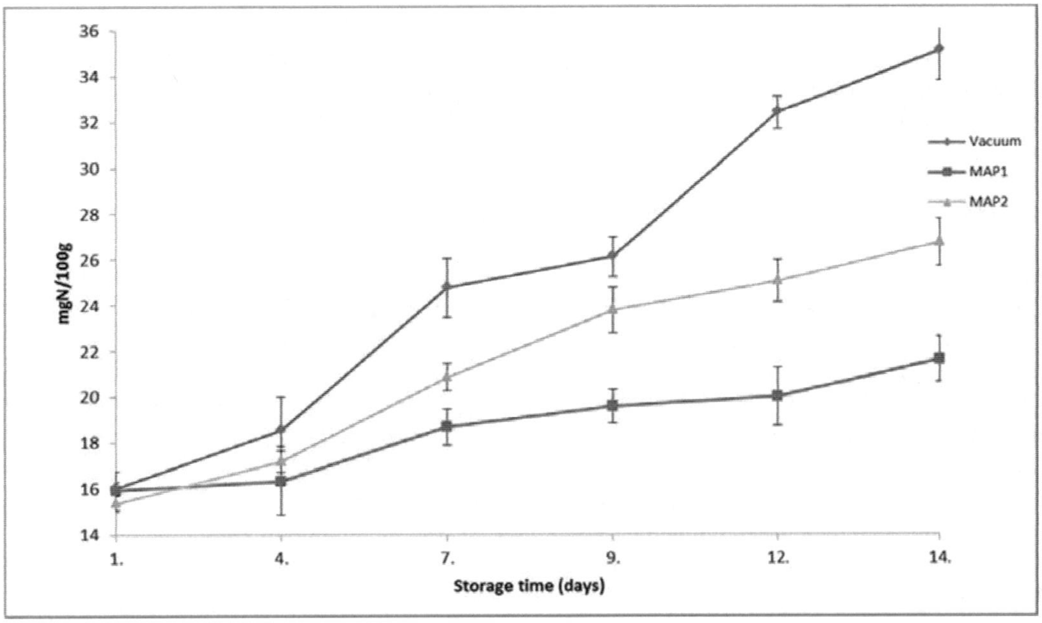

Figure 1. TVB-N value of common carp steaks packaged under different conditions during the storage period.

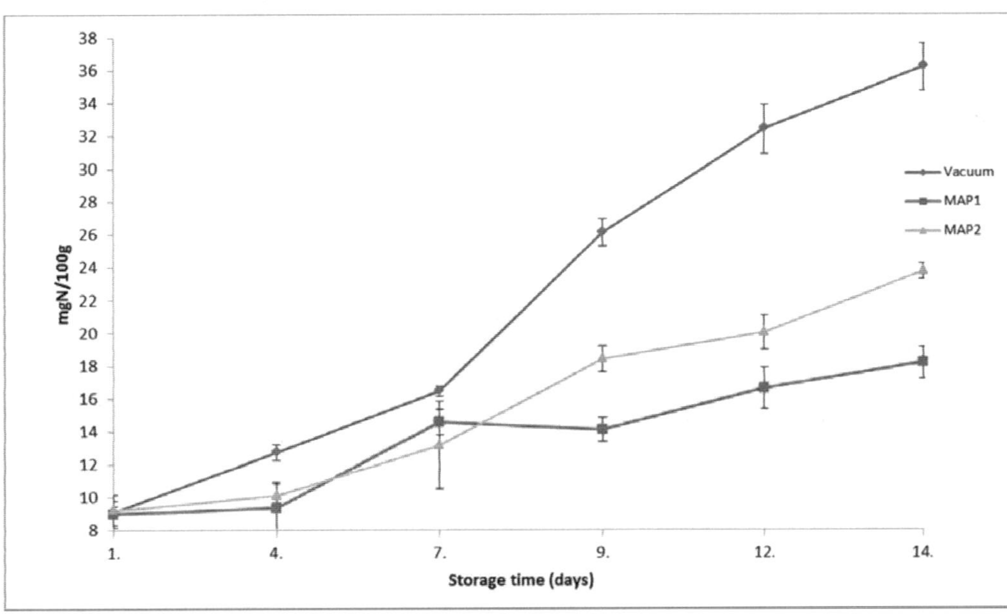

Figure 2. TVB-N value of trout cuts packaged under different conditions during the storage period.

TVB-N levels in fish packaged under MAP 1 changed to lesser extent comparing to those levels in fish packaged under MAP 2 and vacuum. It is interesting to note that up to 7 days of storage, there were no statistically significant differences ($p > 0.05$) in TVBN values between fish packaged in MAP

1 and MAP 2. This may be attributed to the fact that when the total bacteria count of fish flesh is still relatively low, a gas mixture of 60% CO_2 + 40% N_2 is adequate to inhibit the microbial activity, as exhibited by low TVBN values. When microbial populations increase, a higher concentration of CO_2 is required to inhibit the microbial activity and the subsequent spoilage.

According to Masniyom *et al.* [7], a high CO_2 concentration potentially inhibited the growth of mainly Gram negative microorganisms and decreased the deamination capacity of bacteria, resulting in lower production of volatile compounds. The same observations were reported by Milijašević *et al.* [8] and Babić *et al.* [9] for carp and trout stored under MAP, which support the results of the present study.

The TVB-N limit from 25 to 35 mg N/100g has been recommended by some researchers as an indicator for rejecting commercial fresh whole fish and processed fish products [4]. However, no limit for acceptability of TVB-N in common carp and rainbow trout has been established by the Commission Regulation (EC) 2074/2005. Examining the effect of several different gas mixtures on the shelf life of rainbow trout fillets, Gimenez *et al.* [10] found that MAP is very effective in preventing the production of TVB-N. These authors recommended a value of 25 mg N/100g as the highest acceptability level of TVB-N for rainbow trout [10]. In our research, this limit was exceeded in control fish cuts on day 9, while TVB-N values in MAP 1 and MAP remained below this recommended limit. In their research, Ježek and Buhtova [11] recommend 20 mg N/100g in carp meat as the highest acceptable limit for TVB-N. In our current study, this limit was exceeded in control fish cuts (day 7), MAP2 fish cuts (day 9), as well as in MAP1 fish cuts (day 14).

4. Conclusion

Based on the obtained results, it can be concluded that the gas mixture consisting of 90% CO_2 and 10% N_2 was the most suitable for packaging of fresh trout and carp cuts in terms of TVB-N value.

References

[1] Chytiri S, Chouliara I, Savvaidis I N and Kontominas M G 2004 *Food Microbiol.* **21** 157
[2] Debevere J and Boskou G 1996 *Int J. Food Microbiol.* **31** 221
[3] Farber J M 1991 *J. Food Prot.* **54** 58-70
[4] Connell J J 1990 *Control of fish quality* (Oxford: Fishing News Books) p 122
[5] Huss H H 1995 FAO Fisheries technical paper – 348 (Available at: http://www.fao.org/docrep/V7180E/V7180E00.HTM)
[6] Gram L and Huss H H 1996 *Int J. Food Microbiol* **33** 121
[7] Masniyom P Benjakul S and Visessanguan W 2002 *J. Sci. Food Agric.* **82** 873
[8] Milijašević M, Babić J, Baltić MŽ, Spirić A, Velebit B, Borović B and Spirić D 2010 *Tehnologija mesa* **51**(1) 66
[9] Babić J, Dimitrijević M, Milijašević M, Đorđević V, Petronijević R, Grbić S and Spirić A 2014 *Hem. Ind.* **68** 69
[10] Gimenez B P, Roncales P and Beltran J A 2002 *J. Sci Food Agric.* **82** 1154
[11] Ježek F and Buchtová H 2010 *Acta Vet. Brno* **79** 117

Chemical and biotechnological processing of collagen-containing raw materials into functional components of feed suitable for production of high-quality meat from farm animals

M I Baburina, A N Ivankin and I A Stanovova

The V.M.Gorbatov All-Russian Meat Research Institute, Russian Academy of Agricultural Sciences, Moscow, Russian Federation

E-mail: baburina2005@yandex.ru

Abstract. The process of chemical biotechnological processing of collagen-containing raw materials into functional components of feeds for effective pig rearing was studied. Protein components of feeds were obtained as a result of hydrolysis in the presence of lactic acid of the animal collagen from secondary raw materials, which comprised subcutaneous collagen (cuticle), skin and veined mass with tendons from cattle. For comparison, a method is described for preparing protein components of feeds by cultivating *Lactobacillus plantarum*. Analysis of the kinetic data of the conversion of a high-molecular collagen protein to an aminolyte polypeptide mixture showed the advantage of microbiological synthesis in obtaining a protein for feeds. Feed formulations have been developed to include the components obtained, and which result in high quality pork suitable for the production of quality meat products.

1. Introduction

Feeds of animal origin play an important role in providing essential amino acids for maintaining high animal productivity and improving meat quality. A decisive role in the development of the forage reserve is assigned to the creation of new feed additives. Increases in meat production require additional volumes of mixed fodder for fattening animals. However, the need for mixed fodders for animal feed is reduced by increasing their nutritional value [1-4].

At present, much attention has been paid to the rational use of low-value products from slaughter and processing of livestock to produce protein hydrolysates, which are used to produce feed and microbiological media [5-7].

The most promising raw materials for obtaining protein hydrolysates for fodder are collagen-containing wastes that are the most difficult to process into nutrients: beef, including veins, pork skin and collagenous tendons [8]. The yield of such collagen-containing wastes from the slaughter of 1000 head of cattle or pigs of different fatness is, on average, 2-10% by weight for each batch of animals [9].

Rational uses for most meat production wastes have not yet found been found, which, beside material losses, leads to environmental pollution. Therefore, studies aimed at solving the production of high quality feed protein from animal processing waste are quite a current, but complex, problem [10-12].

In connection with the foregoing, our research task was to develop ways to process collagen waste that would preserve the biological value of protein products and ensure high digestibility when fed to animals, in order to obtain high-quality meat.

2. Material and Methods

Collagen products were used: beef hypodermic collagen, bovine mass of cattle tendons, and skin of pigs.

Theoretical and experimental studies were carried out using conventional, standard and novel methods of biochemical, physico-chemical and structural-mechanical analysis [13-15].

Determination of the total nitrogen of the protein was performed with the Kjeltec 8200, Foss (Denmark) analyzer.

Determination of amine nitrogen was carried out by spectrophotometric method using 2,4,6-trinitrobenzenesulfonic acid (TNBS). The degree of hydrolysis was determined as the ratio of amine nitrogen to total nitrogen.

The mass fraction of amino acids was determined on the amino acid analyzer Biotronic 6001 (Germany), using distribution chromatography after acid hydrolysis of proteins.

Free amino acids were determined in products after treatment by adding 10% trichloroacetic acid to precipitate the proteins, neutralizing to pH 2.0, filtering through a Millipore membrane filter with a nominal pore diameter of 0.22 μm, followed by dilution of the filtrate in sample dissolution buffer pH 2.2. Quantitative estimation of the content of individual amino acids was carried out by comparing the areas of peaks on the aminogram calculated using the Winpeak integrating system of Eppendorf-Biotronic (Germany) for the areas of peaks obtained by analyzing a standard mixture of amino acids containing 2.5 μmol of each amino acid in 1 mL of standard solution.

Kinetic measurements during acid hydrolysis were carried out as follows. All substrates were, before hydrolysis, subjected to drying and grinding on a disintegrator. Portions of crushed substrates weighing 1 g were placed in glass ampoules, 4 ml of an acid solution of the required concentration were introduced, purged with argon for 3 minutes, sealed and placed in a thermostat for hydrolysis. Hydrolysis was carried out with solutions of lactic acid at concentrations of 10, 20, 30, 40% for 0.5 - 8 hours at 95-105°C, and then measuring the concentration of nitrogen of the amino end groups of amino acids and lower peptides by formal titration and the content of free amino acids in hydrolysates at the end of the process. The optimum concentration of acid was that at which the maximum degree of hydrolysis of the protein was observed in 6 hours of the process in combination with the minimum possible destruction of labile amino acids [14].

The obtained hydrolysates were used in liquid form with content (%): protein - 18.4; fat - 5.5; ash - 1.6; carbohydrates - 4.5; moisture - 70, amine nitrogen - 365 mg%. In dry form, obtained after spray drying, they were (%): protein - 55.2; fat - 16.5; ash - 4.4; carbohydrates - 13.5; moisture - 10, amine nitrogen - 1033 mg%.

Lactobacillus plantarum ATCC 8014 was used to produce the microbial protein. *L. plantarum* was cultivated in nutrient medium containing (g/L): peptone or lactic acid hydrolyzate of animal collagen - 20, yeast extract - 14, K_2HPO_4 - 6, KH_2PO_4 - 3, NaCl - 5, $MgSO_4$ - 5, initial pH 6.8. The inoculum was grown in Erlenmeyer flasks with a capacity of 0.75 L containing 0.2 L of this medium into which 10 g/L of a sterile glucose solution was added.

A seed culture was produced in flasks at 200 rpm on a shaker at 37°C for 16 h until the optical density exceeded the optical density of uninoculated medium by 6-8 times at 546 nm. The seed culture was used to seed the nutrient medium at amounts of 20% of the volume of 10 L of the Ankum 2M fermenter so that the filling factor was 0.65. Temperature 37°C, sterile air flow 3 L/min.L, stirring speed 200-350 rpm, pH 6.5-6.8 was maintained by adding 25% ammonia solution and 10% sterile glucose solution. The grown cells were separated by centrifuging at 3000 *g* and drying with a spray

IOP Conf. Series: Earth and Environmental Science **85** (2017) 012037 doi:10.1088/1755-1315/85/1/012037

dryer at a low temperature. A protein preparation with a moisture content of 10% and a protein content of 55-65% was obtained.

3. Results and Discussion

As raw materials, high-collagen products from beef and pork carcasses were used. These included the mass of split beef carcass meat with tendons, as well as pork skin. In these high-collagen raw materials, there are no amino acids such as cysteine and tryptophan, and histidine, methionine and tyrosine occur in very small quantities.

The amino acid composition of the raw material showed that, depending on the type and amount of tendons, the total protein content of an inferior amino acid composition is 15-45%. Moreover, the protein content increased with increasing the specific content of the cartilaginous tissue of the tendons, ligaments and other high-collagen tissues. The increased protein content is a plus from the point of view of the potential protein reserve, but this protein cannot be used directly without biochemical processing in feed compositions because of known factors – the stiffness of the cartilaginous tissue and the negative effect on the taste characteristics of the feed.

Since the feedstock can contain 20 to 90% of protein in dry matter, one of the possible (affordable, inexpensive, economical) methods of processing it is hydrolysis with food acid, which allows the hydrolysable collagen proteins to be cleaved with a high degree of conversion [13-15].

Previous studies have shown that the use of proteinases with collagenase activity (crab, microbial or animal origin) allows partial transformation of the raw material to produce protein peptide, amino acid mixtures that have high biological value. The yield of this processing is on average, depending on the type of raw materials, 5-45%; however, a significant fraction of the raw material in the presence of enzymes is not completely processed [14, 16-18].

The collagenous raw material used in the current study contained 3-25% lipids. It is known that in the collagenous tissue of mature animals, the total content of carbohydrates is 0.7-1.3% depending on the location, and in cartilaginous tissues, carbohydrate content can exceed 3% of the mass of the raw material. Moreover, if we consider the fleshy tissues, then the typical composition of carbohydrates is: lactic acid 0.9%, glucose-6-phosphate 0.15%, glycogen 0.1%, glucose 0.05% [2, 14].

In the cartilaginous tissue, the main carbohydrate component is chondroitin sulfates A, B, C, plus hyaluronic acid, and combinations of these compounds in protein complexes. This material can be so strong that in biochemical analysis, part of carbohydrates can collapse with an underestimation of the observed results [15].

Thus, for the production of fodder-functional hydrolysates, an animal feed containing 15-45% protein, 3-25% fat, 1-3% carbohydrate, and the rest (up to 70%) water was used.

It is known that acid hydrolysis leads to the destruction of some amino acids, so it was of interest to evaluate the kinetics of the release process of individual amino acids during hydrolysis for subsequent use as feed. The greatest total yield of free amino acids from a protein derived from veined mass with tendons was observed upon exposure to 20% $CH_3CH(OH)COOH$. Therefore, kinetic studies were carried out using lactic acid of this concentration. The kinetic dependences of the accumulation of most amino acids during hydrolysis at 95, 100, and 105 °C had the form characteristic of pseudo-first order reactions.

The macroconstantity of the reaction rate k_{eff} (sec^{-1}) was found graphically from the equation $ln(P_\infty - P) = lnP_\infty - k_{eff}t$ as the tangent of the slope of the line in the coordinates $\{ln[P_\infty / (P_\infty - P)]\ ,\ t\}$, calculated by the method of least squares, where P is the concentration of the reaction product at time t, g/L; P_∞ is the concentration of the reaction product after completion of the reaction, g/L.

For the kinetic curve of the first-order reaction, the initial period of which was not fixed, the macroconstant reaction rate can be found graphically from the following expressions [14]: ln $ln[P_\infty - P) = ln\ P_\infty - kt$, where P is the concentration of the reaction product at time t, P_∞ is the concentration of the reaction product after completion of the reaction. This expression shows that in the case of a first-order reaction, the absolute value of the effective rate constant does not depend on the units in which the concentrations of reaction products are expressed. Therefore, any physical quantities proportional to the concentrations can be used to calculate the effective rate constant. Such a value in this case is the nitrogen content of amino groups of amino acids and lower peptides, determined by titration. For P_∞, nitrogen of amino groups was taken, measured after 24 hours hydrolysis with 20% lactic acid.

The effective reaction rate constants were found graphically from the tangent of the slope of the line in the coordinates $\{ln[P_\infty / (P_\infty - P)]\ ,\ t\}$ by the least squares method. The activation energy of the process was found from the Arrhenius equation. The values of the total reaction rate constants at 95-105°C were in the interval 1.2-1.8 x 10^{-4} sec^{-1}, and the activation energy of the process was 24.0 kJ/mol.

The study showed that, in the process of hydrolysis of the studied types of raw materials, the accumulation of aspartic and glutamic acids, glycine, alanine, tyrosine and phenylalanine occurred most rapidly. In all cases, destruction of histidine, serine and methionine was noted, and the degree of destruction of the latter was greatest. The kinetic curves of accumulation of cystine as a result of hydrolysis did not have a maximum. Probably, the rate of accumulation of cystine was higher than the rate of its destruction due to the high content of this amino acid.

Consideration of the kinetics of accumulation and destruction of amino acids during acid hydrolysis made it possible to determine the conditions for achieving the highest degree of protein conversion while preserving labile amino acids: temperature 100°C, time 6 h, lactic acid concentration 20%.

The resulting acid hydrolyzate contained insufficient amounts of methionine and cystine, but, however, it had a satisfactory amino acid composition that could be introduced into the feed composition and used to balance its amino acid composition. The resulting acid hydrolysates contained relatively large amounts of aspartic and glutamic acids, while glycine, proline, and alanine were present in sufficient quantities.

According to the literature, it is known that such amino acids (individually and in various combinations), such as glycine, proline, alanine, etc., are known to have the greatest attractant activity. When added to the feed mixture for pigs, for example, feed intake tripled. These facts indicate the important role of these amino acids in the process of intensive growth of agricultural animals [1, 14].

In the feed industry, collagen additives can be used to produce granulated and extruded feeds in order to increase their biological value and the strength of the pellets. The hydrolyzate of protein raw material after pasteurization and drying can be used both as a highly digestible protein supplement for piglets and as a protein base of a nutrient medium for cultivation of probiotic bacteria [16-19].

At present, Russia is paying great attention to the use of new, non-traditional types of raw materials in the composition of feed additives and mixed fodders, the use of which would improve the physiological and ecological status, productivity, preservation and reproduction of livestock and result in high-quality and environmentally safe food.

To develop the technology of ecologically clean, biologically active, new generation feed additives, the following components were selected as initial components: secondary waste from the meat processing industry (collagen hydrolyzate) and waste from oil extraction plants (corn cake). The selected ingredients of vegetable and animal origin are valuable food products, due to their high protein content and the favorable combination of proteins and fats, okys calcium and phosphorus, corresponding to the balance for these feed components. To increase the biological and nutritional value, the feed additive was subjected to hydrobarrothermal treatment.

As the technology of meat processing improves, the amount of waste will decrease, and the yield of food products from raw materials will increase. Therefore, there is no real reason to expect an increase in the rate of production of animal feed. However, contradictions between the needs of the intensive livestock sector and the production of high-grade protein will be exacerbated in the future. An alternative to animal feed may be a protein of microbiological origin.

The main producers of microbiological protein are yeast, bacteria, low and higher fungi and unicellular algae [19]. Cattle require five years to double their protein mass, pigs - 4 months, chickens - 1 month, but for bacteria and yeast, 1-6 hours are sufficient. At the same time, microorganisms differ from animals, as they have high (from 40 to 55% dry weight) protein content, balanced by amino acid composition, and they also contain carbohydrates, lipids, vitamins, macro- and microelements [18]. Bacteria grow much faster than yeast cells, building up biomass and, in addition, bacterial proteins contain more cysteine and methionine, which allows them to be classified as proteins with high biological value.

As a result of the studies, the nutrient medium and the conditions for culturing lactic acid probiotic bacteria were selected, the amount and age of the seed culture, the temperature regime and the duration of cultivation of the crop, the drying parameters of the additive were determined.

A fodder protein supplement of microbial origin obtained from processing animal proteins and used as a source of nitrogen for the cultivation of lactic acid probiotic bacteria contained about 55% protein and 16% fat, and lactic acid bacteria numbers reached 8.0×10^9 cfu/g.

Animals in the experimental group should receive a similar compound feed (control), but with the inclusion of 4.5% of the novel feed additive developed by us instead of fishmeal. According to preliminary calculations, the use of the developed additive of microbial origin will increase the productivity of animals by 8-10% and reduce the per unit feed costs by 6-8%.

4. Conclusion

Kinetic data on the process of transforming animal collagen into a nutrient component for feeds or microbiological media that is of acceptable nutritional value was analyzed. The study showed that the biotechnological method of growing a microbial protein, in terms of the rate of production of protein product, greatly exceeded the traditional method of growing farm animals followed by the subsequent processing of secondary raw materials into a protein product. The research conducted proved it possible to obtain a high-protein feed supplement with probiotic properties for efficient pig growth. Carcasses obtained from the grown piglets fed on this supplement, resulted in pork of the highest quality, eminently suitable for the production of meat products

References

[1] Soloveva A A, Zinina O V, Rebezov M B, Lakeeva M L and Gavrilova E V 2013 *Young Scientist* **5** 102
[2] Lähteinen T, Rinttilä T, Koort J M K, Kant R, Levonen K, Viljanen M J, Björkroth J and Palva A 2015 *Livestock Sci.* **180** 164
[3] Dhillon G S, Kaur S, Oberoi H S, Spier M R and Brar S K 2016 **Protein Byproducts 1** 21
[4] Martínez-Alvarez O, Chamorro S and Brenes A 2015 *Food Res. Int.* **73** 204
[5] Mora L, Reig M and Toldrá F 2014 *Food Res. Int.* **65** 344
[6] Pagán J, Ibarz A, Falguera V and Benítez R 2013 *J. Food Eng.* **119** 655
[7] Martínez-Alvarez O, Chamorro S and Brenes A 2015 *Food Res. Int.* **73** 204
[8] Ferraro V, Anton M, Santé-Lhoutellier V 2016 *Trends Food Sci. Tech.* **51** 65
[9] Petrovic Z, Djordjevic V, Milicevic D, Nastasijevic I and Parunovic N 2015 *Procedia Food Sci.* **5** 235
[10] Djekic I and Tomasevic I 2016 *Trends Food Sci. Tech.* **54** 94
[11] Djekic I 2015 *Procedia Food Sci.* **5** 61
[12] Veeruraj A, Arumugam M, Ajithkumar T and Balasubramanian T 2015 *Food Hydrocolloids* **43** 708
[13] Lisitsyn A B, Ivankin A N and Neklyudov A D 2002 *Methods of practical biotechnology* (Moscow: VNIIMP) p 420
[14] Neklyudov A D and Ivankin A N 2007 *Collagen: production, properties and application* (Moscow: Publishing house MGUL) p 336
[15] Veeruraj A, Arumugam M and Balasubramanian T 2013 *Process Biochem.* **48** 1592
[16] Li Z R, Wang B, Chi C, Zhang Q H, Gong Y, Tang J J, Luo H and Ding G 2013 **Food Hydrocolloids 31** 103
[17] Tamilmozhi S, Veeruraj A, Arumugam M 2013 *Food Res. Int.* **54** 1499

[18] Lähteinen T, Lindholm A, Rinttilä T, Junnikkala S, Kant R, Pietilä T E, Levonen K, Ossowski I, Solano-Aguilar G, Jakava-Viljanen M and Palva A 2014 *Vet. Immunol. Immunopathology* **158** 14

[19] Giang H H, Viet T Q, Ogle B and Lindberg J E 2012 *Livestock Sci.* **143** 132

Meat-based enteral nutrition

O K Derevitskay and A S Dydykin

The V.M.Gorbatov All-Russian Meat Research Institute, Russian Academy of Agricultural Sciences, Moscow, Russian Federation

E-mail: das_tih@mail.ru

Abstract. Enteral nutrition is widely used in hospitals as a means of nutritional support and therapy for different diseases. Enteral nutrition must fulfil the energy needs of the body, be balanced by the nutrient composition and meet patient's nutritional needs. Meat is a source of full-value animal protein, vitamins and minerals. On the basis of this research, recipes and technology for a meat-based enteral nutrition product were developed. The product is a ready-to-eat sterilised mixture in the form of a liquid homogeneous mass, which is of full value in terms of composition and enriched with vitamins and minerals, consists of particles with a size of not more than 0.3 mm and has the modified fat composition and rheological characteristics that are necessary for passage through enteral feeding tubes. The study presents experimental data on the content of the main macro- and micro-nutrients in the developed product. The new product is characterised by a balanced fatty acid composition, which plays an important role in correction of lipid metabolism disorders and protein-energy deficiency, and it is capable of satisfying patients' daily requirements for vitamins and the main macro- and microelements when consuming 1500-2000 ml. Meat-based enteral nutrition can be used in diets as a standard mixture for effective correction of the energy and anabolic requirements of the body and support of the nutritional status of patients, including those with operated stomach syndrome.

1. Introduction

Enteral nutrition is a kind of nutritional therapy whereby nutrients are introduced perorally or through a gastric tube. The use of this nutrition is most topical for surgical and gastroenterological patients and for any disease when the nutrient requirements of patients cannot be satisfied naturally. With that, the nutrition provided must fully meet the body's energy needs, have balanced nutrient composition and fulfil patients' nutritional needs [1]. Adequate protein consumption is necessary to maintain the nitrogen balance during illness. Provision of the optimal qualitative composition of protein (a ratio of essential and non-essential amino acids) in diets is achieved by including both plant and animal proteins, which increases their total biological value. Meat is a source of full-value animal protein, which contains all essential amino acids, some fat- and water-soluble vitamins and easily digestible haeme iron. In addition, inclusion of meat-containing products into a diet allows the extension and diversification of curative nutrition, when much attention is paid to food palatability in the design of diets, especially for patients with appetite disorders and loss of taste sensations with development of psychogenic anorexia.

2. Materials and Methods

An experimental meat-based enteral nutrition product enriched with vitamins and minerals and with a modified fatty acid profile was studied.

The organoleptic and physico-chemical indicators were determined by standard methods, fatty acid composition by gas chromatography, microbiological indicators by conventional methods, vitamins by high performance liquid chromatography, folic acid by competitive enzyme immunoassay using the test system RIDASCREEN®FAST Folsäure (Folic Acid) and minerals by the atomic absorption method. The component particle size was measured on histological preparations of the material using an image analysis system or ocular micrometre. Amino acid scores were calculated as the ratio of amount of amino acids in 1 g of the studied protein to amount of corresponding amino acids in 1 g of ideal protein, expressed in percentages; with that, the FAO/WHO amino acid scale was taken as a reference.

3. Results and Discussion

Researchers at the FGBNU The V.M. Gorbatov All Russia Meat Research Institute have developed recipes and technology for meat-based enteral nutrition products, which are intended for people in the post-operative period, and for conditions associated with appetite loss, mandibulofacial, craniocerebral and burn injuries, as well as chewing or swallowing impairment. The technology is adapted to the conventional conditions of the plants that produce meat-based child nutrition products. The developed product is a ready-to-eat, sterile mixture in the form of a liquid homogeneous mass, which is of full value in terms of composition, consists of particles not more than 0.3 mm in size and has a pleasant, slightly sweet taste and light brown colour. To stabilise the homogenous multicomponent mass and to prevent separation of the liquid phase during storage, we used a complex stabilising additive, which consisted of a mixture of hydrocolloids in an amount that ensured the rheological characteristics necessary for passing through enteral feeding tubes [2]. The total composition of nutrients was based on the recommendations for healthy adults. The nutrient content in a portion (240 g) of the enteral nutrition product and the percentage of the daily requirement of macro- and micro-nutrients are given in tables 1 and 2.

Table 1. The nutrient content in a portion (240 g) of the enteral nutrition product and the percentage of the daily requirement.

Indicator	Content in a portion	% of daily requirement
Energy value, kcal	254.0	10
Protein mass fraction, g	9.6±0.1	12
Fat mass fraction, g	9.6±0.2	10
Carbohydrate mass fraction, g	33.6±0.2	10

Specialised enteral nutrition products must contain the full complex of vitamins and minerals, to avoid their deficiency when this artificial support is used over a long period of time. The mineral and vitamin contents of the developed product meet patients' daily requirements for vitamins and the main macro- and microelements when consuming 1500-2000 ml of product. The composition of the vitamin-mineral premix added to the product was adjusted to take into account results of preliminary studies. These showed losses of individual components in the course of the technological process and during product storage, so the adjusted amount provided a guaranteed level of those vitamins and minerals in the finished product.

Adequate protein intake is necessary to maintain nitrogen balance during illness. Assurance of the optimal qualitative composition of protein (a ratio of essential and non-essential amino acids) in a diet is achieved by including both animal and plant proteins, which increases their total biological value. The protein component in the studied product was a mixture of animal protein (beef with a mass fraction of fatty and connective tissues of not more than 6%) and plant protein (soya), in a ratio that ensured the product's high biological value; the minimal utility coefficient score was 0.99 unit fractions. This compares well with the utility coefficient that characterises the balance of essential amino acids relative to the physiologically necessary norm (FAO/WHO), which is 0.83 unit fractions.

Table 2. The content of selected vitamins and minerals in a portion (240 g) of the product and the percentage of daily requirement.

Mass fraction of vitamins		% of average daily requirement	Mass fraction of minerals		% of average daily requirement
B1, mg	0.36±0.02	24	Sodium (Na) , mg	240.0±36.0	19
B2, mg	0.38±0.03	24	Potassium (K) , mg	420.0±21.0	12
PP, mg	4.3±0.06	24	Calcium (Ca) , mg	192.0±28.8	19
B5 , mg	1.2±0.02	21	Phosphorous (P) , mg	172.8±34.4	21
B6, mg	0.4±0.01	21	Magnesium (Mg) , mg	55.2±11.0	14
B12, µg	0.48±0.01	48	Iron (Fe) , mg	3.8±0.57	26
Bc, mg	0.099 ±0.016	31	Copper (Cu), µg	360.0±72.0	36
H, µg	9.6± 0.9	19	Zinc (Zn) , mg	2.9±0.43	19
C, mg	24.0±4.0	38	Iodine (I), mg	0.030±0.004	21
A, µg	196.8±6.18	24	Manganese (Mn), mg	0.48±0.07	24
D3, µg	1.68±0.07	31	Selenium (Se), µg	13.68±2.74	19
E, mg	3.12±0.07	31	Chloride (Cl), mg	300.0±60.0	13

The fatty acid ratio plays an important role in correction of lipid metabolism disorders and protein-energy deficiency. Among the anti-alimentary factors of meat products is quite a high fat content with a prevalence of saturated fatty acids and cholesterol, which can lead to increasing blood lipid levels. Modification of the fat composition of specialised products *via* enrichment with monounsaturated and polyunsaturated fatty acids plays an important role in correcting lipid metabolism disorders. The fat component in the studied product is a combination of fat from the meat raw material, which contains mainly saturated fatty acids, and a mixture of rapeseed and soya oils, which provides a high level of monounsaturated and polyunsaturated fatty acids. Monounsaturated acids decrease the amount of harmful low density lipoproteins without injuring the protective high density lipoproteins. The oil ratio was selected in such a way as to fulfil the patient's need for essential ω6 and ω3 polyunsaturated fatty acids in a ratio not higher than 5:1. Table 3 presents the fatty acid composition of the meat-based enteral nutrition product. The fat component of this product contains relatively low levels of saturated fatty acids, has significantly higher content of monounsaturated fatty acids and a low ratio of omega-6 to omega-3 fatty acids (up to 4:1).

Table 3. Fatty acid composition of the enteral nutrition product.

Fatty acid	Fatty acid content, %
Lauric 12:0	0.03
Myristic 14:0	0.53
Isopentadecanoic 15:0i	0.04
Anteisopentadecanoic 15:0ai	0.04
Pentadecanoic 15:0	0.11
Palmitic acid 16:0	9.34
Margaric 17:0	0.25
Stearic 18:0	4.69
Arachidic 20:0	0.75
Behenic 22:0	0.34
Total saturated fatty acids	16.12
Myristoleic 14:1	0.17
Pentadecenic 15:1	0.05
Hexadecenic 16:1	0.08
Palmitoleic acid 16:1 9-cis	1.11
Erucic acid 22:1	0.12
Heptadecenic 17:1	0.28
Oleic 18:1	49.36
Elaidic 18:1	0.25
Gondoinic 20:1	0.88
Vaccenic 18:1 11-trans	2.26
Octadecenoic 18:1 11-cis	0.01
Iso-octadecenoic 18:1i	0.14
Total monounsaturated fatty acids	54.71
Linoleic 18 :2	23.47
α-Linolenic acid 18: 3 ω-3	5.54
Iso-Linoleic 18:2i	0.16
Total polyunsaturated fatty acids	29.17

To manufacture the product, meat raw material (beef) was obtained from healthy animals raised without the use of growth promoters, hormones, feed antibiotics or other kinds of unconventional feed materials. This ensured the specified safety requirements of finished product were met. Table 4 presents the indicators that characterise finished product safety.

The product met the requirements of commercial sterility for canned foods of group A in terms of the sanitary and microbiological indicators.

The clinical efficiency and product tolerability were assessed in the clinic at FGBUN's FRC of Nutrition and Biotechnology in patients with operated stomach syndrome [3]. The use of our meat-based enteral nutrition product in the diet of patients with operated stomach syndrome did not cause the occurrence of any adverse effects and was not accompanied by an increase in the intensity of existing complaints. Importantly, consumption of this product was characterised by decreases in the manifestation of heaviness in the epigastrium and decreased nausea, plus improvement of stool consistency in this population of patients. Moreover, it also facilitated an increase in all body composition indicators, and positively affected dynamics of laboratory indicators of protein metabolism among the patients.

Table 4. Safety indicators of the enteral nutrition product.

Indicator	Levels detected in the prepared enteral nutrition product	Maximum allowable levels
Mass concentration of toxic elements, mg/kg:		
Lead	0.026±10%	0.5
Arsenic	< 0.01	0.1
Cadmium	< 0.01	0.05
Mercury	< 0.002	0.03
Antibiotics, mg/kg:		
Levomycetin	not detected (detection limit 0.00008 mg/kg)	not allowed (<0.0003)
Tetracycline group	not detected (detection limit 0.01 mg/kg)	not allowed (<0.01)
Streptomycin	not detected (detection limit 0.1 mg/kg)	not allowed
Bacitracin	not detected (detection limit 0.004 mg/kg)	not allowed (<0.02)
Pesticides, mg/kg:		
Hexachlorocyclohexane (α-, β-, γ-isomers)	not detected	0.1
DDT and its metabolites	not detected	0.1
Radionuclides, specific activity, Bq /kg:		
Cesium-137	0±13	not more than 200
Strontium-90	0±18	-

4. Conclusion

As a result of this study, the recipe and technology for a meat-based enteral nutrition product have been developed. The product has balanced amino acid and fatty acid compositions, and is enriched with vitamins and minerals. Consumption of the product ensured effective correction of the energetic requirements among patients with nutritional status disorder caused by operated stomach syndrome.

References

[1] Manzyuk L V, Saltanov A I, Sel'chuk V Yu, Snegovoy AV and Snigur P V 2007 On the

question of enteral nutrition in oncology *Russian Med. Journal* **25** 1347-51
[2] Dydykin A S, Derevitskaya O K and Zokhabyan P R 2015 Mixtures on the basis of meat raw material for enteral nutrition *Meat Ind.* **7** 25-7
[3] Zainudinov Z M, Isakov V A, Zokhrabyan P R, Dydykin A S, Derevitskaya O K, Aslanova M A and Soldatova N E 2017 Assessment of clinical effectiveness and tolerability of meat containing canned foods for enteral nutrition *Problems of Nutrition* **86**(3) 6-14

Use of artificial intelligence in the production of high quality minced meat

B R Kapovsky, V A Pchelkina, P I Plyasheshnik, A S Dydykin and A A Lazarev

The V.M.Gorbatov All-Russian Meat Research Institute, Russian Academy of Agricultural Sciences, Moscow, Russian Federation

E-mail: system@vniimp.ru

Abstract. A design for an automatic line for minced meat production according to new production technology based on an innovative meat milling method is proposed. This method allows the necessary degree of raw material comminution at the stage of raw material preparation to be obtained, which leads to production intensification due to the traditional meat mass comminution equipment being unnecessary. To ensure consistent quality of the product obtained, the use of on-line automatic control of the technological process for minced meat production is envisaged. This system has been developed using artificial intelligence methods and technologies. The system is trainable during the operation process, adapts to changes in processed raw material characteristics and to external impacts that affect the system operation, and manufactures meat shavings with minimal dispersion of the typical particle size. The control system includes equipment for express analysis of the chemical composition of the minced meat and its temperature after comminution. In this case, the minced meat production process can be controlled strictly as a function of time, which excludes subjective factors for assessing the degree of finished product readiness. This will allow finished meat products with consistent, targeted high quality to be produced.

1. Introduction

Modern trends in the development of meat processing equipment include the use of highly precise methods for meat raw material analysis and high-calibre control of raw meat processing [1]. According to expert estimates, large manufacturers of sausages and meat products in the Russian Federation (RF) will trend towards quality stabilisation and maximal automation of technological processes.

At present, according to different estimates, virtually all Russian meat processors utilise 85-95% imported equipment. In this connection, it is necessary to solve the problem of substituting this expensive, imported meat processing equipment in the light of the requirements of the Food Security Doctrine of the Russian Federation.

Earlier, the V.M. Gorbatov All-Russian Meat Research Institute developed the theoretical aspects for frozen meat raw material processing by a novel milling method: methodology was developed for analytical calculation of performance of the meat cutting process using a mill; a method was proposed for calculating energy expenditure for the meat raw material milling process based on the specific shear force established experimentally; an automatic control system was proposed for the minced meat production process using a process control computer (PCC) that ensures guaranteed high quality of finished products [2].

The 21st century is imposing new requirements on meat and meat product manufacture. To increase enterprise profitability, novel, innovative approaches to the organisation of raw material processing into finished products will be necessary, including mechanisation, robotisation and automation of production. Many manufacturers of food equipment are gradually beginning to develop universal lines for production of certain product types, where raw material is supplied at the entry; with that, processing, transportation and mixing with other ingredients are carried out in automatic mode. At the exit, a finished product or raw material suitable for the following technological processing stages is obtained [3].

The Executive Order of the RF President No. 642 of 01.12.2016 "On the Scientific and Technological Development Strategy of the Russian Federation" states that in the next 10 to 15 years, a transition to advanced intellectual production technologies, effective processing of agricultural produce, and the development of safe and quality food products including functional foods must be considered the priorities for the scientific and technological development of the RF [4]. Therefore, the development of a system for on-line automatic control of the technological process for minced meat production using artificial intelligence methods and technologies is quite topical.

2. Automatic line design

At present, the Institute has been developing an automatic line for minced meat production intended for the manufacture of semi-prepared products and sausages, including products for children's nutrition. The line design is presented in figure 1.

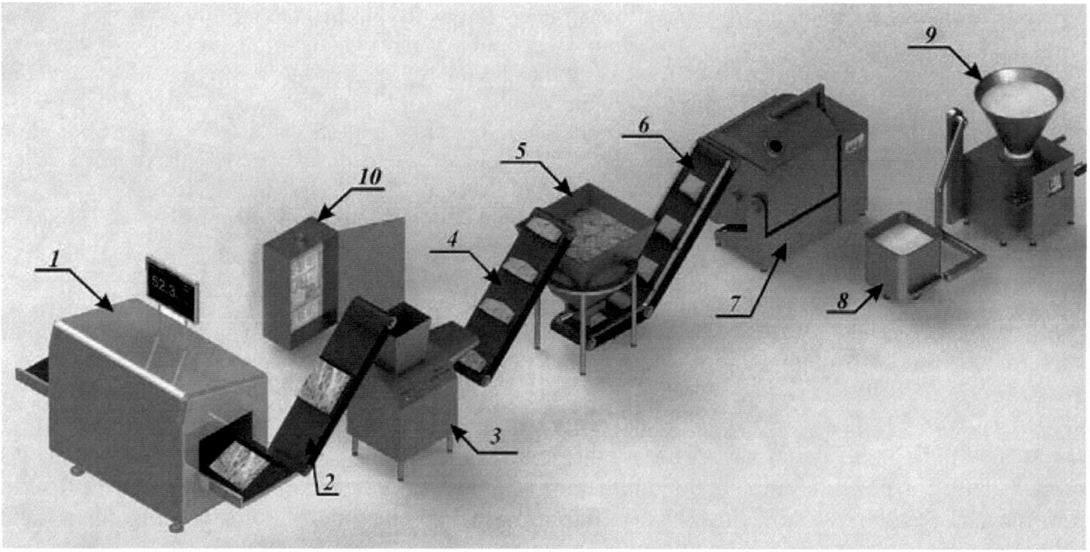

Figure 1. Design of the automatic line for minced meat production: 1 – input control unit for processed raw material; 2, 4, 6 – conveyers; 3 – milling cutter; 5 – hopper for minced raw material; 7 – minced meat mixer; 8 – elevator with a buggy; 9 – stuffer; 10 – process control computer (PCC).

Frozen meat blocks are moved from freezing chambers to the input control unit 1), where unconditioned raw material that contains foreign material inclusions (metallic items, wood, glass, fragments of packaging materials etc.) are identified and rejected. Meat blocks that pass this test are transferred by the conveyer (2) into the milling cutter (3). Minced raw material is carried by the conveyer (4) into the hopper (5), from which it is sent in measured weight quantities by the conveyer (6) to the minced meat mixer (7). After mixing, the formulated mass is transferred into the buggy and loaded by the elevator (8) into the stuffer (9). Then, sausages are subjected to thermal treatment according to a standard technological scheme. The process of minced meat processing is controlled by the PCC (10) that monitors, in real-time, the raw material temperature, the comminution process, the chemical composition (content of water, protein and fat; pH) and the temperature of the minced meat, forming a command for mixing (producing) minced meat with optimal rheological characteristics. When manufacturing a meat mass for homogenised canned foods for children's nutrition, the automatic line is additionally furnished with a heat exchange apparatus and homogeniser instead of a stuffer.

3. Single stage comminution

The proposed design utilises a new technology for meat product manufacture based on an innovative method of raw material comminution. The traditional technology for sausage production involves consecutive comminution of frozen meat blocks using three meat cutting machines: block cutters (comminution to medium-sized particles), grinders (coarse comminution to small-sized particles) and cutters (fine comminution). In the proposed technology, only one machine, a single stage comminutor, finely comminutes the initial raw material. This has the following advantages: 1) costs of minced meat and sausage production are significantly reduced due to exclusion of two expensive meat cutting machines for preliminary raw material comminution (the frozen meat cutter and grinder are unnecessary); 2) each stage of the preliminary comminution using traditional technology is characterised by electricity consumption and meat losses; with the new technology, these losses and consumption are excluded; 3) a reduction in the length of the chain for minced meat production leads to intensification of minced meat/sausage production, making the novel technology more cost effective. To obtain the necessary particle size when producing homogenised canned foods for children's nutrition, the meat mass for these canned foods is traditionally processed in a colloidal mill or micro-cutter or using a dual system of disintegrators after meat raw material is comminuted in a grinder and then is transferred to a homogeniser. Using the novel, single stage comminutor results in a product with the necessary degree of comminution at the early stage of raw material preparation. This leads to intensification of production due to exclusion of the grinder and disintegrators that perform fine comminution of the meat mass for these canned foods. Therefore, the proposed design, utilising the new technology, would be suitable for use in the children's nutrition industry with all the advantages described above for sausage production.

The single stage meat comminutor is furnished with a cutting element in the form of a rotating body (screw) with cutting edges on its flanges [5]. The screw can be made from a set of mills in materials approved for contact with the food environment. When cutting raw material with mills, the contact area of the cutting edges with meat is minimal compared to the area of the side surfaces of cutter knives (the traditional technology of fine comminution); therefore, it is possible to ensure a reduction in the energy expenditure due to friction. By processing meat blocks of typical industrial sizes by the novel milling method in a single stage, the traditional technological chain of meat block processing (frozen meat block cutter – grinder – cutter) is reduced to a minimum, which ensures resource savings. This technological chain reduction leads to a significant (by 2.4 times when the automatic line performance is 2000 kg/hour) decrease in electricity usage, which allows meat processing plants to save financial resources.

4. Intelligent control system (ICS)

In the traditional technology for minced meat production, using batch-operated technological units, finished product quality, to a large extent, depends on the operator. The new technology, realised on the proposed automatic line, envisages complete automation of the technological process with the use of the PCC; that is, control is conducted according to the principle of an unmanned operation based on artificial intelligence. This will allow finished products of guaranteed high quality to be obtained by computer control both of each technological operation and the whole technological process in real-time.

To this end, on the proposed line, the creation of an intelligent control system (ICS) for quality management of minced meat production is planned, the distinctive characteristics of which will be: 1) function in a mode for information interaction with the external environment via sensing elements that measure parameter values to follow the process; 2) the system will be trainable in the process of operation: it accumulates and uses statistical information to increase the precision of predicting the degree of raw material comminution; 3) the system will produce predictions of environment effects, assessing numerical characteristics of such effects.

Control of single stage comminution envisages maintenance of the stipulated degree of raw material comminution in the automatic mode upon stabilisation of the cutting mode parameters in real-time.

In this case, the ICS will operate under the principle of compensating for parameter deviation from a set value (figure 2).

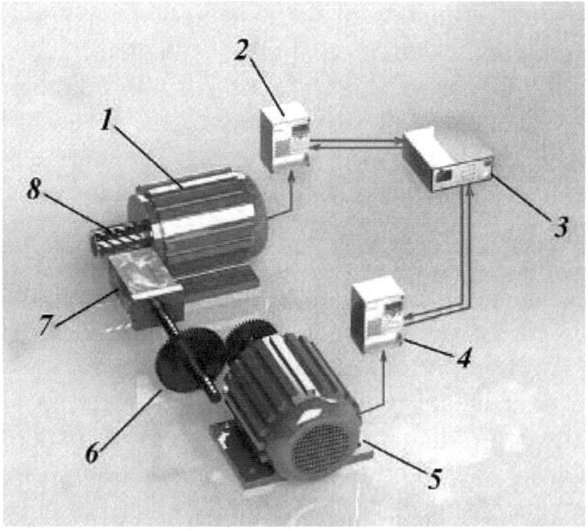

Figure 2. Control of single stage meat block comminution: 1 – electric motor for the driver of the cutting mechanism for raw material; 2, 4 – frequency convertors; 3 – PCC; 5 – electric motor for the driver for the mechanism feeding meat blocks into the comminution zone; 6 – spiral gear; 7 – frozen meat block; 8 – screw (comminutor cutting body).

The PCC calculates the moment of the meat's resistance to comminution by the dynamics of the cutting driver of the comminutor upon fixing the values of the electromagnetic moment of the electric motor of the driver of the cutting mechanism. As a result of this calculation, the PCC memory unit archives the realisation of a random process – the changes in the moment of meat resistance to comminution over time – in the form of a data array. Based on the statistical data obtained, the PCC assesses the correlation function of the process and calculates an estimate of the spectral density of the process by the standard method based on the discrete Fourier transformation. This assumes that a change in the moment of resistance to comminution is a stationary process (in a broad sense, a random process) that has an ergodic property upon stabilisation of the parameters of the raw material cutting regime. The PCC uses the obtained estimate of the spectral density of this random process in a Monte-Carlo statistical computation. Increasing the length of the process realisation in the mode of machine (i.e. modelling) time means the PCC obtains a larger volume of statistical information about the process and, consequently, about changes in the rotational speed of the comminutor cutting shaft on which this process depends. Thus, in addition to the statistical information (obtained when cutting a real meat block) regarding changes in the rotational speed of the cutting shaft under the influence of the moment of resistance to comminution, the PCC will have an additional volume of similar statistical information due to computer modelling of changes in "virtual" meat blocks. This will allow the PCC to calculate point and interval estimates for process changes in the rotational speed of the comminutor cutting shaft during the process of comminution. On the basis of these estimates, the PCC will make the same estimates for changes in the typical size of the meat shavings obtained, with targeted statistical accuracy and reliability using established analytical dependence. The PCC will also calculate the dispersion of these estimates; that is, it will determine a "diffusion" degree for a boundary range of meat shaving sizes, which is important in manufacturing products for children's nutrition. In the process of real comminution of meat blocks, the PCC accumulates and processes statistical information about the comminution process; that is, the PCC is trained in the work regime, improving its prediction of the degree of raw material comminution. As a result of the presented operative algorithm, the PCC will produce statistical information about the degree of comminution in an explicit (numerical) form. The PCC can include equipment for express analysis of the chemical composition of comminuted meat as well as its temperature after comminution. In this case, the following process for minced meat production can be controlled strictly as a function of time, thereby excluding subjective factors in the assessment of finished product readiness.

5. Conclusion

The proposed technology for meat raw material comminution and the designed automatic production line will result in finished minced meat/sausage products with consistent, targeted high quality. Complete automation of the technological process of minced meat production using the ICS opens the door to designing automatic plants – the meat processing plants of the future.

References

[1] Maksimov D A and Zakharov A N 2013 Trends in the development of meat processing equipment *All About Meat* **3** 10-2
[2] Lisitsyn A B, Ivashov V I, Zakharov A N, Kapovsky B R and Kozhevnikova O E 2013 Intelligent quality control system for minced meat *All About Meat* **6** 32-8
[3] Plyasheshnik P I, Kapovsky B R, Glebochev S N and Shikhov S S 2016 Automated lines in the meat sector *Meat Industry* **1** 32-4
[4] Kapovsky B R, Pchelkina V A, Plyasheshnik P I, Lazarev A A and Dydykin A S 2017 Methods of incoming control of meat blocks on the technological lines *Meat Industry* **5** 28-31
[5] Lisitsyn A B, Semenova A A, Kapovsky B R, Kuznetsova T G and Zakharov A N 2017 An innovative method of fine comminution of meat raw material. An alternative to a long-standing tradition *Fleischwirtschaft International* **2** 60-5

Aflatoxin M_1 in processed milk: Occurrence and seasonal variation with an emphasis on risk assessment of human exposure in Serbia

D Milićević, D Spirić, S Janković, B Velebit, T. Radičević, Z Petrović and S Stefanović

Institute of Meat Hygiene and Technology, Kaćanskog 13, Belgrade, Serbia

E-mail: dragan.milicevic@inmes.rs

Abstract. The objectives of this study were to assess aflatoxin M_1 (AFM_1) contamination in processed milk and dairy products, and to estimate the mean daily exposure of the adult Serbian population to AFM1 due to milk consumption. A total of 1734 samples, comprising heat treated cow's milk (n=1233), infant formulae (n=349), milk powder (n=94) and dairy drink (n=58), were analyzed for AFM_1 presence using an Enzyme-Linked Immunosorbent Assay (ELISA) commercial kit. Samples were collected from different regions of Serbia during four seasons each year during 2015 and 2016. The incidences of AFM_1 contamination were 77.8% with a mean level of 0.027±0.03 μg/L (range of <0.005-0.278 μg/L) in samples collected in 2015, and 98.4% with a mean level of 0.039±0.02 μg/L (range of <0.005-0.28 μg/L) in samples collected in 2016. The highest AFM_1 levels were measured in October 2015 (0.278 μg/L) and September 2016 (0.279 μg/L). Based on EU regulation, 214 (17.3%) milk samples exceeded the maximum residue limit (0.05 μg/L). The estimated daily intake (EDI) of AFM_1 during different seasons of year for males and females was in the range of 0.022-0.330 (mean 0.20) ng/kg/bw/day and 0.022-0.30 (mean 0.18) ng/kg/bw/day, respectively. The calculated EDI indicate a public health concern due to the carcinogenic effects of AFM_1.

1. Introduction

Aflatoxin M_1 (AFM_1) is the most significant toxin in milk and dairy products. This compound is the hydroxylated form of the aflatoxin B_1 (AFB_1) that is formed in the liver of lactating animals, and is excreted into the raw milk by dairy cows when animals have been fed with feedstuffs containing AFB_1 [1]. Aflatoxins are potent, naturally occurring carcinogenic mycotoxins with a genotoxic effect and chronic and acute liver toxicity, especially in combination with hepatitis [2]. Since AFM_1 is stable at high temperatures (≥250 °C) and so cannot be removed from milk by heating processes, this toxin can be transferred into the milk products. Thus, AFM_1 contamination can be a substantial public health concern. AFM_1 has been classified as a Group 2 human carcinogen by the International Agency for Research on Cancer [3]. For this reason, and taking into account the significance of milk and milk products in the human diet (especially for children), the maximum allowed levels of AFM_1 are strictly regulated worldwide and range from 0 to 1.0 μg/kg. In Serbia, high incidences of detectable AFM_1 in milk have resulted in periodic changing of the maximum residue levels (MRL) for AFM_1 set by the Serbian Government [4].

Despite the importance that the presence of AFM_1 in milk has for public health concerns, in Serbia, the risks associated with AFM_1 in dairy products have yet to be fully understood. Assessment of risk

IOP Conf. Series: Earth and Environmental Science **85** (2017) 012040 doi:10.1088/1755-1315/85/1/012040

of exposure to AFM_1-contaminated milk and dairy products, is one of the most useful methods to evaluate the severity and probability of liver cancer risk in toxicological studies [5]. Therefore, the aims of the present survey were to determine the incidence of detectable levels of AFM_1 in milk during different seasons, and to estimate the mean daily exposure of the adult Serbian population to AFM1 due to milk consumption.

2. Materials and Methods

2.1. Sample collection and preparation
A total of 1734 samples, comprising heat treated cow's milk (n=1233), infant formulae (n=349), milk powder (n=94) and dairy drink (n=58) were obtained in the period from January 2015 to December 2016 from dairy producers during occasions of self-controls or official controls. Samples were stored at 2-8 °C or frozen at -20 °C until further analysis of AFM_1. Milk samples were prepared according to the ELISA kit manufacturer's instructions. Samples were centrifuged for 10 min at 3500 g at 10 °C. After centrifugation, the upper cream layer was completely removed by aspirating through a Pasteur pipette. Skimmed milk was used directly in the test (100 µl per well).

2.2. Enzyme-linked immunosorbent assay (ELISA) analysis
The ELISA test procedure was performed using the Aflatoxin M1 ELISA kit (Tecna S.r.l., Italy). Preparation of the samples and ELISA test procedure were performed according to the instruction provided by the manufacturer. The detection limit of the method was 0.005 µg/kg. In the case of AFM_1 levels higher than 250 µg/kg, samples were diluted with sample dilution buffer and reanalyzed. Relative standard deviation of reproducibility was 6%. Recovery was 110%.

2.3. Exposure assessment
Estimated daily intake (EDI; ng/kg body weight (bw/day) was calculated through a deterministic method [6]. For exposure calculations, levels of AFM_1 below LOD were assumed to be equal to half the LOD for derived mean concentrations (0.0025 µg/kg) [7]. The daily intake level of milk and mean body weight for males and females was estimated based on data retrieved from Kos et al. [8]. Results were expressed as ng/kg bw/day.

3. Results and Discussion
The occurrence of AFM_1 contamination in heat-treated cow milk and dairy products and EDI due to consuming AFM_1 contaminated milk, collected during four seasons of 2015-2016, are presented in table 1 and figure 1.

3.1. Occurrence of AFM_1
AFM_1 was detected in 90.6% (117 out of 1233) of milk samples at concentrations ranging from 0.005 to 0.28 µg/kg, with a mean level of 0.035±0.029 µg/kg. Individual results by month (data not shown) showed the highest incidence of AFM_1 in 2016 (98.4%) with a mean concentration 0.039±0.02 µg/kg, in comparison to 2015 (77.8%) with a mean concentration 0.027±0.03 µg/kg. However, the range of AFM_1 concentrations between years was similar; 0.005-0.28 µg/kg. The highest concentration, which exceeded the MRL, was determined in winter and spring season (0.279 µg/kg). Statistical analysis of the data showed significant differences between periods of investigation in the level of AFM_1 in heat-treated cow's milk ($p < 0.001$) (table 1).
These differences are presumably due to contamination of feeds as influenced by local weather conditions during pre-harvest and harvest stage. Increased levels of AFM_1 in Serbian milk since 2013 were most probably the consequence of feeding corn contaminated with AFB_1. AFB_1 contamination is prevalent in warm and humid climates and is reported in temperate countries following severe drought, particularly in cases of inadequate storage conditions [9]. Serbia has a continental to moderate continental climate with frequent, heavy rainfall. The temperatures (up to 40 °C) and relative

humidity (up to 80%) are high throughout the year. Under such climate conditions, a high incidence and levels of mycotoxins can be expected [10].

Table 1. The occurrence of AFM1 in different types of dairy products.

Type of milk	Year	N	Incidence n (%)	Concentration of AFM_1 (μg/kg)		Above MRL n (%)	
				Mean ± Sd	Range (min.-max.)	SRB n (%)[1,2]	EU n (%)[1]
Heat treated cow's milk	2015	468	364 (77.8)	0.027±0.03[a]	<0.005-0.278	12 (2.5)	43 (9.1)
	2016	765	753 (98.4)	0.039±0.02[a]	<0.005-0.28	2 (0.2)	171 (22.3)
	Sum of 2015 and 2016	**1233**	**1117 (90.6)**	**0.035±0.029**	**<0.005-0.28**	**14 (1.1)**	**214 (17.3)**
Infant formulae	2015	172	5 (2.9)	0.0086±0.001	<0.005-0.010	0	0
	2016	177	18 (10.2)	0.0012±0.0023	<0.005-0.017	1 (0.5)	1 (0.5)
	Sum of 2015 and 2016	**349**	**23 (6.6)**	**0.011±0.0025**	**<0.005-0.017**	**1 (0.2)**	**1 (0.2)**
Milk powder	2015	68	15 (22.1)	0.021±0.01	<0.005-0.035	0	0
	2016	26	10 (38.4)	0.013±0.008	<0.005-0.027	0	0
	Sum of 2015 and 2016	**94**	**25 (26.6)**	**0.018±0.01**	**<0.005-0.035**	**0**	**0**
Dairy drink	2015	38	7 (18.4)	0.054±0.05	<0.005-0.147	0	3 (7.9)
	2016	20	6 (30.0)	0.010±0.0025	<0.005-0.013	0	0
	Sum of 2015 and 2016	**58**	**13 (22.4)**	**0.034±0.04**	**<0.005-0.147**	**0**	**3 (5.2)**

Means with same letters within the same column are significantly different at $p < 0.001$.
N – total number of analysed samples.
n – number of samples.
MRL 10.05; 2 0.25.

In our study, AFM_1 was found in 23 out of 349 (6.6%) of infant formulae (levels ranged from <LOD to 0.017 μg/kg, mean 0.011±0.0025 μg/kg). The toxin incidence was higher in 2016 (18 out of 177, 10%) than those obtained in 2015 (5 out of 172, 3%). Regarding milk powder, the incidences of AFM_1 contamination in samples collected from 2015 and 2016 were 22% (15 out of 68) and 38.4% (10 out of 26), with concentrations ranging from <LOD to 0.035 μg/kg and <LOD to 0.027 μg/kg, respectively. For the dairy drink samples, AFM_1 was detected in 13 out of 58 (22.4%) samples, with the mean level of 0.034±0.04 μg/kg for these positive samples. The incidences of AFM1 contamination in samples collected in 2015 and 2016 were 7 out of 38 (18%) and 6 out of 30 (30%) with a range of <LOD to 0.147 μg/kg (mean 0.054±0.05 μg/kg) and <LOD to 0.013 μg/kg (mean 0.010±0.0025 μg/kg), respectively.

Based on EU Regulation No 466/2001, the maximum permitted level (MPL) for AFM_1 in milk (raw milk, milk for the manufacture of milk based products and heat-treated milk) is 0.05 μg/L. According to this standard, 9.1% (43/468) in 2015 *vs.* 22.3% (171/765) in 2016, of milk samples had AFM_1 levels that exceeded this limit. However, according to Serbian legislation, only 12 (2.5%) and 2 (0.2%) in 2015 and 2016 of milk samples exceeded the Serbian MRL, respectively. Moreover, in the present study, the AFM_1 concentrations of three dairy drink samples was higher than 0.05 μg/kg. Since 2013, high levels of AFM_1 in milk and dairy products have been detected in Serbia [8,11-16]. Overall, seasonal differences of AFM_1 concentrations in milk and dairy products observed in this study also were recently reported in Serbia [1,16], and Croatia [17,18].

3.2 Exposure assessment - Estimated daily intake (EDI)

The EDI of AFM_1 for the adult Serbian population due to milk consumption during the period of the investigation is presented in figure 1. As shown in figure 1, the EDI in this study was higher during September 2015 to June 2016 than the amount recommended by Kuiper-Goodman (0.2 ng/kg bw/day) [19]. However, our study did not take the contribution of other dairy products into account.

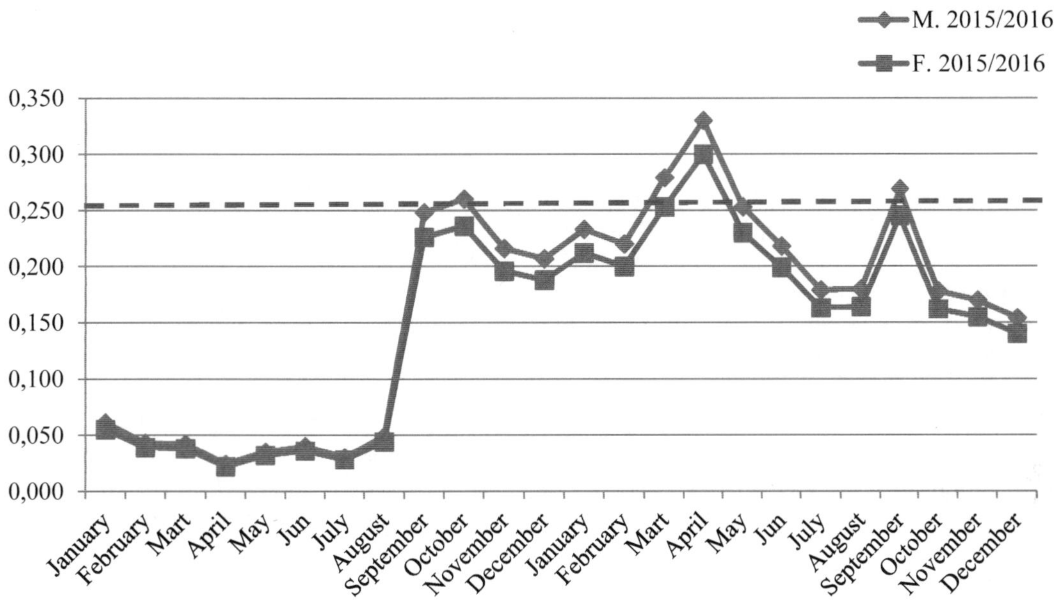

Figure 1. EDI (ng/kg bw/day) for AFM1 in heat-treated milk during seasons 2015-2016 (M-male, F-female).

Considering our results, it could be concluded that consumers are constantly exposed to this toxin through milk consumption. The average calculated EDI in our study was higher than the level reported by Torovic et al. [5], but lower than those reported by Skrbic et al. [11] or Kos et al. [8]. At the international level, on the basis of the mean concentrations of AFM_1 in milk and the milk consumption in the GEMS/Food regional diets [20], the EDI for the European diet was calculated as 0.11, for Latin American 0.058, for Far Eastern 0.20, for Middle Eastern 0.10, and for African diet as 0.002 ng/kg bw/day. Because of the carcinogenic potential of aflatoxins, JECFA [20] concluded that daily exposure, even to <1 ng/kg bw, contributes to the risk of liver cancer.

4. Conclusion

Considering the results obtained in this study, the EDI and potential hazard index, the unsuitably large intakes of AFM_1 by the Serbian population are higher than the recommended permissible intakes by international expert committees. Moreover, integrated prevention strategies at pre-harvest or post-harvest times, including further studies, in an attempt to identify and then control potential influencing factors, are necessary. Also, AFB_1 contamination of animal feeds and milk should be regularly monitored, particularly in regions where milk samples were previously contaminated above the legal limit.

Acknowledgments

This study was supported by project No. TR 31008 funded by the Ministry of Education, Science and Technological Development of the Republic of Serbia. Part of this research was dedicated to COST action TD 1404, supported by European Cooperation in Science and Technology (COST).

References

[1] Flor-Flores M E, Lizarraga E, López de Cerain A and González-Peñas E 2015 Presence of mycotoxins in animal milk: A review *Food Control* **53** 163–76

[2] Milicevic R D, Skrinjar M M and Baltic T 2010 Real and perceived risks for mycotoxin contamination in foods and feeds: Challenges for food safety control *Toxins* **2**(4) 572-92

[3] IARC 2002 IARC monographs on the evaluation of carcinogenic risks to humans *Some traditional herbal medicines, some mycotoxins, napthalene and styrene* (vol. 82) (Lyon: IARC Press) pp 1-556

[4] Serbian Regulation 2015 Amendment on Serbian regulation 2011 Maximum allowed contents of contaminants in food and feed *Official Gazette of the Republic of Serbia* No. 84/15

[5] Torovic L 2015 Aflatoxin M1 in processed milk and infant formulae and corresponding exposure of adult population in Serbia in 2013–2014 *Food Addit. Contam.* Part B **8:4** 235-44 DOI: 10.1080/19393210.2015.1063094

[6] IPCS – International Programme on Chemical Safety 2009 Dietary exposure assessment of chemicals in food. In Principles and methods for the risk assessment of chemicals in food Genève, Switzerland: WHO

[7] WHO GEMS/Food–EURO 1995 Second Workshop on Reliable Evaluation of Low-Level Contamination of Food: report on a workshop in the frame of GEMS/Food-Euro 1995 WHO, Kulmbach

[8] Kos J, Lević J, Đuragić O, Kokić B and Miladinović I 2014 Occurrence and estimation of aflatoxin M1 exposure in milk in Serbia *Food Control* **38** 41–6

[9] Frazzoli C, Gherardi P, Saxena N, Belluzzi G and Mantovani A 2017 The hotspot for (global) one health in primary food production: Aflatoxin M1 in dairy products *Front. Public Health* **4** 294 doi: 10.3389/fpubh.2016.00294

[10] Milicevic D, Nastasijevic I and Petrovic Z 2016 Mycotoxin in the food supply chain—implications for public health program. *J. Environ. Sci. Health C* **34**(4) 293-319 DOI:10.1080/10590501.2016.1236607

[11] Skrbic B, Zivancev J, Antic I and Godula M 2014 Levels of aflatoxin M1 in different types of milk collected in Serbia: assessment of human and animal exposure *Food Control* **40** 113-19

[12] Škrbić B, Antić I and Živančev J 2015 Presence of aflatoxin M1 in white and hard cheese samples from Serbia *Food Control* **50** 111-7

[13] Spirić M D, Stefanović M S, Radičević M T, Đinović Stojanović J M, Janković V V, Velebit B M and Janković D S 2015 Study of aflatoxins incidence in cow feed and milk in Serbia during 2013 *Hem. Ind.* **69**(6) 651–6

[14] Tomasevic I, Petrovic J, Jovetic M, Raicevic S, Milojevic M and Miocinovic J 2015 Two-year survey on the occurrence and seasonal variation of aflatoxin M1 in milk and milk products in Serbia *Food Control* **56** 64-70

[15] Polovinski Horvatovic M, Glamocic D, Jajic I, Krstovic S, Guljaš D and Gjorgjievsk S 2016 Aflatoxin M1 in raw milk in the region of Vojvodina *Mljekarstvo* **66**(3) 239-45

[16] Miocinovic J, Keskic T, Miloradovic Z, Kos A, Tomasevic I and Pudja P 2017 The aflatoxin M1 crisis in the Serbian dairy sector: the year after. *Food Addit. Contam.* Part B: Surveillance Communications **101** 1-4

[17] Bilandzic N, Bozic Đ, Đokic M, Sedak M, Solomun Kolanovic B, Varenina I, Tankovic S and Cvetnic Ž 2014 Seasonal effect on aflatoxin M1 contamination in raw and UHT milk from Croatia *Food Control* **40** 260-4

[18] Bilandžic N, Varenina I, Solomun Kolanovic B, Božic Đ, Đokic M, Sedak M, Tankovic S, Potocnjak D and Cvetnic Ž 2015 Monitoring of aflatoxin M1 in raw milk during four seasons in Croatia. *Food Control* **54** 331-7 doi: 10.1016/j.foodcont.2015.02.015

[19] Kuiper-Goodman T 1990 Uncertainties in the risk assessment of three mycotoxins: aflatoxin, ochratoxin, and zearalenone. *Can. J. Physiol. Pharmacol.* **68**(7) 1017–24
[20] JECFA Joint FAO/WHO Expert Committee on Food Additives 2002 Evaluation of certain mycotoxins in food. Fifty sixth report of the Joint FAO/WHO expert committee on food additives. Geneva: *World Health Organization*

IOP Conf. Series: Earth and Environmental Science **85** (2017) 012041 doi:10.1088/1755-1315/85/1/012041

Changes of pH and peroxide value in carp (*Cyprinus carpio*) cuts packaged in modified atmosphere

M Milijašević, J Babić Milijašević, J Đinović-Stojanović, S Vesković Moračanin and S Lilić

Institute of Meat Hygiene and Technology, Kaćanskog 13, Belgrade, Serbia

E-mail: milan.milijasevic@inmes.rs

Abstract. The aim of our research was to examine the influence of packaging in modified atmosphere on the pH and peroxide value in muscle of common carp (*Cyprinus carpio*), as well as to determine the most suitable gas mixtures for packing of that freshwater species. Three sample groups of carp cuts were investigated. One group of carp cuts was placed on top of flaked ice placed in polystyrene boxes. Two other groups were packaged in modified atmosphere with different gas ratios: 80%O_2+20%CO_2 (MAP 1) and 90%CO_2+10%N_2 (MAP 2). All carp cuts were stored in the same conditions at 3±0.5°C, and on 1, 3, 5, 7, 9, 11, 13, 15. and 17 days of storage, chemical testing was performed. The results obtained indicate that the packaging of common carp under 90%CO_2+10%N_2 slowed proteolytic reaction as well as secondary lipid oxidation.

1. Introduction

The quality of fresh fish is a major concern to industry and consumers. This is a complex concept involving a whole range of factors, which include safety, nutritional quality, availability, convenience and integrity, freshness, eating quality and the obvious physical attributes of the species, size and product type [1].

Fish are highly perishable and prone to vast variations in quality due to differences in species, environmental habitats, feeding habits and action of autolysis enzymes as well as hydrolytic enzymes of microorganisms on the fish muscle [2]. The high water and free amino acid content, and the lower content of connective tissue as compared to other flesh foods lead to the more rapid spoilage of fish. In general, fish have a limited shelf-life in comparison with meat products (veal, lamp, pork, poultry) as a result of the high post mortem pH in the flesh (usually > 6.0), the presence of large amounts of non-protein nitrogen (NPN), the high content of polyunsaturated fatty acids (PUFA), the presence of autolytic enzymes etc.

Deterioration of fish mainly occurs as a result of bacteriological activity leading to loss of quality and subsequent spoilage [3,4]. Microorganisms present on the surface of fish produce a large variety of hydrolytic enzymes, in particular proteases. Endogenous proteases also play an important role in the post mortem degradation of fish muscle protein [5]. These processes lead to a change to the textural and sensory characteristics of fish muscle.

Hydrolytic changes in lipids are the cause of the release of free fatty acids (FFA), which are much more susceptible to oxidative changes. Fish oil contains large amounts of polyunsaturated fatty acids which lead to the initiation of oxidation reactions and the formation of hydroperoxides of fatty acids and other, often toxic, secondary oxidation products. The formation of peroxides (measured by the peroxide value; PV) is considered an indicator of the rate of primary oxidation, while the thiobarbituric acid (TBA) value is an indicator of secondary oxidation [6,7].

The changes in lipids of fish are responsible for the quality deterioration with the extended storage, especially under inappropriate conditions. They involve lipolysis, lipid oxidation, and the interaction of the products of these processes with nonlipid components such as protein. Fish muscles contain an abundance of long chain lipids with a high proportion of polyunsaturated fatty acid that undergoes changes due to oxidation during processing and storage.

The shelf-life of fresh chilled fish is relatively short and at ambient temperatures of 2±2°C it is about 2 to 3 days. The shelf life of fresh chilled fish can be extended by vacuum packaging or modified atmosphere packaging (MAP) [8,9].

The aim of this research was to monitor changes of selected chemical parameters of common carp (*Cyprinus carpio*) steaks packaged in modified atmosphere during the storage at 3±0.5°C and to determine the shelf life of the products.

2. Materials and Methods

2.1. Sample collection

Samples from fourteen common carp (*Cyprinus carpio*) of average body weight of 2.50±0.30 kg were obtained from fishpond where a semi-intensive rearing system was used. Fish were transported live to the fish slaughtering and processing facility, where they were stunned, slaughtered, scaled, and carcasses were cut into steaks 2 cm thick and 220 g average weight. The carp steaks were divided into three groups.

The control fish steak group was placed on top of flaked ice placed in polystyrene boxes with outlets for water drainage. The ice/fish ratio was 3:1 and maintained constant throughout the experiment. The other two sample groups of carp steaks were packaged in modified atmosphere with different gas ratios: MAP1:80%O_2+20%CO_2 and MAP2: 90%CO_2+10%N_2. The machine used for packaging the carp steaks was Variovac (Variovac Primus, Zarrentin, Germany), and material used for packaging was foil OPA/EVOH/PE (oriented polyamide/ethylene vinyl alcohol/polyethylene, Dynopack, Polimoon, Kristiansand, Norway) with low gas permeability (degree of permeability for O_2 – 3.2 cm_3/m_2/day at 23°C, for N^2 – 1 cm^3/m^2/day at 23°C, for CO_2 – 14 cm^3/m^2/day at 23°C and for steam 15 g/m^2/day at 38°C). The ratio of gas:carp steak in the packages was 2:1. All samples were stored in the same conditions at 3±0.5°C and on 1, 3, 5, 7, 9, 11, 13, 15 and 17 days of storage, chemical testing was performed.

2.2. Chemical analysis

Muscle pH was measured by Cyber Scan pH-510 digital pH-meter (EUTECH Instruments, Netherland). PV, expressed in milliequivalents of peroxide oxygen per kilogram of fat, was determined by EN ISO 3960:2009 method.

2.3. Statistical analysis

The mean values and standard deviations were calculated by using column statistics for the six values in each analysed group. Significant differences between groups were calculated by using one-way ANOVA analysis by Tukey's comparative test in the program Microsoft Office Excel (2010). Differences were considered as significant when p value was < 0.05.

3. Results and Discussion

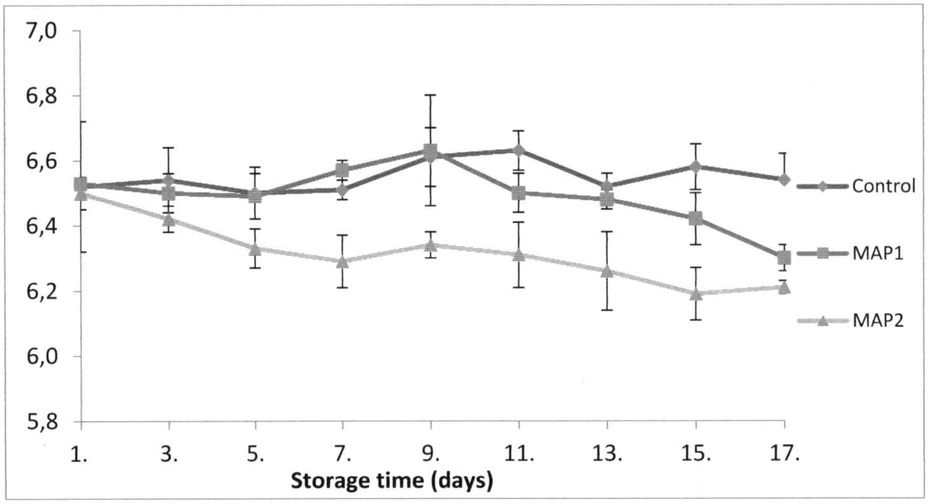

Figure 1. Changes in pH value of common carp steaks packaged under different conditions during the storage period.

Figure 1 shows pH values in common carp steaks depending on the type of packaging and the length of the storage period. In MAP1 carp steaks, significant ($p < 0.01$) a pH increase was observed between day 5 (pH: 6.49±0.03) and day 9 (pH: 6.63±0.06) of the experiment. From then on, the pH value began to decrease and reached 6.30±0.04 on storage day 17. In contrast, a decrease in pH value was detected in carp steaks in MAP2 during the whole experimental period. The lowest pH value of 6.19±0.02 in this experimental group was recorded on storage day 15. pH values of control carp steaks fluctuated during the storage period, and ranged from 6.51±0.09 to 6.63±0.03. Compared to control carp steaks, MAP packaged fish in the 90%CO_2+10%N_2 atmosphere had a lower pH during the entire storage period, while the pH in carp steaks packaged in MAP1 was significantly lower ($p < 0.01$) after 9 days of storage. The mean pH value for the control fish and carp steaks packaged in MAP1 and MAP2 throughout the storage period was 6.55±0.08; 6.49±0.06 and 6.32±0.07, respectively.

The pH values in the muscle tissue of live fish are close to 7.0, but the post mortem pH generally ranges from 6.0 to 7.1, depending on the season of the year, fish species and other factors. Due to the amount of lactic acid produced during glycolysis under anaerobic conditions, the post mortem pH of fish muscle decreased, while the degree of pH reduction influenced quality of fish meat [10].

As shown in figure 1, the lowest pH value was recorded in carp steaks packaged in 90%CO_2+10%N_2. Other authors [4,11,12] also recorded significantly lower pH values in fish samples packaged in modified atmosphere with higher percentage of CO_2; this is explained by dissolution of CO_2 in the fish muscle, which is associated with increase of carbonic acid production.

The moderate increase of the pH in MAP1 carp steaks after five days of storage can be attributed to the higher quantity of basic compounds produced by the activity of fish spoilage bacteria [13], which had favourable growth conditions because of high concentration of O_2 in this gas mixture.

The pH values of common carp muscle in our research as well as differences in pH values under various experimental conditions during storage correspond to the findings of other authors [8,14,15].

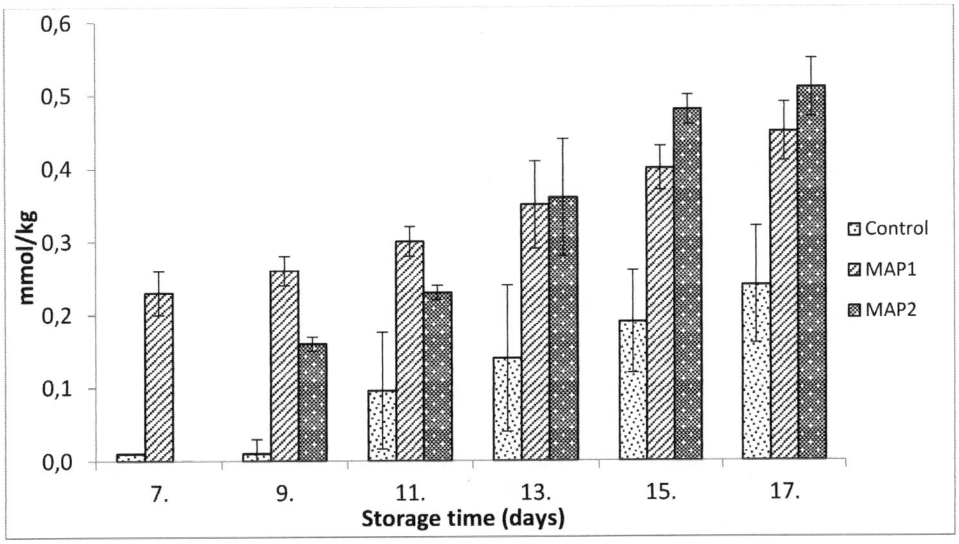

Figure 2. Changes in PV of common carp steaks packaged under different conditions during the storage period

Figure 2 shows the PV in the carp steaks depending on the type of packaging and the length of the storage period. In our research, during the first five days of storage, peroxide was not detected in both unpackaged and packaged fish meat. Later during the study, the PV was lower in control carp steaks than in those packaged in MAP. From day 7 to day 13, PV values were at highest level in carp steaks kept in the oxygen-rich atmosphere (80%). At the end of the study (day 15 and day 17), PV values were higher in fish packaged in the atmosphere without oxygen. As reported by Jayasingh *et al.* [16], lipid oxidation was higher in carp steaks packaged in MAP with 80% O_2 than in control fish, exposed to ambient air, which is in agreement with the results of the present study. As concluded in Ruiz-Capillas and Moral [13], lipid oxidation depends on the synergy effect between CO_2 and O_2. For that reason, lipid oxidation in an atmosphere with 40% O_2 could be more intensive compared with atmosphere with 60% O_2. Fluctuations in PVs that have been recorded in our research are in line with the results of other authors [6], pointing out the fact that PV cannot be considered as suitable indicator of fish muscle freshness.

4. Conclusion

Packaging common carp under 90%CO_2+10%N_2 slowed proteolytic reaction as well as secondary lipid oxidation. According to those indicators, packaging common carp in 90%CO_2+10%N_2 is more suitable compared to packaging in an 80%O_2+20%CO_2 gas mixture.

Acknowledgments
This work was supported by grants from the Ministry of Education, Science and Technological Development of the Republic of Serbia (project No. III 46009).

References

[1] Ashie I N A, Smith J P, Simpson B K 1996 *Critical Reviews in Food Science and Nutrition* **36** 87
[2] Jinadasa B K K K 2014 *Nat. and Sci.* **12**(5) 106
[3] Liston J 1980 *Advances in fish science and technology* p 138
[4] Milijašević M, Babić J, Baltić M Ž, Spirić A ,Velebit B, Borović B and Spirić D 2010 *Teh. mesa* **51**(1) 66
[5] Godiksen H, Morzel M, Hyldig G and Jessen F 2009 *Food Chem.* **113**(4) 889

[6] Ježek F and Buchtová H 2010 *Acta Vet. Brno* **79** 117
[7] Huss H H 1995 *FAO Fisheries technical paper*–348
(Available at: http://www.fao.org/docrep/V7180E/V7180E00.HTM)
[8] Babić J, Dimitrijević M, Milijašević M, Petronijević R, Grbić S, Spirić A 2014 *Hem. Ind.* **68** 69
[9] Ježek F and Buchtova H 2012 *Pol. J. Vet. Sci.* **15**(3) 439
[10] Ashie I N A, Smith J P, Simpson B K 1996 *Crit. Rev. in Food Sci. Nut.* **36**(1-2) 87
[11] Provincial L, Gil M, Guillen E, Alonso V, Roncales P and Beltran J A 2010 *I. J. S. T.* **45** 1828
[12] Stamatis and Arkoudelos J S 2007 *J. Sci. Food and Agr.* **87** 116
[13] Ruiz-Capillas C and Moral A 2001 *Eur. Food Res. Technol.* **212** 413
[14] Masniyom P, Benjama O and Maneesri J 2013 *International Food Research Journal* **20**(3) 1401
[15] Goulas A E, Kontominas M G 2007 *Eur. Food Res. Technol* **224** 545
[16] Jayasingh P, Cornforth D P, Brennand C P, Carpenter C E and Whittier D R 2002 *J. Food Sci.* **9** 3493

IOP Conf. Series: Earth and Environmental Science **85** (2017) 012042 doi:10.1088/1755-1315/85/1/012042

Effect of chard powder on colour and aroma formation in cooked sausages

V V Nasonova and E K Tunieva

The V.M.Gorbatov All-Russian Meat Research Institute, Russian Academy of Agricultural Sciences, Moscow, Russian Federation

E-mail: vvnas@mail.ru

Abstract. The use of nitrate-containing vegetable powders instead of sodium nitrite in meat products requires changes in technological production parameters in order to obtain traditional organoleptic characteristics in the finished products. The aim of this work was to study the effect of chard powder on colour and aroma formation in cooked sausages. Cooked sausage samples were: control with nitrite curing mixture; type 1 sausages with chard powder and ascorbic acid; type 2 sausages with chard powder and sodium ascorbate. To transform nitrate ions contained in the vegetable chard powder to nitrite ions using a denitrifying culture, preliminary thermal treatments were used: 30 and 60 min at 40±2°C, after which the sausages were cooked until a temperature of 72±2°C was achieved. The sausages were stored for 40 days at 0-6°C. When sausage meat was initially held at 40°C for 60 min, a homogenous pink colour formed in the sausages with the vegetable powder. The indicators of lightness, redness and yellowness in cooked sausages as well as the indicators of instrumental odour assessment did not differ significantly ($p>0.05$). The indicators of colour stability during storage were 1.1-3.0% higher in the sausages with the chard powder compared to the control. The mass fraction of sodium nitrite in the experimental sausages was 2.0-2.2 higher than in the control ($p>0.05$). As a result of cooked sausage storage, the differences in the sodium nitrite content in the control and types 1 and 2 sausages were similar. During storage, the mass fraction of sodium nitrite decreased in types 1 and 2 sausages by 55.6 and 54.8%, respectively ($p<0.05$). Cooked sausages with the chard powder contained 2.1-2.4 times more sodium nitrate than did control sausages ($p<0.05$). However, all tested sausage samples complied with legislative requirements in terms of their sodium nitrite and nitrate levels.

1. Introduction

The food additive sodium nitrite E250 is used in the meat industry as a colour fixing agent, and due to its multiple functions (colour formation, preservative and antioxidative abilities, participation in aroma formation), it is practically irreplaceable in meat product manufacture. Taking into consideration the negative effect of excessive amounts of sodium nitrite on human health, doses of this additive in meat product formulations are strictly limited by existing legislation. However, up to now, there are no scientific data on the negative effect on human health of meat products with sodium nitrite, even at increased (by 3-4 times) concentrations. In addition, methods for utilising sodium nitrite in health care to induce improvement in cardiovascular system function are known [1]. However, taking into consideration the negative consumer attitude to products with food additives, the use of alternative sodium nitrite sources is topical. These sources include vegetable powders that contain nitrate, which is transformed into nitrite by the effect of denitrifying bacteria, and nitrite then takes part in the process of colour formation [2-5]. This technology exists in different countries and is interesting in terms of processing vegetables with excessive levels of nitrate and their further use.

For more complete expenditure of sodium nitrite on colour formation and, consequently, maximum effect of its use, on one hand, and the minimum residual content in a finished product, on the other hand, it is possible to use ascorbic acid (E300) and its salts (E301). Ascorbic acid occupies a special place among antioxidants. It is not only a means for retardation of the oxidative process in meat systems but also is an important auxiliary factor for colour formation with sodium nitrite participation. When using nitrate-containing components, it is necessary to use several technological steps, e.g., a preliminary stage of thermal treatment, denitrifying cultures that are necessary for full transformation of nitrate ions into nitrite ions, correct course of the colour formation process and assurance of the minimal residual sodium nitrite content. In this connection, the aim of this work was to study the effect of nitrate-containing chard powder on colour formation in cooked sausages made without the use of the colour fixing agent, sodium nitrite.

2. Materials and Methods

Cooked sausages contained the main raw materials, beef and pork, plus water, salt, spices, sugar and food-grade phosphates. The control sausages were made using traditional technology with a nitrite curing mixture (the mass fraction of sodium nitrite in the mixture was 0.6%) and sodium ascorbate. In type 1 sausages, sodium nitrite and sodium ascorbate were replaced with vegetable chard powder (0.26%) with a sodium nitrate level of 3.0%, ascorbic acid and a denitrifying culture containing *Pediococcus pentosaceus*, *Staphylococcus carnosus*, *Staphylococcus xylosus*, *Lactobacillus sake* Deb. Hanseni. Type 2 sausages differed from type 1 sausages by the use of sodium ascorbate instead of ascorbic acid. The vegetable chard powder was additionally dissolved in a small amount of water with constant mixing before adding into ground meat. The denitrifying culture with a small amount of water was added to meat raw materials with constant mixing. The temperature of prepared ground meat was not higher than 12°C.

To transform the nitrate ions occurring in the vegetable chard powder into nitrite ions with necessary participation of the denitrifying culture, preliminary thermal treatments at 40±2°C for 30 or 60 min was used. Then, sausages were cooked at 80±2°C until they achieved 72±2°C, and cooled to 4±2°C. After that, sausages were stored for 40 days at 0-6°C and relative humidity of 75-78%.

The odour visual fingerprints were determined using a VOCmeter device (electronic nose). Three specimens were taken from each test sausage sample for analysis on the VOCmeter device. To prepare a specimen, a test sausage sample (excluding the superficial layers) was minced and the necessary quantity was placed in a special glass container (vial). Each vial was tightly closed and incubated. After incubation, a needle was injected into each vial for automatic sampling of the analysed gas, which entered the VOC meter device. The readings of the MOS 1-4 sensors were used for visualisation of the experimental results.

Colour characteristics of the sausages were measured in the CIELab system using the spectrocolorimeter "Spectroton" with simultaneous determination of reflection coefficients at 24 fixed wavelengths located in 13 nm intervals in the visible spectral range from 380 to 720 nm with the measurement results mathematically processed using a microprocessor controller built into the measurement unit.

To detect colour stability during storage, a criterion for colour stability (U, %) was used. Colour stability was calculated according to the following equation:

$$U = \left(1 - \left(\frac{|L_1 - L_2|}{3L_1} + \frac{|a_1 - a_2|}{3a_1} + \frac{|b_1 - b_2|}{3b_1}\right)\right) 100$$

where: L_1, L_2 – values of lightness before and after storage;
a_1, a_2 – values of redness before and after storage;
b_1, b_2 – values of yellowness before and after storage.

IOP Conf. Series: Earth and Environmental Science **85** (2017) 012042 doi:10.1088/1755-1315/85/1/012042

The mass fraction of sodium nitrite was determined by the method based on the reaction of nitrite with n-(1-naphthyl) ethylenediamine dihydrochloride and sulphanilamide in a protein-free filtrate and photocolorimetric determination of colour intensity.

The mass fraction of sodium nitrate was determined by the method based on nitrate reduction to nitrite using a cadmium column, photometric measurement of the intensity of colour that was formed in the reaction of sulphanilamide and n-(1-naphthyl)ethylenediamine dihydrochloride with nitrite, determination of an amount of the latter and its conversion to nitrate minus nitrite in the sausage.

3. Results and Discussion

To transform nitrate ions to nitrite ions, the presence of the denitrifying culture for a specified temperature and duration is necessary. The reaction of colour formation is activated during product heating at a temperature higher than 30°C and proceeds up to the product achieving 50°C. With the further increase in temperature up to 60-70°C, nitrosomyoglobin and nitrosohaemoglobin lose their protein partly due to denaturation and change into nitrosomyochromogen and nitrosohaemochromogen that impart a red colour to meat. When nitrite ions are absent, the colour reaction proceeds to the development of metmyoglobin, and the finished product will have a grey colour after thermal treatment, which is unacceptable for the most types of traditional meat products and sausages.

These mechanisms for colour formation were taken into account upon optimisation of the conditions of vegetable powder transformation during holding in the thermal chamber.

Therefore, sausage meat was held at a temperature of 40°C for 30 min or 60 min in order to achieve the uniform colour of cooked sausages. Holding for 30 min was insufficient to achieve a uniform pink colour on the cross-section surfaces of type 2 sausages, which were characterised by the presence of grey spots. In contrast to the use of the vegetable chard powder with sodium ascorbate, its combination with ascorbic acid in type 1 sausages provided a uniform colour after just 30 min at 40°C. The control sausages were characterised by a uniform pink colour.

Holding sausages for 60 min at 40°C produced formation of a uniform colour in sausage of both types 1 and 2 (with the vegetable powder). Thus, an additional holding for 60 min at 40°C is necessary for nitrate transformation when manufacturing cooked sausages with chard powder.

After achieving a temperature of 72±2°C, all sausages attained the characteristic, traditional colour of cooked sausages. Based on the performed research, the additional preliminary stage of thermal treatment was established for cooked sausages to ensure transformation of nitrate ions to nitrite ions with participation of the denitrifying culture: not more than 60 min at 40°C.

As a result of the organoleptic assessment, it was established that during storage, all sausages had similar and acceptable consumer characteristics traditional for these types of meat products.

The instrumental assessment of colour characteristics suggested an absence of significant differences in lightness, redness and yellowness of the control cooked sausage samples and sausages of types 1 and 2 (table 1).

Colour stability in cooked sausages was 97.6-99.4% after 20 days of storage and 95.2-98.2% after 40 days of storage. With that, the experimental sausages were superior to the control in terms of colour stability (by 1.1 and 3.0% for types 1 and 2, respectively) after 40 days of storage.

Table 1. Colour characteristics of sausages.

Sausage samples	Colour characteristics, colour units			Colour stability during storage, %
	L-lightness	a-redness	b-yellowness	
		1 day		
Control	60.7±1.3	10.2±0.6	11.0±0.8	-
Sausage type 1	60.1±1.8	10.9±0.7	12.3±0.4	-
Sausage type 2	59.5±1.2	11.3±0.6	12.0±0.6	-
		10 days		
Control	60.8±0.8	10.0±0.2	11.2±0.4	98.8
Sausage type 1	61.1±1.1	10.7±0.4	12.1±0.5	99.4
Sausage type 2	60.8±1.3	11.0±0.4	12.4±0.6	98.7
		20 days		
Control	60.6±1.3	10.2±0.5	11.8±0.6	97.6
Sausage type 1	60.6±0.5	10.6±0.5	12.3±0.3	99.4
Sausage type 2	59.7±1.7	11.2±0.7	12.4±0.5	98.7
		40 days		
Control	60.9±1.5	9.0±0.5	11.3±0.4	95.2
Sausage type 1	60.9±1.1	10.7±0.4	13.3±0.5	96.3
Sausage type 2	60.5±1.1	10.8±0.3	11.7±0.5	98.2

During storage, all sausages had similar consumer characteristics, traditional for this type of meat product. For more reliable odour assessment, this was assessed instrumentally. The results of the multisensory investigations of cooked sausages are given in figure 1.

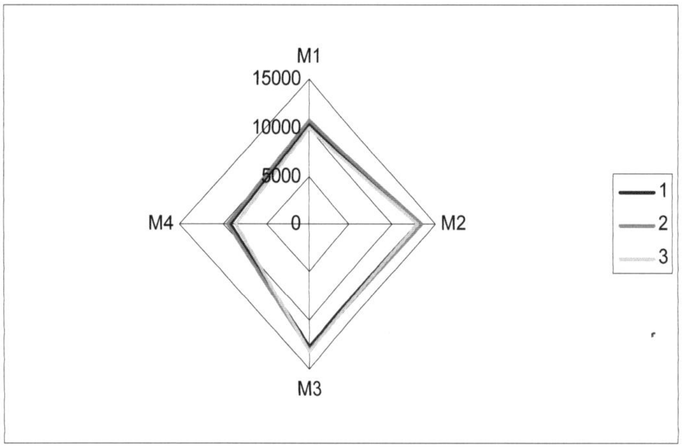

Figure 1. Odour visual fingerprints, obtained in multisensory analysis of cooked sausages after 1 day of storage (1-control, 2-sausage type 1; 3-sausage type 2).

The readings of the sensors M1-M4 that characterise the content of aroma of volatile components in the gas phase of the sausage samples did not have significant differences.

Taking into consideration that the sodium nitrite content in meat products is strictly regulated, we determined the mass fraction of sodium nitrite in sausages during storage (table 2).

Table 2. Sodium nitrite content in cooked sausages

Cooked sausages	Mass fraction of sodium nitrite, %
	1 day
Control	0.00164±0.00020
Sausage type 1	0.00360±0.00032
Sausage type 2	0.00336±0.00036
	20 days
Control	0.00160±0.00015
Sausage type 1	0.00308±0.00024
Sausage type 2	0.00307±0.00026
	40 days
Control	0.00152±0.00009
Sausage type 1	0.00160±0.00018
Sausage type 2	0.00152±0.00012

During storage, all cooked sausage samples complied with Russian legislative requirements for sodium nitrite and nitrate content (not more than 0.005 and 0.025%, respectively).

The mass fraction of sodium nitrite in sausage types 1 and 2 in the process of storage decreased by 55.6 and 54.8% ($p<0.05$), respectively, after 40 days. The nitrite content in the control did not change during sausage storage ($p>0.05$).

After one day of cooked sausage storage, the mass fraction of sodium nitrite in the experimental sausages was 2.0-2.2 times higher than in the control ($p<0.05$). However, during storage, the mass fraction of sodium nitrite in all analysed sausages became similar, which, apparently, can be explained by transformation of formed nitrogen oxide to nitrate under the impact of oxygen (table 3). The sodium nitrate content in the experimental sausages was higher compared to in the control by 2.4 and 2.1 times for types 1 and 2 sausages, respectively ($p<0.05$).

Table 3. Sodium nitrate content in cooked sausages.

Cooked sausage	Mass fraction of sodium nitrate, %
Control	0.00208±0.00021
Sausage type 1	0.00506±0.00042
Sausage type 2	0.00443±0.00032

4. Conclusion

Based on the results of the performed investigations, we established a regime for the preliminary thermal treatment of cooked sausages to transform nitrate ions present in the vegetable chard powder to nitrite ions with participation of an added denitrifying culture: not less than 60 min at 40°C. During storage, the control and experimental sausages had similar and acceptable consumer characteristics traditional for these types of meat products, which was confirmed by organoleptic assessment and instrumental analysis of colour and odour after production and during storage. The residual nitrite and nitrate content in the sausages with the natural, vegetable source of nitrate complied with legislative requirements.

References

[1] Larsen F J, Ekblom B, Sahlin K, Lundberg J O and Weitzberg E 2006 Effects of dietary nitrate on blood pressure in healthy volunteers *N Engl J Med.* **355** 2792–3

[2] Sindelar J J, Cordray J C, Sebranek J G, Love J A and Ahn D U 2007 Effects of vegetable juice powder concentration and storage time on some chemical and sensory quality attributes of uncured, emulsified cooked sausages *.J Food Sci.* **72**(5) 324-32

[3] Magrinya N, Bou R, Rius N, Codony R and Guardiola F 2016 Use of tocopherol extract and

different nitrite sources and starter cultures in the production of organic botifarra catalana, a cooked cured sausage *Food Sci Technol Int.* **22**(3) 221-34

[4] Kostecki A, Bilska A and Danyluk B 2015 Chosen quality parameters of pork sausage produced without curing mixture *Acta Sci. Pol. Technol. Aliment.* **14**(3) 191-8

[5] Eisinaite V, Vinauskiene R, Viskelis P and Leskauskaite D 2016 Effects of Freeze-Dried Vegetable Products on the Technological Process and the Quality of Dry Fermented Sausages *J. Food Sci.* **81**(9) C2175-82

Authentication of meat and meat products vs. detection of animal species in feed – what is the difference?

K Nešić[1], D Stojanović[2] and M Ž Baltić[3]

[1] Institute of Veterinary Medicine of Serbia, Autoput 3, Belgrade, Serbia
[2] Faculty of Medicine, University of Nis, Bulevar Zorana Đinđića 81, Niš, Serbia
[3] Faculty of Veterinary Medicine, University of Belgrade, Bulevar Oslobođenja 18, Belgrade, Serbia

E-mail: ksenija_n@yahoo.com

Abstract. Authenticity of food is an issue that is growing in awareness and concern. Although food adulteration has been present since antiquity, it has broadened to include entire global populations as modern food supply chains have expanded, enriched and become more complex. Different forms of adulteration influence not only the quality of food products, but also may cause harmful health effects. Meat and meat products are often subjected to counterfeiting, mislabelling and similar fraudulent activities, while substitutions of meat ingredients with other animal species is one among many forms of food fraud. Feed is also subject to testing for the presence of different animal species, but as part of the eradication process of transmissible spongiform encephalopathies (TSE). In both food and feed cases, the final goal is consumer protection, which should be provided by quick, precise and specific tools. Several analytical tests have been employed for such needs. This paper provides an overview of authentication of meat and meat products compared with species identification in feed control, highlighting the most prevalent laboratory methods.

1. Introduction

Adulteration in food has been a concern since the beginning of civilization, as it not only influences the quality of food products but also may cause harmful health effects. Adequate testing of food and adulterant detection in various food products are required for value assessment and to assure consumer protection against fraudulent activities [1]. The food protection concept includes food quality, food safety, food fraud and food defense. Although there are some overlaps between these four elements, in general, they are defined as follows: A food quality risk is an economic threat that is unintentional and influences a product's value to consumers. A food safety risk is unintentional contamination of food that causes adverse health consequences. A food fraud risk is economically motivated and intentional, but is not intended to be a public health food threat. A food defense risk is a public health threat that is intentional, such as malicious tampering or terrorism. It is ideologically motivated and makes the food injurious to health. [2].

Food fraud is a public health food risk that is growing in awareness, concern and danger. It has been present since antiquity, as evidence found in Roman amphorae containing fraudulent olive oil and wine shows. But in ancient times, the scale was limited and it covered a small geographic area. Because modern food supply chains have been expanded, enriched and compounded, the risk of food fraud has broadened to include entire global populations [3]. This collective term is used to cover deliberate substitution, addition, tampering, or misrepresentation of food, food ingredients, or food packaging; it also includes false or misleading statements made about a product for economic gain [4]. Most food fraud cases are not harmful, but there are exceptions, like melamine in Chinese skimmed milk powder [5], sudan dyes in spices [6], false labeling of puffer fish as monkfish [7] and the plasticizer di(2-ethylhexyl) phthalate (DEHP) being used as a cheaper substitute for clouding agents in food and beverages [8].

The main areas susceptible to fraud in the meat industry are: 1) the origin of meats and the animal feeding regime (as in the case of certificated regional products, for example); 2) substitutions of meat ingredients with other animal species, tissues, fat or proteins; 3) modifications of the processing methods for producing meat products and 4) additions of non-meat components such as water or additives. Recently, greater numbers of people are concerned about the meat they eat, so accurate labelling is important to inform their choice. Detection of animal species in feed occurs from a completely differently motivation, and it recently became mandatory in the eradication process of transmissible spongiform encephalopathies (TSE). In fact, shortly after the escalation of the bovine spongiform encephalopathy (BSE) crisis in Great Britain in 1986, it was determined that the source of infection was contaminated feed, i.e. infectious ruminant protein processed in meat and bone meal (MBM). One of the most important measures was the introduction of legislation which prevented these nutrients from entering the food chain and the establishment of appropriate analytical control of implementation of such regulations [9].

This paper intends to provide an overview on the issue of authentication of meat and meat products in comparation to species identification in feed control, highlighting the most prevalent laboratory methods. The aim is also to emphasize that simple transfer of protocols between these similar, but not the same matrices is not advisable.

2. Detection of animal species

Meats derived from different animal species are priced differently, and consumers want to know what kind of meat they are purchasing with absolute certainty. Adulteration of meat products poses a serious problem, not only for economic reasons, but also for religious and moral reasons, as well as due to allergies to meat derived from individual animal species. Proper labelling is also important to help fair trade [10]. According to the European Council Regulation (EC) No. 178/2002 28 January 28 2002, laying down the general principles and requirements of food law, food adulteration and misleading consumers is illegal. The same rules for food are implemented in Serbia through national regulations, as well as compliant legislation that applies to animal feed [11]. Some examples of undeclared animal species in meat products are given in table 1, adapted from Ballin [12].

Table 1. Examples of undeclared animal species in meat products.

Investigated product	Country of investigation	% of mislabelling cases (No of analysed samples)
Hamburgers	Mexico	39% (23)
Sausages	Mexico	29% (17)
Meat products	USA	15.9% raw; 22.9% cooked (902)
Meat products	Turkey	22% (100)

Identifying authenticity of meat and meat products is an important issue in food regulatory control for determining fraudulent actions and for assuring accurate food labelling, while detection of animal species in feed belongs to the group of systemic measures for the eradication and prevention of TSEs, so it is a part of the activities within the food safety chain [13]. According to feed legislation [14], feed must be in accordance with the following principles: no use of ruminant processed animal protein (PAP), no use of any PAP in feed intended for ruminants and herbivores, and no intra-species recycling (commonly termed cannibalism). So, as indicated in the TSE Road Map 2 [15], the goal of the European Commission is to continue the review of the measures following a stepwise approach supported by a solid scientific basis while maintaining a high level of food safety, so there must be adequate analytical methods to ensure absolute reliability.

In both food and feed cases, the final goal is consumer protection, which should be provided by quick, precise and specific food and/or feed control and proper identification of animal species. Several analytical methodologies have been employed for such needs, based on anatomical, histological, microscopic, organoleptic, chemical, electrophoretic, chromatographic, or immunological principles. However, due to specificities of their distinctive limitations, many techniques have been surpassed by the DNA-based molecular techniques or determination of proteins/peptides. In recent decades, most of the methods for identifying the species origin in meat and meat products, and especially in feed, are based on polymerase chain reaction (PCR) due to their high specificity and sensitivity and rapid processing time [16].

The authenticity of meat and meat products, as well as feed, includes various aspects and involves a wide range of analytes. Meat substitution is among the most frequent fraudulent and/or accidental situations, and refers to both species and tissue adulteration. Unfortunately, there is no perfect analytical tool that can provide solutions for all the problems in controlling the composition of food and feed. Still, many complementary, different approaches do exist. Techniques based on spectroscopy would be appropriate screening methods for meat and meat products, while more reliable and unambiguous results are achievable by immunoassays, molecular or mass spectrometry-based analysis [17]. Nevertheless, the most common methodologies in species determinations are based on DNA and immunological principles. Genetic methods are the most specific and sensitive for this purpose. However, they require expensive laboratory equipment and a certain degree of expertise. As an alternative, immunological assays can be used to reduce the test time and cost. Among these, ELISA (Enzyme-Linked ImmunoSorbent Assay) has been the most widely used technique in detecting food authenticity because of its specificity, simplicity and sensitivity. Regarding feed control, for all EU Member States and candidate countries, the only officially allowed method, is to implement SOPs and protocols published on the EU Reference Laboratory for Animal Proteins in Feedingstuffs (EURL-AP) website (http://eurl.craw.eu).

3. Enzyme-Linked ImmunoSorbent Assay (ELISA)

ELISA is an immunological technique in which an enzyme, a protein that catalyzes a biochemical reaction, is utilized to detect the presence of an antibody or an antigen in a sample. Two types, the indirect and the sandwich ELISA, are the most commonly used for food authentication. The indirect ELISA uses two antibodies, one specific to the antigen and the other coupled to an enzyme. This second "enzyme-linked" antibody gives the assay its name, and causes a chromogenic or fluorogenic substrate to produce a signal. Sometimes, this second antibody can be linked to a protein such as avidin or streptavidin if the primary antibody is biotin labeled. In the sandwich ELISA, the antigen is bound between two antibodies: the capture antibody and the detection antibody. The detection antibody can be tied to an enzyme or can bind the conjugate (enzyme-linked antibody) that produces the biochemical reaction. This method produces qualitative or quantitative results. Qualitative ELISA detects the presence or absence of an analyte in the sample. The cutoff between positive and negative result is determined by the analyst and is statistically based. In quantitative ELISA, used to detect amounts of the analyte, the optical density or fluorescent units of the sample is interpolated into a standard curve which is typically a serial dilution of the target [18].

Both polyclonal and monoclonal antibodies against muscle and serum animal proteins are used in the method variants for identification of different animal species. Until recently, the great limitation of ELISA methodology has been that the target proteins are sometimes denatured during processing. This has been particularly important for feed control in which the animal proteins are treated by sterilization: by steam pressure of 3 bar, at a temperature of 133°C for 20 minutes, and therefore, the target protein epitope cannot be present in the condition detectable by the antibodies. This limitation has been mainly solved because of the development of antibodies against thermostable proteins [19]. In recent years, advances in immunoassay technology have led to development of lateral flow tests (or dipsticks), which employ the same principles as the ELISA tests, but coat the antibodies and other reagents on a nitrocellulose membrane rather than the inside of test wells or paddles, and they use colloidal gold, dye, or latex bead conjugates to generate a visible signal. The simplicity of both type of tests and the short time required for the analysis make them suitable for screening of a large number of food samples [20].

According to Giovannacci et al. [21], there are limits for species detection in processed meat products, depending on various parameters, such as the fat content, the severity of heat processing, the origin of muscles and the maturation state of meat. Hence, detection limits might be different from one product to another. Although, from a theoretical point of view, ELISA methods, as well DNA methods are quantitative, they mainly give qualitative results. However, Kotoura et al. [22] described a strategy to determine the content of beef meat in different mixed meat products. In contrast to the performance of most ELISA tests for meat authentication, the reliability of immunological determination of ingredients of animal origin in animal feed is still under question [23]. Moreover, some requirements, such as greater sensitivity and better specificity need to be fulfilled [23]. Based on the latest EFSA scientific opinion [24], positive results require confirmation by another method, while false negatives frequently occur.

4. Polymerase Chain Reaction (PCR)

PCR is an efficient way to copy small segments of DNA, to determine animal species. Compared to proteins, DNA has a higher thermal stability, it is present in the majority of cells and potentially enables identical information to be obtained from the same animal, regardless of the tissue of origin. As this technique can theoretically amplify one copy of target DNA, the limit of detection is therefore often lower than observed in protein-based methods. It allows identification of meat species under different processing conditions. However, the variability of DNA at the species and target tissue levels makes DNA-based methods somewhat unsuitable for the quantification of exact percentages of different species in meat and meat products. That is why the majority of published methods are qualitative, although there are some studies that have demonstrated a correlation between meat content (w/w) and signal intensity [25]. Quantitative analysis should be based on real time PCR and results expressed as genome/genome equivalents. The PCR amplified sequence must originate from genomic DNA, while use of mitochondrial DNA is not possible in quantification of species as the amount of mitochondrial DNA in tissue varies. The specific PCR amplified DNA sequence has a large influence on the limit of quantification [26].

Regarding feed control, besides microscopy, the introduction of PCR as an official method and the validation of a PCR assay for the detection of ruminant DNA in feed allowed the re-authorization of non-ruminant PAP in feed for aquaculture animals as of 1 June 2013. The next steps could be the use of poultry PAP for pigs and pig PAP for poultry, but only after validation of adequate analytical tools (e.g. PCR assays) that would allow efficient control. Due to interference of authorized animal ingredients (e.g. fats, blood products, dairy products) with PCR results, additional analytical approaches will probably be needed [27]. In addition to its own research, in 2014, the EURL-AP initiated an international laboratory network to investigate and develop alternative techniques, such as aptamers, mass spectrometry and ELISA. The most promising method is probably mass spectrometry [28]. Recently the identification of proteins and peptide biomarkers allowing the detection of PAPs by mass spectrometry produced very interesting results, but efforts must be continued and the journey to validation and implementation in the control laboratories is long. All official protocols and the versions in force are easily downloadable from the EURL-AP website [29].

Species differentiation of raw materials in industrial food and feed preparation, as well as the detection of animal species in foods and feeds remains a constant challenge. Various PCR protocols have proved adequate for detecting small amounts of DNA in meat samples, by amplifying a target region of template DNA in a rapid and sensitive manner. Real time PCR seems to be the backbone of the future system of TSE prevention in feed control [30].

5. Conclusion

Among the wide range of analytical methods available for determining food authenticity, ELISA has performed well because of its sensitivity, specificity, promptness, cheapness, ease of performance and the low investment in equipment compared to other techniques. Although ELISA limitations for feed control have been shown, and most of the immunoassays for meat speciation are not applicable to feed detection because protein antigens denature during high-temperature rendering, this methodology could allow specific screening in both the meat and feed industries. On the other hand, PCR-based methods are highly specific and sensitive, but generally are unable to distinguish between different tissues of the same species. Also, variability of DNA content on the species and target tissue levels makes DNA-based methods somewhat unsuitable for quantification of different animal species in foods and feeds. For all these reasons, the proteomic approach, identifying different peptide biomarkers, and usually combined with mass spectrometry, is of importance.

It is important to emphasize that differences in the authentication of meat and meat products in comparison with feed do exist and they are related to the complexity of the matrices and processing treatments. Feed control, as food safety issue, is a very well regulated area, and analytical methods and procedures to detect constituents of animal origin are strictly prescribed. These must be fully respected and completely implemented during official controls. Finally, all analytical methods and control strategies, regardless of sample type, have to enforce national and trans-national laws and regulations and to ensure full consumer protection from all possible viewpoints: food fraud, food quality, food safety and food defense.

References

[1] Bansal S, Singh A, Mangal M, Mangal A K and Kumar S 2017 Food adulteration: Sources, health risks, and detection methods *Crit. Rev. Food Sci. Nutr.* **57**(6) 1174-89
[2] Spink J and Moyer D C 2011 Defining the Public Health Threat of Food Fraud *J. Food Sci.* **76**(9) R157-63
[3] Manning L and Soon J M 2016 Food Safety, Food Fraud, and Food Defense: A Fast Evolving Literature *J. Food Sci.* **81**(4) R823-34
[4] Spink J, Helferich O K and Griggs J E 2010 Combating the impact of product counterfeiting *Distribution Bus. Manage. J.* **10** 6
[5] Gossner C M, Schlundt J, Embarek P B, Hird S, Lo-Do-Wong D, Beltran J J O, Teoh K N and Tritscher A 2009 The melamine incident: implications for international food and feed safety *Environ. Health Perspect.* **117**(12) 1803–8
[6] Stiborova M, Martınek V, Rydlova H, Hodek P and Frei E 2002 Sudan I is a potential carcinogen for humans. Evidence for its metabolic activation and detoxication by human recombinant cytochrome P450 1A1 and liver microsomes 1 *Cancer Res.* **62** 5678–84
[7] Cohen N J, Deeds J R, Wong E S, Hanner R H, Yancy H F, White K D, Thompson T M, Wahl M, Pham T D, Guichard F M, Huh I, Austin C, Dizikes G and Gerber S I 2009 Public health response to puffer fish tetrodotoxin poisoning from mislabeled product, *J. Food Protec.* **724** 810–7
[8] Yang J, Hauser R and Goldman R H 2013 Taiwan food scandal: the illegal use of phthalates as a clouding agent and their contribution to maternal exposure *Food Chem. Toxicol.* **58** 362–8
[9] Nesic K and Pavlovic N 2012 Animal health protection through the control of use of animal proteins in feed *Proc. First Int. Symp. Animal Sci.* Belgrade, Serbia pp 912-9
[10] Montowska M and Pospiech E 2011 Authenticity Determination of Meat and Meat Products on the Protein and DNA Basis *Food Rev. Int.* **27** 84-100
Vlachos A, Arvanitoyannis I S and Tserkezou P 2016 An Updated Review of Meat Authenticity Methods and Applications *Crit. Rev. Food Sci. Nutr.* 56, 7, 1061-1096

[11] Regulation (EC) no 178/2002 of the European Parliament and of the Council of 28 January2002 laying down the general principles and requirements of food law, establishing the European Food SafetyAuthority and laying down procedures in matters of food safety *Off. J. Eur. Union*, L31/1, 01.02.2002, 1-24
Zakon o bezbednosti hrane *Sl. glasnik RS* 41/2009
Pravilnik o deklarisanju, označavanju i reklamiranju hrane *Sl. glasnik RS* 19/2017
Pravilnik o uslovima za deklarisanje, označavanje i reklamiranje hrane za životinje *Sl. glasnik RS* 4/2010
[12] Ballin N Z 2010 Authentication of meat and meat products *Meat Sci* 86, 577-587
[13] Nesic K, Pavlovic N, Jakic-Dimic D 2012 Presence of constituents of animal origin in feed *Proc. 6th Cent. Eur. Cong. Food*, Novi Sad, Serbia, 1543-1546
[14] Regulation (EC) No 51/2013 of 16 January 2013 amending Regulation (EC) No 152/2009 as regards the methods of analysis for the determination of constituents of animal origin for the official control of feed. *Off. J. Eur. Union*, 020, 33-43, 2013.
Pravilnik o izmenama i dopunama Pravilnika o utvrđivanju mera ranog otkrivanja i dijagnostike zarazne bolesti transmisivnih spongioformnih encefalopatija, načinu njihovog sprovođenja, kao i merama za sprečavanje širenja, suzbijanje i iskorenjivanje ove zarazne bolesti *Sl. glasnik RS* 33/2016
[15] European Commission EC 2010 The TSE Road map 2. A Strategy paper on Transmissible Spongiform Encephalopathies for 2010-2015. Communication from the Commission to the European Parliament and the Council. http://ec.europa.eu/food/food/biosafety/tse_bse/docs/roadmap_2_en.pdf
[16] Nesic K 2014 PCR technique for detection of meat and bone meal in feed *Proc. XVI Int. Symp. Feed Technology*, Novi Sad, Serbia, 92-96
Kumar A, Kumar R R, Sharma B D, Gokulakrishnan P, Mendiratta S K and Sharma D 2015 Identification of Species Origin of Meat and Meat Products on the DNA Basis: A Review *Crit. Rev. Food Sci. Nutr.* 55, 10, 1340-1351
[17] Sentandreau M A and Sentandreu E 2014 Authenticity of meat products: Tools against fraud *Food Res. Int.* 60, 19-29
[18] Asensio L, Gonzalez I, Garcıa T and Martın R 2008 Determination of food authenticity by enzyme-linked immunosorbent assay (ELISA) *Food Control* 19 (1), 1–8
[19] Van Raamsdonk L W D, Margry R J C F, van Kaathoven R G C, Bremer M G E G 2015 Inter-laboratory validation study of two immunochemical methods for detection of processed ruminant proteins *Food Chem.* 185, 333–339
[20] Nešić K, Pavlović N and Jojić-Maličević Lj 2012 Utvrđivanje prisustva mesno-koštanog brašna poreklom od goveda u hrani za životinje primenom tri različita komercijalna imunohemijska testa *Tehnologija mesa* 53/2, 173-178
[21] Giovannacci I, Guizard C, Carlier M, Duval V, Martin J and Demeulemester C 2004 Species identification of meat products by ELISA *Int. J. Food Sci. Tech.* 39, 863–867
[22] Kotoura S, Murakami-Yamaguchi Y, Kizu K, Nakamura M, Fuchu H, Miake K, Sugiyamaa M and Narita H 2012 Establishment of a sandwich ELISA for the determination of beef content in processed foods by using monoclonal antibodies to myoglobin. *Food Agric. Immunol.* 23, 289–301
[23] Fumiere O, Veys P, Boix A, von Holst C, Baeten V and Berben G 2009 Methods of detection, species identification and quantification of processed animal proteins in feedingstuffs. *Biotechnol. Agron. Soc. Environ.* 13(S) 59-70
[24] EFSA 2011 Panel on Biological Hazards (BIOHAZ), Scientific Opinion on the revision of the quantitative risk assessment (QRA) of the BSE risk posed by processed animal proteins (PAPs), *EFSA J.* 9 (1): 1947
[25] Farag M R, Alagawany M, Abd El-Hack M E, Tiwari R, Dhama K 2015 Identification of different animal species in meat and meat products: trends and advances. *Adv. Anim. Vet.*

Sci. 3(6), 334-346

[26] Ballin N Z, Vogensen F K and Karlsson A H 2009 Species determination – Can we detect and quantify meat adulteration? *Meat Sci.* 83, 165-174

[27] Nesic K, Samanc H, Vujanac I, Prodanovic R, Nesic V, Velebit B and Savic B 2014 Detection of meat and bone meal in cattle feed and ruminal fluid – comparison and combining of microscopy and polymerase chain reaction *Anim. Feed Sci. Technol.* 187, 86-90

[28] Marbaix H, Budinger D, Dieu M, Fumiere O, Gillard N, Delahaut P, Mauro S and Raes M 2016 Identification of Proteins and Peptide Biomarkers for Detecting Banned Processed Animal Proteins (PAPs) in Meat and Bone Meal by Mass Spectrometry. *J. Agric. Food Chem.* 64 (11), 2405-2414

[29] Fumiere O, Veys P, Marien A, Lecrenier MC, Baeten V and Berben G 2016

[30] Doosti A, Dehkordi P G and Rahimi E 2014 Molecular assay to fraud identification of meat products *J. Food Sci. Technol.* 51(1), 148–152

Nesic K, Pavlovic N, Lazic G and Bozic B 2016 Monitoring of feed for the presence of animal proteins in Serbia - decade overview and close perspective *Proc. II Int. Symp. Vet. Med.*, Belgrade, Serbia, 180-188

Investigation of grass carp by-products from a fish farm in Vojvodina

Đ Okanović[1,4], T Tasić[1], Š Kormanjoš[1], P Ikonić[1], B Šojić[2], M Pelić[3] and M Ristić[1]

[1] Institute of Food Technology, University of Novi Sad, Bulevar cara Lazara 1, Novi Sad, Serbia
[2] Faculty of Technology, University of Novi Sad, Bulevar cara Lazara 1, Novi Sad, Serbia
[3] Scientific Veterinary Institute "Novi Sad", 20, Rumenački put 20, Novi Sad, Serbia

[4] E-mail: djordje.okanovic@fins.uns.ac.rs

Abstract. The quantity of by-products obtained during grass carp primary processing and chemical characteristics of internal organs were investigated. The total average weight of by-products was 783.69 g (36.99%) in relation to live body weight which was cca 2118.5 g. The by-product contributing the largest quantity to total live body weight was the head with 458.22 g (21.63% of live body weight), followed by complete internal organs and tail and fins, with weights of 198.03 g or 9.35% and 57.93 g or 2.73%, respectively .The chemical composition of internal organs from the grass carp was mostly water (65.55%), following by crude fats and crude proteins (17.47% and 13.35%, respectively). The low collagen content (13.43% of total crude protein) indicates the high nutritional quality of the protein content from internal organs. Nitrogenous complexes from the internal organs were predominantly proteins. Digestible nitrogen was approximately equal to total nitrogen (89.38%), indicating that all proteins of the internal organs had high biological value. Based on the results obtained, it can be concluded that carp internal organs could be important sources of proteins and fats, and thus, could be used in Serbia as a raw material for feed and technical fat production.

1. Introduction

The aquaculture industry has grown rapidly over the last decade. Likewise, mariculture is expanding worldwide, thereby increasing the demand for feed ingredients to support production [1]. The rapidly growing fish sector directly depends on the aquafeed industry, which in turn largely depends on fish meal as its primary protein source.

Over the last decade, the global supply of fish meal has been limited, and meeting the demands of a growing industry has become challenging. In addition, fish meal proteins experience periodic fluctuations in pricing and availability [2].

Freshwater fish contain high levels of polyunsaturated fatty acids (PUFA), which makes them very important in human nutrition [3]. Since there are a number of biochemical interactions between the n-6 and n-3 series fatty acids, a balanced ratio between these fatty acids in the food is important for the normal functioning of the body in humans, as well as in animals [4]. Consumption of fish meat is increasing, due to its high content of PUFA, amino acids and lipid soluble vitamins which are important ingredients for human health. According to the latest data from FAO [5], the average consumption of fish in Serbia is 5-10 kg per capita per year, which is significantly below the European and global consumption [6].

Common carp is economically the most significant farmed fish species in Serbia [7], and the cyprinids are the most common species in the total world production of freshwater fishes (71.9%, 24.2 million tons in 2010) [5]. Grass carp was imported to the country as a regulator of hydro vegetation. It feeds on higher underwater plants and the meat is of good quality [8].

Manufacturing and development of fish products could increase the amount and contribute to better sales of fish, not only in traditional fish markets, but also in retail stores and supermarkets. However, technological processes, preservation and storage of fish meat differ from those for mammalian meat [9]. For proper manufacturing of fish products, knowledge of the chemical composition and characteristics of raw fish meat is very important in order to apply the most appropriate technology procedures that are adjusted to individual fish species.

Fish processing and new fish product development can produce novel sales of fish, not only in traditional fish markets, but also in all other consumer goods stores [10]. The demands of modern markets are increasingly directed towards processed fish, especially fillets. Larger quantities of edible and non-edible by-products are obtained in industrial conditions of primary fish processing [11].

Fish yield, expressed as the ratio of the weight of the carcass without the head, scales, fins and internal organs and whole fish weight, are essential parameters for all technological operations related to fish processing, since the economy of production is directly dependent on it [12].

By-products of grass carp processing contain valuable nutrients which can be sources for the food, pharmaceutical and feed industries [13]. In order to obtain more complete perception of the quality of animal by-products, it is necessary, in addition to knowledge of basic chemical composition, to obtain complete information on the quality of the most important nutritional components – proteins. However, the high crude protein content of some raw materials is not a guarantee of its high usability, ie. protein digestibility [14].

Inedible by-products obtained during grass carp slaughter belong to the third category of by-products [15], and are significant sources of proteins and fats that are convenient raw materials for processing into proteinaceous feeds for swine and pets.

Due to the increasing industrial grass carp processing and need for complex utilization of the by-products obtained, the aim of this research was to investigate the quantity of by-products and nutritive value of internal organs of grass carp.

2. Materials and Methods

The quantity of by-products and quality of internal organs were monitored during the fish harvesting and processing of grass carp from fish ponds in Vojvodina in industrial conditions.

Grass carp (mean weight approx. 2100 g) from the Ečka Fish Farm were delivered live to the manufacturing plant where they were immediately sacrificed. Scales, gills and viscera, heads (flat transverse incision just behind the gill arch) and the fins were removed with a knife. The following tissues/organs were weighed: fish before cutting, scales, head, tail and fins and total internal organs. The internal organs were not separated because in industrial conditions it is standard procedure to treat them as one tissue mass.

Chemical characteristics of internal organs were determined at the Institute of Food Technology in Novi Sad. The entire internal organs from each fish were individually placed in plastic bags to be used as one sample, labelled and refrigerated at 4°C. Four hours after slaughter, samples were transferred to the chemistry laboratory. All samples (a sample was one set of internal organs from one fish) were ground with a homogenizer, then packed into aluminium foil bags and stored for 24 h at 4°C prior to examination.

The basic chemical composition was assessed by determining moisture [16], total protein [17], hydroxyproline (i.e. the relative content of connective tissue proteins) [18], free fat [19], and total ash [20]. Nitrogen fractions and digestible nitrogen were determined according the AOAC methods for free fat content [21].

Data were statistically evaluated [22] using arithmetic mean ($\overline{X}$), standard deviation (SD) and coefficient of variation (CV).

3. Results and Discussion

The quantities of by-products obtained from carp processing are presented in table 1.

Table 1. Quantity of grass carp by-product, g and %*.

	$\overline{X}$	SD	CV	%
Live body weight	2118.50	98.23	4.64	100.00
Scales	69.51	2.49	3.58	3.28
Head	458.22	31.65	6.91	21.63
Tail and fins	57.93	5.11	8.82	2.73
Total internal organs	198.03	12.22	6.17	9.35
Total by-products	783.69	51.32	6.55	36.99
Fillet	1001.19	75.33	7.52	47.26
Bones and skin	305.85	19.56	6.40	14,.44
Carcass	1307.04	83.14	6.36	61.70

* % according to grass carp live weight.

After cutting off the head, tail and fins and removal of complete internal organs, average carcass weight was 1307.04 g (61.70%) of total live body weight. Routine removal of skin, bones, spine and rib of carcass produced an average fillet weight of 1001.19 g (47.26%). Total average weight of by-products was 783.69 g (36.99%) in relation to live weight which was cca 2118.5 g. The by-product with the largest proportion of total live body weight was the head, with 458.22 g (21.63%), followed by the internal organs and the tails and fins which had weights of 198.03 g (9.35%) and 57.93 g (2.73%), respectively.

The head weight largely depends on the processing method (straight or round cut behind the gills). In the research of Tumbas and Petrović [23], the head, obtained with a circular cut, was 11% of the live body weight. The weight of tail and fins 99.15 g (5.11%) was smaller than the weight of heads. Total internal organs weighed 143.77 g (12.22%). According to Ristić *et al.* (1992), grass carp by-product percentages ranged from head 19.79%, tail and fins 3.09% and total internal organs 9.47%.

Results of the chemical composition of the internal organs are shown in table 2. This raw material, apart from water (65.55%), contained mostly crude fat (17.47%) and then crude proteins (13.35%). The low proportion of collagen (13.43%) in the total crude protein indicates the high nutritional quality of the protein. The fat content in grass carp ranges from 2.3 to 16.8%, while the protein content is less variable and generally is in the range of 14 to 18% [3,24].

As seen in table 2, the nitrogen complex in grass carp internal organs was composed mostly of protein. The high digestibility of the protein (89.38%) indicates the high biological value of the internal organ proteins. Ristić *et al.* [13] studied a set of grass carp internal organs that contained a higher proportion of fat (31.64%), and also had protein with good digestibility (91.57%).

Table 2. Chemical composition of grass carp internal organs, %.

Parameter	$\overline{X}$	SD	CV
Moisture	65.55	13.83	21.10
Crude protein	13.35	1.83	15.94
Relative content of connective tissue proteins	13.43	0.60	4.47
Crude fat	17.47	2.35	13.45
Ash	1.00	0.16	16.00
N-free extract	2.63	0.18	6.84
Non-protein N	0.46	0.08	17.39
Protein digestibility	89.38	3.30	3.69

4. Conclusion

Based on the results obtained in this study, it is possible to conclude the following:

• After cutting off the head, tail and fins and removal of complete internal organs the average carcass weight was 1307.04 g (61.70% of the live weight).

• The total average weight of by-products was 783.69 g or 36.99% in relation to live weight (cca 2118.5 g).

• The head had the largest proportion of weight in relation to live weight, being 458.22 g (21.63% of live weight). Weights of tail and fins were much smaller, being 57.93 g (2.73%). The total internal organs weighed 198.03 g (9.35%).

• Biochemical analyses show that the internal organs, apart from water, contained significant amounts of crude fat (17.47%) and protein (10.68%), making them suitable for feed processing.

• The amount of digestible nitrogen in the internal organs was approximately equal to total nitrogen (89.38%), indicating that all proteins from the internal organs have high biological value.

• Inedible internal organs obtained during carp slaughter could be an important source of fatty acids and could be used as raw material for processing into feeds for use in animal nutrition.

Acknowledgements

Investigations were carried out within the project "The quantity, quality and the possibility utilization of by-products origin from the growing and processing of fish from the fish pond on the territory AP Vojvodina" financially supported by the Municipality Secretary of Science and Technological Development, No. 114-451-480/2016.

References

[1] FAO 2014 State of world fisheries and aquaculture (Available at: http://www.fao.org/3/a i3720e.pdf)

[2] Tacon A G J Metian M 2008 Global overview on the use of fish meal and fish oil in industrially compounded aquafeeds: trends and future prospects *Aquaculture* **285** 146–58

[3] Vladau V, Bud I and Stefan R 2008 Nutritive value of fish meat comparative to some animals meat Bulletin of University of Agricultural Sciences and Veterinary Medicine. *Animal Sciences and Biotechnologies* **65**(1–2) 301–5

[4] Đorđević V, Baltić M, Ćirković M, Kilibarda N, Glamočlija N, Stefanović S and Miščević M 2009 Quantitative and qualitative determination of enrofloxacin residues in fish tissues *Acta Veterinaria Belgrade* **59**(5-6) 579-89

[5] Ćirković M, Ljubojević D, Đorđević V, Novakov N, Petronijević R, Matekalo-Sverak V and Trbović D 2012 The Breed Effect on Productivity and Meat Nutrient Compsition of Fish

Kafkas Üniversitesi Veteriner Fakültesi Dergisi **18**(5) 775-80

[6] Ćirković M, Trbović D and Ljubojević D 2011 Meat quality of fish farmed in polyculture in carp ponds in Republic of Serbia *Tehnologija Mesa* **52** 106-21

[7] FAO 2012 Demand and supply of aquafeed and feed ingredients for farmed fish and crustaceans: trends and future prospects *The State of World Fisheries and Aquaculture* pp 172-81

[8] Vujković G, Jovanović B, Marošan Lj, Manojlović D and Svrzić G 1993 Sporedni proizvodi slatkovodnih riba – potencijalna sirovina za stočnu hranu *Tehnologija mesa* 34(2-3) 99-100

[9] Okanović Đ 2015 Randman ciprinidnih vrsta riba, Monografija autora Ćirković M, Ljubojević D, Novakov N, Đorđević V: "Gajenje i kvalitet mesa šaranskih riba", Poglavlje 15, (Novi Sad: Naučni institut za veterinarstvo) pp 256-66

[10] Baltić Ž M, Kilibarda N and Dimitrijević M 2009 Factors significant for the shelf-life of fish and selected fish products in retail *Tehnologija Mesa* **50**(1-2), 166-176

[11] Ristić M, Filipović S and Sakač M 2002 Liquid protein feedstuff from freshwater fish by-products as a component of animal feed *Roumanian Biotechnological Letters* **7**(3) 729-36

[12] Ljubojević D, Ćirković M, Novakov N, Babić J, Lujić J, Marković T 2012 Faktori koji utiču na randman šaranskih riba *Tehnologija Mesa* **53**(1) 14-9

[13] Ristić M, Kormanjoš Š, Milić M, Ćurčić R and Petrović N 1992 Utvrđivanje količina i kvaliteta sporednih proizvoda koji se dobijaju preradom slatkovodnih riba *Tehnologija Mesa* Beograd, **32**(3) 87-92

[14] Ristić M, Okanović Đ and Sakač M 2011 Karakteristike životinjskih sporednih proizvoda i njihova namena-Monografija (Novi Sad: FINS) pp 1-195

[15] REGULATION (EC) No 2009 1069 of the European Parliament and of the Council

[16] SRPS ISO 1442:1997. Meat and meat products – Determination of the moisture content

[17] SRPS ISO 937:1992. Meat and meat products – Determination total protein content

[18] SRPS ISO 3496:2002, Meat and meat products – Determination hydroxyproline content – the relative content of connective tissue proteins

[19] SRPS ISO 1444:1997. Meat and meat products – Determination of free fat content

[20] SRPS ISO 936:1998. Meat and meat products – Determination total ash content

[21] AOAC 1998. Official methods of analysis (16th ed.). Gaithersburg, MD: Association of Official Analytical Chemists

[22] Statistical Office of the Republic of Serbia 2014: http://webrzs.stat.gov.rs/WebSite/

[23] Tumbas Lj and Petrović M 1978 Uticaj težine amura na randman mesa *Ribarstvo Jugoslavije* **5** 116

[24] Ljubojević D, Ćirković M, Đorđević V, Puvača N, Trbović D, Vukadinov J and Plavša N 2013 Fat quality of marketable fresh water fish species in the Republic of Serbia *Czech Journal of Food Sciences*, **31**, 445-50

Effects of replacing beef fat with pre-emulsified pumpkin seed oil on some quality characteristics of model system chicken meat emulsions

M Serdaroğlu, B Nacak, M Karabıyıkoğlu, M Tepe, I Baykara and Y Kökmen

Food Engineering Department, Engineering Faculty, Ege University, 35100 Bornova, İzmir, Turkey

E-mail: meltem.serdaroglu@ege.edu.tr

Abstract. In this study, the effects of adding pumpkin seed oil (PSO) in water emulsion to model system chicken meat emulsions (MSME) on product quality and oxidative stability were investigated. MSME were produced by replacing 25% (P25) and 50% (P50) of beef fat with PSO-in-water emulsion (PSO/W) while control treatment was prepared with only beef fat. Addition of PSO/W to the formulation resulted in significant differences in chemical composition and pH values of both raw and cooked MSME treatments. The use of PSO/W produced significant improvements to emulsion stability, oxidative stability and cooking yield of MSME. It was determined that the use of PSO/W formulation results in decreased total expressible fluid values and increased cooking yields of the emulsions. It was observed that the highest cooking yield and the lowest total expressible fluid were found in the sample containing 50% PSO/W. It should be a feasible strategy to produce fat-reduced meat products with healthier lipid profiles by using PSO/W.

1. Introduction

Fat has an important role in meat products; it helps emulsion stabilization, improves binding properties, water holding capacity and cooking yields as well as provides sensorial characteristics such as juiciness [1,2]. However, diets with high animal fat contents have been related to increased obesity, cardiovascular disease and coronary heart disease due to their high saturated fatty acid and cholesterol contents [3,4,5]. Thus, the meat industry has begun to work on reformulation strategies to produce healthier meat products by decreasing saturated fatty acids, decreasing cholesterol and adding natural antioxidants. One of these strategies is using formulations with vegetable oils, since they are free of cholesterol and have a higher ratio of unsaturated to saturated fatty acids than animal fats [6,7,8]. However, the use of vegetable oils directly in product formulation can cause technological problems and quality loss in meat products [9]. Therefore, pre-emulsions constitute an innovative approach in low-fat product formulations since they can be used in fat reduction processes, beneficially modifying fatty acid profiles, masking off flavors and improving the sensory properties of products [10,11,12].

Pumpkin seed oil (PSO) is rich in bioactive compounds such as polyunsaturated (linoleic and linolenic) fatty acids, β-carotene, lutein, β-tocopherol, chlorophyll and phytosterols, and it is widely used in salads around the world, especially in Hungary, Slovenia and Austria. Due to its color and foaming problem, PSO cannot be used in manufacturing processed foods or processes such as frying [13].

To the best of our knowledge, no research has been performed regarding utilization of PSO in meat model systems or meat products. Therefore, the objective of this study was to investigate the effect of using PSO in oil-in-water emulsion as a fat replacer on the technological characteristics and oxidative stability of model system chicken meat emulsions.

2. Materials and Methods

The pumpkin seed oil-in-water (PSO/W) emulsion was prepared according to Poyato *et al.* [14] with modifications. The aqueous phase was prepared with the mixture of egg white powder (5 g/100 g) and water (45 g/100 g). The oil phase (50 g/100 g) was added to the aqueous phase after both phases were heated separately to 55°C. After the emulsification process (6000 rpm, Ultra-Turrax® T25basic), the emulsion was cooled to room temperature and PSO/W was kept under refrigeration (4°C) until use. Three different model system chicken meat emulsions (MSME) were prepared following the procedure described by Cofrades *et al.* [15] with modifications (table 1). One contained only beef fat (C), two other MSME were prepared by replacing 25% (P25) and 50% (P50) of beef fat with PSO/W. Chicken breast meat and beef fat were passed through a grinder with a 3-mm plate (Arnica, Turkey). The minced meat was homogenized for 1 min in a kitchen-type mixer (Tchibo, Germany) which was placed in cooling bath (2°C). Fat or PSO/W, half of the ice, plus salt and sodium tri-polyphosphate (STPP) were added and mixed for 1 min. The other half of the ice was added and mixed again for 2 mins. Portions of each emulsion (approximately 25 g) were placed in Falcon tubes (50 ml), which were hermetically sealed then centrifuged at 4500 rpm at 4°C for 1 min to eliminate any air bubbles. Samples were heat-treated for 30 min in a water bath at 70 °C, then cooled to room temperature for further analyses.

Table 1. Formulation (%) of MSME1

Samples[a]	Meat (%)	Beef fat (%)	PSO/W[b] (%)	Water (Ice) (%)
C	68	20	-	10
P25	68	15	5	10
P50	68	10	10	10

Samples contain: 1.5% salt and 0.5% STPP.
[a] Sample denomination: C: %100 beef fat.
P25: 75% beef fat + 25% PSO/W.
P50: 50% beef fat + 50% PSO/W.
[b] Pumpkin seed oil-in-water emulsion.

Moisture, protein and ash contents of raw and cooked MSME treatments were determined by AOAC methods [16]. Fat content was evaluated according to Flynn and Bramblet [17], pH values of emulsions were measured by using a pH-meter (WTW pH 3110 set 2, Germany) equipped with a penetration probe. Emulsion stability, recorded as total expressable fluid (TEF), and water holding capacity (WHC) were determined according to Hughes *et al.* [18]. The weights of meat emulsions before and after cooking were recorded and the cooking yield calculated. Lipid oxidation on days 0, 1 and 4 of storage was evaluated by the TBA method as described by Witte *et al.* [19]. The data were analyzed by one-way ANOVA using the SPSS software version 20. Differences among the means were compared using Duncan's Multiple Range test. A significance level of $p<0.05$ was used for evaluations.

3. Results and Discussion

Table 2. Chemical composition and pH values of raw and cooked MSME treatments.

	Raw MSME					Cooked MSME				
	Moisture (%)	Ash (%)	Fat (%)	Protein (%)	pH	Moisture (%)	Ash (%)	Fat (%)	Protein (%)	pH
C	62.05 ± 0.44^{b}	1.46 ± 0.08	18.77 ± 0.97^{a}	16.29 ± 0.46^{b}	6.06 ± 0.01^{a}	60.23 ± 0.96^{c}	1.51 ± 0.06	15.44 ± 0.70^{ab}	19.77 ± 0.78	6.17 ± 0.02^{b}
P25	64.66 ± 1.20^{ab}	1.53 ± 0.05	16.74 ± 0.65^{b}	16.93 ± 0.93^{ab}	6.02 ± 0.01^{a}	62.54 ± 0.87^{b}	1.41 ± 0.18	16.07 ± 0.64^{a}	19.85 ± 0.26	6.20 ± 0.01^{a}
P50	65.91 ± 0.42^{a}	1.39 ± 0.10	15.04 ± 0.72^{c}	17.56 ± 0.35^{a}	5.96 ± 0.01^{b}	64.28 ± 0.47^{a}	1.48 ± 0.06	15.18 ± 0.47^{b}	19.91 ± 0.21	6.17 ± 0.01^{b}

Data are presented as the mean values of 3 replications ± SD.
^{abc}: Means with different letters in the same column are significantly different ($p \leq 0.05$).

Chemical compositions and pH values of raw and cooked MSME treatments are shown in table 2. Replacing beef fat with PSO/W showed an increasing effect on moisture and protein contents of raw MSME ($p<0.05$), since PSO/W contributed water and protein (egg white powder) to the formulation. The fat content of raw MSME decreased with respect to the incremental addition of PSO/W ($p<0.05$), while no significant differences were observed in ash content ($p>0.05$). Addition of PSO/W decreased the pH values of raw MSME because of lower pH value of PSO [20].

The highest moisture content was found in P50 after cooking ($p<0.05$), probably the result of the lower TEF of this treatment. Protein and ash contents of cooked MSME treatments were similar ($p>0.05$). Addition of PSO/W to the formulation showed significant effects on fat content and pH values of cooked MSME. The highest fat content was found in P25 ($p<0.05$). The fat content of MSME with added PSO/W (P25 and P50) was more or less constant in raw and cooked MSME treatments; this could be the result of higher emulsion stability of these formulations.

Table 3. MSME characteristics.

	WHC (%)	TEF (%)	Cooking Yield (%)
C	93.84 ± 0.20	12.26 ± 0.46^{a}	86.64 ± 1.11^{c}
P25	94.35 ± 0.15	9.98 ± 0.23^{b}	89.47 ± 1.11^{b}
P50	94.37 ± 0.48	7.73 ± 1.59^{c}	92.65 ± 0.75^{a}

Data are presented as the mean values of 3 replications ± SD.
^{abc}: Means with different letters in the same column are significantly different ($p \leq 0.05$).

The aim of adding pre-emulsified fat or oils is having stable characteristic in meat products. WHC, cooking yield and emulsion stability (TEF%) results of MSME are shown in table 3. WHC of the MSME formulations were similar ($p>0.05$). PSO/W addition to formulations showed significant effect on emulsion stability of treatments ($p<0.05$). The highest TEF% values were found in C treatment while replacing beef fat with PSO/W showed a decreasing effect on TEF% values of MSME treatments ($p<0.05$). Cooking yield depends on the ability of the protein matrix to stabilize both fat and water molecules [5]. The lowest cooking yield was observed in C treatment and higher cooking yields were found when beef fat was replaced with PSO/W ($p<0.05$). The higher protein content of

P25 and P50 treatments could be the reason for their higher cooking yields and more stable emulsions, since more protein could have entrapped water and fat molecules in the system. It is well-known that emulsifiers are amphiphilic in their native state in emulsions because of their hydrophilic and hydrophobic interactions [21]. Another reason for the lower TEF% values and higher cooking yields in P25 and P50 formulations may be due to the amphiphilic properties of egg white powder in PSO/W.

Table 4. TBA values of MSME treatments during storage (mg malondialdehyde/kg product).

	Day 0	Day 1	Day 4
C	$0.63{\pm}0.02^{a,X}$	$0.61{\pm}0.02^{a,X}$	$0.69{\pm}0.02^{a,Y}$
P25	$0.52{\pm}0.01^{b,X}$	$0.48{\pm}0.01^{b,Z}$	$0.63{\pm}0.02^{b,X}$
P50	$0.47{\pm}0.02^{c,X}$	$0.49{\pm}0.01^{b,X}$	$0.34{\pm}0.03^{c,Y}$

Data are presented as the mean values of 3 replications ± SD.
^{abc}: Means with different letters in the same column are significantly different ($p \leq 0.05$).
^{XYZ}: Means with different letters in the same row are significantly different ($p \leq 0.05$).

Lipid oxidation can have negative effects on the quality of meat and meat products since they can cause sensory attribute (color, texture, odor and flavor) and nutritional quality changes. TBA values of MSME treatments during 4-days storage are shown in table 4. The oxidative status of MSME is strongly influenced by the type of fat/oil used in the formulation. The lower TBA values were found when beef fat was replaced with PSO/W ($p<0.05$) and this trend was observed during the storage period ($p<0.05$). The lower TBA values in P25 and P50 could be the result of adding pre-emulsion, which provided a protective effect from lipid oxidation and also added an oil with antioxidant properties to the formulations [22]. There was a slight decrement in TBA values of P50 treatment on day 4, probably due to the result of decomposition of malondialdehydes by bacterial processes [23,24] or further oxidation of malondialdehydes to other products [25,26]. Although the highest TBA values were found in C treatment during storage, these oxidation values were below 1.0 (mg malondialdehyde/kg product), which is the accepted limit for rancidity in meat products [27,28].

4. Conclusion

The results of this study indicated that replacement of beef fat with pumpkin seed oil-in-water emulsions significantly affected chemical composition and pH values of raw and cooked MSME treatments. PSO/W addition to the formulations resulted lower total expressible fluid values, higher cooking yield and higher oxidative stability during storage. Our study showed that meat products with a healthier lipid profile could be manufactured by using pumpkin seed oil in pre-emulsions.

References

[1] McAfee A J, McSorley E M, Cuskelly G J, Moss B W, Wallace J M W and Bonham M P 2010 Red meat consumption: An overview of the risks and benefits *Meat Sci.* **84**(1) 1−13
[2] Rodriguez Furlan L T, Padilla A P and Campderros M E 2014 Development of reduced fat minced meats using inulin and bovine plasma proteins as fat replacers *Meat Sci.* **96** 762–8
[3] Özvural E B and Vural H 2008 Utilization of interesterified oil blends in the production of frankfurters *Meat Sci.* **78** 211-6
[4] Sanchez-Zapata E, Daaz-Vela J, Perez-Chabela M., Perez-Alvarez J and Fernandez-Lopez J 2013 Evaluation of the effect of tiger nut fibre as a carrier of unsaturated fatty acids rich oil on the quality of dry-cured sausages *Food Bioprocess Tech* **6** 1181–90.
[5] Kumar Y, Kairam N, Ahmad T and Yadav D N 2016 Physico chemical, microstructural and sensory characteristics of low-fat meat emulsion containing aloe gel as potential fat replacer *Int. J. Food Sci. Tech.* **51**(2) 309–16

[6] Muguerza E, Gimeno O, Ansorena D and Astiasaran I 2004 New formulations for healthier dry fermented sausages: a review *Trends Food Sci. Tech.* **15** 452–7
[7] Choi Y S, Choi J H, Han D J, Kim H Y, Lee M A, Kim H W and Kim C J 2009 Characteristics of low-fat meat emulsion systems with pork fat replaced by vegetable oils and rice bran fiber *Meat Sci.* **82** 266–71
[8] Afshari R, Hosseini H, Khaksar R, Mohammadifar M A, Amiri Z and Komeili R 2015 Investigation of the effects of inulin and b-glucan on the physical and sensory properties of low-fat beef burgers containing vegetable oils: Optimization of formulation using d-optimal mixture design *Food Technol. Biotech.* **53**(4) 436-44
[9] Beriain M J, Gómez I, Petri E, Insausti K and Sarriés M V 2011 The effects of olive oil emulsified alginate on the physico-chemical, sensory, microbial, and fatty acid profiles of low-salt, inulin-enriched sausages *Meat Sci.* **88**(1) 189–97
[10] McClements D J, Decker E A, Park Y and Weiss J 2009 Structural design principles for delivery of bioactive components in nutraceutical and functional *foods Critical Review of Food Science and Nutrition* **49** 577–606
[11] Dickinson E 2011 Double emulsions stabilized by food biopolymers *Food Biophys.* **6** 1–11
[12] Cofrades S, Antoniou I, Solas M T, Herrero A M and Jimenez-Colmenero F 2013 Preparation and impact of multiple (water-in-oil-in-water) emulsions in meat systems *Food Chem.* **141**(1) 338–46
[13] Murkovic M, Hillebrand A, Winkler J, Leitner E and Pfannbauser W 1996 Variability of fatty acid content in pumpkin seeds (*Cucurbita pepo* L.) *Z Lebensm Unters Forsch*, **203** 216-19
[14] Poyato C, Astiasaran I, Barriuso B and Ansorena D 2015 A new polyunsaturated gelled emulsion as replacer of pork back-fat in burger patties: Effect on lipid composition, oxidative stability and sensory acceptability *LWT-Food Sci. Technol.* **62** 1069-75
[15] Cofrades S, López-López I, Solas M T, Bravo L and Jiménez-Colmenero F 2008 Influence of different types and proportions of added edible seaweeds on characteristics of low-salt gel/emulsion meat systems *Meat Sci* **79** 767–76
[16] AOAC 2007 *Official methods of analysis* 18th Ed. AOAC International Association of Official Analytical Chemists Gaithersburg MD USA
[17] Flynn A W and Bramblett V D 1975 Effects of frozen storage cooking method and muscle quality and attributes of pork loins *J Food Sci* **40** 631–33
[18] Hughes E, Cofrades S and Troy D J 1997 Effects of fat level, oat fibre and carrageenan on frankfurters formulated with 5, 12 and 30% fat *Meat Sci.* **45** 273-81
[19] Witte V C, Krause, G F and Bailey M E 1970 A new extraction method for determining 2-thiobarbituric acid values of pork and beef during storage *J. Food Sci.* **35** 582–5
[20] Al-Okbi S Y, Abdou D, Hamed T E, Alaa A, Hosam S, El-Alim A and Mahmoud D 2017 Enhanced prevention of progression of non alcoholic fatty liver to steatohepatitis by incorporating pumpkin seed oil in nanoemulsions *J. Mol. Liq.* **225** 822–32
[21] Viljanen K 2005 Protein oxidation and protein-lipid interactions in different food models in the presence of berry phenolics *Doctoral dissertation* University of Helsinki Department of Applied Chemistry and Microbiology p 36
[22] Öztürk B, Urgu M and Serdaroğlu M 2017 Egg white powder-stabilised multiple (water-in-olive oil-in-water) emulsions as beef fat replacers in model system meat emulsions. *J. Sci. Food Agr.* **97** 2075–83
[23] Casaburi A, Aristoy M C, Cavella S, Di Monaco R, Ercolini D and Toldra F 2007 Biochemical and sensory characteristics of traditional fermented sausages of Vallo di Diano (Southern Italy) as affected by the use of starter cultures *Meat Sci.* **76** 295–307
[24] Fregonesi R P, Portes R G, Aguiar A M M, Figueira L C, Gonclaves C B, Arthur V, Lima C G, Fernandes A M and Trindade M A 2014 Irradiated vacuum-packed lamb meat stored under refrigeration: Microbiology, physicochemical stability and sensory acceptance *Meat Sci.* **97**(2) 151–5

[25] Georgantelis D, Ambrosiadis I, Katikou P, Blekas G and Georgakis S A 2007 Effect of rosemary extract, chitosan and α-tocopherol on microbiological parameters and lipid oxidation of fresh pork sausages stored at 4°C *Meat Sci.* **76** 172–81

[26] Azizkhani M and Tooryan F 2014 Antioxidant and antimicrobial activities of rosemary extract, mint extract and a mixture of tocopherols in beef sausage during storage at 4°C *J. Food Safety* **35** 128–36

[27] Ockerman H W 1976 *Quality control of post mortem muscle and tissue* Vol. 4 Columbus USA Department of Animal Science Ohio State University (Available et: https://kb.osu.edu/.../ QUALITY%20CONTROL%20OF%20POST%20%20MORTERM%20MUSCLE%20TISSUE%20Vol.4.pdf)

[28] Yıldız-Turp G and Serdaroğlu M 2008 Effect of replacing beef fat with hazelnut oil on quality characteristics of sucuk – A Turkish fermented sausage *Meat Sci* **78**(4) 447–54

The influence of smoking in traditional conditions on content of polycyclic aromatic hydrocarbons in *Petrovská klobása*

S Škaljac[1], Lj Petrović[1], M Jokanović[1], V Tomović[1], T Tasić[2], M Ivić[1], B Šojić[1], P Ikonić[2] and N Džinić[1]

[1]Faculty of Technology, University of Novi Sad, Bulevar cara Lazara 1, Novi Sad, Serbia
[2]Institute of Food Technology, University of Novi Sad, Bulevar cara Lazara 1, Novi Sad, Serbia

[1]E-mail: snezanasavatic@gmail.com

Abstract. The aim of this study was to determine the content of 13 polycyclic aromatic hydrocarbons (acenaphthylene, fluorene, phenanthrene, anthracene, pyrene, benz[a]anthracene, chrysene, benzo[b]fluoranthene, benzo[k]fluoranthene, benzo[a]pyrene, indeno[1,2,3-cd]pyrene, dibenz[a,h]anthracene and benzo[ghi]perylene) the from Environmental Protection Agency list (US-EPA PAH) in traditional dry fermented sausage *Petrovská klobása*. Sausages were smoked in traditional conditions and samples for analyses were taken on day 0 of production (0), at the end of drying (T1) and at the end of the storage period (T2).
The highest total content of 13 US-EPA PAHs was determined in sausages at the end of the storage period (73.5 μg/kg). Phenanthrene was the most abundant of the PAHs in all examined sausage samples (0-4.90 μg/kg; T1-18.0 μg/kg and T2-26.3 μg/kg). Benzo[a]pyrene, with a maximum allowed content in smoked meat products of 2 μg/kg (EC No. 835/2011), was below the limit of detection in all examined samples. Also, PAH4 (the total content of benz[a]anthracene, chrysene, benzo[b]fluoranthene and benzo[a]pyrene) with a maximum allowed content in smoked meat products of 12 μg/kg (EC No. 835/2011), was below the limit of detection in all examined sausage samples. According to the results obtained in this study, and in regard to the European regulation on PAHs content, the dry fermented sausage *Petrovská klobása,* smoked in traditional conditions, was safe for consumers.

1. Introduction

Petrovská klobása is traditional dry fermented sausage from Bački Petrovac in the north of Serbia. It is produced in the traditional manner without additives and starter cultures. This product is protected with designation of origin (PDO) according to Serbian legislation because of its specific and distinctive quality. Its intense red colour is one of the essential features which differentiate it from other products of the same type. This colour is formed thanks to the use of red paprika and a unique smoking process (usage of specific types of wood, a mixture of cherry and apricot) [1, 2].

The traditional smoking process gives a special colour, taste and aroma to *Petrovská klobása* [2], but incomplete wood combustion during this process can produce considerable amounts of polycyclic aromatic hydrocarbons – PAHs. PAHs are a group of organic compounds, consisting of two or more condensed aromatic rings, and which may be regarded as potentially genotoxic and carcinogenic to humans [3, 4, 5]. The PAH content in smoked meat products depends on meat products characteristics and factors related to the smoking process such as direct (traditional conditions) or indirect methods (industrial conditions) [6, 7]. Reports which show that concentrations of PAH in traditional smoked meat products can reach high levels have, in recent years, prompted considerable interest in PAH quantification and control [2, 8, 9].

Thus, the aim of this study was to determine the effects of smoking in traditional conditions on the content of 13 PAHs (acenaphthylene - Acy, fluorine - Fln, phenanthrene - Phe, anthracene - Ant, pyrene - Pyr, benz[a] anthracene – BaA, chrysene - CHR, benzo[b]fluoranthene - BbF, benzo[k] fluoranthene - BkF, benzo[a]pyrene - BaP, indeno[1,2,3-cd]pyrene - IcP, dibenz[a,h]anthracene - DhA and benzo[ghi]perylene – BgP) from the Environmental Protection Agency list (US-EPA PAHs) in *Petrovská klobása*.

2. Materials and Methods

2.1. Sausage production

The materials used in this study were dry fermented sausages (*Petrovská klobása*). The basic formulation of sausage mixture for *Petrovská klobása* was lean pork meat (80%) and pork fat (20 %). Other ingredients were added to the pork meat and fat – red hot paprika powder, salt, garlic, caraway, sugar – and then the mixture was mixed mechanically. Raw sausage mixture was stuffed into collagen casings (500 mm long; 55 mm in diameter). After a rest day, sausages were smoked in traditional conditions.

During smoking in traditional conditions, smoke was produced by the combustion of sweet cherry and apricot wood. Sausages came in direct contact with the smoke (direct smoking), with distance of 3 m between fire and sausages. The smoking process lasted for 10 days, with pauses. Temperature and humidity were not controlled, but were influenced by outdoor conditions. Temperature was from 1.8 to 15.6°C (average 8.69°C) and relative humidity was from 54.4 to 95.5% (average 74.4%). The smoking process was followed by drying and ripening, which were conducted in order to reach a sausage moisture content lower than 35% (about 60 days). After drying, sausages were stored under controlled conditions (temperature and relative humidity) in industrial chamber, until day 270.

Samples for analyses were taken at day 0 day of production (control – 0), at the end of drying (T1) and at the end of storage (T2). All determinations were made in three samples.

2.2. PAH determination

The analysis of PAH was carried out on a GC 6890N gas chromatograph coupled to a MS 5975 mass spectrometer (Agilent, Palo Alto, CA, USA). Chromatographic conditions and PAH standard preparation were performed according to Škaljac *et al.* [2].

2.3. Statistical analysis

All data are presented as mean, standard deviation (SD) and range. Data were analysed statistically with one way ANOVA and post-hoc test (DUNCAN'S test). Differences were considered significant at $P < 0.05$. Statistical analysis was conducted using STATISTICA software version 13.2.

3. Results and Discussion

The PAH (μg/kg) contents of *Petrovská klobása* sausages smoked in traditional conditions are presented in table 1. The PAHs determined in raw sausage mixture before smoking (day 0 of production) were Acy, Fln and Phe, while other investigated PAHs were below the limit of detection. Roseiro *et al.* [4] reported that in raw sausage mixture, over 99% of the total PAHs were light PAHs (Naph, Ace, Ant, Fln and Phe). These findings were consistent with the results in our study. Roseiro *et al.* [4] reported higher value of Σ16 US-EPA PAHs (250.33 μg/kg dry matter) in Portuguese raw sausages, compared to values obtained in our study (Σ13 US-EPA PAHs - 9.57 μg/kg; i.e. 21.7 μg/kg dry matter), but Martorell *et al.* [10] reported lower value of Σ16 US-EPA PAHs in fresh meat sausages (1.25 μg/kg).

At the end of the drying period (day 60 of production), higher PAH content was determined in the *Petrovská klobása* compared to those levels determined in raw sausage mixture. The PAH determined in sausages at the end of the drying period were Acy (9.80 μg/kg), Fln (11.1 μg/kg), Phe (18.0 μg/kg), Ant (5.35 μg/kg) and Pyr (2.27 μg/kg), while other investigated PAH were below the limit of detection (table 1). The most abundant PAH in sausage samples was Phe, which was in accordance with the levels in other dry fermented sausages [11]. BaP, as well as PAH4 (BaA, BbF, BaP and CHR), with maximum contents in smoked meat products being set by European Commission Regulation (2 μg/kg and 12 μg/kg, respectively), were below the limit of detection in all examined samples of *Petrovská klobása* at the end of the drying period (table 1). From the point of view of PAH content, the results obtained in this study confirmed the safety of *Petrovská klobása* smoked in traditional conditions.

Table 1. Content of 13 US-EPA polycyclic aromatic hydrocarbons (μg/kg) in *Petrovská klobása* smoked in traditional conditions.

Polycyclic aromatic hydrocarbons		0	T1	T2
Acenaphthylene	Acy	$2.57^a \pm 0.35$	$9.80^b \pm 0.10$	$21.8^c \pm 1.85$
Fluorene	Fln	$2.10^a \pm 0.30$	$11.1^b \pm 0.10$	$13.0^c \pm 0.65$
Phenanthrene	Phe	$4.90^a \pm 0.50$	$18.0^b \pm 1.40$	$26.3^c \pm 2.60$
Anthracene	Ant	nd	$5.35^a \pm 0.55$	$12.5^b \pm 0.40$
Pyrene	Pyr	nd	2.27 ± 0.06	nd
Benz[*a*]anthracene	BaA	nd	nd	nd
Chrysene	BbF	nd	nd	nd
Benzo[*b*]fluoranthene	BkF	nd	nd	nd
Benzo[*k*]fluoranthene	BaP	nd	nd	nd
Benzo[*a*]pyrene	CHR	nd	nd	nd
Indeno[1,2,3-*cd*]pyrene	IcP	nd	nd	nd
Dibenz[*a,h*]anthracene	DhA	nd	nd	nd
Benzo[*ghi*]perylene	BgP	nd	nd	nd
Σ EU PAH4[1]		nd	nd	nd
Σ 13 US-EPA PAHs		$9.57^a \pm 1.15$	$46.52^b \pm 0.80$	$73.50^c \pm 4.20$

In the same row, different letters means that values are significantly different ($P < 0.05$).
nd - not detected
Results are expressed as means ± standard deviations
[1]BaA, CHR, BbF and BaP;

During storage, the PAH content increases because PAH from casings penetrate into the sausages and because of further drying of sausages [2, 6, 12]. On the other hand, during storage, some decrease in PAH content is also expected, caused by light decomposition and interaction with other components of the formulation [13, 14]. Thus, in this study, PAH contents were also determined in *Petrovská klobása* sausages at the end of the storage period (day 270 of production). The content of 13 US-EPA PAHs in *Petrovská klobása* sausages smoked in traditional conditions at the end of the storage period (day 270 of production) are presented in table 1. The PAHs determined in sausage samples were Acy (21.8 μg/kg), Fln (13.0 μg/kg), Phe (26.3 μg/kg) and Ant (12.5 μg/kg). Other investigated PAHs were below the limit of detection. At the end of the storage period, contents of detectable PAHs in sausages were significantly ($P<0.05$) higher than their contents at the end of the drying process. This phenomenon was mainly caused by further drying of sausages during the storage period [2]. Results of this study show that at the end of drying and at the end of storage, the total content of ∑13 US-EPA PAHs was lower compared with content of most other traditional dry fermented sausages [4, 7, 11, 15, 16], but higher compared with Androlla (36.45 μg/kg) and Botillo (29.39 μg/kg) dry fermented sausages from Spain [17]. Those authors found that the content of ∑16 US-EPA PAHs ranged from 36.45 to 2609.81 μg/kg in traditional dry fermented sausages from Portugal, Spain and Serbia [17]. Results presented in our study show that the process of smoking and drying in traditional conditions proceeded correctly and provided low content of PAH in *Petrovská klobása.*

4. Conclusion
BaP, as well as PAH4 were below the limit of detection in all examined samples of *Petrovská klobása* smoked in traditional conditions. According to the results obtained in this study and with regard to the European regulation on PAH content, the dry fermented sausage *Petrovská klobása*, smoked in traditional conditions was safe for consumers.

Acknowledgement
This research was financially supported by the Ministry of Education, Science and Technological Development, Republic of Serbia, project No TR31032

References
[1] Petrović Lj, Džinić N, Tomović V, Ikonić P and Tasić T 2007 Code of practice Registered geographical indication Petrovská klobása *Intellectual Property Office Republic of Serbia*, Decision No. **9652/06 G-03/06**
[2] Škaljac S, Petrović Lj, Tasić T, Ikonić P, Jokanović M, Tomović V, Džinić N, Šojić B, Tjapkin A and Škrbić B 2014 *Food Control* **40** 12-8
[3] Conde F J, Ayala J H, Afonso A M and González V 2005 *J. Agr. Food Chem.* **53** 176–82
[4] Roseiro L C, Gomes A and Santos C 2011 *Food Chem. Toxicol.* **49** 1340–5
[5] Škrbić B, Đurišić-Mladenović N, Mačvanin, N, Tjapkin A and Škaljac S 2014 *Maced. J. Chem. Chem. Eng.* **2** 227-36
[6] Ledesma E, Rendueles M and Díaz M 2016 *Food Control* **60** 64-87
[7] Gomes A, Santos C, Almeida J, Elias M and Roseiro L C 2013 *Food Chem. Toxicol.* **58** 369-74
[8] CAC/RCP 68/2009 2009 Codex Alimentarius Commission (CAC). Code of practice for the reduction of contamination of food with polycyclic aromatic hydrocarbons (PAH) from smoking and direct drying processes (Available at: http://www.codexalimentarius.org/ download/ standards/11257/CXP_068e.pdf)
[9] EC 2002 European Commission, Health and Consumer Protection Directorate-General, Scientific Committee on Food. (2002). Opinion of the Scientific Committee on Food on the Risk to Human Health of Polycyclic Aromatic Hydrocarbons – Occurrence in Food. SCF/CS/CNTM/PAH/29 CAC/RCP 68/2009
[10] Martorell I, Perelló G, Martí-Cid R, Castell V, Llobet J M and Domingo J L 2010 *Environ. Int.* **36** 424–32
[11] Alves P, Alfaia M, Škrbić B, Živannčev J, Fernandes M, Bessa R and Fraquezaa M 2017 *J. Food Compos. Anal.* **59** 124-31
[12] Šimko P 2005 *Mol. Nutr. Food Res.* **49** 637–47
[13] Bartkiene E, Bartkevics V, Mozuriene E, Krungleviciute V, Novoslavskij A, Santini A, Rozentale I, Juodeikiene G and Cizeikiene D 2017 *Food Control* **71** 285-92
[14] Šimko P 1991 *Food Chem.* **40** 293–300
[15] Santos C, Gomes A and Roseiro L C 2011 *Food Chem. Toxicol.* **49** 2343–7
[16] Lorenzo J M, Purrinos L, Bermudez R, Cobas N, Figueiredo M and García-Falcón M C 2011 *Meat Sci.* **89** 105-9

[17] Lorenzo J M, Purrinos L, García-Falcón M C and Franco D 2010 *Meat Sci.* **86** 660-4

IOP Conf. Series: Earth and Environmental Science **85** (2017) 012047 doi:10.1088/1755-1315/85/1/012047

Effect of salting on back fat hydrolysis and oxidation

E K Tunieva, V V Nasonova, I A Stanovova, K I Spiridonov and A A Kurzova

The V.M.Gorbatov All-Russian Meat Research Institute, Russian Academy of Agricultural Sciences, Moscow, Russian Federation

E-mail: lenatk@bk.ru

Abstract. Technological factors significantly affect the rate of hydrolytic and oxidative changes in fat. The aim of the research was to study the effect of sodium chloride on hydrolysis and oxidation of fat raw material, including the impact of thermal treatment. Back fat was minced, sodium chloride was added (in amounts of 0.0, 2.0, 3.5 or 5.0%), then it was thermally treated or not. Determination of the acid value (AV) was carried out by titration with aqueous potassium hydroxide of free fatty acids in the ether-alcohol solution of back fat; the peroxide value (PV) was based on oxidation of iodhydric acid with peroxides contained in fat followed by titration of released iodine with sodium thiosulphate. The thiobarbituric acid value (TBAV) was determined by the development of stained substances due to interaction of fat oxidation products with 2-thiobarbituric acid and measurement of color intensity using a spectrophotometer. Adding 5.0% sodium chloride to back fat led to a 30.1% decrease in AV. Addition of 2.0% sodium chloride inhibited the development of the oxidation products and led to a 17% decrease in the PV and to a 25% decrease in TBAV ($p<0.05$). In the presence of 5.0% sodium chloride, the secondary oxidation products significantly increased by 34.1% ($p<0.05$) and 24.3% ($p<0.05$) on days 1 and 3 of storage, respectively. Thermal treatment mitigated the effect of sodium chloride on the indicators of hydrolytic and oxidative spoilage ($p>0.05$). The results obtained showed an ambiguous effect of sodium chloride on the processes of fat oxidation, depending on dosage and the use of thermal treatment, justifying the necessity to develop approaches that allow reduction of the sodium chloride content in meat products that are not subjected to thermal treatment.

1. Introduction

Non-meat ingredients in formulations that are used for formation of the necessary technological properties of meat and organoleptic characteristics of finished products play an important role in the development of hydrolysis and oxidation products. Table salt is an essential component in all types of meat products and its effect on the oxidative changes in fats is quite considerable. Despite a large number of studies on safety and quality formation in meat products in the process of salting, available literature data regarding the effect of table salt on the mechanism of oxidative processes are rather controversial. According to some data, sodium chloride has an anti-oxidative effect [1] or has no effect on the fat chemical properties [2], while other studies demonstrate the pro-oxidative effects [3-5] of table salt. Inconsistency of data concerning the effect of salt on the chemical changes in lipids, apparently, can be explained by differences in selection of subjects of research, which, as a rule, are various types of meat with different meat tissue content and, consequently, with different water content, as well as finished products produced with various types of technological treatment.

Changes in fat during storage are commonly measured using three parameters. The acid value (AV) of fat characterizes the depth of the hydrolytic degradation of fats and when studying back fat during storage, it is an indicator of the oxidative spoilage along with the peroxide value (PV), which indicates the degree of the fat oxidative spoilage. The thiobarbituric acid value (TBAV) assesses accumulation of secondary fat oxidation products (aldehydes).

An example of a sodium chloride effect is modification of heme of muscle tissue proteins, which catalyzes lipid oxidation. With a complex substrate such as meat or fat and in the process of salting, the number of factors that can have a considerable effect on fat hydrolysis and oxidation is significant, and these processes can occur as a consequence of only indirect effects of sodium chloride on fats. Thus, to fully understand the mechanisms of how salt affects the development of products of oxidative spoilage, it is necessary to individually consider potential factors that influence the character of the table salt effect on fat oxidation, including salt dose, type of thermal treatment and absence/presence of meat tissue. In this connection, the aim of this research was to study the effect of different sodium chloride doses on the hydrolytic and oxidative changes in back fat before and after thermal treatment.

2. Materials and Methods

Samples of pork back fat (2nd class) from 2-year-old Large White pigs were comminuted in a grinder with a 3 mm grind plate, and sodium chloride (0.0, 2.0, 3.5 or 5.0 %) was added. Salted back fat was thermally treated or not, and stored at 4±2°C for 1 and 3 days.

The AV was determined by titration of free fatty acids in an ether-alcohol solution of fat with an aqueous solution of KOH; PV was measured by oxidation of iodhydric acid with peroxides contained in the fat followed by titration of released iodine with sodium thiosulphate. Determination of TBAV was by development of stained substances due to interaction of fat oxidation products with 2-thiobarbituric acid; color intensity was measured spectrophotometrically.

3. Results and Discussion

Sodium chloride concentration had an ambiguous effect on the rate of fat hydrolysis and oxidation product formation (figure 1). Back fat with 2.0% table salt showed a 20.6% decrease in AV by day 3 of storage; when the salt dose was 5.0%, the AV reduced by 30.1% compared to unsalted back fat ($p<0.05$), apparently, due to a water activity decrease caused by the amount of sodium chloride, which facilitated a retardation of hydrolytic changes in the fat. With that, thermal treatment mitigated the effect of table salt on the AV ($p<0.05$) compared to unsalted back fat (figure 2).

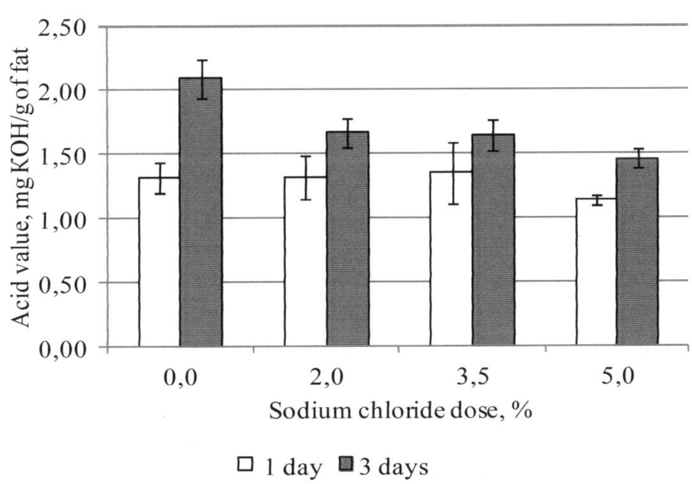

Figure 1. Effect of sodium chloride dose on the acid value of back fat during storage.

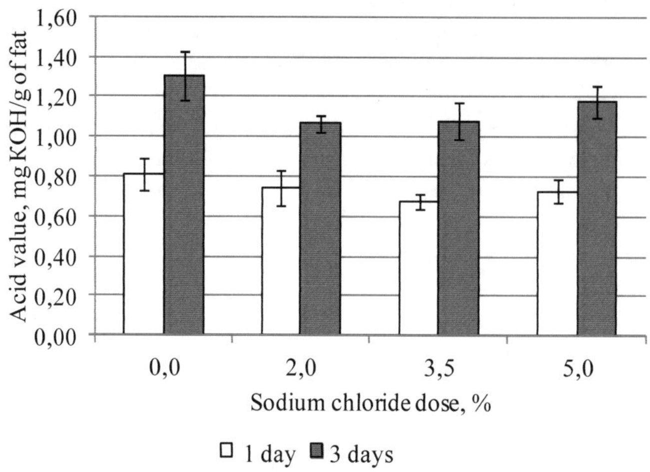

Figure 2. Effect of sodium chloride dose on the acid value of back fat after thermal treatment.

However, the same trends were not observed when PV (figures 3 and 4) or TBAV (figures 5 and 6) were used to assess primary or secondary oxidative spoilage of the salted back fat.

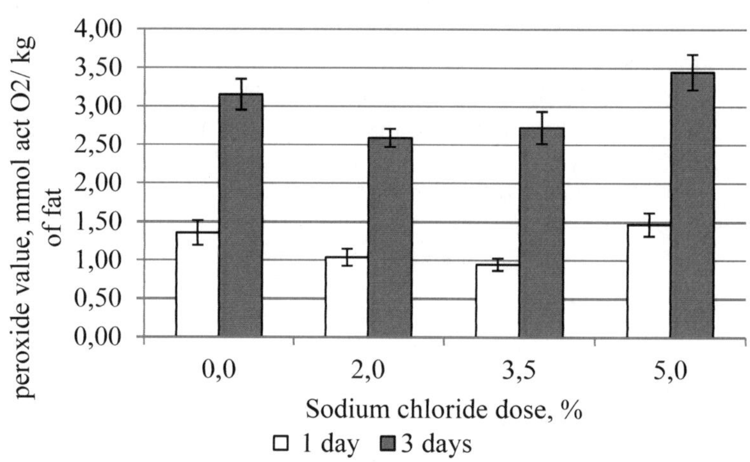

Figure 3. Effect of sodium chloride dose on the peroxide value of back fat during storage.

When 2.0 % table salt was added to back fat, the indicators of oxidative spoilage decreased on day 3 by 17% for PV and by 25 % for TBAV, compared with unsalted back fat.

An increase in the sodium chloride dose up to 3.5 % did not significantly affect the PV ($p>0.05$) compared with control back fat (figure 3). With that, as a result of thermal treatment, the PV (an indicator of oxidative spoilage) decreased by 11.1-31.1 % and 15.1-20.7 % depending on the sodium chloride dose on days 1 and 3 of storage, respectively. Moreover, as a result of thermal treatment, the effect of table salt on the hydrolytic and oxidative changes in the back fat decreased.

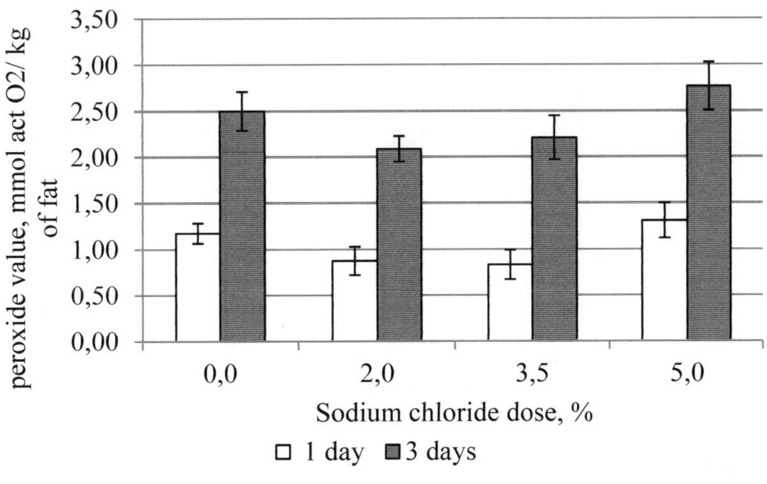

Figure 4. Effect of sodium chloride dose on the peroxide value of back fat after thermal treatment.

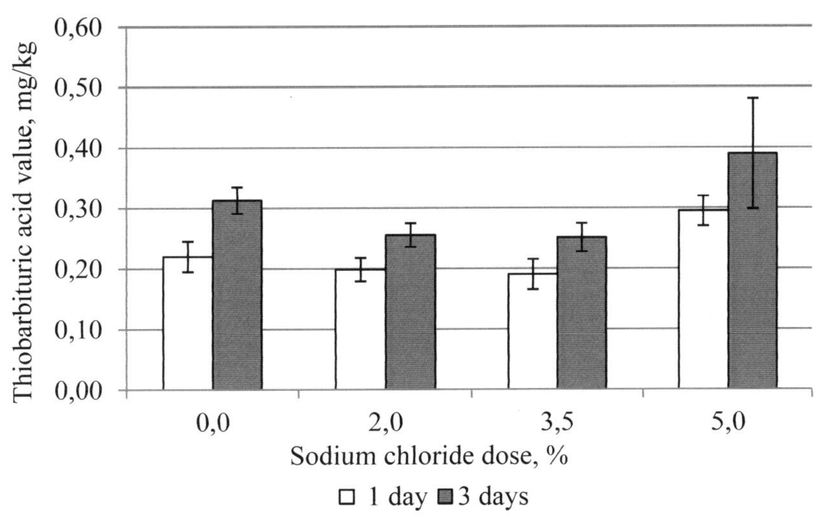

Figure 5. Effect of sodium chloride dose on the thiobarbituric acid value of back fat during storage.

Up to 3.5% salt did not significantly affect the formation of secondary oxidation products in back fat after 1 day of storage ($p>0.05$), as shown by TBAVs measured (figure 5). However, after 3 days of storage, a 25% decrease in the TBAV ($p<0.05$) was observed in the presence of 2.0 % of table salt. An increase in the sodium chloride dose up to 3.5 % did not significantly change the oxidation rate, since the TBAV was similar to that of fat with 2.0% added salt. However, when 5.0% salt was added, a substantial increase in the oxidation products, as measured by TBAV, occurred (34.1 %, $p<0.05$, and 24.3 %, $p<0.05$ on days 1 and 3 of storage, respectively). As a result of the thermal treatment, the TBAV decreased by 5.2-6.3 % ($p<0.05$) and 14.7-24.9 % ($p<0.05$) after 1 and 3 days of storage, respectively.

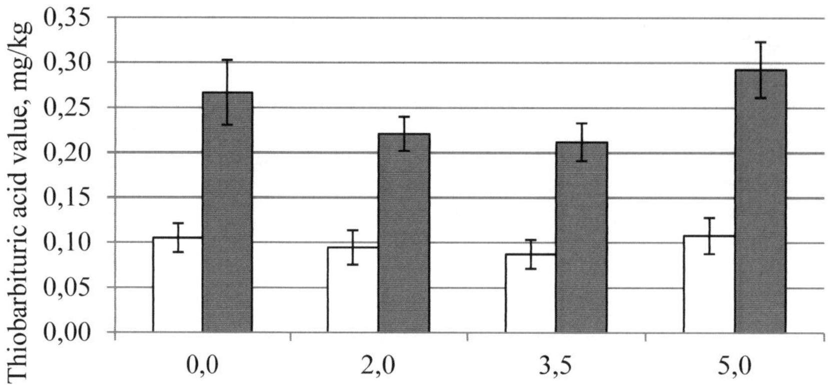

Figure 6. Effect of sodium chloride dose on the thiobarbituric acid value of back fat after thermal treatment.

The data obtained are consistent with the results of a previous study on the effect of salts on the qualitative composition of volatile substances that are formed in back fat in the process of salting. For example, addition of sodium chloride at levels of 2.0 and 3.5% inhibited the oxidative changes in polyunsaturated fatty acids compared to an unsalted sample. However, an increase in the table salt dose up to 5.0% facilitated an increase in the amount of saturated fatty acids formed due to oxidation of unsaturated acids. This tendency was characteristic both for volatile and etherified fatty acids. It was established that back fat salted with 2.0 and 3.5% table salt produced lower levels of secondary oxidation products. However, sodium chloride in higher doses acted as a pro-oxidant, initiating oxidation of fatty acids to form carbonyl compounds as well as the products from fatty acid interaction with ammonia, hydrogen sulfide and other substances that are formed as a result of amino acid breakdown [6].

4. Conclusion

This study has contributed to our understanding of the ambiguous effect of sodium chloride on the hydrolytic and oxidative changes of salted back fat. Based on the results obtained, it can be concluded that addition of up to 2.0% table salt to back fat facilitates retardation of fat's hydrolytic and oxidative changes. However, sodium chloride in high doses acts as a catalytic agent for fat's oxidative changes. With that, thermal treatment mitigated the negative effect of table salt on fat oxidation. Since decreasing the amount of table salt in meat products will lead to deterioration of their taste and technological characteristics, it is necessary to use approaches that lessen these changes. First of all, the anti-oxidative protection of meat raw materials should be increased by the additional use of antioxidants for meat products with a high salt content that are not subjected to thermal treatment (uncooked smoked and air-dried meat products). Secondly, technological recommendations on varying sodium chloride content due to the use of salt substitutes in meat products must be developed.

References

[1] Andres A I, Cava R, Ventanas J, Muriel E and Ruiz J 2004 Lipid oxidative changes throughout the ripening of dry-cured Iberian hams with different salt contents and processing conditions *Food Chem.* **84**(3) 375–81

[2] Calligaris S and NicoliMc 2006 Effect of selected ions from lyotropic series on lipid oxidation rate *Food Chem.* **94**(1) 130-4

[3] Gheisari H R, Moller J K S, Adamsen C E and Skibsted L H 2010 Sodium Chloride or Heme Protein Induced Lipid Oxidation in Raw, Minced Chicken Meat and Beef Czech *J. Food Sci.*

28 364–75
[4] Rhee K S and Ziprin Y A 2001 Pro-oxidative effects of NaCl in microbial growth-controlled and uncontrolled beef and chicken *Meat Sci.* **57** 105–12
[5] Jin G, He L, Zhang J, Yu X, Wang J and Huang F 2012 Effects of temperature and NaCl percentage on lipid oxidation in pork muscle and exploration of the controlling method using response surface methodology (RSM) *Food Chem.* **131**(3) 817–25
[6] Tunieva E K and Ivankin A N 2015 Influence of sodium chloride on the composition of the volatiles formed during oxidation of lard *Vse o Mjase* **2** 27-30 (in Russian)

Importance of medium chain fatty acids in animal nutrition

B Baltić[1], M Starčević[2], J Đorđević[2], B Mrdović[1] and R Marković[2]

[1] Institute of Meat Hygiene and Technology, Kaćanskog 13, 11000 Belgrade
[2] University of Belgrade, Faculty of Veterinary Medicine, Bulevar oslobodjenja 18, 11000 Belgrade, Republic of Serbia

Email: branislav.baltic@inmes.rs

Abstract. Fats in animal and human nutrition are a common subject of research. These studies most often pay attention to particular fat groups (saturated, unsaturated, polyunsaturated fats or fats grouped by the length of their fatty acid chains into short, medium or long chain fatty acids). Medium chain fatty acids (MCFAs) have two main sources: milk and coconut oil. To date, research has shown these acids have positive effects on health, production, feed digestibility and lower body and muscle fats in broilers and swine. MCFAs possess antibacterial, anticoccidial and antiviral effects. Also, it has been proven that these acids act synergistically if they are used together with organic acids, essential oils, or probiotics. Nowadays, commercial MCFA products are available for use in animal nutrition as feed additives.

1. Introduction

In recent years, replacing antibiotics in animal feedstuffs with biologically active substances has become a very current topic. The trend of banning the use of antibiotics, coccidiostats and other medical growth promoters in animal feedstuffs has developed because of fears these compounds contribute to the spread of bacterial resistance. Removing antibiotics from feedstuffs, after they were banned by the European Union (EU) in 2006, has put great pressure on livestock producers to find alternatives to antibiotics. Hence, modern feedstuff production is based on adding bioactive ingredients to feed, which should reduce the need for antibiotics and other medication, and positively influence animal health and welfare. In intensive breeding systems, these ingredients also reduce the negative influence of stress factors from the environment on animal immune systems and on animal production parameters. Consequently, in animal nutrition, more attention is now focused on competitive exclusion, probiotics, prebiotics, antibacterial peptides, yeasts and other additives. These additives also include medium chain fatty acids (MCFAs), which can be used in broiler and piglet nutrition. A variety of commercial MCFA products are already available on the market [1,2].

The structure of fat tissue of monogastric animals (pigs, poultry) is very similar to the fat structure of the feedstuffs on which the animals are fed. This means that source and type of fat in an animal feed can greatly influence the composition of fatty tissue and deposits in the resultant carcasses [3,4,5]. Nowadays, the EU policy of food modification is transferring from large-scale, cheap production to more expensive production, but hopefully with safer final food products. Clearly, the topic of modern animal feed additives is extremely relevant for the health and well-being of human consumers of the resultant meats.

2. Fat and fatty acids in human and animal diets

The term fats and oils, regardless of the animal or vegetable origin of these compounds, is a synonym for lipids that are used in human and animal nutrition. Besides their importance in body energy systems (they are both a source and a storage means of energy), they induce better resorption of liposoluble vitamins, slow down food passage through the intestine and, therefore, enable better exploitation of the food. They also increase the efficiency of energy consumption, as well as acceptability of the food. The energy value of the fats and oils in food and feed depends on numerous contributors, such as length of the carbon chain, specific organization of saturated and unsaturated fatty acids in the glycerol molecule, structure of free fatty acids, structure of the food/feed, amounts and types of triglycerides added to food/feed, and the intestinal flora, species, sex and age of the human/animal [6].

Fatty acids are variable by the chain length and degree of saturation. Chain length can vary from 2 to 40 carbon atoms, but the most common fatty acids contain 12 to 22 C atoms. Fatty acids are commonly categorized as short chain (SCFA; up to 4 C), medium chain (MCFA; from 6 to 12 C) and long chain (LCFA; more then 12 C). Additionally, they are categorized by the number of double bonds they contain; saturated fatty acids (SFA) have no double bonds in their fatty acid chain; monounsaturated fatty acids (MUFA) have one double bond in their fatty acid chain, and; polyunsaturated fatty acids (PUFA) have more than one double bond in their fatty acid chain. MCFA are a group of fatty acids with 6 to 12 C atoms obtained from edible fats such as coconut oil and milk fat by lipid fraction separation. Commercial MCFA products usually have 8 to 10 C atoms [7].

Differences in fat structure between MCFA and LCFA influences molecule size and water solubility, and also can result in differences during digestion, absorption and transportation of these fats in tissues and organs. Long term usage of MCFA in livestock feedstuffs, because of differences in the energy ratio between MCFA and LCFA and differing effects the two fat types have on intake, transport and efficiency of utilization, can influence the energy balance in animals [8,9]. Among the differences observed between MCFA and LCFA in animal feeds, these two types of fatty acids, according to their chain length and degree of saturation, must pass different metabolic pathways.

Fatty acid metabolic differences between the two fat types begin in the digestive tract, where MFCA is absorbed more efficiently then LCFA. Pigs can partly absorb MCFA via their stomachs. Medium chain triglycerides from MCFAs (MCTs) can be absorbed by intestinal enterocytes, after which they are hydrolyzed by the microsomal lipase. Compared with LCFA, MCFAs undergo simpler degradation to fatty acid and glycerol with help of pancreas lipase, after which they are absorbed via the portal circulation and transported to the liver where they undergo fast oxidation. LCFAs, as chylomicrons, are transported by the lymphatic system [10]. However, MCFAs are mostly saturated fatty acids, so they have great oxygenic stability, much higher so than LCFAs. This difference in the structure between LCFAs and MCFAs is reflected in the biological processes utilizing them. There are also differences in ketogenic and lipogenic capacity for these two groups of fatty acids. MCFA enter the mitochondria independently of the carnitine transport system and undergo preferential oxidation. Variations in ketogenic and lipogenic capacity also exist [8,11].

Alongside their energy value, bioactive fatty acids such as MFCAs, linoleic acid (C18:2) and linolenic acid (18:3) (these latter two fatty acids are categorized as essential fatty acids) have significant influence on human and animal health [12]. The most frequently mentioned characteristic of MCFAs is their ability to reduce obesity and fat tissue. This effect first was proven in rats, pigs and broilers [13,14,15]. The antidiabetic influence of MCFAs on humans with diabetes type 2 has only rarely been discussed [16]. MCFAs as sources of energy are poorly stored in subcutaneous fat tissue of humans, rats, broilers and pigs. In broilers and especially in pigs, this has special meaning because carcasses with less fat have higher commercial value, and this kind of meat is more in demand by consumers. In broiler meat, the presence of C6:0 has not been proven, but the content of C10:0 was higher than of C8:0. The content of C10:0 and C8:0 in chicken leg muscle and breast meat increased with increasing percentage of this fatty acid in feed for broilers [16].

In modern pig production, piglets are weaned and start to eat feed when they are between 15 and 28 days old. Feed consumption data indicated that digestive tracts of such young piglets are not yet developed, so they cannot use all the nutrition available from feed [17]. This low intake causes poor energy intake which negatively influences growth and development of the piglets. Digestibility of LCFAs is decreased by 65% to 80% in these young piglets compared with older animals, as result of low pancreas and intestinal lipase activity, which causes insufficient fat absorption from the digestive system. Also, in this period of life, piglets cannot synthesize enough carnitine, which has an important negative influence on transportation of LCFAs into mitochondria for energy production [18]. Moreover, unsaturated bonds in LCFAs can be damaged by free radicals, which can cause cascade damage to endogenous lipids and peroxide occurrence [19]. However, MFCA is a good supplement for weaned piglets. Individually, MCFAs have an unacceptable odor, and so their triacylglycerols (medium chain triacylglycerols; MCTs) have usually been added to feed for animals [20]. Effects MCTs, or MCFA, or mixtures of two or all four important MCFAs (caproic acid (C6), caprylic acid (C8), capric acid (C10) and lauric acid (C12)) as supplements for growing piglets depend on the amount added. The life stage when piglets are fed MCFA also influences piglet growth [21,22].

Broilers, in their first days of life, have limited fat absorption and limited digestion capability, because of their low amounts of synthesized bile salt and lipase. In young individuals, fat digestibility is 6% less then in adult birds [23], but the young animals have far better ability to digest PUFA [22]. MCFAs have specific nutritive value and metabolic and antibacterial effects in poultry [24,25,26]. In egg production, usage of MCFA gave better results, eggs had stronger shells, and greater egg protein quality (larger Haugh units), higher content of calcium, and reductions in the number of *Escherichia coli* bacteria [27] resulted.

Pigs which have been fed with feed supplemented with MCFAs had better growth performance [8,28]. Digestibility of MCTs is much higher (98.5 instead 93.4%) then long chain triacylglycerols (LCTs) in weaned piglets. With two to five weeks of receiving MCTs, the digestive system of weaned piglets is fully functional for normal digestion and exploitation of nutritive components and energy from feed [8]. These authors showed that MCTs have a positive effect on production results, feed digestibility, blood plasma metabolites and antioxidant capacity of weaned piglets [8]. Digestibility of proteins and higher concentrations of protein in blood plasma especially stand out [8].

3. MCFA and microbiota of animal digestive tract

The role of SCFAs and MCFAs in control of infection and maintenance of health and integrity of digestive tract was examined in broilers [28] and other animals [29]. Fatty acids are generally inhibitory to microorganisms, but different fatty acids have different minimum inhibitory concentrations (MIC), depending on the type of fatty acid, type of microorganism and environmental pH. Low pH increases the concentration of dissociated SCFA, which in that conformation, can pass into the bacterial cells where the intercellular pH is higher. This higher pH dissociates the SCFA, so intracellular pH decreases and subsequently changes the bacterial cell's metabolism [29].

MCFAs produce a strong antibacterial effect due to the anionic part of the molecule, but how much effect is due to change of the bacterial pH and how much is due to influence on the metabolic level of the bacteria is not yet known. The anionic part of fatty acids changes the physico-chemical characteristics of the digestive tract environment in which the microorganisms exist, and influences the expression of microorganism and host genes. The molecular basis of the mechanism by which MCFA decrease numbers of *Salmonella* spp., *Clostridium* spp. and *E. coli* in the digestive tract is still unclear, but it has been proven that MCFA decreases the number of intracellular lymphocytes in epithelium cells of the digestive tract [29]. Additionally, the anticoccidial properties of MCFAs have been proven [30,31] by studying high quality coconut oil (enhanced virgin coconut oil – EVCO) that contains MCFAs and their proper monoglycerides [31]. These materials also inhibited growth of both Gram-positive and Gram-negative bacteria and the yeast, *Candida albicans* [31].

The structure of intestinal microbiota has important influences on the animal's ability to digest feed properly, causing positive, negative and neutral effects [32]. Modification of gastrointestinal microbiota reduced colonization of the animals with pathogenic microorganisms, which can have a big influence on intestinal wall structure [33]. For example, Uni et al. [34] established that stress (delayed access to feed after hatching, exposure to bacteria, etc.) is related to changes in intestinal morphology, such as shorter intestinal villus cells and depth crypts. The immune system of newly-hatched chickens, especially the mucosal immune system, requires oral intake of feed for full and fast development.

Hermans at al. [35] reported that usage of MCFA emulsion in drinking water for broilers decreases the number of *Campylobacter* in their digestive tracts, and posited that this may especially be due to the lowered possibility that water acted as a source of these pathogens. MCFA are new generation additives, and can be used for *Salmonella* control. Broilers fed with MCFAs and organic acids performed better then control broilers, because they had better feed digestibly and lower numbers of *Salmonella enterica* Enteritidis in the cecum [36]. Immersell et al. [37] determined that usage of MCFA significantly reduced the level of *S. enterica* Enteritidis colonization when broilers were infected the third day after adding MCFA in feed. Zeiger et al. [38] used lauric acid in broiler feed, and found statistically significantly reduced numbers of *Campylobacter* on the carcasses.

4. MCFA and synergistic effects with other feed additives

Adding SCFA and MCFA in feed has numerous positive effects on animal health that can be seen when health of the animals is endangered, especially by compromised integrity of digestive tract. Mathis et al. [39] proved that the combination of organic acids and MCFA significantly reduced clinical symptoms of diseases in artificial necrotic enteritis of broilers. Almo et al. [39] showed in broilers infected with viral malabsorption syndrome (MAS) that adding SCFA and MCFA together in feed increased broiler growth and resulted bigger broiler weight at the end of the production cycle. How this directly impacts viruses is not known, but it is considered that SCFA and MCFA together have a synergistic effect on bacteria, whereby MCFA damages microorganisms' cell walls, thus allowing SCFA access into the bacterial cytoplasm to produce an antibacterial effect [40]. Fat, protein and crude fiber digestibility can be improved if MCFA is used with selected organic acids (propionic, fumaric). This combination increases intestinal villi and crypt depths but does not affect villus width, so the ratio between villus height and crypt depth is more favorable. Use of MCFA (C6 and C8) and plant extracts (thymol, cinnamon oil, eucalyptus oil) in turkey nutrition increased bird weight after 15 weeks of fattening and did not influence total feed intake, but caused better feed conversion [41]. Adding MCFA to animal feed can be used in parallel with probiotics (e.g. *Enterococcus faecium*), giving weaning piglets better productivity, increased feed digestibility and changed biochemical blood profile (cholesterol, triglyceride, glucose, creatinine) [42]. The authors stressed that MCFA and probiotic together had a synergistic effect on piglet nutrition, and this effect should be an effective replacement for antibiotics [42]. Zeits et al. [43] used a different ratio of C12 and C14 in broiler nutrition. They did not measure any differences in microbiota, intestine morphology, or fat or cholesterol content in meat or liver, but adding MCFA had a positive effect on feed conversion and pectoral weight of broilers [43]. Using SCFA and MCFA in parallel in broiler nutrition had a positive effect on health and production results (live weight, weight of pectoral muscle, growth, feed intake and digestibility) [43]. The parallel use of MCFA and coccidiostats positively influenced production results in broilers (weight at the end of the production cycle, body weight, feed intake, growth, conversion) [44].

Acknowledgment
This paper was supported by Ministry of Education, Science and Technological development, Republic of Serbia, through the funding of Project No 31034.

References

[1] Baltić Ž M, Marković R and Đorđević V 2011 Nutrition and meat quality *Tehnologija mesa* **52** pp 154-159
[2] Šefer D, Marković R, Nedeljković-Trailović J, Petrujkić B, Radulović S and Grdović S 2015

Primena biotehnologije u ishrani životinja, The application of biotechnology in animal nutrition *Veterinarski Glasnik* **69** pp 127-137

[3] Baião N C and Lara L J C 2005 Oil and fat in broiler nutrition *Rev. Bras. Cienc. Avic.* **7** pp 129-141

[4] Đorđević V, Đorđević J, Baltić Ž, Laudanović M, Teodorović V, Bošković M, Peurača Mile and Marković R 2016 Effect of sunflower, linseed and soybean meal in pig diet on chemical composition, fatty acid profile of meat and backfat, and its oxidative stability *Acta Veterinaria* **66** pp 359-372

[5] Đorđević J, Pantić S, Bošković M, Marković R, Dokmanović-Starčević M, Baltić T, Laudanović M and Baltić Ž M 2015 Uticaj konjugovane linolne kiseline na masnokiselinski sastav i senzorne osobine suvog svinjskog vrata *Tehnologija mesa* **56** pp109-120

[6] Ratnayake W N and Galli C 2009 Fat and fatty acid terminology, methods of analysis and fat digestion and metabolism: a background review paper *Ann. Nutr. Metab.* **55** pp 8-43

[7] Babayan V K 1987 Medium chain triglycerides and structured lipids *Lipids* **22** pp 417-420

[8] Li Y, Zhang H, Yang L, Zhang L and Wang T 2015 Effect of medium-chain triglycerides on growth performance, nutrient digestibility, plasma metabolites and antioxidant capacity in weanling pigs *Animal Nutrition* **1** pp 12-18

[9] Hanczakowska E 2017 The use of medium chain fatty acids in piglet feeding–a review *Ann. Anim. Sci.* DOI: 10.1515/aoas-2016-0099

[10] Ooyama K, Kojima K, Aoyama T and Takeuchi H 2009 Decrease of food intake in rats after ingestion of medium-chain triacylglycerol *J. Nutr. Sci. Vitaminol.* **55** pp 423-427

[11] Papamandjaris A A, MacDougall D E, and Hones P J H 1998 Medium chain fatty acid metabolism and energy expenditure: obesity treatment implications *Life sci.* **62** pp 1203-1215

[12] Aluko R E 2012 *Functional foods and nutraceuticals* p. 22 New York: Springer.

[13] Takeuchi H, Noguchi O, Sekine S, Kobayashi A and Aoyama T 2006 Lower weight gain and higher expression and blood levels of adiponectin in rats fed medium-chain TAG compared with long-chain TAG. *Lipids* **41** pp 207-212

[14] Newport M J, Storry J E and Tuckley B 1979 Artificial rearing of pigs: 7*. Medium chain triglycerides as a dietary source of energy and their effect on live-weight gain, feed: gain ratio, carcass composition and blood lipids *Br. J. Nutr.* **41** pp 85-93

[15] Mabayo R T, Furuse M, Kita K and Okumura J 1993 Improvement of dietary protein utilisation in chicks by medium chain triglyceride *Br. Poult. Sci.* **34** pp121-130

[16] Chu H C and Chiang S H 2017 Deposition of Dietary Bioactive Fatty Acids in Tissues of Broiler Chickens *Int. J. Poult. Sci.* **54** pp 173-178

[17] Wu G 1998 Intestinal mucosal amino acid catabolism *J. Nutr.* **128** pp 1249-1252

[18] Cera K R, Mahan D C and Reinhart G A 1988 Weekly digestibilities of diets supplemented with corn oil, lard or tallow by weanling swine *J Anim Sci* **66** pp 1430-1437

[19] Fritz I B and Yue K T 1963 Long-chain carnitine acyltransferase and the role of acylcarnitine derivatives in the catalytic increase of fatty acid oxidation induced by carnitine *J. Lipid Res.* **4** pp 279-288

[20] Oprean L, Iancu R, Stan R and Traşcă C 2011 Comparison between types of feeding on goat milk composition *JASB* **44** pp 76-79

[21] Devi S M and Kim I H 2014 Effect of medium chain fatty acids (MCFA) and probiotic (*Enterococcus faecium*) supplementation on the growth performance, digestibility and blood profiles in weanling pigs *Veterinarni Medicina* **59**

[22] Hanczakowska E, Świątkiewicz M, Natonek-Wiśniewska M and Okoń K 2016 Medium chain fatty acids (MCFA) and/or probiotic Enterococcus faecium as a feed supplement for piglets *Livest Sci* **192** pp 1-7

[23] Freitas E R, Sakomura N K, Neme R and Santos A L D 2005 Valor energético do óleo ácido de soja para aves *Pesq. agropec. bras.* pp 241-246

[24] Wiseman J 1984 Assessment of the digestible and metabolizable energy of fats for non-ruminants. In: Wiseman J. Fats in animal nutrition. London: Butterwoths 1984. pp 277-297

[25] Skřivanová E, Molatová Z, Skřivanová V and Marounek M 2009 Inhibitory activity of rabbit milk and medium-chain fatty acids against enteropathogenic *Escherichia coli* O128 *Vet. Microbiol.* **135** pp 358-362

[26] Batovska D I, Todorova T, Tsvetkova V and Najdenski H M 2009 Antibacterial study of the medium chain fatty acids and their 1-monoglycerides: individual effects and synergistic relationships *Pol. J. Microbiol.* **58** pp 43-47

[27] Lee S I, Kim H S and Kim I 2015 Microencapsulated organic acid blend with MCFAs can be used as an alternative to antibiotics for laying hens *Turk. J. Vet. Anim. Sci.* **39** pp 520-527

[28] Hong S M, Hwang J H and Kim I H 2012 Effect of medium-chain triglyceride (MCT) on growth performance, nutrient digestibility, blood characteristics in weanling pigs *AJAS* **25** pp 1003-1008

[29] Deschepper K, Gantois I, Maertens L and Van Meenen E 2013 A balanced mixture of medium chain fatty acids improves zootechnical performances and slaughter results of broilers *In Proceedings of the 19th European Symposium of Nutrition* PPV 10

[30] Price K L, Lin X, Van Heugten E, Odle R, Willis G and Odle J 2013 Diet physical form, fatty acid chain length, and emulsification alter fat utilization and growth of newly weaned pigs *J. Anim. Sci.* **91** pp 783-792

[31] Lückstädt C and Mellor S 2011 The use of organic acids in animal nutrition, with special focus on dietary potassium diformate under European and Austral-Asian conditions *Recent Adv Anim Nutr Aust* **18** pp 123-130

[32] Lensing M, Van der Klis J D, Fabri T, Cazemier A and Else A J 2010 Efficacy of a lactylate on production performance and intestinal health of broilers during a subclinical *Clostridium perfringens* infection *Poult Sci* **89**(11) pp 2401-2409

[33] Sato H, Nitana A, Kurosawa T and Oikawa S 2004 Anticoccidial efficacy of medium-chain triglycerides (MCT) in calves *J Vet Med Sci* **66**(12) pp 1583-1585

[34] Garrido M N, Skjervheim M, Oppegaard H and Sørum H 2004 Acidified litter benefits the intestinal flora balance of broiler chickens *Appl Environ Microbiol* **70**(9) pp 5208-5213

[35] Uni Z, Smirnov A and Sklan D 2003 Pre-and posthatch development of goblet cells in the broiler small intestine: effect of delayed access to feed *Poult Sci* **82**(2) pp 320-327

[36] Hermans D, Martel A, Garmyn A, Verlinden M, Heyndrickx M, Gantois I, Haesebrouck F and Pasmans F 2012 Application of medium-chain fatty acids in drinking water increases Campylobacter jejuni colonization threshold in broiler chicks *Poult Sci* **91**(7) pp 1733-1738

[37] Chotikatum S, Kramomthong I and Angkanaporn K 2009 Effects of medium chain fatty acids, organic acids and fructo-oligosaccharide on cecal *Salmonella* Enteritidis colonization and intestinal parameters of broilers *THAI J VET MED* **39**(3) pp 245-258

[38] Van Immerseel F, De Buck J, Boyen F, Bohez L, Pasmans F, Volf J, Sevick M, Rychlik I, Haesebrouck F and Ducatelle R 2004 Medium-chain fatty acids decrease colonization and invasion through *hilA* suppression shortly after infection of chickens with *Salmonella enterica* serovar Enteritidis *Appl Environ Microbiol* **70**(6) pp 3582-3587

[39] Zeiger K, Popp J, Becker A, Hankel J, Visscher C, Klein G and Meemken D 2017 Lauric acid as feed additive–An approach to reducing *Campylobacter* spp. in broiler meat *PloS one* **12**(4) e0175693

[40] Mathis G F, Van Dam J T P, Corujo F and Hofacre C L 2005 Effect of an organic acids and medium-chain fatty acids containing product in feed on the course of artificial Necrotic Enteritis infection in broiler chickens In *Proceedings of the 15th European Symposium on poultry nutrition, Balatonfüred, Hungary, 25-29 September, 2005* pp 372-374 World's Poultry Science Association (WPSA)

[41] Del Alamo A G, De Los Mozos J, Van Dam J T P and De Ayala P P 2007 The use of short and medium chain fatty acids as an alternative to antibiotic growth promoters in broilers infected

with malabsorption syndrome. In *Proceedings of the 16th European Symposium on Poultry Nutrition Strasbourg France* pp 317-320

[42] Lipiński K, Mazur M, Makowski Z, Makowska A, Antoszkiewicz Z and Kaliniewicz J 2016 The effectiveness of the preparation medium-chain fatty acids (MCFA) and a herbal product on the growth performance of turkeys *Pol J Natur Sci* **31**(1) pp 47-57

[43] Zeitz J O, Fennhoff J, Kluge H, Stangl G I and Eder K 2015 Effects of dietary fats rich in lauric and myristic acid on performance, intestinal morphology, gut microbes, and meat quality in broilers *Poult Sci* **94**(10) pp 2404-2413

[44] Baltić B, Marković R, Šefer D, Glišić M, Hermans D and De Laet M 2016 The effect of adding a mixture of medium chain fatty acid-"aromabiotic" in the diet on broiler performance *Krmiva* **57**(2) pp 57-62

IOP Conf. Series: Earth and Environmental Science 85 (2017) 012049 doi:10.1088/1755-1315/85/1/012049

The interactive effects of transportation and lairage time on welfare indicators, carcass and meat quality traits in slaughter pigs

N Čobanović[1], D Vasilev[1], M Dimitrijević[1], V Teodorović[1], N Parunović[2], N Betić[2] and N Karabasil[1]
[1]Department of Food Hygiene and Technology, Faculty of Veterinary Medicine, University of Belgrade, Bulevar oslobodjenja 18, 11000 Belgrade, Serbia
[2]Institute of Meat Hygiene and Technology, Kaćanskog 13, 11000 Belgrade, Serbia

E-mail: cobanovic.nikola@gmail.com

Abstract. This study assessed the effects of transportation and lairage time and their interaction on welfare, carcass and meat quality traits in slaughter pigs under commercial conditions. The study was conducted on 120 pigs with a live weight of approximately 115 kg and about six months old. A complete blood picture was measured in pigs to assess pre-slaughter stress. Also, nine different carcass quality parameters including live weight, hot and cold carcass weights, cooling loss, dressing percentage, backfat thickness, meatiness and skin lesions score were measured. The pH and temperature measurements were performed 45 minutes post-mortem. The results showed that short transportation time and slaughtering without lairaging and long transportation time and overnight lairaging negatively influenced the hematological parameters, which meant that the animal welfare was seriously compromised under these pre-slaughter conditions. Long transportation time and overnight lairaging reduced live and carcass weights and increased the incidence of skin lesions on the carcass and DFD pork. In addition, short transportation time and slaughtering without lairaging caused a significant deterioration in pork quality. It can be concluded that, from the standpoint of animal welfare, carcass and meat quality, the above-mentioned pre-slaughter conditions are not recommended to the farmers and/or pork producers.

1. Introduction

Pre-slaughter treatment is the major research topic in animal science since it is closely connected to animal welfare, as well as carcass and meat quality. Assessment of pig welfare status on the day of slaughter is of paramount importance to ensure high pork quality while maintaining the well-being of the animals [1]. On the day of slaughter, several practices have considerable effects on pig welfare and subsequently on carcass and pork quality, such as feeding, slaughter weight, gender, handling during loading or unloading, mixing pigs from different pens, stocking density and time of transportation and lairaging, feed and water withdrawal and the season of slaughter [1-6]. Research findings showed that transportation and lairage time are the two key factors of all pre-slaughter practices that can adversely affect the animal welfare, carcass and pork quality [7-11]. It has been reported that longer transportation (>3 h) results in a higher risk for producing dark, firm and dry (DFD) pork, while pigs subjected to short transportation (<15 min) are more prone to produce pale, soft and exudative (PSE) pork [8]. Furthermore, lairaging shorter than one hour is known to result in a higher occurrence of PSE meat, while when pigs were subjected to overnight lairaging, carcass weight, backfat thickness and meat temperature were reduced [6]. Also, overnight lairaging poses a higher risk for developing DFD pork and skin lesions on the carcass [11]. Therefore, the aim of this study was to determine the effects of transportation and lairage time and their interaction on welfare indicators, carcass and meat quality traits under commercial conditions.

2. Materials and Methods

2.1. Animals, pre-slaughter handling and slaughter procedure

A total of 120 slaughter pigs with average live weight (LW) of approximately 115 kg and about six months old was evaluated. All the animals were of the same breed (Yorkshire x Landrace crossbreeds) and originated from the same farm. The pigs were raised in a finishing piggery on a completely-slotted floor, in groups of 20 animals per pen with an average stocking density of 1 m^2 per pig. During the fattening period, food and water were available *ad libitum* at nipple drinkers and food dispensers. Before transportation, feed and water were not withdrawn. Once loading was finished, the pigs were transported to the same slaughterhouse for less than one hour or more than two hours. Since the information was collected under commercial transport conditions scheduled to pick up animals at several production sites every day, transportation time differed between groups of pigs. Depending on the dynamics of slaughter, the pigs were slaughtered immediately after unloading, or they were held in a lairage overnight and slaughtered the following morning. Lairage density was 0.65 m^2 per pig. During the period of lairaging, water was provided, but no food. At the end of the lairaging period, the pigs were head-only electrically stunned in batches of six animals without restraining. Following bleeding, the carcasses were processed using conventional practice.

2.2. Welfare indicators

Immediately after the onset of bleeding, blood samples were collected from each pig. They were kept refrigerated (4°C) until being processed immediately on arrival at the laboratory. The vacutainers (2 mL) coated with EDTA were used to measure hematological parameters including white blood cells (WBC), lymphocytes (LYM), middle-sized cells (monocytes, eosinophils, basophils) (MID), neutrophils (NEUT), red blood cells (RBC), red cell distribution width (RDW), hemoglobin (HGB), hematocrit (HCT), mean corpuscular volume (MCV), mean corpuscular hemoglobin (MCH), mean corpuscular hemoglobin concentration (MCHC) and platelet count (PLT). The proportions of lymphocytes (LYM%), middle-sized cells (MID%) and neutrophils (NEUT%) were calculated as a percentage of leukocyte concentration on the same device. The indicators of the hematological profile were analyzed by an automatic hematological analyzer Abacus junior vet (Diatron MI PLC, Hungary). Blood glucose (GLUC) levels were measured using a handheld devices analyzer (Gluco Sure Auto Code, ApexBio, Taiwan).

2.3. Carcass and meat quality analyses

The carcasses were weighed immediately after splitting and final washing to obtain the hot carcass weight (HCW), and re-weighed 24 hours after chilling at 4°C to determine the weight of the cooled carcass (CCW). Cooling loss (CL) was calculated based on the difference between the hot and cold carcass weights, expressed as a percentage of the hot carcass weight. The dressing percentage (DP) was calculated as: (hot carcass weight ÷ live weight) x 100. Carcass backfat thickness was measured with a metal ruler at two points (between the 13th and 15th dorsal vertebrae and over M. *gluteus medius*). Meatiness (in percentages) (M) was calculated according to the Official Gazette [12] based on hot carcass weight and the sum of carcass fat thickness on the back (FTB) and at the sacrum (FTS). Skin lesions (SLC) on the left side of pigs' carcasses were visually assessed 45 minutes post-mortem according to the Welfare Quality® protocol [13]. The carcasses were divided into the following regions: i) ears; ii) front part of the carcass (from the head to the end of the shoulder); iii) middle part of the carcass (from the end of the shoulder to the rear part of the carcass); iv) rear part of the carcass; and v) limbs (from the accessory digit upwards). Each region of carcass was scored based on a three-point scale: 0) no visible skin lesions, or only one skin lesion bigger than 2 cm or skin blemishes smaller than 1 cm; 1) between two and 10 skin blemishes bigger than 2 cm; and 2) any wound penetrated into muscles or more than 10 skin blemishes larger than 2 cm. The scoring of the five regions of the carcass was combined into one scoring as follows: 0) all carcass regions with a score of 0; 1) at least one carcass region with a score 1; and 2) at least one carcass region with a score 2. The pH and temperature of the M. *longissimus dorsi* were measured 45 minutes after slaughter using a pH-meter Testo 205 (Testo AG, Lenzkirch, Germany). Pork quality classes (PSE, normal meat, DFD meat) were determined according to Adzitey and Nurul [14] using pH_{45} value. The carcasses showing pH_{45} values lower than 6.0 were classified as PSE meat, while the carcasses showing pH_{45} values higher than 6.4 were classified as DFD meat. The carcasses with pH_{45} between 6.0 and 6.4 were classified as normal pork quality.

2.4. Statistical analysis

Statistical analysis of the results was conducted with SPSS software version 23.00 for Windows. The pigs were divided into two groups with regard to transportation time: short transportation (<1 h) and (n=61) and long transportation (>2 h) (n=59). According to lairage time, the pigs were allocated to two groups: the group of pigs slaughtered immediately after unloading (n=60) and the group of pigs slaughtered after overnight lairaging (n=60). Two-way ANOVA with Tukey's multiple comparison test was performed to test the effect of transportation and lairage time, and their interaction on the welfare indicators, carcass and meat quality traits. Data were described by descriptive statistical parameters as the mean value and pooled standard error of means (SEM). The distribution of pork quality classes in relation to the transportation and lairage time were determined by Chi-square test. A value of $P<0.05$ was considered significant.

3. Results and Discussions

The effects of transportation and lairage time and their interaction on welfare indicators in slaughter pigs can be seen in Table 1. Two-way interaction ($P<0.05$) between transportation and lairage time affected red blood cell count and hemoglobin concentration, so that, these parameters increasing markedly when transportation and lairage time decreased. Thus, pigs subjected to short transportation and slaughtered without rest had the highest red blood cell count and hemoglobin concentration ($P<0.05$). Alterations in red blood cell count, hemoglobin concentrations and hematocrit indicate a situation of adaptation or resistance, in which pigs, after being exposed to the stressful factors, need plenty of oxygen-carrying capacity of blood [2]. Therefore, when pigs encounter stress, the spleen, as a reservoir of red blood cells, contracts and releases erythrocytes into the circulatory system, which provides the muscle mass with a large number of oxygenated erythrocytes, allowing the animal increased physical activity [15-16].

Two-way interaction ($P<0.05$) between transportation and lairage time was found in white cell parameters, indicating that the number of leukocytes and lymphocytes increased as transportation and lairage time decreased, reaching the maximum values in pigs subjected to short transportation and slaughtered immediately after unloading (Table 1). Lymphocytosis and neutrophilia in stressed animals are associated with changes in leukocyte trafficking and release from the bone marrow after the endogenous secretion of epinephrine or corticosteroids from the adrenal glands [2]. In acutely stressed animals, epinephrine-induced changes are seen within minutes after the release, whereby these effects are characterized by a transient elevation in white blood cells with lymphocytosis [8-9]. In addition, as a result of significant two-way interaction ($P<0.05$) between transportation and lairage time, the highest blood glucose levels were found in pigs subjected to short transportation and slaughtered without rest (Table 1), which may be explained by catecholamine-mediated glycogenolysis as a response to acute stressors such as loading, transportation, unloading and short lairaging [17-18]. In contrast, when animals are exposed to chronic stress, high white blood cells count with neutrophilia and eosinopenia induced by corticosteroids is evident a few hours after the secretion [8-9]. Therefore, a typical response to corticosteroids with increased number of leukocytes and neutrophils, as a result of significant two-way interaction (P<0.05) between transportation and lairage time, can be seen in pigs subjected to long transportation time and slaughtered after overnight lairaging (Table 1). In the present research, the results of the analyzed welfare parameters suggest that pigs experienced intense acute stress during shorter transport periods and slaughter immediately after unloading. On the other hand, pigs subjected to longer transport periods in combination with prolonged lairage time were under chronic stress. Therefore, it may be argued that in both cases, animal welfare was seriously compromised.

Table 1. Mean values (±pooled SEM) of welfare indicators in slaughter pigs according to transportation time and lairage time (n=120)

	Short transportation time		Long transportation time		Pooled SEM	TT	LT	TT x LT
Lairage time	Immediate slaughter	Overnight slaughter	Immediate slaughter	Overnight slaughter		Significance		
Number of pigs	41	20	19	40				
WBC (10^9/L)	23.44^a	18.57^b	15.87^b	23.02^a	1.96	NS	NS	*
LYM (10^9/L)	16.83^a	12.72^b	10.74^b	13.35^b	1.51	*	NS	*
MID (10^9/L)	0.22	0.16	0.19	0.22	0.32	NS	NS	NS
NEUT (10^9/L)	6.37^a	5.70^a	4.95^a	9.45^b	1.48	NS	*	*
LYM (%)	69.78^a	70.14^a	68.62^{ab}	59.40^b	5.42	*	NS	NS
MID (%)	0.99	0.83	1.27	0.98	0.57	NS	NS	NS
NEUT (%)	27.55^a	29.03^a	30.11^a	39.61^b	4.81	*	*	*
RBC (10^{12}/L)	8.48^a	7.04^b	7.30^b	6.92^b	0.37	*	*	*
RDW (%)	20.38	20.83	20.24	20.79	0.50	NS	NS	NS
HGB (g/L)	149.07^a	136.75^b	137.11^b	129.53^b	6.16	*	*	*
HCT (%)	41.63^a	40.07	39.55	38.39^b	1.41	*	NS	NS
MCV (fl)	49.66	50.15	49.89	50.48	1.34	NS	NS	NS
MCH (pg)	18.05	17.86	18.13	17.93	0.41	NS	NS	NS
MCHC (g/L)	359.90	358.35	360.05	360.80	3.65	NS	NS	NS
PLT (10^9/L)	228.93	264.55	239.95	256.70	46.88	NS	NS	NS
GLUC (mmol/L)	9.65^a	6.73^b	5.87^{bc}	4.44^c	0.94	*	*	*

TT – significance of transportation time; LT – significance of lairage time; TT x LT – significance of the interaction between transportation and lairage time.
* Statistical significance at (P<0.05); NS: not significant (P>0.05)
- Different letters in the same row indicate a significant difference at $P<0.05$ ($^{a\text{-}c}$)

The effects of transportation and lairage time and their interaction on carcass and meat quality in slaughter pigs are shown in Table 2. Two-way interaction (P<0.05) between transportation and lairage time affected carcass quality traits, indicating that live, hot and cold carcass weights decreased, as transportation and lairage time increased, reaching the minimum values in pigs subjected to long transportation time and slaughtered after overnight lairaging (Table 2). The leading causes of slaughter and carcass weight losses are the decrease of the gastrointestinal tract weight and bladder content during feed and water withdrawal [19]. It has been reported that slaughter weight decreased by 100 g per hour over the 24 h fasting period [20]. Accordingly, it could be considered that the synergistic effect of feed and water deprivation, longer transportation duration and prolonged lairaging increased slaughter weight and carcass loss [19]. Also, pigs after long transportation and overnight lairaging had the highest (P<0.05) pH_{45} value, skin lesion score and incidence of DFD meat. This can be attributed to the fact that the combined effects of long-distance transportation and prolonged lairaging cause muscle fatigue and breakdown of glycogen, which increases the tendency towards DFD meat [7-11]. Likewise, it is well known that extended lairaging (overnight to >24 h), stimulates fighting behavior and subsequently increasing the occurrence of skin lesions on the carcass [11].

Two-way interaction (P<0.05) between transportation and lairage time showed that the lowest pH_{45} value, the highest T_{45} value and incidence of PSE meat were found in pigs subjected to short transportation time and slaughtered without rest (Table 2). Pigs that underwent short transportation and lairaging would not have sufficient time to recuperate from these initial stressors, and, thus, this practice is not recommended, because animals are exhausted and agitated [21-22]. Moreover, such pre-slaughter treatment results in an increase in muscle temperature (+1°C) and lactic acid just prior to slaughter, thus increasing a tendency towards PSE meat [5-11].

Table 2. Mean values (±pooled SEM) of carcass and meat quality traits in slaughter pigs according to transportation and lairage time (n=120)

	Short transportation time		Long transportation time		Pooled SEM	TT	LT	TT x LT
Lairage time	Immediate slaughter	Overnight slaughter	Immediate slaughter	Overnight slaughter		Significance		
Number of pigs	41	20	19	40				
Carcass quality								
LW (kg)	115.60^a	116.40^a	116.60^a	111.40^b	2.30	NS	NS	*
HCW (kg)	94.22^a	95.60^a	95.98^a	91.09^b	2.04	NS	NS	*
CCW (kg)	91.25^a	92.98^a	93.68^a	88.34^b	2.04	NS	NS	*
CL (%)	3.15	2.75	2.37	3.02	0.51	NS	NS	NS
CD (%)	81.49	82.14	82.35	81.69	0.58	NS	NS	NS
FTB (mm)	18.88	16.70	17.95	19.50	3.13	NS	NS	NS
FTS (mm)	36.73	32.85	35.00	42.88	8.40	NS	NS	NS
M (%)	40.00	41.32	40.56	38.40	2.28	NS	NS	NS
SLC	0.71	0.30	0.63	1.23	0.32	*	NS	*
Meat quality								
pH_{45}	6.02^a	6.21^b	6.21^b	6.35^c	0.07	*	*	*
T_{45} (°C)	40.14	39.36	38.86	39.17	0.30	*	NS	*
Pork quality classes (%)								
PSE	51.22^a	20.00^b	21.05^b	7.50^b	-	-	-	-
Normal	48.78	65.00	68.42	50.00	-	-	-	-
DFD	0.00^a	15.00^b	10.53^b	42.50^c	-	-	-	-

TT – significance of transportation time; LT – significance of lairage time; TT x LT – significance of the interaction between transportation and lairage time. PSE meat: $pH_{45}< 6$; normal meat: pH_{45} between 6.0 and 6.4; DFD meat: $pH_{45}> 6.4$.
* Statistical significance at ($P<0.05$); NS: not significant ($P>0.05$)
- Different letters in the same row indicate a significant difference at $P<0.05$ ($^{a-c}$)

4. Conclusion

The results showed that short transportation time and slaughtering without lairaging, as well as long transportation time and overnight lairaging, seriously compromised animal welfare, carcass and pork quality, and, therefore, the above-mentioned pre-slaughter conditions are not recommended to the farmers and/or pork producers.

Acknowledgment
This paper was supported by Ministry of Education, Science and Technological Development, Republic of Serbia, Project "Selected biological hazards to the safety/quality of food of animal origin and the control measures from farm to consumer" (No. 31034).

References

[1] Muchenje V and Ndou S P 2011 How pig pre-slaughter welfare affects pork quality and the pig industry *News and Industry Article, South African Pork Producers Organization (SAPPO)* 1-44.
[2] Chai J, Xiong Q, Zhang C X, Miao W, Li F E, Zheng R, PengJ and Jiang S W 2010 Effect of pre-slaughter transport plant on blood constituents and meat quality in halothane genotype of NN Large White × Landrace pigs *Livestock science* **127** 211-217.
[3] Šefer D, Marković R, Nedeljković-Trailović J, Petrujkić B, Radulović S and Grdović S 2015 The application of biotechnology in animal nutrition *Veterinarski Glasnik* **69** 127-137.
[4] Đorđević V, Đorđević J, Baltić Ž M, Laudanović M, Teodorović V, Bošković M, Peurača M and Marković R 2016 Effect of sunflower, linseed and soybean meal in pig diet on chemical

composition, fatty acid profile of meat and backfat, and its oxidative stability *Acta Veterinaria-Beograd* **66** 359-372.

[5] Čobanović N, Karabasil N, Stajković S, Ilić N, Suvajdžić B, Petrović M and Teodorović V 2016 The influence of pre-mortem conditions on pale, soft and exudative (PSE) and dark, firm and dry (DFD) pork meat *Acta Veterinaria-Beograd* **66** 172-186.

[6] Čobanović N, Bošković M, Vasilev D, Dimitrijević M, Parunović N, Djordjević J and Karabasil N 2016 Effects of various pre-slaughter conditions on pig carcasses and meat quality in a low-input slaughter facility *South African Journal of Animal Science* **46** 380-390.

[7] Gispert M, Faucitano L, Oliver M A, Guàrdia M D, Coll C, Siggens K, Harvey K, Diestre A 2000 A survey of pre-slaughter conditions, halothane gene frequency, and carcass and meat quality in five Spanish pig commercial abattoirs *Meat science* **55** 97–101.

[8] Pérez M P, Palacio J, Santolaria M P, Aceña M C, Chacón G, Gascón M, Calvo J H, Zaragoza P, Beltran J A and Garcıa-Belenguer S 2002 Effect of transport time on welfare and meat quality in pigs *Meat science* **61** 425-433.

[9] Pérez M P, Palacio J, Santolaria M P, del Carmen Aceņa M, Chacón G, Verde M T, Calvo J H, Zaragoza P, Gascón M and García-Belenguer S 2002 Influence of lairage time on some welfare and meat quality parameters in pigs *Veterinary Research* **33** 239-250.

[10] Zhen S, Liu Y, Li X, Ge K, Chen H, Li C and Ren F 2013 Effects of lairage time on welfare indicators, energy metabolism and meat quality of pigs in Beijing *Meat science* **93** 287-291.

[11] Faucitano L 2010 Invited review: Effects of lairage and slaughter conditions on animal welfare and pork quality *Canadian Journal of Animal Science* **90** 461–469.

[12] Official gazette 1985 Rules for the classification of pig carcasses (*Official Gazette SFRJ, No 2/85, 12/85, 24/86*).

[13] Welfare Quality® Consortium 2009 Welfare quality assessment protocol for pigs, 2009 *Welfare Quality Consortium, Lelystad, The Netherlands*.

[14] Adzitey F and Nurul H 2011 Pale soft exudative (PSE) and dark firm dry (DFD) meats: Causes and measures to reduce these incidences-a mini review *International Food Research Journal* **18** 11-20.

[15] Pereira T L, Corassa A, Komiyama C M, Araújo C V and Kataoka A 2015 The effect of transport density and gender on stress indicators and carcass and meat quality in pigs *Spanish Journal of Agricultural Research* **13** 1-11.

[16] Choe J H, Choi M H, Ryu Y C. Lim K S, Lee E A, Kang J H, Hong K C, Lee S K, Kim Y T, Moon S S and Lee K W 2015 Correlations among various blood parameters at exsanguination and their relationships to pork quality traits *Animal Production Science* **55** 672-679.

[17] Bryer P J, Sutherland M A, Davis B L, Smith J F and McGlone J J 2011 The effect transport and space allowance on the physiology of breeding age gilts *Livestock science* **137** 58-65.

[18] Sutherland M A, Bryer P J. Davis B L, Smith J F and McGlone J J 2012 The combined effects of transport and food and water deprivation on the physiology of breeding age gilts *Livestock Science* **144** 124-131.

[19] Lambertini L, Vignola G, Badiani A, Zaghini G and Formigoni A 2006 The effect of journey time and stocking density during transport on carcass and meat quality in rabbits *Meat science* **72** 641-646.

[20] Dalla Costa F A, Devillers N, da Costa M P and Faucitano L 2016 Effects of applying preslaughter feed withdrawal at the abattoir on behaviour, blood parameters and meat quality in pigs *Meat science* **119** 89-94.

[21] Hoffman L C and Fisher P 2010 Comparison of the effects of different transport conditions and lairage times in a Mediterranean climate in South Africa on the meat quality of commercially crossbred Large white × Landrace pigs *Journal of the South African Veterinary Association* **81** 219-223.

[22] Panella-Riera N, Gispert M, Gil M, Soler J, Tibau J, Oliver M A, Velarde A and Fàbrega E

2012 Effect of feed deprivation and lairage time on carcass and meat quality traits on pigs under minimal stressful conditions *Livestock Science* **146** 29-37.

IOP Conf. Series: Earth and Environmental Science **85** (2017) 012050 doi:10.1088/1755-1315/85/1/012050

Distribution of mercury in leg muscle and liver of game birds from Serbia

S Janković[1], D Nikolić[1], S Stefanović[1], T Radičević[1], J Đinović-Stojanović[1], D Spirić[1] and S Tanković[2]

[1] Institute of Meat Hygiene and Technology, Kacanskog 13, 11000 Beograd, Serbia
[2] Ministry of Foreign Trade and Economic Relations of Bosnia and Herzegovina, Veterinary Office of Bosnia and Herzegovina, Marsala Tita 9a/II, 71000 Sarajevo, Bosnia and Herzegovina

Email: sasa.jankovic@inmes.rs

Abstract. The purpose of this study was to determine the distribution of Hg levels in leg muscle and liver of game birds collected within the Serbian National residue monitoring program from 2013 to 2016. Hg levels in samples (n=464) of: pheasants (n=182), mallard (n=25), Eurasian jay (n=7), partridges (n=5) and woodcocks (n=8) were determined by ICP-MS. The highest mean Hg levels were observed in leg muscle samples of woodcocks (0.071 mg/kg) and mallard (0.059 mg/kg). The lowest mean Hg level in liver was determined in partridges (0.008 mg/kg) while the highest was in pheasants (0.262 mg/kg) and mallard (0.161 mg/kg). Statistical analysis showed significantly differences between Hg levels in liver of woodcocks and mallard, as well as between them and livers of other analysed game birds. During the four years (2013-2016), 87.5% of leg muscle and 50% of woodcock livers had Hg levels that exceeded the MRL, while in mallard muscle and liver those percentages were 36% and 40%, respectively.

1. Introduction

Meat is good source of protein, fat, vitamins and some essential elements and makes up an important part of the food we eat [1]. Beside wide range of nutrients, meat may also carry certain toxic substances [2]. Heavy metals are toxic, and still one of the global environmental problems posing health risk to man and wildlife [3]. Heavy metals enter the environment by natural sources (volcanic activity, erosions and weathering) and by human activities (mining, fossil fuels combustion, industrial emissions, direct application of fertilizers and fungicides). Mercury (Hg) is a persistent global pollutant existing in three forms – elemental, inorganic and organic mercury compounds. The inorganic form of mercury can be transformed to methylated form, methylmercury, by microorganisms and this form is mostly present in fish [4,5,6]. Exposure to mercury causes permanent damage to the brain, kidneys, and developing foetus [7,8]. According to International Agency for Research on Cancer (IARC) methylmercury is classified as possibly carcinogenic for humans (2B).

Game meat is considered and valued for its exceptional taste than for its nutritive properties. However, as game animals are a part of the terrestrial soil-plant-animal food chain, they have been reported containing higher toxic element levels than animals from farms [9]. While toxic element levels in muscle are generally low, liver and kidney accumulate higher concentrations of these elements [10]. This is especially expressed in birds, as their metabolism is more rapid compared to other species of animals. So, game birds are considered a wildlife type that is a suitable, representative, potential bioindicator of pollution of the environmental.

Since the meat of bird game is, therefore, a bioindicator of potential pollution in the environment and is consumed by humans, the objective of this study was to examine the levels of Hg in leg muscle and liver of different game bird species and to identify possible differences in Hg levels between species.

2. Materials and Methods

Levels of Hg were measured in leg muscle and liver of game birds (n=464). Pheasants (n=182), mallard (n=25), Eurasian jay (n=7), partridges (n=5) and woodcocks (n=8) were acquired during regular hunting seasons within the Serbian National residue monitoring program from 2013 to 2016.

Frozen samples were thawed at +4°C for a day before analysis and then homogenized. An amount, approximately 0.5 g of homogenized tissue was transferred into a teflon vessel with 5 mL nitric acid (67% Trace Metal Grade, Fisher Scientific, Bishop, UK) and 1.5 mL hydrogen peroxide (30% analytical grade, Sigma-Aldrich, St. Louis, MA, USA) for microwave digestion. The microwave (Start D, Milestone, Sorisole, Italy) program consisted of three steps: 5 min from room temperature to 180°C, 10 min hold at 180°C, 20 min ventilation. After cooling, the digested sample solutions were quantitatively transferred into disposable flasks and diluted to 100 mL with deionized water from a water purification system (Purelab DV35, ELGA, Buckinghamshire, UK).

Inductively coupled plasma mass spectrometry (ICP-MS), (iCap Q mass spectrometer, Thermo Scientific, Bremen, Germany), was used for analysis of the ^{202}Hg isotope. Five-point calibration curve (including zero) was constructed for the qualitative analysis for Hg isotope of the samples. Multielement internal standard (^{6}Li, ^{45}Sc—10 ng/mL; ^{71}Ga, ^{89}Y, ^{209}Bi—2 ng/mL) was introduced online by an additional line through the peristaltic pump.

The quality of the analytical process was verified by analysis of the certified reference material NIST 1577c (Gaithersburg, MD, USA). Reference material was prepared as samples using microwave digestion. Replicate analyses were in the range of certified values.

Statistical analysis of experimental data was performed using software Minitab 16 Statistical Software. One-way analysis of variance – ANOVA and Tukey's HSD test were applied for comparison of Hg levels between leg muscles as well as between livers from different game birds.

3. Results and Discussion

The results of Hg levels in leg muscle and liver of the analysed samples are presented in Table 1 and Table 2, respectively. For calculation, when the levels of Hg were below the limit of detection (LOD, LOD=0.001 mg/kg), that value was assumed to be equal to one half of the LOD (1/2 LOD).

Table 1. Hg levels in leg muscle of game birds from Serbia

	n1	min-max	Mean ± SD	n2
Pheasants	182	< LOD-0.030	0.001±0.002^{a}	
Mallard	25	0.005-0.220	0.059±0.058^{a}	9
Eurasian jay	7	< LOD-0.012	0.008±0.004^{a}	
Partridges	5	< LOD-0.013	0.003±0.006^{a}	
Woodcocks	8	0.021-0.130	0.071±0.036^{a}	7

n1 – number of samples
n2 – number of non-compliamt samples
a-b Different superscripts within the same column indicate significant differences of means according to Tukey's HSD test ($p < 0.05$)

Table 2. Hg levels in liver of game birds from Serbia

	n1	min-max	Mean ± SD	n2
Pheasants	182	< LOD-0.039	0.004±0.007^{c}	
Mallard	25	0.015-0.484	0.175±0.161^{a}	10

Eurasian jay	7	< LOD-0.035	0.046±0.014^{c}	
Partridges	5	< LOD-0.036	0.008±0.016^{c}	
Woodcocks	8	0.071-0.212	0.132±0.060^{b}	4

n1 – number of samples
n2 – number of non-compliamt samples
a-b-c Different superscripts within the same column indicate significant differences of means according to Tukey's HSD test ($p < 0.05$)

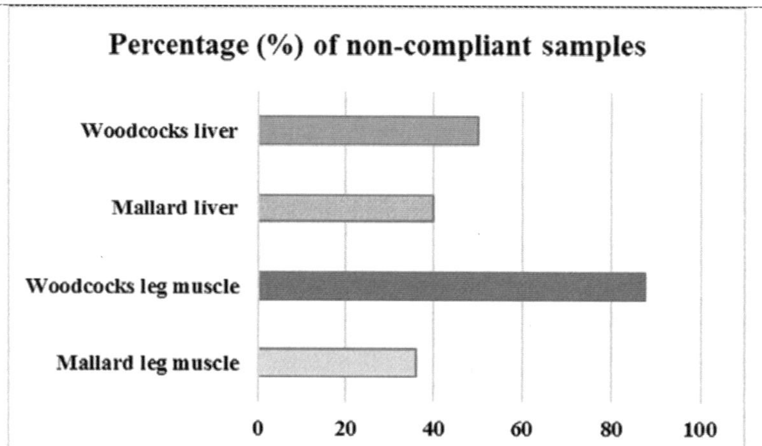

Figure 1. Percentage of non-compliant leg muscle or liver of mallard and woodcocks with regard to Hg levels.

The levels of Hg in leg muscle were within the range < 0.001-0.220 mg/kg. The highest mean Hg levels were established in leg muscle samples of woodcocks (0.071 mg/kg) and mallard (0.059 mg/kg), while liver samples of other game birds contained very low Hg levels, often close to the LOD (Table 1). These could be a consequence of the characteristic feed of mallard and woodcocks, which is rich in crustaceans, worms and fish, food with high levels of Hg. Statistical analysis did not show any significant differences in Hg levels between leg muscle samples of game birds. National legislation [11] does not prescribe maximum residue levels (MRL) for Hg in game tissue. Considering that, in this study, the MRL for Hg in poultry tissue (muscle, liver) is used as the MRL for game. MRL for muscle is 0.030 mg/kg [11]. Nine leg muscle samples of woodcocks and seven of mallard exceeded this level. So, during 2013-2016, 87.5 and 36% of analysed leg samples of woodcocks and mallard, respectively, were non-compliant (Figure 1).

While Hg levels in leg muscle were generally low, liver accumulates higher levels [12, 13, 14]. The lowest mean Hg levels established in liver samples were from partridges (0.008 mg/kg) while the highest was in the liver from pheasants (0.262 mg/kg) and mallard (0.161 mg/kg). Significant differences were established between Hg levels in liver of mallard and woodcocks. Also, Hg levels in liver of mallard and woodcocks significantly differ from the other analysed livers of game birds. National legislation [11] established 0.100 mg/kg as MRL for Hg in game liver. Some liver samples of woodcocks (n=4) and mallard (n=10) exceeded this MRL level (Figure 1). The highest established Hg levels were in liver of mallard (0.484 mg/kg) and woodcocks (0.212 mg/kg).

4. Conclusion

Game meat, including meat of game birds, is commonly consumed by hunters and their families. Meat and offal of game have been recognized as food items with elevated content of toxic elements thus raising concern about exposure to these elements even in a small part of the population. So, periodic control and investigation of Hg and other toxic elements in game meat and offal are needed to assess the safety of these meats with respect to human health and to widen the knowledge base for this type of meat.

References

[1] Akan J C, Abdu F I, Irahman O A, Chiroma Y A 2010 *Res. J. Appl. Sci. Eng. Technol.* **2** 743
[2] Khalafalla F, Ali F, Schwagele F, Abd-El Wahab M 2011 *Vet. Ital.* **47** 351
[3] Walker C H, Hopkin S P, Sibly R M, Peakall D B 2006 Principles of ecotoxicology. Taylor & Francis: Boca Raton, FL, 3
[4] Merritt K A, Amirbahman A 2008 *Limnol. Oceanogr.* **53** 1064
[5] Ersoy B, Celik M 2010 *Food Chem. Toxicol.* **48** 1377
[6] Saei-Dehkordi S S, Fallah A A, Nematollah A 2010 *Food Chem. Toxicol.* **48** 2945
[7] WHO (World Health Organization) 1990 International program on chemical safety, Environmental Health Criteria, 101: Methyl mercury, Geneva, Switzerland, 1-144
[8] Hightower J M, Moore D 2003 *Environ. Health Perspect.* **111** 604
[9] Forte G, Bocca B 2007 *Food Chem.* **105** 1591
[10] López-Alonso M, Miranda M, Castillo C, Hernández J, García-Vaquero M, Benedito J L 2007 *Food Add. Contam.* **24** 943
[11] Official gazette of RS 29/2014 2014
[12] Toman R, Massányi P, Lukáč N, Ducsay L, Golian J 2005 *Ecotoxicol. Environ. Saf.* **62** 112
[13] Massányi P, Toman R, Uhrín V, Renon P 1995 *Ital. J. Food Sci.* **7** 311
[14] Massányi P, Uhrín V 1996 *J. Environ. Sci. Health.* **A32** 1459

IOP Conf. Series: Earth and Environmental Science **85** (2017) 012051 doi:10.1088/1755-1315/85/1/012051

Listeria monocytogenes presence during fermentation, drying and storage of Petrovská klobása sausage

V Janković, R Mitrović, B Lakićević, B Velebit and T Baltić

Institute of Meat Hygiene and Technology, Kaćanskog 13, Belgrade 11000, Serbia

Email: vesna.jankovic@inmes.rs

Abstract. The majority of human listeriosis cases appear to be caused by consumption of ready-to-eat (RTE) foods contaminated at the time of consumption with high levels of *Listeria monocytogenes*. Although strategies to prevent growth of *L. monocytogenes* in RTE products are critical for reducing the incidence of human listeriosis, this pathogen is highly difficult to control in fermented sausage processing environments due to its high tolerance to low pH and high salt concentration. The aims of the present study were to investigate the occurrence, presence and elimination of *L. monocytogenes* in Petrovská klobása sausage during processing, fermentation, drying and storage. *L. monocytogenes*, which was detected at the beginning of the production cycle, disappeared before day 30. The pathogen decline was much faster in those sausages which were dried in controlled, industrial conditions than in those dried applying the traditional, household technique.

1. Introduction

Dry fermented sausages are characterized by their relatively longer shelf-life, which is brought about by the production of lactic acid in the fermentation process (pH<4.5-5) and low water activity (<0.90) of the final product [1]. Traditionally, fermented sausages are made using lactic acid bacteria (LAB) and Gram-positive catalase positive cocci, in particular coagulase-negative staphylococci naturally present in the meat, or with the inoculation of starter cultures at the chopping step. The mixture is then filled into natural or artificial casings, left to ferment and then dried [2].

Petrovská klobása is a traditional and autochthonous fermented pork meat product, which is a part of gastronomic heritage of Slovaks in Vojvodina, and which is produced in a traditional way in rural households in the Municipality of Bački Petrovac. In rural households, this sausage is at the end of November and during December. Petrovská klobása is made by mixing partly cooled (chilled for cca 4 h after slaughter) or cold (chilled for cca 24 h after slaughter) medium-chopped lean pork and fat (up to 10 mm) with addition of powdered red hot spicy paprika, salt, crushed garlic, caraway and sugar. A well-mixed filling, which is prepared within 15-30 minutes by using a unique technique of manual mixing with kneading and overturning, is stuffed into natural casings consisting of the rear part of pig intestines (rectum), forming units 35-45 cm long and 4.5-5.0 cm in diameter. After stuffing, the sausages are left to drain for a while and then they are smoked by a cold process for about 10-15 days with pauses, using specific kinds of wood (cherry wood in particular). When the smoking process is finished, the sausage is kept in a dry and well-ventilated place to dry and ripen, until it achieves an optimum quality, which takes about four months [3,4].

Fermented sausages contaminated with *L. monocytogenes* have rarely been implicated in critical listeriosis outbreaks [5]. Insufficiently dried sausages can have water activity levels close to 0.92-0.94 [6], and *L. monocytogenes* is able to survive during sausage fermentation, overcoming the hurdles encountered during the manufacturing process. In general, the contamination levels at the end of ripening are always lower than 100 CFU/g [7], because *L. monocytogenes* cannot compete with the prevailing LAB. Only without competitive microflora is *L. monocytogenes* able to multiply and reach high levels of contamination (higher than 1000 CFU/g), representing a major public health concern [8-10].

The aims of the present study were to investigate the occurrence, presence and elimination of *L. monocytogenes* in Petrovská klobása sausage during processing, fermentation, drying and storage in natural and artificial casings, and in household and industrial conditions.

2. Materials and Methods

2.1. Preparation of Petrovská klobása

Petrovská klobása dry fermented sausages were manufactured from a mixture of lean minced pork (80%) and pig fat (20%) obtained from carcasses of Large White cross breed animals. Meat was either hot (4 h-chilled meat) or cooled (24 h-chilled meat). After grinding the meat and the fat to a size of about 10 mm (with adjustable plate holder diameter set), raw materials were mixed with seasonings (red hot paprika powder, salt, raw garlic paste, caraway and sucrose) for about 10 min. The seasoned batter was immediately stuffed into collagen casings (two types of collagen casings – one artificial, and one natural pig intestine 500 mm long and 55 mm in diameter), and raw sausages were entirely processed in a traditional smoking/drying room or under industrial conditions during 120 days, then stored until 270 days.

2.2. Samples

Samples produced from the hot meat in a traditional way in the household – natural casing (A1) and A2 (artificial casing), samples produced from the cooled meat, smoked and dried in a household – natural casing (B1), samples produced from the cooled meat, smoked and dried in a household – artificial casing (B2), samples produced from the cooled meat, smoked and dried at the processing plant – natural casing (B3) and samples produced from the cooled meat, smoked and dried at the processing plant – artificial casing (B4) were taken before stuffing (at day 0) and during processing (on days 2,4, 6, 9, 12, 15, 30, 45, 60, 90, 120,150, 210 and 270).

2.3. Microbiological methods

Each sample was tested according to SRPS ISO 11290-1 and 2.

3. Results and Discussion

Results of testing are presented in figures 1, 2 and 3. The sausages contained detectable low levels of *L. monocytogenes*, which occurred naturally in the sausage mixture at the start of the study.

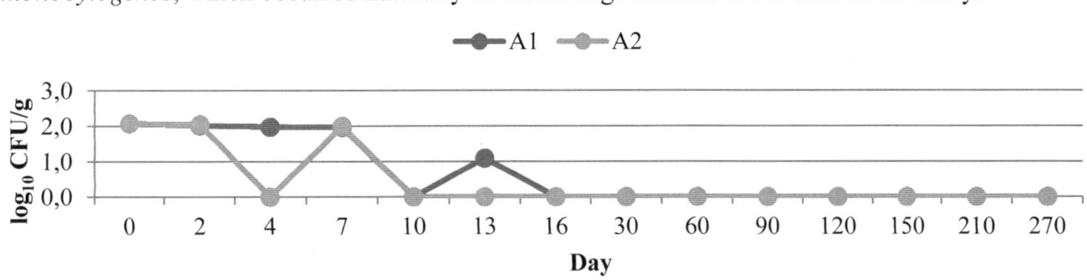

Figure 1. Decline of *Listeria monocytogenes* during smoking, fermentation, drying and storage of Petrovská klobása in groups A1 (hot meat/natural casing/household production) and A2 (hot meat/artificial casing/household production).

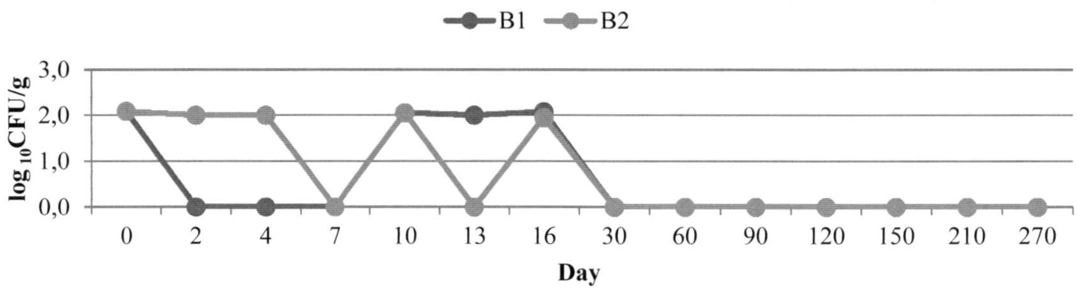

Figure 2. Decline of *Listeria monocytogenes* during smoking, fermentation, drying and storage of Petrovská klobása in groups B1 (cooled meat/natural casing/household production) and B2 (cooled meat/artificial casing/household production).

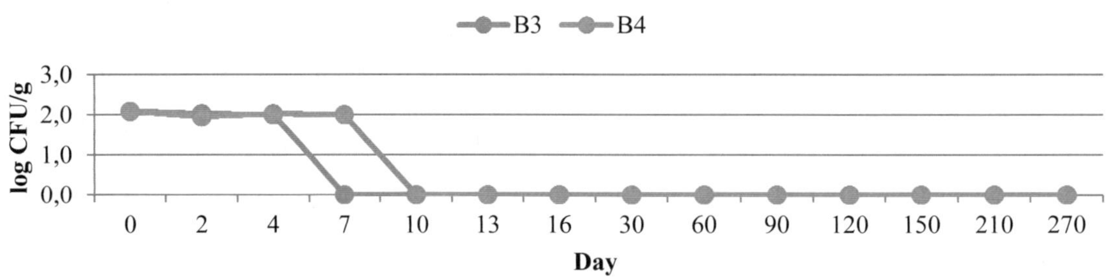

Figure 3. Decline of *Listeria monocytogenes* during smoking, fermentation, drying and storage of Petrovská klobása in groups B3 (cooled meat/natural casing/industrial production) and B4 (cooled meat/artificial casing/industrial production).

The presence of *L. monocytogenes* in A1 (figure 1) was detected on days 0, 2, 4, 6 and 12, and in A2 (figure 1) on days 0, 2 and 4. In B1 and B2 sausages, *L. monocytogenes* was found up to 15 days (figure 2). In B3 and B4 sausages (figure 3), the presence of *L. monocytogenes* was detected until days 4 (B3) and 6 (B4). Generally, with respect to the meat chilling time, filling and drying process (traditional and controlled conditions), *L. monocytogenes* disappeared much faster in sausages which were dried in controlled, industrial conditions than in those dried applying the traditional household technique. The results are in accordance with the results obtained by [11,12,13]. Generally, the declines in pathogen presence can be explained by the fact that the preservatives and protective microbiota have a significant impact on the survival and growth of *L. monocytogenes*. The organism is inhibited in fermented sausages by sequential steps: the "hurdle technology" concept includes several sequential hurdles, essential at different stages of the fermentation or ripening process [14]. These include lowering of pH by fermenting sugars to mainly lactic acid, lowering of water activity by salting, drying by evaporating water, inhibiting growth of aerobic bacteria by creating an anaerobic environment, inhibiting microbial growth by addition of nitrate or nitrite, and inhibiting surface growth by smoking or by addition of specific moulds. Together, these hurdles generally lead to a shelf-stable product [15]. These hurdles are essential in different steps of the fermentation or ripening process and lead to stable and safe final products [16].

4. Conclusion

According to the results, *L. monocytogenes*, which was detected at the beginning of the production cycle, disappeared before day 30. The pathogen decline was much faster in those sausages which were dried in controlled, industrial conditions than in those dried applying the traditional, household technique. Petrovská klobása sausage may be contaminated by *L. monocytogenes* at several stages. The raw materials may be contaminated from the slaughterhouse environment, during the production process or by contact with contaminated unprocessed raw materials, unclean surfaces or people or in the post-processing stages.In order to prevent growth of *L. monocytogenes* in dry fermented sausages, good manufacturing practices, correct sampling schemes, adequate cleaning and disinfection procedures and HACCP principles have to be applied. The use adequate hurdles can minimize the potential for growth of *L. monocytogenes*.

References

[1] Jankovic V 2013 *Determination of the changes in health safety parameters of Petrovská klobása during production in traditional and controlled conditions* PhD thesis (Faculty of Technology University of Novi Sad Serbia) p 102

[2] Meloni D 2015 Presence of *Listeria monocytogenes* in Mediterranean-style dry fermented sausages *Foods* **4** 34–50

[3] Tasić T 2012 *Producing of Biogenic amines in traditional dry fermented sausage Petrovská klobása, during safety and quality standardization process* PhD thesis (Faculty of Technology University of Novi Sad Serbia) p 93

[4] Janković V, Petrović Lj, Lakićević B, Matekalo Sverak V, Lilić S and Mitrović R 2013. Determination of typical house flora during production process of the Petrovská klobása. *Afr. J. of Microbiol. Res.* 7 **32**, 4130–37

[5] European Food Safety Authority (EFSA) European Centre for Disease Prevention and Control (ECDC) 2014 The European Union Summary Report on trends and sources of zoonoses, zoonotic agents and food-borne outbreaks in 2012. *EFSA J.* **12** 3547

[6] Meloni D 2014 Focusing on the main morphological and physiological characteristics of the food-borne pathogen *Listeria monocytogenes J. Vet. Sci. Res.* **1** 1–2

[7] Farber J M and Peterkin P 1991 *Listeria monocytogenes*, a food-borne pathogen *Microbiol. Mol. Biol. Rev.* **55** 476–571

[8] Mc Lauchlin J, Mitchell R T, Smerdon W J and Jewell K 2004 *Listeria monocytogenes* and listeriosis: A review of hazard characterization for use in microbiological risk assessment of foods *Int. J. Food Microbiol* **92** 15–33

[9] Thévenot D, Delignette-Muller M L, Christieans S, Leroy S, Kodjo A and Vernozy-Roland C 2006 Serological and molecular ecology of *Listeria monocytogenes* isolates collected from 13 French pork meat salting-curing plants and their products *Int. J. Food Microbiol.* **112** 153–161

[10] Meloni D, Consolati S G, Mazza R, Mureddu A, Fois F, Piras F and Mazzette R 2014 Presence and molecular characterization of the major serovars of *Listeria monocytogenes* in ten Sardinian fermented sausage processing plants *Meat Sci* **97** 443–50

[11] Holck A, Lars A, McLeod A, T M Rode and Even H 2017 Health and safety considerations of fermented sausages *Journal of Food Quality* **2017**

[12] Lakićević B, Velebit B, Janković V, Spirić D, Baltić T, Mitrović R and Babić J 2014 Taq Man Real Time PCR detection of *Listeria monocytogenes*: a study of enrichment incubation time affecting sensitivity in experimental dry fermented sausages *Tehnologija Mesa* 55 **1** p 26

[13] Janković V, Petrović Lj, Džinić N, Lakićević B, Šarić Lj, Škaljac S, Šojić B and Jokanović M 2014 Microbial ecosystem of processing units during production process of Petrovská klobása *Tehnologija Mesa* 55 **2** 123–29

[14] Meloni D, Piras F, Mureddu A, Mazza R, Nucera D and Mazzette R 2012 Sources of *Listeria monocytogenes* contamination in traditional fermented sausages processing plants in Italy *Ital.*

IOP Conf. Series: Earth and Environmental Science **85** (2017) 012051 doi:10.1088/1755-1315/85/1/012051

J. Food Sci **24** 214–22

[15] Barbuti S and Parolari G 2002 Validation of manufacturing process to control pathogenic bacteria in typical dry fermented products *Meat Sci.* **62** 323–29

[16] Leistner L and Gould G W 2002 *Hurdle Technologies: Combination Treatments for Food Stability, Safety and Quality* (Kluwer; New York, NY, USA) 1-21

Whole genome sequencing: an efficient approach to ensuring food safety

B Lakicevic[1], I Nastasijevic[1] and M Dimitrijevic[2]

[1]Institute of Meat Hygiene and Technology, Kaćanskog 13, 11000 Belgrade, Serbia
[2]Faculty of Veterinary Medicine, Bulevar oslobođenja 18, 11000 Belgrade, Serbia

E-mail: brankica.lakicevic@inmes.rs

Abstract. Whole genome sequencing is an effective, powerful tool that can be applied to a wide range of public health and food safety applications. A major difference between WGS and the traditional typing techniques is that WGS allows all genes to be included in the analysis, instead of a well-defined subset of genes or variable intergenic regions. Also, the use of WGS can facilitate the understanding of contamination/colonization routes of foodborne pathogens within the food production environment, and can also afford efficient tracking of pathogens' entry routes and distribution from farm-to-consumer. Tracking foodborne pathogens in the food processing-distribution-retail-consumer continuum is of the utmost importance for facilitation of outbreak investigations and rapid action in controlling/preventing foodborne outbreaks. Therefore, WGS likely will replace most of the numerous workflows used in public health laboratories to characterize foodborne pathogens into one consolidated, efficient workflow.

1. Introduction

Food safety is a global concern, and consumers have the right to safe and nutritious food. However, the estimated burden of foodborne diseases at global level, i.e. 600 million foodborne illnesses and 420,000 deaths from 31 major food safety hazards in 2010 [1] as well as at EU level, with around 400,000 confirmed human illnesses and 463 deaths [2] from 5 major food safety hazards (causing listeriosis, salmonellosis, campylobacteriosis, STEC infections, yersiniosis) and the related social and economic costs (hospitalization, loss of income, employment and market access) remain unacceptably high [3]. Tracking foodborne pathogens in the food processing – distribution – retail – consumer continuum is of utmost importance for facilitation of outbreak investigation and rapid action in controlling/preventing foodborne disease outbreaks. Whole genome sequencing (WGS), using next generation sequencing (NGS) technology, has the ability to sequence entire genomes [4] and provides rapid, comprehensive information and characterization of microorganisms with a level of precision not previously possible. WGS, now internationally recognized as the new revolution in microbiology food safety, is a technology that likely will transform public health microbiology in a few years. Field applications of WGS have garnered success in multiple contexts in recent years: from transmission studies and outbreak tracking of pathogens in hospitals to traceback investigations and source attribution in foodborne outbreaks at both national and multinational levels [5]. Therefore, WGS likely will replace most of the numerous workflows used in public health laboratories to characterize foodborne pathogens into one consolidated workflow [6].

2. Traditional techniques for distinguishing bacteria

At present, several different techniques are used to identify and characterize bacteria, and to distinguish between different strains of a species. The method of choice varies depending on the species in question and on precisely what information is required. Some of the commonly used methods are:

- Serotyping – different bacterial strains have different sets of molecules coating them; antibodies are used to determine which molecules are present on the outer surface.
- Phage typing – distinguishes between strains based on their susceptibility to infection by various types of bacteriophage (a group of viruses).
- Pulsed Field Gel Electrophoresis (PFGE) – the bacterial genome is chopped up at specific points, to create fragments that are then electrophoretically separated on a gel according to their size. Different strains have different patterns of fragments.
- Multilocus sequence typing (MLST) – sequencing 400-500 base pair fragments of DNA at seven different conserved genes allows small variations within a species to be detected. Quite time consuming and costly, but can be highly discriminatory if the genes are correctly chosen.
- Multilocus variable number tandem repeat analysis (MLVA) – particular regions of DNA are very repetitive, and the number of repeats of a sequence in a particular region varies between different bacterial strains. A few such areas are analysed to determine how many repeats there are in each. MLVA is faster and easier to perform than MLST, but there are issues with reproducibility and validation.

All of the genotyping methods mentioned above have their limitations and ideally the entire genome of a microbial isolate would be sequenced to provide definitive typing. Until quite recently, this would have been very costly and taken years to complete, but the advent of NGS technology, such as the Ion TorrentTM platform from Life Technologies and the Illumina sequencing system, make WGS within days a practical option. Now that high-throughput bench-top NGS instruments like Life Technologies' Ion PGMTM and the Illumina NextSeq 500 are available, WGS is within the reach of smaller clinical microbiology and research laboratories. The costs of NGS are falling, and an entire bacterial genome can now be sequenced for about the same price as MLST typing using conventional PCR/Sanger sequencing. Otherwise, WGS in principle offers one simplified approach, easily reproducible across different laboratories [7]. In addition, WGS is different from current typing methods because it requires – in addition to the actual laboratory work – substantial data processing, storage and analysis to extract useful information from the large amount of generated data. Collaboration and data sharing between organizations and countries is required due to the international dimension of food and waterborne bacterial (FWD) pathogens and food trade in particular. National public health reference laboratories form the first line of such collaboration, as they are usually the first to have sufficient information, i.e. the microbiological typing data, to allow linkage of cases at national level and subsequent detection of human clusters of outbreaks [8].

3. *Listeria monocytogenes* and whole genome sequencing

Listeria monocytogenes is a predominantly foodborne pathogen capable of causing a range of clinical illnesses, including invasive disease such as bacteraemia and meningoencephalitis in humans [9], and is commonly monitored by public health facilities for the emergence of outbreaks [10]. A number of serotypes of *L. monocytogenes* can be isolated from environmental and food sources, but most outbreaks of human disease are due to serotypes 1/2a, 1/2b and 4b [11]. It has been confirmed that listeriosis carries one of the highest hospitalization rates among known foodborne pathogens – up to 91% [12, 13]. Altogether, 1,642 listeriosis cases in humans were reported in the European Union (EU) in 2012 with high fatality rate (17.8%) among the confirmed cases; this was a 10.5% increase of reported listeriosis cases compared with 2011 [14]. *L. monocytogenes* was found in 10.3% of fishery products, 2.1% of heat-treated meat products and 0.5% of soft and semi-soft cheeses collected from supermarkets and shops across the EU, from January 2010 to January 2012. Fermented sausages contaminated with *L. monocytogenes* have rarely been implicated in critical listeriosis outbreaks [14]. WGS has emerged as a powerful technology for the comparison of isolates in outbreak analysis [15], because the *Listeria* genome is fairly small and relatively easy to sequence and analyse [16]. With WGS, *Listeria* outbreaks can be detected when as few as two people have fallen ill. Determining that the same strain of *Listeria* is making people ill is an indication that these illnesses may have come from the same source – for example, the same contaminated food processing facility. By combining real time WGS with data from patients about the foods they ate and data about *Listeria* in foods, public health officials can: detect more cluster (possible outbreaks) of *Listeria* infections, link cases of *Listeria* to a likely source, identify unrecognized sources of *Listeria* and stop *Listeria* outbreaks while they are still small.

4. Food Safety and WGS

WGS is a significant new tool in the area of food safety, including foodborne disease surveillance, food inspection (testing) and monitoring, outbreak detection and investigation (including food attribution studies) and food technology developments. In comparison to the plethora of molecular identification and characterization technologies available to date, WGS is conceptually speaking, quite simple, regardless of the platform used. WGS is universal for all organisms and provides virtually the entire genome, which facilitates targeted exchange and comparison of its data. The results are not only useful for food monitoring, disease surveillance, and outbreak investigation and response, but also for addressing broader questions that are critical for food safety improvements and preventive measures, through source tracking, source attribution and the identification of transmission pathways [17, 18]. Since the data comprise the genetic code, WGS results can be used for more than one purpose simultaneously – such as identification, subtyping, virulence marker detection, antimicrobial resistance (AMR) predictions, and genome-wide association studies. In food monitoring, WGS is used as forensic evidence for source tracking and to inform regulatory action. Since food is a global commodity, global use of this common technology facilitates sharing and collaboration across sectors, and greatly increases the availability of contextual data when interpreting results and recommending regulatory actions on a scientific basis. It is, however, important to emphasize that WGS cannot stand alone. It is just one source of information among the complex systems that comprise the whole food supply chain. The technology requires that clinical, food and environmental isolates/samples from routine testing, inspection and surveillance, and associated data are made available, and that infrastructure is in place to utilize the data for regulatory food safety and public health action. Thus, the implementation of WGS should be accompanied by the establishment of an integrated national food control system and relevant food safety programmes that assimilate information from different sources [17].

5. Conclusion

WGS is poised to become standard methodology in food safety for identification and characterization of foodborne pathogens, including antimicrobial resistant organisms, with a level of precision not previously possible. WGS provides the most extensive analysis for isolate comparison and is superior to current typing methods for pathogens. Also, WGS as an incredibly powerful technique is less labour-intensive and more cost-effective than current typing methods for surveillance. It is already being utilised by Public Health England (PHE) and the Food Standards Agency (FSA) to aid outbreak investigation; but it has a multitude of other potential applications in relation to food.

References

[1] World Health Organization 2015 The World Health Organization estimates of the global burden of foodborne diseases (available

at:http://apps.who.int/iris/bitstream/10665/199350/1/9789241565165_eng.pdf?ua=1)

[2] European Food Safety Authority 2015 The European Union summary report on trends and sources of zoonoses, zoonotic agents and food-borne outbreaks in 2014 *EFSA J.* **13** 4329

[3] Nastasijević I, Lakićević B and Velebit B 2017 Whole genome sequencing application in food safety management and public health protection *11th Serbian Microbiologists Congress MICROMED,* 11-13 may 2017, Belgrade, ed D Obradovic p. 133-34

[4] Hossieni H 2016 Whole genome sequencing: an efficient approach in food safety management system *Nutr. Food Sci. Res.* **3** 1-2

[5] Hill A A, Crotta M, Wall B, Good L, O'Brien S J and Guitian J 2017 Towards an integrated food safety surveillance system: a simulation study to explore the potential of combining genomic and epidemiological metadata R. Soc. Open Sci. **4** DOI:10.1098/rsos.160721

[6] European Food Safety Authority Scientific Colloquium Summary Report 2014 Use of Whole Genome Sequencing (WGS) of food – borne pathogens for Public Health Protection (available at: http://onlinelibrary.wiley.com/doi/10.2903/sp.efsa.2015.EN-743/pdf)

[7] Food Standards Agency 2016 Science report on whole genome sequencing of foodborne pathogens (available at: https://www.food.gov.uk/sites/default/files/csa-whole-genome-seq-reportv2.pdf)

[8] European Centre for Disease Prevention and Control 2015 Expert Opinion on the introduction of next-generation typing methods for food - and waterborne diseases in the EU and EEA (available at: http://ecdc.europa.eu/en/publications/Publications/food-and-waterborne-diseases-next-generation-typing-methods.pdf)

[9] Lakicevic B, Velebit B, Jankovic V, Spiric D, Baltic T, Mitrovic R and Babic J 2014 Taq Man Real Time PCR detection of *Listeria monocytogenes*: a study of enrichment incubation time affecting sensitivity in experimental dry fermented sausages *Tehnologija mesa* **1** 60-5

[10] Hamon M, Bierne H and Cossart P 2006 *Listeria monocytogenes*: a multifaceted model. *Nat. Rev. Microbiol.* **4** 423-34

[11] Orsi R H, den Bakker H C and Wiedmann M 2011 *Listeria monocytogenes* lineages: genomics, evolution, ecology, and phenotypic characteristics. *Int. J. Med. Microbiol.* **301** 79-96

[12] Denny J and McLauchlin J 2008 Human *Listeria monocytogenes* infections in Europe – an opportunity for improved European Surveillance *Euro. Surveill.* **13** 8082

[13] Scallan E, Hoekstra R M, Angulo F J, Tauxe R V, Widdowson M A, Roy S L and Griffin P M 2011 Foodborne illness acquired in the United States–major pathogens. *Emerg. Infect. Dis.* **17** 7-15

[14] European Food Safety Authority 2014 The European Union summary report on trends and sources of zoonoses, zoonotic agents and food-borne outbreaks in 2012 *EFSA J.* **12** 3547

[15] Kwong J C, Mercoulia K, Tomita T, Easton M, Li H Y, Bulach D M, Stinear T O, Seemann T and Howden B P 2016 Prospective Whole-Genome Sequencing enhances national surveillance of *Listeria monocytogenes*. *J. Clin. Microbiol.* **54** 333-42

[16] Smidt P G 2013. *Listeria* Sequencing in Real time (available at: https://www.aphl.org/conferences/proceedings/Documents/2013/InFORM/025a%20-%20Gerner-Smidt%20final.pdf)

[17] Food and Agriculture Organization of the United Nations 2016 Applications of Whole Genome Sequencing in food safety management *Technical background paper* (available at: http://www.fao.org/3/a-i5619e.pdf)

[18] Nastasijevic I, Milanov D, Velebit B, Djordjevic V, Swift C, Painset A and Lakicevic B 2017 Tracking of *Listeria monocytogenes* in meat establishment using Whole Genome sequencing as a food safety management tool: A proof of concept *I. J. Food Microbiol.* **257** 157-64

Sodium and potassium content and their ratio in meatballs in tomato sauce produced with lower amounts of sodium

S Lilić[1], D Nikolić[1], Z Pejkovski[2], B Velebit[1], B Lakićević[1], V Koričanac[1] and D Vranić[1]

[1] Institute of Meat Hygiene and Technology, Kaćanskog 13, Belgrade, Republic of Serbia

[2] Faculty of Agricultural Sciences and Food, Bul. Aleksandar Makedonski bb, Skopje, FYR Macedonia

E-mail: slobodan.lilic@inmes.rs

Abstract. The goal of this study was to examine the possibility of partial replacement of sodium chloride with potassium chloride and ammonium chloride, with the target of achieving less sodium content in meatballs and tomato sauce as well as achieving a better Na:K ratio. The trial consisted of five groups. In the control group of meatballs and sauce, only sodium chloride was added. In group 1, half of the sodium chloride was replaced with potassium chloride related to control group while in group 2 one third of the sodium chloride was replaced with potassium chloride. In group 3, one third of the sodium chloride was replaced with ammonium chloride, and in group 4, sodium chloride was reduced to half the amount in the control group, and 1 g (0.25%) of ammonium chloride was also added. All products were acceptable according to sensory analyses. The largest reductions of sodium content were 44.64%, achieved in meatballs from group 1 and 50.62% in tomato sauce from group 4 in relation to meatballs and tomato sauce from control group. The highest Na:K ratio was calculated in meatballs and tomato sauce from control group, 2.88 and 4.39, respectively. The best Na:K ratio was in meatballs and tomato sauce from group 1, 0.60 and 0.92, respectively, in which half of sodium chloride was replaced with potassium chloride. However, in meatballs and tomato sauce from group 4, with only half the amount of sodium chloride related to control group, the Na:K ratio was worse because in these products, potassium chloride was not added.

1. Introduction

Excessive dietary sodium intake is linked with negative health influences, particularly with essential hypertension and consequential cardiovascular disorders. Intake of sodium via food, in many cases, exceeds the recommendation of the World Health Organization (WHO), and because of that, there are many directives issued from WHO concerning how to reduce sodium content in food. The meat industry is often a target of investigations connected with this topic. There are many studies about reducing sodium chloride and sodium content in meat products, which is done mostly by partial replacement of sodium chloride with other chloride salts. The meat industry is an important producer of ready-to-eat meals prepared or cooked in advance, with no further cooking or preparation required before being eaten. These ready-to-eat meals have become an important choice for modern consumers, with respect to the fast lifestyle of modern societies.

There are not much literature data on reducing the sodium content in ready-to-eat meals. Many investigations are focused on reducing sodium content in cooked meat products, but the real challenge is sodium reduction in meat products that are not thermally treated (dry fermented sausages, dry ham). Since it is the only salt with a clearly salty taste, in food, sodium chloride cannot be totally replaced with other salts. However, it can be partially replaced, and to this purpose, potassium chloride and less often, other chloride salts are used [1]. Besides potassium chloride, magnesium and calcium salts and ascorbates are most commonly used as replacers [2].

The need to reduce sodium in meat products and generally in food will be an aim of the food industry in the future; fast food chains will also have to address this issue, even if people think the amount of salt consumed via fast food is not so large [3]. Nonetheless, salt replacers present a difficult problem because of their degradation of desirable sensory characteristics, including texture and, of course, salty taste [4].

The WHO recommendation for dietary sodium intake is less than 2000 mg daily [5]. In the US, the recommendation is less than 2300 mg daily, but an adequate amount is 1500 mg daily [5]. The WHO guideline for dietary potassium intake is more than 3510 mg daily [6], but in the case of the US and Canada, this is set at 4700 mg daily for adults [7]. According to these recommendations, the Na:K ratio would be less than 0.49 (2300 mg Na daily and 4700 mg K daily) for healthy individuals.

Scientific data suggest that the dietary Na:K ratio is more associated with an increased risk of hypertension and cardiovascular diseases.

Accordingly, the goal of this study was to examine the possibility of partial replacement of sodium chloride with potassium chloride and ammonium chloride with the target of achieving less sodium content in meatballs and tomato sauce, as well as achieving a better Na:K ratio.

2. Materials and methods

The trial consisted of five groups. In the control group of meatballs and sauce, only sodium chloride was added. In group 1, half of the sodium chloride was replaced with potassium chloride while in group 2 one third of the sodium chloride was replaced with potassium chloride related to the control group. In group 3, one third of the sodium chloride was replaced with ammonium chloride and in group 4, sodium chloride was reduced by one half compared with the control, while 1 g (0.25%) of ammonium chloride was also added.

2.1. Meatball preparation

Meatballs were prepared from minced pork leg meat (grind plate 3 mm) purchased from a local market. Meat was well mixed with the ingredients presented in Table 1 to achieve optimal consistency to form into round shapes. Prepared meatballs were briefly fried in a thin layer of sunflower oil.

Table 1 Composition of meatballs, g

Group	Minced pork (leg)	Sodium chloride	Potassium chloride	Ammonium chloride	Ground garlic
Control	400	6.00	-	-	2.00
1	400	3.00	3.00	-	2.00
2	400	4.00	2.00	-	2.00
3	400	4.00	-	2.00	2.00
4	400	3.00	-	1.00	2.00

2.2. Tomato sauce preparation

Sauces were prepared from tomato juice (Tomatino classic, Polimark, Serbia) and the ingredients presented in Table 2. A sauce was prepared from flour fried in sunflower oil for about 1 minute, and after that, water, tomato juice, salt/salt mixture and sugar were added. The sauces were simmered for 10 minutes.

Table 2 Composition of tomato sauce, g

Group	Tomato juice	Water	Sunflower oil	Flour	Sugar	Sodium chloride	Potassium chloride	Ammonium chloride
Control	400	400	6.00	6.00	6.00	6.00	-	-
1	400	400	3.00	3.00	3.00	3.00	3.00	-
2	400	400	4.00	4.00	4.00	4.00	2.00	-
3	400	400	4.00	4.00	4.00	4.00	-	2.00
4	400	400	3.00	3.00	3.00	3.00	-	1.00

2.3. Meal preparation

Meatballs were cooked in prepared tomato sauce for 45 minutes. Half an hour after cooking, the product was presented to sensory assessors for evaluation and sent to determine the sodium and potassium content.

2.4. Determination of Na and K

Aliquots of approximately 0.3 g were transferred into Teflon vessels and 5 mL nitric acid (p.a. Sigma-Aldrich, St. Louis, MA, USA) and 1.5mL hydrogen peroxide (30%, p.a., Merck & Co., Whitehouse Station, New Jersey, USA.) were added. The sample solutions were quantitatively transferred into disposable flasks and diluted, then digested using a microwave program that consisted of three steps as follows: 5 min from room temperature to 180°C, 10 min hold 180°C, 20 min vent. After cooling at room temperature, digests were diluted to 100 mL with deionized water (ELGA).

The analysis was performed by inductively-coupled plasma mass spectrometry (ICP-MS). Measurements were performed using the instrument "iCap Q" (Thermo Scientific, Bremen, Germany), equipped with collision cell and operating in kinetic energy discrimination (KED) mode. The following isotopes were measured: ^{39}K and ^{23}Na.

Torch position, ion optics and detector settings were adjusted daily using tuning solution (Thermo Scientific Tune B), in order to optimize measurements and minimize possible interferences. For the qualitative analysis of the samples, a five-point calibration curve (including zero) was constructed for each isotope in the concentration range of 0.1 – 2.0 mg/L. An additional line of the peristaltic pump was used for on-line introduction of multi-element internal standard (^{6}Li, ^{45}Sc – 10 ng/mL; ^{71}Ga, ^{89}Y, ^{209}Bi – 2 ng/mL) covering wide mass range. Concentrations of each measured isotope were corrected for response factors of both higher and lower mass internal standard using interpolation method.

The quality of the analytical process was controlled by the analysis of the standard reference material (NIST SRM 1577c). Measured concentrations were within the range of the certified values for all isotopes.

3. Results and discussion

All samples of meatballs and tomato sauce as well as whole meals were sensorially acceptable [6].

Results of sodium and potassium content in meatballs and tomato sauce are presented in Table 3. Sodium content in meatballs and tomato sauce was directly influenced by added sodium chloride. The highest sodium content was determined in meatballs and tomato sauce from control group (8537.13 mg/kg and 12750.89 mg/kg) and the lowest was in meatballs and tomato sauce from group 4 (4901.47 mg/kg and 6296.02 mg/kg). The highest sodium content in tomato sauce from the control group was also the result of high sodium content in tomato juice, which was found to be 5410.03 mg/kg. The largest reductions of sodium content were 44.64% achieved in meatballs from group 1 and 50.62% in tomato sauce from group 4 (in the relation to meatballs and tomato sauce from control group). The smallest reduction of sodium was in group 3 meatballs as well as in tomato sauce (26.71% and 24.29%, respectively), in which only one third of sodium chloride was replaced with ammonium chloride

Table 3. Sodium and potassium content in meatballs and tomato sauce, mg/kg

Group	Meatballs		Tomato sauce	
	Sodium	Potassium	Sodium	Potassium
Control	8537.13	2960.59	12750.89	2902.90
1	4726.03	7866.11	7470.57	8156.76
2	5908.05	5654.13	8384.73	6363.66
3	6256.80	3319.82	9653.45	2725.61
4	4901.47	3120.96	6296.02	2102.97

Sodium:potassium ratios (Na:K) of the meatballs and tomato sauce are presented in Table 4. The largest ratio was calculated in meatballs and tomato sauce from control group, 2.88 and 4.39, respectively. The best Na:K ratio was in meatballs and tomato sauce from group 1, 0.60 and 0.92, respectively, in which one half of sodium chloride was replaced with potassium chloride. However, in meatballs and tomato sauce from group 4, with one half the amount of sodium chloride related to control group, the Na:K ratio was worse because in these products, potassium chloride was not added.

Table 4. Sodium:potassium (Na:K) ratios in meatballs and tomato sauce

Group	Meatballs	Tomato sauce
Control	2.88	4.39
1	0.60	0.92
2	1.04	1.32
3	1.88	3.54
4	1.57	2.99

4. Conclusion

All the prepared meatballs and tomato sauce as well as whole meals were sensorially acceptable.

The lowest sodium content was in meatballs and tomato sauce from group 4 (4901.47 mg/kg and 6296.02 mg/kg).

The largest reductions of sodium content were 44.64% achieved in meatballs from group 1 and 50.62% in tomato sauce from group 4, in relation to meatballs and tomato sauce from control group.

The best Na:K ratio was in meatballs and tomato sauce from group 1, 0.60 and 0.92, respectively, in which one half of the sodium chloride was replaced with potassium chloride.

Acknowledgements

This study was funded by grants TR 31083 and III 46009 from the Ministry of Education, Science and Technological Development, Republic of Serbia.

References

[1] Guàrdia M D, Guerrero L, Gelabert J, Gou P and Arnau J 2006 *Meat Sci.* **73** 484-90
[2] Ruusunen M and Puolanne E 2005 *Meat Sci.* **70** 531-41
[3] Moran A J, Ramirez M and Block J P 2017 *Appetite* **113** 155-61
[4] Kamenik J, Saláková A, Vyskočilová V, Pechová A and Haruštiaková D 2017 *Meat Sci.* In press, doi: 10.1016/j.meatsci.2017.05.010.
[5] World Health Organization 2012 *Guideline: Sodium intake for adults and children* Geneva ed

WHO Library Cataloguing-in-Publication Data
[6] World Health Organization 2012 *Guideline: Potassium intake for adults and children* Geneva ed WHO Library Cataloguing-in-Publication Data
[7] Food and Nutrition Board 2005 *Dietary reference intakes for water, potassium, sodium, chloride and sulfate* Washington (DC): National Academy Press
[8] Perez V and Chang E T 2014 *Adv. Nutr.* **5** 712-41
[9] Lilić S, Karan D, Jovanović J, Babić J, Borović B, Stefanović S, Rašeta M 2017 *Meat technology* 58 47-51

Colour and fat content as intrinsic cues for consumers attitudes towards meat product quality

M Ristić[1], K Troeger[1], J Đinović-Stojanović[2], N Knežević[3] and M Damnjanović[3]

[1] BAFF, Former Federal Centre for Meat Research, Kulmbach, Germany (retired researchers)
[2] Institute of Meat Hygiene and Technology, Kaćanskog 13, 11000 Belgrade, Serbia
[3] Zlatiborac Company d.o.o., Mojkovačka 58, 11136 Belgrade, Serbia

E-mail: ristic.rc-kulmbach@t-online.de

Abstract. The aim of this study was to evaluate consumers' attitudes towards sensory properties of chicken, royal and beef salami, meat products from Zlatiborac Meat Company. Sensory evaluation was performed by Serbian consumers (n=1018) in four retail stores (Delhaize) in Belgrade. Consumers were asked for their preference concerning the colour and fat of three selected salami and then completed questionnaire of socio-demographic information including eating behaviour. Selected smoked meat products were evaluated in the DLG Test Center Food, Germany. Consumers, at all education levels and in all age groups, evaluated colour as good and fat as sufficient with a significantly ($p<0.05$) higher percentage in comparison with other offered answers. All smoked products passed the DLG tests and received "DLG award winner" medals in Gold (73%) or Silver (27%).

1. Introduction

Consumers' evaluation of meat and meat products is becoming the critical issue for the meat industry because it has direct influence on its profitability. Many studies point out that consumers' perspectives are complex and include many factors in determination of meat quality and its acceptance [1-3]. Consumers first evaluate and then purchase. Meat must look good to consumers before satisfying their palate when they decide to buy it. Appearance has a great influence on how meat is valued by the consumers. The quality and width of the sensory evaluation were significantly influenced by traditional eating habits [4]. Response of the consumer is based on sensory properties of the meat and meat products [5], but it is also associated with other factors, such as the habits of a particular region, nationality, age and gender, religion, education, socio-economic status, psychological motives (symbolism of food, advertising, brand name, etc.) and physiological motives (thirst, hunger, deficit, health condition, etc.) [3,6].

Consumer perception on meat has traditionally been largely based on intrinsic cues like colour of the meat, the visible fat and fat. Colour and colour stability are the most important attributes of meat quality and various commercial approaches have been used to meat consumers' expectation [7]. Colour of meat and meat products is the first impression that consumers have of any meat product. It is the single most important factor of meat products, and it influences consumer buying decision and affects their perception of the freshness of the product. Colour and colour stability are the most important attributes of meat quality and various commercial approaches have been used to meet consumers' expectation [8]. From the consumers' point of view, fat content is not a good predictor of meat quality and, in the research area of healthier meat products, a possible trend is to replace high energy density fat in formulations with substances providing less energy than fat [9]. According to German Agricultural Society (Deutsche Landwirtschafts-Gesellschaft, DLG), [10] the most important characteristic of meat products during sensory evaluation are appearance, colour, colour maintenance, composition and taste, and so evaluations of these characteristics are multiplied by three and have the most influence on the final assessment of the products. Figure 1 shows the most common deficiencies for international smoked meat evaluated by DLG experts [11]. The main deficiencies were related to their external preparation, colour, consistency, door and taste.

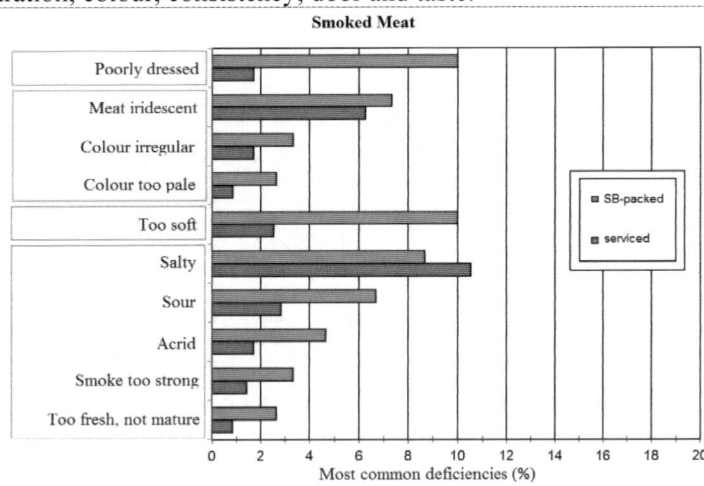

Figure 1. Most common deficiencies for smoked meat evaluated by DLG experts (%), [11].

For the meat industry, it is very important to fully understand consumers' cues because they have a direct influence on profitability. Thus, science and innovation play an important role between the industry and consumer concerns and expectations [12] in order to enhance these cues in existing or new products. To the best of our knowledge, consumer acceptance data is very limited for all meat products produced in Serbia. This pilot study aimed to investigate consumers' attitudes and acceptability towards colour and fat of three types of salami (chicken, royal and beef) of Serbian origin. On the other hand, the results obtained were compared with results of sensory evaluation of selected smoked meat products from Serbia, evaluated by DLG experts in Germany in March 2017.

2. Materials and Methods

Chicken, royal and beef salami were prepared in meat industry (Zlatiborac, Mačkat, Serbia) by using the in-house production protocol. To conduct comparative evaluation of these products, an equal number of samples of chicken, royal and beef salami were manufactured. Chicken salami consisted of chicken breast meat (72.3%) and solid fat (22.8%); royal salami consisted of pork (81%) and solid fat (17%); beef salami consisted of beef (71.8%) and beef tallow (24%). The origin and market name of the products were unknown to the consumers.

Consumer testing was performed in four large retail stores (Delhaize) in February 2017, in Belgrade. A total of 1018 consumers were participated in six days during the period from 10 a.m. to 4 p.m. Consumers were males (37%) and females (63%) older than 18 years of age. For each type of meat product, consumers were asked to express their perception of colour and fat. They were asked to consume all types of salami and answer the following questions: (1) Rate the colour (the offered answers were: good, too pale, too dark); (2) Rate the fat content (the offered answers were: sufficient, too fatty). The authors prepared a questionnaire, modifying a version of DLG-5-points-scheme [10].

IOP Conf. Series: Earth and Environmental Science **85** (2017) 012054 doi:10.1088/1755-1315/85/1/012054

The questionnaire, covering general socio-demographic information related to ages and education levels including eating behaviour, was then put to the participants.

The DLG [10] is the oldest food testing institution in Europe. Around 30000 foodstuffs from Germany and abroad are evaluated every year in the DLG Test Center Food by applying the DLG Quality Tests which are accordance with scientific quality standards [13] and an expert report [14]. Sensory evaluation of eleven smoked meat products from Zlatiborac Meat Company was conducted by DLG experts in Germany, in March 2017 and assessed in accordance with the DLG-5-points-scheme (DLG Quality Test for Ham & Sausage). The DLG-5-points-scheme is a descriptive sensory analysis which included visual (appearance/exterior), haptic (consistency/texture), olfactory (odour) and gustative (taste) criteria of the meat products. Meat products that pass the DLG tests receive a "DLG award winner" medal in Gold, Silver or Bronze. DLG medals are ambassadors for good taste and high quality foods. A product wins a DLG Award in Gold if it is free of faults in the sensory test and satisfies all further quality parameters (5.00 points). If DLG points are between 4.60-4.99 and 4.10-4.59 points, products win the DLG Award in Silver and DLG Award in Bronze, respectively.

Statistical evaluation of the data was conducted by applying ANOVA analysis [15].

3. Results and discussion

3.1. Consumers' evaluation

Results for consumers' attitudes in relation to education and age are shown in Tables 1 and 2.

Table 1. Consumers' attitudes (%) in relation to education
(The questions were: rate the colour and fat content).

Education	Perception of	The offered answers	Salami type Chicken	 Royal	 Beef
BS-Basic School		*Good*	96.4	98.8	72.3
	Colour	*Too pale*	3.6	1.2	-
		Too dark	-	-	27.7
	Fat	*Sufficient*	98.8	78.3	95.2
		Fatty	1.2	21.7	4.8
HS-High / secondary School		*Good*	96.5	99.0	92.2
	Colour	*Too pale*	2.9	0.6	-
		Too dark	0.6	0.4	7.8
	Fat	*Sufficient*	87.7	84.8	96.3
		Fatty	12.3	15.2	3.7
BD-Bachelor Degree		*Good*	96.9	97.9	89.8
	Colour	*Too pale*	2.6	1.9	0.7
		Too dark	0.5	0.2	9.5
	Fat	*Sufficient*	90.3	83.6	96.4
		Fatty	9.7	16.4	3.6

Table 1 shows that consumers of all education levels evaluated colour of Royal salami with the highest percentage of the best offered answers, among the three salami types (chicken, royal and beef). Consumers with Basic School evaluated fat content of chicken salami as sufficient (98.8%) with the highest percentage in comparison with royal (78.3%) and beef (95.2%) salamis. On the other hand, participants with High secondary School and Bachelor degree evaluated fat of beef salami as sufficient with the highest percentage (96.3% and 96.4%, respectively). Noticeably, 27.7% of consumers with basic school education evaluated colour of beef salami as too dark, while consumers with higher education levels evaluated colour as too dark with around 3-3.5-fold lower percentages (HS – 7.8% and BD – 9.5%). Consumers, at all education levels, evaluated colour as good and fat as sufficient with significantly ($p<0.05$) higher percentage in comparison with other offered answers.

The youngest consumers (18-29 years), (Table 2) evaluated colour of chicken and royal salami as good (97.6%), while fat of chicken and beef salami was evaluated as sufficient (95.1%) with the highest percentage. Older consumers, in the remaining four age groups (Table 2), evaluated royal salami as salami with good colour (98%-100%) and fat of beef salami as sufficient (94.1%-98.4%) with the highest percentage. Around twice the percentage of the oldest age group of participants evaluated fat content of chicken and royal salami as fatty (11.3% and 23.2, respectively) in comparison to the youngest participants (4.9% and 11.6%, respectively). Age (i.e. experience) and interest of older population in healthier lifestyles could be a reason for these results. Consumers, in all age groups, evaluated colour as good and fat as sufficient with significantly ($p<0.05$) higher percentage in comparison with other offered answers.

Table 2. Consumers' attitudes (%) in relation to age
(The questions were: rate the colour and fat content).

Age (years)	Perception of	The offered answers	Salami type Chicken	Royal	Beef
18-29		*Good*	97.6	97.6	90.2
	Colour	*Too pale*	1.8	2.4	0.6
		Too dark	0.6	-	9.1
	Fat	*Sufficient*	95.1	88.4	95.1
		Fatty	4.9	11.6	4.9
30-39		*Good*	96.7	99.0	88.5
	Colour	*Too pale*	2.4	1.0	-
		Too dark	1.0	-	11.5
	Fat	*Sufficient*	88.5	84.7	96.7
		Fatty	11.5	15.3	3.3
40-49		*Good*	96.9	98.0	90.6
	Colour	*Too pale*	2.7	1.2	0.8
		Too dark	0.4	0.8	8.6
	Fat	*Sufficient*	89.0	85.9	98.4
		Fatty	11.0	14.1	1.6
50-59		*Good*	94.6	98.2	87.8
	Colour	*Too pale*	5.0	1.4	-
		Too dark	0.5	0.5	12.2
	Fat	*Sufficient*	88.3	82.4	94.1
		Fatty	11.7	17.6	5.9
60 and older		*Good*	98.2	100.0	91.1
	Colour	*Too pale*	1.8	-	-
		Too dark	-	-	8.9
	Fat	*Sufficient*	88.7	76.8	96.4
		Fatty	11.3	23.2	3.6

3.2. DLG evaluation

As in previous years (2009-2016), in March 2017, Zlatiborac Meat Company analysed smoked meat products in the DLG Test Center Food. All products passed the DLG tests and receive "DLG award winner" medals in Gold (73%) or Silver (27%). The deficiencies of smoked meat products win the DLG Award in Silver were marked as "dry edge formation", "not regularly cut off" and "appearance of cut unclear". All these deficiencies were related to their appearance but not to their quality. In Dederer and Hillgärtner's study [16], Zlatiborac's meat products, as the only Meat Company from Serbia, were assigned as non-EU members (as well as products from South Korea, Japan and Thailand), with good DLG evaluations. The DLG evaluation and its medals presented in this study and earlier [17] strongly indicate that Zlatiborac Meat Industry has good marketing opportunities for placement of products on the EU market.

4. Conclusion

This study has contributed to display transparency to the public; it made it possible to gather reliable data on the acceptance and attitudes of consumers towards specified sensory properties of the meat products examined. As the results showed, consumers were satisfied with colour and fat of chicken, royal and beef salami. Also, it was observed that older consumers were likely more aware of health aspects of food products they purchase, which was reflected in their sensory evaluation of meat products. By providing direct transparency from producer to consumers, the meat industry will be able to effectively satisfy consumer needs.

Acknowledgment

The authors would like to thank Mr Dejan Djurdjevic from Zlatiborac Meat Company, Serbia, for his help in sensory evaluation of the meat products.

References

[1] Grunert K G, Bredahl L, Bruns K 2004. *Meat Sci.* **66** 259
[2] Troy D J, Kerry J P 2010. *Meat Sci.* **86** 214
[3] Henchion M M, McCarthy M, Resconi V C 2017. *Meat Sci.* **128** 1
[4] Popov-Raljic J 1999. *Tehnologija i kvalitet gotove hrane*. Tehnoloski fakultet, Novi Sad
[5] Ristic M 2010. *Fleischwirtschaft* **90** 99
[6] Joksimovic J 1977. *Osnovi kontrole i upravljanja kvalitetom u proizvodnji hrane*. Privredni Pregled, Beograd
[7] Renerre M 1990. *Int. J. Food Sci. Technol.* **25** 613
[8] Wideman N, O'Bryan C A, Crandall P G 2016. *Worlds Poult. Sci.* J. **72** 353
[9] Varga-Visi E, Toxanbayeva B 2017. *Acta Aliment.* **46** 181
[10] www.dlg.org, German Agricultural Society
[11] Lautenschläger R, Thumel H, Hillgärtner K 2016. *Fleischwirtschaft* **10**, 47
[12] Grunet KG 2006. *Meat Sci.* **74** 149
[13] DIN EN ISO/IEC 2015 **17065**
[14] DIN EN ISO/IEC 2015 **17024**
[15] www.vassarstats.net
[16] Dederer I, Hillgärtner K 2017. *Fleischwirtschaft* **2** 51
[17] Đinovic-Stojanovic J, Troeger K, Ristic M, Knezevic N, Damnjanovic M 2017 *J. Agr. Sci. Food Tech.* **3** 22

The effect of essential oil from sage (*Salvia officinalis* L.) herbal dust (food industry by-product) on the microbiological stability of fresh pork sausages

B Šojić[1], P Ikonić[2], B Pavlić[1], Z Zeković[1], V Tomović[1], S Kocić-Tanackov[1], N Džinić[1], S Škaljac[1], M Ivić[1], M Jokanović[1] and T Tasić[2]

[1]University of Novi Sad, Faculty of Technology Novi Sad, Bulevar cara Lazara 1, 21000 Novi Sad, Serbia
[2]University of Novi Sad, Institute of Food Technology, Bulevar cara Lazara 1, 21000 Novi Sad, Serbia

E-mail: sojic@tf.uns.ac.rs

Abstract. The effect of essential oil obtained from sage (*Salvia officinalis* L.) herbal dust (a food industry by-product) (SEO), on the pH value, microbiological stability and sensory properties of fresh pork sausages prepared without chemical additives was evaluated during 8 days of aerobic storage at 3±1°C. The addition of SEO significantly ($p<0.05$) reduced the microbial growth in fresh pork sausages. Moreover, SEO added at a level of 0.05 μL/g had no negative effect on sensory properties of this meat product. Hence, the results of this study showed significant antimicrobial activity of SEO obtained from sage filter tea processing by-products and the potential for utilising SEO in fresh pork sausages in order to enhance their stability and safety.

1. Introduction

Fresh pork sausages are among the most common and the most popular processed meat products all around the world [1]. Due to their very high water activity, relatively high fat content, comminuted structure of raw materials, high total number of microorganisms and lack of thermal processing, these products are characterized by a short shelf life. Spoilage of these products can be caused by microbial contamination and lipid oxidation, leading to development of unacceptable sensory characteristics and even foodborne diseases. Hence, the inhibition of microbial growth and delay of lipid oxidation are primary goals that can significantly extend shelf life [1,2]. In order to reduce oxidative changes and to prevent bacterial growth, several synthetic food additives are regularly used by meat processors. However, in recent years due to increasing consumer awareness about potentially toxic effects and health issues, the use of nitrites and synthetic antioxidants (butylated hydroxytoluene – BHT; butylated hydroxyanisole – BHA; tertiary butylhydroquinone – TBHQ) has decreased, while demand for natural additives has rapidly increased [1,3,4,5]. Many natural antioxidants also exhibit antimicrobial activity, and thus, have the advantage of being readily accepted by both consumers and meat processors [1,6]. However, natural antioxidants are often more expensive and less effective than synthetic ones. Consequently, increasing attention is recently being paid to the extraction of antioxidants from agro-food industry by-products [7,8]. The aim of this study was to determine the antimicrobial effect of sage essential oil (SEO), obtained from sage tea processing by-products, when utilised in fresh pork sausage during refrigerated storage.

2. Materials and Methods

2.1. Plant material

Sage (*Salvia officinalis* L.) originated from Montenegro and was kindly donated by a local filter tea factory producing herbal teas (Fructus DOO, Bačka Palanka, Serbia). Sage herbal dust was obtained as by-product in the filter tea factory while all processing steps and applied unit operations were described elsewhere [9]. Plant material had a mean particle size of <0.315 mm and moisture content of 7.24±0.05%. Conventional (hydrodistillation) and novel (supercritical fluid extraction – SFE) extraction techniques were used for recovery of SEO. The official procedure from the fourth edition of the Yugoslavian Pharmacia was applied for hydrodistillation. Briefly, 20 g of sage herbal dust and 400 mL of water were mixed in a 1 L round flask connected with Clevenger-type apparatus. Hydrodistillation was performed for 2 h and essential oil was separated from the aqueous phase after determination of its yield (%).

2.2. Preparation of fresh pork sausage

Fresh pork sausages (of the type called in Serbian *Petrovská klobása*) were produced in the meat processing pilot plant within the Institute of Food Technology Novi Sad (FINS). Pork shoulder and back fat were minced using an electric meat grinder with an 8 mm grinding plate. The sausage batter was obtained by mixing minced pork (80%) and back fat (20%) with salt (1.80%), sweet paprika powder (1.00%), red hot paprika powder (0.70%), caraway (0.20%) and garlic paste (0.07%), in a meat mixer. The amount of seasonings was calculated in relation to minced meat and back fat weight. The resulting mixture was divided into four batches. SEO was added separately to three batches, at concentrations of 0.05 μL/g (SEO1), 0.075 μL/g (SEO2) and 0.1 μL/g (SEO3). The remaining batch was used as the control (C). All batches were stuffed into natural casings (pig small intestines; Ø ≈ 32 mm). Sausages were stored at 3±1°C in the dark for 8 days.

2.3. Samples

Samples taken at distinct periods of storage comprised three randomly selected sausages from each batch after 0, 2, 4, 6 and 8 days. Analyses were carried out on the day of sampling, and were conducted in duplicate for each sample.

2.4. pH determination

The pH of samples was measured using the portable pH meter Testo 205 (Testo AG, USA) equipped with a combined penetration tip with temperature probe. The pH meter was calibrated before the readings using two buffer solutions (pH=4.00±0.05 and pH = 7.00±0.01 at 20±2 °C).

2.5. Microbiological analysis

Microbiological analyses were performed on three samples from each group of the fresh pork sausages in duplicate. Twenty grams of samples were homogenized for 10 minutes at 200 rpm (Unimax 1010, Heidolph, Germany) in 180 mL 1 g/L buffered peptone water (Merck, Darmstadt) and then serial decimal dilutions were prepared (up to 10^{-3}). Pour plates were prepared with 1 mL of each dilution in separate sterile Petri plates, and with appropriate media depending on the type of microorganism studied. The following microbial analyses were performed: total number of aerobic mesophilic bacteria (TBC), *Salmonella spp*., *Escherichia coli*, and *Listeria monocytogenes* [4]. Results were expressed as log cfu/g.

2.6. Sensory analysis

Sensory evaluation of the investigated sausages was performed by ranking tests [10]. Sausages were taken from refrigerated storage (3°C) and baked in an oven at 165°C for 1h**.** The sausages were served warm (50-60°C) and analysed by a panel of 70 consumers. Consumers were asked the following: "Please rank the samples in numerical order according to your preference" (from like extremely – 1 to dislike extremely – 7). Consumers between 19 and 65 years old were students and staff members of the Faculty of Technology, Novi Sad.

2.7. Statistical analysis
Statistical analysis was carried out using STATISTICA 12.0 (StatSoft, Inc., Tulsa, OK, USA). All data were presented as means with their standard deviations (mean±SD). Variance analysis (ANOVA) was performed, with a confidence interval of 95% ($p<0.05$). Means were compared by Fisher's LSD test.

3. Results and Discussion
The effect of SEO on the pH value of fresh pork sausages is shown in Table 1. At the beginning of storage, pH values ranged from 5.60 to 5.66 in the sausages.

Table 1. Effect of different concentrations of sage essential oil (SEO) on pH value of fresh pork sausage during storage

Storage day	C	SEO1	SEO2	SEO3
0	$5.62±0.04^{Aba}$	$5.60±0.05^{Ba}$	$5.60±0.01^{Ba}$	$5.66±0.02^{Aa}$
2	$5.54±0.03^{Cb}$	$5.58±0.03^{Bca}$	$5.62±0.01^{Ab}$	$5.61±0.02^{ABa}$
4	$5.47±0.02^{Cc}$	$5.52±0.01^{Bb}$	$5.55±0.03^{Abbc}$	$5.57±0.04^{Ab}$
6	$5.46±0.01^{Cc}$	$5.45±0.01^{BCc}$	$5.53±0.03^{Acd}$	$5.47±0.01^{Bc}$
8	$5.40±0.02^{Bd}$	$5.40±0.01^{Bd}$	$5.46±0.06^{Ad}$	$5.44±0.03^{ABc}$

Values with different letters $^{(A-C)}$ in the same row are significantly different ($p<0.05$); Values with different letters $^{(a-c)}$ in the same column are significantly different ($p<0.05$).
SEO levels in sausages: 0.05 μL/g (SEO1), 0.075 μL/g (SEO2) and 0.1 μL/g (SEO3).

The results obtained corresponded very well with literature data for this type of sausage [11]. A significant drop of pH was registered in each sausage group during 8 days of refrigerated storage. Most probably, this was result of growth and metabolic activity of lactic acid bacteria, as was previously reported by a number of authors [6, 11]. The microbiological profile of fresh pork sausage during 8 days of storage under refrigeration is shown in Table 2.

Table 2. Effect of different concentrations of sage essential oil (SEO) on total bacterial count (log cfu/g) of fresh pork sausage during storage

Storage day	C	SEO1	SEO2	SEO3
0	$6.90±0.02^{Ac}$	$5.74±0.01^{Ce}$	$5.59±0.01^{De}$	$6.03±0.05^{Bc}$
2	$6.10±0.01^{Ae}$	$6.00±0.02^{Bd}$	$5.88±0.01^{Cd}$	$5.80±0.02^{Dd}$
4	$6.63±0.03^{Ad}$	$6.38±0.02^{Bc}$	$6.01±0.01^{Dc}$	$6.23±0.01^{Cb}$
6	$7.15±0.03^{Ab}$	$6.84±0.02^{Cb}$	$6.73±0.02^{Db}$	$6.98±0.02^{Ba}$
8	$7.66±0.04^{Aa}$	$7.29±0.01^{Ba}$	$7.15±0.01^{Ca}$	$7.10±0.02^{Da}$

Values with different letters $^{(A-E)}$ in the same row are significantly different ($p<0.05$); Values with different letters $^{(a-c)}$ in the same column are significantly different ($p<0.05$)
SEO levels in sausages: 0.05 μL/g (SEO1), 0.075 μL/g (SEO2) and 0.1 μL/g (SEO3).

The addition of SEO significantly ($p<0.05$) reduced the TBC. Probably, it was the consequence of sage antimicrobial properties [11]. The initial TBC ranged from 5.59 log cfu/g (SEO2) to 6.90 log cfu/g (C). As expected, for all sausages, the TBC significantly ($p<0.05$) increased during 8 days of storage. At the end of storage, TBC was significantly ($p<0.05$) different between the sausages, being in the order as follows: C>SEO1>SEO2>SEO3. None of the three analysed foodborne pathogenic bacteria (Salmonella spp., E. coli, or L. monocytogenes) were detected, neither in control nor in sausages with added SEO.

The sensory attribute of preference of C and SEO fresh pork sausages are shown in Figure 1.

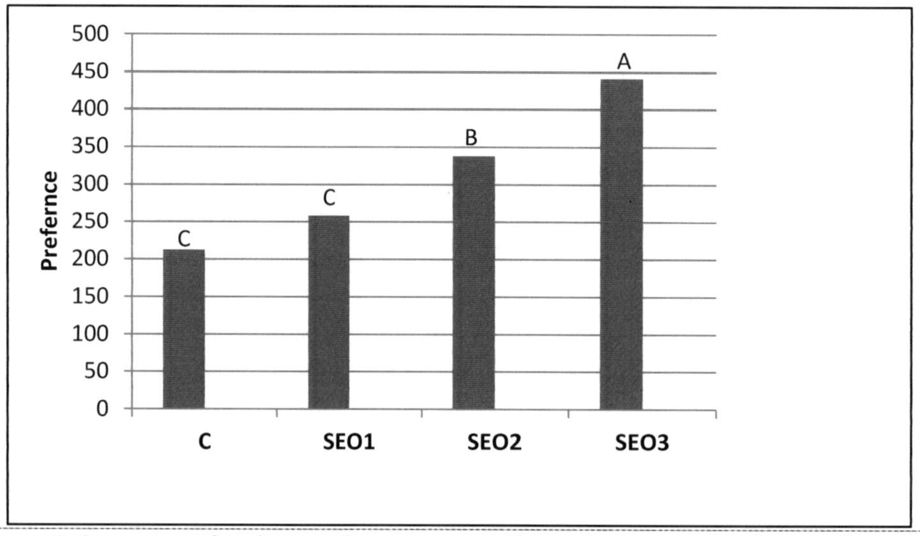

Figure 1. Summary of rank sums for sensory attribute of consumer preference for control sausages and sausages with added SEO

Values with different letters (A-C) are significantly different ($p<0.05$)
SEO levels in sausages: 0.05 μL/g (SEO1), 0.075 μL/g (SEO2) and 0.1 μL/g (SEO3)

SEO addition to fresh pork sausage at the concentration of 0.05 μL/g had no negative effect on the sensory attribute of preference. This was in accordance with literature data [12].

4. Conclusion

The results of this study showed that the utilisation of SEO retarded microbial growth in fresh pork sausages. Use of SEO (at 0.05 μL/g) did not have a negative effect on the sensory attribute of preference by consumers, compared with control sausages. The overall results show that essential oil obtained from sage herbal dust (a food industry by-product) could be successfully applied in the formulation of fresh pork sausages as an antimicrobial agent.

Acknowledgement

The research in this paper was financed by the Provincial Secretariat for Higher Education and Scientific Research, Autonomous Province of Vojvodina, Republic of Serbia No 142-451-3626/2016-01. Also, these results are part of the project No. TR 31032, which is financially supported by the Ministry of Education and Science of the Republic of Serbia.

References

[1] Georgantelis D, Ambrosiadis I, Katikou P, Blekas G and Georgakis S A 2007 *Meat Sci.* **76** 172–181.

[2] da Silveira S M, Luciano F B, Fronza N, Cunha Jr A, Scheuermann G N and Vieira C R W 2014 *LWT - Food Sci. Technol.* **59** 86–93.

[3] Sebranek J G, Sewalt V J H, Robbins K L and Houser T A 2005 *Meat Sci.* **69** 289–296.
[4] Šojić B, Tomović V, Kocić-Tanackov S, Škaljac S, Ikonić P, Džinić N, Živković N, Jokanović M, Tasić T and Kravić S 2015 *Food Control* **54** 282–286.
[5] Šojić B, Tomović V, Jokanović M, Ikonić P, Džinić N, Kocić-Tanackov S, Popović Lj, Tasić T, Savanović J and Živković Šojić N 2017 *Czech J. Food Sci.* **35** 189–193.
[6] Jayawardana B C, Liyanage R, Lalantha N, Iddamalgoda S and Weththasinghe P 2011 *LWT - Food Sci. Technol.* **64** 1204–1208.
[7] Mandic A I, Dilas S M, Cetkovic G S, Canadanovic Brunet J M and Tumbas V T 2008 *Int. J. Food. Prop.* **11** 713–726.
[8] Lorenzo J M, Sineiro J, Amado I R and Franco D 2014 *Meat Sci.* **96** 526–534.
[9] Pavlić B, Vidović S, Vladić J, Radosavljević R, Cindrić M and Zeković Z 2016 *J. Supercrit. Fluids* **116** 36–45.
[10] Lawless H T and Heymann H 2010 Second Edition, Springer, New York, USA.
[11] Sharma H, Mendiratta S K, Agrawal R K, Gurunathan K, Kumar S and Singh T P 2017 *LWT - Food Sci. Technol.* **81** 118–127.
[12] Zhang L, Lin Y H, Leng X J, Huang M and Zhou G H 2013 *Meat Sci.* **95** 145–150.

IOP Conf. Series: Earth and Environmental Science **85** (2017) 012056 doi:10.1088/1755-1315/85/1/012056

Determination of calcium content in mechanically separated meat

A Tasić, J Kureljušić, K Nešić, N Rokvić, M Vićentijević, M Radović and B Pisinov

Institute of Veterinary Medicine of Serbia, Vojvode Toze 14, Belgrade, Republic of Serbia

E-mail: alekstasic79@gmail.com

Abstract. One of the most important control parameters for mechanically separated meat (MSM) is the calcium content, which indicates the presence of bone. The aim of this study was to examine the calcium content in three types of locally-produced MSM from differing animal species. Calcium was quantified in the examined samples of MSM, using indirect titration, and three differing napthol indicators (eriochrome black T, hydroxy naphthol blue and hydroxy naphthol blue disodium salt) were used to track the end point of titration. Each of the three different indicators showed a recovery of more than 90% when examining a certified reference material, and so all three could be successfully used to determine the amount of calcium in MSM. The calcium levels in the MSM samples examined were in the range of 0.050 to 0.100%, with that of beef MSM being the highest. Using a previously published binary logistic regression model for classifying meats as MSM, this corresponded to probabilities of between 0.69 and 0.94, confirming that the three types of MSM examined could indeed be properly classified as MSM.

1. Introduction

Mechanically separated meat (MSM) or mechanically recovered meat are generic terms used to describe residual meat that is separated using machinery from animal bones or carcasses from which most of the meat has previously been removed manually. Mechanical separation involves grinding meat and bones together, as well as passing the mixture through micro sieves or slits to remove bone particles [1,2]. This process enables the use of most of the meat remaining on the bones, which would otherwise be difficult or unprofitable to use. The resulting MSM has the appearance of finely chopped meat. Lipid oxidation is a major factor in reducing the quality and acceptability of meat products, those MSM with a high degree of unsaturation and fat content are more susceptible to lipid oxidation [3]. Meat quality changes can be also perceived by changes in taste, color, texture, nutritional value and the production of potentially toxic compounds [4]. While the amount of calcium is a quality parameter that does not change over time, but the amount above the permissible limit is forbidden in MSM. The chemical composition of MSM is variable due to the natural variation within and between animal species, diet regimens, age of slaughtered animals, meat pieces, bones, pre-treatment of bone (freezing), and machine type in the production process [5]. Poor installation of equipment can also lead to the production of an unacceptable particle size, which affects the quality of products that contain MSM. During the mechanical separation of the bones from the meat, it is inevitable that some bone particles pass into the MSM. These particles contain high levels of calcium, which is reflected in higher calcium levels in MSM than in the meat [6]. Calcium increases during calcification processes, it varies with bone type, as well as with species, feeding or age of the slaughtered animals.

Determination of calcium content in MSM is an indicator of the amount of bone in the product, and can be used to control the yield of mechanical separation processes [7]. Calcium makes up about 37% of the content of bone ash, so that the content of calcium and ash can be used equally to estimate bone content. The calcium content is frequently used as one of the criteria to identify MSM, although the starting material can also affect the amount of calcium in MSM [8]. Testing and control of MSM is important because the product is used in the composition of some sausages, salami and dry soups, in amounts up to about 24% [11]. With increasing use of MSM in such products, besides standard determination of nutritional value, it is also important to control the amount of calcium present.

Chemical methods for the determination of mineral matter or ash in MSM are indirect methods for determining the percentage of bone in the product, but at the same time they are used as a parameter for recognizing different types of MSM. Different methods are used to determine calcium. Calcium is determined using the atomic absorption spectrophotometry method (ISO 6869: 2000) and the AOAC method, using inductively coupled plasma-optic emission spectrometry [9]. Determination of calcium in a food matrix can also be performed using a standard titration method. The method especially standardised for the determination of calcium in MSM is the AOAC Official Method [10]. The advantages of this method are that it is simple and quick, but also it does not require the use of expensive instrumentation.

The aim of this study was to determine the amount of calcium present in different types of locally-produced MSM using the AOAC titration method [10] and compare the results with a previously-published regression analysis for classifying product as MSM [7]. In addition, three different napthol indicators were used to determine the calcium content of a reference MSM.

2. Materials and Methods

The amount of calcium was determined in MSMs (beef, chicken and turkey meat) which had been produced for further use in food products in Serbia. The certified reference material used was beef MSM (DRRR, Germany). All chemicals used were p.a. purities. Indirect titration with the disodium salt of ethylenediaminetetraacetic acid was used, and the titrations were conducted using a calcium carbonate solution. Amounts of MSM (10 g) were digested in a solution of hydrochloric acid, the solution cooled and filtered through filter paper. The aliquoted digests were diluted volumetrically with distilled water and potassium hydroxide and potassium cyanide were added; the final pH was 12.5±0.2. The solutions were titrated, with the titration end point being the appearance of violet color for the indicator hydroxyinatol blue, but two other indicators were used in parallel: disodium hydroxyphenol blue (end point a purple colour) and eriochrome black T (end point a red colour).

The limit of detection (LOD) and limit of quantification (LOQ) were calculated from measurements using blank samples. Standard deviations were calculated using ten independent measurements of blank samples. LOD was defined as 3× the standard deviation. The LOQ was defined as 10× the standard deviation.

3. Results and Discussion

Table 1 shows the mean calcium content of different MSM matrices after titration with different indicators.

Table 1. Calcium content (mg/kg) measured in different types of MSM (10 replications)

Replication	Beef	Chicken	Turkey
1	890	750	680
2	995	780	620
3	895	680	690
4	890	700	700

5	1000	760	850
6	990	900	860
7	1000	790	800
8	980	500	780
9	990	550	760
10	900	800	770
Mean value (mg/kg)	953	721	790

The mean calcium content for beef MSM was 953 mg/kg, or in the range from 0.089 to 0.100%, and this was the highest calcium level detected in the MSMs examined. The quantities of calcium measured in chicken and turkey MSM ranged from 0.05 to 0.09%; and 0.062 to 0.086%, respectively.

In order to validate the method and determine the recovery, the calcium content in the certified reference material was determined using the three different indicators, and the results obtained are shown in Table 2. The certified reference material had a calcium content of 830 mg/kg (ppm), as stated by the manufacturer. Among the three indicators, Eriochrome black T indicator produced the greatest accuracy when used as a titration indicator (Table 2). However, it should be noted the pH range of this indicator is 7.5 to 10.5, which is not entirely suitable for this analysis. The disodium salt of hydroxy naphthol blue, on the other hand, more satisfies the pH range for the determination, i.e. the pH range of this indicator is from 12 to 13. In spite of that, this indicator recovered slightly lower calcium levels in the reference material (Table 2).

Table 2. Calcium levels (mg/kg) in certified reference material detected by titration using three different indicators, and calculated means, standard deviations, relative standard deviations (RSD) and biases

Replication	Eriochrome black T	Hydroxy naphthol blue	Hydroxy naphthol blue disodium salt
1	850	850	820
2	860	870	855
3	820	830	890
4	845	880	845
5	865	900	865
6	870	840	865
7	865	880	870
8	885	910	890
9	835	860	855
10	830	820	845
Mean value (mg/kg)	853	864	860
Standard deviation (mg/kg)	20.17	29.51	21.21
RSD, %	2.37	3.42	2.47
Bias, %	2.77	4.10	3.61

The indicators used in the study are all naphthols, and showed good agreement and high accuracy in the determination of calcium (Table 2). Recoveries were greater than 90%, which confirms that all three tested indicators could be successfully used to determine the amount of calcium in MSM.

It is important to emphasize that pH has a high impact on the titration reaction and the formation of a complex compound. Also, cyanide ions that mask the presence of other ions in solution and enable better determination are of great importance in determining calcium levels.

The LOD for any analytical procedure, the point at which analysis is just feasible, is determined by a statistical approach based on measuring replicate blank (negative) samples or by an empirical approach, consisting of measuring progressively more dilute concentrations of analyte. The LOQ, or concentration at which quantitative results can be reported with a high degree of confidence, is likewise determined by either approach. Table 3 shows the values obtained for the detection limit and limit of quantification using the blank sample and all three types of indicators.

Table 3. Calcium levels (mg/kg) in blank material detected by titration using three different indicators, and calculated means, standard deviations, LODs and LOQs

Replication	Eriochrome black T	Hydroxy naphthol blue	Hydroxy naphthol blue disodium salt
1	8	24	12
2	16	35	15
3	0	12	30
4	24	16	25
5	20	18	36
6	10	20	24
7	12	22	8
8	22	30	16
9	23	35	26
10	28	10	14
Mean value (mg/kg)	16.3	22.2	20.6
Standard deviation (mg/kg)	8.69	8.85	8.91
LOD (mg/kg)	26.07	26.55	26.73
LOQ (mg/kg)	86.90	88.50	89.10

In the EU, the maximum calcium level for the so-called low pressure MSM is 100 mg/100 g (1000 ppm). Therefore, machinery should be adjusted so as not to exceed this limit. In addition to the percentage of calcium, bone particles and their size are also of great importance, because large particles might cause a gritty texture and potential dental problems. Therefore, bone particle size is regulated in places like the United States, where 90% of the bone particles cannot exceed 0.5 mm and no particle should be larger than 0.85 mm. A high bone content, such as we detected in our beef MSM, means that the pressure used in the deboning process was too high or that the meat-to-bone ratio was too low.

The European Food Safety Authority (EFSA, Italy) conducted a binary logistic regression analysis to identify the probability of a product being classified as MSM based on calcium content, among other things. This analysis showed calcium was the most appropriate indicator for classifying a product as MSM [7]. According to that model, a calcium content corresponding to probabilities from 0.05 to 0.99 qualify a product to be classified as MSM [7]. Our results for calcium content in the examined MSM were in the range of 50 to 100 mg/100 g, and from the binary logistic regression model, this corresponds to probabilities of between 0.69 to 0.94, indicating that the products should be classified as MSM. This confirms that the three types of MSM examined can indeed be classified as MSM, because they were in the given range of probability of the model.

4. Conclusion

This work showed that all three types of tested indicators can be used during the examination of the presence of calcium in MSM. The highest accuracy was obtained using the Eriochrome black T indicator, and it achieved the smallest LOD and LOQ among the three indicators. It is important to note that the detection limits and quantification limits for calcium testing obtained by using the different indicators did not differ much. From the presented results obtained by determining the presence of calcium in commercial MSM, it can be concluded that all examined MSM products had a calcium content below the limit prescribed by Serbian legislation.

Acknowledgment

This paper was supported by Ministry of Education, Science and Technological Development, Republic of Serbia, through the funding of Project No III 46009.

References

[1] Christensen M *et al* 2015 Development of an objective method to perform quality classification of comminuted poultry meat *61st International Congress of Meat Science and Technology*, Clermont-Ferrand, France

[2] Meltem Serdaroúlu M and Turp G Y 2005 Effects of deboning methods on chemical composition and some properties of beef and turkey meat *Turk J. Vet. Anim. Sci.* **29** 797-802

[3] Jin S K *et al* 2014 The effects of mechanically deboned chicken hydrolysates on the characteristics of imitation crab stick *Korean J. Food Sci. An.* **34** 192-9

[4] Bigolin J *et al* 2013 Lipid oxidation in mechanically deboned chicken meat: effect of the addition of different agents *Food and Nutrition Sciences* **4** 219-23

[5] Sözen U and Hecer C 2013 Potential risks of mechanically separated poultry meat technology *Akademik Gıda* V **11** (1) 59-63

[6] Crosland A R *et al* 1995 Investigation of methods to detect mechanically recovered poultry meat in meat products – I: Chemical composition *Meat Science* **40** 289-302

[7] EFSA 2013 Scientific Opinion on the public health risks related to mechanically separated meat (MSM) derived from poultry and swine *EFSA Journal* **11** (3) 3137

[8] Field R A 2000 Ash and calcium as measures of bone in meat and bone mixteres *Meat Sci.* **55** 255-64

[9] AOAC Official Method 984.27:1986 *Calcium, Copper, Iron, Magnesium, Manganese, Phosphorus, Potassium, Sodium, and Zinc and infant formula. Inductively coupled plasma emission spectroscopic method*

[10] AOAC Official Method 983.19:1985 *Calcium in Mechanically Separated Poultry and Beef* Titrimetric Method

[11] Trindade M A *et al* 2004 Mechanically separated meat of broiler breeder and white layer spent hens *Sci. Agric. (Piracicaba, Braz.),* **61** 234-9

Monitoring process hygiene in Serbian retail establishments

S Vesković Moračanin, T Baltić and L Milojević

Institute of Meat Hygiene and Technology, Kaćanskog 13, Belgrade, Serbia

Email: tatjana.baltic@gmail.com

Abstract. The present study was conducted to estimate the effectiveness of sanitary procedures on food contact surfaces and food handlers' hands in Serbian retail establishments. For that purpose, a total of 970 samples from food contact surfaces and 525 samples from workers' hands were microbiologically analyzed. Results of total aerobic plate count and total *Enterobacteriaceae* count showed that the implemented washing and disinfection procedures, as a part of HACCP plans, were not effective enough in most retail facilities. Constant and intensive education of employees on proper implementation of sanitation procedures are needed in order to ensure food safety in the retail market.

1. Introduction

Food safety is primarily achieved through a preventive approach such as the implementation of a food safety management system based on the principles of Hazard Analysis and Critical Control Point (HACCP) and good hygiene practice (GHP). Good cleaning practices are prerequisites to the implementation of an HACCP system and are essential for the production of safe food. Food contact surfaces are considered a major concern for food service facilities in controlling the spread of food-borne pathogens because cross contamination via food contact surfaces has been identified as a significant risk factor [1]. Likewise, improvement of the hand hygiene of workers who come into direct contact with food during retail sale of food is of great importance in prevention of food-borne diseases [2]. Food contact surfaces include food containers, utensils, plates, cooking kettle, cutting boards, slicers, knives, steel pallets and spatulas, stainless steel gastronorms and plastic vessels for food distribution [3].

Pathogens are often found attached to the food contact surfaces of processing equipment, such as knives, cutting boards and the conveyor belts of slicing machines and these are typically made of stainless steel, plastic and polyester urethane (PSU), respectively. As microbiological indicators of surface hygiene, the aerobic plate count and total *Enterobacteriaceae* count are the most frequently examined microbiological parameters [4]. In the past two decades, serious outbreaks of food-borne disease were caused by *Listeria monocytogenes*, a pathogen frequently found in retail delicatessens [5]. In addition, *L. monocytogenes* can form biofilms, and thus, successfully persist in retail and/or food processing establishments, which makes this pathogen a significant concern for the food industry and a serious threat to public health. Especially on porous surfaces (such as damaged or old plastic conveyor belts) or in the niches where it is difficult to access, the bacteria can persist and form biofilms [6], after which, routine cleaning may not eliminate them. Thus, it is of great importance to implement sanitary procedures to ensure thorough removal of impurities and minimizing the number of microorganisms on the surfaces to a minimum acceptable level. Therefore, to ensure the safety of the finished food product, it is necessary to control the cleanliness of the environmental surfaces in processing and retail levels, as well as of the hands of workers who manipulate food.

The aim of this study was to estimate the effectiveness of sanitary procedures on food contact surfaces and food handlers' hands in Serbian retail establishments.

2. Materials and Methods

During a one year period (from July 1, 2016 to June 30, 2017), an assessment of the process hygiene was carried out in retail facilities in Serbia. A total of 524 retail facilities were included in this study, where 970 food contact surfaces and 525 food handlers' hands were investigated for microbiological parameters of process hygiene. Depending on size and capacities, retail facilities were classified into categories (A, B and C), where facilities of A category were the largest. The number of samples in each facility was adjusted according to those criteria (Table 1).

Table 1. Number of swabs sampled for the estimation of process hygiene in different categories of retail facilities

Category of retail facilities	Number of facilities	Number and type of swab samples	
		Food contact surfaces	Food handlers' hands
A	128	384	128
B	191	382	191
C	205	204	206
Total	524	970	525

2.1. Swab samples

Swab samples from the food contact surfaces or food handlers' hands were taken after cleaning, washing and disinfection procedures. Sampling was conducted according to the standard method [7]. On the sampling day, swabs were transported to the laboratory in a hand-held refrigerator and analyzed within 24 h.

2.2. Microbiological examinations

Depending on the surface from which the swab was taken, microbiological analyses were performed. Namely, aerobic plate count, total *Enterobacteriaceae* count and the presence of *L. monocytogenes* were analyzed in samples from the surface of the equipment and tools, such as slicing machines, cutting boards, knives or hatchets. The surfaces of food handlers' hands were tested for aerobic plate count and total *Enterobacteriaceae* count. The microbiological examinations were performed following standard SRPS ISO methods: aerobic plate count [8]; total *Enterobacteriaceae* count [9] and detection of *L. monocytogenes* [10]. Results of the microbiological analyses were expressed as number of bacteria per cm^2 (CFU/cm^2) and number of bacteria per swab (CFU/swab), for swabs taken from the food contact surfaces and food handlers' hands, respectively (data not shown).

2.3. Evaluation of microbiological results

The assessment of the obtained results of microbiological contamination was carried out in accordance with the limit values set by the self-control plans of each retail food business operator (Table 2). These compliance criteria were selected because they were practical, achievable and verifiable for the evaluation of hygiene and sanitation program in the food industry.

Table 2. Microbiological criteria–limit values contained in the self-control plans of the food business operators

Microorganisms	Surfaces	
	Equipment and tools	Food handlers' hands
Aerobic plate count	≤ 10 CFU/cm²	≤ 1000 CFU/swab
Enterobacteriaceae	≤ 1 CFU/cm²	≤ 10 CFU/swab
L. monocytogenes	Absence in 100 cm²	

3. Results and Discussion

Serbia, within its process of legal harmonization with the EU, recently reorganized its food safety system to comply with European Union (EU) regulations according to *acquis communautaire* [11]. Among the legislative changes, a new Food Safety Law was introduced in 2009 [12] and a Veterinary Law in 2005 [13]. This mandatory law requires implementation of a food safety system based on HACCP principles for all subjects in the food chain, except primary production [14].

As hygiene indicator organisms, aerobic plate count and *Enterobacteriaceae* count in samples from the surfaces were investigated. The limit that distinguishes dirty (or unsatisfactory) from clean (or satisfactory) food contact surface is not defined by current Serbian or EU regulations. Therefore, food business operators engaged in the production, processing, and distribution of food are obliged to define the limit values for their own process hygiene parameters [15].

During the one year period, in 524 retail facilities (128 –A category, 191 – B category, and 205 – C category), 1495 swabs were examined: 970 from food contact surfaces (slicing machines, cutting boards and knives or hatchets) and 525 swabs from food handlers' hands (Table 1). A high standard of hygiene in the working environment, in particular on food contact surfaces, equipment and facilities, is a fundamental requisite for the prevention of microbial contamination of food [16]. Traditional microbiological analyses (aerobic plate count and *Enterobacteriaceae*) are generally used to evaluate the effectiveness of sanitation operating procedures. Results of analyzed data for aerobic plate count and total *Enterobacteriaceae*, in samples from the food contact surfaces, are presented in Table 3.

Table 3. Results of the process hygiene estimation in different categories of retail facilities

Category of retail facility	No[a]	Non-compliant		No. of swabs	Non-compliant		Finding/Frequency (%)					
		n[b]	%		n[c]	%	APC[d]	%	ENT[e]	%	TBC + ENT[f]	%
A	128	91	71.09	384	177	46.09	127	71.75	9	5.08	41	23.17
B	191	109	57.07	382	146	38.22	118	80.82	1	0.7	27	18.48
C	205	108	52.68	204	84	41.18	69	82.14	1	1.19	14	16.67

[a]Number of retail facilities
[b]Number of all non-compliant retail facilities
[c]Number of non-compliant swabs
[d]Aerobic plate count
[e] *Enterobacteriaceae*
[f]Aerobic plate count + *Enterobacteriaceae*

The results showed that 41.96% of the total swabs tested did not fulfill the process hygiene criteria, whereby increased aerobic plate counts were, for the most part, the reason for the non-compliant results in all facility categories (A – 71.75%, B – 80.82%, and C – 82.14%). These findings should alert food business operators and trigger them to properly apply effective sanitary procedures. During the examinations, *L. monocytogenes* was detected in only one sample from the surface of a plastic cutting board (data not shown). Lakićević and Nastasijević [5] stated that risk mitigation strategies for *L. monocytogenes* should be based on integrated control along the food chain continuum, from farm to retail establishment, in order to ensure food safety, especially safety of ready-to-eat foods. Inadequate cleaning and disinfection of food contact surfaces can result not only in the reduction of the shelf-life of food, but also in the possible presence of pathogens, particularly problematic for those with a low minimum infective dose [17].

Table 4. Microbiological status of the food handlers' hands

Category of retail facility	No of swabs	Non-compliant		Finding/Frequency (%)					
		N[a]	%	APC[b]	%	ENT[c]	%	APC+ ENT[d]	%
A	128	45	35.15	30	66.67	-		15	33.33
B	191	51	26.70	32	62.74	-		19	37.25
C	206	69	33.49	59	85.50	-		10	14.49

[a] Number of non-compliant swabs
[b] Aerobic plate count
[c] *Enterobacteriaceae*
[d] Aerobic plate count + *Enterobacteriaceae*

The results of microbiological examinations of swabs from the food handlers' hands (presented in Table 4) are worrying. Of 524 swabs, 169 (31%) were non-compliant in relation to the set limits (A – 35.15%, B – 26.70%, and C – 33.49%). These findings are very close to those in 2012 conducted by Rašeta *et al.* [2]. Workers who manipulate food must be trained, informed and competent in food handling. Constant attention and compliance with the norms of GHP within HACCP are of great importance, since it is apparent that when workers become less attentive, it reflects immediately on hygiene results. Green and Selman [16] found that non-compliant results were obtained in cases when workers did not pay enough attention to the execution of work procedures/operations for hand washing. Results of the present study confirm a similar situation in Serbian retail establishments. Protection of hands with gloves is considered an effective way of preventing the transmission of bacteria to food [17,18].

4. Conclusion

The results show that the washing and disinfection procedures are not effective enough in most retail facilities. This relatively large study has shown that in all facilities where food is manipulated, even if HACCP has been introduced, constant and intensive education of employees on the proper implementation of sanitation procedures must be employed. Definitely, only proper sanitation will help to reduce the presence of microorganisms, some potentially pathogenic, which can negatively affect the microbiological status of the foodstuffs.

Acknowledgement

The results presented in this paper are part of Project III No 46009, funded by the Ministry of Education, Science and Technological Development of the Republic of Serbia.

References

[1] Cosby C M, Costello C A, Morris W C, Haughton B, Devereaux M J, Harte F and Davidson P M 2008 *Appl. Environ. Microb.* **74** 6918

[2] Rašeta M, Sverak Matekalo V, Đorđević V, Vranić V, Branković Lazić I, Grbić Z, Grubić M and Lončina J 2012 *Proc. Int. Conf. Biological Food Safety and Quality BFSQ,* 4-5 Oct. 2012,Belgrade, ed S Bunčić (Novi Sad: Faculty of Agriculture, University of Novi Sad) 171
[3] Lahou E, Jacxsens L, Verbunt E and Uyttendaele M 2015 *Food Control* **49** 75
[4] Aarnisalo K, Tallavaara K, Wirtanen G, Maijala R and Raaska L 2006 *Food Control* **17** 1001
[5] Lakićević B and Nastasijević I 2017 *Food Rev. Int.* **33** 247
[6] Poulsen L V 1999 *Lebensm. Wiss. Technol.* **32** 321
[7] SRPS ISO 18593: 2010 Microbiology of food and animal feeding stuffs-Horizontal methods for sampling techniques from surfaces using contact plates and swabs
[8] SRPS ISO 4833-1:2014 Microbiology of the food chain-Horizontal method for the enumeration of microorganisms Colony count at 30 degrees C by the pour plate technique
[9] SRPS ISO 21528-2:2004 Microbiology of food and animal feeding stuffs-Horizontal methods for the detection and enumeration of Enterobacteriaceae-Part 2: Colony-count method
[10] SRPS EN ISO 11290-1:2010 Microbiology of food and animal feeding stuffs-Horizontal method for the detection and enumeration of *Listeria monocytogenes*-Part 1: Detection method
[11] Smigic N, Rajkovic A, Djekic I and Tomic A 2015 *Brit. Food J.* **117** 1 94
[12] Food safety law 2009 Official Gazette of the Republic of Serbia No. 41/2009
[13] Veterinary Law 2005 Official Gazette of the Republic of Serbia No. 91/2005
[14] Tomasevic I, Kuzmanović J, Anđelković A, Saračević M, Stojanović M and Djekic I 2016 *Meat Sci.* **114** 54
[15] Commission Regulation (EC) No. 2073/2005 of 15 November 2005 on microbiological criteria for foodstuffs *O. J.* L 338/1
[16] Green L and Selman C 2005 *Food Prot. Trends* **25** 981
[17] Michaels B, Keller C, Blevins M, Paoli G, Ruthman T, Todd E and Griffith C 2004 *Food Serv. Tech.* **4** 31
[18] Montville R, Chen Y and Schaffner D 2002 *Int. J. Food Microbiol.* **73** 305

IOP Conf. Series: Earth and Environmental Science **85** (2017) 012058 doi:10.1088/1755-1315/85/1/012058

Physicochemical properties of honey from Serbia in the period 2014-2016

D Vranić, R Petronijević, J Đinović Stojanović, V Korićanac, J Babić Milijašević and M Milijašević

Institute of Meat Hygiene and Technology, Kaćanskog 13, 11000 Belgrade, Serbia

E-mail: danijela.vranic@inmes.rs

Abstract. Honey is a viscous, aromatic, sweet food that is consumed and enjoyed by people around the world due to its unique nutritional and medicinal properties. The physicochemical parameters of natural honeys, such as moisture, reducing sugars, sucrose, hydroxymethylfurfural (HMF), free acidity, diastase activity, water-insoluble content and electrical conductivity are strictly defined and constitute the quality indicators which characterize individual honey varieties. The present study shows results of honey quality investigation from various regions of Serbia, which were evaluated according to the Serbian Regulation. A total of 372 honey samples (132 acacia honey, 221 blossom honey and 19 honeydew honey) were obtained from the Serbian market. All applied methods were performed according to the Harmonized Methods of the International Honey Commission. Summarizing the results presented, the most important parameters for detecting honey that was non-compliant with the regulation were HMF content along with diastase activity and sugar content. Results show that in 2014 and 2015, a great number of honey samples were of insufficient quality to satisfy regulatory requirements. In 2016, the situation on the Serbian honey market improved and became more under control.

1. Introduction

Honey is a viscous, aromatic, sweet food that is consumed and enjoyed by people around the world due to its unique nutritional and medicinal properties [1]. According to [2], the annual production of honey in Serbia has increased. About 8554 tonnes, 4383 tonnes and 12263 tonnes were produced in 2013, 2014 and 2015, respectively. Average consumption of honey per year is estimated to be only 0.85 kg per capita [3]. All quality criteria of honey are described by law, including the quality control of honey, and rules on the conditions for the production and marketing of honey [4]. The physicochemical parameters of natural honeys, such as moisture, reducing sugars, sucrose, hydroxymethylfurfural (HMF), free acidity, diastase activity (DN), water-insoluble content and electrical conductivity are strictly defined and constitute the quality indicators which characterise individual honey varieties [5]. The beneficial properties of honey alongside with limited availability have increased demand and the high price of honey has produced heightened interest in its adulteration [6]. The most common forms of honey tampering are the addition of cheap sweeteners (such as cane sugar or refined beet sugar, corn syrup, high fructose or maltose syrup) and honeybees fed with sucrose [7].

Table 1 shows the identity and minimum quality requirements for honey (physicochemical parameters) regarding to the Regulations in Europe and Serbia [4, 5]. The present study shows results of honey quality investigation from various regions of Republic of Serbia. Chemical and physical properties of honey are evaluated according to the Serbian Regulation [5].

Table 1. Regulatory values of chemical and physicochemical parameters for honey [4, 5]

Chemical and physicochemical parameters	Composition criteria
Moisture	In general: Not more than 20% Heather and baker's honey: Not more than 23% Baker's honey from heather: Not more than 25%
Free acidity	In general: Not more than 50meq/kg Baker's honey: Not more than 80meq/kg
Diastase activity	In general, except baker's honey: Not less than 8 Not less than 3*
HMF	In general, except baker's honey: Not more than 40mg/kg Not more than 80 mg/kg**
Glucose and fructose content	Blossom honey 60≤ Honeydew honey, blends of honeydew honey with blossom honey45%≤
Sucrose	In general: Not more than 5% False acacia, alfalfa, Menzies Banskia, French honeysuckle, red gum, leatherwood, Citrus spp.: Not more than 10% Lavender, borage: Not more than 10%
Electrical conductivity	Honey not listed below, and blends of these honeys: Not more than 0.8 mS/cm Honeydew and chestnut honey and blends of those except with those listed below: Not less than 0.8 mS/cm; exceptions: strawberry tree, bell heather, lime, ling heather, manuka or jelly bush, tea tree
Water-insoluble content	In general: Not more than 0.1% Pressed honey: Not more than 0.5%

*honeys with low natural enzyme content (e.g. citrus honeys) and an HMF content of not more than 15 mg/kg.
**honeys of declared origin from regions with tropical climate and blends of these honeys

2. Materials and methods

2.1. Honey samples

The total of 372 honey samples (132 acacia honey, 221 blossom honey and 19 honeydew honey) were obtained from different regions from the Serbian market. The sample distribution by the year was: 2014: 21 samples of acacia honey, 39 samples of blossom honey and 4 samples of honeydew honey; 2015: 35 samples of acacia honey, 74 samples of blossom honey and 9 samples of honeydew honey; 2016: 76 samples of acacia honey, 108 samples of blossom honey and 6 samples of honeydew honey. All samples were stored at ambient temperature prior to analysis.

2.2. Methods of physical and chemical analysis

All applied methods were performed according to the Harmonized Methods of the International Honey Commission [8]. The moisture content (%) was determined from the refractive index of the honey by reference to a standard table. Free acidity was determined by titration to pH 8.30 and expressed as milliequivalents/kg (meq/kg). Electrical conductivity (mS/cm) was performed using a conductivity meter at 20°C in a 20% (dry matter basis) solution of honey samples prepared with ultrapure water. Content of insoluble matter (%) is defined as that material found by the procedure to be insoluble in water. Determination of sugars (glucose, fructose and sucrose), (%) were performed with a Waters 2690 high-performance liquid chromatograph equipped with a refractive index (RI) detector (Waters model 2414). Duplicate injections were performed and average peak areas were used for the peak quantification. Glucose, fructose and sucrose purity ≥99.5 % (Sigma–Aldrich) were used as standards to determine the sugar content of honey. Quantification was performed according to the external standard method on peak areas. Determination of DN after Schade was performed and results are expressed in Göthe units per gram of honey. The concentration of 5-(hydroxymethyl)-furan-2-carbaldehyde (HMF) was determined using reverse phase HPLC equipped with UV detection and result is expressed in mg/kg.

2.3. Statistical analysis
For statistical evaluation of data and graphical expression of results Microsoft Excel with Data Analysis Tool Pack from MS Office was used.

3. Results and discussion
Analysis of the water content showed that all samples were in accordance with the regulation, and the maximum value of 20% was not exceeded [4,5]. The results obtained were similar compared to data reported in studies of 201 honey samples originating from the entire territory of Serbia which was performed during 2009 (average moisture content ranged from 16.12% in acacia honey samples to 17.98% in sunflower honey samples), [9]. In 187 evaluated honeys harvested in Northwest Spain, the average water content was 17.6% [10], and in 39 pine honey samples in Greece, water content was between 10.50% and 20.50% [11].

The water-insoluble content in all honey samples were similar and ranged from 0.01% to 0.04% and all these values were lower than the permitted limit (at most 0.1%). The water-insoluble content is an indicator of purity of honey, because it is defined as that matter which is not soluble in water, such as pollen, honeycomb and mineral particles, etc. [12]. Similar results for the water-insoluble content in honey were reported in the study of acacia and linden honey samples in Croatia [13].

The mean free acidity values of honey ranged from the lowest 10.82 meq/kg (2014), 10.87 meq/kg (2015) and 8.23 meq/kg (2016) for acacia honey; 17.44 meq/kg (2014), 14.65 meq/kg (2015) and 16.46 meq/kg (2016) for blossom honey to the highest 26.03 meq/kg (2014), 18.53 meq/kg (2015) and 23.59 meq/kg (2016) meq/kg for honeydew honey. Free acidity values of all examined honeys were below the legal limit (lower than 50 meq/kg). Free acidity is an important parameter which is characterised by the presence of organic acids in equilibrium with lactones, internal esters and some inorganic ions such as phosphates, sulphates and chlorides [14]. Although higher values for free acidity can be indicative of fermentation of sugars into organic acids [6] and related to the deterioration of honey, variation in free acidity among different honeys can be explained by blossom origin, the presence of different organic acids or some inorganic ions, geographical origin or harvest season [15,16]. The results obtained in the current study agree with data in the literature [9,17,18,19].

Diastases (α-and b-amylases) are enzymes naturally present in honey. Diastase content depends on the blossom and geographical origins of the honey. They are sensitive to heat (thermolabile) and consequently are able to indicate overheating of the product and the degree of preservation [20]. Similar to HMF content, the activity of diastase (DN) can be used as an indicator of aging and increased temperature, because the activity of this enzyme can be reduced during storage or when the product is subjected to heating above 60°C [18]. The current law stipulates a minimum DN value of 8.00 Göthe units. However, honeys with naturally lower DN tolerate a minimum of 3 Göthe units if honeys have up to 15 mg/ kg of HMF [5]. Hence, where DN is low, it is essential that this honey sample contain a maximum of 15 mg/kg of HMF, because as DN is low, it is necessary to prove that honey has not undergone heat treatment or prolonged storage [6]. The lowest mean DN values were in acacia honey samples (in 2014, 2015 and 2016 were 13.05; 8.86 and 11.60, respectively), while the highest mean DN values were in samples of honeydew honey (18.72; 15.63 and 23.50, respectively). The mean DN values for during 2014, 2015 and 2016 in blossom honeys were 14,30, 11.86 and 16.96, respectively. In 2014, analysis showed non-compliant DN levels in 13 acacia honey samples (61.90%), 16 blossom honey (42.11%) and 1 honeydew honey (25%); in 2015: 20 samples of acacia (57.14%), 41 blossom (56.16%) and 8 honeydew honey samples (88.89%); in 2016: 13 acacia honey samples (17.11%) and 10 blossom honey (9.26%). All these honey samples contained low diastase activity (DN), which were below a minimum value of 8.00. According to the results of a study in which the bees were fed with commercial glucose [21], bees should not be fed glucose in excessive amounts, as this could promote an enzyme deficiency (especially diastase) among enzymes which convert glucose and fructose. Honeys with lower enzyme content are produced from young nectars in early spring and a low enzyme concentration is caused by a low concentration of nectar and higher sugar content with reduced activity of the bees during their growth [22].

The range of HMF content in acacia honeys was 3.07-140.06 mg/kg (2014); 2.11-387.43 mg/kg (2015) and 0.57-211.35 mg/kg (2016), while HMF in blossom honeys was 1.15-234.15 mg/kg (2014); 0.58-392.81 mg/kg (2015) and 0.18-265.59 mg/kg (2016). In analysed honeydew honeys, except the control in 2015 (range 21.31-363.13 mg/kg), all HMF levels were lower than the permitted limit (max. 40 mg/kg), and so the honeys can be considered as fresh honeys (2014: range 3.84-33.06 mg/kg and 2016:1.57-32.81 mg/kg). The HMF results obtained, in general, were higher than reported in literature for honeys from Spain (6.80 mg/kg), [23], from Turkey (2.52 mg/kg), [24] and from Argentina (8.98 mg/kg), [19]. The Regulations [4,5] set a maximum HMF value of 40.00 mg/kg for honey mixtures or processed honey and a maximum value of 80.00 mg kg/kg if the honey and honey blends contain a declared origin from regions with a tropical climate. Among the acacia honey samples analysed, 21 of them had HMF values above the maximum stipulated [5], 2014: 9 samples (42.86%), 2015: 7 (21.21%) and 2016: 5 (6.58%). The number of blossom honeys with HMF values above the maximum permitted value (51) was higher (2014: 10 samples (25.64%), 2015: 34 (45.95%) and 2016: 7 (6.48%) than the number of acacia honeys that exceeded the limit. For honeydew honeys, only 7 samples from 2015 had HMF values above the maximum permitted value (although this was 77.78% of these honeys in that year). HMF is formed by the decomposition of monosaccharides when honey is heated or stored for a long time [6]. High HMF content in honeys can also indicate falsification by adding invert syrup, because HMF can be produced by heating sugars in the presence of an acid to the inversion of sucrose [18,25].

Since 2015, as a result of harmonization of national legislation with that of the EU, measurement of the electrical conductivity of honey has been expanded from only honeydew honey [26] to the other types of honey with few exceptions [5]. The electrical conductivity is often used in the quality control of honey to distinguish blossom honey from honeydew honey [11]. The Serbian Official Regulation on quality of honey [5] recommends a maximum value of 0.8 mS/cm for acacia and blossom honey and minimum value of 0.8 mS/cm for honeydew honey. Results of the electrical conductivity of honeydew samples, analysed during 2014 and 2015 varied in the range 0.40-2.27 mS/cm (mean 1.15 mS/cm) and 0.12-1.10 mS/cm (mean 0.33 mS/cm), respectively. Mean values and ranges of conductivity for examined honeys in 2016 were 0.16 mS/cm (0.06-0.52) mS/cm; 0.45 mS/cm (0.06-2.05) mS/cm and 0.94 mS/cm (0.46-1.83) mS/cm for acacia, blossom and honeydew samples, respectively. Two samples of honeydew honey (50%) in 2014 and eight samples (88.89%) in 2015 had values below the minimum value (1.00 mS/cm), [26], while during 2016, 2 samples of blossom honey (2.35%) and 2 samples of honeydew honey (33.33%) had values higher than 0.80 mS/cm and lower than 0.80 mS/cm for blossom and honeydew honey, respectively, as defined in the current national Regulation [5].

Monosaccharides make up about 75% of the sugars found in honey, along with 10-15% disaccharides and small amounts of other sugars [6]. Sugar composition depends mainly on the honey's botanical and geographical origin, and is affected by climate, processing and storage [15,27]. In 2014, the ranges of reducing sugars expressed as sum of glucose and fructose, and sucrose contents in acacia honeys were 24.54-71.56% and not detected (n.d.)-22.67%, respectively; in 2015, 23.97-79.13 and 0.99-26.23% and, in 2016, 33.42-76.23 and 0.96-30.72%, respectively, while the range of these parameters in blossom honeys were 24.50-79.96% and n.d.-28.18%, respectively in 2014; 20.60-84.47% and 0.98-48.18% , respectively in 2015 and 36.42-82.80 and 0.63-29.81%, respectively in 2016. The results for these quality parameters for honeydew honey were 49.97–69.93% and 1.85-6.13%, respectively in 2014; 33.87-66.36% and 1.11-22.34% in 2015 and 58.03-65.62% and 2.47-4.98%, respectively in 2016. According to results in 2014, a total of 11 honey samples were non-compliant for lower content of reducing sugars and for higher content of sucrose than legally allowed, which was 52.38% of all examined acacia honeys; 14 samples (35.90%) of blossom honey (reducing

IOP Conf. Series: Earth and Environmental Science **85** (2017) 012058 doi:10.1088/1755-1315/85/1/012058

sugars) and 5 blosson honey samples (12.82%) (sucrose), plus 2 samples (50%) (reducing sugars) of honeydew honey were non-compliant; in 2015 there were 18 non-compliant acacia honeys (51.43%) (lower content of reducing sugars) and 20 (57.14%) (higher content of sucrose); for blossom honey, 33 samples (44.59%), (reducing sugars) and 24 samples (32.43%) (sucrose) were non-compliant, and 8 honeydew samples (88.89%) (reducing sugars) and 6 samples (66.67%) (sucrose) were non-compliant; in 2016, there were 10 non-compliant acacia honey samples (13.16%) (lower contents of reducing sugars) and 5 (6.58%), (higher content of sucrose); among blossom honey, 8 samples (7.41%) (reducing sugars) and 5 samples (4.63%) (sucrose) were non-compliant. The results obtained for reducing sugars and sucrose contents in all examined samples of honeydew honey during 2016 were in accordance with composition criteria defined in the current national regulation [5].

Summarizing the results presented, the most important parameters for detecting honey that was non-compliant with the law were HMF content along with diastase activity and sugar content. Moisture, water-insoluble matter and free acidity alone were not relevant factors for evaluation of honey quality. Previous articles on the same topic were of local significance and did not cover the entire market of Serbia, or they used insufficient parameters for checking compliance with regulations [17,28]. The number of non-compliant honey samples by year and type is shown on Figure 1. Control of honey in 2014 and 2015 showed that the quality of honey on the Serbian market was seriously affected during this period. These results might indicate possible adulteration [29]. The drastic decrease of non-compliant honey among all honey types in 2016 is noticeable. That is the most likely consequence of tightening control by government authorities.

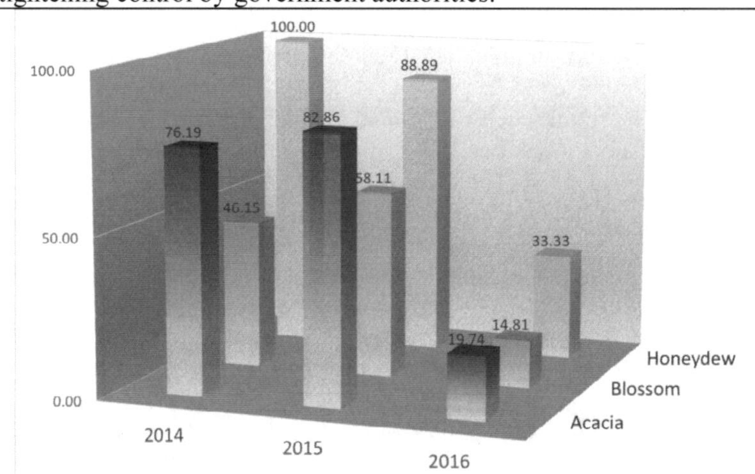

Figure 1. Distribution (%) of non-compliant honey samples by year and type

4. Conclusion

Quality control of honey from the entire Serbian market during the period 2014-2016 showed that the most relevant parameters for detecting non-compliant honeys were determination of HMF, sugar content and diastase activity. Electrical conductivity was relevant only for determining whether honeydew honey conformed with regulation, but free acidity, moisture and water-insoluble matter were of no significance for determining compliance of honey with legislation.

Results show that in 2014 and 2015, a great number of honey samples were of insufficient quality to satisfy regulatory requirements. In 2015, for example, the number of non-compliant honeys culminated with almost 83% of acacia honey, 58% of blossom honey and 89% of honeydew honey being non-compliant. After government action early in 2016, the situation on the Serbian honey market improved and became more under control. The number of non-compliant honey samples decreased to under 20% for both acacia and blossom honey. In interpreting the results for the non-compliant samples of honeydew honey, it should be noted that relatively few of these were sampled and, therefore, the trend must be taken with the reserve.

Serbia exports honey, especially rare and valuable honey like acacia, sunflower and linden honey. Adulteration and degradation of the Serbian honey quality has a significant impact on the public health and economy of the country. Therefore, it is necessary to emphasise the importance of honey quality control over the whole territory of Serbia. Positive results of the government action on enhancing the quality control of honey show that this control must be continued and maintained.

Acknowledgments
This work was supported by grants from the Ministry of Education, Science and Technological Development of the Republic of Serbia (project no. III 46009).

References

[1] El-Haskoury R, Kriaa W, Lyoussi B, Makni M 2017. *J. Food Drug Anal.* doi: http://dx.doi.org/10.1016/j.jfda.2016.11.016
[2] Stat. Yearbook Rep. Serbia 2016. **49** 237
[3] Stat. Off. Rep. Serbia 2006-2011. *Bull.* **71**
[4] Council directive 2001/110/EC 2002. *OJEC* **L10** 47
[5] *Off. Gazette RS* 101/2015
[6] da Silva PM, Gauche C, Gonzaga LV, Oliveira CAC, Fett R 2016. *Food Chem.* **196** 309
[7] Puscas A, Hosu A, Cimpoiu C 2013. *J. Chromatogr. A* **1272** 132
[8] http://www.bee-hexagon.net/en/network.htm
[9] Lazarevic KB, Andrić F, Trifkovič J, Tešić Ž, Milojković-Opsenica D 2012. *Food Chem.* **132** 2060
[10] Escuredo O, Míguez M, Fernández-González M, Seijo MC 2013. *Food Chem.* **138** 851
[11] Karabagias IK, Badeka A, Kontakos S, Karabournioti S, Kontominas MG 2014. *Food Chem.* **146** 548
[12] D'Arcy BR 2007. High-power Ultrasound to Control of Honey Crystallization. Corporation, RIAD, Ed. 07/145 Australian Government, Australia
[13] Tucak Z, Beuk D, Jurnić V, Tušek T, Knežević SV, Tolušić Y 2009. *Coll. Antroplol.* **33** 11
[14] Moreira RFA, Maria CAB, Pietroluongo M, Trugo LC 2007. *Food Chem.* **104** 1236
[15] Tornuk F, Karaman S, Ozturk I, Toker OS, Tastemur B, Sagdic O, Dogan M, Kayacier A 2013. *Ind. Crops Prod.* **46** 124
[16] Alves A, Ramos A, Gonçalves MM, Bernardo M, Mendes B 2013. *J. Food Comp. Anal.* **30** 130
[17] Milosevic M, Pavlovic Z, Sobajic S, Djordjevic B, Djuric S 2013. *Hrana & Ishrana* **54** 19
[18] Yücel Y, Sultanoglu P 2013. *Food Biosci.* **1** 16
[19] Cantarelli MA, Pellerano RG, Marchevsky EJ, Camina JM 2008. *J. Argent. Chem. Soc.* **96** 33
[20] Ahmed M, Djebli N, Aissat S, Khiati B, Meslem A, Bacha S 2013. *Asian Pac. J. Trop. Biomed.* **3** 816
[21] Guler A, Kocaokutgen H, Garipoglu AV, Onder H, Ekinci D, Biyik S 2014. *Food Chem.* **155** 155
[22] Vorlova L, Pridal A 2002. *Acta Univ. Agric. Silvic. Mendel. Brun.* **5** 57
[23] Terrab A, Diez MJ, Heredia FJ 2002. *Food Chem.* **79** 373
[24] Turhan K 2007. *Fresen. Environ. Bull.* **16** 45
[25] Capuano E, Fogliano V 2011. *Food Sci. Technol.* **44** 793
[26] Off. Gazete SCG 45/2003
[27] Escuredo O, Dobre I, Fernández-González M, Seijo MC 2014. *Food Chem.* **149** 84
[28] Prica N, Živkov-Baloš M, Jakšić S, Mihaljev Ž, Kartalović B, Babić J, Savić S 2014. *Arhiv Vet. Med.* **7** 99
[29] Bogdanov S, Martin P 2002. *Mitt. Geb. Lebensmittelunters. Hyg.* **93** 232

IOP Conf. Series: Earth and Environmental Science **85** (2017) 012059 doi:10.1088/1755-1315/85/1/012059

Effect of vegetable oils on fatty acid composition and cholesterol content of chicken frankfurters

D Belichovska[1], Z Pejkovski[2], K Belichovska[2], Z Uzunoska[3] and A Silovska-Nikolova[4]

[1] Faculty of Environmental Resources Management – Food Safety, MIT University, Treta Makedonska Brigada 66a, 1000 Skopje, Republic of Macedonia
[2] Faculty of Agricultural Sciences and Food, Ss. Cyril and Methodius University, Aleksandar Makedonski bb, 1000 Skopje, Republic of Macedonia
[3] Faculty of Technology and Technical Sciences, St. Clement of Ohrid University, Dimitar Vlahov bb, 1400 Veles, Republic of Macedonia
[4] Quality Control, Rimes MS Group, 40 1632, 1000 Skopje, Republic of Macedonia

E-mail: daniela.belichovska@gmail.com

Abstract. To study the effect of pork adipose tissue substitution with vegetable oils in chicken frankfurters, six frankfurter formulations were produced: control; with pork backfat; with olive oil; with rapeseed oil; with sunflower oil; with palm oil, and; with a mixture of 12% rapeseed oil and 8% palm oil. Fatty acid composition and cholesterol content and some oxides thereof were determined in the final products. The use of vegetable oils resulted in improvement of the fatty acid composition and nutritional of frankfurters. Frankfurters with vegetable oils contained significantly less cholesterol and some of its oxides, compared to the frankfurters with pork fat. The formulation with palm oil had the least favourable fatty acid composition. The use of 12% rapeseed oil improved the ratio of fatty acids in frankfurters with a mixture of rapeseed and palm oils. Complete pork fat replacement with vegetable oils in chicken frankfurter production is technologically possible. The mixture of 12% rapeseed oil and 8% palm oil is a good alternative to pork fat from health aspects. Further research is needed to find the most appropriate mixture of vegetable oils, which will produce frankfurters with good sensory characteristics, a more desirable fatty acid ratio and high nutritional value.

1 Introduction

Meats and processed meats are associated with nutrients and nutritional profiles that are often considered negative, including high levels of saturated fatty acids, cholesterol, sodium and high fat and caloric contents [1]. Animal fats have long been utilized in processing meat products. However, the use of animal fat in meat products has emerged as a topic of increasing concern to consumers [2]. Diets rich in animal fat are associated with several types of obesity, hypertension, cardiovascular and coronary heart diseases [3].

In recent years, consumer demands for healthier meat and meat products with reduced levels of fat, cholesterol, decreased contents of sodium chloride and nitrite, improved composition of fatty acid profile and incorporated health-enhancing ingredients are rapidly increasing worldwide [4]. Modern eating habits have changed towards a diet high in saturated fatty acids and unbalanced in polyunsaturated fatty acids, with a high n-6/n-3 ratio [5]. However, this ratio should be ideally <5 [5].

Frankfurters are non-fermented, emulsion type cooked sausages [6], and are one of the most popular traditional meat products in the world [7]. However, their consumption likely has negative health effects regarding the amounts and types of animal fats that they contain [8].

The potential health risks associated with the consumption of high fat foods has led to development of new formulations to modify traditional food products so they contain less fat [9]. The substitution of animal fat with vegetable oils has been suggested to improve the fatty acid profile and to decrease the cholesterol levels of meat products. Vegetable oils have positive effects on the cardiovascular system [3].

Therefore, the objective of the present investigation was to examine the effects of pork backfat substitution with several vegetable oils and fats on fatty acid composition, cholesterol content and some oxides thereof.

2 Materials and Methods

Frankfurters made from chicken (boneless breast and thigh) were used in the present research. Six different frankfurters were produced, compositions of which are given in table 1. The control frankfurters were prepared with pork backfat (Po). The other five types of frankfurters were prepared with olive oil (O); rapeseed oil (R); sunflower oil (S); palm oil (Pa), and a mixture of 12 % rapeseed oil and 8 % palm oil (Mi).

Table 1. Frankfurter formulation (%).

Ingredients	Po	O	R	S	Pa	Mi
Chicken breasts without skin	23	23	23	23	23	23
Chicken thighs without skin	23	23	23	23	23	23
Pork backfat	20					
Olive oil		20				
Rapeseed oil			20			12
Sunflower oil				20		
Palm oil					20	8
Nitrite curing salt	1.7	1.7	1.7	1.7	1.7	1.7
Ice	32	32	32	32	32	32
Sodium tripolyphosphate	0.2	0.2	0.2	0.2	0.2	0.2
Mixture of spices	0.2	0.2	0.2	0.2	0.2	0.2
Sodium isoascorbate	0.075	0.075	0.075	0.075	0.075	0.0705
Soy protein isolate	2	2	2	2	2	2

Artificial collagen edible casings, 23 mm in diameter (Naturin GmbH, Weinheim, Germany), were used for stuffing the batter. Heat treatment was performed in smokehouse at 78°C until an internal temperature of 72°C was achieved.

Fatty acids in the frankfurters were detected based on their methyl ester content according to the method described by Park and Goins [10]. Separation of fatty acid methyl esters was performed on a gas chromatograph (Agilent Technologies 6890) with a flame-ionization detector and capillary column type HP-88 (100 m x 0.25 mm x 0.20 μm).

The cholesterol content of the frankfurters was determined by the modified method according to Ubhayasekera et al. [11]. After saponification of cholesterol with the basic reagent, cleaning of the sample was carried out by the SPE procedure using Strata Si-1 columns (Phenomemex 8B-5 ∅12-HBJ). The cholesterol content was determined by high pressure liquid chromatography.

For determination of cholesterol oxides, the method by Ubhayasekera et al. [11] was used. Once cholesterol was separated, the sample was cleaned and SPE (Strata Si-1, Phenomemex) was performed. Cholesterol oxides were determined by gas chromatography with capillary column HP-5ms (30 m x 0.32 mm x 0.25 µm).

The program package SAS/STAT [12] was used for statistical processing of data and significance, means were separated using Duncan's test.

3 Results and Discussion

3.1 Fatty acid profile of frankfurters

The fatty acid composition of frankfurters is presented in table 2. The highest content of saturated fatty acids (SFA) (46.23%) was observed in frankfurters with palm oil (Pa), which was even higher than the SFA content in frankfurters with pork fat (41.73%) (Po). The lowest SFA content (12%) was observed in frankfurters with rapeseed oil (R). The frankfurters with olive oil (O), sunflower oil (S) and mixture of 12% rapeseed oil and 8% palm oil (Mi) had lower contents of SFA as well, compared to the control treatment (Po). The differences between frankfurter formulations were statistically significant ($P<0.001$). It is clear that the type of vegetable oil used affected the fatty acid composition of the frankfurters. Among the SFAs, the most common was palmitic acid, with the highest content (39.93%) in Pa frankfurters and the lowest (8.08%) in S frankfurters. Stearic acid was found at percentages among SFA from 2.88% (in R frankfurters) to 13.46% (Po). The differences between frankfurters were statistically significant ($P<0.001$).

Table 2. Fatty acid composition of frankfurters (mean±standard deviation).

% of total fatty acids	Treatments					
	Po	O	R	S	Pa	Mi
C10:0	0.09 ± 0.01^{a}	0.01 ± 0.00^{c}	0.01 ± 0.00^{c}	0.01 ± 0.00^{c}	0.02 ± 0.00^{b}	0.01 ± 0.00
C12:0	0.09 ± 0.00^{b}	0.01 ± 0.00^{d}	0.01 ± 0.00^{d}	0.01 ± 0.00^{e}	0.16 ± 0.00^{a}	0.08 ± 0.00
C14:0	1.57 ± 0.11^{a}	0.22 ± 0.03^{d}	0.11 ± 0.02^{d}	0.11 ± 0.01^{d}	1.13 ± 0.24^{b}	0.50 ± 0.07
C16:0	25.98 ± 0.73	13.81 ± 0.14	8.16 ± 0.28^{e}	8.08 ± 0.30^{e}	39.93 ± 0.58^{a}	21.90 ± 0.3
C16:1*n*-9c	2.42 ± 0.06^{a}	1.24 ± 0.01^{b}	0.45 ± 0.02^{dc}	0.32 ± 0.00^{d}	0.61 ± 0.19^{c}	0.48 ± 0.05
C18:0	13.46 ± 0.09^{a}	3.38 ± 0.09^{d}	2.88 ± 0.08^{e}	4.04 ± 0.03^{c}	4.50 ± 0.25^{b}	3.51 ± 0.11
C18:1*n*-7c	0.18 ± 0.02^{a}	0.02 ± 0.04^{d}	0.05 ± 0.00^{cb}	0.04 ± 0.00^{cd}	0.08 ± 0.00^{b}	0.06 ± 0.00
C18:1*n*-9c	36.22 ± 0.26^{e}	67.17 ± 0.04^{a}	46.24 ± 0.15^{b}	23.51 ± 0.20^{f}	37.27 ± 0.12	42.97 ± 0.4
C18:2c9t12	0.00 ± 0.00^{d}	0.02 ± 0.00^{c}	0.03 ± 0.01^{b}	0.00 ± 0.00^{d}	0.04 ± 0.00^{a}	0.03 ± 0.00
C18:2c9c1	13.85 ± 0.50	10.54 ± 0.15^{e}	32.73 ± 0.10^{b}	61.20 ± 0.52^{a}	11.02 ± 0.38^{e}	23.71 ± 0.1
C18:3*n*-6	0.05 ± 0.00^{a}	0.02 ± 0.00^{cd}	0.02 ± 0.00^{cb}	0.01 ± 0.00^{d}	0.02 ± 0.00^{b}	0.02 ± 0.00
C18:3*n*-3	0.24 ± 0.03^{c}	0.39 ± 0.03^{b}	0.44 ± 0.00^{a}	0.23 ± 0.01^{c}	0.38 ± 0.04^{b}	0.39 ± 0.01
C20:4*n*-6	0.64 ± 0.12^{a}	0.24 ± 0.02^{c}	0.21 ± 0.02^{c}	0.18 ± 0.02^{c}	0.45 ± 0.09^{b}	0.23 ± 0.04
C20:5*n*-3	0.38 ± 0.26^{b}	0.33 ± 0.01^{cb}	0.00 ± 0.00^{d}	0.02 ± 0.00^{d}	0.82 ± 0.01^{a}	0.16 ± 0.01
C22:5*n*-3	0.12 ± 0.00^{a}	0.02 ± 0.01^{d}	0.03 ± 0.00^{c}	0.03 ± 0.00^{dc}	0.05 ± 0.01^{b}	0.03 ± 0.00
C22:6*n*-3	0.58 ± 0.38^{ba}	0.40 ± 0.02^{bc}	0.05 ± 0.00^{d}	0.05 ± 0.00^{d}	0.73 ± 0.05^{a}	0.21 ± 0.01
ΣSFA	41.73 ± 0.64	17.82 ± 0.03	12.00 ± 0.24^{f}	13.14 ± 0.29^{e}	46.23 ± 0.11^{a}	26.62 ± 0.4
ΣMUFA	41.68 ± 0.28	70.04 ± 0.08^{a}	54.16 ± 0.19^{b}	24.88 ± 0.20^{f}	39.93 ± 0.27^{e}	48.26 ± 0.3
ΣPUFA	16.59 ± 0.40	12.14 ± 0.05^{f}	33.84 ± 0.10^{b}	61.98 ± 0.49^{a}	13.84 ± 0.23^{e}	25.12 ± 0.1
n-6	14.59 ± 0.36	10.80 ± 0.13^{f}	32.97 ± 0.10^{b}	61.42 ± 0.50^{a}	11.51 ± 0.30^{e}	24.03 ± 0.1
n-3	1.32 ± 0.68^{b}	1.14 ± 0.06^{b}	0.52 ± 0.01^{c}	0.33 ± 0.02^{c}	1.97 ± 0.09^{a}	0.80 ± 0.04
n-6/*n*-3	12.90 ± 5.36	9.46 ± 0.63^{d}	63.78 ± 1.23^{b}	185.88 ± 12.74	5.85 ± 0.39^{d}	30.02 ± 1.4
PUFA/SFA	0.40 ± 0.02^{e}	0.68 ± 0.00^{d}	2.82 ± 0.06^{b}	4.72 ± 0.14^{a}	0.30 ± 0.00^{e}	0.94 ± 0.02

Trans FA	0.28±0.02^{b}	0.15±0.01^{d}	0.36±0.01^{a}	0.09±0.00^{e}	0.10±0.01^{e}	0.23±0.01
IA	0.56±0.01^{b}	0.18±0.00^{d}	0.10±0.00^{e}	0.10±0.00^{e}	0.83±0.01^{a}	0.33±0.01

$^{a-f}$– Means within a row with different superscript letters are significantly different (P<0.001).
Po – Pork backfat (control); O – Olive oil; R – Rapeseed oil; S – Sunflower oil; Pa – Palm oil; Mi – Mixture of 12% rapeseed oil and 8% palm oil.
SFA – Saturated fatty acids; MUFA – Monounsaturated fatty acids; PUFA – Polyunsaturated fatty acids; IA – Atherogenic index.

Frankfurters produced with sunflower oil (S) had convincingly the lowest content (24.88%) of monounsaturated fatty acids (MUFA). That was due to the low content of these acids in the sunflower oil itself. The highest content of these acids was found in frankfurters with olive oil (O) (70.04%). There were significant differences (P<0.001) between the amounts of MUFA in the different frankfurters. The most abundant MUFA was the *cis*-form of oleic acid, which ranged from 23.51% (S frankfurters) to 67.17% (M frankfurters). In terms of the total percentage of all the fatty acids, it was also the most common fatty acid in the frankfurters.

Polyunsaturated fatty acids (PUFA) were the most plentiful (61.98%) in the frankfurters produced with sunflower oil (S). This was a result of the high content of PUFA in sunflower oil. A somewhat high PUFA content (33.84%) was also measured in the frankfurters with rapeseed oil (R). Therefore the frankfurters with Mi, the composition of which also contained rapeseed oil, were characterized by a slightly higher content of PUFA than the other frankfurters, with the exception of the frankfurters with S and R. Frankfurters produced with olive oil (M) had the lowest content of PUFA (12.14%). Frankfurters with Pa had the highest content of n-3 fatty acids (1.97%), and those with S had the lowest (0.33%). Differences between frankfurters were significant (P<0.001).

The ratio of n-6 to n-3 fatty acids is dependent on the type of oil used in the frankfurters. It can be concluded that only frankfurters with Pa (5.85) and O (9.46) had a favourable n-6/n-3 ratio. Other frankfurters had unfavourable n-6/n-3 ratios. Frankfurters with S had the most unfavourable n-6/n-3 ratio (185.88).

The lowest value of PUFA/SFA ratio (0.30) was found in the frankfurters with Pa (table 2). This value is inconvenient because the PUFA/SFA ratio was lower than 0.5. An unfavourable PUFA/SFA ratio (0.4) was also found in the Po frankfurters. The highest PUFA/SFA ratio (4.72) was found in frankfurters with S. Other frankfurter formulations also had favourable PUFA/SFA ratios which were higher than 0.5. Differences between frankfurters were statistically significant (P<0.001).

The highest content of *trans*-fatty acids was found in the rapeseed oil formulation (0.36%) and the lowest (0.09%) in the sunflower oil formulation. In control frankfurters (Po), *trans*-fatty acids comprised 0.28% of the fatty acids. Frankfurters produced with beef fat contained more *trans*-fatty acids while frankfurters with sunflower oil contained less [13]. Frankfurters formulated with olive oil had lower *trans*-fatty acids content than control frankfurters with pork fat [14].

The value of the atherogenic index (IA) is also dependent on the type of oil used in frankfurter production (table 2). Values of IA higher than 0.5 are considered undesirable. Unfavourable IA values were determined in the palm oil treatment (0.83) and pork fat treatment (0.56). The remaining treatments had a favourable IA values. The most desirable IA value (0.10) was found in the frankfurters containing rapeseed (R) or sunflower (S) oils. There were significant differences (P<0.001) among the frankfurter formulations.

The results of the present investigation showed that the oils or fats used significantly affected the fatty acid composition and the nutritional value of frankfurters. Other researchers also concluded that by replacing animal fats with vegetable oils in the beef frankfurter production, the nutritional content of product was improved due to alteration of the fatty acid composition [15]. Frankfurters produced with sunflower oil, unlike those produced with beef fat, were characterized by a better fatty acid composition [13,16]. Generally, from a nutritional point of view, it can be concluded that frankfurters produced with vegetable oils are far more favourable than those produced with pork fat, because they contain less saturated and more unsaturated fatty acids.

3.2 Cholesterol content and cholesterol oxides in the frankfurters

The cholesterol content of the frankfurters ranged from 37.96 to 48.41 mg/100 g (table 3). In frankfurters produced with vegetable oils, the cholesterol content was lower compared to those with pork fat. Differences between frankfurters were statistically significant (P<0.001). This means that pork fat significantly increased the cholesterol levels in the frankfurters. This stems from the fact that plant cells do not contain cholesterol, as opposed to animal cells that do contain cholesterol.

Table 3. Content of cholesterol content and some cholesterol oxides in the frankfurters (mean±standard deviation).

Parameters	Treatments					
(mg/100g)	Po	O	R	S	Pa	Mi
Cholesterol	48.41 ± 0.89^{a}	38.90 ± 0.38^{bc}	37.96 ± 0.64^{d}	39.24 ± 0.17^{bc}	38.66 ± 0.22^{c}	39.48 ± 0.06^{b}
7b-hydroxy cholesterol	1.37 ± 0.37^{a}	0.78 ± 0.06^{b}	1.55 ± 0.76^{a}	0.33 ± 0.13^{c}	1.59 ± 0.12^{a}	1.22 ± 0.13^{a}
5a-hydroxy cholesterol	0.00 ± 0.00^{a}	0.00 ± 0.00^{a}	0.00 ± 0.00^{a}	0.00 ± 0.00^{a}	0.00 ± 0.00^{a}	0.00 ± 0.00^{a}
20a-hydroxy cholesterol	1.73 ± 0.11^{ab}	1.79 ± 0.31^{a}	1.55 ± 0.13^{b}	0.51 ± 0.30^{c}	0.09 ± 0.02^{d}	0.19 ± 0.03^{d}
25a-hydroxy cholesterol	0.03 ± 0.02^{a}	0.00 ± 0.00^{b}	0.00 ± 0.00^{b}	$0.00 \pm 0.00b^{b}$	0.00 ± 0.00^{b}	0.00 ± 0.00^{b}

$^{a\text{-}d}$ – Means within a row with different superscript letters are significantly different (P<0.001).
Po – Pork backfat (control); O – Olive oil; R – Rapeseed oil; S – Sunflower oil; Pa – Palm oil; Mi – Mixture of 12% rapeseed oil and 8% palm oil.

In general, the cholesterol content in all six frankfurter formulations was low. The relatively low cholesterol content was due to the use of chicken thighs and breasts without skin, by which a good part of the cholesterol in the poultry meats was eliminated. Lower fat content in the raw material results in lower fat and cholesterol contents in the finished products [17]. The lowest cholesterol content (37.96 mg/100 g) was measured in the frankfurters with rapeseed oil (R). Frankfurters produced with pork fat (Po) had the highest content (48.41 mg/100 g) of cholesterol, which was expected. The cholesterol content of commercial chicken frankfurters amounts to 100 mg/100 g [18]. In another study, vegetable oil and rice bran fibre-containing reduced-fat frankfurters had significantly lower cholesterol contents than the control frankfurters with pork fat [14].

From table 3, it is clear there was no 5a-hydroxy cholesterol in frankfurters (levels were below the limit of detection). 25a-hydroxy cholesterol was present in trace amounts (0.03 mg/100 g) only in control frankfurters (Po). 7b-hydroxy cholesterol and 20a-hydroxy cholesterol were present in all frankfurter formulations. The content of 7b-hydroxy cholesterol ranged from 0.33 to 1.59 mg/100 g, while the content of 20a-hydroxy cholesterol in frankfurters ranged from 0.09 to 1.79 mg/100 g. There were significant differences (P<0.001) between frankfurters.

4 Conclusion

Total pork fat replacement with vegetable oils is possible in chicken frankfurter production. In general, frankfurters produced with vegetable oils had significantly more favourable fatty acid composition and nutritional value (PUFA/SFA ratio, n-6/n-3 ratio and IA value).

The cholesterol content and some cholesterol oxides in frankfurters with vegetable oils were significantly (P <0.001) lower in comparison with pork fat frankfurters.

From a nutritional health aspect, frankfurters produced with palm oil had the least favourable fatty acid composition due to the unfavourable fatty acid composition of palm oil. However, a mixture of

12% rapeseed oil and 8% palm oil is a good alternative to pork fat. The use of palm oil improved the sensory characteristics of frankfurters (data not shown), while rapeseed oil has improved the fatty acid composition and the nutritional value of the products.

In order to improve the fatty acid composition and nutritional value of frankfurters produced with a similar mixture of rapeseed and palm oils, further research is needed. A better mixture of vegetable oils will contribute to the preservation of sensory characteristics while improving the fatty acid composition and nutritional value of chicken frankfurters.

References

[1] Whitney E N and Rolfes S R 2002 *Understanding nutrition* (Ninth edn Belmont, CA: Wadsworth)
[2] Dzudie T, Scher J, Tchiégang C and Hardy J 2005 Effect of fat sources on the physico-chemical nutritional and textural properties of beef sausages *J. Food Technol.* **3** 220-5
[3] Özvural E B and Vural H 2008 Utilization of interesterified oil blends in the production of frankfurters *Meat Sci.* **78** 211-6
[4] Zhang W Xiao S Samaraweera H, Lee E J and Ahn D U 2010 Improving functional value of meat products *Meat Sci.* **86** 15–31
[5] Rondelly S G, Martinez O and Garcia P T 2004 Effects of different dietary lipids on the fatty acid composition of broiler abdominal fat *Braz. J. Poult. Sci.* **6** 171-5
[6] Estévez M, Morcuende D and Cava R 2006 Extensively reared Iberian pigs versus intensively reared white pigs for the manufacture of frankfurters *Meat Sci.* **72** 356-64
[7] Ayo J, Carballo J, Solas M T and Jimenez-Colmenero F 2008 Physicochemical and sensory properties of healther frankfurters as affected by walnut and fat content *Food Chem.* **107** 1547–52
[8] Bloukas J G and Paneras E D 1993 Substituting olive oil for pork backfat affects quality of low-fat frankfurters *J. Food Sci.* **58** 705-9
[9] Mendoza E, Garcia M L, Casas C and Selgas M D 2001 Inulin as fat substitute in low fat, dry fermented sausages. *Meat Sci.* **57** 387-93
[10] Park P W and Goins R E 1994 In situ preparation of fatty acid methyl ester for analysis of fatty acids compositions in food *J. Food Sci.* **59** 1262-66
[11] Ubhayasekera S J K A, Verleyen T and Dutta P C 2004 Evaluation of GC and GC-MS methods for the analysis of cholesterol oxidation product *Food Chem.* **84** 149-57
[12] SAS/STAT 1999 Version 8.01. Cary, SAS Institute Inc. Software.
[13] Yilmaz I 2004 Quality characteristics and fatty acid composition of turkish type frankfurter made with sunflower oil addition *Fleischwirtschaft International* **1** 52-4
[14] Choi Y S, Choi Y H, Han D J, Kim H J, Lee M A, Jeong J Y Chung H J and Cheon J K 2010 Effects of replacing pork back fat with vegetable oils and rice bran fiber on the quality of reduced-fat frankfurters *Meat Sci.* **84** 557-63
[15] Vural H, Javidipour I and Ozbas O O 2004 Effects of interesterified vegetable oils and sugarbeet fiber on the quality of frankfurters *Meat Sci.* **67** 65-72
[16] Yilmaz I, Şimşek I and Işikli M 2002 Fatty acid composition and quality characteristics of low-fat cooked sausages made with beef and chicken meat, tomato juice and sunflower oil *Meat Sci.* **62** 253-8
[17] Polak T 2000 Specifična problematika zmanjšanja maščob in holesterola v predelavi mesa klavnih živali, perutnine in rib. V: Meso in mesnine za kakovostno prehrano, 2. posvet o vlogi in pomenu mesa v normalni-zdravi in dietni prehrani Portorož Ljubljana Biotehniška fakulteta Oddelek za živilstvo p79–88
[18] Popov-Raljić J 1999 *Tehnologija i kvalitet gotove hrane* (Izdavač: Tehnološki fakultet Novi Sad)

IOP Conf. Series: Earth and Environmental Science **85** (2017) 012060 doi:10.1088/1755-1315/85/1/012060

Meat product based on porcine hearts and aortas ameliorates serum lipid profile and inflammation in hyperlipidemic rats

I M Chernukha, E A Kotenkova and L V Fedulova

The V.M.Gorbatov All-Russian Meat Research Institute, Russian Academy of Agricultural Sciences, Moscow, Russian Federation

E-mail: lazovlena92@yandex.ru

Abstract. The biological effect of porcine hearts and aortas in a hyperlipidemic rat model was confirmed. Porcine heart and aorta mixture in a 3:1 ratio was blended, canned and sterilized at 115°C and 0.23 Mpa for 40 min. Administration of experimental meat product to the animal model decreased total cholesterol, triglycerides and cholesterol low density lipoproteins by 31.8% ($P<0.05$), 28.2%, and 21.6% ($P<0.05$), respectively, compared to those of hyperlipidemic control rats, as well significantly reducing the serum atherogenic index by 41.3% ($P<0.05$) in rats fed the experimental meat product compared with hyperlipidemic control rats. Normalization of white blood cell populations was also detected. Monocyte and granulocyte counts in blood of rats fed the meat product decreased by 71.1% ($P<0.05$) and 57.6% ($P<0.05$) compared to those of the hyperlipidemic control animals. The granulocyte/leucocyte ratio was also reduced by an average of 38.6% ($P<0.05$) in rats fed the meat product compared with hyperlipidemic control rats. The data confirmed the hypolipidemic action of the sterilized meat product. Normalization of white blood cell populations led us to hypothesize an anti-inflammatory action of the new meat product, which, therefore, could be recommended as a part of maintenance therapy for people with lipid disorders or atherosclerosis.

1 Introduction

Novel technologies are actively sought for implementation in the food industry, and special attention is being paid to the development of functional and specialized products. An abundance of publications highlights the results of studies aimed at recipe modification by adding essential nutrients as well as ingredients of vegetable or animal origin [1-3]. Peptides that naturally occur in raw meats or that form during enzymatic hydrolysis or technical processing have become an area of intensive research, especially in the last decade. In numerous studies, it was shown that these peptides have characteristic antioxidant, hypotensive and anticancer, etc. action [4-8].

Modern scientific approaches, particularly proteomics, have enabled confirmation that the proteome of each organ or tissue contains constitutive structural and functional proteins, characterized by specific proteins and peptides, which are involved in organ/tissue function and support the normal physiological state [9]. In this regard, the study of meat by-products as sources of bioactive sequences promoting normalization of metabolic disturbance is a challenging scientific task, as is development of specialized and functional products utilizing bioactive substances. Previously, a significant decrease of total cholesterol, triglycerides and atherogenic fractions of lipoproteins was shown in hyperlipidemic rats which received native tissues of cattle and pigs hearts and the aortas [10], while diets with porcine hearts and aortas demonstrated the greatest efficiency. Proteomic analysis revealed several specific proteins, including apolipoprotein A-1, peroxiredoxin-1, galectin-1, and heat shock proteins in porcine aortas. Fatty acid-binding protein was detected in heart. It was also found that these bioactive substances are destroyed during sterilization process, except fatty acid-binding protein [11]. Nevertheless, we can assume that tissue-specific proteins can decompose into some peptides which can also possess biological action similar to that of native tissues. Therefore, the aim of this study was to assess the influence of the biological effect of meat product produced from porcine hearts and aortas on rat serum lipid profile and white blood cell count.

2 Materials and Methods

Meat product was produced using a ZAO Yoshkar-Olinskiy Myasokombinat. Porcine hearts were chopped into pieces of 2-3 mm and then salted for 12 h. Aortas were chopped into pieces of 2-3 mm and minced in cutter at 3000 rpm for 2-3 min. Heart emulsion was quantitatively added in a 3:1 ratio to minced aortas and the mixture was then additionally homogenized at 3000 rpm for 6-8 min. Homogenate was packed in cans and sterilized at 115°C, 0.23 MPa for 40 min to produce the meat product.

Thirty male Wistar rats (380±20 g) approximately 1 year old were housed in conventional standard conditions; water and feed were available *ad libitum*. Animals were randomly divided into 3 groups: group 1 – negative control (n=10); group 2 – hyperlipidemic control (n=10), and; group 3 – experimental (hyperlipidemic+experimental diet; n=10). Group 1 (negative control) were fed standard chow (Labkorm, Russia) *ad libitum* during the study. A rat model of alimentary hyperlipidemia was developed in group 2 and 3 rats by adding cholesterol (2.0-10.0%) and fat (10.0-25.5%) to the standard diet, and each animal received *per os* vitamin D2 injection (35,000 IU/kg b.w.). After modeling, rats in group 2 (hyperlipidemic control) were fed with standard chow, while group 3 rats received meat product (8 g/kg b.w.) mixed with standard chow. All diets were equally balanced according main nutrients: protein, fat, minerals, etc., and were fed to the rats for 42 days.

Forty-two days after rats commenced the experimental diet, they were euthanized in a VETtech chamber according to the animal welfare rules, and blood samples for biochemical and flow cytometry analysis were taken. Total cholesterol (TC), triglyceride (TG), cholesterol low-density lipoproteins (CL LDL) and cholesterol high-density lipoproteins (CL HDL) were measured in rat serum on an automatic analyzer BioChem FC-360 (HTI, USA) according to the manufacturer's instructions for measurement kits (HTI, USA). Atherogenic index was calculated as AI = (TC-CL HDL)/ CL HDL. White blood cell (WBC), lymphocytes (LYM), granulocytes (GRA) and monocytes (MON) in blood were measured on the Guava EasyCyte cytometer (Merck Millipore, Germany).

STATISTICA 10.0 software was used. Significant differences were tested using two-way analysis of variance (ANOVA), followed by Newman-Keul's test. Differences with P-values less than 0.05 were considered as statistically significant.

3 Results and Discussion

It was shown that long-term consumption of a cholesterol- and animal fat-rich diet led to increases of TC, TG and atherogenic fractions of lipoproteins in rat serum. On day 42, rat serum TC, TG and CL LDL in the group 2 hyperlipidemic rats exceeded those in group 1 by 35.8% ($P<0.05$), 17.0% and 15.5%, respectively. In serum of group 3 rats (fed the experimental meat product), TC, TG and CL LDL decreased by 31.8% ($P<0.05$), 28.2%, and 21.6% ($P<0.05$), respectively, compared to those of the hyperlipidemic control rats (table 1).

Table 1. Rat serum lipid profile.

	Group 1	Group 2	Group 3
CL, mmol/L	2.18±0.12	2.96±0.08*	2.02±0.19#
TG, mmol/L	1.76±0.27	2.06±0.33	1.48±0.18
CL LDL, mmol/L	0.84±0.05	0.97±0.02	0.76±0.11#
CL HDL, mmol/L	0.85±0.03	0.88±0.06	0.81±0.03
AI	1.58±0.09	2.52±0.12*	1.48±0.22#

Group 1 = negative control rats.
Group 2 = hyperlipidemic control rats.
Group 3 = hyperlipidemic rats receiving meat product.
Data are shown as mean ± standard error.
*Significantly different compared with group 1.
#Significantly different compared with group 2.

Redistribution of lipoproteins fractions resulted in a significant increase of serum AI in group 2 rats compared to that of negative control rats by 59.5% ($P<0.05$), while in group 3 rats on the experimental diet, AI was lower than in hyperlipidemic control rats by 41.3% ($P<0.05$) (table 1).

Not only serum lipid profile, but also hematologic parameters have prognostic importance in cardiovascular disease. Blood is a carrier of metabolic products from and to the organs and tissues and is affected by the clinical status of the tissue environment [12]. Thus, the relation between atherosclerosis progression and hematologic parameters is not well defined, but it was determined that progression rate of coronary atherosclerosis was shown to be significantly higher in patients with high GRA/LYM ratios [13-15], as well as in those with increased MON counts [15,16] which is of great importance taking into consideration that monocytes are crucial cells in the genesis of atherosclerotic lesions, as they stick to endothelium, which results in cardiovascular disease [17].

The WBC count increased by 24.1% in blood of control group 2 rats, mainly due to MON and GRA counts, which increased by 2.1 ($P<0.05$) and 2.4 times ($P<0.05$), respectively, compared with negative control rats. MON and GRA counts in blood of group 3 rats, treated with experimental meat product, decreased by 71.1% ($P<0.05$) and 57.6% ($P<0.05$) compared to hyperlipidemic control animals (group 2) (figure1).

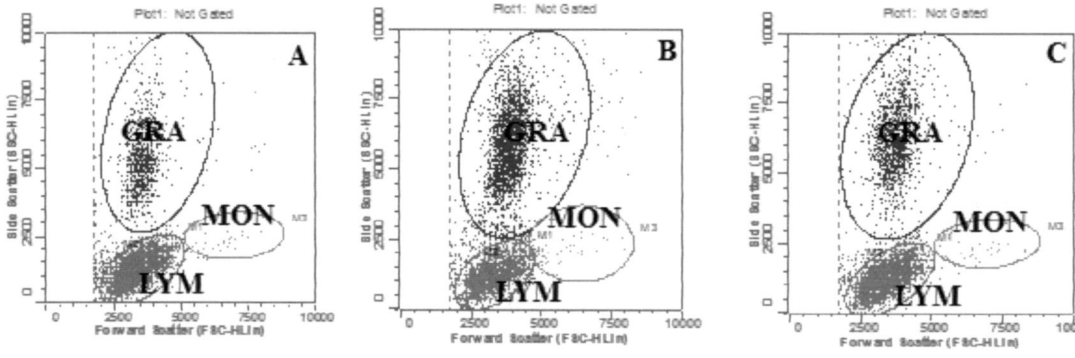

Figure 1. WBC, differentiated into LYM, GRA and MON in the three groups of rats. A – group 1 (negative control), B – group 2 (hyperlipidemic control), C – group 3 (experimental meat product).

The mean GRA\LYM ratio in group 2 rats was 0.57±0.03, which was 2.5 times ($P<0.05$) greater than that of the negative control rats (group 1), while in blood of animals treated with the meat product (group 3) for 42 days, the GRA\LYM ratio was 0.35±0.03; this was 38.6% lower than that of the hyperlipidemic control rats (figure 2).

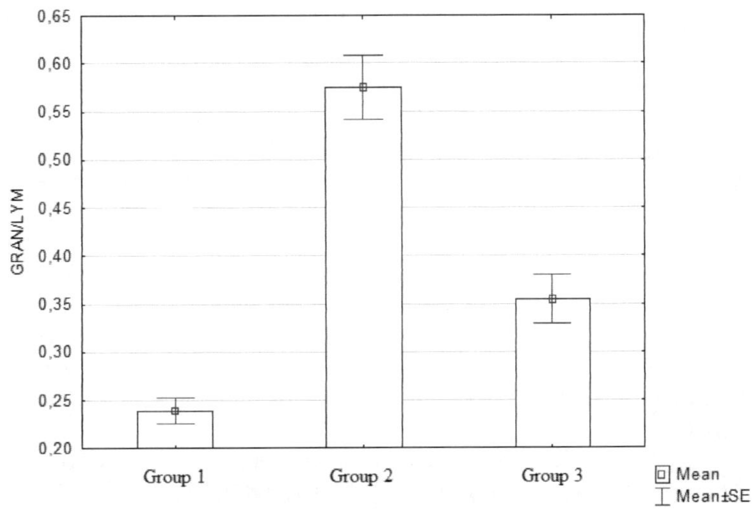

Figure 2. GRA\LYM ratio in rat blood.

The results of this study confirm the hypolipidemic action of the prepared porcine meat product, despite the fact we previously demonstrated decomposition of functional proteins during sterilization. Moreover, normalization of WBC populations was also detected, showing hematopoietic changes due to the anti-inflammatory action of the experimental meat product made from porcine harts and aortas. The developed meat product could be recommended as a part of maintenance treatment for people with lipid disorders or atherosclerosis.

Acknowledgement
This work was supported by the Russian Science Foundation (project No. 16-16-10073).

References

[1] Arihara K 2006 Strategies for designing novel functional meat products *Meat Sci.* **74** 219-29
[2] Hui Y H 2012 *Handbook of Meat and Meat Processing* (USA: Taylor & Francis Group) p 1000
[3] Olmedilla-Alonso B, Jiménez-Colmenero F and Sánchez-Muniz F J 2013 Development and assessment of healthy properties of meat and meat products designed as functional foods *Meat Sci.* **95** 919-30
[4] Bauchart C, Re´mond D, Chambon C, Patureau M P, Savary A I, Reyne`s C and Morzel M 2006 Small peptides (<5 kDa) found in ready-to-eat beef meat *Meat Sci.* **74** 658-66
[5] Bhat Z F, Kumar S and Bhat H F 2015 Bioactive peptides of animal origin: a review *J Food Sci Technol.* **52** 5377–5392
[6] Lafarga T and Hayes M 2014 Bioactive peptides from meat muscle and by-products: generation, functionality and application as functional ingredients *Meat sci.* **98** 227–39
[7] Mine Y and Shahidi F 2006 *Nutraceutical Science and Technology, Book 4: Nutraceutical Proteins and Peptides in Health and Disease* (USA: Taylor & Francis Group) 688 p
[8] Udenigwe C C and Howard A 2013 Meat proteome as source of functional biopeptides *Food Res Int* **54** 1021-32
[9] Fagerberg L et al. 2014 Analysis of the human tissue-specific expression by genome-wide integration of transcriptomics and antibody-based proteomics *Mol Cell Proteomics.* **13** 397-406
[10] Chernukha I M, Fedulova L V and Kotenkova EA 2014 Study on the molecular-biological principles of hypolipidemic activity of cardiac and aortic tissues of pigs and cattle *Archivos Latinoamericanos de Producción Animal* **22** 594-7
[11] Chernukha I M, Fedulova L V, Kotenkova E A, Shishkin S S, Kovalyov L I, Mashentseva N G

and Klabukova D L 2016 Influence of heat treatment on tissue specific proteins in the Sus Scrofa cardiac muscle and aorta *Russian Journal of Biopharmaceuticals* **8** 38-44

[12] Muriithi N J, Maina G S, Maina M B, Kiambi M J, Kelvin J K, Umar A, John M K, Ann N W, Piero N M and Eliud N N M 2015 Determination of hematological effects of methanolic leaf extract of *Vernonia lasiopus* in normal mice *J Blood Lymph* **5** 139

[13] Kalay N, Dogdu O, Koc F, Yarlioglues M, Ardic I, Akpek M, Cicek D, Oguzhan A, Ergin A and Kaya M G 2012 Hematologic parameters and angiographic progression of coronary atherosclerosis *Angiology.* **63** 213-7

[14] Kaya M G 2013 Inflammation and coronary artery disease: as a new biomarker neutrophil/lymphocyte ratio *Turk Kardiyol Dern Ars.* **41** **1**91-2

[15] Swirski F K and Robbins C S 2013 Neutrophils usher monocytes into sites of inflammation *Circ Res.* **112** 744-5

[16] Woollard K J and Geissmann F 2010 Monocytes in atherosclerosis: subsets and functions *Nat. Rev. Cardiol.* **7** 77–86

[17] Arika W M, Nyamai D W, Musila M N, Ngugi M P and Njagi E N M 2016 Hematological markers of in vivo toxicity *J Hematol Thrombo Dis* **4** 236

The relationship between the carcass characteristics and meat composition of young Simmental beef cattle

J Ćirić[1], M Lukić[2], S Radulović[1], J Janjić[1], N Glamočlija[1], R Marković[1] and M Ž Baltić[1]

[1] Faculty of Veterinary Medicine, University of Belgrade, Bulevar Oslobođenja 18, Belgrade, Serbia
[2] Institute of Meat Hygiene and Technology, Kaćanskog 13, Belgrade, Serbia

Email: 1310jecko@gmail.com

Abstract. The objective was to study the relationships between the carcass characteristics and meat composition of young Simmental beef, classified with regard to conformation and degree of fatness scores, and total lipid content, depending on gender. For this purpose, 90 animals (60 male and 30 female Simmental beef cattle) were analysed. The results of the study showed that gender affected carcass measurement scores and chemical composition of meat through its important effect on overall animal fatness. Referring to correlations, male carcass conformation score was negatively related to slaughter weight, total lipid content and fatness score. On the other hand, slaughter weight, hot and cold carcass weight, dressing percentage and carcass conformation was positively related to fatness score, all of them being significant. However, female carcass conformation score was positively related to slaughter weight, total lipid content and fatness score. Hot and cold carcass weights of female Simmental beef cattle were positively correlated to slaughter weight, total lipid content and carcass conformation score. Carcass conformation score and fatness score were affected by gender of young Simmental beef cattle.

1. Introduction

The quality of the beef is affected by many factors, including gender, feeding, animal handling, slaughter of animal, genotype of animals [1-3]. In the European Union, beef carcass classification for conformation and fatness play important point in marketing within and between countries [4]. Nowadays, the food industry prefers to buy steers because they have carcasses with higher fat deposits as indicated by fat thickness and marbling [5,6]. Therefore, meat price in the market is positively related to carcass conformation [7]. Many studies [8-10] have shown the relationships among production factors and beef carcass characteristics. The objectives of this study were to determine the relationship between carcass characteristics and meat composition of young Simmental beef cattle, using the European beef carcass grading system.

2. Materials and Methods

We analysed the carcasses from 60 male and 30 female Simmental beef cattle. Bulls were fasted 18 h before slaughter. Final live weights were recorded. Cattle were slaughtered at a commercial slaughterhouse. After slaughter, hot and cold carcass weights were obtained. Dressing proportions were calculated as the ratio of cold carcass weight to final live weight.

The carcasses were divided between the 12th and 13th rib interface into forequarters and hindquarters [11]. Carcass were classified using the SEUROP classification scales for conformation (S-superior; E-excellent; U-very good; R-good; O-fair; P-poor) scoring from 18 for S+ to 1 for P-. For the fatness classification (1-low; 2-slight; 3-average; 4-high; 5-very high) the score was 15 for 5+ and 1 of 1-. Moisture content was calibrated by reference to an oven drying method [12], fat content by Soxhlet extraction [13] and protein content according ISO [14]. Total cholesterol (mg/100 g of muscle) was determined according to Maraschiello et al. [15].

Statistical analyses of the results were conducted using software GraphPad Prism version 7.0 for Windows (GraphPad Software, San Diego California USA, www.graphpad.com). All parameters were described by descriptive statistics (mean, standard deviation, minimum and maximum value). Pearson's correlation was used to determine relationships among carcass characteristics and carcass weight, total lipids, conformation score and fatness score.

3. Results and Discussion

Characterization of the experimental population is presented in Table 1. In this study, slaughter weight ranged from 403.8 to 715.6 kg (males) and 417.3 to 662.8 kg (female), which is in accordance with results of other authors [3, 16, 17].

The weights of male yearlings (Domestic Simmental) ranged from 499 kg to 604 kg, while females of the same breed ranged from 430 kg to 481 kg [18]. According to Dokmanovic et al. [19], the average weight of yearlings was 533 kg for male and 421 kg for female animals. The average hot carcass weight and cold carcass weight were similar to those in other studies [20,3]. The dressing percentage was between 53.20% and 61.40% (male), 53.30% and 58.30% (female). Our results for dressing percentages accorded with those of Waritthitham et al. [21] and Sanudoa et al. [22]. Drca [18] reported that male Domestic Simmental cattle in Serbia had dressing percentages between 54.20% and 55.40%, while females achieved between 53% and 55.40%.

Average weights of forequarters and hindquarters of males and females are shown in table 1. Males had higher forequarter and hindquarter weights than female cattle. Carcass conformation score and carcass fatness score were affected by gender. On average, males had leaner carcass than the carcasses of females.

Table 1. Mean, standard deviation (SD), minimum (Min) and maximum (Max) value of carcass characteristics of young Simmental beef cattle (male and female).

	Male				Female			
	Mean	SD	Min	Max	Mean	SD	Min	Max
Slaughter weight (kg)	583.9	70.95	403.8	715.6	541.5	55.76	417.3	662.8
Hot carcass weight (kg)	329.9	50.28	218.0	429.6	305.6	31.82	240.4	385.8
Cold carcass weight (kg)	322.1	50.05	212.8	421.9	296.5	31.28	234.5	374.3
Dressing percentage (%)[a]	56.32	2.32	53.20	61.40	56.42	1.36	53.30	58.30
Forequarter weight (kg)	176.1	28.40	112.0	234.1	161.5	16.08	128.2	202.6
Hindquarter weight (kg)	146.1	22.13	100.8	187.8	135.0	15.53	106.3	171.7
Conformation score[b]	13.91	1.38	12.00	16.00	15.39	0.92	14.00	16.00
Fatness score[c]	9.72	2.17	6.00	15.00	12.88	1.42	10.50	15.00

[a] carcass weight x 100/live weight.
[b] – 1 = P (poorest) to 18 = S+ (best).

[c] – 1= (leanest) to 15 = 5+ (fattest).

The chemical composition in this study is represented by several factors: moisture, ash, crude protein, total lipids and total cholesterol. Chemical composition of the *Longissimus* muscle (table 2) revealed that the differences in moisture and total lipids content were due to gender. Generally, males had lower total lipid content than did females. Minchin et al. [23] suggest that the higher percentage for total lipids in cows is due to their high deposition of fat. The low percentage of total lipids in bulls is explained by testosterone; this hormone is related to the higher capacity for muscle growth in bulls and their lower capacity for fat deposition [24]. Variations in moisture percentage occur when there is a variation in lipid percentage in *Longissimus* muscle [6,25]. Some authors [6,25] reported crude protein percentage in *Longissimus* muscle varying between 21% and 24%. Thus, nutrition and gender can alter crude protein percentage in *Longissimus* muscle of bovines.

Table 2. Chemical composition (mean, standard deviation (SD), minimum (Min) and maximum (Max) values) of *Longissimus* muscle of young Simmental beef (male and female).

	Male				Female			
	Mean	SD	Min	Max	Mean	SD	Min	Max
Moisture (%)	75.72	1.69	71.45	78.78	72.44	0.85	70.66	74.01
Ash (%)	0.93	0.14	0.69	1.64	0.89	0.06	0.79	1.02
Crude protein (%)	21.05	0.75	19.16	22.92	21.01	0.87	18.38	22.26
Total lipids (%)	2.23	1.50	0.50	6.14	5.71	1.47	3.42	9.34
Total cholesterol (mg/100 g of muscle)	49.05	6.12	40.15	62.96	58.83	13.50	43.99	84.12

Correlations between carcass characteristics and slaughter weight, total lipids, conformation score and fatness score of young male Simmental beef cattle are presented in table 3. Slaughter weight was positively correlated to fatness score, but negatively to conformation score. Hot and cold carcass weight had the strongest correlations with slaughter weight. These relationships are consistent with those reported in other studies [26-28]. As carcass weight increased, carcass dimensions (forequarter weight and hindquarter weight) increased. Carcass conformation score was negatively correlated to slaughter weight, total lipid content and carcass fatness score. On the other hand, fatness score was positively correlated to slaughter weight and total lipid content. Not surprisingly, the carcass fatness score most strongly associated with total lipid content. Similar results were presented by Indurain et al. [4].

Table 3. Correlation coefficients (r) between carcass characteristics with slaughter weight, total lipids, conformation score and fatness score of young male Simmental beef.

	SW[c]	TL[d]	CS[b]	FS[e]
Slaughter weight (kg)	-	0.002	-0.087	0.408**
Hot carcass weight (kg)	0.810	0.308	0.014	0.374**
Cold carcass weight (kg)	0.816	0.317	0.008	0.380**
Dressing percentage (%)[a]	0.682***	0.267	-0.018	0.368**
Forequarter weight (kg)	0.819	0.341	-0.001	0.357**
Hindquarter weight (kg)	0.793	0.254	0.022	0.399**
Conformation score[b]	-0.049	-0.123	-	-0.060
Fatness score[c]	0.293*	0.349*	-0.060	-

[a] carcass weight x 100/live weight.
[b] Conformation score – 1 = P (poorest) to 18 = S+ (best).
[c] Slaughter weight (kg).
[d] Total lipids (%).
[e] Fatness score – 1= (leanest) to 15 = 5+ (fattest).
*P<0.05; **P<0.01; ***P<0.001.

Table 4 shows correlations between carcass characteristics with slaughter weight, total lipids, conformation score and fatness score of young female Simmental beef cattle. The female slaughter weight was moderately correlated with total lipid content. Similarly, hot carcass weight and cold carcass weight had moderate correlations with total lipid content. This is because female cattle deposit more fat than males [6]. Males produce higher slaughter weight carcasses than female cattle. This greater growth of male in comparison with female cattle seems to be due to the higher production of anabolic hormones by the testicles [29]. Dressing percentage was negatively correlated to slaughter weight, total lipids, conformation score and fatness score. Carcass fatness score was positively correlated with total lipid content and as a consequence, had a similar relationship with the conformation score. These relationships agree with results reported earlier [26,28].

Table 4. Correlation coefficients (r) between carcass characteristics with slaughter weight, total lipids, conformation score and fatness score of young female Simmental beef cattle.

	SW[d]	TL[e]	CS	FS
Slaughter weight (kg)	-	0.537	-0.002	-0.105
Hot carcass weight (kg)	0.176	0.604	0.109	-0.115
Cold carcass weight (kg)	0.971	0.596	0.097	-0.115
Dressing percentage (%)[a]	-0.071	-0.064	-0.097	-0.234
Forequarter weight (kg)	0.959	0.500	0.055	-0.119
Hindquarter weight (kg)	0.963	0.648	0.139	-0.108
Conformation score[b]	0.185	0.143	-	0.094
Fatness score[c]	-0.054	0.313	0.094	-

[a] carcass weight x 100/live weight.
[b] CS Conformation score – 1 = P (poorest) to 18 = S+ (best).
[c] FS Fatness score – 1= (leanest) to 15 = 5+ (fattest).
[d] SW Slaughter weight (kg).
[e] TL Total lipids (%).
*P<0.05; **P<0.01; ***P<0.001.

Acknowledgements
This paper was supported by the Ministry of Education, Science and Technological Development of the Republic of Serbia, Project "Selected Biological Hazards to the Safety/Quality of Food of Animal Origin and the Control Measures from Farm to Consumer" (TR 31034) and HERD, Agriculture: Project "Research, education and knowledge transfer promoting entrepreneurship in sustainable use of pastureland/grazing" 09-1548 (332160UÅ) (Norway).

References
[1] Smith S B, Gill C A, Lunt D K and Brooks M A 2009 Regulation of fat and fatty acid composition in beef cattle *Asian-Aust. J. Anim. Sci.* **22** 9 1225–1233
[2] Mach N, Bach A, Velarde A and Devant M 2008 Association between animal, transportation, slaughterhouse practices, and meat pH in beef *Meat Sci.* **78** 3 232-238
[3] Lukic M, Ivanovic J, Starcevic M, Djordjevic J, Markovic R and Baltic M 2016 Carcass performance of Simmental and Holstein Friesian beef cattle in Serbia *Meat Technol* **57** 2 95-101
[4] Indurain G, Carr T R, Goñi M V, Insausti K and Beriain M J 2009 The relationship of carcass measurements to carcass composition and intramuscular fat in Spanish beef *Meat Sci* **82** 2 155-161
[5] Moreira F B, Souza N E, Matsushita M, Prado I N and Nascimento W G 2003 Evaluation of carcass characteristics and meat chemical composition of *Bos indicus×Bos taurus* crossbred steers finished in pasture systems *Braz. Arch. Biol. Techn.* **46** 609-616
[6] Prado I N, Oliveira A N, Rotta P P, Perotto D, Prado I N, Silva R R, Souza N E and Moletta J L 2009 Chemical and fatty acid composition of *Longissimus* muscle of crossbred bulls finished in feedlot *Asian-Aust. J. Anim. Sci.* **22** 1054-1059
[7] Osoro K, Martínez A and Castro P 2003 Desarrollo de Sistemas Eficientes de Producción de Carne de Calidad en Zonas Bajas KRK Ediciones (SERIDA), Oviedo, Spain
[8] Maltin C A, Lobley G E, Grant C M, Miller L A, Kyle D J, Horgan G W, Matthews K R and Sinclair K D 2001 Factors influencing beef eating quality 2. Effects of nutritional regimen and genotype on muscle fiber characteristics *Anim. Sci.* **72** 279-287
[9] Moloney A P, Fallon R J, Mooney M T and Troy D J 2004 The quality of meat and fatness of bulls offered ad libitum concentrates, indoors or at pasture *Livest Prod Sci* **87** 2 271-276
[10] Nogalski Z, Wroński M, Wielgosz-Groth Z, Purwin C, Sobczuk-Szul M, Mochol M and Pogorzelska P 2013 The effect of carcass conformation class (Europe system) on the slaughter quality of young crossbred beef bulls and Holstein-Friesians/Porównanie wartości rzeźnej buhajków mieszańców mięsnych i holsztyno-fryzów w zależności od klasy uformowania w systemie EUROP. *Ann. Anim. Sci.* **13** 1 121-131
[11] Baltic M 1994 *Kontrola namirnica* Institut za higijenu i tehnologiju mesa Beograd
[12] ISO 1997 *Determination of moisture content* ISO 1442:1997 standard In: International standards meat and meat products Geneva Switzerland: International Organization for Standardization
[13] ISO 1992 *Determination of total fat content* ISO 1443:1992 standard In: International standards meat and meat products Geneva Switzerland: International Organization for Standardization
[14] ISO 1992 *Determination of nitrogen content* ISO 937:1992 standard. In: International standards meat and meat products Geneva Switzerland: International Organization for Standardization
[15] Maraschiello C, Diaz I, Regueiro J AG 1996 Determination of cholesterol in fat and muscle of pig by HPLC and capillary gas chromatography with solvent venting injection *J. High Res. Chromatog.* **19** 165–168
[16] Zapletal D, Chládek G, Subrt J 2009 Breed variation in the chemical and fatty acid. Breed variation in the chemical and fatty acid compositions of the longissimus dorsi muscle in Czech Fleckvieh and Montbeliarde cattle *Livest Sci* **123** 28–33.
[17] Piasentier E et al. 2009 Fatty acid composition and sensory properties of Italian Simmental beef

as affected by gene frequency of Montbeliarde origin *Meat Sci* **83** 543–50

[18] Drca D 2009 Examination of lean meat in beef cattle carcasses in slaughterhouse in Cajetina University of Belgrade, Serbia

[19] Dokmanovic M, Lukic M, Baltic M Z, Ivanovic J, Markovic R, Grbic S and Glamoclija N. 2014 Analysis of beef production volume in Serbia from 1985 to 2011 *Tehnologija Mesa* **55** 1 73–80

[20] Kamieniecki H, Wójcik J, Pilarczyk R, Lachowicz K, Sobczak M, Grzesiak W, Błaszczyk P 2009 Growth and carcass performance of bull calves born from Hereford, Simmental and Charolais cows sired by Charolais bulls *Czech J Anim Sci.* **54** 2 47–54

[21] Waritthitham A, Lambertz C, Langholz H J, Wicke M, Gauly M 2010 Assessment of beef production from Brahman x Thai native and Charolais x Thai native crossbred bulls slaughtered at different weights I: Growth performance and carcass quality. *Meat Sci* **85** 191–195

[22] Sanudoa C, Maciea E S, Olletaa J L, Villarroel M, Paneaa B, Albert P 2004 The effects of slaughter weight, breed type and ageing time on beef meat quality using two different texture devices *Meat Sci* **66** 925–932

[23] Minchin W, Buckley F, Kenny D A, Monahan F J, Shalloo L, O'Donovan M 2009 Effect of grass silage and concentrate based finishing strategies on cull dairy cow performance, carcass and meat quality characteristics *Meat Sci* **81** 1 93–101

[24] Rule D C, MacNeil M D, Short R E 1997 Influence of sire growth potential, time on feed, and growing-finishing strategy on cholesterol and fatty acids of ground carcass and *Longissimus* muscle of beef steers *J. Anim. Sci.* **75** 1525-1533

[25] Rotta P P, Prado I N, Prado R M, Moletta J L, Silva R R, Perotto D 2009 Carcass characteristics and chemical composition of the *Longissimus* muscle of Nellore, Caracu and Holstein-Friesian bulls finished in a feedlot *Asian-Aust. J. Anim. Sci.* **22** 598-604

[26] Greiner S P, Rouse G H, Wilson D E, Cundiff L V, Wheeler T L 2003 Prediction of retail product weight and percentage using ultrasound and carcass measurements in beef cattle *J. Anim. Sci.* **81** 7 1736-1742

[27] Realini C E, Williams R E, Pringle T D, Bertrand J K 2001 *Gluteus medius* and rump fat depths as additional live animal ultrasound measurements for predicting retail product and trimmable fat in beef carcasses *J. Anim. Sci.* **79** 6 1378-1385

[28] Tait R G, Wilson D E, Rouse G H 2005 Prediction of retail product and trimmable fat yields from the four primal cuts in beef cattle using ultrasound or carcass data *J. Anim. Sci.* **83** 6 1353-1360

[29] Lee C Y, Henricks D M, Skelley G C, Grimes L W 1990 Growth and hormonal response of intact and castrate male cattle to trenbolone acetate and estradiol *J. Anim. Sci.* **68** 2682-2689

IOP Conf. Series: Earth and Environmental Science **85** (2017) 012062 doi:10.1088/1755-1315/85/1/012062

Distribution of manganese and selenium in four different pork cuts commercially available in the Serbian market

J Đinović-Stojanović, D Nikolić, S Janković, D Vranić, M Milijašević and J Babić-Milijašević

Institute of Meat Hygiene and Technology, Kaćanskog 13, 11000 Belgrade, Serbia

E-mail: jasna.djinovic@inmes.rs

Abstract. This study aimed to provide information on levels of Mn and Se in four different pork cuts (loin, neck, hind leg and shoulder) commercially available on the Serbian market, with a view to providing information on dietary intakes of metals associated with the consumption of these meat cuts. In total, for 50 pork cuts, the levels of Mn and Se were determined by inductively-coupled plasma mass spectrometry (ICP-MS). The following ranges of Mn and Se were found (mg kg^{-1}) in loin 0.055-0.130 and 0.074-0.365, in neck 0.014-0.365 and 0.045-0.196, in hind leg 0.032-0.099 and 0.066-0.123, in shoulder 0.012-0.290 and 0.027-0.515, respectively. The highest mean levels were obtained for Mn (0.124 mg kg^{-1}) in shoulder and for Se (0.209 mg kg^{-1}) in loin. The Estimated Daily Intake (EDI) of essential elements through consumption of 114.1 g mammalian meat/person/day was below 1% of Adequate Intake (AI) for Mn and between 18.9% and 43.2% of Recommended Dietary Allowance (RDA) for Se.

1. Introduction

Meat and meat products are an excellent source of the major essential nutrients such as hydrophilic vitamins, high quality proteins and minerals etc. Red meat is the ideal dietary source of vitamin B12 and several minerals, while poultry supplies an important amount of niacin [1]. The levels of essential elements (e.g. Zn, Cu, Fe, Mn, Se) in meat depend on meat type [2]. These elements are critically important for life, are necessary for adequate physiological function and should be available through dietary intake. Insufficient intake of essential elements can cause improper metabolic function and can cause fatigue, poor growth, anaemia, chronic diseases and ultimately, death [3]. Essential elements act as enzyme cofactors, organic molecule stabilizers, participants in redox reactions etc [4].

The latest available data according to the UK Nutrient Databank [5], maintained by the Food Standards Agency, contains extensive information on the nutrient content of foods commonly consumed and is published by McCance and Widdowson's 'The Composition of Foods' (CoF) book series [6]. This databank contains Mn and Se levels for more than 3000 items of food and alcoholic beverages (table 1). It shows that meat and meat products are an average source of Mn and Se in the presented food sub-groups (table 1).

Table 1. Amounts of Mn and Se in different food sub-groups (adapted from McCance and Widdowson, 2015).

Food sub-groups	Mn	Se
	Range (min-max)	
	[mg/100g]	[μg/100g]
Cereals and cereal products	0.13-1.91	3-10
Milk and milk products	0.01-0.50	1-11
Vegetables	0.10-3.10	1-21
Nuts and seeds	1.30-1.80	1-32
Fruits	0.1-0.8	1-8
Fish and fish products	0.01-0.54	12-43
Meat and meat products	0.01-0.27	2-15
Beverages	0.01-4.70	1-5
Sugars, preserves and snacks	0.01-3.30	1-4
Soups, sauces and miscellaneous foods	0.03-0.57	1-2
Alcoholic beverages	0.01-0.05	--

Meat and meat products contribute a significant part of the human diet in Serbia. Globally, pork is the most widely consumed meat (15.8 kg/capita/year), followed by poultry (13.6 kg/capita/year), beef (9.6 kg/capita/year) and finally sheep and goat meat (1.9 kg/capita/year) [7]. Meat consumption statistics vary among and within countries. The World Health Organization (WHO), through the Global Environment Monitoring System – Food Contamination Monitoring and Assessment Programme (GEMS/Food) cluster diets 2012 [8], gives data for the daily intake of 383 different food items from 183 countries. According to that data, the daily intake of meat for adult Serbian population is 114.1 g of mammalian species [8].

The main objectives of this study were to: (1) analyse and compare the levels of Mn and Se in four different pork cuts from the Serbian market; (2) determine which pork cuts contained the highest levels of each element; (3) estimate the daily intake of analysed essential elements derived from oral consumption of pork cuts.

2. Materials and Methods

2.1 Sample collection

A total of 50 different pork cuts (loin, neck, hind leg and shoulder) commercially available in Serbia were collected during one year (from September 2014 to August 2015). After collection, meats were labelled and stored in polyethylene bags and frozen at -18°C prior to analysis.

After acid mineralization of homogenized pork cuts, microwave digestion (Digestion System: Milestone, Sorisole, Italy) was used for sample preparation. Analysis of Mn and Se was performed by inductively coupled plasma mass spectrometry (ICP-MS), (iCap Q mass spectrometer, Thermo Scientific, Bremen, Germany). The most abundant isotopes were used for quantification. The accuracy of the analysis was verified by analysing the certified reference material NIST SRM 1577c (bovine liver, Gaithersburg, MD, USA).

2.2 Statistical analysis

Statistical analysis of experimental data was performed using software Statistica 10.0 (StatSoft Inc., Tulsa, OK, USA). Analysis of variance (ANOVA) and Tukey's HSD comparison of the means of results were used for analysing variations.

2.3 Estimated daily intake

The Estimated Daily Intake (EDI) of Mn and Se through consumption of analysed pork cuts was calculated. The following equation was used:

$$EDI\ (mg\ day^{-1}) = C_{elements} \times DC_{pork\ cut}$$

where $C_{elements}$ is the concentration of element (mg kg^{-1}) detected in pork cut and $DC_{pork\ cut}$ is the average per capita daily consumption of mammalian meat (114.1 g), [8].

3. Results and Discussion

The results obtained for the two elements in four different pork cuts were compared with the literature data. However, for most of the literature data, exact specifications of the samples analysed were not available. Table 2 shows grouped results for Mn and Se levels in pork cuts (SD – standard deviation, n – number of samples).

Table 2. Mn and Se levels (mean ± SD*) of selected pork cuts.

	n**	Levels (mg kg^{-1}) Mn	Se
Pork cuts			
1 Loin	10	0.088±0.026	0.209±0.109
2 Neck	14	0.089±0.056	0.120±0.051
3 Hind leg	6	0.064±0.024	0.091±0.023
4 Shoulder	20	0.124±0.078	0.150±0.126

The lowest average Mn and Se levels were found in hind leg while the highest average Mn and Se levels were observed in shoulder and loin, respectively. Post-hoc Tukey's HSD test showed that there were no statistically significant differences (at $p<0.05$ level) in the Mn and Se levels in various pork cuts (figure 1).

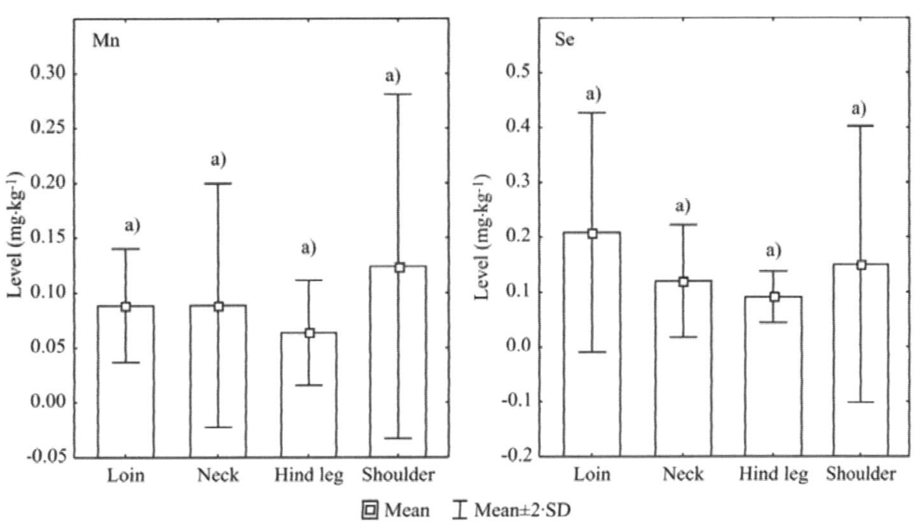

Figure 1. Levels of Mn and Se pork cuts. Data are presented as mean ± 2SD. Different letters in the bars of each element indicate significant differences of means between types of canned fish, according to Tukey's HSD test (p<0.05).

The Mn levels determined in pork cuts (0.064-0.124 mg kg^{-1}) were more than two to almost five-fold higher than those reported by Tomović et al. [9] in pigs from different genetic lines (Mn: 0.025 ± 0.004 mg/kg) but lower than levels given Batista et al. [2] in pork (0.214 mg kg^{-1}). Se levels in the analysed pork cuts (0.091-0.209 mg kg^{-1}) were higher than reported by Jablonska et al. [10] in 24 selected meat (ranging from 0.007 to 0.161 mg kg^{-1}), and Fajt et al. [11], in pork samples from different herds in Czech Republic (0.087 mg kg^{-1}). On the other hand, Se levels in pork reported by Batista et al. [2], (0.461 mg kg^{-1}) were higher than those obtained in our study.

Table 3 shows the EDI of Mn and Se, based on the average Serbian per capita consumption of mammalian meat. Estimates are for adults up to 19 years old. The results are expressed as % of the Adequate Intake (AI) and the Recommended Dietary Allowance (RDA) published by Institute of Medicine [12,13]. The results obtained showed that the contribution was dependent on the type of pork cut. Overall, pork cuts were estimated to provide between 0.32 and 0.79 % of the AI for Mn and between 18.9 and 43.2 % of the RDA for Se. The results from this study showed that estimated pork cut consumption levels (for loin, neck, hind leg and shoulder) can be considered as important dietary sources of Se. On the other hand, other food types are clearly necessary to provide adequate dietary levels of Mn for Serbian populations.

Table 3. Estimated daily intakes (EDI) of essential elements based on the average Serbian per capita consumption of mammalian species (114.1 g person^{-1} day^{-1}) [8].

			EDI (expressed as % of the AI[a] or % of the RDA[b])							
Elements	Males	Females	Loin		Neck		Hind leg		Shoulder	
	AI (mg day^{-1})		M	F	M	F	M	F	M	F
Mn	2.3	1.8	0.44	0.56	0.44	0.56	0.32	0.41	0.62	0.79
	RDA (μg day^{-1})									
Se	55	55	43.2	43.2	24.9	24.9	18.9	18.9	31.1	31.1

[a]AI – Adequate intake (AI) for males (M) and females (F) up to 19 years old [12].
[b]RDA – Recommended Dietary Allowance (RDA) for males (M) and females (F) up to 19 years old [13].

4. Conclusion

This study revealed that pork cuts can be considered as important dietary sources of Se, but other food types are necessary to provide adequate dietary levels of Mn for Serbian population. The results showed there were no significant differences in the levels of Mn and Se among the analysed pork cuts. Mn was the most abundant in shoulder, while Se prevailed in loin. Periodic control of meat and meat products is necessary to provide more data on essential as well as on toxic elements to ensure the quality and safety of meat and meat products.

Acknowledgments

This work was supported by grants from the Ministry of Education, Science and Technological Development of the Republic of Serbia (project no. III 46009).

References

[1] Marangoni F, Corsello G, Cricelli C, Ferrara N, Ghiselli A, Lucchin L and Poli A 2015 *Food Nutr. Res.* **59** 27606
[2] Batista B L, Grotto D, Hornos Carneiro M F and Barbosa F Jr 2012 *J. Toxicol. Env. Heal. A.* **75** 1269
[3] FAO/WHO 2002 *Report of a joint Food and Agriculture Organization of the United Nations/World Health Organization expert consultation* Bangkok Thailand.
[4] Fleet J C, Replogle R and Salt D E 2011 *J. Nutr.* **141** 520
[5] UK Nutrient Databank. http://www.fooddatabanks.ifr.ac.uk/
[6] McCance R A and Widdowson E W 2015 *The 7th edition of McCance and Widdowson's the Composition of Foods* (London: Public Health England, Wellington House) p 1-550

[7] FAOSTAT. Food and Agriculture Organization of the United Nations 2014. http://faostat.fao.org/
[8] www.who.int/foodsafety/chem/gems/en
[9] Tomović M V, Petrović S Lj, Tomović S M, Kevrešan S Ž and Džinić R N 2011 *Food Chem.* **124** 342
[10] Jablonska E, Gromadzinska J, Klos A, Bertrandt J, Skibniewska K, Darago A and Wasowicz W 2013 *J. Food Comp. Anal.* **31** 259
[11] Fajt Z, Drabek J, Steinhauser L and Svobodova Z 2009 *Neuro Endocr. Lett.* **30** 17
[12] Institute of Medicine 2000 *Dietary Reference Intakes for Vitamin C, Vitamin E, Selenium, and Carotenoids* (Washington DC: The National Academies Press)
[13] Institute of Medicine 2001 *Dietary Reference Intakes for Vitamin A, Vitamin K, Arsenic, Boron, Chromium, Copper, Iodine, Iron, Manganese, Molybdenum, Nickel, Silicon, Vanadium, and Zinc* (Washington, DC: The National Academies Press)

IOP Conf. Series: Earth and Environmental Science **85** (2017) 012063 doi:10.1088/1755-1315/85/1/012063

A comparison between Warner-Bratzler shear force measurement and texture profile analysis of meat and meat products: a review

S Novaković and I Tomašević

Animal Source Food Technology Department, Faculty of Agriculture, University of Belgrade, Nemanjina 6, Belgrade, Republic of Serbia.

E-mail: sasa.novakovic@agrif.bg.ac.rs

Abstract: Texture is one of the most important characteristics of meat and we can explain it as the human physiological–psychological awareness of a number of rheological and other properties of foods and their relations. In this paper, we discuss instrumental measurement of texture by Warner-Bratzler shear force (WBSF) and texture profile analysis (TPA). The conditions for using the device are detailed in WBSF measurements, and the influence of different parameters on the execution of the method and final results are shown. After that, the main disadvantages are reflected in the non-standardized method. Also, we introduce basic texture parameters which connect and separate TPA and WBSF methods and mention contemporary methods with their main advantage.

1. Introduction

Meat texture is a feature that can be defined by certain homogeneous properties which are detected by human senses relating to vision, hearing, somesthesis and kinesthesis [1]. These properties are perceived as hardness/firmness, gumminess, resilience, cohesiveness, springiness, adhesiveness, and viscosity. However, the textural properties of meat are often adapted by food processing, where the aim is often to make the structure of meat, and food in general, more delicate and easier to chew. The methods used for texture assessment can be separated into three groups: sensory, instrumental and indirect methods. Instrumental methods of texture assessment frequently apply mechanical analyses, measuring the food resistance, as the opposing force of the food is more solid than the strength of gravity. Since the applied power is beyond the strength of the tested sample, the sample is frequently ruined in this procedure. Therefore, the mechanical test of texture measurement is typically destructive [12]. The Warner-Bratzler Shear Force (WBSF) test and texture profile analysis are classic instrumental methods for estimation of meat tenderness (toughness). The study of Dar and Light [13] pointed out the key role of the texture when it comes to food quality identification by consumers, and also the influence of consumer attitudes. However, in this review, only the instrumental methods for analyzing meat texture will be explored.

2. Instrumental measurement of food texture

Several intrinsic and extrinsic features affect meat quality characteristics, including the trait of tenderness. These factors are separated into pre-slaughter factors and post-slaughter ones. In addition to animal stress, pre-slaughter factors include species, genotype, nutrition and age of the animal that typically affects weight and fatness of the carcass. Post-slaughter features refer to methods of stimulation, scalding, dangling, ageing, etc. Tenderness is an attribute of meat, and food in general, which is measured as a sensory characteristic. Moreover, as well as juiciness, it contributes to the mouth-feel [14], and combined with texture, juiciness and taste, it makes up the whole sense of quality as perceived by customers. Since meat is processed in consumers' mouth, its thermo-mechanical neutral features are of great importance when it comes to the mouth-feel, as well as the perception of smell and moisture. Texture measurement can be assessed by different instrumental methods. Puncture, compression, shear and tension are the main and generally used procedures for evaluating texture, giving values of force, deformation, slope and area [2].

3. Warner-Bratzler Shear Force (WBSF)

The most frequently applied instrumental procedure for assessing meat tenderness that has been used since 1930s is the WBSF test [3]. This test measures the maximum force (N) as a function of knife movement (mm) and the compression to shear (cut off) a sample of meat (MPa). The result of this measurement shows the hardness (toughness) of meat [13]. The term shear refers to sliding of meat parallel to the plane of contact, with the applied force tangential to the segment. Nevertheless, this word is commonly used in food technology to attribute any cutting action which splits a product into two fragments.

In the WBSF method, different devices for analysis can be used with a particular head or blade attached to them. These include machines such as Texture Analyzers [4], Instron devices or other common test devices [5]. Therefore, WBSF is performed either by a unique machine or by some other automatic device with the WBSF blade mounted to it. In the examination, a blade cuts through the meat samples so that shearing is perpendicular to the longitudinal positioning of the muscle fibers [3].

It was previously stated that exact requirements should be provided for WBSF test, both regarding the meat sample and the device used. Numerous studies have been carried out testing dissimilar modifications. One such study was conducted by Voisey and Larmond [15], who observed the effect of the changing angle of the cutting edges of the blade. They came to the conclusion that if the angle of the blade extends from 30° to about 70°, it increases the shear force. On the other hand, widening of the angle over this point does not lead to more increase in shear force. Separately from this survey, the above-mentioned researchers also studied different blade thicknesses and the width between the blade and the anvil [15]. They also concluded that changes in the test performance rate caused noteworthy variations in the rupture force as well as other evaluated parameters. They proved that alternations in the rate travel of the anvil did not have an important influence on the increase of the correlation between the receptive tenderness rating and the WBSF rating.

Voisey and Larmond [15] studied differences in the features of the blades produced by various manufacturers. These differences included the blade thickness, the angle of the hole, the clearance between the head and the anvil, etc. They came to the conclusion that it was necessary to standardize the Warner-Bratzler blade dimensions and specifications in order to avoid getting inconsistent results from various laboratories which all claimed to have used a 'Warner-Bratzler' blade. The original blade was made of stainless steel. On the other hand, the modern Warner-Bratzler blades are made of aluminum alloy which is not as resistant to wear as stainless steel and therefore probably suffers changes in the dimensions more quickly than stainless steel.

The meat samples must be uniformly round and of the same diameter for the WBSF test. Specifically, beef samples and other animals' large muscles in general are supposed to be cut cylindrically with an internal diameter of either 0.5 or 1 inch (1.27 or 2.54 cm). On the other hand, smaller muscles are, without cutting, put into the triangular hole of the blade. Afterwards, the sample is sheared into two pieces. The newly obtained surface cross-section is measured and included as a correction in the WBSF calculation. This cross-section area can be evaluated by pressing the surface on a piece of filter paper, marking the line around it and later measuring it by planimeter.

Nowadays, the interrelation between the diameter of the cross-section of noncylindrical samples and the WBSF is still not evident, although there has been research on this subject. Kastner and Henrickson

[6] tested cooked pork chops and discovered a nonlinear correlation between diameter and WBSF. Nevertheless, the results change when the data is recalculated according to cross-section, and the relationship looks to be linear, meaning that the shear force linearly corresponds to the cross-section area. Pool and Klose [16] found comparable results with cooked turkey meat. They noted that the force was proportional to diameter. Other researchers used different samples and equipment but the results are variable and are not clear enough to draw evident conclusions, except that the sample diameter should be uniform for each individual study. Naturally, this makes comparisons between different institutions and machines/protocols difficult. Wheeler et al. [7] tested how sampling, cooking and coring influence WBSF values for beef, and compared the shear evaluations of five institutions. They established the necessity of standardized procedures in order to accomplish consistent results for WBSF tests on cooked beef.

Numerous requirements must be fulfilled when it comes to the automatic testing machine, as well as the blades used. They must be V-notch blades made for the WBSF machine that meet the precise specifications such as the thickness, the bevel on the cutting edge, etc. Warner-Bratzler shear blade specifications are: (1) thickness of 1.1684 mm (0.046 inches); (2) V-notched (60° angle) cutting blade; (3) cutting edge beveled to a half-round; (4) angle of V rounded to a quarter-round of a 2.363 mm diameter circle; (5) spacers providing 2.0828 mm gap for the cutting blade to slide through.

Meat must also be standardized by cooking and chilling overnight to 2-5°C. After chilling, the meat is firm enough to be adequately cored. If this standard chilling step is not used, then the meat should undergo some other procedure to provide consistent temperature, and hence, uniform diameter cores. The width should be the same for each round core; 1.27 cm (0.5 inches). The cores must be removed parallel to the longitudinal direction of the muscle fibers which provides for them to be sheared perpendicular to the muscle fiber orientation.

The automatic testing machines should be used at the crosshead speed of 200 to 250 mm/minute. Any other shear tests which are not carried out according to these specifications (for example using a different blade or a blade not appropriately beveled) or on samples with unfulfilled requirements must not be called WBSF tests.

4. Texture profile analysis (TPA)

TPA is a procedure invented in 1963 by a group of scientists at General Foods Corporation. Originally, the procedure was designed to be conducted on a specific instrument known as the General Foods Texturometer (GFT), and it was available to anyone who had access to this instrument. In 1968, the method was modified and adjusted by Bourne in order to function on an Instron Universal Testing Machine (IUTM) [8]. His adjustments changed the experimental protocol, but at the same time, he managed to overcome some instrumental difficulties of its predecessor.

The main issues with the GFT performance were deformation of food samples and unreliable instrumental readings. The device was built as a human jaw, thus reproducing the process of mastication. In the procedure, the engaged power was in a sinuosity mode and chewing mimicry was achieved by motions of a lever with a plunger set on it. However, as the plunger moved towards the plate and mimicked about 42 bites every minute, it also deformed the food sample. The deformations were uneven due to the lever rotation and different influences of the plunger [8]. The direction of pressure changed as the lever swept through its arc. In addition to these issues, another problem with the GFT was the fact that instrumental interpretations were not solely based on deformation and stresses resulting from the food, because there was some flexibility in the construction of the strain gauges attached to the lever, which were used for measuring the stresses.

The main indicators of TPA analysis can be divided into primary and secondary (Table 1).

Table 1. Primary and secondary characteristics of meat texture [17].

Parameter	Sensorial definition	Instrumental definition
Primary		

characteristics		
Hardness	Force obligatory to compress a food between molars. Definite as power needed to reach given deformation	Peak power of the first compression cycle
Springiness	Proportion at which a deformed material goes back to its unreformed state after deforming power is removed	Height that the food recuperates during the time that elapses between the end of the first chew and the start of the second chew
Adhesiveness	The effort needed to overwhelm the attractive forces between the superficies of the food and the superficies of other constituents with which the food derives into interaction (e.g. tongue, teeth). Work obligatory to pull food away after a superficial	The negative part for the first chew, representing the effort needed to pull compressing sound away after sample
Cohesiveness	The force of internal bonds compensates the body of the produce (superior the value the superior the cohesiveness)	The proportion of positive energy throughout the second to that of the first compression sequence (descending strokes only)
Secondary characteristics		
Brittleness (Fracture force)	Power at which a material fractures. Connected to the primary parameters of hardness and cohesiveness, where fragile materials have low cohesiveness. Not all foods rupture and thus value may tell to hardness if only single peak is current. Inelastic foods are never adhesive	The first important break in the first compression round
Gumminess	Energy obligatory to crumble a semi-solid food produce to a state prepared for swallowing. Connected to foods with low hardness height	Calculated parameter: Produce of Hardness x Cohesiveness
Chewiness	Energy obligatory to chew a solid food to a state where it is prepared for swallowing. Characteristic is problematic to quantify exactly due to difficulties of mastication (shear, penetration)	Calculated Parameter: Produce of Gumminess x Springiness (basically primary parameters of Hardness x Cohesiveness x Springiness)

The TPA test imitates the chewing process similar to the one in the human mouth, and its performance speed is equivalent to that of the human jaw. Many studies aimed to check the human bite speed and calculated it to be between 33 and 66 mm/s. Nevertheless, it was proved that sensory correlations with tests are greater if the speeds are higher. If TPA parameters are applied to different types of food, the significance of standardization and protocol for the procedure used must be cited.

Barbut et al. [9] presented variations in sample length (L) from 10 to 20 mm, diameter (D) from 13 to 73 mm, and D/L ratio from 1 to 4. Furthermore, the compression ratio varied from 50 to 85% and compression speed from 5 to 200 mm/min. The effects of varying D/L, speed and compression rate on beef wieners were studied by the same authors [9]. A decrease in D/L resulted in a decrease of hardness, cohesiveness and gumminess, and an increase in springiness and chewiness.

Increasing the compression rate causes reduced springiness, cohesiveness, gumminess and chewiness. At the same deformation rate, a shorter sample is actually deformed at a higher strain rate and, consequently, should exhibit higher stress than a longer sample under the same strain. Thus, TPA parameters are comparable only when the tests are performed by a standard procedure [9]. The values obtained for a ground salami meat product and a whole muscle corned beef product resulted in recommending the following test parameters: D/L = 1.5; compression ratio = 75%; and speed rate = 1-2 cm/min. Using these standard conditions will allow direct comparison of data from different laboratories/institutions and reduce confusion and mistakes that result from selecting inappropriate parameters [9].

5. Comparison of instrumental methods for texture evaluation

Ruiz de Huidobro [10] studied the relationship of the WBSF test and TPA to the sensory features of beef, which, of course, can be related to its texture characteristics. Overall, the authors found the TPA test to be more suitable for beef texture assessments. The TPA assessment predicted sensory hardness better than the WBSF examination. WBSF and the sensory rating of chewiness were related, showing the decline in the course of aging. However, the receptive juiciness did not fluctuate significantly with aging. On the other hand, the WBSF test had the highest coefficient of variability (27.5%). Overall, as measures of toughness, the WBSF and TPA tests were positively correlated.

TPA seems more convenient for predicting sensory texture of meat than the WBSF method, provided the study is on raw meat. When cooked meat is examined, the WBSF method is better, although it is not a very accurate predictor of meat texture [10].

Conclusion

The evaluation of texture and structure measurements for meat and meat products is significant in quality control for meat industry. This review discussed the main instrumental methods, WBSF and TPA, used to measure meat texture. Both of them are useful for instrumental measurement of meat texture, with greater importance for TPA in raw beef texture evaluations. With all these findings, it is expected that TPA will be used much more for this purpose than WBSF in the future.

Acknowledgement

This work was performed within the National Project number TR46009, supported by the Ministry of Education, Science and Technology, Republic of Serbia.

References

[1] Hagen O, Solberg C, Sirnes E and Johnston I A 2007 Biochemical and structural factors contributing to seasonal variation in the texture of farmed Atlantic halibut (*Hippoglossus hippoglossus* L.) flesh *J. of Agri. Food Chem.* **55** 5803-8.

[2] Cheng J H, Dai Q, Sun D W, Zeng X A, Liu D and Pu H B 2013 Applications of non-destructive spectroscopic techniques for fish quality and safety evaluation and inspection *Trends Food Sci. Tech.* **34** 18-31.

[3] Destefanis G, Brugiapaglia A, Barge M T and Molin E Dal 2008 Relationship between beef consumer tenderness perception and Warner-Bratzler shear force *Meat Sci.* **78** 153-6.

[4] Cai J, Chen Q, Wan X and Zhao J 2011 Determination of total volatile basic nitrogen (TVB-N) content and Warner-Bratzler shear force (WBSF) in pork using Fourier transform near infrared (FT-NIR) spectroscopy *Food Chem.* **126** 1354-60.

[5] Yancey J W S, Apple J K, Meullenet J F and Sawyer J T 2010 Consumer responses for tenderness and overall impression can be predicted by visible and near-infrared spectroscopy, Meullenet–Owens razor shear, and Warner-Bratzler shear force *Meat Sci.* **85** 487-92

[6] Kastner C L and Henrickson R L 1969 Providing uniform meat cores for mechanical shear force measurement *J. Food Sci.* **34** 603-5
[7] Wheeler T L, Shackelford S D and Koohmaraie M 1996 Sampling, cooking, and coring effects on Warner-Bratzler shear force values in beef *J. Anim. Sci.* **74** 1553-62
[8] Bourne, M C 1968 Texture profile of ripening pears *J. Food Sci.* **33** 223-6
[9] Barbut S 2016 *The Science of Poultry and Meat Processing* (Guelph: University of Guelf)
[10] de Huidobro F R, Miguel E, Blázquez B and Onega E 2005 A comparison between two methods (Warner-Bratzler and texture profile analysis) for testing either raw meat or cooked meat *Meat Sci.* **69** 527-36
[11] Caine W R, Aalhus J L, Best D R, Dugan M E and Jeremiah L E 2003 Relationship of texture profile analysis and Warner-Bratzler shear force with sensory characteristics of beef rib steaks *Meat Sci.* **64** 333-9
[12] Saláková A 2012 *Instrumental Measurement of Texture and Color of Meat and Meat Products* vol 02/2012 (Brno: Maso International)
[13] Dar Y and Light J 2014 *Food Texture Design and Optimization* (Chichester: Wiley)
[14] Lawless H T and Heymann H 2010 *Sensory Evaluation of Food: Principles and Practices* (New York: Springer) 179-201.
[15] Voisey P W and Larmond L 1974 Examination of factors affecting the performance of Warner-Bratzler meat shear test *Canadian Institute of Food Science and Technology Journal* **7**
[16] Pool M F and Klose A A 1969 The relation of force to sample dimensions in objective measurement of tenderness of poultry meat *J. Food Sci.* **34** 524-6
[17] Coelho G M, Weschenfelder A V, Meinert E M, Amboni R D and Beirão L H 2007 Effects of starch properties on textural characteristics of fish burgers: sensory and instrumental approaches *Boletim do Centro de Pesquisa de Processamento de Alimentos* **25** (Curitiba: Universidade Federal do Parana)

IOP Conf. Series: Earth and Environmental Science **85** (2017) 012064 doi:10.1088/1755-1315/85/1/012064

Conventional and unconventional extraction methods applied to the plant, *Thymus serpyllum* L

D Đukić[1], P Mašković[1], S Vesković Moračanin[2], V Kurćubić[1], M Milijašević[2] and J Babić[2]

[1] University of Kragujevac, Faculty of Agronomy, Department of Food Technology, Cara Dušana 34, Čačak, Serbia
[2] Institute of Meat Hygiene and Technology, Kaćanskog 13, Belgrade, Serbia

E-mail: afdekanat@kg.ac.rs

Abstract. This study deals with the application of two conventional and three non-conventional extraction approaches for isolation of bioactive compounds from the plant *Thymus serpyllum* L. The extracts obtained were tested regarding their chemical profile (content of phenolics, flavonoids, condensed tannins, gallotannins and anthocyanins) and antioxidant activities. Subcritical water extract of *Thymus serpyllum* L. generally had the highest concentrations of the chemical bioactive compounds examined and the best antioxidant properties.

1. Introduction

Phenolic compounds, which are the product of secondary metabolism of plants, are one the most investigated class of natural products due to their wide range of biological activities, such as antioxidant, cytotoxic, antimicrobial, anti-inflammatory, antiulcer, antispasmodic, antiviral and many other activities [1, 2]. There are over 8000 compounds which belong to one of the following groups, among others: simple phenolics, phenolic acids, stilbenes, flavonoids, coumarins and tannins [3, 4].

For isolation of these and other compounds from their natural sources, many different approaches can be applied. Among them are conventional extraction techniques such as maceration and Soxhlet extraction and non-conventional techniques such as ultrasound-assisted, microwave-assisted and subcritical water extraction techniques [5, 6]. Every technique possesses certain advantages and disadvantages, and they are usually combined to obtain improved results.

There are only two studies which reported the useful properties of the plant *Thymus serpyllum* L. [7, 8]. The aim of this study was to apply conventional and non-conventional extraction techniques for isolation of biologically active compounds from *Thymus serpyllum* L., to investigate the biological activity of the extracts obtained and to establish their chemical profiles. Results were compared in order to evaluate the efficiency of the applied extraction techniques and the activities of the extracts obtained.

2. Materials and methods

2.1. Thymus serpyllum L

Thymus serpyllum L. (wild thyme) is a perennial shrub, native to areas of northern and central Europe, belongs to the family *Lamiaceae*. Due to its pharmacological properties, the essential oil of wild thyme, a plant used in traditional medicine, is an important natural resource for the pharmaceutical industry. In addition, it can be a source of natural antioxidants, nutritional supplements, or components of functional foods in the food industry. The chemical composition and yield of the essential oil of *Thymus serpyllum* are considered to be affected by geographic region, the development stage of the plant, the harvest season, habitat, and climatic conditions [9].

The aim of this study was to examine the use of different extraction methods, one conventional and one non-conventional, on the composition and quality of essential oil from *Thymus serpyllum* L. collected in Central Serbia.

2.2. Extractions

2.2.1. Conventional method. Soxhlet extraction was conducted in the following manner: plant material (75.0 g) was crushed and homogenized into small 3–5mm pieces by a cylinder crusher and placed in the Soxhlet apparatus. Extraction was carried out for eight hours using 96% ethanol as a solvent (600 mL). The Soxhlet extract (SE) obtained was filtered through filter paper (Whatman – qualitative filter paper Grade 1, Sigma-Aldrich) and evaporated by a rotary evaporator (Devarot, Slovenia) under vacuum and dried at 60ºC to the constant weight. The dried extracts were stored in a dark glass bottle at 4°C to prevent oxidative damage.

Maceration was conducted using the following procedure: plant samples (10.0 g) were extracted using 96% ethanol (300 mL) as a solvent. The extraction process was carried out under laboratory conditions at 22°C in a sheltered, dry place for seven days, with occasional shaking to improve the maceration process. After seven days, maceration extract (ME) was filtered through filter paper and concentrated to dry mass by a rotary evaporator under vacuum and dried at 60°C to the constant weight. The dried extracts were stored in a dark glass bottle at 4°C to prevent oxidative damage.

2.2.2. Non-conventional extractions. Ultrasound-assisted extraction (UAE) was performed in ultrasonic water bath (EUP540A, Euinstruments, France). A sample (5.0 g) was placed in volumetric flask and 100 mL of solvent (96% ethanol) was added. The mixture was sonicated for 30 min at a frequency of 40 kHz and ultrasound power of 90% (216 W).

Microwave-assisted extraction (MAE) was performed in a domestic microwave oven, which was previously modified for this purpose. Extraction was conducted using the same sample weight, solvent volume and extraction time. The extraction procedure program was as follows: one min pre-heating at 160 W; one min pre-heating at 320 W and 30 min extraction at 600 W.

Subcritical water extraction (SWE) was performed in a previously described home-made extractor system [10]. In all experimental runs, 5.0 g of plant material was mixed with 100 mL of double-distilled water. Extraction was performed at a pressure of 40 bars and at 140°C for 30 min. Agitation was assured by vibrational movements of vessel platform at a frequency of 3 Hz. After extractions, process vessels were immediately cooled in a flow-through water-bath at 20°C.

2.3. Spectrophotometric assays

Total phenolics (TPC) and total flavonoids (TFC) contents were determined using the previously described methods [11]. Condensed tannins (CT) were determined according to a previously described method which relies on the precipitation of proanthocyanidins with formaldehyde, while gallotannins (GA) were determined using the described potassium iodate assay [12]. Anthocyanins were determined according the previously described procedure [13] using pH single and differential methods. Antioxidant activities of the obtained extracts were determined using the following, previously described assays: total antioxidant capacity [14], lipid peroxidation assay [15], hydroxyl radical scavenging activity [16] and DPPH radical scavenging activity [17] with slight modification [18].

2.4. Statistical analysis
Statistical analysis was carried out using Statistica 6.0. (StatSoft Inc, Tulsa, US). All extractions were performed at least in triplicate unless specified otherwise. Results are presented as a value ± standard deviation (SD). Significance levels were defined at $p < 0.05$.

3. Results and Discussion

3.1. Chemical profile of Thymus serpyllum L. extracts
Results for TPC, TFC, CT, GA and TAC obtained using spectrophotometric assays are presented in table 1. According to the results, the highest contents of all compound classes was observed in SCW extract, while the lowest level of compounds was in SE extract. The amount of TPC obtained in SCW was about 50% higher than the corresponding value obtained by SE, while this percentage was slightly lower in the case of TFC, CT, GA and TAC. Results obtained for MAE and UAE extracts were slightly lower than those for SCW (about 10–20%), while MAC extract in some cases also achieved satisfactory results. The reasons for such diversity among the investigated extracts could be different mechanisms of thermal and mass transfers, as well as different solubility of compounds in the medium.

Antioxidant activity was established using four different assays: total antioxidant capacity, inhibition of lipid peroxidation, hydroxy radical scavenging and DPPH scavenging activities, the results of which are presented in table 2.

Table 1. Chemical profile of *Thymus serpyllum L.* extracts obtained by spectrophotometric assays.

Extract	TPC (mg GAE/g)[a]	TFC (mg RU/g)	CT (mg GAE/g)	GA (mg GAE/g)	TAC (mg C_3G/g)
SE	95.86 ± 0.49	16.18 ± 0.37	53.85 ±0.44	16.13 ± 0.94	102.05 ± 0.40
MAC	115.14 ± 0.26	19.57 ± 0.58	55.19 ±0.37	18.74 ± 0.96	111.63 ± 0.73
UAE	124.42 ± 0.87	20.55 ± 0.17	58.65 ± 0.54	19.19 ± 0.42	122.56 ± 0.19
MAE	133.56 ± 0.19	21.81 ± 0.60	59.70 ±0.19	20.36 ± 0.77	129.72 ± 0.54
SWE	141.12 ±0.23	23.24 ± 0.18	61.53 ± 0.81	23.56 ± 0.64	130.32 ± 0.28

[a] Results are mean values ± SD from three extractions.
SE – Soxhlet Extraction.
MAC – Macerate Extraction.
UAE – Ultrasound-Assisted Extraction.
MAE – Microwave-Assisted Extraction.
SWE – Subcritical Water Extraction.
TPC – total phenolics.
TFC – total flavonoids
CT – condensed tannins.
GA – gallotannins.
TAC – total anthocyanins.

Table 2. Antioxidant activity of *Thymus serpyllum L.* extracts obtained by spectrophotometric assays.

Extract	TA (mg AA/G)[a]	ILP_{50}[b] (mg/mL)	OH_{50}[b] (mg/mL)	IC_{50}[b] (mg/mL)
SE	111.34 ± 0.43	30.35±0.42	32.51±0.88	45.68±1.01
MAC	118.92±0.48	29.73±0.08	29.15±0.93	36.83±0.98
UAE	133.71±0.29	27.57±0.44	25.83±0.77	35.47±0.69
MAE	158.09±0.82	26.56±0.81	23.41±0.81	29,60±0.99
SWE	170.32±0.87	20.71±0.45	19.63±0.94	22.73±0.53

[a] Results are mean values ± SD from three extractions.
[b] ILP_{50}, OH_{50} and IC_{50} values were determined by non linear regression analysis.
SE – Soholot Extraction.
MAC – Macerated Extraction.
UAE – Ultrasound-Assissted Extraction.
MAE – Microwave-Assissted Extraction.
SWE – Subcritical Water Extraction.
TA – Total antioxidant activity.
ILP_{50} – lipid peroxidation activity.
OH_{50} – hydroxy radical scavenging activity.
IC_{50} – DPPH radical scavenging activity.

The highest antioxidant activity of all four tested antioxidant types was observed in SWE, while the lowest was detected in SE. MAE and UAE expressed similar lipid peroxidation activities, hydroxy radical scavenging and DPPH scavenging activities, while these extracts differed in total antioxidant capacity (MAE had more TA capacity). On the other hand, similar activity between MAC and SE was observed in the case of total antioxidant capacity, lipid peroxidation test and hydroxy radical scavenging, while DPPH activity was higher in SE.

4. Conclusion

The present study showed that *Thymus serpyllum* L. possesses good potential to be used as a source of biologically active compounds. Based on the presented results, the prepared extracts exhibited high antioxidant activity. Subcritical water extract generally showed the best properties, but other non-conventional techniques demonstrated satisfactory results. All in all, the results obtained in this study encourage further and deeper investigation of this plant together with the new fields and possibilities for its application.

Acknowledgments

The results presented in this paper are part of Project III No 46009 and TR 31057 funded by the Ministry of Education and Science of Serbia.

References

[1] Ćetković G S, Čanadanović-Brunet J M, Djilas S M, Tumbas V T, Markov S L and Cvetković D D 2007 *Int J Mol Sci* **8** 1013
[2] Roleira F M F, Tavares-da-Silva E J, Varela C L, Costa S C, Silva T, Garrido J and Borges F 2015 *Food Chem* **183** 235
[3] Croteau R, Kutchan T M and Lewis N G 2000 *Natural products (Secondary metabolites)*, in: Buchanan B B et al (Eds.) Biochemistry & Molecular Biology of Plants (American Society of Plants Biologists Rockville) p 1250
[4] Dewick P M 2009 *Medicinal Natural Products*: *A Biosynthetic Approach* (John Wiley & Sons Ltd Chichester, UK)
[5] Azmir J, Zaidul I S M, Rahman M M, Sharif K M, Mohamed A, Sahena F, Jahurul M H A,

Ghafoor K, Norulaini N A N and Omar A K M 2013 *J Food Eng* **117** 426
[6] Teo C C, Tan S N, Yong J W H, Hew C S and Ong E S 2010 *J Chromatogr A* **1217** 2484
[7] Guendouze-Bouchefa N, Madani K, Chibane M, Boulekbache-Makhlouf L, Hauchard D, Kiendrebeogo M, Stévigny C, Okusa P N and Duez P 2015 *Ind Crops Prod* **70** 459
[8] Pavlović R D, Lakušić B, Došlov-Kokoruš Z and Kovačević N 2009 *Pharmazie* **64** 656
[9] Jarić S, Mitrović M and Pavlović P 2015 Review of Ethnobotanical, Phytochemical, and Pharmacological Study of Thymus serpyllum L. *Evid Based Complement Alternat Med*
[10] Cvetanović A, Švarc-Gajić J, Gašić U, Tešić Ž, Zengin G, Zeković Z and Đurović S 2017 *J Supercrit Fluids* **120** 32
[11] Brighente I M C, Dias M, Verdi L G and Pizzolatti M G 2007 *Pharm Biol* **45** 156
[12] Vermerris W and Nicholson R 2006 *Phenolic Compound Biochemistry* (Springer (Dordrecht) The Netherlands) p 276
[13] Vulic J, Tumbas V, Savatovic S, Djilas S, Cetkovic G and Canadanovic-Brunet J 2011 *Acta Period Technol* **42** 271
[14] Prieto P, Pineda M and Aguilar M 1999 *Anal Biochem* **269** 337
[15] Hsu C K, Chiang B H, Chen Y S, Yang J H and Liu C L 2008 *Food Chem* **108** 633
[16] Hinneburg I, Dorman D H J and Hiltunen R 2006 *Food Chemistry* **97** 122
[17] Takao T, Kitatani F, Watanabe N, Yagi A and Sakata K 1994 *Biosci Biotechnol Biochem* **58** 1780
[18] Kumarasamy Y, Byres M, Cox P J. Jaspars M, Nahar L and Sarker S D 2007 *Phyther Res* **21** 615

IOP Conf. Series: Earth and Environmental Science **85** (2017) 012065 doi:10.1088/1755-1315/85/1/012065

Carcass quality traits of three different pig genotypes, White Mangulica, Duroc × White Mangulica and Large White pigs, reared under intensive conditions and slaughtered at 150 kg live weight

M Ivić[1], V Tomović[1], R Šević[2], M Jokanović[1], S Škaljac[1], N Džinić[1], B Šojić[1], T Tasić[3] and P Ikonić[3]

[1] University of Novi Sad, Faculty of Technology Novi Sad, Bulevar cara Lazara 1, 21000 Novi Sad, Serbia
[2] Group Univerexport Bačka ad, Joint-Stock Company, Meat Industry, Novosadski put 10, 21400 Bačka Palanka, Serbia
[3] University of Novi Sad, Institute for Food Technology, Bulevar cara Lazara 1, 21000 Novi Sad, Serbia

E-mail: ramajana1988@gmail.com

Abstract. The effect of breed on carcass composition was studied for autochthonous purebred White Mangulica (WM), crossbred Duroc x White Mangulica (DWM) and purebred Large White (LW) pigs. Pigs were slaughtered at a target body weight of about 150 kg. After slaughter, carcass yield, backfat thickness, the thickness of the lumbar muscle and chilling loss were measured and calculated. WM pigs had the highest percentage of carcass yield, while DWM produced an intermediate carcass yield, between those of the pure breeds. The backfat thickness was highest in WM pigs, compared to DWM and LW pigs (67, 41 and 27 mm, respectively, $P < 0.001$). WM and LW pigs had respectively the lowest and the highest thickness of the lumbar muscle (62 and 72 mm), with DWM pigs at an intermediate position (69 mm). As regards chilling loss, WM and DWM pigs showed better results than LW pigs (1.74, 1.75 and 1.92 %, respectively). Overall, evidence of additive genetic effects was present for all investigated parameters, with crosses showing intermediate values between pure breeds.

1. Introduction

Mangulica is one of three primitive autochthonous pig breeds in Serbia and is considered endangered. In Serbia there are three varieties: White (Blond), Swallow Belly and Red. This breed emerged in Hungary as a result of crossing several Hungarian aboriginal pig breeds which disappeared or were altered by the end of the nineteenth century by crossing with the Serbian Šumadinka pig [1].

However, in the middle of the last century, the population decreased and nearly disappeared because of the carcass composition, and poorer reproductive and growing performance than modern commercial breeds. Mangulica meat belongs to the group of fat-rich meats [2]. Its average total mass consists of 65% - 70% fat tissue and 30% - 35% meat [3]. Its fresh meat is darker, more succulent and softer than the meat from other pig breeds. Its odour is stronger. The tenderness of Mangulica pork is much higher than that obtained from any commercial pig breed [2].

The Mangulica breed is very resistant when kept in extensive conditions (free-range) on pasture, foraging on acorns, tubers, rhizomes and roots, it is well adapted to continental climate conditions and resistant to some pig diseases [1,2]. Today, the Mangulica breed is reared in an organized system in several herds in Serbia and is the most numerous indigenous pig breed [4]. Recently, the interest in this autochthonous breed has increased, not only for the purpose of gene preservation but also for production of meat products manufactured in traditional ways, which are much appreciated by consumers.

Crossbreeding is extensively used in pig production to increase the total efficiency of pig production and to improve the quality of the meat. Crossing with the Duroc pig breed is often used to improve the productivity of the autochthonous animals without greatly affecting their hardiness or reducing the level of intramuscular fat [5,6], because Duroc is notable in having a high muscle lipid (marbling fat) content relative to subcutaneous fat compared with other modern breeds [1,5-7].

This paper reports on carcass quality characteristics of White Mangulica pigs, both pure and crossbred with Duroc, reared under intensive farming conditions (system). Additionally, one of the tasks was to compare determined carcass quality traits of White Mangulica and its crosses with other European autochthonous breeds and their crosses with modern pig breeds.

2. Materials and Methods

A total of 60 pigs (female and castrated male) were studied: 20 purebred White Mangulica (WM) pigs, 20 crossbred pigs (Duroc sires crossed with White Mangulica dams - DWM) and 20 purebred Large White (LW) pigs. All animals were raised on a modern farm and slaughtered in a modern slaughterhouse in Serbia according to national legislation which is mainly harmonized with EU legislation. All animals were reared under the same environmental and production regime. Pigs were fed the same commercial diets and were slaughtered at a target body weight of about 150 kg. All pigs had *ad libitum* access to feed and water. Feed was withdrawn 12 h before slaughter, but water was freely available. On the slaughter day, pigs were individually weighed and transported to the slaughterhouse. The pigs were held in lairage for 2 h, with free access to water. All the animals were slaughtered and dressed in three days (20 pigs on each day), using standard commercial procedures [8].

All measurements were performed on the right side of the carcass. Carcass measurements were performed on the day of slaughter. Backfat thickness with the skin was measured in millimetres at the thinnest point over *M. gluteus medius* (S value). The thickness of the lumbar muscle in millimetres was measured as the shortest distance from the front (cranial) end of the *M. gluteus medius* to the upper (dorsal) edge of the spinal canal (M value). Chilling loss was expressed as the percentage of weight loss, measuring the mass of one carcass side after slaughter and at the end of the chilling process (24 h *post mortem*).

All data are presented as mean, standard deviation and range. Data were analysed statistically with one way ANOVA and post-hoc test (DUNCAN'S test). The software package STATISTICA 12 was used (StatSoft, 2015) for analysis.

3. Results and Discussion

The results of the determination of carcass quality traits (carcass yield, S value, M value and chilling loss) of three different pig genotypes (WM, DWM and LW) are presented in Table 1. The highest average carcass yield was found in WM pigs (84.7%), slightly lower average carcass yield was found in DWM pigs (84.0%), while LW pigs had the lowest average carcass yield (83.6%), but differences between the examined groups were not significant, indicating that the genotype did not affect carcass yield. The individual values of carcass yield ranged from 83.0 to 85.8% (WM), from 82.5 to 85.5% (DWM) and from 82.1 to 84.7% (LW).

Table 1. Carcass quality traits of three different pig genotypes (White Mangulica-WM, Duroc x White Mangulica-DWM and Large White pigs-LW).

Parameter	WM	DWM	LW	*P* value
Carcass yield (%)	84.7±1.0	84.0±0.9	83.6±0.9	0.055
Range of carcass yield	83.0–85.8	82.5–85.5	82.1–84.7	
S (mm)	$67\pm7^{a,o,x}$	$41\pm4^{b,p,y}$	$27\pm4^{c,q,z}$	<0.001
Range of S	55–80	35–46	22–35	
M (mm)	$62\pm3^{b,p,y}$	$69\pm6^{a,o,xy}$	$72\pm5^{a,o,x}$	<0.001
Range of M	55–65	60–80	65–80	
Chilling loss (%)	1.74 ± 0.10^{b}	1.75 ± 0.16^{b}	1.92 ± 0.20^{a}	0.031
Range of chilling loss	1.65–1.92	1.44–2.00	1.38–2.05	

^{abc} Means with different letters in the same row indicate significant differences at P<0.05.
^{opq} Means with different letters in the same row indicate significant differences at P<0.01.
^{xyz} Means with different letters in the same row indicate significant differences at P<0.001.

By comparing the obtained values for the carcass yield with the results of other authors who have also comparatively examined the carcass quality (carcass yield) of autochthonous breeds and/or their crosses with modern breeds, it can be concluded that in this study, higher average carcass yields were observed [9,10,11,12,13,14,15,16,17,18,19]. In this investigation, the slaughter of all pigs was carried out at a body weight of 150 kg, or at an average age of 244 (LW), 308 (DWM) and 532 (WM) days. However, according to a number of similar studies, increase in pigs' body weight (age increase) leads to a significant increase in pigs' carcass yield [16,20-23].

The highest average S value was found in WM pigs (67 mm), a lower average S value was found in DWM pigs (41 mm), while the lowest S value was found in LW pigs (27 mm). The individual S values ranged from 55 to 80 mm (WM), from 35 to 46 mm (DWM) and from 22 to 35 mm (LW). Differences in the average S values were statistically significant ($P < 0.001$) among all three examined pig genotypes, which indicates that the pig genotype affected the backfat thickness measured at the thinnest point over *M. gluteus medius* (S value). Comparable measurements for backfat thickness were found by several other authors [9,10,24,25] who have also comparatively examined the quality of carcasses of autochthonous breeds and/or their crosses with modern breeds and/or modern breeds, and reported the same trend. However, some authors [11,19,26,27] reported no significant difference in the backfat thickness between pigs of autochthonous breeds and their crosses with modern breeds of pigs.

Further, analysing the S value, it can be concluded that the S value varies greatly depending on the genotype [15,18,23,25,28,29]. Generally, native or autochthonous breeds have greater ability for fat deposition in comparison with modern breeds [10,21,30,31]. In similar studies, the highest backfat thickness over the *M. gluteus medius* at the thinnest point was found in the Iberian Retinto pigs (65.3 mm) reared under intensive production systems and whose pre-slaughter age was 231 days and the pre-slaughter weight was 144.4 kg [24], while in the case of Corsican x LW pigs, reared under the extensive conditions, fed with a concentrated diet and slaughtered with body weight of 141 kg, the highest backfat thickness over the *M. gluteus medius* was 48 mm [11]. The lowest backfat thickness over the *M. gluteus medius* was determined in LW pigs reared indoors and fed a commercial mixture (26.1 mm) [9,10].

Regarding the M value, the highest average value was found in LW pigs (72 mm), slightly lower average M value was found in DWM pigs (69 mm) and the lowest average M value was determined in WM pigs (62 mm). The individual M values ranged from 65 to 80 mm (LW), from 60 to 80 mm (DWM) and from 55 to 65 mm (WM). The average M value in LW and DWM pigs was statistically significantly higher ($P < 0.01$) in comparison with average M value in WM pigs. The results are in agreement with those of other studies. Other authors have also measured the thickness of *M. longissimus thoracis et lumborum*, ranging between 3rd and 4th lumbar vertebra [31] and at the last rib [32] in similar pig genotypes. Their results showed that the thickness of *M. longissimus thoracis et lumborum* was significantly higher in modern pigs compared to autochthonous pigs or in comparison with the crossbred pigs (modern x autochthonous pigs). There is also a relationship between the *M. longissimus thoracis et lumborum* surface (the loin eye area) and pig genotype [33]. An increase in the age and/or weight of similar pig genotypes leads to an increase in the area of the loin eye muscle [12,21,34,35].

Furthermore, the highest average chilling loss was found in LW pigs (1.92%), while in DWM and WM pigs, slightly lower average chilling loss was detected (1.75 and 1.74%, respectively). The values of chilling loss for LW pigs ranged from 1.38 to 2.05%, for DWM pigs from 1.44 to 2.00% and for WM pigs from 1.65 to 1.92%. The average chilling loss of LW pigs was statistically significantly higher ($P = 0.031$) compared to the average chilling loss of DWM and WM pigs, indicating that pig genotype affected the chilling loss. The average chilling loss found in carcasses of the modern pig (LW) is characteristic for a commercial carcass chilling process for commercial pigs [8].

4. Summary

Overall, additive genetic effects were evident for carcass traits, with crosses showing intermediate values between pure breeds. The performance of the DWM crossed pigs confirmed the theory according to which additive genetic effects mainly control morphology and body composition. Our study indicates significant differences between breeds for backfat thickness with the skin at the thinnest point over *M. gluteus medius*, the thickness of the lumbar muscle and chilling loss. No significant differences were observed for carcass yield.

Acknowledgement
This research was financially supported by the Ministry of Education, Science and Technological Development, Republic of Serbia, project TR31032.

References

[1] Tomovic V, Šević R, Jokanović M, Šojić B, Škaljac S, Tasić T, Ikonić P, Lušnic-Polak M, Polak T and Demšar L 2016 *Arch. Anim. Breed.* **59** 401–15
[2] Parunović N, Petrović M, Matekalo-Sverak V, Radović Č and Stanišić N 2013 *S. Afr. J. Anim. Sci.* **43** 123–36
[3] Egerszegi I, Rátky J, Solti L and Brüssow K 2003 *Arch. Tierz.* **46** 245–56
[4] Petrović M, Radović Č, Parunović N, Radojković D and Savić R 2012 *Biotechnol. Anim. Husb.* **28** 303–11
[5] Edwards S A 2005 *Livest. Prod. Sci.* **94** 5 –14
[6] Pugliese C and Sirtori F 2012 *Meat Sci.* **90**, 511–18
[7] Wood J D, Enser M, Fisher A V, Nute G R, Sheard P R, Richardson R I, Hughes S I and Whittington F M 2008 *Meat Sci.* **78** 343–58
[8] Tomović V, Petrović Lj and Džinić N R 2008 *Meat Sci.* **80** 1188–93
[9] Franci O, Campodoni G, Bozzi R, Pugliese C, Acciaioli A and Gandini G 2003 *Ital. J. Anim. Sci.* **2** 59–65
[10] Franci O, Bozzi R, Pugliese C, Acciaioli A, Campodoni G and Gandini G 2005 *Meat Sci.* **69** 545–50
[11] Coutron-Gambotti C, Gandemer G and Casabianca F 1998 *Meat Sci.* **50** 163–74

[12] Senčić Đ, Bukvić Z, Antunović Z and Šperanda M 2005 Poljoprivreda **11** 43–48
[13] Morcuende D, Estévez M, Ramírez M and Cava R 2007 *Meat Sci.* **76** 86–94
[14] Ramírez R and Cava R 2007 *Meat Sci.* **75** 388–396
[15] Salvatori G, Filetti F, Di Cesare C, Maiorano G, Pilla F and Oriani G 2008 *Meat Sci.* **80** 623–31
[16] Serrano M P, Valencia D G, Fuentetaja A, Lázaro R and Mateos G G 2008 *Meat Sci.* **80** 1122–28
[17] Cordero G, Isabel B, Menoyo D, Daza A, Morales J, Piñeiro C and Lopez-Bote C J 2010 *Span. J. Agric. Res.* **8** 962–70
[18] Robina A, Viguera J, Perez-Palacios T, Mayoral A I, Vivo J M, Guillen M T and Ruiz J 2013 *Span. J. Agric. Res.* **11** 1057–67
[19] Franco D, Vazquez J A and Lorenzo J M 2014 *Meat Sci.* **96** 195–202
[20] Mayoral I, Dorado M, Guillén M T, Robina A, Vivo J M, Vazquez C and Ruiz J 1999 *Meat Sci.* **52** 315–24
[21] Renaudeau D, Hilaire M and Mourot J 2005 *Anim. Res.* **54** 43–54
[22] Maiorano G, Cavone C, Paolone K, Pilla F, Gambacorta M and Manchisi A 2007 *Ital. J. Anim. Sci.* **6** 698–700
[23] Franco D, Carballo J, Bermñudez R and Lorenzo J M 2016 *Ann. Anim. Sci.* **16** 259–73
[24] Serrano M P, Valencia D G, Nieto M, Lázaro R and Mateos G G 2008 *Meat Sci.* **78** 420–28
[25] Sirtori F, Crovetti A, Zilio D M, Pugliese C, Acciaioli A, Campodoni G, Bozzi R and Franci O 2011 *Ital. J. Anim. Sci.* **10** 188–93
[26] Galián M, Peinado B, Martínez C, Periago M J, Ros G and Poto A 2007 *Anim. Sci. J.* **78** 659–67
[27] Poto A, Galián M and Peinado B 2007 *Livest. Sci.* **111** 96–103
[28] Wood J D, Nute G R, Richardson R I, Whittington F M, Southwood O, Plastow G, Mansbridge R, da Costa N and Chang K C 2004 *Meat Sci.* **67** 651–67
[29] Fortina R, Barbera S, Lussiana C, Mimosi A, Tassone S, Rossi A and Zanardi E 2005 *Meat Sci.* **71** 713–18
[30] Serra X, Gil F, Pérez-Enciso M, Oliver M A, Vázquez J M, Gispert M, Díaz I, Moreno F, Latorre R and Noguera J L 1998 *Livest. Prod. Sci.* **56** 215–23
[31] Maiorano G, Gambacorta M, Tavaniello S, D'Andrea M, Stefanon B and Pilla F 2013 *Ital. J. Anim. Sci.* **12** 426–31
[32] Furman M, Malovrh Š, Levart A and Kovač M 2010 *Arch. Tierz.* **53** 73–84
[33] Čandek-Potokar M, Žlender B and Bonneau M 1998 *Ann. Zootech.* **47** 3–16
[34] Alfonso L, Mourot J, Insausti K, Mendizabal J A and Arana A 2005 *Anim. Res.* **54** 33–42
[35] Wojtysiak D and Połtowicz K 2014 *Meat Sci.* **97** 395–403

IOP Conf. Series: Earth and Environmental Science **85** (2017) 012066 doi:10.1088/1755-1315/85/1/012066

Comparison of analytical methods for the determination of histamine in reference canned fish samples

S Jakšić[1], M Ž Baloš[1], Ž Mihaljev[1], J Prodanov Radulović[1] and K Nešić[2]

[1] Scientific Veterinary Institute "Novi Sad", Rumenački put 20, 21000 Novi Sad, Serbia

[2] Institute of Veterinary Medicine of Serbia, Vojvode Toze 14, Belgrade, Serbia

E-mail: sandra @niv.ns.ac.rs

Abstract. Two screening methods for histamine in canned fish, an enzymatic test and a competitive direct enzyme-linked immunosorbent assay (CD-ELISA), were compared with the reversed-phase liquid chromatography (RP-HPLC) standard method. For enzymatic and CD-ELISA methods, determination was conducted according to producers' manuals. For RP-HPLC, histamine was derivatized with dansyl-chloride, followed by RP-HPLC and diode array detection. Results of analysis of canned fish, supplied as reference samples for proficiency testing, showed good agreement when histamine was present at higher concentrations (above 100 mg kg^{-1}). At a lower level (16.95 mg kg^{-1}), the enzymatic test produced some higher results. Generally, analysis of four reference samples according to CD-ELISA and RP-HPLC showed good agreement for histamine determination (r=0.977 in concentration range 16.95–216 mg kg^{-1}) The results show that the applied enzymatic test and CD-ELISA appeared to be suitable screening methods for the determination of histamine in canned fish.

1. Introduction

Histamine is a product of the decarboxylation of histidine by microbiological histidine decarboxylase, caused by the growth of certain bacteria in protein-rich food like fish, cheese and wine [1]. The amount of histamine formed depends on the bacterial species, the temperature and the time of exposure [2,3]. Histamine is a causative agent of scombroid poisoning or histamine fish poisoning. The consumption of food containing significant concentration of histamine can cause symptoms similar to those associated to seafood allergies [4]. Histamine can be present mainly in *Scombridae* (Tuna, Mackerel) and *Clupeidae* (Herring, Sardine), in species which contain a high level of free histidine [5].

The critical dose of oral histamine has been estimated to be in the range of 100–200 mg kg^{-1}, and the EC regulation [6] stated for histamine that nine samples must be taken from each batch. These samples must fulfil the following requirements:

- the mean value must not exceed 100 mg kg^{-1};
- two samples may have a value of more than 100 mg kg^{-1} but less than 200 mg kg^{-1};
- no sample may have a value exceeding 200 mg kg^{-1} [6].

A wide variety of procedures for the determination of histamine and biogenic amines have been published: colorimetric methods, thin layer chromatography (TLC) methods, enzymatic methods, immuno-enzymatic methods, and flow injection analysis with fluorimetric detection [7]. However,

IOP Conf. Series: Earth and Environmental Science **85** (2017) 012066 doi:10.1088/1755-1315/85/1/012066

precise and reliable quantitative analysis for histamine and biogenic amines are chromatographic techniques, such as gas chromatography (GC), high performance liquid chromatography (HPLC), and high performance thin layer chromatography (HPTLC), as well as capillary electrophoresis [8,9]. Those allow the simultaneous analysis of histamine and other biogenic amines in fish and fishery products. These methods usually required extensive sample cleanup, and expensive equipment as well as trained staff. To enhance food control measurements, rapid, easy, and simple analytical methods for this compound are usually used as screening methods. In the case of contaminated samples, the use of confirmatory methods is necessary. In the EU regulation, it is specified that examinations must be carried out in accordance with reliable, scientifically recognized reference methods, such as HPLC. The reference method for histamine in Europe is the International Standard Organisation (ISO) method [10].

The aim of this paper was evaluation and comparison of histamine determination in four reference materials by the standard HPLC method, and by screening methods, an enzymatic test and competitive direct enzyme-linked immunosorbent assay (CD-ELISA).

2. Material and Methods

Studies were performed using canned fish samples intended for proficiency testing (PT). Proficiency testing provider, FAPAS, processes these samples under strict quality control procedures to ensure a homogenous reference material. Since numerous laboratories participate in those schemes, samples are considered as reference materials and are used for evaluation and comparison of screening and confirmatory methods for histamine determination. Four samples were used, with different histamine concentrations (table 1). All determinations in the study were performed in triplicate.

2.1. Screening methods

Two screening procedures were used in our study: Ridascreen® Histamine enzymatic test and Veratox® for Histamine CD-ELISA.

Ridascreen® Histamine (enzymatic) (Art. No. 1605, R-Biopharm, Germany) is an enzymatic test in microtiter plate format for the quantitative determination of histamine in fresh fish, canned fish, fish meal, wine, cheese and milk. The test was performed according to the manufacturer's manual. The basis of this test is an enzymatic reaction. The microtiterplate is coated with a reagent (electron carrier) and a dye. Histamine-dehydrogenase catalyzes the oxidation of histamine to imidacetaldehyde in presence of an electron carrier and a dye. The formation of dye is measured at 450 nm and is proportional to the histamine concentration. Calculation of histamine concentrations used linear regression; the dilution factor of the samples was taken into account. Since the quantitation range of the test is 2–100 mg kg^{-1}, samples with histamine contents higher than 100 mg kg^{-1} were further diluted with water and re-tested.

Veratox® for Histamine (AOAC-RI #070703 approved method, Product No. 9505, Neogen, USA) is used for the quantitative analysis of histamine in scombroid species of fish, such as tuna, bluefish and mahi-mahi. The test is a competitive direct ELISA. Free histamine in the sample and controls competes with enzyme-labeled histamine (conjugate) for the antibody-binding sites. After a wash step, substrate reacts with the bound enzyme conjugate to produce blue color that is measured at 630 nm. Since the range of quantitation was from 2.5 mg kg^{-1} to 50 mg kg^{-1}, samples with histamine contents higher than 50 mg kg^{-1} were further diluted with water and re-tested.

The color intensity in the microtiter wells was measured photometrically using a Thermo Multiskan FC photometer (Thermo Scientific). Special software, the Rida®Soft Win (Art. No. Z9999, R-Biopharm, Germany), was used to evaluate the results.

2.2. RP-HPLC method

Determination of histamine by HPLC was done according to the standard method [10]. Extraction was performed by mixing the sample with perchloric acid. Precolumn derivatization was performed using dansyl-chloride. Histamine was separated from other biogenic amines by HPLC Dionex UltiMate

IOP Conf. Series: Earth and Environmental Science **85** (2017) 012066 doi:10.1088/1755-1315/85/1/012066

3000 Series (Thermo Scientific, Germany) on SupelCosil™ LC-18-DB column (250 x 4.6 mm id, particle size 5 μm), by gradient elution, using acetonitrile and water as components of mobile phase, and UV detection (254 nm). The system was controlled by Chromeleon® 7 software (Thermo Scientific). Histamine concentration (mg kg^{-1}) was calculated from the area ratio of histamine and internal standard (1,7-diaminoheptane), using a matrix matched calibration curve covering the concentration range from 0 to 500 mg kg^{-1}.

2.3. Statistical Analysis
The difference between the PT declared histamine content in samples and the values obtained using different methods was analyzed using the *t* test. The difference between the histamine content obtained by the screening methods and the HPLC method for tested contaminated samples of canned fish was analyzed using linear regression analysis (PAST, Version 2.12, Oslo, Norway).

3. Results and Discussion
Results of histamine determination in all analyzed reference samples are presented in table 1. As can be seen from table 1, results for all samples except for FAPAS Proficiency test 27161 obtained by both screening methods were in the acceptance range. However, the result for sample with low histamine concentration (16.96 mg kg^{-1}), obtained by enzymatic method (21.33 mg kg^{-1}) was some higher than the acceptable value (13.4–20.5 mg kg^{-1}). We noted, on analyzing data from all participants in the proficiency scheme report for FAPAS 27161, that laboratories with screening methods obtained a higher average value (19.16 mg kg^{-1}) than the declared value. Contrary to this, statistical analysis (*t* test) showed that there was no significant difference between the contents of histamine obtained by the enzymatic test and declared content at the 0.05 level (p=0.074) in our study. Also, there was no significant difference in either case between the determined and declared content at the 0.05 level for all analyzed samples.

Generally, the screening methods gave some higher recovery (98–126%) compared to the HPLC method (99–111%), while intermediate precision (day-to-day variability) was in a similar range (3.8-15.5 and 5.5-14.7%).

Linear regression analysis (figure 1) showed good agreement between screening methods and the standard method in all analyzed samples (r=0.977).

The results obtained are similar to literature data. CD-ELISA and HPLC methods showed good agreement (R=0.969; concentration range 0.7–420 mg/kg) for histamine analysis in commercial soybean paste, suggesting that the CD-ELISA can be used as a rapid indicator for biogenic amines, including histamine [11]. Analysis of 50 commercial cheeses according to CD-ELISA and RP-HPLC also showed good agreement for histamine (r=0.979; concentration range 2–1800 mg kg^{-1}) [12].

Although commercial test kits are generally used for determining histamine in fresh and canned fish and fish meal, information on their performance and application to traditional fish products, which differ in product properties, showed different correlations with HPLC [13]. Therefore, new commercial test kits should be evaluated against the approved analytical method before being applied to new types of products.

IOP Conf. Series: Earth and Environmental Science **85** (2017) 012066 doi:10.1088/1755-1315/85/1/012066

Table 1. Results of histamine determination in reference samples using different methods (N=3).

Sample (Histamine in canned fish)	PT declared value (acceptance range) mg kg^{-1}	Method	
FAPAS Proficiency test 27149	137 (116–158)	Enzymatic test	HPLC
Average (mg kg^{-1})		148.2	138.2
SD (mg kg^{-1})		5.6	16.7
RSD (%)		3.8	12.1
Recovery (%)		108	101
p [a]		0.074	0.915
FAPAS Proficiency test 27161	16.95 (13.4–20.5)	Enzymatic test	HPLC
Average (mg kg^{-1})		21.33	16.84
SD (mg kg^{-1})		2.10	2.47
RSD (%)		9.8	14.7
Recovery (%)		126	99
p [a]		0.068	0.944
FAPAS Proficiency test 27176	216 (186–247)	Enzymatic test	HPLC
Average (mg kg^{-1})		200.1	237.0
SD (mg kg^{-1})		11.6	13.3
RSD (%)		5.5	5.5
Recovery (%)		98	111
p [a]		0.659	0.089
FAPAS Proficiency test 27197	174 (148–199)	ELISA	HPLC
Average (mg kg^{-1})		183.5	165.3
SD (mg kg^{-1})		28.4	9.0
RSD (%)		15.5	5.4
Recovery (%)		105	95
p [a]		0.620	0.235

t test value (*p* values) comparison of data obtained using different methods and declared content.

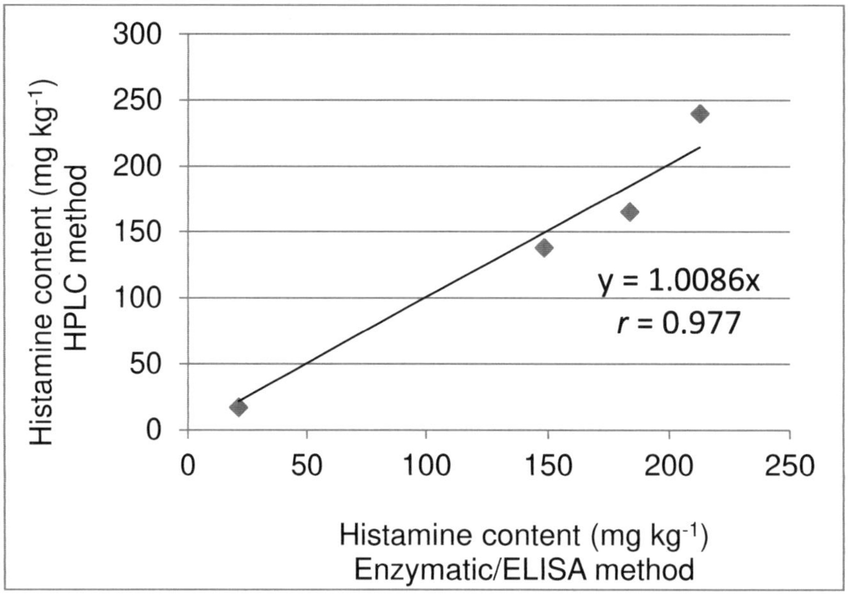

Figure 1. Correlation of results obtained by screening and standard method.

4. Conclusion

To ensure food safety with respect to histamine, it is preferable to use a rapid, simple and cheap method for screening. Results of histamine determination in reference samples showed that the enzymatic test and CD-ELISA give reliable and accurate values, especially for determination of critical high histamine concentrations (100–200 mg kg^{-1}). However it is necessary to validate any screening method and its reliability. According to the EC [6], it is important to confirm any high concentration of histamine in potentially positive samples by the reference method. Since quality control and consumer safety is of great importance from the aspect of histamine contamination, there is a constant challenge to develop new, fast, and reliable methods for different types of samples.

Acknowledgments

The work was financially supported by the Ministry of Education, Science and Technological Development of the Republic of Serbia (Project No 031071).

References

[1] Watanabe T, Taguchi Y, Maeyama K and Wada H 1991 *Handbook of Experimental Pharmacology* (B Uvnäs Berlin: Springer) p 145-163
[2] Guizani N, Al-Busaidy M A, Al-Belushi I M, Mothershaw A and Rahman M S 2005 *Food Res. Int.* 38 215
[3] Hongpattarakere T, Buntin N and Nuylert A 2016 *J. Food Sci. Technol.* **53** 245
[4] Petrović J, Babić J, Jakšić S, Kartalović B, Ljubojević D and Ćirković M 2016 *J. Food Prot.* **79** 90
[5] Prester Lj 2011 *Food Addit. Contam. A* **28** 1547
[6] COMMISSION REGULATION (EC) No 2073/2005
[7] Etienne M 2006 SEAFOOD plus–Traceability–Valid–Methods for chemical quality assessment-Methodology for histamine and biogenic amines analysis (Nantes: Ifremer) 1-20
[8] Bedia E F 2013 *Trends Anal. Chem.* **52** 239
[9] Őnal A, Kepekci T S E and Őnal C 2013 *Food Chem.* **138** 509

[10] ISO 19343:2017 Microbiology of the food chain - Detection and quantification of histamine in fish and fishery products - HPLC method
[11] Kim T K, Lee J I, Kim J H, Mah J H, Heang H J and Kim Y W 2011 *Food Sci. Biotech.* **20** 1747
[12] Aygun O, Schneider E, Scheuer R, Usleber E, Gareis M and Martlbauer E 1999 *J. Agric. Food Chem.* **47** 1961
[13] Kőse S, Kaklikkaya N, Koral S, Tufan B, Buruk K C and Azdin F 2011 *Food Chem.* **125** 1490

IOP Conf. Series: Earth and Environmental Science **85** (2017) 012067 doi:10.1088/1755-1315/85/1/012067

Presence of sulphites in different types of partly processed meat products prepared for grilling

V Koričanac, D Vranić, D Trbović, R Petronijević and N Parunović

Institute of Meat Hygiene and Technology, Kaćanskog 13, 11000 Belgrade, Serbia

E-mail: vladimir.koricanac@inmes.rs

Abstract. In the period January 2016 to May 2017, the presence and levels of sulfite were examined in 270 samples of hamburger, sausage (various types), pljeskavica (Serbian-style meat patties of various types) and ćevapi or ćevapčići (grill kebabs) from the Serbian market. Some (12.59%) of these partly processed meat products contained sulfites, expressed as SO_2, at levels above 10 mg/kg, and so did not meet requirements laid down in the National Regulation. In the remainder of the meat products (87.41%), sulfite contents were below 10 mg/kg, which is considered as "not detected". By groups, 100% of hamburgers, 91.76% of sausages and 90.48% of pljeskavica met requirements of National Regulation. The meat product group with the biggest percentage of non-compliant meat products in which sulfites were detected was the ćevapi or ćevapčići – 18.10% of them contained sulfites. All in all, most of the partly processed meat products from the Serbian market met the National Regulation regarding sulfite content, and they were safe for consumption. Nonetheless, the high percentage of ćevapi or ćevapčići that contained sulfites leads us to conclude that regular and periodic control is necessary and one of the most important steps in ensuring safe and quality meat products for consumers.

1. Introduction

Sulfites are compounds that contain the sulfite ion (SO_3^{2-}). Sulfur dioxide (SO_2) has been used since ancient times for its cleansing, disinfecting and purifying properties. In addition, sulfites have a number of technological uses, for example, as antioxidants, bleaching agents, flour treatment agents and preservatives. Sulfites are permitted in various foods such as wine, cordials, dried fruit and vegetables. They are used in the food industry to maintain food color, to prolong shelf life, and to prevent microbial growth [1]. Sulfites are present in some foods, naturally or as additives, but they are not used in most fresh foods. According to the U.S. Food and Drug Administration (FDA), about 1% of population are sensitive to sulfites, and so that is why the same body requires sulfites to be declared when used as an ingredient in the food, when used as a processing aid, or when present in an ingredient used in the food. That applies where the concentration of total SO_2 is equal or over 10 mg/kg (ppm) [2]. Below this limit, the amount of sulfites is considered as insignificant, as this is limit of detection, and food is considered as sulfite free.

Sulfites can cause allergy like reactions (intolerances), most commonly asthma symptoms in those with underlying asthma, sometimes allergic rhinitis-like reactions, occasionally urticaria (hives) and very rarely, anaphylaxis (severe allergic reaction). Wheezing is the most common reaction. The mechanism by which reactions occur is unclear. SO_2 gas is an irritant, and so reflex contraction of the airways from inhaling it is one possible explanation. This mechanism might explain the rapid onset of symptoms when drinking liquids like beer or wine, when SO_2 gas is inhaled during the swallowing process. Some people with asthma and who react to sulfites have a partial deficiency of the enzyme sulfite oxidase, which helps to break down SO_2. Not many people have positive skin allergy tests to sulfites, indicating true (IgE-mediated) allergy [3]. Sensitivity to sulfites in food is dependent on how much a person is exposed to SO_2 or sulfites from all sources. The pathogenesis of adverse reactions to sulfites has not been clearly documented but it is unlikely that sulfite reactions are allergic and immunity-mediated or produce anaphylactic reactions. Labeling of foods containing sulfite at concentrations of 10 mg/kg or more is required in the European Union, although the threshold for sensitivity reactions could be even lower [4].

The toxicity of sulfites is generally low. Evaluations by the Scientific Committee for Food (SCF) and by The Joint FAO/WHO Expert Committee on Food Additives (JECFA) have led to the conclusion that, for most consumers, sulfites in foods are of low health concern, although single, large oral doses of sulfites can produce gastrointestinal disturbances [1].

Sulfites are used as food additives for a variety of applications. In most countries, sodium and potassium sulfate ($NaHSO_3$, $KHSO_3$), sodium and potassium meta-bisulfite ($Na_2S_2O_5$, $K_2S_2O_5$), and sodium sulfite (Na_2SO_3) are allowed, along with sulfur dioxide gas (SO_2). In some countries, potassium sulfite (K_2SO_3) and sulfurous acid (H_2SO_3) are allowed. The U.S. has a long-standing restriction on use of sulfites in meats, but this restriction does not exist in all other countries [2].

In Serbia, sulfites in food are regulated by National Regulation [5]. Accordingly, sulfites are not allowed in partly processed meat products, except in some rare cases and types of foods (breakfast sausage with 6% grain or burger meat with 4% vegetables and/or grain). Similarly, to the FDA, all values under 10 ppm are considered as insignificant.

The European Standard [6] specifies a distillation method for the determination of the sulfite content, expressed as SO_2, in foodstuffs in which the content of sulfite is at least 10 mg/kg. The method is applicable in the presence of other volatile sulfur compounds. It is not applicable to cabbage, dried garlic, dried onions, ginger, leeks and soy proteins. It has been shown that the analysis of isolated soy protein leads to false positive results [6].

The principle of this method is based on measuring free sulfite plus a reproducible portion of bound sulfites (such as carbonyl addition products) in foods. The test portion is heated with refluxing solution of hydrochloric acid to convert sulfite to SO_2. A stream of nitrogen is introduced below the surface of the refluxing solution to sweep SO_2 through a water-cooled condenser and, via a bubbler attached to the condenser, into hydrogen peroxide solution, where SO_2 is oxidized to sulfuric acid. The generated sulfuric acid is titrated with standardized sodium hydroxide solution. The sulfite content is directly related to the generated sulfuric acid [6].

The aim of this study was to determine levels of sulfites in retail meat products commonly consumed in Serbia.

2. Materials and Methods

In the period January 2016 to May 2017, 270 samples of different types of partly processed meat products prepared for grilling (6 hamburgers, 85 sausages, 63 pljeskavica (Serbian-style meat patties of various types) and 85 ćevapi or ćevapčići (grill kebabs)) were examined for SO_2 content. Samples were collected from the Serbian market, and were produced mostly by domestic producers.

The presence of sulfites was determined according to the standard procedure [6] and expressed as SO_2 in mg/kg (ppm).

All chemicals used for analysis of the presence of sulfites in partly processed meat products prepared for grilling were of analytical grade and were used as received without any further purification. The results analysis and graphical presentation of their distribution was performed using Microsoft Office Excel 2016.

3. Results and Discussion

The results of determination of sulfites in the examined partly processed meat products prepared for grilling are shown in table 1. Also, distributions of the results, by groups and in all meat products examined, are graphically presented in figures 1-5.

Table 1. Number of samples of partly processed meat products prepared for grilling examined for the presence of SO_2 , for the period January 2016 – May 2017.

Type of samples	Not detected (<10 mg/kg SO_2)	Detected (>10 mg/kg SO_2)
Hamburger (6)	6	0
Sausage, various types (85)	78	7
Pljeskavica, various types (63)	57	6
Ćevapi or ćevapčići (116)	95	21
All samples (270)	236	34

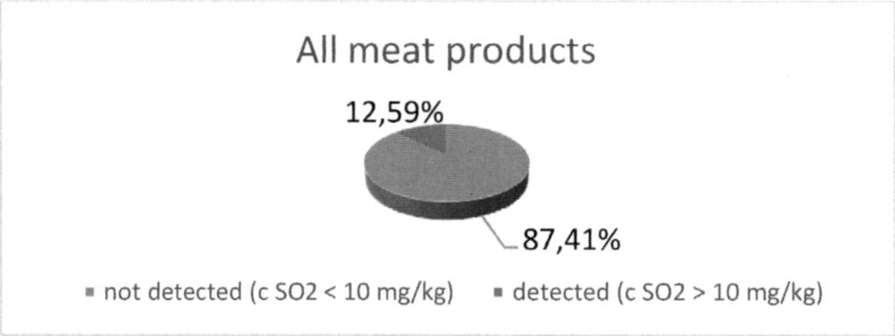

Figure 1. Percentage of all meat products examined where the presence of sulfites was detected or not detected.

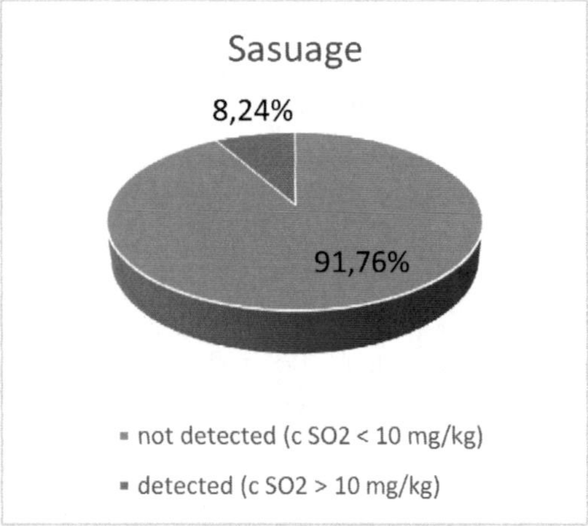

Figure 2. Percentage of sausages where the presence of sulfites was detected or not detected.

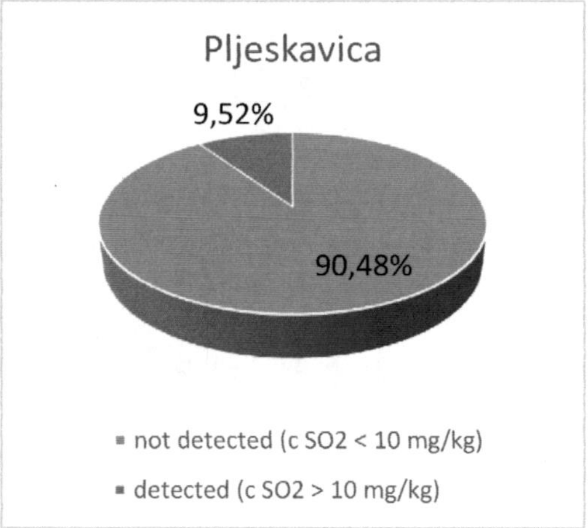

Figure 3. Percentage of pljeskavica where the presence of sulfites was detected or not detected.

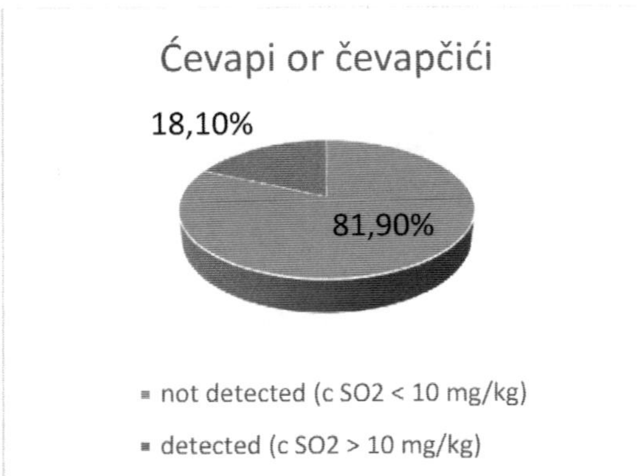

Figure 4. Percentage of ćevapi or ćevapčići where the presence of sulfites was detected or not detected.

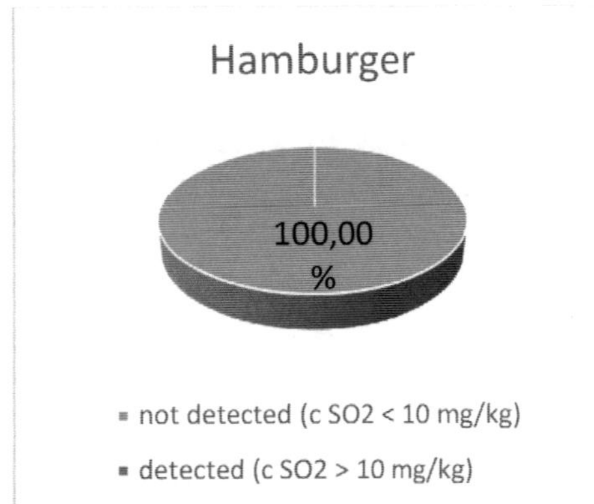

Figure 5. Percentage of hamburger where the presence of sulfites was detected or not detected.

4. Conclusion

Most of the tested meat products, 87.41% of them, met the National Regulation regarding the presence of sulfites, did not contain sulfites at detectable levels, and were safe for consumption. Hamburgers were the most safe, as 100% of them were without detectable sulfites, although we note that just six hamburgers were examined. The meat product group that had the highest percentage of products with sulfites detected was the group of ćevapi or ćevapčići, where 18.10% of examined products did not meet requirements laid down in National Regulation.

All in all, most of products across the selected meat groups were safe for consumption. However, the relatively high number of partly processed meat products in which sulfites were detected leads us to conclude that periodic and regular control by the responsible authority is a necessary step in ensuring the safety and fitness for consumption of these products.

References

[1] Keng-Wen L, Dennis P H, Hui-Ying H, Chiu-Hu W, Shih-Pei N and Min-P Ling 2016 *Toxicology Reports* **3** 544-551
[2] http://farrp.unl.edu/sulfitesinternatl
[3] https://www.allergy.org.au/images/pcc/ASCIA_PCC_Sulfite_sensitivity_2014
[4] European Food Safety Authority, Opinion of the Scientific Panel on dietetic products, nutrition and allergies on a request from the Commission relating to the evaluation of allergenic foods for labelling purposes 2004 *EFSA J.* **32** 1–197
[5] Pravilnik o prehrambenim aditivima, *Sl. Glasnik* RS br. 63/2013
[6] SRPS EN 1988-1:2004 Foodstuffs - Determination of sulfite - Part 1: Optimized Monier-Williams method

Antimicrobial compounds of porcine mucosa

E A Kotenkova, E A Lukinova and L V Fedulova

The V.M.Gorbatov All-Russian Meat Research Institute, Russian Academy of Agricultural Sciences, Moscow, Russian Federation

E-mail: lazovlena92@yandex.ru

Abstract. The aim of the study was to investigate porcine oral cavity mucosa (OCM), nasal cavity mucosa (NCM), rectal mucosa (RM) and tongue mucosa (TM) as sources of antimicrobial compounds. Ultrafiltrates with MW >30 kDa, MW 5-30 kDa and MW <5 kDa were obtained. All ultrafiltrates had antimicrobial activity against *Escherichia coli* and *Proteus vulgaris*. NCM ultrafiltrates revealed the highest antibacterial activity in respect to negative control: for the fraction with MW >30 kDa, the zone of microbial growth inhibition was 7.5 mm, for the MW<5 kDa fraction, it was 7 mm, and for MW 5-30 kDa fraction, it was 4.5 mm. No significant differences were found in high molecular weight proteomic profile, while qualitative and quantitative differences were observed in the medium and low molecular weight areas, especially in OCM and NCM. HPLC showed 221 tissue-specific peptides in OCM, 156 in NCM, 225 in RM, but only 5 in TM. The results observed confirmed porcine mucous tissues as a good source of antimicrobial compounds, which could be an actual alternative for reduction of microbial spoilage of foods.

1. Introduction

Analysis of scientific publications has shown there is intensive investigation of bioactive substances derived from animal tissues. The urgent need for alternative ways to combat pathogenic microorganisms that are rapidly acquiring resistance has led to comprehensive study of substances with antimicrobial action, especially in the last decade, while antimicrobial peptides have been known for more than 60 years. Thus, currently, about 2,000 antimicrobial compounds are found in tissues of different animal species.

Nowadays, a wide range of publications are dedicated to isolation and study of substances with antimicrobial action from microorganisms [1], cultured and wild plants (barley [2]), secretions from arthropods (scorpion venom [3,4]) and mollusks (oyster slime [5]), skin of amphibians, reptiles and fishes (cod [6], sturgeon [7], salamander [8], frogs [9,10]), blood cells of mammals and birds (chicken thrombocytes [11], leukocytes of goats [12], foxes [13], elk [14], cattle [15], etc.). For example, one of the largest databases supported by the University of Trieste lists more than 800 antimicrobial peptides (and some proteins) from various species such as amoeba, plants, penguins, people, etc. Various attempts have been made at classification of these compounds. Commonly, peptides are classified according to secondary structure: linear α-helical peptides (e.g., LL-37), peptides with β-strands linked by disulfide bridges (e.g., defensins), loop peptides (e.g., bactenecin), and those with a high content of specific amino acids (e.g., histatins) [16]. Antimicrobial substances are active against both gram-negative and gram-positive bacteria, and also fungi, viruses and protozoa. The mechanism of action is based on the destruction of microbial membrane due to integrity damage or interaction with certain areas of membranes. First, peptide and negatively charged membrane of the microorganism are electrostatically attracted, then the membrane is disrupted due to pore formation, through which ions and other cell components exit the microbal cell [17].

More than 500 antimicrobial proteins and peptides were found in tissues of mammals and are classified into histatins, cathelicidins and defensins. This later category is the most widely studied, and it is subdivided into alpha-defensins (found mainly in neutrophils and Paneth cells), beta-defensins (localized in leukocytes and epithelial cells) and theta-defensins (the least studied, found in some primates) [18]. Initially, defensins in mammals were described in 1956 by Robert C. Skarnes, and W. Dennis Watson as leukins and James G. Hirsch as phagocytins produced by rabbit leukocytes. H. I. Zeya and John K. Spitznagel related open substances to one molecular family, which they identified as cationic antimicrobial proteins. Only in 1985, did Michael E. Selste et al. give them their modern name – defensins [19].

Nevertheless, despite the ready availability and low cost of farm animal by-products, the question of their use as a source of substances with antimicrobial action is not enough in focus. However, the intensive development of proteomics in recent years has made it possible to study the composition of such substances. The results confirm that proteome and peptidome of barrier tissues can be significantly changed, mainly due to biosynthesis of proteins and peptides that are involved in cellular processes, metabolism, and immune protection. Moreover, overexpression of such compounds continues for some time after inflammation subsides. In addition, the particular combination of antimicrobial agents produced varies considerably depending on the type of pathogen that caused the inflammatory response. In this regard, the study of antimicrobial proteins and peptides contained in the mucous membranes of cattle and pigs is highly relevant due to their border position and, as a result, intensive contact with a wide range of biological agents (pathogenic and opportunistic microorganisms, viruses, fungi).

In summary, it is possible that meat by-products could contain significant resources of antimicrobial proteins and peptides, and their biological activity is of primary worth for study in developing alternative approaches to food processing technology, with the aim of prolonging food shelf life.

2. Materials and Methods

The objects of study were extracts and ultrafiltrates of porcine oral cavity mucosa (OCM), nasal cavity mucosa (NCM), rectal mucosa (RM) and tongue mucosa (TM). Extraction was carried out on laboratory dispersing equipment (Labotex, Russia) with 0.9% NaCl for 24 hours, ratio 1:2, at a temperature of 4-5 °C, 300 rpm. Extract was separated by centrifuging using a CM-6M (ELMI, Latvia) at 3500 rpm for 8 min, and then ultrafiltrated on PES membranes (MWCO 5 and 30kDa) by tangential filtration on a VivaFlow 200 system (Sartorius, Germany).

Antibacterial activity against *Escherichia coli* and *Proteus vulgaris* was determined by the disc-diffusion method. Cell cultures of *E. coli* and *P. vulgaris* were seeded on agar surfaces in Petri dishes, then paper discs moistened with ultrafiltrates (concentration: 0.1 g/ml 0.05 g/ml, 0.025 g/ml and 0.012 g/ml) were placed on the agar surfaces and incubated at 37°C. Zones of microbial growth inhibition were measured after 20 and 40 hours.

One-dimensional electrophoresis of extracts was carried out according to the Laemmli method in the 15% gradient SDS-PAGE with standards purchased from Fermentas (Fermentas, Lithuania).

Analysis of the peptide profile was carried out using high performance liquid chromatography (HPLC) with mass spectrometer (liquid chromatograph AGILENT 1200 C with a mass selective detector, AGILENT 6410, USA).

3. Results and Discussion

All ultrafiltrates of porcine ORM, NCM, RM and TM possessed antimicrobial activity against *E. coli* and *P. vulgaris*. NCM ultrafiltrates revealed the highest antibacterial activity: for the fraction with MW >30 kDa, the microbial growth inhibition zone was 7.5 mm, for the MW<5 kDa fraction, it was 7 mm, and for the MW 5-30 kDa fraction, it was 4.5 mm. Zones of microbial growth inhibition of ORM and TM ultrafiltrates (MW>30 kDa) were 4 mm.

Proteomic studies showed that the greatest number of low and high molecular weight proteins were detected in ORM and NCM, (about 30 bands), while in TM had 28 bands, and RM had 23 bands (figure 1).

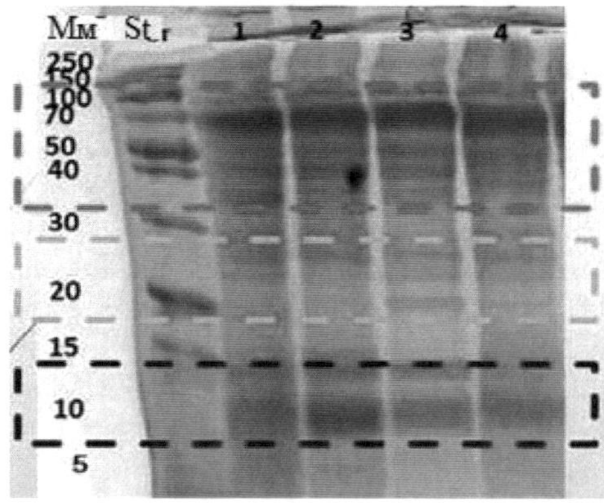

Figure 1. One-dimensional electrophoresis of porcine extracts. St – Molecular weight marker (250, 150, 100, 70, 50, 40, 30, 20, 15, 10 and 5 kDa). Lanes: 1 – TM, 2 – OCM, 3 – NCM, 4 – RM. Dotted lines show locations of: purple – high-molecular fraction (Mm>30 kDa), green – medium molecular fraction (Mm 5-30 kDa), blue – low molecular fraction (Mm <5 kDa).

According to the UniProt Protein DataBase, we can assume that tissue-specific proteins could be present in the investigated porcine extracts but were hidden in bands of the major proteins and the intensity of bands was formed both by major and minor (tissue specific) proteins. Thus, TM and NCM both contained a band of 15-17 kDa, which may correspond to protegrin-1,2,3 (16-17 kDa) with bactericidal activity against *E. coli, Listeria* and *Candida albicans*, while a band in the NCM probably corresponds to an antibacterial protein PR-39 (19 kDa) with antimicrobial activity against both *E. coli* and *Bacillus megaterium*. TM and OCM may contain the antibacterial peptide AP 3910 (4 kDa), while in both OCM and NCM, in the region of 8-10 kDa, AP 2 (9 kDa) and hepcidin (9 kDa) appear as likely candidates. These compounds are also characterized by a high antibacterial activity [10].

HPLC results showed that OCM contained 221 tissue-specific peptides, NCM contained 156, RM contained 225, while in TM, only 5 such peptides occurred (figure 2).

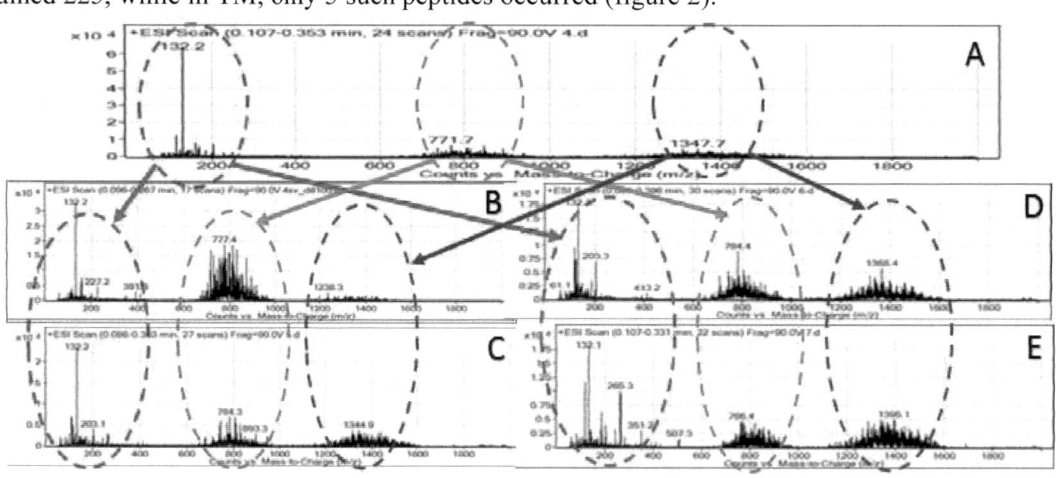

Figure 2. Peptide profiles of extracts. A – porcine muscle tissue, B – TM, C – NCM, D – OCM, D – RM. Dotted lines show ranges of: purple – 1100-1600 Da, green – 600-100 Da, red – 100-400 Da.

It was revealed that high molecular weight ultrafiltrates of porcine NCM and TM possessed the highest antibacterial activity. Bacterial growth inhibition zones were as large as 7.0 mm in diameter. Proteomic studies did not reveal significant differences in the high molecular range, while qualitative and quantitative differences were observed in the medium and low molecular weight areas, especially in OCM and NCM. The largest number of tissue-specific peptides was observed in RM and OCM (225 and 221, respectively).

The observed results confirm the investigated porcine mucosa tissues to be a good source of antimicrobial compounds, which could be suitable as actual alternative compounds to reduce microbial spoilage of foods.

Acknowledgement
This work was supported by the Russian Science Foundation (project No. 17-76-10033).

References

[1] Hassan M, Kjos M, Nes I F, Diep D B and Lotfipour F 2012 Natural antimicrobial peptides from bacteria: characteristics and potential applications to fight against antibiotic resistance *J Appl Microbiol.* **113** 723-36
[2] Bamdad F, Sun X, Guan L L and Chen L 2015 Preparation and characterization of antimicrobial cationized peptides from barley (*Hordeum vulgare* L.) proteins *LWT - Food Sci Technol.* **63** 29-36
[3] Bea R de L, Petraglia A F and Johnson L E 2015 Synthesis, antimicrobial activity and toxicity of analogs of the scorpion venom BmKn peptides *Toxicon.* **101**
[4] Zhong J, Zeng X C, Zeng X, Nie Y, Zhang L, Wu S and Bao A 2017 Transcriptomic analysis of the venom glands from the scorpion *Hadogenes troglodytes* revealed unique and extremely high diversity of the venom peptides *J Proteomics.* **150** 40-62
[5] Pales Espinosa E, Koller A and Allam B 2016 Proteomic characterization of mucosal secretions in the eastern oyster, *Crassostrea virginica J Proteomics.* **132** 63-76
[6] McDonald M et al. 2015 Structure-function relationships in histidine-rich antimicrobial peptides from Atlantic cod *Biochim Biophys Acta.* **1848** 1451-61
[7] Shamova O V, Orlov D , Balandin S V, Shramova E I, Tsvetkova E V, Panteleev P V, Leonova Yu F, Tagaev A A, Kokryakov V N and Ovchinnikova T V 2014 Acipensins – novel antimicrobial peptides from leukocytes of the Russian sturgeon *Acipenser gueldenstaedtii Acta Naturae* **6** 99–109
[8] Geng X et al. 2015 Proteomic analysis of the skin of Chinese giant salamander (*Andrias davidianus*) *J Proteomics.* **119** 196-208
[9] Conlon J M, Mechkarska M, Radosavljevic G, Attoub S, King J D, Lukic M L and McClean S 2014 A family of antimicrobial and immunomodulatory peptides related to the frenatins from skin secretions of the Orinoco lime frog *Sphaenorhynchus lacteus* (Hylidae) *Peptides.* **56** 132-40
[10] Holden W M, Reinert L K, Hanlon S M, Parris M J and Rollins-Smith LA 2015 Development of antimicrobial peptide defenses of southern leopard frogs, *Rana sphenocephala*, against the pathogenic chytrid fungus, *Batrachochytrium dendrobatidis Dev Comp Immunol.* **48** 65-75
[11] Sycheva M V 2016 The effect of antimicrobial peptides from platelets of *Gallus gallus* on antibiotic sensitivity of microorganisms *БОНЦ УрО РАН* **1** 7
[12] Shamova O V, Orlov D S, Zharkova M S, Balandin S V, Yamschikova E V, Knappe D, Hoffmann R, Kokryakov V N and Ovchinnikova TV 2016 Minibactenecins ChBac7.Nα and ChBac7. Nβ – antimicrobial peptides from leukocytes of the goat *Capra hircus Acta Naturae.* **8** 136-146
[13] Ilina E I, Berlov M N, Dubrovsky Ya A, Bogomolova E G and Kokryakov V N 2013 Antimicrobial peptide from leukocytes of *Vulpes vulpes* red fox *Vestnik of Saint Petersburg University* **2** 56-63

[14] Yuhnev V A, Shartukova M A, Lugovkina N V, Kokryakov V N and Shamova O V 2014 Search of novel antimicrobial peptides of the cathelicidins and defensins families in moose (*Alces alces*) leucocytes *Vestnik of Saint Petersburg University* **1** 115-31

[15] Wan J, Lia Y, Chena D, Yua B, Chenb G, Zhenga P, Maoa X, Yua J and He J 2016 Recombinant plectasin elicits similar improvements in the performance and intestinal mucosa growth and activity in weaned pigs as an antibiotic *Animal Feed Sciand Technol.* **211** 216-26

[16] McDermott A M 2004 Defensins and other antimicrobial peptides at the ocular surface *Ocul Surf.* 2004 **2** 229-47

[17] Bahar A A and Ren D 2013 Antimicrobial peptides *Pharmaceuticals (Basel).* **6** 1543–75

[18] Tecle T, Tripathi S and Hartshorn K L 2010 Review: Defensins and cathelicidins in lung immunity *Innate Immun.* **16** 1 51-9

[19] Abaturov A Ye 2011 Cationic antimicrobial peptides of non-specific respiratory protection: defensins and cathelicidins. defensins – molecules undergoing renaissance (Part 1) *Child`s health* **7** 161-71

Isolation and detection of *Listeria monocytogenes* in poultry meat by standard culture methods and PCR

J Kureljušić [1], N Rokvić[1], N Jezdimirović[1], B Kureljušić[1], B Pisinov[1] and N Karabasil[2]

[1] Institute of Veterinary Medicine of Serbia, Vojvode Toze 14, 11000 Belgrade, Serbia
[2] University of Belgrade, Faculty of Veterinary Medicine, Bulevar oslobodjenja 18, 11000 Belgrade, Serbia

E-mail: jasnakureljusic@yahoo.com

Abstract. *Listeria* is the genus of a bacteria found in soil and water and some animals, including poultry and cattle. It can be present in raw milk and food made from raw milk. It can also live in food processing plants and contaminate a variety of processed meats. Microscopically, *Listeria* species appear as small, Gram-positive rods, which are sometimes arranged in short chains. In direct smears, they can be coccoid, so they can be mistaken for streptococci. Longer cells can resemble corynebacteria. Flagella are produced at room temperature but not at 37°C. Haemolytic activity on blood agar has been used as a marker to distinguish *Listeria monocytogenes* among other *Listeria* species, but it is not an absolutely definitive criterion. Further biochemical characterization is necessary to distinguish between the different *Listeria* species. The objective of this study was to detect, isolate and identify *Listeria monocytogenes* from poultry meat. Within a period of six months from January to June 2017, a total of 15 samples were collected. Three samples were positive for the presence of *Listeria monocytogenes*. Biochemical and microbiological tests as well as PCR technique using specific primers were used to confirm *L. Monocytogenes* in the samples.

1. Introduction

Listeria monocytogenes is recognized as an important foodborne pathogen in many industrialized countries. The consumption of food contaminated by *L. Monocytogenes* has been identified as the main transmission route for this pathogen in both humans and animals. In humans, listeriosis is a rare but serious illness that can lead to abortion or serious cases of meningitis or encephalitis, and even death [1]. Cases are observed especially in vulnerable and immunocompromised humans such as newborn infants, pregnant women, cancer or AIDS patients and the elderly. Because of the high fatality rate (20–30%), listeriosis ranks among the most frequent causes of death due to foodborne illnesses [2,3]. Since 2006, increasing numbers of listeriosis cases have been observed in several European Union countries, including France, predominantly in people of more than 60 years of age [4,5]. *L. Monocytogenes* is widely disseminated in the environment (soil, surface water, plants, and infected animals). The ubiquitous character of the pathogen inevitably results in the contamination of numerous food products (such as milk and dairy products, raw vegetables, meat and meat products and seafood). Poultry, poultry products, eggs and egg products have rarely been reported to be involved in *L. Monocytogenes* outbreaks [6]. In most studies, the contamination of poultry meat occurs during the slaughtering and processing phases [7,8,9,10,11]. Very few studies have reported an incidence of *L. Monocytogenes* at the farm level [12].

2. Materials and Methods

Isolation of *L. monocytogenes* was performed according to the ISO standard method [13]. A food sample of 25 g was added to 225 ml of half strength Fraser broth which is used as primary enrichment media, to obtain a 1:10 sample dilution. All samples were homogenized 30- 60 seconds and incubated at 30°C for 24 h ± 2 h. Pre-enrichment media may contain selective agents to inhibit the growth of competing microorganisms, but usually at lower concentrations than those used in selective enrichments.

From this primary enrichment, 0.1 ml was inoculated into 10 mL of Fraser Broth which is secondary enrichment medium, and incubated for 48 h at 37°C. A loopful of the Fraser Broth enrichment culture was streaked on the surface of different selective agar, ALOA agar and second Palcam agar, and incubated in an incubator at 37 or 35°C for 48 ± 2 h. The selective enrichment culture is inoculated on to two selective agar media and incubated at 37 or 35°C for 48 ± 2 h. Selective agar were observed for suspected colonies at 24 to 48 h of incubation. Characteristic colonies were purified on TSYEA for identification.

Two individual colonies of each suspect isolate were suspended in 50 µl of DNA/RNA free water and heated at 95°C for 5 minutes. To demonstrate the *L. monocytogenes* genome, a commercial kit (TopTaq Master Mix Kit, Qiagen®, Germany) and primers (Lip1 gatacagaaacatcggttggc and Lip2 gtgtaacttgatgccatcagg) and the thermal protocol described by Jofre *et al.*, [14] were used. 10 µl of the PCR product was analysed by agar gel electrophoresis in 2% gel, with addition of ethidium bromide and visualized in UV transilluminator.

3. Results and Discussion

From a total of 15 poultry meat samples that were tested, 3 were positive for presence of *L. monocytogenes* and all 3 isolates were confirmed by PCR. Pure cultures were isolated from all three suspect poultry meat samples, and were shown by PCR to contain the marker for the *L. monocytogenes* genome (figure 1).

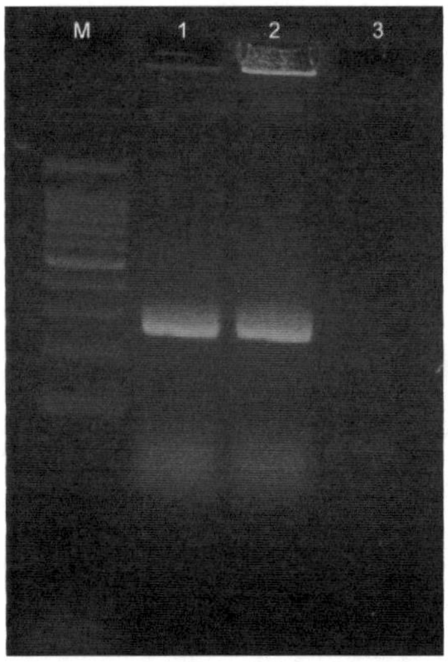

Figure 1. Agarose gel electrophoresis. M - molecular marker (100 bp), 1. *L. monocytogenes* positive sample (274 bp), 2. positive control, 3. negative control.

Listeria is not particularly resistant to disinfectants. In order to obtain optimal disinfection, it is thus necessary to respect cleaning and disinfection procedures. Moreover, the ability of *L. monocytogenes* to develop biofilms on a variety of surfaces makes disinfection treatment difficult [15].

The best way to avoid contamination would be to prevent biofilm formation in food industries by frequently disinfecting and cleaning surfaces [16]. This procedure should be sufficient to remove cells not yet strongly adhered, although any such cleaning can fail and not remove mature biofilms. To fulfil this purpose, the use of appropriate disinfectants is essential, and other strategies have been tested with this aim. Some examples are the use of ozone or acidic water, usually considered eco-friendly biocides, as they do not leave chemical residuals [17]. Besides these products, natural compounds extracted from bacteria or aromatic plants cultures and some GRAS (generally recognized as safe) ingredients have also been evaluated to eradicate biofilms [18,19].

4. Conclusion

L. monocytogenes is considered as a highly pathogenic bacterium that, world-wide, contaminates a wide range of food products, and has a high mortality rate in infected patients. Our investigation showed *L. monocytogenes* was present in Serbian poultry meat. The hygienic status of the slaughterhouse and sanitary practices observed at the farm could be relevant for the *Listeria* status of poultry meat, although this data is not reported here. These factors have already been reported in studies related to *Salmonella* and *Campylobacter* in poultry flocks, but this is the first time this idea is presented for *L. monocytogenes* in Serbia.

Acknowledgment

This paper was supported by Ministry of Education, Science and Technological development, Republic of Serbia, through the funding of Project No III 46009.

References

[1] Ahmed S S, Tayeb B A, Ameen A M, Merza S M and Sharif Y H M 2016 Isolation and detection of *Listeria monocytogenes* in minced meat, frozen chicken and cheese in Duhok Province, Kurdistan Region by using RT-PCR *Journal of Food & Industrial Microbiology* **2**1

[2] Vaillant Vet al. 2005 Foodborne infections in France *Foodborne Pathogens Dis.* **2** 221–232

[3] EFSA 2009 The Community Summary Report on trends and sources of zoonoses and zoonotic agents in the European Union in 2007 *EFSA J* 223

[4] Denny J, McLauchlin J, 2008 Human Listeria monocytogenes infections in Europe – an opportunity for improved European surveillance *Euro Surveill* **13** 1–5

[5] Goulet V, Hedberg C, Le Monnier A and De Valk H 2008 Increasing incidence of listeriosis in France and other European countries *Emerg. Infect. Dis.* **14** 734–40

[6] Gottlieb S L et al. 2006 Multistate outbreak of listeriosis linked to turkey deli meat and subsequent changes in US regulatory policy *Clin. Infect. Dis* **42**(1) 29-36

[7] Lawrence L M and Gilmour A 1994 Incidence of *Listeria* spp. and *Listeria monocytogenes* in a poultry processing environment and in poultry products and their rapid confirmation by multiplex PCR *Appl. Environ. Microbiol.* **60** 4600–04

[8] Chasseignaux E, Gérault P, Toquin M T, Salvat G, Colin P and Ermel G 2002 Ecology of *Listeria monocytogenes* in the environment of raw poultry meat and raw pork meat processing plants *FEMS Microbiol. Lett.* **210** 271–75

[9] Rorvik L M, Aase B, Alvestad T and Caugant D A 2003 Molecular epidemiological survey of *Listeria monocytogenes* in broilers and poultry products *J. Appl. Microbiol.* **94** 633–40

[10] Barbalho T C F, Almeida P F, Almeida R C C and Hofer E 2005 Prevalence of *Listeria* spp. at a poultry processing plant in Brazil and a phage test for rapid confirmation of suspect colonies. *Food Control* **16** 211–16

[11] Loura C A C, Almeida R C C and Almeida PF 2005 The incidence and level of *Listeria* spp. and *Listeria monocytogenes* contamination in processed poultry at a poultry processing plant *J. Food Saf.* **25** 19–29

[12] Chemaly M, Toquin M T, Le Nôtre Y and Fravalo P 2008 Prevalence of *Listeria*

IOP Conf. Series: Earth and Environmental Science **85** (2017) 012069 doi:10.1088/1755-1315/85/1/012069

monocytogenes in poultry production in France *J. Food Prot.* **71** 1996–2000
[13] ISO 2010 11290-1 Microbiology of food and animal feeding stuffs — Horizontal method for the detection and enumeration of *Listeria monocytogenes* — Part 1: Detection method
[14] Jofre A, Martin B, Garriga M, Hugas M, Pla M, Rodriguez-Lazaro D and Aymerich T 2005 Simultaneous detection of *Listeria monocytogenes* and *Salmonella* by multiplex PCR in cooked ham *Food Microbiology* **22** 109–15
[15] Aury K, Le Bouquin S, Toquin M T, Huneau-Salaün A, Le Nôtre Y, Allain V, Petetin I, Fravalo and P Chemaly M 2001 Risk factors for *Listeria monocytogenes* contamination in French laying hens and broiler flocks *Prev Vet Med.* **98** 271-8
[16] Casarin L S and Tondo E C 2015 Adhesion and Biofilm Formation of *Listeria monocytogenes* on Food Contact Surfaces: General Aspects and Control Measures In: *Listeria monocytogenes: Incidence, Growth Behavior and Control* Nova Science Publishers: Nova Iorque.
[17] Ashraf M A, Ullah S, Ahmad L, Qureshi A K, Balkhair K S and Abdur Rehamn M 2014 Green biocides, a promising technology: current and future applications to industry and industrial processes *J. Sci. Food Agric.* **94** 348-403
[18] Neyret C, Herry J M, Meylheuc T and Dubois Brissonnet F 2014 Plant derived compounds as natural antimicrobials to control paper mill biofilms *J Ind Microbiol Biotechnol.* **41** 87-96
[19] Bridier A, Sanchez-Vizuete P, Guilbaud M, Piard J C, Naital M and Briandet R 2015 Biofilm associated persistence of food borne pathogens *Food Microbiol.* **45** 167-178

Distribution of cadmium in leg muscle and liver of game birds from Serbia

D Nikolić, J Đinović-Stojanović, S Stefanović, T Radičević, D Trbović, D Spirić and S Janković

Institute of Meat Hygiene and Technology, Kaćanskog 13, 11000 Beograd, Serbia

E-mail: dragica.nikolic@inmes.rs

Abstract. The aim of this study was to present the distribution of cadmium (Cd) levels in leg muscle and liver of game birds. Samples (n=464) of: pheasants (n=182), mallards (n=25), Eurasian jay (n=7), partridges (n=5), woodcocks (n=8) and common quail (n=5) were collected during regular hunting seasons within the Serbian National Residue Monitoring Program from 2013 to 2016. Analysis of Cd was performed by ICP-MS. In all liver samples, Cd levels were above the limit of detection (LOD=0.001 mg/kg) while in 66.4% of muscle samples, Cd was detected. Statistical analysis showed significant differences between Cd levels in leg muscle and liver of woodcocks and others game birds. The highest mean Cd level was observed in muscle samples of woodcocks (0.042 mg/kg). The lowest mean Cd levels in liver were observed in common quails (0.130 mg/kg) and mallards (0.160 mg/kg) while the highest levels were measured in woodcocks (1.247 mg/kg) and pheasants (0.262 mg/kg). During four years of the Serbian National Residue Monitoring Program, leg muscle samples of woodcocks (n=3), liver samples of pheasants (n=23), woodcocks (n=6) and mallards (n=3) exceeded the maximum residue limit (MRL).

1. Introduction

Environmental pollution with heavy metals is world-wide problem [1]. Human activities, technological and industrial development all influence heavy metal contamination in the environment. Cadmium (Cd) occurs on agricultural land as a contaminant of phosphorous fertilizers and can be found in sewage sludge, surface waters and plants. Cd is a ubiquitous and toxic heavy metal, recognized as a potential health threat to human and wildlife species as it is not biodegradable and accumulates in living organisms [2, 3]. According to the International Agency for Research on Cancer, Cd and its compounds are classified as carcinogenic to humans [4].

Game and game birds are free-living wildlife. As they freely choose food and are the part of soil-plant-animal chain, Cd contamination in free-living game could be fair bioindicator of environmental pollution [5]. Accumulation of Cd in tissues of game, including game birds is studied almost world-wide [6-9].

Scientific information about Cd levels in tissue of game birds from Serbia is scarce. The aim of this study was to determine and compare distribution of Cd in leg muscle and liver of different species of game birds.

2. Materials and Methods

Levels of Cd were measured in leg muscle and liver of game birds (n=464). Pheasants (n=182), mallards (n=25), Eurasian jay (n=7), partridges (n=5), woodcocks (n=8) and common quail (n=5) were acquired during regular hunting seasons and analysed within the Serbian National Residue Monitoring Program from 2013 to 2016.

Frozen samples were thawed at 4°C the day before the analysis and then homogenized. An amount, of approximately 0.5 g of homogenized tissue was transferred into a teflon vessel with 5 mL of nitric acid (67% Trace Metal Grade, Fisher Scientific, Bishop, UK) and 1.5 mL of hydrogen peroxide (30% analytical grade, Sigma-Aldrich, St. Louis, MA, USA) for microwave digestion. The microwave oven (Start D, Milestone, Sorisole, Italy) program consisted of three steps: 5 min from room temperature to 180°C, 10 min hold at 180°C and 20 min ventilation. After cooling, the digested sample solutions were quantitatively transferred into polypropylene volumetric flasks and diluted to 100 mL with deionized water obtained from a water purification system (Purelab DV35, ELGA, Buckinghamshire, UK).

Inductively coupled plasma mass spectrometry (ICP-MS), (iCap Q mass spectrometer, Thermo Scientific, Bremen, Germany), was used for analysis of the 111Cd isotope. A five-point calibration curve (including zero) was constructed for the qualitative analysis of the samples. Multielement internal standard (6Li, 45Sc—10 ng/mL; 71Ga, 89Y, 209Bi—2 ng/mL) was introduced online by an additional line through the peristaltic pump.

The quality of the analytical process was verified by analysis of the certified reference material NIST 1577c (Gaithersburg, MD, USA). Reference material was prepared in the same way as samples using microwave digestion. Obtained concentrations were in the range of certified values.

Statistical analysis of experimental data was performed using Minitab 16 Statistical software. One-way analysis of variance – ANOVA and Tukey's HSD test were applied for comparison of Cd levels between different leg muscles samples as well as between livers from different game birds. For the purpose of calculation, when the levels of Cd were below the limit of detection (LOD, LOD=0.001 mg/kg), that value was assumed to be equal to 1/2 LOD.

3. Results and Discussion

The results of Cd levels in leg muscle and liver of the analysed samples are presented in table 1 and table 2, respectively.

Table 1. Cd levels in leg muscle of game birds from Serbia.

	n1	min-max	Mean ± SD	n2
Pheasants	182	< LOD-0.049	0.006 ± 0.008^{a}	
Mallards	25	< LOD-0.026	0.005 ± 0.006^{a}	
Eurasian jay	7	0.003-0.010	0.005 ± 0.003^{a}	
Partridges	5	< LOD-0.008	0.004 ± 0.003^{a}	
Woodcocks	8	0.018-0.065	0.042 ± 0.019^{b}	3
Common quail	5	< LOD-0.028	0.008 ± 0.012^{a}	

n1 – number of samples.
n2 – number of non-compliant samples.
$^{a-b}$ Different superscripts within the same column indicate significant differences of means according to Tukey's HSD test ($p < 0.05$).

Table 2. Cd levels in liver of game birds from Serbia.

	n1	min-max	Mean ± SD	n2
Pheasants	182	0.014-1.162	0.262±0.257[a]	23
Mallards	25	0.005-0.746	0.186±0.166[a]	2
Eurasian jay	7	0.084-0.277	0.174±0.086[a]	
Partridges	5	0.084-0.292	0.160±0.078[a]	
Woodcocks	8	0.213-3.204	1.247±0.989[b]	6
Common quail	5	0.005-0.467	0.130±0.190[a]	

n1 – number of samples.
n2 – number of non-compliant samples.
[a–b] Different superscripts within the same column indicate significant differences of means according to Tukey's HSD test ($p < 0.05$).

The levels of Cd in leg muscle were from < 0.001 to 0.042 mg/kg. The highest percentages of detected Cd levels in leg muscle were detected in woodcocks, partridges and common quails (figure 1). The highest mean Cd level was observed in leg muscle of woodcocks (0.042 mg/kg) while in other game birds, Cd was found only at very low levels, often close to the LOD (table 1). Statistical analysis showed significant differences between Cd levels in leg muscle of woodcocks and others game birds. According to the national legislation [10] there is no maximum residue level (MRL) for Cd in game tissue, therefore MRL for Cd in poultry tissue (muscle, liver) was used instead. The MRL for Cd in poultry muscle is 0.050 mg/kg [10]. Three leg muscle samples of woodcocks analysed during four years (2013-2016) exceeded this level i.e. those samples were non-compliant.

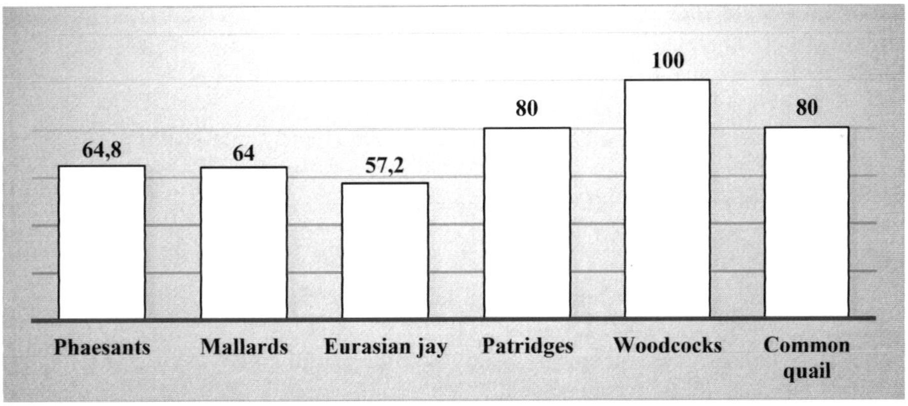

Figure 1. Percentage (%) of leg muscle samples of game birds with detected Cd levels.

While Cd levels in leg muscle was mostly low, liver accumulated higher levels [8, 11, 12] resulting in detectable levels of Cd in all liver samples examined. The lowest mean Cd levels in liver were established in livers from common quail (0.130 mg/kg) and mallards (0.160 mg/kg) while the highest were in livers from woodcocks (1.247 mg/kg) and pheasants (0.262 mg/kg). Significant differences were established between Cd levels in liver of woodcocks and others game birds. National legislation [10] established 0.500 mg/kg as the MRL for Cd in game liver. Some livers of pheasants (n=23), mallards (n=2) and woodcocks (n=6) exceeded this level. The highest measured Cd levels were in liver of mallard (3.204 mg/kg) and pheasant (1.162 mg/kg).

To the best of our knowledge, most of the literature on the subject of Cd levels in tissue of game birds is for pheasants and mallards. According to Petrovic and Jankovic [9], two liver samples (0.70

and 0.95 mg/kg) of pheasants collected during hunting seasons 2004 and 2005 in Serbia exceeded the MRL. The mean Cd level in leg muscle of pheasants from the current study was lower while in liver, it was higher compared to the mean Cd levels reported by Koréneková et al. [13] (0.019 and 0.024 mg/kg, respectively). Szymczyk and Zalewski [14] reported lower mean Cd level in pheasants' livers (range from 0.130 to 0.180 mg/kg) in birds living near a non-ferrous metallurgy area as well as Cd levels among their non-compliant samples (1.121 mg/kg) than we obtained in this study. Also, Cd levels in liver of mallards were examined in Szymczyk's and Zalewski's [14] study. The authors reported Cd levels in liver of analysed mallards in the range of 0.014 to 0.394 mg/kg, which was lower compared to our data, while higher levels of Cd in liver were found in mallards from the polluted region of Ślask (average 1.274 mg/kg) [14].

4. Conclusion

During the Serbian National Residue Monitoring Program from 2013 to 2016, 7% of leg muscle and liver samples of game birds were non-compliant for Cd levels. Hunters and members of their households commonly eat game meat, thus intaking higher amounts of Cd than the average population. It is assumed that the amount of consumed game meat could pose a health risk even in this small part of the population. Therefore, continual control of Cd and other toxic elements in game meat is needed in order to assess the safety of these tissues with respect to human health.

References

[1] Walker C H, Hopkin S P, Sibly R M and Peakall D B 2006 *Principles of Ecotoxicology* (Boca Raton, FL: Taylor & Francis)
[2] Larison J R, Likens G E, Fitzpatrick J W and Crock J G 2000 *Nature* **406** 181
[3] Chan D Y, Fry N, Waisberg M, Black W D and Hale B A 2004 J. *Toxicol. Environ. Health* **67** 397
[4] IARC International agency for research on cancer 2012 (Lyon, France)
[5] Alexander J, Benford D, Cockburn A, Cravedi J, Dogliotti E and Domenico A 2009 *EFSA J.* **980** 1
[6] Kramárová M et al. 2005 *Bull. Vet. Inst. Pulawy* **49** 465
[7] Mochiyuki M, Hondo R, Kumon K, Sasaki R, Matsuba H and Ueda F 2002 *Environ. Monit. Assess.* **73** 229
[8] Toman R, Massányi P, Lukáč N, Ducsay L and Golian J 2005 *Ecotoxicol. Environ. Saf.* **62** 112
[9] Petrovic Z and Jankovic S 2008 *Tehnologija mesa* **49** 36
[10] Official gazette of RS 29/2014
[11] Massányi P, Toman R, Uhrín V and Renon P 1995 *Ital. J. Food Sci.* **7** 311
[12] Massányi P and Uhrín V 1996 *J. Environ. Sci. Health.* **A32** 1459
[13] Koréneková B, Skalická M, Kožárová I, Nagy J, Máté D and Nad P 2008 *Slovak J. Anim. Sci.* **41** 184
[14] Szymczyk K and Zalewski K 2003 *Pol. J. Environ. Stud.* **12** 381

IOP Conf. Series: Earth and Environmental Science **85** (2017) 012071 doi:10.1088/1755-1315/85/1/012071

Use of olive oil-in-water gelled emulsions in model turkey breast emulsions

M Serdaroğlu and B Öztürk

Ege University Engineering Faculty Food Engineering 35100 Bornova-Izmir/Turkey

E-mail: burcu.ozturk@ege.edu.tr

Abstract. Today, gelled emulsion systems offer a novel possibility in lipid modification of meat products. In this study, we aimed to investigate the quality characteristics of model turkey emulsions that were prepared with olive oil-in-water gelled emulsion (GE) as partial or total beef fat replacer. The results indicated that while most of the GE treatments showed equivalent emulsion characteristics in terms of emulsion stability, water-holding capacity and cook yield, utilization of 100% GE as the lipid source could increase total expressible fluid of the model turkey emulsion and thus negatively affect the quality. Utilization of GE was effective in total fat reduction, as the model turkey emulsions formulated with more than 50% GE had significantly lower fat content compared to full-beef fat control model emulsion. However, beef fat replacement with GE produced considerable changes in colour parameters. Finally, it was concluded that utilization of GE as a partial beef fat replacer has good potential to enhance stability and reduce total fat in turkey meat emulsion products.

1. Introduction

Although the consumption of poultry meat products has been continuously increasing in the last decades, consumers currently associate further processed meat products as a high animal fat-containing, unhealthy food source. Therefore, an important goal for the meat industry is to suggest novel lipid modification strategies. Incorporation of gelled emulsion (GE) in poultry product formulations is one of the new approaches in lipid modification that ensures both the product yield and healthier composition. An emulsion gel is defined as "an emulsion with a gel-like network structure and solid-like mechanical properties" [1]. GEs mimic the functional and sensory characteristics of animal fat used in most of the widely-consumed meat products [2]. Olive oil has long been known as a functional vegetable oil that is a rich source of mono- and polyunsaturated fatty acids, providing beneficial health impacts [3]. In this study, we aimed to research the effects of olive oil-in-water GE as beef fat replacers in model turkey emulsions, with particular regard to emulsion stability parameters and final quality.

2. Materials and Methods

Fresh boneless, post-rigor turkey breast muscles (*Pectoralis major*), beef fat (BF), olive oil and other ingredients were purchased from the local market. For preparation of GEs, the oil phase (50 g/100 g emulsion) containing polyglycerol polyricinoleate (PGPR) as surfactant (6.4 g/100 g oil), was added to the aqueous phase containing 3 g gelatine/100 g emulsion and 9 g inulin/100 g emulsion, and homogenized. Both phases were previously heated separately to 55°C on a hot plate stirrer. After the homogenization process (6000 rpm, Ultra-Turrax® T25, UK), the emulsion was cooled to room temperature.

The GE was kept for 12 h at 4°C until being used in meat emulsions [4]. Model system turkey emulsions were produced according to Öztürk *et al.* [5]. Each treatment was prepared to initially contain 20% total lipid. Control (C) group consisted of 100% BF, whereas GE treatments were formulated by replacing 30% (G1), 50% (G2), 70% (G3) or 100% (G4) of BF. Turkey breast muscles and beef fat were separately minced through a 3 mm plate grinder (Arnica, Turkey). Minced meat was homogenised for 1 min using a kitchen-type mixer (Tchibo, Germany) that was placed in ice bath. After that, BF (control) and/or GE, half of the ice, sodium chloride (1%), sodium tripolyphosphate (0.5%) and sodium nitrite (0.015%) were added and mixed for 1 min. The other half of the ice was then added and mixing was continued again for 2 min.

Emulsions were placed in hermetically sealed centrifuge tubes and centrifuged at 3100 × g at 4°C for 1 min to eliminate any air bubbles. These were then heat-treated in a 70°C water bath for 30 min and cooled to room temperature. Turkey model emulsions were stored in sealed tubes at 4°C prior to analysis. Emulsion stability (ES) as total expressible fluid (TEF) and expressible fat (EFAT) [6], water-holding capacity (WHC) [6], and cook yield (CY) [7] were analysed to evaluate turkey model emulsion characteristics. pH was measured by using a pH-meter equipped with a penetration probe. Total moisture, protein and ash analyses were carried out according to AOAC [8]. Total lipid content was evaluated according to Flynn and Bramblett [9]. Colour (L*, a*, b*) of the emulsions was measured with a portable colorimeter (Konica Minolta, Japan). Data was analysed by ANOVA and Duncan's Post-Hoc tests using the SPSS software.

3. Results and Discussion

TEF and EFAT, WHC and CY results are presented in figure 1a, 1b and 1c, respectively. TEF was recorded as being between 6.23-10.65% in the various model turkey emulsions. G4 emulsions had the highest TEF value among the formulations ($P<0.05$), whereas G1 and G2 emulsions had similar TEF compared to C emulsions.

This result showed that increased concentrations of GE might have an undesired impact on emulsion stability but concentrations of 30% and 50% should favour stability. EFAT values were between 6.73-9.08%, where the lowest value was measured in G1 group, indicating that this emulsion had good stability in terms of both expressible fluid and fat.

The other groups had similar EFAT compared to the control. WHC of the model turkey emulsions were in the range of 54.29-63.67% and all the GE model turkey emulsions showed similar WHC to C emulsions. In GE treatments, G2 group had higher WHC than G4 ($P<0.05$). Thus, the total fat replacement seemed to decrease WHC. The CY measured was between 94.93-96.83%, where GE model turkey emulsions had similar values to C emulsion, in accordance with WHC results. Similar to our results, Serdaroğlu *et al.* [4] reported that total beef fat replacement with olive oil GEs in model beef emulsions could negatively affect emulsion characteristics, but could show equivalent stability when replacement level was up to 50%.

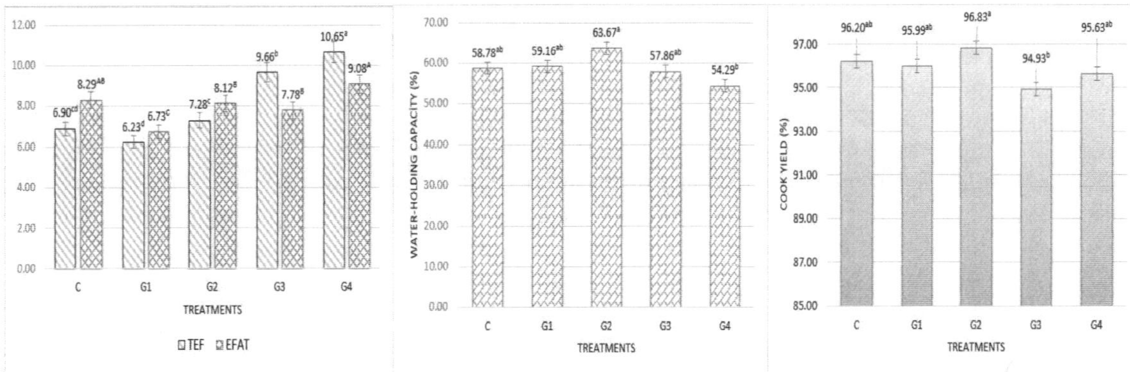

(a) (b) (c)

Figure 1. (a) Emulsion stability (TEF & EFAT), (b) WHC and (c) CY of the treatments. Different letters indicate significant difference (*P*<0.05). Standard deviation of the means was recorded between 0.21-1.81, 0.06-0.89, 1.13-2.74 and 0.03-0.84 for TEF, EFAT, WHC and CY, respectively.

Chemical composition and pH values of the model turkey emulsions are presented in table 1. Total moisture, protein, lipid and ash content were between 62.62-66.01, 16.45-17.82, 10.45-16.64 and 1.98-2.21, respectively. No significant differences were obtained in moisture and protein content of the model turkey emulsions. Similar moisture contents could be attributed to the similar WHCs of the model turkey emulsions.

In addition, since added GE was not a source of protein, the total protein content of the model turkey emulsions did not differ. A significant lipid reduction was achieved in the model turkey emulsions formulated with more than 50% GE (*P*<0.05). This result showed that incorporating GEs in turkey product formulations could offer a favourable advantage, promoting healthier composition of the meat products.

The lowest lipid content was recorded in G4 emulsions (*P*<0.05), whilst G2 and G3 emulsions had similar lipid content. Ash content of G4 emulsions was also significantly lower than other treatments (*P*<0.05). pH values of the treatments range between 5.99-6.06. G2 and G3 emulsions had similar pH value compared to C, but pH value was higher in G4 and lower in G1 than in C emulsions (*P*<0.05). Therefore, an average level of GE should be useful to maintain the pH value.

Table 1. Chemical composition and pH values of model turkey emulsions.

Treatments	Moisture (%)	Protein (%)	Lipid (%)	Ash (%)	pH
C	63.26 ± 1.02	17.55 ± 1.41	16.64^{a} ± 0.38	2.21^{a} ± 0.07	6.02^{b} ± 0.01
G1	62.62 ± 0.65	17.40 ± 0.31	16.41^{a} ± 0.45	2.18^{a} ± 0.03	5.99^{c} ± 0.01
G2	65.55 ± 1.26	16.45 ± 1.06	13.74^{b} ± 0.14	2.16^{a} ± 0.03	6.01^{b} ± 0.01
G3	66.01 ± 1.27	17.12 ± 1.00	12.72^{b} ± 0.50	2.13^{a} ± 0.09	6.01^{b} ± 0.01
G4	65.73 ± 1.12	17.82 ± 0.70	10.45^{c} ± 0.36	1.98^{b} ± 0.03	6.06^{a} ± 0.02

a, b, c: Different letters indicate significant difference (*P*<0.05).
Data is presented as mean values ± standard deviation.

Colour parameters of model turkey emulsions are shown in figure 2. L*, a* and b* values were between 74.43-77.79, 2.92-5.30 and 8.18-11.43, and replacement of beef fat with GEs led to significant differences in all of the colour parameters (*P*<0.05). The lowest L* and b* and the highest a* values were recorded in C emulsions (*P*<0.05). Thus, utilization of GE resulted in increased L* and b* and decreased a* values (*P*<0.05). G4 emulsions had the highest L* value (*P*<0.05) while L* values were similar in G1, G2 and G3 emulsions, meaning that GE concentrations more than 70%

could produce a larger change in L* values. In GE treatments, a* values were decreased with increased GE concentrations ($P<0.05$), except in G2 and G3 emulsions that had similar a* values. In b* values, an increasing trend was recorded with increased concentrations of GE ($P<0.05$), probably due to the natural yellow-greenish colour of olive oil used in GE formulations. Similar to our study, Serdaroğlu *et al.* [4] reported that addition of GE in model beef systems increased lightness and yellowness but reduced redness compared to control samples. It should be noted that since instrumental colour parameters are not certainly in a linear relationship with colour acceptability, the visual satisfaction of the consumers should be evaluated in products formulated with GEs.

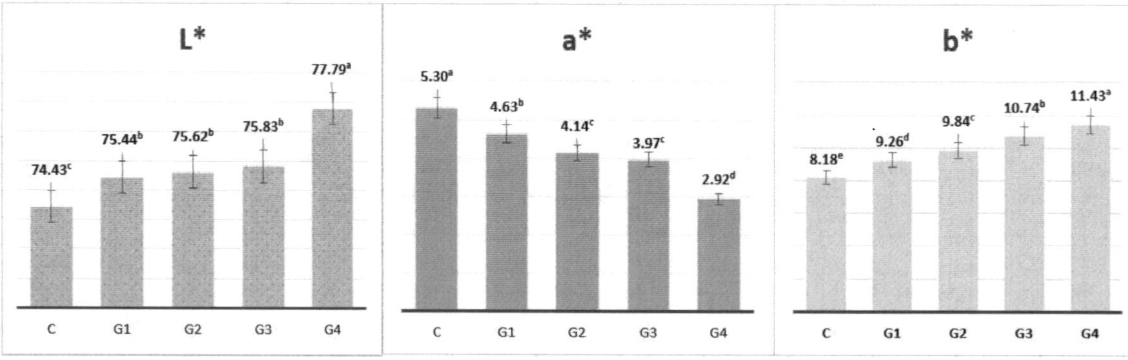

Figure 2. L*, a* and b* values of the model turkey emulsions. Different letters indicate significant difference ($P<0.05$). Standard deviation of the means was recorded between 0.19-1.15, 0.30-0.83 and 0.05-1.16 for L*, a* and b*, respectively.

4. Conclusion

The results of our study indicated that incorporating GE into model turkey meat emulsions presented favourable effects in terms of emulsion stability and fat reduction. Utilization of GE resulted in turkey meat emulsions with equivalent emulsion stability parameters to full-beef fat turkey meat emulsions, and thus, the use of GE offers technological advantages. However, increased concentrations of GE could have a negative impact on emulsion stability parameters in terms of expressible fluid, although no significant changes were observed in water-holding capacity or cooking yields. Reductions in total fat were obtained in model turkey emulsions that were produced with more than 50% GE as beef fat replacer, which is key to development of healthier product formulations. However, the colour parameters were highly affected by GE incorporation. Further study should be performed regarding the adoption of GE in various meat products.

References

[1] Dickinson E 2013 *J. Sci. Food Agr.* **93** 710
[2] Poyato C, Ansorena D, Berasategi I, Navarro-Blasco I and Astiasarán I 2014 *Meat Sci.* **98** 615
[3] Lurueña-Martínez M A, Vivar-Quintana A M and Revilla I 2004 *Meat Sci.* **68** 383
[4] Serdaroğlu M, Nacak B, Karabıyıkoğlu M and Keser G 2016 *Korean J. Food Sci. An.* **36** 744
[5] Öztürk B, Urgu M and Serdaroğlu M 2017 *J. Sci. Food Agr.* **97** 2075
[6] Hughes E, Cofrades S and Troy D J 2004 *Meat Sci.* **45** 273
[7] Murphy E W, Criner P E and Grey B C 1975 *J Agr Food Chem.* **23** 1153
[8] AOAC 2012 *Official Methods of Analysis* (Gaithersburg, MD, USA)
[9] Flynn A W, Bramblett V D 1975 *J Food Sci.* **40** 631

IOP Conf. Series: Earth and Environmental Science **85** (2017) 012072 doi:10.1088/1755-1315/85/1/012072

Consumer perception and acceptance of pork and chicken sausage

M Ristić[1], K Troeger[1], J Đinović-Stojanović[2], N Knežević[3] and M Damnjanović[3]

[1] BAFF, Former Federal Centre for Meat Research, Kulmbach, Germany (retired researchers)
[2] Institute of Meat Hygiene and Technology, Kaćanskog 13, 11000 Belgrade, Serbia
[3] Zlatiborac Company d.o.o., Mojkovačka 58, 11136 Belgrade, Serbia

E-mail: ristic.rc-kulmbach@t-online.de

Abstract. This study was performed to evaluate consumers' perception and acceptance of selected pork and chicken sausage (budim and chicken sausages, respectively) from Zlatiborac Meat Company. Sensory evaluation was performed by Serbian consumers (n=1157) in three retail stores in Belgrade. Consumers were asked for their preference for taste, salt content and smoke of two sausages and to recognize the kind of meat which was used to make these meat products. Consumers evaluated taste, salt content and smoke flavor of budim and chicken sausages with the highest percentage of the best offered answer. Between 47-55%, 72-76% and 82-84% of consumers evaluated the taste of sausages as good, the salt content as well-balanced and the smoke flavor as balanced, respectively. Tukey's HSD test was applied to analyze variations of male and female perception and acceptance of analyzed sausages.

1. Introduction

Meat and meat products have played a very important role in the daily diet of consumers over the world. They are highly valued sources of protein, iron, zinc and vitamins etc. On the other hand, some scientific research emphasized the strong relationship between high red meat consumption and an increased risk of cancers, diabetes, and cardiovascular diseases [1,2]. Despite that, global meat production is projected to more than double from 229 million tones (1999/2001) to 465 million tones by 2050 [3]. Meat consumption statistics vary among and within countries and depends on many factors, such as socio-economics, ethics, cultural, religious beliefs and tradition [4].

Sensory evaluation of meat and meat products is not an easy task for consumers. Consumer trust often leads to consumer acceptance of products. The relationship between consumer perception of quality and meat industry has a direct influence on profitability. Also, it is well documented that consumer reactions to food scandals result in economic implications [5]. Thus, scientific knowledge and technology innovation play an important role and can help industry to respond to consumer expectations [6]. Modern consumers are becoming increasingly interested in health and safety of food with good sensory properties [7,8].

This study was carried out to elaborate consumers' perception and acceptance of two sausages. The main objectives of this study were to evaluate pork and chicken sausage (budim and chicken sausage,

IOP Conf. Series: Earth and Environmental Science **85** (2017) 012072 doi:10.1088/1755-1315/85/1/012072

respectively) by consumers in three retail stores in Belgrade, Serbia, and determine the main deficiencies of those meat products in order to improve their quality.

2. Materials and Methods

Consumer testing was performed in three large retail stores (Delhaize, DIS and Mercator S) in January 2016, in Belgrade. A total of 1157 consumers were participated in one day during the period from 10 a.m. to 4 p.m. Consumers were males (45%) and females (45%) older than 18 years of age.

The questionnaire started with general questions about the consumers, referred to their age, number of family members, education levels and shopping habits. Thereafter respondents were asked to evaluate budim and chicken sausages produced by Zlatiborac Meat Company, but the origin and market name of the products were unknown to the consumers. Budim sausage (in Serbian *budimska kobasica*) consists of pork (57.5%), beef (19.2%) and solid fat (19.2%). Chicken sausage (in Serbian *pileća čajna kobasica*) consists of chicken drumstick meat (69.2%), pork (9.7%) and solid fat (19.5%). Consumers were asked the following questions: (1) Rate the taste (The offered answers were: good, satisfactory, unsatisfactory); (2) Rate the salt content (The offered answers were: balanced, not salty enough, too salty); (3) Rate the smoke flavor (The offered answers were: balanced, not strong enough, too strong); (4) Which type of meat has been processed? (The offered answers were: pork, beef, poultry and mixture).

Statistical analysis of experimental data was performed using software Statistica 10.0 (StatSoft Inc., Tulsa, OK, USA). Tukey's HSD test for comparison of consumers answers were used to analyze variations of male and female perception of budim and chicken sausages.

3. Results and discussion

Results concerning some demographic parameters of consumer are shown in figures 1 and 2. Consumers (n=1157) were males (45%) and females (45%) older than 18 years of age.

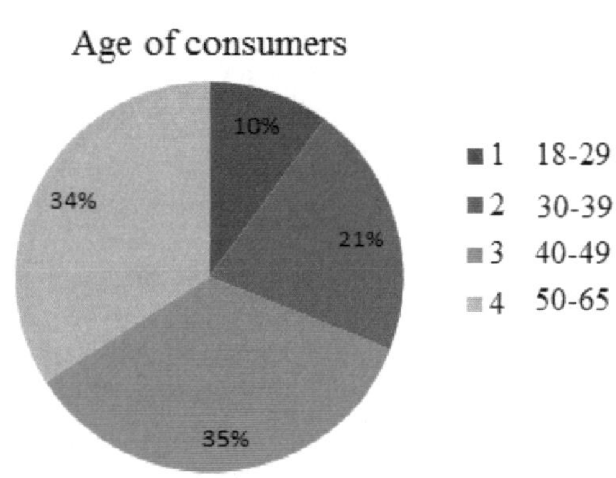

Figure 1. Age of consumers.

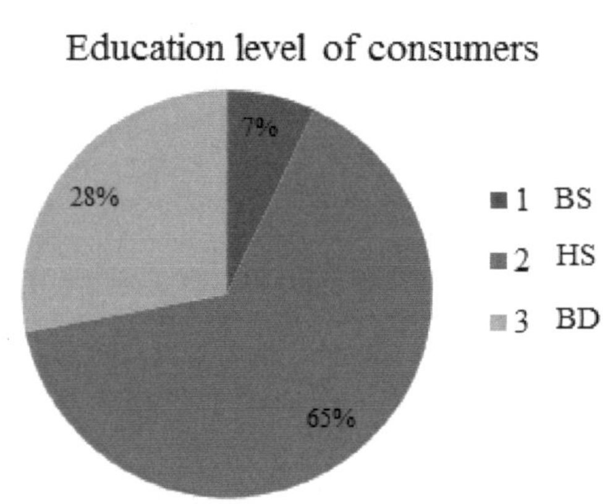

Figure 2. Education level of consumers' (BS–Basic School; HS–High/secondary School; BD–Bachelor Degree.

The highest percentages of consumers were between 40 and 49 (34.5%) and between 50 and 65 years old (34.4%), and more than 60% of them had completed high school. The average test person was male, aged 40-49, with a secondary school education, living in a family of four.

The results of sensory evaluation of budim and chicken sausage by consumers are shown in figures 3 and 4.

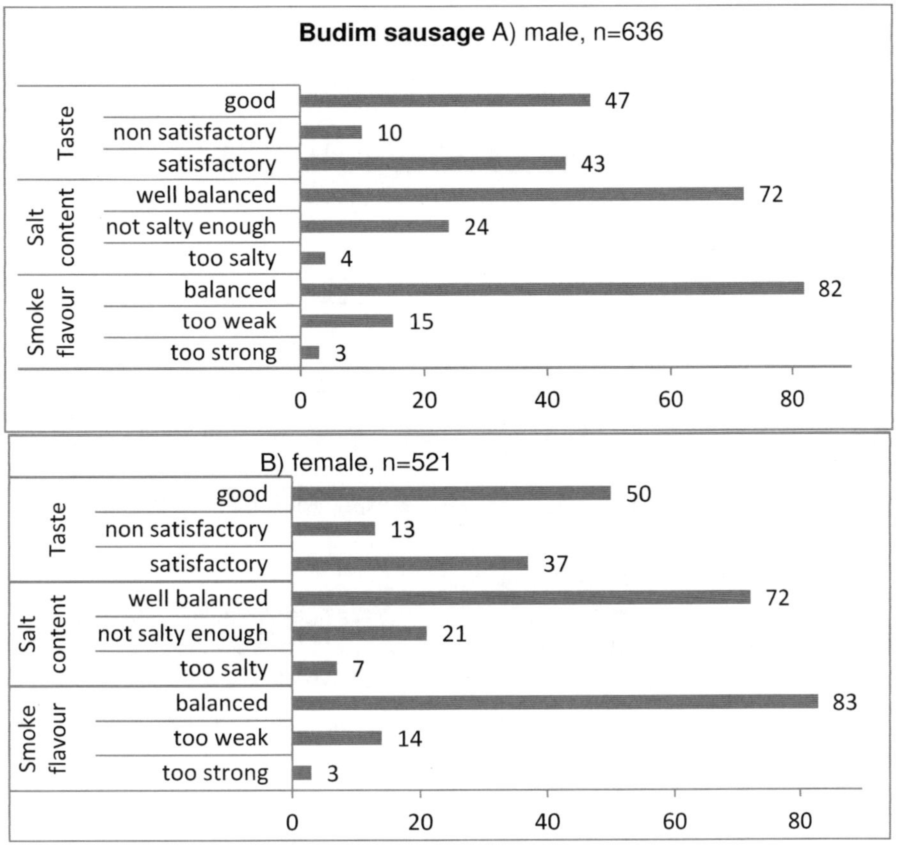

Figure 3. Sensory evaluation of budim sausage by consumers (%).

Sensory evaluation of budim sausage (figure 3) showed that only 47-50% of consumers evaluated the taste as good while 10-13% of them were not satisfied with the taste. More than 70% and 80% of consumers evaluated salt content and smoke flavor as well-balanced and balanced, respectively. In the case of chicken sausage more than 50 % of consumers evaluated the taste as good; around 75% of consumers rated the salt content as well-balanced; more than 80% of consumers rated the smoke flavor as balanced (figure 4). Among the possible answer categories, the highest percentage of consumers were not satisfied with salt content of both sausages, since 21-24% and 18-22% of consumers evaluated salt content of budim and chicken sausage, respectively, as not salty enough. Post-hoc Tukey's HSD test showed that there were no statistically significant differences (at $p<0.05$ level) in male and female answers concerning rating the taste, salt content or smoke flavor of budim and chicken sausages.

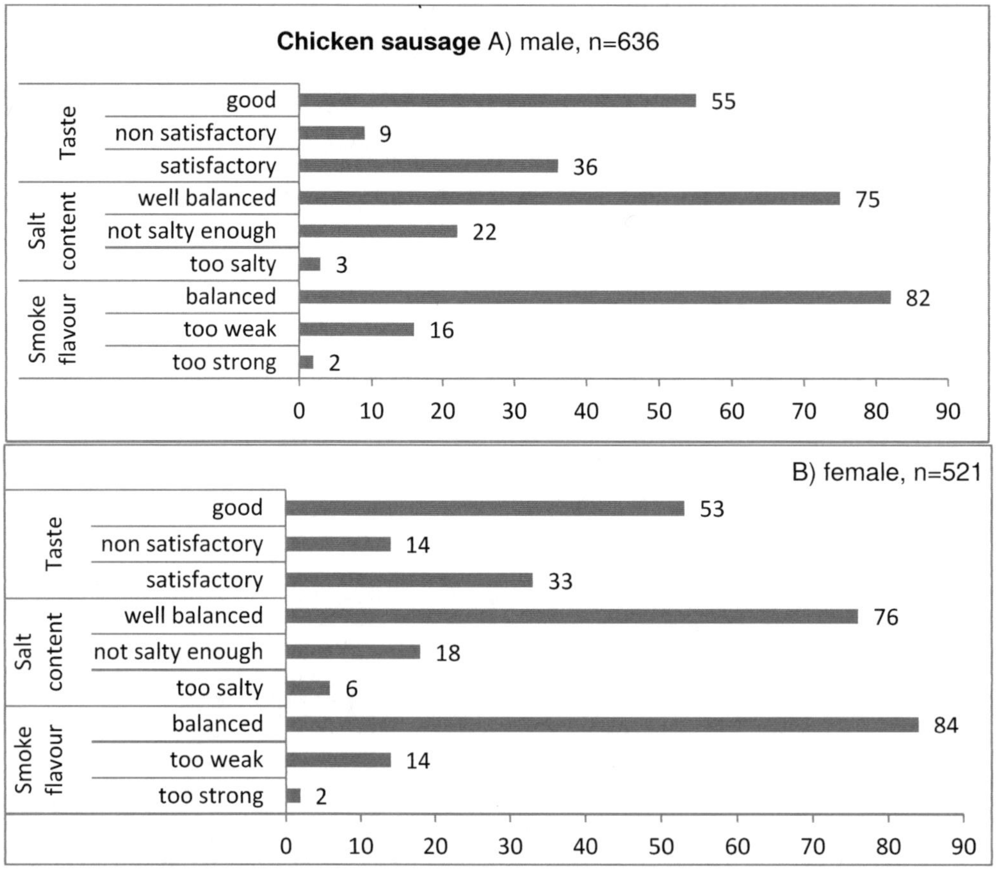

Figure 4. Sensory evaluation of chicken sausage by consumers (%).

Consumers were also asked to identify which animal species they thought the meat incorporated in the sausages was from (table 1). Only 57% and 38% of consumers recognized that the budim and chicken sausages were prepared from pork and mixed (beef and pork) meat, respectively.

Table 1. Consumers' answers [%] to the question: What kind of meat was used to make this sausage? Answers in bold and italic are correct answers.

Type of meat	Sausage	
	Budim	Chicken
Beef	6	5
Mixed (beef and pork)	27	***38***
Pork	***57***	28
Chicken	10	29

The data obtained from this study could be compared with the data from DLG (German Agricultural Society) evaluation in Germany in 2016 [9], where meat company Zlatiborac participated. In the DLG Test Center Food, in 2016, 1577 international meat products were evaluated in accordance with the DLG-5-points-scheme. It is a descriptive sensory analysis with scales on the basis of assessment by experts, which included visual (appearance/exterior), haptic (consistence/texture), olfactory (odor) and gustative (taste) criteria of the meat products. The main deficiencies of the international meat products were related to their external appearance, consistency, odor and taste [9]. They were denoted as "sinew component too high" (11%), "surface film" (7%), "too soft" (5%), "sour" (5%), etc. The main deficiencies of meat products from Serbian origin [10] were "sinew component too high", "smoke too strong", "sour" and "salty". However, beside mentioned deficiencies, all meat products from Serbia pass the DLG tests and received a "DLG award winner".

It is clear that Serbian consumers and DLG experts had different perceptions of Serbian meat products in some aspects of sensory evaluation. For example, DLG experts evaluated some products as "salty" and "smoke was too strong" [10]. However, although consumers in Belgrade evaluated taste, salt content and smoke flavor of budim and chicken sausages with the highest percentage of the best offered answer, the main deficiencies of analyzed meat products were evaluated as "not salty enough" (18-24%) and smoke flavor was "too weak" (14-16%) (figures 3 and 4). Serbian consumers' responses concerning analyzed meat products could be associated with the fact that salty and highly-smoked meat products are traditional in the country.

4. Conclusion

Sensory evaluation of budim and chicken sausages produced by Zlatiborac Meat Company showed that consumers were satisfied with the taste, salt content and smoke flavor of the products. Their responses concerning the analyzed meat products could be associated with traditional meat production in Serbia. Post-hoc Tukey's HSD test showed that there were no statistically significant differences (at $p<0.05$ level) in male and female answers concerning sensory evaluation of budim and chicken sausages. The obtained data could be useful cues for industry to enhance some parameters of meat product quality in order to place them on the EU market.

Acknowledgments

The authors would like to thank Mr Dejan Djurdjevic from Zlatiborac Meat Company, Serbia, for his help in sensory evaluation of the meat products.

References

[1] McAfee A J, McSorley E M, Cuskelly G J, Moss B W, Wallac E J M W and Bonham M 2010 *Meat Sci.* **84** 1
[2] Alisson-Silva F, Kawanishi K and Varki A 2016 *Mol. Aspects Med.* **51** 16
[3] Steinfeld H, Gerber P, Wassenaar T, Castel V, Rosales M and De Haan C 2006 Livestock's long shadow: Environmental issues and options, Rome, Italy
[4] Naylor R W, Droms C M and Haws K L 2009 *JPP&M* **28** 221
[5] Rieger J, Kuhlgatz C and Anders S 2016 *Food Policy* **64** 82
[6] Troy D J and Kerry J P 2010 *Meat Sci.* **86** 214
[7] Olewinik-Mikolajewska A, Guzek D, Glabaska D and Gutkowska K 2016 *J. Sens. Stud.* **31** 193
[8] Chamhuri N and Batt P J 2015 *Brit. Food J.* **117** 1168
[9] Lautenschläger R, Thumel H and Hillgärtner K 2016 *Fleischwirtschaft* **96** 10 47
[10] Đinovic-Stojanovic J, Troeger K, Ristic M, Knezevic N and Damnjanovic M 2017 *J. Agr. Sci. Food Tech.* **3** 15

Changes in total viable count and TVB-N content in marinated chicken breast fillets during storage

T Baltić[1], J Ćirić[2], B Velebit[1], R Petronijević[1], B Lakićević[1], V Đorđević[1] and V Janković[1]

[1]Institute of Meat Hygiene and Technology, Kaćanskog 13, Belgrade, Serbia
[2]Faculty of Veterinary Medicine, University of Belgrade, Bulevar Oslobođenja 18, Belgrade, Serbia

E-mail: tatjana.baltic@inmes.rs

Abstract. Marination is a popular technique for enhancing meat properties. Depending on the marinade type and ingredients added, marination can improve sensory, chemical and microbiological quality of meat products. In this study, the total viable count and total volatile basic nitrogen (TVB-N) content in marinated chicken breast fillets were investigated. The possible correlation between bacterial growth and formation of TVB-N was also tested. Chicken breast fillets were immersed in a solution of table salt (as a control) orthree different marinades,which consisted of table salt, sodium tripolyphosphate and/or sodium citrate, and stored in air for nine days at 4±1°C. Analyses of the total viable count and TVB-N were performed on days0, 3, 6 and 9 day of storage. The total viable count gradually increased in all examined groups, and statistically significant differences ($p<0.01$; $p<0.05$) between treatments on days0, 3 and 6 day of storage were established. TVB-N values in marinated chicken were significantly higher ($p<0.01$; $p<0.05$) compared to the control. Using the multiple linear regression, a positive correlation between total viable count and formation of TVB-N in chicken marinated with sodium citrate was established ($p<0.05$), while the intensity of TVB-N formation was lowest in chicken marinated with sodium tripolyphosphate.

1. Introduction

Poultry meat has become a mass consumer product throughout the world: in every region, in countries with very different levels of development and in diverse forms. Furthermore, as an important source of proteins, poultry meat has a high biological value and it has been frequently recommended for its nutritious low-fat content [1]. In the last few decades, consumers' demand for convenience foods has resulted in an expansion of the processed meat and poultry industry [2].

The food industry constantly seeks new ways to add value to products, and one example is the marination of raw meat to add flavour and extend shelf-life [3]. Marination also improves the tenderness, juiciness, flavour, colour, and cooking yield of meat and poultry [4]. Depending on marinade type, marination might affect chemical and sensory attributes of the meat, as well as the microbiological safety of the final meat product. For the marination of poultry meat, alkaline marinades are often used. A typical marinade solution for commercial chicken products is made of 90% water, 6% table salt (sodium chloride), and 4% sodium tripolyphosphate (STPP) [5]. However, thequantity of phosphate used in meat products is limited by legislation [6], and more often the phosphates are combined with other additives in order to increase the water binding capacity of the meat. Sodium citrate has been utilized as a phosphate replacer to enhance water-holding properties of the meat [7]. Although these additives are not considered as antimicrobials, studies of the antimicrobial activity of sodium tripolyphosphate and other polyphosphates on food contaminants have been conducted suggesting their possible use as antimicrobials [8-10]. According to standard NF-V01-003 [11], total viable counts in marinated chicken meat should not exceed the maximum recommended limit of 5.7 log CFU/g of meat.

In addition to microbial counts, some chemical parameters could indicate meat freshness. The total volatile basic nitrogen (TVB-N) content is a chemical indicator of the meat quality, and it is associated with the amino acid decarboxylase activity of microorganisms during storage. The above-mentioned standard [11] recommends that TVB-N value in marinated chicken meat should not exceed 60 mgN/100g of the meat, while Balamatsia *et al.* [12] suggested a value of 40 mg N/100 g as an upper limit for fresh poultry meat.

This study was conducted to investigate the changes in a total viable count and TVB-N content in marinated chicken breast fillets during storage. Therefore, the addition of STPP and sodium citrate to chicken breast fillets as a means of enhancing sensory quality, and the effect of various concentrations of STPP (1 and 2%) and sodium citrate (1 and 2%), on the behaviour of total viable count and formation of TVB-N on days0, 3, 6and 9of storage at 4.0±1°C were investigated. The possible correlation between bacterial growth and formation of TVB-N was also tested.

2. Materials and Methods

Deboned and skinless chicken breast meat (*m. pectoralis major*) was purchased in a local poultry abattoir. In the laboratory, meat was aseptically filleted into pieces weighing approximately 0.1 kg and divided into four groups.

For marinade, table salt with 99 to 99.5% sodium chloride (Solana dd Tuzla, Bosnia and Herzegovina), a "Carfosel Genius"commercial mixture of sodium tripolyphosphate (Prayon, Belgium), tri-sodium citrate dihydrate ($C_6H_5Na_3O_7 \times 2H_2O$) (Merck, Germany) and drinking water were used. Three different marinades and a solution of 6% table salt as a control (group I) were prepared and kept refrigerated 24 hours at 4±1°C before use. The first marinade consisted of water, 6% NaCl and 2% phosphate (group II), the second consisted of water, 6% NaCl and 2% citrate (group III) and the third one consisted of water, 6% NaCl, 1% phosphate and 1% citrate (group IV). Chicken breast fillets were immersed in marinade solutions or in the control solution of table salt in the ratio of 1:2 (100g of meat/200 mL of marinade or table salt solution). Marination process lasted 4h at 4±1°C, with stirring at times. After 4 h of immersing, the fillets were drained, packed in sterile Stomacher bags (Stomacher 400 Classic Bags, Vicor, Serbia) and stored 9 days in air at 4±1°C.

The total viable count was determined according to the standard ISO method [13]. Results of the microbial count determinations are presented as log CFU/g of meat.

Determination of TVB-N content in marinated chicken breast fillets during storage was performed according to the method suggested by Goulas and Kontominas [14]. Results of the TVB-N content in marinated chicken breast fillets are presented as mg N/100 g of meat.

All measurements were conducted in triplicate for each sample. All parameters were described by means and standard deviation, in format X±SD. Statistical analysis of the results was conducted using GraphPad Prism version 6.00 for Windows (GraphPad Software, San Diego, CA, USA, www.graphpad.com). One-way ANOVA and post hoc Tukey's test were performed to assess the significance of differences among various groups. Values of $p<0.05$ and $p<0.01$ were considered significant. Multiple linear regression (MLR) was performed in JMP 10 Software (SAS Institute Inc., Cary, NC).

3. Results and Discussion

The total viable count is a microbiological parameter commonly used to determine hygiene status of the meat and contributes shelf-life of meat and meat products. The total viable counts in marinated chicken breast fillets during 9 days of storage are presented in Table 1. During the storage period, the total viable counts gradually increased in all examined groups of marinated chicken, and significant differences ($p<0.01$; $p<0.05$) were found between groups on days 0, 3and 6 of storage. At the beginning of the study (day 0), the initial total viable count was the lowest in group II marinated chicken, and the highest in group III marinated chicken.

Table 1. Mean total viable counts (log CFU/g) in marinated chicken breast fillets during storage

Marinade group	Day of storage			
	0	3	6	9
I (control)	4.11 ± 0.44^{a}	5.33 ± 0.51^{A}	7.03 ± 0.24^{A}	8.86 ± 0.51
II	3.68 ± 0.35^{A}	5.87 ± 0.48^{B}	7.15 ± 0.27^{Ba}	8.70 ± 0.43
III	4.73 ± 0.31^{aB}	5.66 ± 0.33^{C}	6.44 ± 0.18^{AB}	8.60 ± 0.41
IV	3.82 ± 0.30^{AB}	6.03 ± 0.67^{ABC}	6.74 ± 0.18^{a}	8.95 ± 0.30

Within a row, means with common superscript letter are significantly different $^{A-C}$ $p<0.01$; ^{a} $p<0.05$

Group I – water and 6% NaCl

Group II – water, 6% NaCl and 2% phosphate

Group II – water, 6% NaCl and 2% citrate

Group IV – water, 6% NaCl, 1% phosphate and 1% citrate

At the end of the study (day 9), the total viable counts in all tested groups reached values above 8 log CFU/g, and no significant differences ($p>0.05$) were found between marinade treatments. These results are in agreement with findings from Susiluoto *et al.* [15] and Samoui *et al.* [16]. According to standardNF-V01-003, the maximum recommended limit for a total viable count in marinated chicken is 5.7 log CFU/g [11], which, in our study, was exceeded on day 3(marinade groups II and IV) and day 6(marinade groups I, II, III, IV).

For estimating meat freshness or shelf-life, TVB-N as a product of microbial decarboxylase activity was investigated. TVB-N values in marinated chicken were significantly higher ($p<0.01$; $p<0.05$) compared to those in the control, but were below the recommended value of 40 mg N/100 g of meat until day 9. The average TVB-N content in marinated chicken breast fillets during the storage period of 9 days are presented in Table 2.

Table 2. Average of total volatile basic nitrogen (TVB-N) in marinated chicken fillets during storage (mg N/100 g)

Marinade group	Day of storage			
	0	3	6	9
I (control)	20.44 ± 1.35^{AC}	20.21 ± 0.21^{BCD}	19.83 ± 2.61^{B}	37.57 ± 0.41^{Aab}
II	22.82 ± 0.39^{aC}	23.94 ± 0.88^{AC}	30.64 ± 3.45^{ABC}	52.78 ± 6.19^{a}
III	20.44 ± 1.35^{aAB}	26.11 ± 1.30^{ABa}	23.24 ± 4.81^{A}	57.11 ± 10.22^{A}
IV	21.75 ± 0.23^{B}	24.64 ± 0.53^{aD}	23.29 ± 1.08^{C}	51.75 ± 9.05^{b}

Within a row, means with common superscript letter are significantly different $^{A-D}$ $p<0.01$; $^{a-b}$ $p<0.05$

Group I – water and 6% NaCl

Group II – water, 6% NaCl and 2% phosphate

Group II – water, 6% NaCl and 2% citrate

Group IV – water, 6% NaCl, 1% phosphate and 1% citrate

On day 0 of storage, the lowest TVB-N value was in control group chicken (group I) and the highest was in group III marinated chicken. By day 6, TVB-N values in all tested groups were below the recommended value of 40 mg N/100 g. On day 9day of storage, TVB-N values in marinated chicken exceeded the recommended value and were statistically significant higher ($p<0.01$or $p<0.05$) compared to those of the control group where the TVB-N value was below 40 mg N/100 g. Patsias *et al.* [17] found that the initial TVB-N value of 12 mg/100g sharply increased in chilled chicken fillets stored in air, resulting in high TVB-N values (49 g/100g) after 9 days of storage. In the present study, increases in the total viable count were followed by subsequent increases of TVB-N content in marinated chicken breast fillets, but this was not the case in control group chicken fillets. Khalafalla *et al.* [18] stated a clear relationship between the microbiological quality of broiler chicken breasts and level of TVB-N formation. The MLR results we obtained showed that the most significant ($p<0.05$) increase in TVB-N occurred in chicken breast fillets treated with sodium citrate (marinade groups III and IV) where positive correlations between total viable count and TVB-N was established ($p<0.05$). Although changes in total viable count in marinade group II showed a similar pattern as in other experimental groups, MRL analysis showed a negative correlation between these two parameters (microbial counts and TVB-N) (Table 3).

Table 3. MLR results of total viable count vs. TVB-N

Marinade group	Estimate	Std Error	t Ratio	t Ratio	Prob>|t|
I (control)	-0.59	1.16	-0.51		0.6159
II	-4.88	1.48	-3.31		0.0037*
III	7.36	1.34	5.50		<0.0001*
IV	3.04	1.39	2.19		0.0409*

* significance level of $p<0.05$
Group I – water and 6% NaCl
Group II – water, 6% NaCl and 2% phosphate
Group II – water, 6% NaCl and 2% citrate
Group IV – water, 6% NaCl, 1% phosphate and 1% citrate

4. Conclusion

Sodium tripolyphosphate and sodium citrate, alone or in combination, did not inhibit microbial growth in marinated chicken breast fillets during storage. MLR results indicated that sodium tripolyphosphate slows down TVB-N formation. Moreover, it seems that sodium citrate potentiates the growth of aerobic bacteria and has a positive effect on TVB-N formation in marinated chicken breast fillets.

Acknowledgement

This paper was supported by the Ministry of Education, Science and Technological Development, Republic of Serbia, Project "Selected biological hazards to the safety/quality of food of animal origin and the control measures from farm to consumer" (TR 31034).

References

[1] Mantilla S P S, Santos E B, de Freitas M Q, de Carvalho Vital H, Mano S B and Franc R M 2012 Refrigerated poultry breast fillets packed in modified atmosphere and irradiated: bacteriological evaluation, shelf life and sensory acceptance *B. J. Microbiol.* **43** 1385–92
[2] Baltić T, Baltić Ž M, Mišić D, Ivanović J, Janjić J, Bošković M and Dokmanović M 2015 Influence of marination on *Salmonella* spp. growth in broiler breast fillets *Acta Vet. Beograd* **65** 417–28
[3] Rhoades J, Kargiotou C, Katsanidis E and Koutsoumanis K P 2013 Use of marination for controlling *Salmonella enterica* and *Listeria monocytogenes* in raw beef *Food Microbiol.* **36** 248–53
[4] Guerrero-Legarreta I and Hui Y H 2010 *Handbook of Poultry Science and Technology* vol 1ed I Guerrero-Legarreta (New York: Wiley-Interscience) p 524
[5] Smith DP and Acton JC 2001 Marination, cooking, and curing of poultry products *Poultry meat*

processing ed AR Sams (Boca Raton: CRC Press, Taylor & Francis Group) p 257

[6] Rulebook on quality of minced meat, semi-processed and processed meat products 2015 Official Gazette of the Republic of Serbia No. 94/2015

[7] Alvarado C and McKee S 2007 Marination to improve functional properties and safety of poultry *Meat. J. Appl. Poult. Res.* **16** 113–20

[8] Lee Y L, Cesario T, Owens J, Shanbrom E and Thrupp L D 2002 Antibacterial activity of citrate and acetate *Nutrition* **18** 665–6

[9] Vareltzis K, Soultos N, Koidis P, Ambrosiadis J and Genigeorgis C 1997 Antimicrobial effects of sodium tripolyphosphate against bacteria attached to the surface of chicken carcasses *LWT-Food Sci. Technol.* **30** 665–9

[10] Baltić T, Mitrović R, Todorović E, Ivanović J, Lončina J and Baltić M Ž 2012 The influence of vacuum-packaging and marination on the behaviour of total *Enterobacteriaceae* on chicken breast fillets *Proc. Int. Conf., Biological Food Safety and Quality BFSQ,* 4-5 Oct. 2012,Belgrade, ed S Bunčić (Novi Sad: Faculty of Agriculture, University of Novi Sad) p 95

[11] Association Française de Normalisation–AFNOR 2004 NF-V01-003 Hygiene and safety foods – Validation of the microbiological shelf life – Perishable and cooled foods

[12] Balamatsia C C, Patsias A, Kontominas M G and Savvaidis I N 2007 Possible role of volatile amines as quality-indicating metabolites in modified atmosphere-packaged chicken fillets: Correlation with microbiological and sensory attributes *Food Chem.* **104** 1622–8

[13] International Organization for Standardization–ISO 2013 ISO 4833-1:2013 Microbiology of the food chain – Horizontal method for the enumeration of microorganisms – Part 1: Colony count at 30 °C by the pour plate technique

[14] Goulas A and Kontominas M G 2007 Effect of modified atmosphere packaging and vacuum packaging on the shelflife of refrigerated chub mackerel (*Scomber japonicus*): biochemical and sensory attributes *Eur. Food Res. Technol.* **224** 545–53

[15] Susiluoto T, Korkeala H and Björkroth J 2002 *Leuconostoc gasicomitatum* is the dominating lactic acid bacterium in retail modified-atmosphere-packaged marinated broiler meat strip products on sell by day *Int. J. Food Microbiol.* **80** 89–97

[16] Samoui S, Hlima HB, Salah RB and Ghorbel R 2011 Effects of sodium lactate and lactic acid on chemical, microbiological and sensory characteristics of marinated chicken *Afr. J. Biotechnol.* **10** 11317–26

[17] Patsias A, Badeka A V, Savvaidis I N and Kontominas M G 2008 Combined effect of freeze chilling and MAP on quality parameters of raw chicken fillets *Food Microbiol.* **25** 575–81

[18] Khalafalla F A, Ali F H M and Hassan A H A 2016 Quality improvement of broiler chicken breasts by nisin and lactic acid *J. World Poult. Res.* **6** 37–47

Effect of starter cultures on survival of *Listeria monocytogenes* in Čajna sausage

M Bošković[1], V Tadić [1], J Đorđević [1], M Glišić[1], B Lakićević[2], M Dimitrijević[1] and M Ž Baltić[1]

[1] Faculty of Veterinary Medicine, University of Belgrade, Bulevar oslobođenja Belgrade, Serbia
[2] Institute of Meat Hygiene and Technology, Kacanskog 13, Belgrade, Serbia

E-mail: marijaboskovic116@gmail.com

Abstract. The aim of the study was to evaluate the survival of *Listeria monocytogenes* during the production of Čajna sausage with short maturation time. Sausage batter was inoculated with three different serotypes 4b and serotype 1/2a of *L. monocytogenes*. Control sausages were without any starter culture added; the second batch was inoculated with strains of *Lactobacillus sakei*, *Staphylococcus carnosus* and *Staphylococcus xylosus*, and the third batch was inoculated with strains *of Debaryomyces hansenii*, *Lactobacillus sakei*, *Pediococcus acidilactici*, *Pediococcus pentosaceus*, *Staphylococcus carnosus* and *Staphylococcus xylosus*. After 18 days of ripening, *L. monocytogenes* was not detected in any of the sausages, but during this fermentation and drying, the numbers of this pathogen was lower in the sausages inoculated with starter cultures.

1. Introduction

Čajna sausage is a traditional dry, fermented meat product, widely produced and consumed in Serbia [1]. Sausages are prepared by mixing ground meat, fat with combinations of spices, flavorings, salt, sugar, additives, and frequently, starter cultures [2]. During dry sausage manufacture, the fermentation and ripening processes lead to pH and water activity (a_w) decreases, which remains the main manner of achieving the safety of this type of fermented product [3].

Raw meat for sausage production and the final product tend to be consumed without prior cooking, which is why fermented sausages are a ready-to-eat product [2,4]. Thus, even if dry, fermented sausages are generally recognized as microbiologically safe, when initial contamination of the raw materials is high or there is insufficient control, the safety of these products can become compromised [5]. In past years, several outbreaks of food-borne illness associated with fermented meats have been reported and *Listeria monocytogenes* have been often detected in finished fermented sausages [6].

According to a USDA risk assessment [7] dry, fermented sausages are classified in a group designated as moderate risk for *L. monocytogenes* presence. Due to its ubiquitous nature, *L. monocytogenes* can contaminate meat and meat products during slaughter and processing operations, including cutting, slicing and packing [2,8]. Furthermore, this pathogen can persist and grow at low pH values, at low water activity, and at refrigeration temperatures [8,9] and pose a serious risk to human health.

Insufficient data is available on the survival of *L. monocytogenes* in Čajna sausages even though the product is widely consumed. Therefore, the aim of present study was to evaluate if *L. monocytogenes* is able to survive during the production of short maturation Čajna sausages produced with autochthonous and commercial starter cultures.

IOP Conf. Series: Earth and Environmental Science **85** (2017) 012074 doi:10.1088/1755-1315/85/1/012074

2. Materials and methods

2.1. Inoculum preparation

L. monocytogenes serotype 4b ATCC 19115, serotype 4b NCTC 11994, and serotype 4b and 1/2a were previously isolated from smoked herring and smoked salmon, respectively. All isolates were resuscitated twice in Brain Heart Infusion (BHI) broth (Oxoid, UK) and incubated at 37°C for 24 h. Then bacterial cultures were mixed in approximately equal proportions to produce the cocktail inoculum, and an appropriate volume of the inoculum was added to sausage batter to produce approximately 10^6 CFU of *L. monocytogenes* per gram of batter.

2.2. Sausage preparation and inoculation of pathogens

Čajna sausage was prepared from a mixture of lean pork (75%) and pig fat (25%) obtained from carcasses of Yorkshire x Landrace crossbreed pigs. Sausage batter was prepared by grinding frozen lean meat and fat tissue to 8 mm size and mixing with 2.1% salt containing 0.6% sodium nitrite and the spice mixture, Čajna nova (Prima commerce, Serbia) (4 g/kg) in a commercial meat processing plant and transported under refrigerated conditions to the faculty workshop. The batter for fermented sausage was inoculated with the cocktail of *L. monocytogenes* strains. After mixing, the batter was divided into three batches: the first batch was control (C) without any starter culture added, the second batch (EI) was inoculated with strains of *Lactobacillus sakei, Staphylococcus carnosus* and *Staphylococcus xylosus* (Biostart Sprint, RAPS GMBH, Obertrum, Austria), and the third (EII) was inoculated with strains of *Debaryomyces hansenii, Lactobacillus sakei, Pediococcus acidilactici, Pediococcus pentosaceus, Staphylococcus carnosus* and *Staphylococcus xylosus* (BACTOFERM™ B-LC-007, The Craft Butchers Pantry, US). The prepared mixtures were stuffed into 34mm-diameter collagen casing, to produce 30 sausages of approximately 450 g each for each group. Čajna sausage is a rapidly fermented sausage, and the ripening process (fermentation and drying) for the sausages lasted 18 days. First, the sausages were cold smoked for 8h during 3 days at 21-23°C and 80-85% relative humidity and, then were dried for 18 days in a climate chamber at 17°C with a relative humidity of 75%.

2.3. Microbiological analyses

Samples were analyzed for *L, monocytogenes*, lactic acid bacteria (LAB) count and total *Enterobacteriaceae* count on day 0 and on days 3, 7, 14 and 18 of storage. For bacterial enumeration, 10g of sausage samples were added to 90 ml of Buffered Peptone Water (BPW) (Merck, Germany) and homogenized in a Stomacher blender (Stomacher 400 Circulator, Seward, UK) for 2 min. Serial decimal dilutions were prepared and 0.1 ml or 1 ml of appropriately diluted suspension was inoculated directly on the surface of the appropriate media for enumeration of the different bacteria. *L. monocytogenes* was enumerated on Agar Listeria acc. to Ottaviani and Agosti (ALOA, Oxoid) and plates were incubated for 24-48h at 37°C according to ISO 11290-2:1998. LAB were enumerated on MRS (Merck, Germany) following incubation at 30°C for 72h according to ISO 15214:1998 and *Enterobacteriaceae* were enumerated on Violet Red Bile Glucose Agar (VRBGA, Merck, Germany) after incubation at 37°C for 24h according to ISO 21528-2:2004. After an appropriate period of incubation for each type of bacteria, plates were examined visually for typical colony types and morphological characteristics associated with each growth medium, number of colonies was counted, and results were recorded as colony forming units per g (CFU/g).

2.4. pH and a_w values

The pH was measured using a pH meter, TESTO 205 (Lenzkirch, Germany). Water activity (a_w) was measured with an aqualab water activity meter series 3 TE (Decagon Devices Inc., USA) according to the manufacturer's instructions.

2.5. Statistical analysis

Statistical analysis of the results was conducted using the software GraphPad Prism version 6.00 for Windows (GraphPad Software, San Diego, California USA, www.graphpad.com). The results were expressed as mean±standard deviation. The effects of different starter cultures during ripening period were appraised by one-factor analysis of variance (ANOVA) with Tukey's multiple comparison test at 95% confidence level (difference considered significant if $P<0.05$).

3. Results

L. monocytogenes counts ranged between 6.30 and 6.35 log CFU/g at the beginning of the study without significant differences between sausages (P>0.05). Slight but non-significant decreases (0.13-0.17 log CFU/g) were noted within the first three days of fermentation. Then, *L. monocytogenes* counts decreased significantly in all sausage types and were below the detection limit on day 18. From day 7 until day 14, the decrease of *L. monocytogenes* was significantly ($P<$ 0.05) slower in the sausages without starter cultures added. No significant differences were found in *L. monocytogenes* numbers among treatments with starter cultures added, except on day 14 when significantly lower numbers of this pathogen were measured in the sausages inoculated with strains of *Debaryomyces hansenii, Lactobacillus sakei, Pediococcus acidilactici, Pediococcus pentosaceus, Staphylococcus carnosus* and *Staphylococcus xylosus.* Even so, at the end of the ripening process, *L. monocytogenes* was below the detection limit in all types of sausages, regardless of whether or not they were inoculated with the starter cultures.

Table 1. *L. monocytogenes* counts (log CFU/g) during the fermentation and drying of Čajna sausage.

	0	3	7	14	18
C	6.35±0.05 ^{A}	6.17±0.04 ^{A}	4.90±0.05^{A}	2.78±0.02^{A}	nd
EI	6.33±0.03 ^{A}	6.20±0.02 ^{A}	4.50±0.06^{B}	2.32±0.02^{B}	nd
EII	6.30±0.02 ^{A}	6.16±0.01 ^{A}	4.35±0.04^{B}	2.02±0.03^{C}	nd

Within a column, means with a different letter are significantly different: A-C, P<0.05; nd-below detection limit
C – control sausages
EI – sausages with Biostart Sprint commercial starter
EII – sausages with BACTOFERM™ B-LC-007 commercial starter

The initial number of LAB was 4.52 log CFU/g in control sausages and 5.09 and 5.37 log CFU/g in sausages with starter cultures added (Table 2). As expected, LAB counts increased rapidly in all sausages during processing. During the ripening and fermentation process, in control sausages, LAB counts increased by 3.42 log CFU/g, while in the sausages with starter cultures added, the measured increase was 3.74 and 4 log CFU/g. On day 14, LAB had reached their final population level and slightly, but not statistically significant ($P>0.05$) decreased on day 18.

Table 2. LAB counts (log CFU/g) during the fermentation and drying of Čajna sausage.

	0	3	7	14	18
C	4.52±0.09^{A}	6.06±0.07^{A}	7.23±0.36^{A}	8.13±0.05^{A}	7.94±0.15^{A}
EI	5.37±0.07^{B}	7.33±0.06^{B}	8.96±0.20^{B}	9.19±0.07^{B}	9.11±0.06^{B}
EII	5.09±0.07^{B}	6.97±0.09^{C}	8.82±0.05^{B}	9.14±0.07^{B}	9.04±0.06^{B}

Within a column, means with a different letter are significantly different: A-B, P<0.05
C – control sausages
EI – sausages with Biostart Sprint commercial starter
EII – sausages with BACTOFERM™ B-LC-007 commercial starter

Initial *Enterobacteriaceae* counts ranged between 4.44 and 4.47 log CFU/g (Table 3). The numbers of *Enterobacteriaceae* did not change significantly during the first 3 days of fermentation but decreased thereafter in all sausages and were below the detection limit at the end of the ripening process.

Table 3. *Enterobacteriaceae* counts (log CFU/g) during the fermentation and drying of Čajna sausage.

	0	3	7	14	18
C	4.44±0.07^{A}	4.36±0.05^{A}	2.91±0.06^{A}	1.91±0.04^{A}	nd
EI	4.48±0.05^{A}	4.34±0.04^{A}	2.80±0.05^{A}	1.36±0.03^{B}	nd
EII	4.47±0.06^{A}	4.38±0.05^{A}	2.87±0.05^{A}	1.59±0.06^{C}	nd

Within a column, means with a different letter are significantly different: A-C, P<0.05; nd-below detection limit
C – control sausages
EI – sausages with Biostart Sprint commercial starter
EII – sausages with BACTOFERM™ B-LC-007 commercial starter

Changes in pH of the sausages during fermentation and drying are presented in Table 4. During this period, the pH decreased in all sausages, indicating a normal sausage making process, but decreases occurred significantly ($P < 0.05$) faster in sausages inoculated with the starter cultures than in sausages without starter culture added. Decreases in the pH values by 0.77-0.89 coincided with the increases in LAB counts.

Table 4. pH during the fermentation and drying of Čajna sausage.

	0	3	7	14	18
C	6.05±0.07^{A}	5.86±0.06^{A}	5.51±0.06^{A}	5.34±0.03^{A}	5.28±0.03^{A}
EI	6.03±0.08^{A}	5.78±0.08^{A}	5.42±0.04^{B}	5.20±0.03^{B}	5.18±0.02^{B}
EII	6.05±0.05^{A}	5.81±0.05^{A}	5.41±0.03^{B}	5.19±0.02^{B}	5.16±0.03^{B}

Within a column, means with a different letter are significantly different P<0.05
C – control sausages
EI – sausages with Biostart Sprint commercial starter
EII – sausages with BACTOFERM™ B-LC-007 commercial starter

The addition of 2.1% salt produced an initial a_w of about 0.968 in the initial sausage batters (Table 5). During fermentation and drying, a_w decreased in all sausages to levels of 0.909-0.912, and decreases were significant from day 7. Statistical analysis did not show significant differences (P>0.05) between the sausages inoculated with starter culture and the control sausages.

Table 5. Water activity (a_w) during the fermentation and drying of Čajna sausage.

	0	3	7	14	18
C	0.968±0.001^{A}	0.961±0.001^{A}	0.948±0.004^{A}	0.928±0.001^{A}	0.909±0.001^{A}
EI	0.968±0.001^{A}	0.963±0.001^{A}	0.952±0.001^{A}	0.929±0.001^{A}	0.912±0.001^{A}
EII	0.968±0.001^{A}	0.962±0.002^{A}	0.952±0.001^{A}	0.929±0.001^{A}	0.912±0.002^{A}

Within a column, means with a different letter are significantly different P<0.05
C – control sausages
EI – sausages with Biostart Sprint commercial starter
EII – sausages with BACTOFERM™ B-LC-007 commercial starter

4. Discussion

Serotypes 1/2a, 1/2b, and 1/2c of *L. monocytogenes* are frequently isolated from food products and serotypes 1/2a, 1/2b, and 4b cause 95% of the human cases of listeriosis [10,11], which is the reason these serotypes were used in the present study.

The final a_w and pH values and the concentration of salts, nitrites, and spices should suppress or inhibit the growth of pathogenic microorganisms in the fermented sausages, but *L. monocytogenes* could survive these conditions [12]. This is why bacteriocin-producing starters are often added. The addition of starter cultures is of particular importance for the microbial stability of quick-ripened fermented sausages, which are not greatly dried [13], like Čajna sausage. *Lb. sakei* presented in both commercial starter cultures can produce bacteriocins that showed broad inhibitory activity against pathogenic microorganisms including *L. monocytogenes* [14,15]. The results from the present study indicated that *Lb. sakei* as starter cultures for sausages played an important role in the control of *L. monocytogenes*. Furthermore, *Pediococcus acidilactici* pediocin-producing strain was added to one of the commercial starter cultures, and *L. monocytogenes* counts in the EII sausages inoculated with this starter culture were the lowest. Nonetheless, *L. monocytogenes* was inhibited, not only in the sausages with starter cultures, but also in non-inoculated sausages. This can be attributed to the fact that during spontaneous fermentation, LAB rapidly dominated the total microbiota in the present study. Autochthonous LAB are recognized as good competitors and exhibit a bioprotective or inhibitory effect on fortuitous microbiota as a result of the competition for nutrients and/or of the production of bacteriocins or other antagonistic compounds such as organic acids, hydrogen peroxide and enzymes [16].

Furthermore, Mataragas et al. [3] suggested that inactivation of *L. monocytogenes* is observed when pH and a_w values are within the range that does not support growth of the pathogen (pH≤5.0 and a_w≤0.94). In the present study, final a_w values were in within this limit. The pH decreased faster in sausages inoculated with the starter cultures than in sausages without starter culture added. This result suggests that the starter cultures played an active role in the fermentation process. Even so, on day 18, the pH still was above pH 5. pH and a_w are important factors for *L. monocytogenes* inactivation in fermented sausages; however, fermentation and drying temperatures have a significant role. It is suggested that temperatures at or above 20°C are needed, especially during the first 48 h of fermentation, for rapid inactivation of *L. monocytogenes* [6,17]. The significant reduction of this pathogen in the present study could be attributed to the high temperatures used during first three days of fermentation.

In accordance with present results Drosinos et al. [18] reported that *L. monocytogenes* was rapidly inactivated, decreasing by 3-4 log CFU/g during 28 days and by 4-5 log CFU/g in in fermented products from Croatia and Bosnia and Herzegovina. Contrary to results from present study, Ducic et al. [5] reported only reductions of 0.8 and 0.5 log CFU/g *L. monocytogenes* in fermented sausages without starters, even though the pathogen was inoculated at similar levels to those we used. Differences in the results could be explained by characteristics of isolates used and the production process.

A low *Enterobacteriaceae* count is crucial to obtaining high-quality hygienic sausages [19]. In the present study, the initial number of this bacterial group was relatively high. This could be attributed to the poor hygienic status of incoming raw materials or the processing environment in the meat processing plant where stuffing for sausages was prepared [5]. Nonetheless, the number of *Enterobacteriaceae* decreased during storage and was below 2 log CFU/g in all sausages at the end of the rapid fermentation. These results indicate the inhibitory effect of the inoculated probiotic cultures but also the acidification which was observed in control sausages as result of the activity of autochthonous LAB against *Enterobacteriaceae*.

Acknowledgment

This paper was supported by Ministry of Education, Science and Technological Development, Republic of Serbia, through the funding of Project No 31034.

References

[1] Džinić N, Ivić M, Jokanović M, Šojić B, Škaljac S and Tomović V 2016 Chemical, Color, Texture and Sensory Properties of Čajna Kobasica, a Dry Fermented Sausage. *Quality of life*, **13** 1-2

[2] Rajkovic A, Tomasevic I, De Meulenaer B and Devlieghere F 2017 The effect of pulsed UV light on *Escherichia coli* O157: H7, *Listeria monocytogenes*, *Salmonella Typhimurium*, *Staphylococcus aureus* and staphylococcal enterotoxin A on sliced fermented salami and its chemical quality *Food Control* **73** 829-37

[3] Mataragas M, Rantsiou K, Alessandria V and Cocolin L 2015 Estimating the non-thermal inactivation of *Listeria monocytogenes* in fermented sausages relative to temperature, pH and water activity *Meat sci.* **100** 171-78

[4] Rahman M S 2007 *Handbook of food preservation*. CRC press.

[5] Ducic M, Klisara N, Markov S, Blagojevic B, Vidakovic A and Buncic S 2016 The fate and pasteurization-based inactivation of *Escherichia coli* O157, *Salmonella Typhimurium* and *Listeria monocytogenes* in dry, fermented sausages *Food Control* **59** 400-06

[6] Mataragas M, Bellio A, Rovetto F, Astegiano S, Decastelli L and Cocolin L 2015. Risk-based control of food-borne pathogens *Listeria monocytogenes* and *Salmonella enterica* in the Italian fermented sausages Cacciatore and Felino *Meat sci.* **103** 39-45

[7] U.S. Department of Agriculture (USDA) 2003 Food Safety and Inspection Service (FSIS), Food and Drug Administration (FDA) Quantitative assessment of the relative risk to public health from food-borne *Listeria monocytogenes* among selected categories of ready-to-eat foods USDA Washington, D.C.

[8] Colak H, Hampikyan H, Ulusoy B and Bingol E B 2007 Presence of *Listeria monocytogenes* in Turkish style fermented sausage (sucuk) *Food Control* **18** 30-32

[9] Erkmen O 2008 Modeling the effects of sucuk production technique on *Listeria monocytogenes*, aerobic bacteria and lactic acid bacteria during ripening and storage *Food. Bioprod. Process.* **86** 220-26

[10] Swaminathan B and Gerner-Smidt P 2007 The epidemiology of human listeriosis *Microbes Infect.* **10** 1236-43

[11] Martín B, Perich A, Gómez D, Yangüela J, Rodríguez A, Garriga M and Aymerich T 2014 Diversity and distribution of *Listeria monocytogenes* in meat processing plants *Food Microbiol.* **44** 119-27

[12] Benkerroum N, Daoudi A, Hamraoui T, Ghalfi H, Thiry C, Duroy M and Thonart P 2005 Lyophilized preparations of bacteriocinogenic *Lactobacillus curvatus* and *Lactococcus lactis* subsp. *lactis* as potential protective adjuncts to control *Listeria monocytogenes* in dry-fermented sausages *J. Appl. Microbiol.* **98** 56-63

[13] Meloni D 2015 Presence of *Listeria monocytogenes* in Mediterranean-style dry fermented sausages *Foods* **1** 34-50

[14] Gao Y, Li D and Liu X 2014 Bacteriocin-producing *Lactobacillus sakei* C2 as starter culture in fermented sausages *Food control* **35** 1-6

[15] Pérez-Ibarreche M, Castellano P, Leclercq A and Vignolo G 2016 Control of *Listeria monocytogenes* biofilms on industrial surfaces by the bacteriocin-producing *Lactobacillus sakei* CRL1862 *FEMS microbiology letters* **363** 118

[16] Pragalaki T, Bloukas J G and Kotzekidou P 2013 Inhibition of *Listeria monocytogenes* and *Escherichia coli* O157: H7 in liquid broth medium and during processing of fermented sausage using autochthonous starter cultures *Meat sci.* **95** 458-64

[17] Gounadaki A, Skandamis P, Drosinos E H and Nychas G J E 2005 Survival of *Listeria monocytogenes* during the fermentation and ripening process of sausages. In *Proceedings of the 1st Hellenic Symposium in Food Biotechnology and Technology. Association of Greek Chemists and Hellenic Association of Chemical Engineers, Athens* pp. 436-441

[18] Drosinos E H, Mataragas M, Vesković-Moračanin S, Gasparik-Reichardt J, Hadžiosmanović M

and Alagić D 2006 Quantifying nonthermal inactivation of *Listeria monocytogenes* in European fermented sausages using bacteriocinogenic lactic acid bacteria or their bacteriocins: a case study for risk assessment *J. Food. Prot.* **69** 2648-63

[19] Rubio R, Aymerich T, Bover-Cid S, Guàrdia M D, Arnau J and Garriga M 2013 Probiotic strains *Lactobacillus plantarum* 299V and *Lactobacillus rhamnosus* GG as starter cultures for fermented sausages *Food Sci. Technol.* **54** 51-56

Some risk factors that affect contamination of mussels (*Mytilus galloprovincialis*) from the Bay of Kotor, Montenegro

N Grković[1], B Velebit[2], V Teodorović[1], N Karabasil[1], D Vasilev[1], V Đorđević[2] and M Dimitrijević[1]

[1] Faculty of Veterinary Medicine, University of Belgrade, Bulevar Oslobodjenja 18, 11000 Belgrade, Serbia
[2] Institute of Meat Hygiene and Technology, Kacanskog 13, 11040 Belgrade, Serbia

E-mail: nevena.ilic@vet.bg.ac.rs

Abstract. Pollution and contamination of the Bay of Kotor ecosystem arise from both anthropogenic sources and natural weathering. In recent decades, a need has arisen for regular control of marine organisms, which are used in human nutrition, because the entire bay is constantly and increasingly exposed to negative anthropogenic impact. Molluscs, including mussels (*Mytilus galloprovincialis*), can be involved in foodborne disease. They are filter feeding organisms, able to retain and concentrate in their bodies the bacteria, parasites, viruses and biotoxins of marine algae present in their external environment. A structured field study was undertaken in the Bay of Kotor, Montenegro, in order to investigate plausible influence of environmental factors, like rainfall and temperature, on the variability of *Escherichia coli* and norovirus (NoV). This study focuses on human-derived pathogens that are abundant in sewage-related sources. We proved the negative correlation between outside temperature and the number of *E.coli* and the presents of Norovirus in Bay of Kotor mussel. We used this data from the sampling site to discuss options to better manage the risk of contamination of shellfish. From the aspect of food safety, an upgrade of monitoring plans in the future could lead to obtaining safer products.

1. Introduction

With the increase in the consumption of seafood in recent years, marine mussels (*Mytilus galloprovincialis*) have become commercially more important seafood species worldwide. Also they are available throughout the year, and are reasonably tolerant to environmental change and pollution [1]. In recent decades, growing urbanization and industrialization led to extensive settlement of the Montenegrin coast, and caused pollution of the environment of the Bay of Kotor, especially the sea water and organisms that live in it. Wastes from various industries, shipyards, hotels and hospitals near the bay discharge into the sea and are a constant source of pollution of the aquatic environment.

Although seawater temperature is a good predictor of the temporal degree of contamination, especially of noroviruses, within a site it clearly cannot, on its own, predict the overall degree of contamination occurring. For this, the role of other variables was investigated.

Pollutant material, which can be rich in protozoa, viruses or bacteria, will originate from a variety of point sources such as sewage discharges and rainfall. Near the Bay of Kotor lies Crkvice, the area with the highest average precipitation in Europe. This water brings a large amount of suspended particles into the bay that affect the ecological conditions in seawater. The amount of pollutants that come into the sea due to large amounts of precipitation is not inappreciable [2].

Current legislation specifies that an evaluation of the sources and types of faecal contamination impacting shellfish production areas combined with monitoring of faecal indicator organisms (*Escherichia coli*) in shellfish flesh is undertaken to provide an indication of the risk of contamination with bacterial and viral pathogens [3]. Food safety agencies are trying to mitigate the risk of pathogens in aquatic environments, but still there is no comprehensive indicator system for their monitoring [4].

This study was undertaken to investigate the significance of some of environmental factors on the microbiological contamination of shellfish, as determined by *E. coli* concentration in the shellfish and the presence of norovirus. The potential for rapid diagnosis, high throughput, as well as sensitivity and quantitative nature of the real-time PCR test all make it highly suitable to replace virus isolation for both surveillance and diagnostic and research purposes [5]. The object was to identify whether such factors need to be taken into account when assessing the potential effects of proposed sewage discharges on commercial shellfish.

2. Materials and Methods

We used an existing database of positive samples of NoV (genogroups I and II) and *E.coli* (reported as most probable number/100 g) quantified in mussels (*Mytilus galloprovincialis*) from 6 sampling points within 6 production areas around the coast of the Bay of Kotor was used in this study. The research was done in a one-year period, from July 2015 to July 2016, on a monthly basis. The NoV and *E. coli* dataset analysed has been previously reported by Ilić et al. [6].

Levels of *E. coli* and NoV were quantified using appropriate ISO methods for most probable number [7] and quantitative Real-Time reverse transcription (RT)-PCR [8], respectively.

Information on potential risk factors includes rainfall, outside temperature and average daily quantity of waste water assumed to influence NoV and *E.coli* contamination. The outside temperature was measured at the moment of mussel sampling, and precipitation was measured seven days earlier. Sewage discharges impacting shellfish beds was from the database of consented water company and private sewage discharges to controlled waters. The dataset (rainfall and temperature) was produced by Environment Agency.

We investigated the relationships between microbiological parameters (NoV and *E. coli*) in mussels and the risk factors using Spearman's correlation. This coefficient is a statistical measure of the strength of the relationship between pairs of data and is denoted by r.

3. Results and Discussion

The results of environmental monitoring showed that the Bay of Kotor experienced variable, seasonal rainfall in harvesting areas that were investigated. The rainfall fluctuated between 64.1 mm in August and 122 mm in September (average=93.05 mm) (Fig. 1). The outside temperature, during all monitored seasons, varied between -0.4±1.0°C in January and 20.0±1.0°C in August (Fig. 2). All production areas were commercially harvested and classified under Regulation (EC) No 854/2004.

Analysis of correlation between rainfall in Bay of Kotor and the prevalence of norovirus in marine mussels, harvested in the same location, showed that there was a poor positive correlation ($r=0.47$) which was not statistically significant ($p>0.05$). Also, analysis of correlation between rainfall and the number of *E. coli* in mussels showed there was a mean positive correlation (r=0.57), which was not statistically significant (p>0.05) (Fig. 3).

The results of our research do not agree with the results of Campos et.al. [9], who proved that the predictive environmental factor for *E. coli* contamination in mussels in England and Wales was rainfall. In the current study in the Bay of Kotor, rainfall was not found to be associated with NoV contamination in the oysters. At some sites, *E. coli* levels have been found to reduce during high levels of rainfall possibly due to either a suspension of shellfish filtration activity at low salinities or to the dilution of impacting sewage plumes [10], and it is possible that a similar process also influenced the levels of NoV in shellfish.

There was a poor negative correlation between seasonal occurrence of NoV and outside temperature (r=-0.35), which was not statistically significant (p>0.05). Statistically significant (p<0.01) and medium negative correlation (r=-0.77) was obtained between outside temperature and the number of *E. coli* in the mussels (Fig. 4).

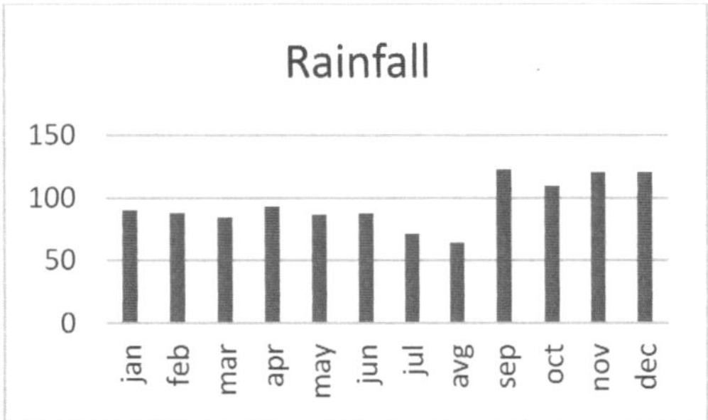

Figure 1. Seasonal variation of rainfall

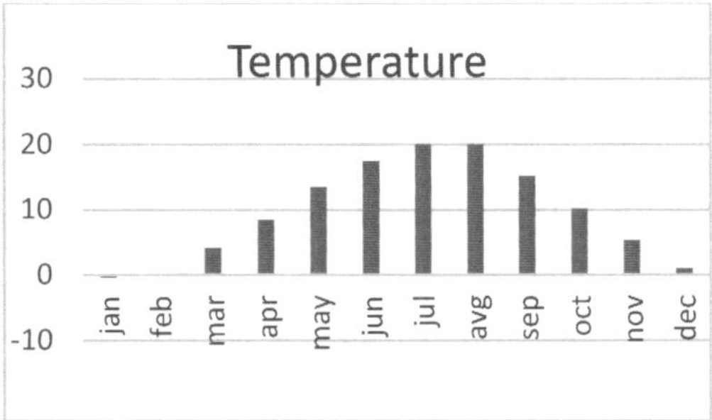

Figure 2. Seasonal variation of temperature

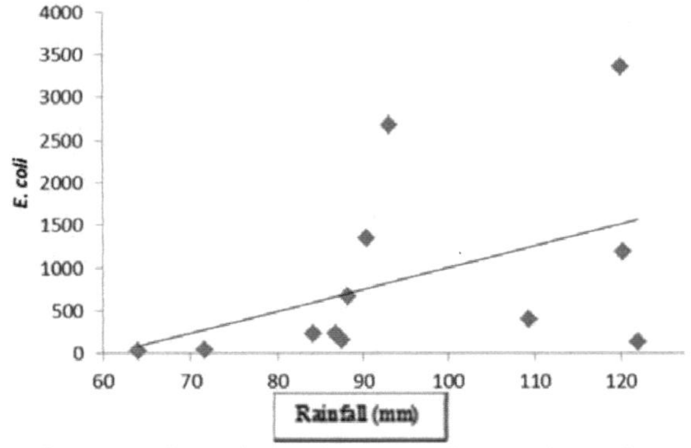

Figure 3. Correlation between the number of *E.coli* (MPN/ 100g shellfish) in Bay of Kotor mussels and rainfall during four seasons. (n=12, Spearman |r|=0.57, p>0.05)

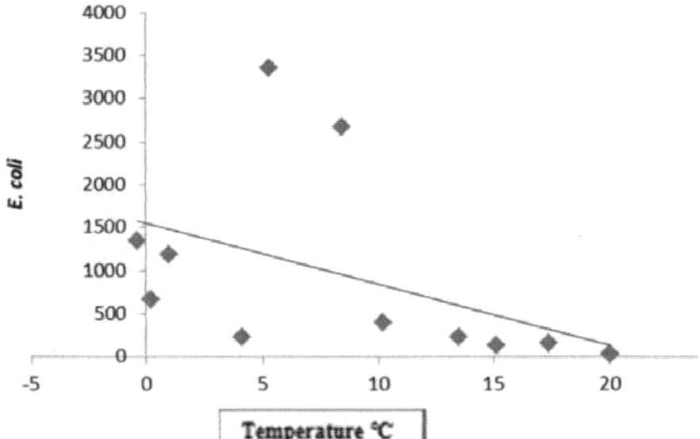

Figure 4. Correlation between the number of *E.coli* (MPN/100g shellfish) in Bay of Kotor mussels and outside temperature during four seasons.
(n=12, Spearman |r|=-0.77, p<0.01)

Overall, the predictive environmental factor for *E.coli* contamination in the mussels was outside temperature, while the predictive factor for total NoV (GI+GII) in the mussels was water temperature [11].

During summer months in the Bay of Kotor, the average daily quantity of sewage discharged was 2600 m^3/day (wastewater from households and from industry), which is much more than average daily quantity of sewage discharged in winter, when it was 2000 m^3/day.

4. Conclusions

From the results shown above, we concluded that, in the harvesting areas, the sources and abundance of human faecal pollution from sewage discharges were the main factor influencing the risk of contamination in mussels. However, it is also apparent that additional site-specific factors can further influence the extent of impact of this contamination on the shellfish.

We also observed that, in the catchments studied, the environmental factors driving the abundance of *E. coli* in shellfish do not operate in a similar way to those driving the abundance of NoV. The strength and significance of the risk factors varies between NoV and the statutory indicator of faecal pollution (*E. coli*).

Therefore, a distinct set of measures is required to manage the risk of NoV contamination of mussels. For all pathogens that can potentially be found in seawater, sewer overflows that catch sewerage water, produce the biggest risk of contamination, as they discharge crude, untreated sewage into the environment**.**

The results obtained in this study should contribute to the introduction of shellfish monitoring programmes, which would be improved with acquisition of all environmental parameters data, including demographic, hydrometric, climatic and pollution source characteristics and Real-Time qRT-PCR tests on cultured areas. Control measures for pathogens in mussels should focus on reducing volumes of sewage contamination in production areas.

References

[1] Stankovic S, Jović M, Milanov R and Joksimović D, Trace elements concentrations (Zn, Cu, Pb, Cd, As and Hg) in the Mediterranean mussel (*Mytilus galloprovincialis*) and evaluation of mussel quality and possible human health risk from cultivated and wild sites of the southeastern Adriatic Sea, Montenegro. *J. Serb. Chem. Soc.*, 2011, **76**:1725-1737

[2] Mehta A V and Yang S, Precipitation climatology over Mediterranean Basin from ten years of TRMM measurements, *Advances in Geosciences. Copernicus Publications*. 2008, **17**: 87–91

[3] European Parliament and Council of the European Union. Regulation (EC) No 854/2004 of the European Parliament and of the Council of 29 April 2004 laying down specific rules for the organisation of official controls on products of animal origin intended for human consumption

[4] European Food Safety Authority. Norovirus (NoV) in oysters: Methods, limits and control options. *EFSA Journal*, 2012, 10, 2500.

[5] Milic N , Radalj A, Nisavic J, Standard and molecular methods in the diagnostics of infections caused by equine herpesviruses 1 and 4, *Veterinarski Glasnik,* 2017, OnLine-First Issue 00, Pages: 2-2, https://doi.org/10.2298/VETGL170227002M

[6] Ilic N, Velebit B, Teodorovic V, Djordjevic V, Karabasil N, Vasilev D, Djuric S, Adzic B, Dimitrijevic M, Influence of environmental conditions on norovirus presence in mussels harvested in Montenegro. *Food Environ Virol*. 2017. DOI 10.1007/s12560-017-9298-0

[7] EN ISO/TS 16649-3:2005. *Microbiology of Food and Animal Feeding Stuffs — Horizontal Method for the Enumeration of Beta-glucuronidase-positive Escherichia coli — Part 3: Most Probable Number Technique Using 5-Bromo-4-chloro-3-indolyl-beta-d-glucuronide*. International Organization for Standardization

[8] EN ISO/TS 15216-1:2013. *Microbiology of Food and Animal Feed — Horizontal Method for Determination of Hepatitis A Virus and Norovirus in Food Using Real-time RT-PCR — Part 1: Method for Quantification.* International Organization for Standardization
[9] Campos C J A , Acornley R, Morgan O C, Kershaw S. Trends in the levels of *Escherichia coli* in commercially harvested bivalve shellfish from England and Wales, 1999–2008, *Mar Pollut Bull.* 2013, **67(1-2)**:223-7.
[10] Lee R J, Morgan O C. Environmental factors influencing the microbiological contamination of commercially harvested shellfish. *Water Sci. Technol*. 2003, **47 (3)** : 65–70.
[11] European Food Safety Authority. The European Union summary report on trends and sources of zoonoses, zoonotic agents and food-borne outbreaks in 2014. *EFSA Journal*, 2015, 13(12).

Characterization of biofilms produced by *Escherichia coli* O157 isolated from cattle hides

L Milojević, B Velebit, T Baltić, A Nikolić, R Mitrović and V Đorđević

Institute of Meat Hygiene and Technology, Kaćanskog 13, Belgrade, Serbia

Email: lazar.milojevic@inmes.rs

Abstract. This study aimed to investigate possibility *E. coli* O157 from cattle hides to produced biofilms. We had 28 suspect primoisolates and 17 were confirmed to be *E. coli* O157. Biofilm production test showed that more than 50% of this isolates did not produce biofilm. From the other half of the isolates, 5 of them were weakly adherent, 3 were moderately adherent. Since *E. coli* O157 are one of the main foodborne hazards in meat processing industry and the discovery that some of them can produce moderately adherent biofilms, request necessity of strict implementation of HACCP procedures to prevent further expansion this pathogen.

1. Introduction

Escherichia coli is a Gram-negative rod-shaped enteric bacterium, and their species are commonly classified by their virulence properties, mechanisms of pathogenicity, clinical syndromes, and O and H serotypes [1]. Following this, *E. coli* O157:H7 is classified by its serotype, O157:H7, because it expresses the 157th somatic (O) antigen and the 7th flagellar (H) antigen[2]. *E. coli* O157:H7 produces Shiga toxins 1 and 2 (Stx1 and Stx 2) as important virulence factors. This bacterium is significant for public health, it causes disorders that are characterized as with hemorrhagic colitis, bloody diarrhea and hemolytic uremic syndrome (HUS) with renal tubular damage [3]. This pathogen has a very low infective dose and it is sufficient low than 50 organisms to cause disease [4]. The most common mode of transmission *E. coli* O157:H7 infection is through food and water, but it is also possible to transfer from person to person [5], but the most important source of this pathogen is the cattle since *E. coli* O157:H7 naturally colonizes the intestinal tract of about 1% of healthy cattle [6]. Beef carcasses can become contaminated by fecal during slaughtering and processing [7].

E. coli is one of many bacteria that can switch between planktonic form and biofilm form. Several reasons can explain the need for bacteria to create biofilm, in this way bacteria can avoid being washed away by water flow or, cells in biofilms are about 1000 times more resistant than their planktonic [8]. *E. coli* O157:H7 has shown the ability to attach, colonize, and form biofilms on a variety of surfaces [9]. Biofilms that are attached to food contact surfaces, such as stainless steel, polyvinyl chloride, and polyurethane. To create a biofilm, the bacteria must be close enough to the surface. At about 10–20 nm distance from the surface, the negative charges on the bacterial surface are repelled by negative charges on most environmental surfaces. This reflection transcends van der Waals forces, and thanks to the fimbriae and flagella provide mechanical attachment to the surface [10]. Biofilm formation can be described in three stages: attachment, maturation and dispersion. The attachment step is categorize as a two-stage process: initial reversible attachment and irreversible attachment. The irreversible attached biofilm can tolerate stronger physical or chemical forces [11]. During the maturation process, more biofilm scaffolds, such as proteins, DNA, polysaccharides, etc. are secreted into the biofilm by the entrapped bacteria. After maturation, the dispersion step is followed, which is also very important for the biofilm life cycle. Biofilms disperse because of myriads of factors, such as lack of nutrients, intense competition, outgrown population and dispersal could occur in the whole or just a part of it. Biofilm consists of the structure of the Extracellular Polymeric Substance (EPS). Matrix of biofilms is composed of one or more of extracellular polysaccharides, DNA and proteins. This exopolysaccharides are synthesized extracellularly or intracellularly and secreted into the outside, they serve as scaffolds for other carbohydrates, proteins, nucleic acids and lipids to adhere. Some of the most carbohydrates are mannose, galactose and glucose, followed by N-acetyl-glucosamine, galacturonic acid, arabinose, rhamnose and xylose and most of them are not biofilm specific, but their production can increase as a result of a stress response, as colonic acid production in *Escherichia coli* [12].

Since *E. coli* O157:H7 is significant for public health and the most important source of the pathogen are beef carcasses, the goal of this study was to detect the prevalence *E. coli* O157 of these samples to create biofilms.

2. Material and Methods

DNA Extraction and rfbE gene amplification

The 1.5 ml microcentrifuge tubes containing 1 mL of cell suspension (3,1 × 10^9 CFU/mL) were centrifuged for 10 min at 10.000× *g*. Supernatant was then discarded and DNA extracted from cell pellet using the DNeasy Blood and Tissue kit (Qiagen, Germany) following manufacturer's instructions for Gram-negative bacteria.

DNA concentration was measured using a Biophotometer spectrophotometer (Eppendorf, Germany) and its integrity checked by visualization on 1.2% agarose gels. Then, samples were stored at −20°C before analyses.

A qPCR assay targeting the *E. coli* O157:H7 rfbE gene was performed using *rfb*E forward (5'-TTT CAC ACT TAT TGG ATG GTCTCAA-3') and *rfb*E reverse primer (5'-CGA TGA GTT TAT CTG CAA GGT GAT -3'). The 20 µL volume reaction mixture contained 2×Brilliance SYBR Green qPCR buffer (Agilent, USA), 500 nm of each PCR primer, and 2 µL of DNA template. After a 3 min denaturation at 95°C, the qPCR mixtures were subjected to 40 cycles of amplification at 95°Cfor 10 sec, 60°C for 20 sec, using a qPCR thermal cycler (MX3005P, Agilent, USA).

Biofilm characterization

Biofilm formation on microtiter plate was performed following the methodology of Christensen's et al., (1985), by inoculating the microtiter wells with bacterial strains [13]. The strains were grown aerobically in blood-heart-infusion (BHI) broth at 37°C for 24 hours. After incubation, the OD of the cultures was adjusted to 0.5 at 600nm. In 96 well-microtiter plates, 200µL of cell suspension diluted to 1:100 was added. BHI broth, *E. coli* O157 biofilm-positive and biofilm-negative strains were used as controls. The plates were incubated at 37°Cfor 24 hours. After incubation, the growth medium was discarded and microtiter plate wells were washed twice with 200 µL of 0.85 % NaCl. The plates were vigorously shaken in order to remove all non-adherent bacteria. The remaining attached bacteria were fixed by subsequent drying in incubator at60°C for 60 minutes.

The bound cells were stained with 200 µL of 0.1 % (v/v) crystal violet (CV) solution for 15 minutes at room temperature. Excess dye was removed by washing each well (3×) with 200 µL of 0.85 % NaCl. The quantification of attached cells was performed by adding 200 µL of 33% (v/v) glacial acetic acid as a CV solvent. O.D at 595 nm of dissolved CV was measured in a microtiter plate reader. Biofilm assays were performed in triplicate.

All strains were classified into the following categories: non-adherent, weakly, moderately, or strongly adherent, based upon the ODs of bacterial biofilms. We defined the cut-off OD for the microtiter-plate test as three standard deviations above the mean OD of the negative control.

3. Results and Discussion

Out of 28 tested suspect primoisolates (using mTSB broth with novobiocin and cefixime-tellurite sorbitol MacConkey agar), 17 were confirmed to be *E. coli* O157 (60.7%) Figure 1.

These samples were subsequently tested on capability for biofilm production.

Figure 1. Dissociation curves of O157 positive strains

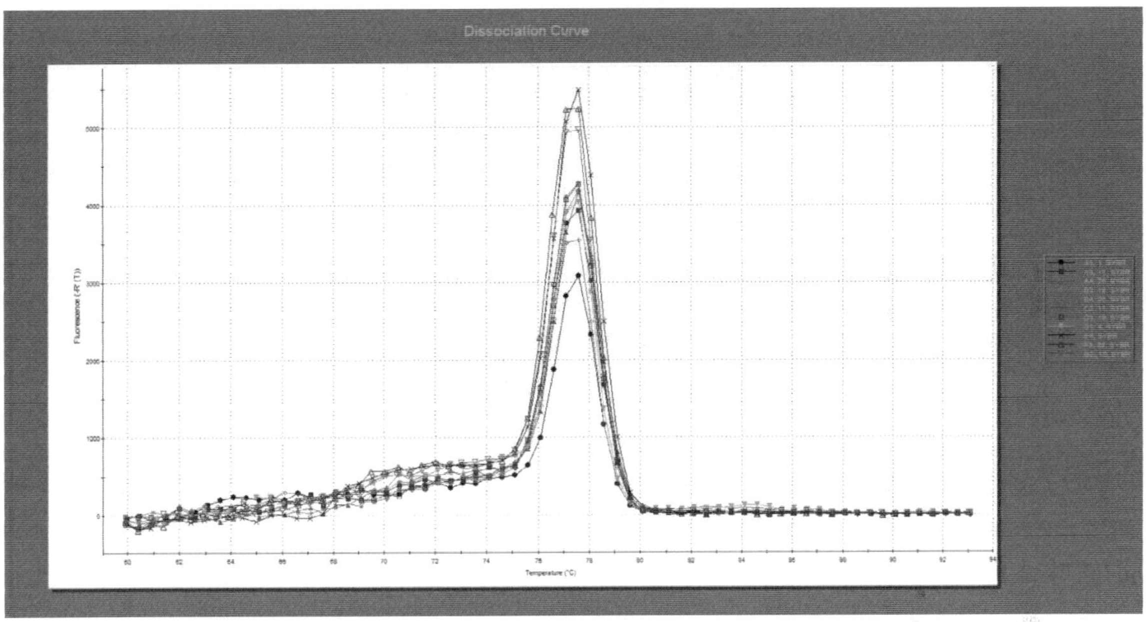

Biofilm production test indicated that more than 50% of the isolates did not produce biofilm (Figure 2). However, within the other half of the isolates, 5 of them were weakly adherent, 3 were moderately adherent, while none of the isolates exhibited particularly high capability for biofilm production. In his research [14], Dourou confirmed the ability of *E. coli* O157 to form biofilms, and showed that the formation of biofilms can depend on a number of factors such as temperature or surface. Narisawa [15] came to research results that 2,1% and 0,5% *E. coli* form biofilms, which is approximately equal to the prevalence that we found in our study. Al-Shabib [16] found that of the 10 tested strains, four strains formed strong biofilm while other formed moderate to week biofilm which is much higher prevalence than ours. In contrast, Biscola [17] has published the results in which 5 of 18 *E. coli* O157 formed biofilms, which is very similar to our results.

Figure 2. Biofilm characterization produced by *Escherichia coli* O157

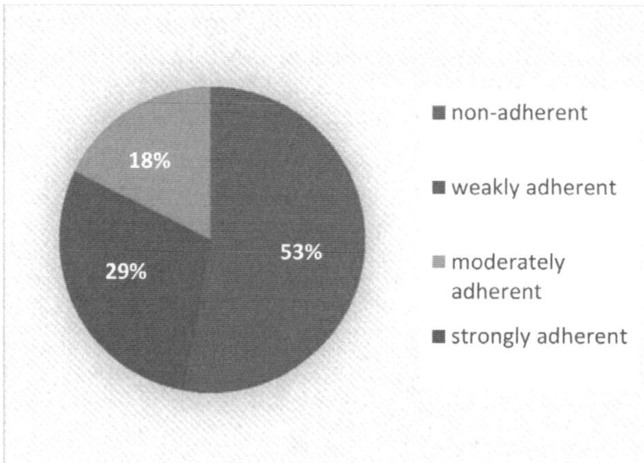

4. Conclusion

E. coli O157 which produce moderately adherent biofilms on cattle hides are one of the main foodborne hazards in meat processing industry. For sure, these findings require necessity of strict implementation of HACCP procedures during exenteration and dehiding of cattle in order to decrease the number of it and prevent further dissemination of this pathogen.

References

[1] Montville TJ and Matthews KR 2005 *Food Microbiology: an Introduction*(Washington (DC): ASM Press) p 380

[2] Mead PS, Slutsker L, Dietz V, McCaig LF, Bresee JS, Shapiro C, Griffin PM and Tauxe RV 1999 Food-related illness and death in the United States. *Emerg. Infect. Dis.***5**607-25

[3] Mohawk KL, Melton-Celsa AR, Zangari T, Carroll EE and O'Brien AD 2010 Pathogenesis of Escherichia coli O157:H7 strain 86-24 following oral infection of BALB/c mice with an intact commensal flora *Microb. Pathog.***48** 131-42

[4] Tilden J, Young W, McNamara AM, Custer C, Boesel B and Lambert-Fair MA 1996 A new route of transmission of Escherichia coli: infection from dry fermented salami Am J Public Health **86** 1142-5

[5] Besser RE, Griffin PM and Slutsker L.1999 Escherichia coli O157:H7 gastroenteritis and the hemolytic uremic syndrome: an emerging infectious disease Annu. Rev. Med.**50** 355-67

[6] Moxley RA 2004 Escherichia coli O157:H7: and update on intestinal colonization and virulence mechanisms Anim. Health Res. Rev.**5**(1) 15-33

[7] Rekow CL, Brashears MM, Brooks JC, Loneragan GH, Gragg SE and Miller MF 2011 Implementation of targeted interventions to control Escherichia coli O157:H7 in a commercial abattoir *Meat Sci.***87** 361-5

[8] Jefferson KK 2004 What drives bacteria to produce abiofilm? FEMS Microbiol. Lett. **236**(2) 163–73

[9] Uhlich GA, Cooke PH and Solomon EB 2006 Analysis of the red-dry-rough phenotypeof an *Escherichia coli* O157:H7 strain and its role in biofilm formation and resistance to antibacterial agentsAppl. Environ. Microbiol.**72** 2564-72

[10] Palmer J, Flint S and Brooks J 2007 Bacterial cell attachment, thebeginning of a biofilm. *J. Ind. Microbiol. Biotechnol.* **34**(9) 577–88

[11] Sutherland I 2001 Biofilm exopolysaccharides: a strong andsticky framework *Microbiology* **147**(1) 3–9

[12] Prigent-Combaret C, Vidal O, Dorel C and Lejeune P 1999 Abiotic surface sensing and biofilm-dependent regulation ofgene expression in Escherichia coli. *J. Bacteriol.***181**(19) 5993–6002

[13] Christensen GD, Simpson WA, Younger JJ, Baddour LM, Barrett, FF, Melton DM and Beachey EH 1985 Adherence of coagulase-negative staphylococci to plastic tissue culture plates: a quantitative model for the adherence of staphylococci to medical devices *Journal of clinical microbiology* **22**(6) 996-1006

[14] Dourou D, Beauchamp CS, Yoon Y, Geornaras I,Belk KE, Smith GC,Nychas GE and Sofos JN 2011 Attachment and biofilm formation by Escherichia coli O157:H7 at different temperatures, on various food-contact surfaces encountered in beef processing *International Journal of Food Microbiology* **149**(3) 262-8

[15] Narisawa N, Furukawa S, Ogihara H and Yamasaki M 2005 Estimation of the biofilm formation of Escherichia coli K-12 by the cell number *Journal of Bioscience and Bioengineering* **99** 78-80

[16] Al-Shabib NA, Husain FM, Ahmad I, Khan MS, Khan RA and Khan JM 2017 Rutin inhibits mono and multi-species biofilm formation by foodborne drug resistant Escherichia coli and Staphylococcus aureus *Food Control* **79** 325-32

[17] Biscola FT, Abe CM and Guth BEC 2011 Determination of adhesin gene sequences in, and biofilm formation by, O157 and non-O157 Shiga Toxin-Producing Escherichia coli strains isolated from different sources Applied and Environmental Microbiology **77**(7) 2201-8

Antimicrobial resistance among Salmonella enterica serovar Infantis from broiler carcasses in Serbia

A Nikolić[1], T Baltić[1], B Velebit[1], M Babić[2], L Milojević[1], V Đorđević[1]

[1] Institute of Meat Hygiene and Technology, 13 Kaćanskog st., Belgrade, Serbia
[2] Faculty of Veterinary Medicine, University of Belgrade, Bulevar Oslobođenja 18, Belgrade, Serbia

E-mail: aleksandra.nikolic@inmes.rs

Abstract. This study aimed to investigate antimicrobial resistance of *Salmonella* Infantis isolates from poultry carcasses in Serbia. A total of 48 Salmonella isolates were examined for antimicrobial resistance. A panel of 10 antibiotics was selected for testing. Isolates showed resistance to sulfamethoxazole, ceftazidime and cefotaxime (100%). However, the highest number of *Salmonella* Infantis isolates were sensitive to chloramphenicol. The usage of antibiotics in food producing animals could result in antimicrobial resistance pathogenic bacteria especially *Salmonella* spp. in poultry, which may be transmitted to humans through the food chain and increase risk of treatment failures.

1. Introduction

Poultry meat is essential for a complete diet, since it contains high amounts of highly digestible proteins, unsaturated lipids (mainly found in the skin and easily removed), B-group vitamins (mainly thiamin, vitamin B6, and pantothenic acid), and minerals (like iron, zinc, and copper) which makes it highly acceptable for consumption by consumers of all ages [1, 2]. However, poultry meat is involved in the transmission of several food-borne pathogens, which are distributed across the whole production chain, from chicken birth to the final product [3]. *Salmonella* food poisoning is one of the most common and widely distributed diseases in the world [4], whereby outbreaks are usually connected with consumption of contaminated poultry meat. Perhaps more concerning is the fact that multidrug resistance (MDR), is being increasingly detected among numerous *Salmonella* serotypes recovered from animals and humans worldwide [5, 6]. The recovery of antimicrobial-resistant *Salmonella* in foods of animal origin has raised concerns that loss of antibiotic efficiency may compromise the treatment of human salmonellosis, because antimicrobial-resistant strains appear to be more often associated with severe disease [7, 8]. It has been reported that *Salmonella* exhibiting lower susceptibility to several antibiotics such as fluoroquinolones (e.g., ciprofloxacin) and extended-spectrum cephalosporins (e.g., ceftiofur and ceftriaxone), which are important in treatment of human salmonellosis, in both adults and children, respectively [9, 10, 11]. Some authors have confirmed the presence of a *Salmonella enterica* subspecies *enterica* serovar Infantis (6, 7, r, 1, 5) in all 17 samples, and also same serovars isolated in the case of diseased humans [12]. Therefore, the aim of this study was to investigate the antimicrobial resistance of *Salmonella* Infantis isolated from broiler carcasses in Serbia.

2. Materials and Methods

The study included a total of 48 *Salmonella* Infantis isolates, where 39 were originating from the broiler meat and 9 isolates were isolated from the skin of the broiler neck. The isolates were revitalized in 10 mL of BHI broth (Lab, UK), and incubated for 24h at 37°C. After incubation period, *Salmonella* suspensions were transferred by sterile swabs on Mueller-Hinton agar, followed by antibiotic discs application (automatic applicator, Oxoid, UK), and incubated for 24 hours at 37°C. After 24 hours of incubation the growth inhibition zones were measured, and the results were interpreted according to Clinical and Laboratory Standards Institute (CLSI) 2006[] recommendations as sensitive, intermediate sensitive and resistant. The following antibiotic discs (Oxoid, UK) were used: ampicillin 10 mg, cefotaxime/clavulanic acid (cephalosporin) 30 mg, ceftazidime 10 mg + 30 mg, chloramphenicol 30 mg, ciprofloxacin (quinolone) 5 mg, gentamicin (aminoglycoside antibiotic) 30 mg, tetracycline 30 mg, nalidixic acid (quinolone) 30 mg, trimethoprim/sulfamethoxazole (inhibitors of folic acid) 30 µg.

3. Results and Discussion

From a total of 48 *Salmonella* Infantis isolates from poultry carcasses, all were resistant to sulfamethoxazole, ceftazidime and cefotaxime. However, the highest number of *Salmonella* spp. isolates were sensitive to chloramphenicol. Antimicrobial resistance (intermediate sensitive and sensitive) of *Salmonella* Infantis isolated from broiler carcasses is shown in Figure 1.

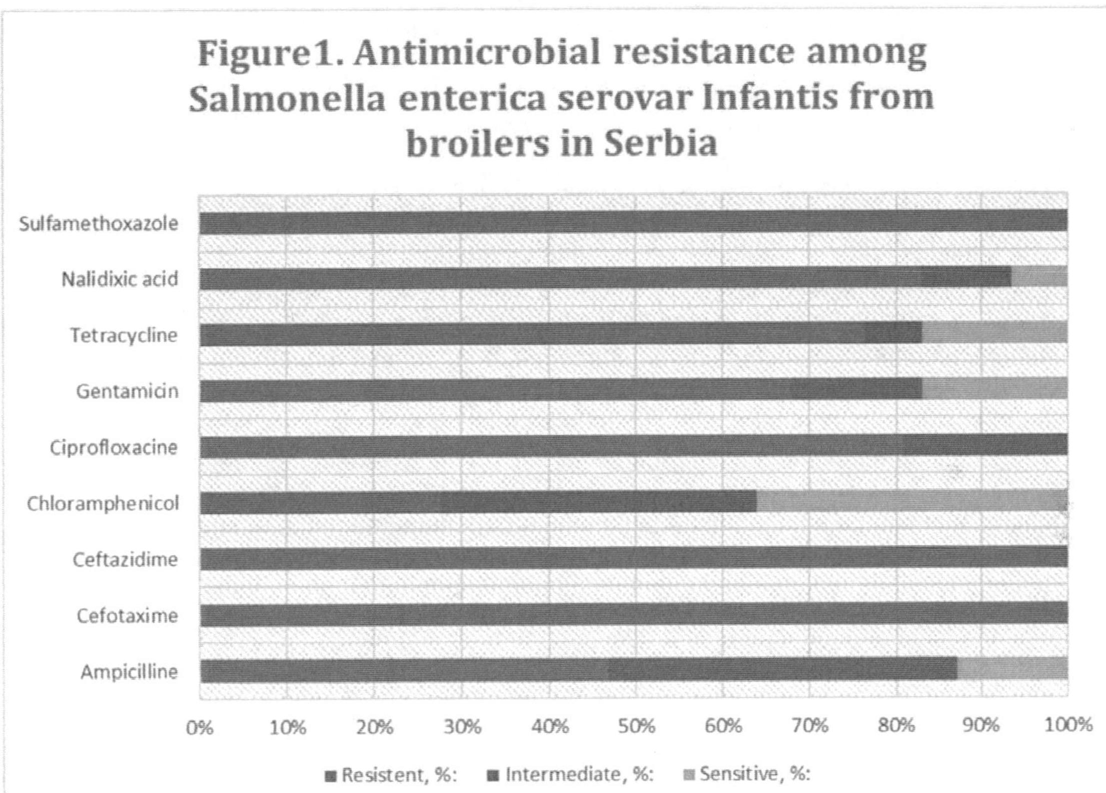

The usage of antibiotics in food producing animals could result in antimicrobial resistance pathogenic bacteria especially *Salmonella* spp. in poultry, which may be transmitted to humans through the food chain and increase risk of treatment failures. Due to the massive use of *S.* Enteritidis and *S.* Typhimurium vaccine as part of *Salmonella* the eradication programs, probably lead to the high prevalence of *S. Infantis* in the poultry [13]. The presence of *S.* Infantis in broiler meat in the recent years is constantly increasing [12]. *Salmonella* spp. isolated from poultry carcasses showed resistance to ampicillin and nalidixic acid (95.5%), tetracycline (91%) and cefotaxime (68.25%) [12]. However, our results show resistance to the same antibiotics, with the highest resistance to cefotaxime (100%) and the lowest resistance to ampicillin (47%).

Some authors compared the results of *Salmonella* antimicrobial resistance in 2007 to 2013-2014, where they found that resistance was significant different [13]. In 2007, resistance was most frequently to amoxicillin and sulfametoxazol [13], which coincides with our results, where resistance to sulfamethoxazole was 100%. During 2013 and 2014, we detected significant resistance of *S.* Infantis to nalidixic acid and tetracycline [13], also in accordance with our results (tetracycline 78%, nalidixic acid 82%). This indicates that *Salmonella* have a tendency to behave clonally as has been confirmed by genotype and resistotype [14].

References

[1] Nucleo de Estudos e Pesquisas em Alimentação 2011. Tabela brasileira de composição de alimentos. 4th. Campinas: UNICAMP p 161

[2] Marangoni F, Corsello G, Cricelli C, Ferrara N, Ghiselli A, Lucchin L and Poli A 2015 Role of poultry meat in a balanced diet aimed at maintaining health and wellbeing: an Italian consensus document. *Food & nutrition research* **59** (1) 27606

[3] Franco RM, do Nascimento ER, de Carvalho LR and de Azevedo MDGG 2014 *Salmonella* spp. detection in chicken meat and cross-contamination in an industrial kitchen. *African Journal of Microbiology Research* **8** (11) 1130-1139

[4] Gomez T M, Motarjemi Y, Miyagawa S, Käferstein F K and Stöhr K 1997. Foodborne salmonellosis. World health statistics quarterly. *Rapport trimestriel de statistiques sanitaires mondiales* **50** (1-2) 81-89

[5] White D G, Zhao S, Sudler R, Ayers S, Friedman S, Chen S, McDermott P F, McDermott S, Wagner D D and Meng J 2001. The isolation of antibiotic-resistant *Salmonella* from retail ground meats *New England journal of medicine* **345** (16) 1147-1154.

[6] Zhao S, McDermott PF, Friedman S, Abbott J, Ayers S, Glenn A, Hall-Robinson E, Hubert SK, Harbottle H, Walker RD, Chiller TM and White DG 2006 Antimicrobial resistance and genetic relatedness among *Salmonella* from retail foods of animal origin: NARMS retail meat surveillance *Foodborne Pathogens & Disease* **3** (1) 106-117

[7] Helms M, Vastrup P, Gerner-Smidt, P and Mølbak K 2002. Excess mortality associated with antimicrobial drug-resistant Salmonella Typhimurium *Emerging infectious diseases* **8** (5) 490

[8] Varma J K, Mølbak K, Barrett T J, Beebe J L, Jones T F, Rabatsky-Ehr T, Smith KE, Vugia D J, Chang H G and Angulo F J 2005 Antimicrobial-resistant nontyphoidal *Salmonella* is associated with excess bloodstream infections and hospitalizations *The Journal of infectious diseases* **191** (4) 554-561

[9] Gupta A, Fontana J, Crowe C, Bolstorff B, Stout A, Duyne S V, Hoekstra M P, Whichard J M, Barrett T J and Angulo F J 2003 Emergence of multidrug-resistant *Salmonella enterica* serotype Newport infections resistant to expanded-spectrum cephalosporins in the United States *The Journal of infectious diseases* **188** (11) 1707-1716

[10] Mølbak K, Baggesen D L, Aarestrup F M, Ebbesen J M, Engberg J, Frydendahl K, Gerner-Smidt P, Petersen A M and Wegener H C 1999 An outbreak of multidrug-resistant, quinolone-resistant *Salmonella enterica* serotype Typhimurium DT104 *New England Journal of Medicine* **341** (19) 1420-1425

[11] Zhao S, McDermott P F, Friedman S, Qaiyumi S, Abbott J, Kiessling C, Ayers S, Singh R, Hubert S, Sofos J and White D G 2006 Characterization of antimicrobial-resistant *Salmonella* isolated from imported foods *Journal of food protection* **69** (3) 500-507

[12] Rašeta M, Teodorović V, Bunčić O, Katić V, Branković LI, Polaček V, Vidanović D 2014 Antibiotic resistance and molecular studies on Salmonella Enterica subspecies Enterica serovar Infantis isolated in human cases and broiler carcasses *Acta Veterinaria-Beograd* **64** (2), 257-268

[13] Petrovica J, Babic J, Stojanova I, Velhnera M 2015 The change in antimicrobial resistance profile of meat chain-associated Salmonella in Serbia *Procedia Food Science* **5** 231 – 234

[14] Velhner M, Potkonjak D, Stojanovic D, Mitevski D, Stojanov I, Petrovic J. 2013 Rezistencija salmonela na antibakterijske lekove i mere kontrole u zivinarskoj proizvodnji. *Veterinarski glasnik* **67** (1-2):87-96.

IOP Conf. Series: Earth and Environmental Science **85** (2017) 012078 doi:10.1088/1755-1315/85/1/012078

Allergen labelling in meat, dairy and cereal products from the Serbian market

D Spirić[1], D Nikolić[1], J Ćirić[2], S Janković[1], S Stefanović[1], V Janković[1] and V Teodorović[2]

[1] Institute of Meat Hygiene and Technology, Kaćanskog 13, Belgrade, Serbia
[2] Faculty of Veterinary Medicine, Bulevar oslobođenja 18, Belgrade, Serbia

E-mail: danka.spiric@inmes.rs

Abstract. Allergens in food are a great health risk, because of the ratio of severity of problems compared to small amounts of ingested allergen. Since 2014, Serbian producers and importers of food have been obliged to declare allergens from the list of *Codex Alimentarius* on the product packaging. Surveillance of different meat, diary, and cereal product took place in 2016, with aim of checking if the Serbian regulatory requirements for labelling of allergens in food are being fulfilled. Out of 68 different meat products, 20 were not labelled for allergens. Thirty-six labels of various dairy products were examined revealing that allergen information was included on 27 of them. Only one of eight examined cereal products did not have allergen labelling.

1. Introduction

Allergens in foodstuffs are a major risk for the sensitive part of the population. The incidence of food allergy is increasing, according to recent combined studies, in children and in adults [1]. An increase in the incidence of food-related allergies can be associated with an increase in the number of data from different studies, as well as an increase in consumers' awareness of health problems themselves. Nevertheless, probable or clinically confirmed findings led to a severe allergy rate of 3.08% among children in the United States [2]. The prevalence of allergies caused by foods in the European population was calculated for milk, eggs, wheat, soybean, peanut, nuts and shellfish, in turn; 0.2% (0.2-0.3), 0.2% (0.2-0.3), 0.1% (0.01-0.2), 0.3% (0.1-0.4), 0.2% (0.2-0.3), 0.5% (0.08-0.8), 0.1% (0.02-0.2), and 0.1% (0.06-0.3) [3]. This study [3] covers the most common food allergens that are defined by the list contained in the *Codex Alimentarius*, and they are also contained in Serbian regulatory requirements [4]. Avoiding the use of allergens in the diet is the most effective way of preventing food-related allergic reactions. One problem occurs when foods have basic compositions that do not contain allergens, but these can still occur in traces, due to cross-contamination. To avoid the exposure of the vulnerable part of the population to allergens, manufacturers are required to properly declare allergens listed in the Serbian regulatory requirements [4]. Applicable European Union (EU) regulation [5] prescribes the declaration of allergens to ensure the safety of the food chain from packaging, export, distribution, wholesale, retail and catering production, as well as inspectors' responsibilities for the application of these regulations. The EU lists 14 allergens that, in addition to precautionary measures from an existing Good Manufacturing Practice, are required to be included on the product declaration [5]. Previous regulations relate to originally packaged food, while catering, public kitchens, restaurants, and the like can rely on voluntary information (oral or written) on allergens in the meal.

Cereals containing gluten are to be listed on the declaration, with a bold font. Gluten-free labelling is voluntary, but if groceries are voluntarily labelled as gluten-free, the foodstuffs must meet the requirements of the regulations [6,7]. The labelling of crustaceans or shellfish as an allergenic group is mandatory, regardless of the species name: e.g. shrimps (crabs); oysters (shellfish). Similarly, fish products/ingredients are labelled as fish, regardless of the type: e.g. trout (fish). Eggs, regardless of the type of bird from which they originate, are listed on the declaration as an allergen. Soya is listed as

IOP Conf. Series: Earth and Environmental Science **85** (2017) 012078 doi:10.1088/1755-1315/85/1/012078

either soybean, soybean meal or soy/soya, and it is included in parentheses if the product is mentioned: e.g. Tofu (soya). Declaring peanuts as nuts should be avoided, but the name *peanuts* should be specified. In Serbia, dairy products are a special case. According to article 23 [4], ingredients do not need to be listed on butter, cheese, plain unadulterated yoghurt etc., since apart from bacterial cultures and enzymes, they largely consist of milk. For other foods, when there is no specific note that milk components are allergenic ingredients in a food, e.g. lactose, whey and casein, milk must be listed in brackets in the ingredient list e.g. casein (milk). Allergenic nuts from the list are almond, nut, cashew nut, pecan, hazelnut, pistachio nuts, macadamia, Queensland walnuts and their products, and these must be listed by species in the ingredient list. Chestnut and coconut contain the word *nuts*, but are not listed as allergens by *Codex Alimentarius*. Sesame and mustard must be listed in the ingredient list both as raw materials and as a product, e.g. Tahini (sesame). Sulphites in an amount greater than 10 mg/kg (mg/L) are listed as an allergen, but are also included as an ingredient e.g. Sulphur dioxide.

Amendment of older Serbian regulations [7] concerning labelling and declarations of foodstuffs resulted in new Serbian regulatory requirements [4]. The new requirements [4] have common elements mandatory for the declaration of packaged foods: the name under which the food is sold, a list of ingredients, the net quantity, expiry date/use by date or best before date, storage conditions, batch or lot marking, name and head office of the manufacturer (food business operator), which produces/packs the food, instruction for use of the food, the category of food quality, if prescribed, and the content of alcohol in drinks with greater than 1.2% v/v alcohol. In addition to these common requirements, there is now a clearly defined requirement [4] that on the food declaration, ingredients that could cause allergies and/or intolerances are listed.

The aim of this market research was to determine if the Serbian regulatory requirements with respect to allergen labelling of retail foods was being fulfilled.

2. Materials and methods

Allergen declarations were examined using photographs of samples from "infolab" database of sample analysis of Institute of Meat Hygiene and Technology. Samples were divided into 3 groups: meat and meat products; dairy products and cereal based products. Each group of samples was classified according to method of giving allergen information on the packaging. Percentage of each labelling style was determined regarding total number of samples in the group.

3. Results and discussion

Allergen declarations for those allergens listed in Serbian regulatory requirements [4] were studied. Labels from 112 meat, milk and cereal products from different manufacturers were examined and the ingredient lists and allergen declarations were studied.

3.1 Meat and meat products

Among retail meat products, 14 kulen (a type of spicy, dry-cured raw meat sausage) declarations, 15 hot-dog declarations, 22 canned meat declarations and 17 declarations for ready-to-eat groceries produced by various domestic and foreign manufacturers were examined (table 1). Two producers of kulen included allergen information (a note) about the possible presence of soy and gluten, neither soy nor gluten were listed in the ingredient list. One kulen producer included the allergen information "without gluten and allergens", while other manufacturers did not indicate the possibility of the presence of allergens either in the ingredient list or in a special note.

IOP Conf. Series: Earth and Environmental Science **85** (2017) 012078 doi:10.1088/1755-1315/85/1/012078

Table 1. Number and percentage of different meat products with various forms of allergen labelling.

Method of giving allergen information on the packaging	Kulen	Hot-dog	Canned meat	Ready-to-eat meal
Allergen information listed in a separate note	2 (14%)	7 (47%)	6 (27%)	1 (6%)
Emphasized font in Ingredient list	/	4 (27%)	16 (73%)	7 (41%)
Other method of labelling allergens	1 (7%)	3 (20%)	5 (23%)	4 (23%)
No labelling of allergens	13 (93%)	2 (13%)	3 (14%)	5 (29%)

Seven out of fifteen local hot-dog manufacturers listed a special note giving allergen information on the presence of soybean allergens (7), gluten (2), milk (1), while the only EU producer mentioned lactose, celery, mustard, soybean, wheat and peanuts in their allergenic information note. Three manufacturers who included a note for allergen information and, in a bold font, also listed the allergens in the ingredient list. Four producers who did not have an allergen info note indicated allergens using bold font in the ingredient list (soy protein and wheat protein). The two manufacturers did not list allergenic substances either in the ingredient list or in a special note. The sixteen producers of pressed hams emphasized allergens in the ingredient list using a bold font, as the only way of labelling allergens . Six declarations, in addition to allergens being in the listed ingredients, also had specific allergen information notes on the presence of soybeans (5) and milk (1). Five manufacturers did not declare allergens as ingredients, but had allergenic information about the possible presence of milk proteins. Three manufacturers did not have any of the regulated allergens either as ingredients or traces. Ready-to-eat meat dishes can contain vegetables, spices and additives. The seven producers labelled the allergenic substances with a bold font in the ingredient lists. Only one producer included a separate note for allergens. Four manufacturers used bold font to highlight allergens in the ingredient list, and had special allergen information in a separate note. The remaining five manufacturers did not list allergens on the product packaging at all.

3.2 Dairy products

Eight cheese declarations from various manufacturers were examined, as well as seven declarations from yogurt and 21 from ice-cream (table 2). Three cheese producers included milk in the ingredient list using a bold font. One manufacturer also stated in a special note that the product contained milk. The remaining five producers did not mention milk on the ingredient label or in a separate note (which is in accordance with Serbian regulation for these types of products).

Table 2. Number and percentage of dairy products with various forms of allergen labelling.

Product	Cheese	Yoghurt	Ice-cream
Allergen information listed in a separate note	1 (12%)	3 (43%)	4 (19%)
Emphasized font in Ingredient list	3 (37%)	6 (75%)	16 (76%)
Other method of labelling allergens	/	/	/
No labelling of allergens	5 (62%)	3 (43%)	1(5%)

Sixteen ice-cream declarations, from different producers, declared allergens by using a bold font in the ingredient list; milk (11), soybean (5), gluten (3), hazelnut, egg (2), peanuts, sesame (1). Gluten was printed in bold font as biscuit, gluten or starch. Four ice-cream declarations did not have information on allergens printed with an accented font, while they did have a special note indicating the product could contain soybeans (3), peanuts (3), and nuts (2). One ice-cream did not include any allergen information on the declaration.

3.3 Cereal products
Eight samples of frozen dough were examined from different manufacturers. Four frozen doughs had ingredient lists with allergenic ingredients emphasised by text font, while two products had a special note citing possible allergens. Three frozen doughs also had an allergen information note, but the allergens were not emphasized by the font used in the ingredient lists. Only one frozen dough did not have either allergen marked with a special font in the ingredient list, nor a special note about the allergen content of the product.

The Serbian regulatory requirement states that allergens must be listed in the ingredients when it cannot be undoubtedly concluded that the raw material(s) used are not allergen(s) [4]. However, it is not necessary to specify allergenic components when the product is made exclusively of one main ingredient, e.g. unadulterated plain yoghurt, which contains milk and bacterial cultures. Therefore, a high percentage of Serbian retail yoghurts and cheeses did not have allergenic milk declared on their labels, since this is not mandatory according to local law. Elsewhere, five percent of chocolate sampled had declared allergens in a special note [8]. Studies on the declaration of food allergens in Europe give special attention to allergy-preventive labelling. Most surveillance studies include labelling and testing for the presence of allergens in preventively-labelled products. When a study of allergy-preventive milk labelling [9] was carried out, milk in amounts greater than 1 mg/serving were detected in products with two different types of preventive declarations: in 60% of the products listing milk as a minor ingredient, and in all the tested foods listing milk fat as a minor ingredient. A study involving the declaration of peanuts as an allergen [10] detected a lower quantity of peanuts in products where it was listed as a possible ingredient (7.3%) than when peanuts was a component to a lesser degree (33%). In a study that included absence of milk, egg and peanut allergen labelling for groceries from large commercial chains, 2.6% of studied groceries had egg allergen that was not declared, while 3% of the groceries contained undeclared milk allergen but none of the examined groceries had undeclared peanut traces [11]. Seventeen percent of products examined contained advisory labelling [12] on the U.S. market, comparing to 23% (table 1) and 22% (table 2) of Serbian retail products.

References:

[1] Gendel S M, 2014 *Food Allergen Risk Management in the United States and Canada Risk Management for Food Allergy* pp 201–6
[2] Gupta R S, Springston E E, Warrier M R, Smith B, Kumar R, Wang X, Holl J R and Pongracic, J, 2011 *Journal of Allergy and Clinical Immunology* **127** p 33
[3] Nwaru B I, Hickstein L, Panesar S S, Roberts G, Muraro A and Sheikh A, 2014 *EAACI Food Allergy and Anaphylaxis Guidelines Group Allergy* **69** pp 992-1007.
[4] Rulebook on Declarations, Labeling and Advertising of Food *Official Gazette of the Republic of Serbia* No. 85/2013 and 101/2013)
[5] EU Food Information for Consumers Regulation (No. 1169/2011) (EU FIC) and Food Information Regulation (FIR) 2014(SI 2014/1855)
[6] Rulebook on the health safety of dietary products *Official Gazette of the Republic of Serbia*", No. 45/2010, 27/2011, 50/2012, 21/2015 and 75/2015)
[7] *Commission Regulation No*. 41/2009

[8] Van Hengel A J 2007 *Trends in Food Science & Technology* **18** pp 96-100
[9] Crotty M P and Taylor S L 2010 *J Allergy Clin Immunol* **125** pp 935-7
[10] Hefle S L, Furlong T J, Niemann L, Lemon-Mule H, Sicherer S and Taylor S L 2007 *J Allergy Clin Immunol* **120** pp171-6
[11] Ford L S, Taylor S L, Pacenza R, Niemann L M, Lambrecht D M and Sicherer S H 2010 *J Allergy Clin Immunol* **126** pp 384-5
[12] Pieretti M M, Chung D, Pacenza R, Slotkin T and Sicherer S H 2009 *J Allergy Clin Immunol* **124** pp 337-3

IOP Conf. Series: Earth and Environmental Science **85** (2017) 012079 doi:10.1088/1755-1315/85/1/012079

The characterization of lactic acid bacteria isolated during the traditional production of Užička sausage

B Borović, B Velebit, S Vesković, B Lakićević and T Baltić

Institute of Meat Hygiene and Technology, Kaćanskog 13, Belgrade, Serbia

E-mail: branka.borovic@inmes.rs

Abstract. Užička sausage is a traditionally fermented dry sausage that is produced in western Serbia. It is made of beef and pork with the addition of solid fat and natural spices. The whole manufacturing process lasted for 21 days. The goal of this study was to create a collection of lactic acid bacteria isolated during the ripening and identify them using molecular methods. A total of 50 isolates from different stages of ripening (fermentation and drying) were identified by molecular methods. *Leuconococcus mesenteroides*, *Lactobacillus brevis*, and *Lactobacillus sakei* were the predominant microorganisms in Užička sausage.

1. Introduction

Modern manufacturing procedures employ technologies for contemporary meat production that are significantly faster than traditional procedures. However, the resulting meat products, especially dried meat and dry fermented sausages, often fail to acquire the sensory characteristics typical for traditionally manufactured products. Basic differences occur due to usage of smoke produced in smoke generators, while traditional procedures employ smoke obtained from open fire beds. Next, neither starter cultures, nor glucono delta-lactone are used in traditional procedures. Ripening (fermentation and drying) processes, rather, are influenced by proteolytic and lipolytic enzymes from muscle and fat tissue as well as the epiphytic microbiota present. The dynamics of these processes, which are mainly directed by activity of the microbiota, are not under control in traditional manufacturing, and therefore, dry fermented sausages manufactured in traditional ways often fail to meet uniform quality standards.

The production of fermented sausages is an area of meat processing that has been the subject of very intensive scientific research in recent decades. Due to the desirable and recognizable sensory properties of traditionally produced dry sausages, demand for them is constantly growing. The production of these products is still based on local customs and traditional production methods.

The microorganisms that are most often responsible for these transformations are lactic acid bacteria (LAB), coagulase negative cocci (CNC) and yeasts [1]. LAB play an essential role in the production of fermented meat products. The products of their metabolism affect the process of ripening, development of desired sensory and nutritive characteristics of the products and, at the same time, inhibit the development of undesired microbiota [2]. They contribute to the development of flavor, color and texture of meat products. Reduction of pH as a result of lactic acid synthesis by LAB is an important effect of fermentation, which ensures the safety, stability and shelf life of meat products [3,4,5]. Traditional fermented sausages with a specific geographical origin have unique sensory characteristics and are generally of high quality [6].

There is growing interest in being able to isolate and describe the LAB population in traditional fermented foods, especially with the advancement of methods in molecular microbiology [7,8,9]. Molecular methods are all reliable, simple to interpret and their application avoids subjective interpretation by the user.In order to protect the traditional approach to sausage manufacturing, it is essential to understand the microbial diversity and to select autochthonous starter cultures that can be used in the production of innovative foods with a geographical origin [8].LAB that have been isolated from fermented sausages belong to the genera *Lactobacillus*, *Weissella*, *Leuconostoc*, *Pediococcus*, *Enterococcus* and *Lactococcus* [8,9,10]. Among them, lactobacilli are most frequently isolated from sausages produced with different technologies, especially the species *Lactobacillus sakei*, *Lactobacillus curvatus* and *Lactobacillus plantarum*[3,9,10].Other lactobacilli that canbe found in lower numbers are *Lactobacillus brevis*, *Lactobacillus buchneri* and *Lactobacillus paracasei* [11,12].

One of the traditionally fermented dry sausages in Serbia is Užička sausage, which is produced in western Serbia. The goal of this study was to create a collection of LAB isolated during the ripening of Užička sausage and complete their molecular identification.

2. Materials and Methods

2.1 Užička sausage

Užička sausage was traditionally produced in an artisanal household in Zlatibor district, Serbia. Užička sausage was composed of beef, pork and fat in the %ratio of 50:20:20. Meat and fat were ground to the size of 3 mm and mixed with nitrite curing salt (2.5%), sucrose (0.33%) and spice mixture (0.25%; composed of sweet and hot red peppers, black pepper and garlic). Prepared stuffing was filled into small beef intestine casings (diameter of 40 mm) to form sausages. Sausages were cold smoked for 4 days. The whole manufacturing process (smoking, fermentation and drying) lasted for 21 days.

2.2 Microbiological investigation

Sausage samples for microbiology examinations were taken on days0, 2, 4, 7, 14 and 21. The experiment was repeated three times. Three replicate sausage samples were collected on each sampling day and used for analysis. Each sausage sample weighing 25 g was homogenized in 225 mL of MRD (Oxoid, UK) in a stomacher (AES, France) for 90 s. Serial dilutions (10-fold) were plated onto MRS agar (Oxoid, UK) in duplicate, and incubated for 48 h at 30°C under microaerophilic conditions. From each plate, single colonies were randomly picked and streaked on fresh agar plates in order to obtain pure cultures. Suspect LAB isolates from MRS agar were examined by Gram straining and catalase reaction. A total of 50 Gram-positive and catalase-negative isolates were further identified and characterized by molecular methods. Total DNA from LAB was extracted from a single colony by using the DNeasy Blood and Tissue Kit (Qiagen GmbH, Germany) according to the manufacturer's protocol for Gram-positive bacteria. PCR was performed in a final volume of 50 μL containing 1× PCR buffer (10× PCR buffer: 500 mM KCl, 100 mM Tris-HCl, 0.8% Nonidet P40), 2.5 mM MgCl2, 10 μM dNTP, 200 nM of each primer, 1 U of Taq polymerase (Fermentas, Lithuania) and 100 ng of DNA template. DNA in PCR tubes was amplified in a thermal cycler (Techne, UK) using primers P1V1 (GCGGCGTGCCTAATACATGC) and P4V3 (ATCTACGCATTTCACCGCTAC), complementary to the V1-V3 region of the 16S rRNA, by 5 min at 95°C, 35 cycles of 1 min at 95°C, 1 min at 42°C, 2 min at 72°C and the final extension of 5 min at 72°C. PCR products were purified by QIAquick PCR purification kit (Qiagen, Germany) and sent for sequencing to IIT Biotech (Bielefeld, Germany). The BLAST algorithm was used to determine the most related sequence relatives in the NCBI nucleotide sequence database (http://blast.ncbi.nim.nih.gov).

IOP Conf. Series: Earth and Environmental Science **85** (2017) 012079 doi:10.1088/1755-1315/85/1/012079

3. Results and Discussion

Table 1. LAB species identified by molecular methods during production of Užička sausage using traditional techniques.

Day of ripening	Strain identification by 16S rRNA gene sequencing
0	*Ln. mesenteroides subsp. mesenteroides* (7 isolates) *Lb. curvatus* *Lb. brevis* *Pediococcus parvulus* *Lc. lactis subsp. cremoris* (2 isolates)
2	*Ln. mesenteroides subsp. mesenteroides* (2 isolates) *Ln. inhae* *Weissella hellenica* *Lb. brevis* (2 isolates)
4	*Lb. sakei* (3 isolates) *Lb. brevis* (4 isolates) *Weissella hellenica*
7	*Ln. mesenteroides subsp. mesenteroides* (2 isolates) *Pediococcus pentosaceus* *Lb curvatus* *Lb. brevis* (2 isolates)
14	*Ln. mesenteroides subsp. mesenteroides* (5 isolates) *Lb. sakei* *Weissella hellenica*
21	*Lb. brevis* (2 isolates) *Lb. sakei subsp. sakei* (4 isolates) *Ln. mesenteroides subsp. mesenteroides* (5 isolates)

The most dominant microorganisms in Užička sausage were *Leuconostoc mesenteroides*, *Lb. brevis*, and *Lb. sakei* (Table 1). Besides them, *Lactococcuslactis subsp.cremoris*, *Lb.curvatus*, *Weissella hellenica*, *Pediococcus pentosaceus, Pediococcus parvulus* and *Leuconostoc inhae* were also identified, although in lower levels.

The prevalence of *Ln.mesenteroides* found in Užička sausage is not in agreement with findings of authors who studied traditionally fermented sausages [9,10].However, it was in accordance with results obtained for another type of sausage produced in Serbia, Petrovská klobása[13],in which *Ln.mesenteroides* and *Lactobacillus* constituted the majority of the microbiota. Some of the *Leuconostoc* isolates from fermented sausages play an important role in sausage flavor development and could also exhibit strong antimicrobial activity [14,15]. However, the role of *Leuconostoc* species in the sausage ripening is not yet well understood. The finding of high levels of *Lb.brevis*is in agreement with other authors[16,17], while the predominant presence of *Lb.sakei* is in accordance with other authors [3,9,10,18]. These species are the most adapted of the lactobacilli to the fermented sausage environment [9].

4. Conclusion

Results of this study showed that during the ripening of Užička sausage, *Ln. mesenteroides*, *Lb.brevis,* and *Lb. sakei* were the most dominant species. At the beginning of the production, *Ln. mesenteroides* prevailed. From day 4 until the end of fermentation, the dominant lactobacilli were *Lb.brevis* and *Lb. sakei*. Isolates of LAB from this study could be used as starter cultures in the production of traditional dry sausages.

Acknowledgement

This research was funded by grants TR 20127 and III 46009 from the Ministry of Education, Science and Technological Development, Republic of Serbia.

References

[1] Hutkins R W 2006 Microorganisms and Metabolism *Microbiology and technology of fermented foods* (Oxford UK: Blackwell Publishing Professional) pp 15-66.

[2] Veskovic-Morachanin S 2010 *Tehnologija mesa* **51** 83-94

[3] Fontana C, Cocconcelli P S and Vignolo G 2005 *Int. J. Food. Microbiology.* **103** 131-42

[4] Frece J, Kos B, Beganović J, Vuković S and Šušković J 2005 *World J. Microb. Biot.* **21** 1401-8

[5] Bonomo M G, Ricciardi A, Zotta T, Parente E, and Salzano G 2008 *Meat Sci.* **80** 1238-48

[6] Raseta M, Veskovic-Moracanin S, Borovic B, Karan D, Vranic D, Trbovic D and Lilic S 2010 *Tehnologija mesa* **51** 45-51

[7] Cocolin L, Rantsiou K, Iacumin L, Urso R, Cantoni C and Comi G 2004 *Appl. Environ. Microbiol.* **70** 1883-94

[8] Rantsiou K, Drosinos E, Gialitaki M, Metaxopoulos I, Comi G and Cocolin L 2006 *Int. J. Food Microbiology* **112** 215-22

[9] Urso R, Comi G and Cocolin L 2006 *Syst. Appl. Microbiol.* **29** 671-90

[10] Comi G, Urso R, Iacumin L, Rantsiou K, Cattaneo P, Cantoni C and Cocolin L 2005 *Meat Sci.* **69** 381-92

[11] Aymerich T, Martín B, Garriga M and Hugas M 2003 *Appl. Environ. Microbiol.* **69**4 583-94

[12] Papamanoli E, Tzanetakis N, Litopoulou-Tzanetaki E and Kotzekidou P 2003 *Meat Sci.* **65** 859-67

[13] Danilovic B, Jokovic N, Petrovic Lj, Veljovic K, Tolinacki M and Savic D 2011 *Meat Sci.* **88** 668-74

[14] Lee J Y, Kim C J and Kunz B 2006 *Meat Sci.* **72** 437-45

[15] Xiraphi N, Georgalaki M, Rantsiou K, Cocolin L, Tsakalidou E and Drosinos E H 2008 *Meat Sci.* **80** 194-203

[16] Kozachinski L, Drosinos E, Chaklovica F, Cocolin L, Gasparik-Reichardt J and Veskovic S, 2008 *Food Technol. Biotechnol.* **46** 93-106

[17] Benito M, Serradilla M J, Ruiz-Moyano S, Martin A, Pérez-Nevado F and Cordoba G 2008 *Meat Sci.* **80** 656-61

[18] Cocolin L, Dolci P, Rantsiou K, Urso R, Cantoni C and Comi G, 2009 *Meat Sci.* **82** 125-32

Coccidiostats in unmedicated feedingstuffs for poultry

T Radičević, S Janković, S Stefanović, D Nikolić, J Đinović-Stojanović and D Spirić

Institute of Meat Hygiene and Technology, Kaćanskog 13, 11000 Belgrade, Serbia

E-mail: tatjana.radicevic@inmes.rs

Abstract. Coccidiostats are compounds that are widely used as feed additives to prevent and treat coccidiosis, a contagious disease affecting mainly poultry, and which is associated with warm and humid conditions, as can be found on poultry farms. In Serbia and in the EU, specific coccidiostats are authorized as poultry feed additives. A wide range of these products is available for prevention (as additives) and treatment of coccidiosis (as veterinary medicinal products). The aim of this study is to present findings of residues of coccidiostats in unmedicated feed for chickens for fattening and laying hens as possible causes for coccidiostat residues in liver and eggs. The reasons for these compounds occurring in animal tissues and primary products of animal origin could be an inappropriate withdrawal period after the last administration of medicated feed or cross-contamination of unmedicated feed during the production on the same production line as medicated feedingstuffs, because of inadequate cleaning procedures and/or hygiene practices.

1. Introduction

Coccidiosis is a parasitic disease affecting livestock, especially poultry. The disease is caused by protozoan parasites of the genus *Eimeria*. In warm and humid environments, it causes intestinal lesions and it is highly contagious and spreads from one animal to another by contact with infected feces. Even in the presence of high sanitary standards and good farm management, coccidiosis can occur with serious impact on animal health and welfare and possible high mortality rates. The economic damage caused by coccidiosis should also be taken into consideration [1]. Effective coccidiostats are, at the moment, indispensable to protect the health and welfare of poultry and other animals against coccidiosis.

The use of coccidiostats in the EU is regulated by Commission Regulation 2003/1831/EC [2] on additives for use in animal nutrition. This Regulation lays down rules for authorization of feed additives and classifies feed additives into five categories, namely technological, sensory, nutritional, and zootechnical additives and coccidiostats or histomonostats. Currently eleven coccidiostats are authorized as feed additives in accordance with regulation [2] – decoquinate, diclazuril, halofuginone, lasalocid, maduramicin, monensin, narasin, nicarbazin, robenidine, salinomycin and semduramicin. Each coccidiostat can be used at a prescribed concentration and during a certain time interval for broilers and young chickens but not for laying hens. Despite the requirements set for companies that produce feeds (cleaning of equipment and production line, especially between medicated and non-medicated feed) [3], under practical conditions during production, unavoidable carry-over occurs when a certain percentage of a feed batch containing coccidiostats remains in the production circuit and contaminates the subsequent feed batch. This unavoidable carry-over or cross-contamination can result in the exposure of non-target animal species with potential health risks for animals, as well as the potential residues in foods derived from these species such as eggs. Maximum levels of unavoidable carry-over of certain coccidiostats is regulated by Commission Directive 2002/32/EC on undesirable substances in animal feed [4], Annex I section VII – authorized feed additives in non target feed following unavoidable carry-over, and subsequent Directives and Regulations amending Annex I (Directive 2009/8/EC, Regulations 574/2011 and 744/2012). A carry-over rate of 1% is allowed for feed used during the period before slaughter (withdrawal feed), for other feed to which no coccidiostats are added and for non-target feed for continuous food producing animals (laying hens or dairy cows). Maximum levels for the presence of coccidiostats in food resulting from the unavoidable carry-over of these substances in non-target feed are regulated by the EU [5,6].

In Serbia, the use of these same eleven coccidiostats as feed additives and unavoidable carry-over in non-target feed is regulated by Regulation on quality of feed in article 89 (permitted substances with a coccidiostatic action which can be added to the mixtures and premixtures) and article 99 (maximum permitted harmful substances – authorised feed additives in non target feed following unavoidable carry-over) [7]. Table 1 presents the maximum concentrations of active substances allowed in non-target feed in Serbia given an unavoidable maximum level of carry-over of 1% of the coccidiostats.

Table 1. Maximum levels of unavoidable carry-over (1%) of coccidiostats in non-target feed

Coccidiostat	Maximum levels of unavoidable carry-over (1%) (mg of active substance/kg of feed)
Narasin	0.70
Lasalocid	1.25
Maduramicin	0.05
Semduramicin	0.25
Salinomycin	0.70
Monensin	1.25
Nicarbazin (DNC)	1.25
Diclazuril	0.01
Robenidine	0.70
Decoquinate	0.40
Halofuginone	0.03

Chickens for fattening receive coccidiostats through feed during their entire life, except during the period before slaughter – the withdrawal period (this is 3-5 days for the different coccidiostats). Coccidiostats are added to their feed at different levels: 100-125 mg/kg (monensin), 75-125 mg/kg (lasalocid), 60-70 mg/kg (salinomycin), 5-6 mg/kg (maduramicin), 40-50 mg/kg (narasin), 40-50 mg/kg (nicarbazin), 30-40 mg/kg (robenidine), 1 mg/kg (diclazuril) [7]. A withdrawal period before slaughter is required in order to avoid residues of coccidiostats in edible tissues, although there is no risk to consumer health from ingestion of coccidiostats residues in tissues of animals exposed to feed cross-contaminated up to a level of 10% [8-10]. None of the coccidiostats are licensed for use in egg laying hens and eggs should be free of coccidiostats, but residues of coccidiostats in eggs have been detected in Europe at levels from 0.3 to <40 μg/kg [11-13], and in Serbia within the National Residue Monitoring Programme, at levels from 1 to <20 μg/kg. The reported findings of coccidiostats residues can be attributed to the cross-contamination of unmedicated feed at feed production mills, although illegal use of the drugs cannot be discarded.

2. Materials and Methods

Diclazuril (DICL), robenidine (ROBN), narasin (NAR), nicarbazin (DNC), monensin (MON), salinomycin (SAL), maduramicin (MAD) and lasalocid (LAS) were all purchased from Sigma-Aldrich (St. Louis, USA). Water, methanol, acetonitrile and N,N-dimethylformamide were all HPLC grade and also purchased from Sigma-Aldrich. Formic acid LC grade was from Merck (Merck KGaA, Darmstadt, Germany). Individual stock solutions at a concentration of 1.00 mg/mL were prepared in methanol for all compounds except for narasin and nicarbazin, which were dissolved in dimethylformamide (DMF); all were stored at -20°C. Working standard solution (a mixture of analytes) was prepared in acetonitrile by diluting stock solutions into a range that equated to the carryover levels in feed and stored at 4°C.

Coccidiostats were analyzed with Waters ACQUITY connected to a TQD mass spectrometer (Waters, Miliford, MA, USA). The instrument was controlled by MassLynks software version 4.1.The analytical column used for separation was Kinetex 100 x 2.1 mm, 2.6μ C18 100A with UltraGuard cartridge (Phenomenex, Torrance, CA, USA). The oven temperature was set at 45°C. The chromatographic separation was achieved in gradient mode using water acidified with 0.1% formic acid (mobile phase A) and acetonitrile acidified with 0.1% formic acid (mobile phase B) at a flow rate of 0.55 mL/min. Electrospray ionization (ESI) was used with both positive and negative mode, with following parameters: capillary voltage 3.5kV, source temperature 130°C, desolvation temperature 400°C, desolvation and cone gas 900 and 60 L/h, respectively. Argon was used as collision gas. The precursor and products ions for each analyte, cone voltages and collision energies are presented in Table 2.

Feeds labeled as withdrawal feed or feed for laying hens, from 2014 to May 2017, were used in this study. We also analyzed different batches of rinsing feed, produced when switching from production of medicated feed to unmedicated feed, provided to us by mill owners. Before analysis, feeds were ground using a laboratory mill IKA A11 Basic (IKA Werke, Germany). A 5.0 g portion of each feed was individually weighed into polypropylene jars with caps. Acetonitrile 25 mL was added and the jars were shaken on a horizontal shaker IKA Yellow line (IKA Werke, Germany) for 60 mins. The extracted feeds were filtered individually through nylon 0.22μm syringe filter into HPLC vials.
Quantification was carried out using matrix extracted calibrations curves at four levels. With every batch of feeds examined, blank feed samples were fortified at four different levels with mixed working standard solution and submitted to the full extraction procedure.

Table 2. Mass spectrometry parameters for quantification of coccidiostats in poultry feeds

Compound	Precursor ion (m/z)	Product ions (m/z)	Cone voltage (V)	Collision energies (eV)	Ionization mode
Narasin	787.30	431.30	70	50	ES+
		531.40	70	45	
Lasalocid	613.20	359.30	70	32	ES+
		377.40	70	35	
Maduramicin	934.8	629.50	36	28	ES+
		647.50	36	20	
Salinomycin	773.50	265.10	70	50	ES+
		431.10	70	50	
Monensin	693.00	461.10	70	60	ES+
		479.10	70	60	
Nicarbazin	301.00	107.10	50	40	ES-
		137.00	50	40	
Diclazuril	405.00	334.00	33	21	ES-
	406.90	336.00	33	19	
Robenidine		138.00	35	30	
	333.90	155.00	35	20	ES+
		178.00	54	30	

3. Results and Discussion

Results obtained from analyzed rinsing feed proved that production of medicated and unmedicated feed on the same production line led to unavoidable carry over, i.e. cross-contamination. According to the legislation, SAL can be added to feed for chickens for fattening at concentrations of between 60-70 mg/kg [7]. Based on these concentrations, from up to 4% to less than 1% of the added SAL was left in the production lines after several batches of rinsing feed, intended to clean the lines, were processed. Similar results were obtained for other coccidiostats. Concentrations of residual coccidiostats in the feeds produced after the rinsing feeds were processed are shown in Figure 1. As can be seen, only after several batches of rinsing feed were processed did the concentrations of coccidiostats fall below maximum levels of unavoidable carry-over (1%).

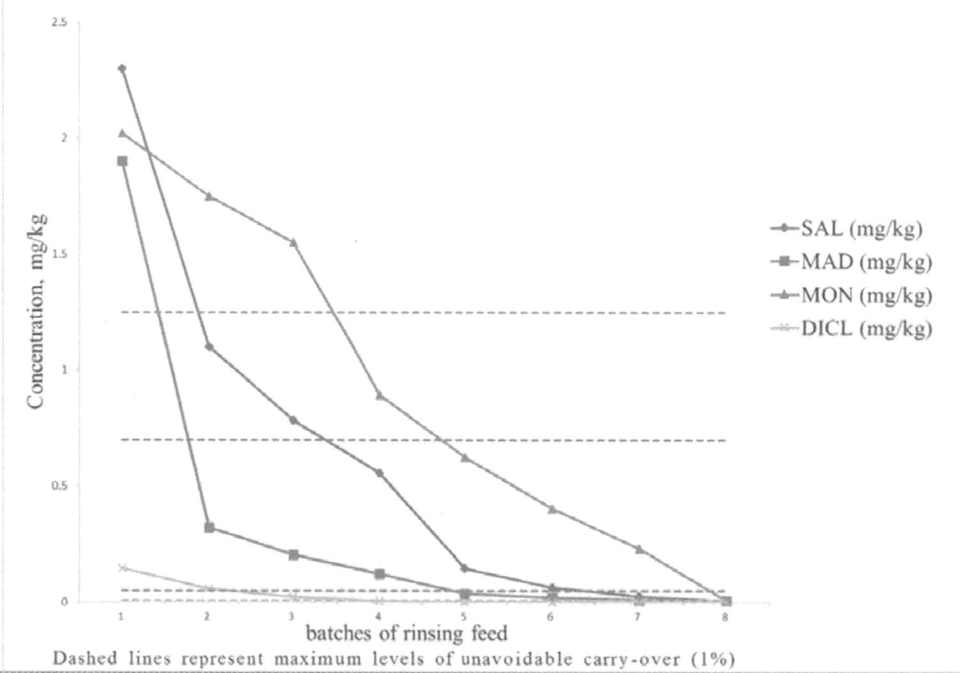

Figure 1. Concentration of different coccidiostats (mg/kg feed) in rinsing feed.

Concentrations of different coccidiostats found in analyzed feed labeled as withdrawal feed or feed for laying hens are presented in Table 3 and Table 4, respectively. For withdrawal feeds, the highest concentration of the coccidiostats SAL, MAD, LAS, DICL and ROBN corresponded to the doses added to starter feed for chickens for fattening [7]. Feeds that had concentrations of coccidiostats above maximum levels of unavoidable carry-over (1%), were declared non-compliant. Among the feeds, 13.7% contained non-compliant concentrations of SAL (Table 3). However, the majority of feeds had concentrations of coccidiostats below maximum levels of unavoidable carry-over (1%) and in 20-25% of feeds, concentrations of coccidiostats were below the limit of detection (LOD). The most disturbing fact was that a large number of feeds (80%) contained concentrations below maximum levels of unavoidable carry-over (1%) of several (2-4) different coccidiostats.

Table 3. Concentration of different coccidiostats (mg/kg feed) in withdrawal feed

Compound	n 1	min – max	n 2	n 3
Salinomycin SAL	124	0.005 – 60	74	17 (13.7%)
Maduramicin MAD	124	0.005 – 6	15	8 (6.5%)
Narasin NAR	124	0.005 – 12	40	4 (3.2%)
Lasalocid LAS	124	0.005 – 116	14	2 (1.6%)
Diclazuril DICL	124	0.005 – 0.920	5	1 (0.8%)
Robenidine ROBN	124	0.005 – 27.6	2	2 (1.6%)
Nicarbazin DNC	124	0.010 – 6.2	18	3 (2.4%)

n 1 – number of feeds; n 2 - number of feeds with coccidiostat level above LOD and below maximum levels of unavoidable carry-over (1%); n 3 - number of feeds with coccidiostat level above maximum levels of unavoidable carry-over (1%)

In feed for laying hens, in which the use of coccidiostats is not licensed, of 133 feeds analyzed, MAD was above maximum levels of unavoidable carry-over (1%) in 13 (9.8%) of feeds (Table 4). Bodi et al. [14] investigated carry-over of MAD from feed into eggs of laying hens. Feeding the hens a diet containing 50μg/kg MAD (the maximum level of unavoidable carry-over 1%), resulted in concentrations of MAD in eggs up to 11 μg/kg. They calculated that the carry-over rate from feed into eggs was 8% for MAD. Kennedy at al. [12] reported the carry-over of LAS from medicated to unmedicated feed during manufacture. LAS was found at levels up to 1mg/kg, level high enough to result in residues in eggs.

Table 4. Concentration of different coccidiostats (mg/kg feed) in feed for laying hens

Compound	n 1	min – max	n 2	n 3
Salinomycin SAL	133	0.005 – 7.3	81	3 (2.3%)
Maduramicin MAD	133	0.005 – 0.832	19	13 (9.8%)
Narasin NAR	133	0.005 – 0.769	13	1 (0.75%)
Lasalocid LAS	133	0.005 – 0.486	17	0
Diclazuril DICL	133	0.005 – 0.660	1	1 (0.75%)
Robenidine ROBN	133	0.005 – 0.125	8	0
Monensin MON	133	0.005 – 3.6	9	1 (0.75%)

n 1 – number of feeds; n 2 - number of feeds with coccidiostat level above LOD and below maximum levels of unavoidable carry-over (1%); n 3 - number (%) of feeds with coccidiostat level above maximum levels of unavoidable carry-over (1%)

4. Conclusion

The obtained results indicate that more attention should be paid to the feed production process in order to avoid cross-contamination from medicated to non-medicated feed. Besides the measures usually taken after any non-compliant findings of coccidiostats in non-medicated feed, adequate education of feed producers is highly recommended. This must be focused on better understanding of rules for production and hygiene practices and implementation of control measures in the use of coccidiostats, including closer cooperation between producers and competent authorities responsible for the residue surveillance program and official control.

Acknowledgments
This work was supported by grants from the Ministry of Education, Science and Technological Development of the Republic of Serbia (Project no. III 46009).

References

[1] Mortier L, Daeseleire E, Van Peteghem C 2005 *J. Chromatogr.* **820** 261
[2] Commission Regulation 2003/1831/EC of The European Parliament and of the Council **L 268** 29
[3] Commission Regulation 2005/183/EC of The European Parliament and of the Council **L 35** 1
[4] Commission Directive 2002/32/EC of The European Parliament and of the Council **L 140** 10
[5] Commission Regulation 2009/124/EC of The European Parliament and of the Council **L 40** 1
[6] Commission Regulation 2012/610/EC of The European Parliament and of the Council **L 178** 1
[7] Official Gazette RS No. 4/2010, 113/2012, 27/2014, 54/2017
[8] EFSA – European Food Safety Authority, EFSA *J* 2007 **552** 1
[9] EFSA – European Food Safety Authority EFSA *J* 2008 **593** 1
[10] EFSA – European Food Safety Authority EFSA *J* 2008 **594** 1
[11] Kennedy DG, Hughes PJ, Blanchflower WJ 1998 *Food Addit. Contam.* **15** 535
[12] Kennedy DG, Blanchflower WJ, Hughes PJ, McCaughey 1996 *Food Addit. Contam.* **13** 787
[13] Rokka M, Earola S, Perttila U 2005 *Mol. Nutr. Food Res.* **49** 38
[14] Bodi D, Fry H, Schafft H, Lahrssen-Wiederholt M, Preiss-Weigert A 2012 *J. Agric. Food Chem.* **60** 6946

IOP Conf. Series: Earth and Environmental Science **85** (2017) 012081 doi:10.1088/1755-1315/85/1/012081

Verification of rapid method for estimation of added food colorant type in boiled sausages based on measurement of cross section color

J Jovanović, R B Petronijević, M Lukić, D Karan, N Parunović and I Branković-Lazić

Institute of Meat Hygiene and Technology, Kaćanskog 13, 11000 Belgrade, Serbia

E-mail: radivoj.petronijevic@inmes.rs

Abstract. During the previous development of a chemometric method for estimating the amount of added colorant in meat products, it was noticed that the natural colorant most commonly added to boiled sausages, E 120, has different CIE-LAB behavior compared to artificial colors that are used for the same purpose. This has opened the possibility of transforming the developed method into a method for identifying the addition of natural or synthetic colorants in boiled sausages based on the measurement of the color of the cross-section. After recalibration of the CIE-LAB method using linear discriminant analysis, verification was performed on 76 boiled sausages, of either frankfurters or Parisian sausage types. The accuracy and reliability of the classification was confirmed by comparison with the standard HPLC method. Results showed that the LDA + CIE-LAB method can be applied with high accuracy, 93.42 %, to estimate food color type in boiled sausages. Natural orange colors can give false positive results. Pigments from spice mixtures had no significant effect on CIE-LAB results.

1. Introduction

Technological processes and components used in manufacturing have a major influence on sensory properties of meat products, and therefore, on acceptance by consumers [1]. Meat products have a characteristic color, which could make them acceptable or not at first glance. The color of meat products can be adjusted by addition of natural and/or synthetic dyes. This addition not only provides a uniform color of the products from batch to batch, but also allows the replacement of high-quality components with ones of lower quality, fraud and deception of the consumer. Usage of synthetic and some natural food dyes can have adverse reactions on human health, especially in children and sensitive adults [2-5]. Hence, their use in meat products is regulated [6-8].

Color measurement of meat and meat products using the CIE (Commission Internationale de l'Eclairage) L*a*b* system (CIE-LAB) is in common use, along with sensory evaluation [9-10]. Instrumental color measurement is independent of auditors' subjective impressions. The widespread application of CIE-LAB measurement can be explained due to the uniform distribution of color and similarity to color perception of the human eye. The CIE-LAB method along with chemometrics has been applied to predict color properties of food during accelerated shelf-life studies, recipe changes, package and storage conditions and so forth [11-14]. CIE-LAB measurements cannot be applied for direct determination of added food colorants in meat products without chemometrics, because other

IOP Conf. Series: Earth and Environmental Science **85** (2017) 012081 doi:10.1088/1755-1315/85/1/012081

coloring materials, like paprika and chili from spice mixtures, and the color of the constituent meat, contribute to the overall color [15]. Reliability of measurements depends on color homogeneity of the analyzed surface, and in the case of meat products, freshly-sectioned surfaces are required. Measurement of cross-sectional color of fine-grained products like frankfurters and Parisian sausage gives accurate and repeatable results.

It is not necessary to emphasize the need for rapid methods and techniques for food color monitoring throughout the entire food supply chain, from the manufacturers to retailers. In order to contribute to this topic, the goal of this study was to investigate the possibility of re-purposing a previously developed method [15] to distinguish the addition of synthetic or natural food colorants in boiled sausages with even cross-sectional color. After initial adjustment, the method was verified on retail products, and results of the CIE-LAB method were compared with results of HPLC method [15].

2. Materials and methods

Reagents, meat products for development and calibration of the CIE-LAB method, sensory and HPLC analysis, and CIE-LAB measurements were performed according to the procedures previously described [15]. Meat products for verification (76 boiled sausages, either frankfurters or Parisian sausages) were obtained from importers, exporters or retailers.

2.1. Statistical analysis

Techniques and methods previously described were used for data analysis [15]. Principal component analysis (PCA) and linear discriminant analysis (LDA) [16] were used to determine the existence of data grouping based on CIE-LAB measurement as well as to produce the model for determination of the colorant type in boiled sausages. MS Office Excel and JMP 10 Statistical Discovery (SAS Institute, SAD) were used for data processing and all calculations.

3. Results and discussion

Ten fine-grained boiled sausage products with different food colorants at different concentration levels produced in an industrial facility [15] were used as a training set. Sixty results for CIE-LAB measurements were used for analysis. First of all, PCA was applied to determine grouping either by type or colorant concentration level. Graphically represented results of PCA are shown in Figure 1.

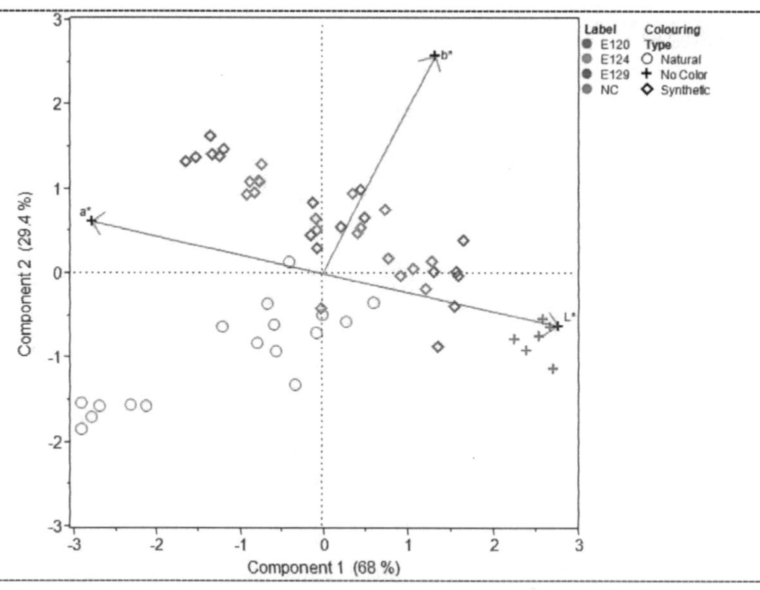

Figure 1. PCA results for the training set of ten boiled sausage products

Principal components 1 (PC1) and 2 (PC2) explained 89.4 % of total variance. Results showed that the main factors for grouping results by dye concentration level were the amount of red (a*) and lightness (L*). As the dye concentration increased, a* also increased, while L* values decreased. For sausages without added colorant (NC, No Colorant, orange cross in Figure 1) the luminance value (lightness, L*) was the most relevant grouping factor. Synthetic colorants (E 124 and E 129) were separated from natural dye (E 120) by b*, the yellowness value. Values of b* increased with increasing concentrations of synthetic dye, while increases in the amount of added natural colorant decreased b* values. Values of L* and a* showed more significant differences with concentration changes of natural than of synthetic dyes. In conclusion, PCA results for the training set indicated that sausages could be separated into three groups according to type of added food colorant – with no added colorant, with synthetic colorant and with natural colorant.

Re-purposing the CIE-LAB method to identify the type of colorant added to boiled sausages with homogenous section color involved LDA of the training set data. LDA was selected because it is suitable for calibration with categorical output variables. Results of LDA are shown in Table 1.

Table 1. LDA results for the training set of ten boiled sausage products

Counts: Actual Rows by Predicted Columns	Natural	No Color	Synthetic
Natural	17	0	1
No Color	0	6	0
Synthetic	1	2	33

Actual groups in the training set are given by rows, and predicted by columns in Table 1. Only 5 of the 60 measurements resulted in a misclassification. In the natural group, 1 was misclassified; in the group with no colorant, 0 were misclassified, and; 3 were misclassified in the group with synthetic colorants. The modified CIE-LAB method produced 91.67 % accurate prediction in the training set of boiled sausages. By groups, the percent of accurate prediction was as follows: 94.44 % in the natural colorants group, 100 % in the group without added colorant, and 91.67 % in the synthetic colorants group.

The verification set consisted of 76 samples of boiled sausages. Sensory audit of cross-section color was performed simultaneously, along with CIE-LAB measurements. Food dye content and type were determined and confirmed by the HPLC-DAD method [15]. Results of colorimetric, sensory and chromatographic determinations were compared and applied for estimation of CIE-LAB method validity (Table 2).

Table 2. Verification results for 76 boiled sausages from retail, export or import using CIE colorimetry + LDA to predict the type of colorant added to the foods

No	Type[a]	CIE colorimetry			Sensory	Probability (%)			Estimated dye (CIE)	Dye type (HPLC)	Correctness of classification
		L*	a*	b*		NC[b]	Nat[c]	Syn[d]			
1	FR	67.60	15.63	16.12	3.08	0.00	0.00	100.00	Syn	E 120	-
2	PS	74.71	14.30	13.78	4.92	100.00	0.00	0.00	NC	ND	+
3	PS	75.79	13.08	11.73	4.67	100.00	0.00	0.00	NC	ND	+
4	PS	66.25	16.21	20.73	2.83	0.00	0.00	100.00	Syn	E 129	+
5	PS	68.11	15.14	17.63	3.00	0.00	0.00	100.00	Syn	E 129	+
6	PS	72.78	8.83	13.67	3.25	80.18	11.87	7.95	NC	ND	+
7	PS	63.97	19.29	14.92	3.25	0.00	64.63	35.37	Nat	E 120	+
8	PS	69.01	15.61	12.88	4.83	0.00	99.91	0.09	Nat	E 120	+
9	PS	72.24	14.77	11.96	4.92	87.14	12.85	0.01	NC	ND	+
10	FR	68.63	16.57	11.98	3.58	0.00	100.00	0.00	Nat	E 120	+
11	FR	66.96	15.71	13.65	3.00	0.00	99.76	0.24	Nat	E 120	+
12	FR	68.41	14.31	13.90	2.50	0.00	94.10	5.90	Nat	E 120	+
13	FR	69.48	15.26	14.97	3.25	0.00	0.08	99.92	Syn	E 129	+

IOP Conf. Series: Earth and Environmental Science **85** (2017) 012081 doi:10.1088/1755-1315/85/1/012081

No	Type[a]	CIE colorimetry			Sensory	Probability (%)			Estimated dye (CIE)	Dye type (HPLC)	Correctness of classification
		L*	a*	b*		NC[b]	Nat[c]	Syn[d]			
14	FR	68.02	15.14	15.57	2.25	0.00	0.03	99.97	Syn	E 110	+
15	PS	66.83	17.01	12.56	4.63	0.00	100.00	0.00	Nat	E 120	+
16	PS	71.35	14.87	13.04	4.75	22.23	68.70	9.07	Nat	E 120	+
17	PS	68.92	14.78	17.12	3.38	0.00	0.00	100.00	Syn	E 129	+
18	FR	62.53	19.83	14.50	3.38	0.00	99.67	0.33	Nat	E 120	+
19	PS	63.60	19.16	15.68	3.00	0.00	3.02	96.98	Syn	E 110	+
20	FR	67.84	15.86	14.32	4.00	0.00	45.62	54.38	Syn	E 100	-
21	FR	68.82	15.45	14.74	4.13	0.00	1.13	98.87	Syn	E 100	-
22	PS	63.91	16.20	13.63	3.00	0.00	100.00	0.00	Nat	E 120	+
23	PS	65.44	18.26	16.45	3.50	0.00	0.00	100.00	Syn	E 110	+
24	FR	61.03	18.94	12.44	4.00	0.00	100.00	0.00	Nat	E 120	+
25	FR	66.83	17.41	12.76	4.00	0.00	100.00	0.00	Nat	E 120	+
26	FR	63.31	18.91	12.49	4.50	0.00	100.00	0.00	Nat	E 120	+
27	FR	63.95	18.43	10.52	3.50	0.00	100.00	0.00	Nat	E 120	+
28	PS	60.61	18.27	12.34	3.50	0.00	100.00	0.00	Nat	E 120	+
29	FR	59.57	20.55	13.48	3.50	0.00	100.00	0.00	Nat	E 120	+
30	PS	61.47	19.67	12.51	3.50	0.00	100.00	0.00	Nat	E 120	+
31	PS	66.60	18.00	11.42	2.50	0.00	100.00	0.00	Nat	E 120	+
32	PS	63.65	17.24	13.96	3.50	0.00	99.99	0.01	Nat	E 120	+
33	PS	66.22	17.15	14.44	3.00	0.00	75.17	24.83	Nat	E 120	+
34	FR	62.76	15.79	16.00	2.50	0.00	28.47	71.53	Syn	E 110	+
35	PS	63.94	18.70	17.01	3.00	0.00	0.00	100.00	Syn	E 100	-
36	PS	67.17	17.86	12.88	3.50	0.00	99.98	0.02	Nat	E 120	+
37	PS	68.95	13.43	12.32	3.50	0.00	100.00	0.00	Nat	E 120	+
38	PS	72.31	13.90	14.32	-	93.96	0.00	6.03	NC	ND[e]	+
39	PS	63.48	16.31	13.20	-	0.00	100.00	0.00	Nat	E 120	+
40	PS	66.63	15.94	13.01	4.00	0.00	100.00	0.00	Nat	E 120	+
41	PS	60.58	19.56	12.40	4.00	0.00	100.00	0.00	Nat	E 120	+
42	FR	65.47	12.23	13.89	4.00	0.00	100.00	0.00	Nat	E 120	+
43	FR	65.91	13.01	13.81	4.50	0.00	99.99	0.01	Nat	E 120	+
44	FR	64.35	14.10	13.85	3.50	0.00	100.00	0.00	Nat	E 120	+
45	PS	65.70	15.36	13.38	4.00	0.00	100.00	0.00	Nat	E 120	+
46	PS	66.23	17.24	13.76	3.50	0.00	99.56	0.44	Nat	E 120	+
47	PS	63.68	17.11	13.24	3.00	0.00	100.00	0.00	Nat	E 120	+
48	PS	65.34	16.67	14.03	3.50	0.00	99.73	0.27	Nat	E 120	+
49	PS	69.24	14.00	13.79	4.00	0.00	89.36	10.64	Nat	E 120	+
50	PS	65.91	13.01	13.81	3.50	0.00	99.99	0.01	Nat	E 120	+
51	PS	63.69	17.08	14.14	3.50	0.00	99.97	0.03	Nat	E 120	+
52	PS	61.03	17.93	14.19	4.00	0.00	100.00	0.00	Nat	E 120	+
53	FR	66.92	16.54	13.58	4.50	0.00	99.70	0.30	Nat	E 120	+
54	FR	60.83	19.89	13.11	4.00	0.00	100.00	0.00	Nat	E 120	+
55	PS	60.58	19.71	12.49	4.00	0.00	100.00	0.00	Nat	E 120	+
56	PS	63.69	17.08	14.14	3.50	0.00	99.97	0.03	Nat	E 120	+
57	FR	65.47	12.23	13.89	3.50	0.00	100.00	0.00	Nat	E 120	+
58	PS	61.03	17.93	14.19	3.00	0.00	100.00	0.00	Nat	E 120	+
59	FR	65.91	13.01	13.81	4.50	0.00	99.99	0.01	Nat	E 120	+
60	FR	64.35	14.10	13.85	4.00	0.00	100.00	0.00	Nat	E 120	+
61	PS	65.70	15.36	13.38	4.00	0.00	100.00	0.00	Nat	E 120	+
62	PS	66.23	17.24	13.76	3.50	0.00	99.56	0.44	Nat	E 120	+
63	PS	63.68	17.00	13.24	3.50	0.00	100.00	0.00	Nat	E 120	+
64	PS	65.34	16.67	14.03	4.50	0.00	99.73	0.27	Nat	E 120	+
65	PS	69.24	13.97	13.79	4.00	0.00	89.64	10.36	Nat	E 120	+
66	PS	65.39	16.33	12.96	4.00	0.00	100.00	0.00	Nat	E 120	+
67	PS	67.62	15.86	15.44	4.00	0.00	0.09	99.91	Syn	E 110	+
68	FR	60.63	19.98	12.76	4.00	0.00	100.00	0.00	Nat	E 120	+
69	PS	64.28	15.34	15.26	3.00	0.00	78.18	21.82	Nat	E 120	+
70	PS	61.04	16.80	15.41	3.50	0.00	99.56	0.44	Nat	E 120	+
71	FR	66.91	16.35	13.67	4.50	0.00	99.56	0.44	Nat	E 120	+
72	PS	61.47	16.02	15.44	3.50	0.00	99.34	0.66	Nat	E 120	+

No	Type[a]	CIE colorimetry			Sensory	Probability (%)			Estimated dye (CIE)	Dye type (HPLC)	Correctness of classification
		L*	a*	b*		NC[b]	Nat[c]	Syn[d]			
73	PS	68.32	15.09	13.18	3.50	0.00	99.90	0.10	Nat	E 120	+
74	PS	65.89	15.23	14.68	3.50	0.00	87.41	12.59	Nat	E 120	+
75	PS	68.28	13.43	12.48	3.50	0.00	100.00	0.00	Nat	E 120	+
76	PS	66.54	14.90	15.45	3.00	0.00	1.72	98.28	Syn	E 100	-

[a] FR = frankfurter; PS = Parisian sausage
[b] NC = no colorant
[c] Nat = natural colorant
[d] Syn = synthetic colorant
[e] ND = none detected

Verification of the method for estimating the type of food dye added to boiled sausages based on CIE-LAB colorimetry measurements was conducted by comparison of correctly classified vs. misclassified sausages. The overall estimated accuracy of verification was 93.42 %. Five of 76 sausages were misclassified, of which four were confirmed by HPLC-DAD as containing the orange natural food dye, curcumin (E 100). Considering that the b* value was the main discriminant factor for differentiation of synthetic and natural food colorants used in the training set, and that in orange food colorants, the amount of yellow (b*) is significantly higher than in red food colorants, it is easy to deduce why the use of E100 gave false positive results for synthetic dyes. In the case of the other orange colorant detected in the sausages (Sunset yellow FCF, E 110), it was correctly classified since it is a synthetic colorant.

Similarly to the training set, no sausages without added food colorant were misclassified in this verification of 76 sausages. The absence of misclassified samples in the group of sausages with no added color points to the fact that pigments from spice mixtures had no significant effect on CIE-LAB determination.

4. Conclusion

CIE-LAB measurement is a fast, nondestructive method for *in-situ* determination of product color characteristics and it can be applied, with chemometrics, to determine the amount of colorants in meat products. The fine-grained boiled sausages studied, with even cross-sectional color, gave repeatable CIE-LAB measurement results [15].

The previously developed method for estimation of added food dye amount based on multiple linear regression (MLR) and CIE-LAB measurements was successfully transformed into a method for estimation of added colorant type (natural, synthetic, or no colorant) by substitution of color concentration with color type and use of LDA in place of MLR.

Verification results showed that the LDA + CIE-LAB method could be applied with high accuracy to estimate food colorant type in boiled sausages. However, natural orange colorants can be falsely categorized as synthetic colorants. Sausages with added natural red colorant, except for one sample, were all correctly classified, as were sausages without added color.

Pigments from spice mixtures had no significant effect on CIE-LAB results.

This CIE-LAB method can be used as a complementary method alongside HPLC determinations and sensory analysis for rapid estimation of the type of added colorant in boiled sausages as an important control factor in food supply chain.

References

[1] Troy D J and Kerry J P 2010 *Meat Sci.* **86** 214-226
[2] *EFSA Journal* 2015 **13**(4):4073
[3] *EFSA Journal* 2015 **13**(2):4007
[4] *EFSA Journal* 2014 **12**(7):3765
[5] *EFSA Journal* 2013 **11**(6):3234
[6] *OJ* L 354 2008 16-33

[7] *Sl. glasnik RS* 2013 **63**
[8] US Food and Drug Administration 2010 Food Additives *Federal Food, Drug, and Cosmetic Act*
[9] Zhang Y, Luo S and Wang W 2014 *J Food Process Pres* **38** 1259-1268
[10] Vasilev D, Vuković I, Tomović V, Jokanović M, Vasiljević N, Milanović-Stevanović M, Tubić M 2009 *Tehnologija mesa* **50** 342-350
[11] Sheridan C, O'Farrell M, Lewis E, Flanagan C, Kerry J and Jackman N 2007 *J Opt. A-Pure Appl Op* **9** S32–S39
[12] Korifi R, Le Dréau Y, Antinelli J F, Valls R and Dupuya N 2013 *Talanta* **104** 58–66
[13] Rodríguez-Pulido F J, Gordillo B, González-Miret M L and Heredia F J 2013 *Comput Electron Agr* **99** 108–115
[14] Wu D and Sun D W 2013 *Trends Food Sci Tech* **29** 5-20.
[15] Petronijević R B, Matekalo-Sverak V F, Spirić A T, Babić J A, Milijašević M P, Trbović D K and Vuković I K 2014 *Hem Ind* **68** 781-791
[16] Miller J N and Miller J C 2005 *Statistics and chemometrics for analytical chemistry* (Harlow: Pearson Education Ltd.)

Incidence of *Salmonella* Infantis in poultry meat and products and the resistance of isolates to antimicrobials

V Kalaba[1], B Golić[1], Ž Sladojević[1] and D Kalaba[2]

[1] P I Veterinary Institute of the Republic of Srpska "Dr Vaso Butozan" Banja Luka, Branka Radicevica 18, 78000 Banja Luka, Republic of Srpska
[2] Faculty of Medicine, Department of Pharmacy, Save Mrkalja 14, 78000 Banja Luka, Republic of Srpska

E-mail: vesna.kalaba@yahoo.com

Abstract. Globalisation, climate change, changes in eating habits and the food industry, modern animal husbandry and market demands often have a negative impact on quality assurance, food safety and animal health. After the eradication of some zoonotic diseases that previously often jeopardized the human population, today in developed countries, the focus is mainly on the control of zoonoses transmitted by food. *Salmonella* is one of the most common pathogens that can be transmitted from animals to humans, and its reservoirs are poultry, cattle and pigs, so one transmission route to humans is from contaminated food of animal origin. Multidrug-resistant isolates of *Salmonella*, which can transfer their resistance genes to other microorganisms, are considered a serious threat to public health. Control of *Salmonella* primarily depends on a good monitoring system and knowledge of the presence of serovars and strains in an epizootiological area. During the first nine months of 2016, 1321 samples of poultry meat and products were examined, among which 108 harboured *Salmonella*. Altogether, 29 of the 108 isolates (26.85%) were *Salmonella* Infantis. For all 29 *S.* Infantis isolates, antimicrobial resistance was tested by the disc diffusion method. The isolates showed 100% resistance to amoxicillin, and nalidixic acid.

1. Introduction

Bacteria of the genus *Salmonella* belong to the family *Enterobacteriaceae*. The *Salmonella* genus includes over 2,570 different serotypes. All species of this genus are pathogenic to humans and cause different types of illnesses, from intestinal infections such as diarrhoea to generalized infections which are life-threatening such as typhoid fever. Infectious diseases in humans are mainly caused by *Salmonella enterica* subspecies *enterica* (*S.*), serotype Typhi or Paratyphi. Other serotypes mostly cause intestinal infections, but can also cause septicaemia, and often the existence of long-term carrier status.

Apart from in humans, *Salmonella* can cause disease in many animal species, and human disease is commonly associated with the consumption of food of animal origin. Humans can become infected through the faecal-oral route, by consuming different kinds of contaminated food and water or through contact with animals. *Salmonella enterica* subsp. *enterica* serotype Enteritidis (*S.* Enteritidis), *S.* Typhimurium, *S.* Infantis and *S.* Hadar are ubiquitous serotypes with a wide range of hosts. They rarely cause systemic disease in healthy adults, but they are able to colonize gastrointestinal tract of many animal species. Thanks to frequent colonization and a high level of excretion in faeces of

animals reared for human consumption, ubiquitous serotypes can enter the food chain and cause salmonellosis cases in humans [1].
The ubiquity of *Salmonella* and its prevalence in the natural environment, on farms and in the food chain, as well as its adaptability to many animal reservoirs and routes of transmission, makes epidemiological studies of these bacteria very complex. For most epidemics among humans, the source of infection can be detected relatively easily, because they are usually foodborne, caused by one serotype and mainly of limited duration, but hospital epidemics are exceptions. When animals are bred for human consumption on farms, sources, routes and methods of transmission are usually more complex, especially as multiple strains and serotypes of *Salmonella* can persist concurrently on farms for a long time. Hence, the eradication of *Salmonella* on farms is a challenge. The first step in preventing the spread of infection between animals and humans is monitoring the movement of individual strains of *Salmonella* in both populations. For this purpose, it is not enough to identify isolates just to the level of *Salmonella* serotypes, because not all isolates within each serotype are identical. The divisions at different levels within the serotypes have significance in epidemiological studies and at the local level, in the context of supervision of these pathogens.

The role of poultry in the epidemiology of human *Salmonella* infections is recognized to be due to the development of intensive poultry production worldwide. Infected poultry is one of the most important reservoirs of *Salmonella* that are transmitted to humans through the food chain. In the past, the main motive for the control of salmonellosis in poultry was to reduce production losses. Today, the interest in protecting public health, plus political and consumer pressures, have resulted in the prevention of foodborne salmonellosis being a top priority for poultry producers [2]. The presence of *Salmonella* in poultry (compared to other domestic food-source animals), means poultry products have far greater significance as potential sources of infection for humans. Therefore, monitoring the presence and movement of *Salmonella* in poultry flocks is the first step in the control of this zoonosis, which later in the poultry chain is continued by microbiological control of poultry meat, eggs and their products [3].

Paratyphoid infections in poultry are caused by motile *Salmonella* serotypes, of which the most common causes are *S.* Enteritidis and *S.* Typhimurium. These organisms can infect a variety of hosts, including humans. In poultry, paratyphoid infections are generally asymptomatic, while in humans they cause symptoms of food poisoning [4].

In Serbia, foodborne salmonellosis outbreaks were linked commonly with cakes (22.2%), eggs (8.9%) or fried chicken (4.4%). However, according to information on the presence of *Salmonella* in some types of foods in Belgrade in the last twenty years, *Salmonella* was more likely to be isolated from chicken (49.2%), but was detected significantly less often in cakes and pies (8.8%) [5].

S. Enteritidis is, together with *S.* Typhimurium, a leading cause of salmonellosis in humans in the world. However, according to available data, in recent years, the number of isolates of *S.* Enteritidis and *S.* Typhimurium has decreased by half, but a constant increase in the number of isolates of *S.* Infantis has been recorded with its main sources being pigs, poultry and animal food [6]. The *S.* Infantis presence in human and animal food, as well as in animals, has led to this being one of the ten most frequently identified *Salmonella* serotypes in the European Union (EU) [6]. The European Food Standards Agency (EFSA) [6] stated that, according to results from the European Centre for Disease Control (ECDC), for the period from 2012 to 2014, *S.* Infantis was the fourth most common *Salmonella* serovar in the EU. Furthermore, it was revealed that *S.* Infantis contains isolates of several different clones which are present not only in food, but also in humans [7].

Human infection caused by *S.* Infantis *via* food is increasingly being monitored worldwide. *S.* Infantis is also increasing among human isolates in Europe (Germany, Hungary, Netherlands, Finland), but the rise of this serovar is also noticed in countries around the world (Australia, New Zealand, Russia, Argentina, Brazil, Canada, Japan [6,8,9,10]. In Hungary, in recent years, there has been an increased occurrence of *S.* Infantis not only in poultry production, but also in humans [11,12]. Likewise in Germany, *S.* Infantis is the third most common *Salmonella* serovar in recent years in humans, and poultry and pork meat are major sources of infection [7,13].

According to available literature, the Republic of Srpska and neighbouring countries have not conducted studies that include isolates from poultry, people and food of poultry origin, in an attempt to investigate their possible clonal association as evidence of circulating strains in the poultry food chain. However, available data from Croatia, where *S.* Infantis has been monitored for several years in broiler flocks, showed isolations of *S.* Infantis are increasing in absolute terms, but also in relation to other pathogenic strains of *Salmonella* that are being monitored [14].

Many scientific studies in various countries have recognized poultry meat as a potential source of *S.* Infantis, and the presence of this organism as the predominant *Salmonella* serovar in poultry in Israel was associated with an increase in the number of people with diabetes during 2007-2009 [9,15]. Also, over a period of five years (2004-2009), 76 *S.* Infantis isolates were obtained from broilers from nine countries in Central and Eastern Europe [10,16]. A significant increase (up 162%) of human infection by *S.* Infantis was reported in the United States in 2014, compared to the previous reporting period (2011- 2013) [17].

EFSA and ECDC [18], reporting on antimicrobial resistance among pathogens and indicator bacteria in humans, animals and food, stated that *S.* Infantis showed resistance to more than 90% of the tested antimicrobials. The greatest resistance (including to ciprofloxacin, streptomycin, sulphonamides and tetracyclines) was shown by isolates from broiler meat and hens. In countries where vaccination against *S.* Enteritidis and *S.* Typhimurium is carried out with good results, it can be assumed that this could allow other *Salmonella* to enter the poultry at farm level; *S.* Infantis is a potential candidate for this [7].

The Republic of Srpska's Policy on Microbiological Criteria for Foods [19] requires serotyping of *S.* Enteritidis and *S.* Typhimurium only when isolates are from fresh broiler chicken meat, or from broilers, hens, or breeding or fattening flocks of turkeys, while all other *Salmonella* isolations are recorded as contamination with *Salmonella* spp., and do not require further subtyping.

Good farming and good hygienic practices (GFP and GHP) are a set of measures that are applied to control *Salmonella* spp. [20]. The basis of food production with minimal contamination with *Salmonella* is good manufacturing practice as well as the implementation of hazard analysis and critical control points (HACCP) from preparation to delivery [21].

The aim of this study was to determine the overall incidence of *S.* Infantis in poultry meat and poultry meat products, frozen and fresh in the Republic of Srpska, as well as to determine the resistance of the *S.* Infantis isolates to antimicrobial drugs.

2. Materials and methods

Microbiological testing of fresh and frozen poultry meat was carried out from January to October 2016 as part of the normal business activities in the Laboratories for Microbiology of Food, Animal Feed and Water, Veterinary Institute of the Republic of Srpska "Dr Vaso Butozan". In total, 1321 samples were tested including: 39 grilled chickens, 244 chicken fillets, 54 chicken wings, 24 chicken meat samples, 136 fresh neck skins, 234 frozen neck skins, 40 livers, 265 samples of mechanically deboned meat, 56 thighs together with drumsticks, 14 marinated meats, 6 chicken breasts with skin and 215 eggs. All samples were tested for *Salmonella* according to the method ISO 6579/Cor 2:2010 [22].

Further biochemical identification and serological typing was conducted on suspected colonies grown on solid media. Biochemical reactions (triple sugar iron, urea, lysine decarboxylase test, ONPG, indole production and Voges-Proskauer) were used for the biochemical identification which confirmed the physiological characteristics of the genus *Salmonella*. Serological typing of *Salmonella* isolates was by the Kauffman-White-Le Minor scheme [23], with polyvalent and monovalent antisera. Subtyping of isolates was conducted using the antisera O6, O7, H1;r, and H2;5 for *S.* Infantis, and O9, H1;g, and H1;m for *S.* Enteritidis.

Disk diffusion according to Kirby-Bauer [25] was used for testing the susceptibility of the *S.* Infantis isolates to antimicrobial drugs. The tested isolates were first streaked on trypticase soy agar (TSA) and incubated for 24 h at 37°C. *Salmonella* suspensions were prepared in saline from the grown colonies corresponding to the density of 0.5 McFarland standard. *Salmonella* suspensions were

IOP Conf. Series: Earth and Environmental Science **85** (2017) 012082 doi:10.1088/1755-1315/85/1/012082

swabbed onto Mueller Hinton agar with sterile swabs (HiMedia, India), and then commercial antibiotic discs (Liofilchem, Italy) were placed on the plates. The *S.* Infantis isolates were examined for resistance to: streptomycin, at the level of 10 μg, cefotaxime 30 μg, cephaclor 30 mg, cephalexin 30 mg, ceftazidime 30 mg, ceftriaxone 30 mg, kanamycin 30 mg, pipemidic acid 20 mg, amoxicillin 30μg and nalidixic acid 30 μg. After 24 h of incubation at 37°C, inhibition zones were measured. Isolates that produced intermediate inhibition zones were considered resistant.

3. Results and discussion

In the period from 1 January to 1 September 2016, 1321 samples of poultry meat and products were tested for the presence of *Salmonella*. A total of 108 (8.18%) of the samples contained *Salmonella*. Biochemical characterization confirmed their belonging to the genus *Salmonella*. Of the 108 *Salmonella* isolates recovered, 6 (5.55%) were *S. E*nteritidis, and 29 (26.85%) were *S.* Infantis, according to the antisera used. Table 1 shows the number of isolates of *S.* Infantis, *S.* Enteritidis and total *Salmonella* according to the number and type of samples.

Table 1. The incidence of *S.* Infantis, *S.* Enteritidis and *Salmonella* spp. according to poultry meat and product type.

Sample	Sample size			*S.* Infantis		*S.* Enteritidis		*Salmonella* spp.	
	N	+	%	+	%	+	%	+	%
Grilled chicken	39	1	2.56	1	100.00	0	0.00	0	0.00
Chicken fillet	244	24	9.83	5	20.83	0	0.00	19	79.16
Chicken wings	54	9	16.66	5	55.55	1	11.11	3	33.33
Chicken meat	24	5	20.83	1	20.00	0	0.00	4	80.00
Fresh neck skins	130	16	12.30	3	14.28	0	0.00	13	81.25
Frozen neck skins	234	21	8.97	3	14.28	5	23.80	13	61.90
Liver	40	2	5.00	1	50.00	0	0.00	1	50.00
Mechanically deboned meat	265	15	5.66	4	26.66	0	0.00	11	73.33
Drumsticks together with thighs	56	9	16.07	6	66.66	0	0.00	3	33.33
Marinated meat	14	0	0.00	0	0.00	0	0.00	0	0.00
Chicken breasts with skin	6	4	66.66	0	0.00	0	0.00	4	100.00
Eggs	215	2	0.93	0	0.00	0	0.00	2	100.00
Total	**1321**	**108**	**8.18**	**29**	**26.85**	**6**	**5.55**	**73**	**67.59**

The most frequently *Salmonella*-positive types of poultry were chicken breasts with skin (66.66%), chicken meat (20.83%), chicken wings (16.66%), drumsticks together with thighs (16.07%) and fresh neck skin (12.30%).

S. Infantis, compared to the total number of isolated *Salmonella*, was predominant in drumsticks together with thighs (66.66%) and chicken wings (55.55%).

Table 2 shows the results of testing the resistance the *S.* Infantis isolates to antimicrobials by the disc diffusion method.

Table 2. Antimicrobial resistance of *S.* Infantis isolates from poultry meat and products.

Antimicrobial	Number of isolates tested	**Number of resistant isolates**	%	Number of sensitive isolates	%
Streptomycin	26	**9**	34.61	17	65.38
Cefotaxime	29	**16**	55.17	13	44.82
Cefaclor	29	**12**	41.37	17	58.62
Cephalexin	29	**9**	31.03	20	68.96
Ceftazidime	29	**20**	68.96	9	31.03
Ceftriaxone	29	**0**	0.00	29	100.00
Kanamycin	29	**0**	0.00	29	100.00
Pipemidic acid	29	**15**	51.72	14	48.27
Amoxicillin	26	**26**	100.00	0	0.00
Nalidixic acid	29	**29**	100.00	0	0.00

The *S.* Infantis isolates showed resistance to: amoxicillin and nalidixic acid (100%), ceftazidime (68.96%), cefotaxime (55.17%), pipemidic acid (51.72%), chloramphenicol (41.37%), streptomycin (33.33%) and cephalexin (31.03%), but not to ceftriaxone or kanamycin. These results coincide with the results of other researchers who have studied this issue [26].

For 20 (68.97%) of the *S.* Infantis isolates, multiple resistance was found to four or more antimicrobial drugs. The results are shown in table 3.

Table 3. Multidrug-resistant *S.* Infantis (number of isolates) according to poultry product sources.

Antimicrobials	Grilled chicken (1)	Chicken wings (2)	Chicken fillet (5)	Fresh skins (3)	Drumsticks together with thighs (6)	Frozen neck skin (3)	Mechanically deboned meat (4)	Liver (1)	Chicke n meat (1)
Streptomycin	NT[a]	**R**[b]	S[c]	**R**	**R**	S	S	NT	NT
Amoxicillin	NT	**R**	S	S	S	NT	**R**	NT	NT
Cefotaxime	**R**	S	**R**	**R**	**R**	S	S	S	**R**
Kanamycin	S	S	S	S	S	S	S	S	S
Cefahlor	S	**R**	S	**R**	S	S	**R**	S	S
Cephalexin	S	**R**	S		S	S	**R**	S	S
Ceftazidime	**R**	**R**	S	**R**	**R**	S	**R**	S	**R**
Pipemidic acid	**R**	**R**	**R**	**R**	**R**	S	**R**	**R**	**R**
Nalidixic acid	**R**	**R**	**R**	**R**	**R**	S	**R**	**R**	**R**
Ceftriaxone	S	S	S	S	S	S	S	S	S
Total	**4**	7	**3**	**6**	**5**	**0**	**6**	**2**	**4**

[a] NT – not tested
[b] R – resistant
[c] S – sensitive

Research into resistance to antimicrobial drugs among foodborne microorganisms that cause disease is very topical. The level of antimicrobial resistance in some *Salmonella* is 100%, e.g. resistance to tetracycline, chloramphenicol, streptomycin, and sulphonamides in *S.* Typhimurium DT104 isolates. The occurrence of so-called related resistance, such as that in some *S.* Typhimurium DT104, is of particular importance since the use of one antimicrobial drug selectively acts on all resistance genes present, and thus, the isolates have a selective advantage in the case of administration of ampicillin, chloramphenicol, streptomycin, sulphonamides or tetracyclines [25]. The selection of resistant isolates among bacterial pathogens whose are primary reservoirs are animals is achieved by selective pressure based on unjustified and excessive use of antimicrobials (or antimicrobial growth promoters) in the treatment or prevention of animal diseases.

Many different *Salmonella* isolates exhibit multiple resistances to streptomycin, kanamycin, sulphonamides, tetracyclines and some ß-lactams (penicillins and cephalosporins). It has been shown that the plasmids that carry resistance to several antimicrobials are common among *Salmonella* and that these genes are often grouped in integrons. Integrons in *Salmonella* can carry genes for resistance to chloramphenicol, sulphonamides, tetracyclines and streptomycin. The usual resistance genes, class one integrons, which are common in many bacterial species, were also found in *Salmonella* [27]. Altogether, 11% of human *Salmonella* isolates in the US carry antimicrobial resistance genes, and some isolates have acquired resistance to gentamicin and the third generation of cephalosporins [8].

Since there are no results of previous surveillance of poultry meat or products in the markets in the Republic of Srpska, we clearly cannot comment on changes in the presence of *Salmonella* in poultry meat and products, but the finding of 8.18% *Salmonella*-positive samples is still a warning. Among the poultry isolates, 26.83% were *S.* Infantis, which is worrying, since it confirms the significant exposure of the human population in the Republic of Srpska to this pathogen. The occurrence of antimicrobial resistant, and in particular multiple-resistant, isolates of *S.* Infantis is also alarming.

The presence of *S.* Infantis in poultry meat could be a result of poor manufacturing and hygiene practices in slaughterhouses and manufacturing plants, but it could also be a consequence of the implementation of vaccination against *S.* Enteritidis and *S.* Typhimurium, which some researchers have already confirmed [7,28]. Some experts define the generally increasing incidence of *S.* Infantis as the emergence of a new pathogenic microorganism causing infectious disease in humans, and which has occurred in a new host; alternatively, the incidence of the pathogen could be significantly increasing as a result of long-term changes in the epidemiology of *S.* Infantis [29].

4. Conclusions

Based on the results, the following conclusions can be made:

1. In the first nine months of 2016, the incidence *Salmonella* in samples of poultry meat and meat products was 8.18%, of which 5.55% were *S.* Enteritidis and 26.85% were *S.* Infantis.
2. *S.* Infantis isolates showed 100% resistance to amoxicillin and nalidixic acid. Multiple resistances to four or more antimicrobials were found in 68.97% of the *S.* Infantis poultry isolates.
3. Having in mind that the sources, routes and methods of transmission of *Salmonella* in the food chain are inextricably linked to the risks of infection and illness in humans, epidemiological studies should include monitoring and typing of these pathogens in humans, and in animals intended for human consumption, in all phases of food production process and distribution.
4. Poultry meat contaminated with *S.* Infantis poses a risk to human health, particularly children, the elderly and people with immunodeficiency.
5. Raising consumer awareness regarding methods of preparing food, especially poultry meat and meat products (heat treatment destroys *S.* Infantis) is required, but also the importance of good hygiene practice for the food business operators, in order to avoid cross-contamination, must constantly be stressed.

References

[1] Van Hoorebeke S, Van Immerseel F, Haesebrouck F, Ducatelle R and Dewulf J 2011 The influence of the housing system on *Salmonella* infections in laying hens: a review *Zoonoses Public Hlth* **58** 304-11

[2] WrayC and Wray A 2000 *Salmonella in domestic animals* (CABI *Publishing*, UK)

[3] Orlić D, Vidić B, Suvajdžić L and Stojanov I 2001 Salmoneloza živine, mere kontrole u cilju obezbeđenja zdravstveno ispravnih jaja i mesa za ishranu. *Zbornik rezimea*, 1. Međunarodni simpozijum „Hrana u 21. veku“ (Naučni institut za ratarstvo i povrtarstvo) 254

[4] Breytenbach J H, 2004 *Salmonella* Control in poultry *Intervet International* 12

[5] Pavlović N, Slavica M, Branislava Z and Purtić Kljajić 2014 Hrana izvor zaražavanja u epidemijama salmoneloza – istraživanje jačine dokaza 4. Simpozijum bezbednost i kvalitet namirnica animalnog porekla *Zbornik radova* Beograd 6-19
[6] EFSA/ECDC 2015 European Food Safety Authority, European Centre for Disease Prevention and Control. The European Union Summary Report on Trends and Sources of Zoonoses, Zoonotic Agents and Food-borne Outbreaks in 2014. *EFSA Journal* **13** 4329
[7] Miller T, R. Prager, W Rabsch, K Fehlhaber and M Voss 2010 Epidemiological relationship between *Salmonella* Infantis isolates of human and broiler origin *Lohman Information* **45** 27-31
[8] CDC 2012 Number and incidence of *Salmonella* infection by serotype 2011http://www.cdc.gov/foodnet/data/trendts/tsables
[9] Gal-Mor O, L Valinsky, M Weinberger, S Guy, J Jale, Y I Schorr, A Raisfeld, V Agmon and I. Nissan 2010 Multidrug resistant *Salmonella enterica* serovar Infantis in Israel *Emerg Infect Dis* **16** 1754-57
[10] Rahmani M, S M Peighambari, C A Svendsen, L M Cavaco, Y Agres and R S Hendriksen 2013 Molecular clonality and antimicrobial resistance in *Salmonella* enterica serovars Enteritidis and Infantis from broilers in three northern regions of Iran *BMC Vet Res***9** 66
[11] Nógrády N, G et al. 2008 Prevalence and characterization of *Salmonella* Infantis isolates originating from different points of the broiler chicken-human food chain in Hungary *Int J Food Microbiol* **127** 162-167
[12] Nógrády N A, Toth A. Kostyak J and Paszti B 2007 Emergence of multidrug-resistant clones of *Salmonella* Infantis in broiler chickens and humans in Hungary *Antimicrob Chemother* **60** pp 645-8
[13] Miller T, P G Braun, Fehlhaber K, Prager Y, Pfeifer and Rabsch W 2013 Typing of *Salmonella* enterica serovar Infantis from 51 outbreaks in Germany between 1974 and 2009 by a novel phage-typing scheme *Epidemiol Infect* **142** 75-83
[14] Jurinović L, Šimpraga B, Krstulović and Sokolović M 2015 Pojavnost bakterije *Salmonella* Infantis u tovnih pilića u Hrvatskoj u razdoblju od 2010 do 2014 godine, *Zbornik radova, XI simpozija Peradarski dani* 39-43
[15] Aviv G, K. Tsyba N, Steck M, Salmon-Divon A, Cornelius G, Rahav G, Grassl A, and Gal-Mor O 2013 A unique megaplasmid contributes to stress tolerance and pathogenicity of an emergent *Salmonella* enterica serovar Infantis strain *Environ Microbiol* **16** 977-994
[16] Nógrády N, Kiraly, M, Davies R and Nagy B 2012 Multidrug resistant clones of *Salmonella* Infantis of broiler origin in Europe *Int J Food Microbiol* **157** 108–112.
[17] CDC, Centers for Disease Control and Prevention 2015 Preliminary Incidence and Trends of rends of Infection with Pathogens Transmitted Commonly Through Food — Foodborne Diseases Active Surveillance Network
[18] Miller T, Prager M, Rabsch W, Fehlhaber K and M Voss 2010 Epidemiological relationship between *Salmonella* Infantis isolates of human and broiler origin *Lohman Information* **45** 27-31
[19] EFSA/ECDC 2015b Scientific Report of EFSA and ECDC EU Summary Report on antimicrobial resistance in zoonotic and indicator bacteria from humans, animals and food in 2013 *EFSA Journal* **13** 4036.
[20] Pravilnik o mikrobiološkim kriterijumima za hranu Sl. glasnik Republike Srpske, br 109/12
[21] EFSA (European Food Safety Authority) 2004 Opinion of the Scientific Panel on Biological Hazards on a request from the Commission related to the use of antimicrobials for the control of the *Salmonella* in poultry *EFSA Journal* **115**, 1-83
[22] Cooke B C 2002 The industrial production of safe animal feeds in Europe. Food Assurance in the Pre-harvest Phase Wageningen Academic **2** 71-86.
[23] BAS EN ISO 6579/Cor2:2010 Horizontalna metoda za utvrđivanje prisustva *Salmonella* vrsta
[24] Grimont P A D and Weill F X 2007 *Antigenic Formulae of the Salmonella Serovars* (WHO

Collaborating Centre for Reference and Research on *Salmonella* World Health Organisation & Institut Pasteur)

[25] Kirby-Bauer A 1996 Antimicrobial sensitivity testing by agar diffusion method *J Clin Pathol,* **44** 493.

[26] Bulajić S, Mijačević M and Kalaba V 2011 Nove spoznaje o uzročnicima alimentarnih obolenja, Veterinarski žurnal Republike Srpske *Veterinary Journal of Republic of Srpska* **11**, 164-171

[27] Rašeta M, Teodorović V, Bunčić O, Katić V, Lazić Branković I, Polaček V and Vidaković D. 2014 Antibiotska rezistencija i molekularna ispitivanja *Salmonella* enterica subspecies enterica serovar Infantis izolovanih u slučajevima oboljenja ljudi i sa trupova brojlera *Acta Veterinaria* **64** 257-268

[28] Rao S, Maddox C W, Hoien-Dalen P, Lanka S and Weigel RM 2008 diagnostic Accuracy of Class 1 Integron PSR method in detection of antibiotic resistance in *Salmonella* isolated from swine production systems *J Clin Microbiol* **46** 916-920.

[29] RabschW B, M Hargis, R M Tsolis, R A, Kingsley, Hinz H, Tschape H and Baumler A 2000, Competitive exclusion of *Salmonella* Enteritidis by *Salmonella* Gallinarum in poultry. *Emerg Infect Dis* **6** 443-448

[30] Cleaveland S, D T Haydon and Taylor 2007 Overviews of pathogen emergence: Which pathogens emerge, when and why? *Curr Top Microbiol Immunol* **315** 85-111

IOP Conf. Series: Earth and Environmental Science **85** (2017) 012083 doi:10.1088/1755-1315/85/1/012083

Antimicrobial-resistant bacteria in wild game in Slovenia

M Križman, A Kirbiš and U Jamnikar-Ciglenečki

Institute of Food Safety, Feed and Environment, Veterinary Faculty, University of Ljubljana, Gerbiceva 60, Ljubljana, Slovenia.

E-mail: manja.krizman@vf.uni-lj.si

Abstract. Wildlife is usually not exposed to clinically-used antimicrobial agents but can acquire antimicrobial resistance throughout contact with humans, domesticated animals and environments. Samples of faeces from intestines (80 in total) were collected from roe deer (52), wild boars (11), chamois (10) red deer (6) and moufflon (1). After culture on ChromID extended spectrum β-lactamase (ESBL) plates to select for growth of ESBL-producing bacteria, 25 samples produced bacterial colonies for further study. Six species of bacteria were identified from the 25 samples: *Stenotrophomonas maltophilia*, *Serratia fonticola*, *Stenotrophomonas nitritireducens*, *Enterococcus faecium*, *Enterococcus faecalis* and *Escherichia coli*. Two ESBL enzymes were amplified from group TEM and three from group CTX-M-1. Undercooked game meat and salami can be a source of resistant bacteria when animals are not eviscerated properly.

1. Introduction

The mere occurrence of antimicrobial resistance and corresponding resistance genes in the environment is an ancient phenomenon, due to the simple fact that most of the antimicrobial substances currently in use are based on natural compounds produced by soil bacteria like streptomycetes [1]. The function of these antibiotics was presumably more related to microbial competition for ecological niches, than the role they play today in clinical settings [2,3,4].

While various bacterial species are important in terms of multiresistance and nosocomial infections in humans and veterinary medicine, we consider extended-spectrum β-lactamase (ESBL)-producing Gram-negative bacteria like *Escherichia coli* as being key indicator pathogens to trace the evolution of multiresistant bacteria in the environment and wildlife. These multiresistant bacteria have also made their way into livestock farming and companion animals [5,6,7]. Although so far it is not clear how ESBL-*E. coli* make their way into the natural environment, it seems unlikely that pathogens isolated from wildlife have acquired resistance through new parallel mutations in respective genes. Horizontal transfer of resistance genes from clinical isolates or the intake of already resistant bacteria from human waste, sewage, or domesticated animal manure might be more probable [8,3]. Antimicrobial-resistant *E. coli* isolates originating from wildlife species were reported for the first time at the beginning of the 1980s from Japanese wild birds [9,10,11]. However, the detection of ESBL-*E. coli* from wild boar was reported from Portugal and Czech Republic [12,13].

Much of the Slovenian countryside (58.3%) is covered with forest, which is why wild game presence near farms is not unusual. In 2015, according to the Republic of Slovenia Statistical Office, hunters killed 33,668 roe deer, 8,367 wild boars, 6,064 red deer and 2,302 chamois, among other wild game species. In Slovenia, game meat is treated as high quality meat and is often served as a speciality in restaurants in the form of meat dishes and dried game meat products. For these reasons, in our study we analysed 80 samples of intestinal content of game animals for the presence of antimicrobial-resistant bacteria and resistance patterns.

2. Material and methods

2.1. Sample collection

In 2014 and 2015, a survey throughout Slovenia was performed to screen certain game animals as a potential source of enteric viruses as well as antibiotic-resistant bacteria. In total, 80 samples of game animal intestinal content were collected from five different wildlife species, comprising 52 samples from roe deer (*Capreolus capreolus*), 11 from wild boars (*Sus scrofa*), 10 from chamois (*Rupicapra rupicapra*), 6 from red deer (*Cervus elaphus*) and 1 from moufflon (*Ovis musimon*). The age of the game animals was from 5 months to 10 years and they were culled by five Slovenian hunting families between July 2014 and March 2015. Samples of intestinal content of each animal were collected by hunters after culling. The lower part of each intestine was placed in a sterile plastic bag. Samples were stored at -20°C and, as soon as possible, sent to the Veterinary Faculty where they were stored below -60°C until use.

2.2. Bacterial isolation and identification

Suspensions (10%) of the intestinal contents were prepared in RPMI-1640 (Thermo Fisher Scientific, Carlsbad, CA, USA) and centrifuged for 10 min at 1000 *g*. Volumes (0.1 ml) of each supernatant were added to 0.9 ml of peptone water (Buffered Peptone Water, Biolife, Italy) and mixed on a vibromix (Tehtnica, Železniki, Slovenia). Mixtures were then incubated for enrichment at 37°C for 16-20 h. Subsequently, enriched mixtures were inoculated, using 10 µl sterile disposable sampling loops, onto selective ChromID ESBL agar plates (BioMerieux, Marcy l'Etoile, France) and incubated for 24 h at 37°C. Representative colonies from all plates showing bacterial growth were inoculated onto blood agar (Blood agar base No.2, Oxoid, Hampshire, United Kingdom, supplemented with 5% ovine blood) and incubated for 24 h at 37°C. The next day, bacterial isolates were confirmed by matrix-assisted laser desorption/ionization, time of flight (MALDI-TOF, Bruker Daltonics, Bremen, Germany) and mass spectrometry [14].

2.3. Characterization of β-lactamases

β-lactamase resistance genes, bla_{TEM}, bla_{SHV} and $bla_{CTX\text{-}M}$, were screened by PCR [15,16]. The multiplex PCR method described by Woodford [16] did not work properly; therefore, we used the same primers in five singleplex reactions for each $bla_{CTX\text{-}M}$ group (CTX-M-1, CTX-M-2, CTX-M-8, CTX-M-9, CTX-M-25), respectively. The PCR products were analysed using electrophoresis on 2.0% agarose gels.

3. Results

3.1. Bacterial isolation and identification

Among 80 samples of intestinal contents, 25 (31.3%) exhibited bacterial growth on ChromID ESBL plates. Most positive samples belonged to roe deer 16 (61.5%), followed by wild boar 7 (25.9%) and red deer 2 (7.4 %). According to MALDI-TOF and mass spectrometry, bacterial cultures on ChromID ESBL agar plates belonged to six different bacterial species *Stenotrophomonas maltophilia*, *Serratia fonticola*, *Stenotrophomonas nitritireducens*, *Enterococcus faecium*, *Enterococcus faecalis* and *E. coli*.

3.2. Detection of ESBL-producing bacteria and characterisation of ESBL

Among the 25 positive bacterial isolates on ChromID ESBL plates, two (2.5%) were positive for the ESBL enzyme group, TEM. Both isolates originated from samples collected from wild boars; one isolate was identified as *Enterococcus faecium* and the other as *Enterococcus faecalis*. Both intestinal content samples came from the same hunting family, based in the town of Škofja Loka. In Figure 1, typical positive PCR products are shown.

Bacteria producing ESBL from the ESBL enzyme group CTX-M-1 were detected in 4 out of 80 (5.0%) samples of intestinal contents. All isolates were obtained from intestinal contents of roe deer. Two isolates were identified as *Serratia fonticola,* one as *Stenotrophomonas maltophilia* and one as *Stenotrophomonas nitritireducens.* Three of these intestinal content samples were collected by the hunters from Škofja Loka while one sample came from a hunting family from Polhov Gradec (Table 1). Both hunting families hunt in adjoining territories, on the terrain of central and Gorenjska regions.

Table 1 Overview of the detected ESBL-producing isolates.

Isolate number	Animal species	Appearance of isolates on ChromID ESBL plate	MALDI –TOF identification	ESBL enzyme group	Hunting family
17/14	Roe deer	Blue colonies	*Serratia fonticola*	CTX-M-1	Škofja Loka
18/14	Roe deer	Blue colonies	*Serratia fonticola*	CTX-M-1	Škofja Loka
20/14	Roe deer	Yellow colonies	*Stenotrophomonas nitritireducens*	CTX-M-1	Škofja Loka
30/14	Wild boar	White-blue colonies	*Enterococcus faecium*	TEM	Škofja Loka
34/14	Wild boar	White-blue colonies	*Enterococcus faecalis*	TEM	Škofja Loka
67/14	Roe deer	White colonies	*Stenotrophomonas maltophilia*	CTX-M-1	Polhov Gradec

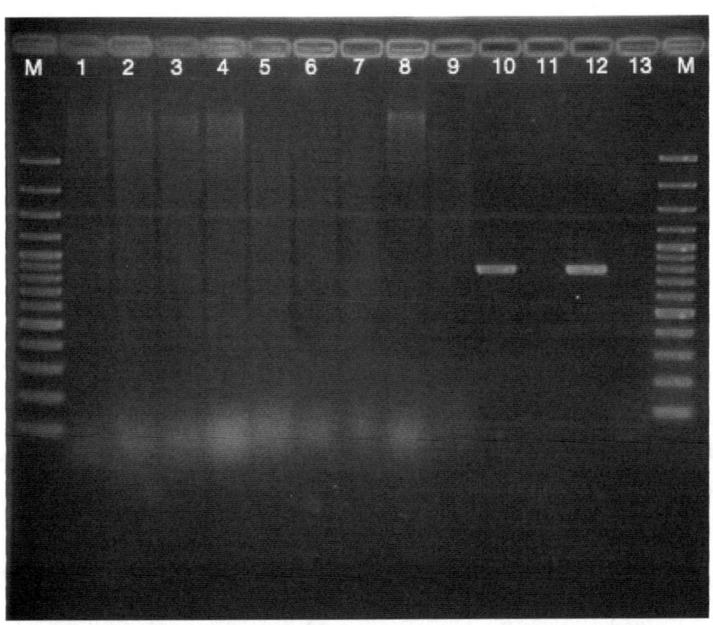

Figure 1. Agarose gel electrophoresis of amplified DNA of *bla*$_{TEM}$ gene of *Serratia fonticola* on 2% agarose gel. (line 10 (30/14) and 12 (34/14) amplified *bla*$_{TEM}$); DNA marker contains bands (from bottom to the top) of 100bp, 200bp, 300bp, 400bp, 500bp, 600bp, 700bp, 800bp, 900bp, 1000bp, 1200bp, 1500bp, 2000bp and 3000bp.

4. Discussion

The results show that ESBL-producing bacteria are present in wild animals. However, they reveal that carriage of these multiresistant bacteria is not widespread among wild game in Slovenia. We cultivated 25 (31.3%) multiresistant bacterial isolates that grew on ChromID ESBL agar plates, and only in five isolates was a resistance gene amplified. The reasons could be that the resistance is chromosomal in origin or some new genes are involved. Most of our multiresistant isolates were common soil bacteria; however, we isolated also two strains of *S. fonticola*. Some species of the genus *Serratia* have medical significance as opportunistic pathogens [17,18]. In addition, we isolated *E. faecium*, which can be commensal in the human intestine but it can also be pathogenic, causing diseases such as neonatal meningitis or endocarditis. From the genus *Enterococcus*, we also isolated *E. faecalis*. *E. faecalis* is a commensal bacterium inhabiting the gastrointestinal tracts of humans and other mammals. However, it can cause life-threatening infections in humans, especially in the nosocomial (hospital) environment, where the naturally high levels of antibiotic resistance found in *E. faecalis* contribute to its pathogenicity [19].

The results show that undercooked game meat and dried game meat products could be a source of multiresistant bacteria if evisceration of game is not conducted properly in the field. In addition, a further study on antimicrobial-resistant bacteria in game meat should be performed to determine the true prevalence of multiresistant bacteria with greater certainty.

5. Conclusion

As previously suggested, thorough spatial and temporal studies of antimicrobial resistance in different natural habitats are warranted [20,21] to fully understand the importance of wildlife as a source of antimicrobial drug resistance.

In our study, we only determined the ESBL enzyme group. Therefore, the next step is to perform nucleotide sequencing from these PCR products from the 25 bacterial isolates to determine the nature of their antibiotic resistance genes.

Acknowledgements

This work was supported by the Slovenian Ministry for Higher Education, Science and Technology (Research program P4-0092). The authors thank Rok Drešar, Janez Hafner, Samo Kumar, Nejc Nastran, Jože Podmeninšek and Gašper Pustovrh for providing the samples.

References

[1] D`Costa V M, et al 2011 Antibiotic resistance is ancient. *Nature* **477** 457
[2] Martinez J L 2009a Environmental pollution by antibiotics and by antibiotic resistance determinants. *Environ. Pollut.* **157** 2893
[3] Martinez J L 2009b The role of natural environments in the evolution of the resistance traits in pathogenic bacteria. *Proc. Biol. Sci.* **276** 2521
[4] Allen H K, Donato J, Wang H H, Cloud-Hansen K A, Davies J and Handelsman J 2010 Call of the wild: antibiotic resistance genes in natural environments. *Nat. Rev. Microbiol.* **8** 251
[5] Smet A, Martel A, Persoons D, Dewulf J, Heyndrickx M, Herman L, Haesebrouck F and Butaye 2010 Broad-spectrum beta-lactamases among *Enterobacteriaceae* of animal origin:

molecular aspects, mobility and impact on public health. *FEMS Microbiol. Rev.* **34** 95

[6] Ewers C, Grobbel M, Bethe A, Wieler L H and Guenther S 2011 Extended-spectrum beta-lactamases-producing gram-negative bacteria in companion animals: action is clearly warranted! *Berl. Munch. Tierarztl. Wochenschr.* **124** 10

[7] Wieler L H, Ewers C, Guenther S, Walther B and Lübke-Becker A 2011 methicillin-resistant staphylococci (MRS) and extended spectrum-beta lactamases (ESBL)-producing *Enterobacteriaceae* in companion animals: nosocomial infections as one reason for the rising prevalence of these potential zoonotic pathogens in clinical samples. *Int. J. Med. Microbiol.* **301** 635

[8] Kummerer K 2009 Antibiotics in the aquatic environment – a review – part I. *Chemosphere* **75** 417

[9] Sato G, Oka C, Asagi M and Ishiguro N 1978 Detection of conjugative R plasmids conferring chloramphenicol resistance in *Escherichia coli* isolated from domestic and feral pigeons and crows. *Zentralbl. Bakteriol. Orig.* **241** 407

[10] Kanai H, Hashimoto H and Mitsuhashi S 1981 Drug resistance and conjugative R-plasmids in *Escherichia coli* strains isolated from wild birds (Japanese tree sparrows, green pheasants and bamboo partridges). *Jpn. Poult. Sci.* **18** 234

[11] Tsubokura M, Matsumoto A, Otsuki K, Animas S B and Sanekata T 1995 Drug resistance and conjugative R plasmids in *Escherichia coli* strains isolated from migratory waterfowl. *J. Wildl. Dis.* **31** 352

[12] Poeta P, et al., 2009 Wild boars as reservoirs of extended-spectrum beta-lactamases (ESBL) producing *Escherichia coli* of different phylogenetic groups. *J. Basic Microbiol.* **49** 584

[13] Literak I, Dolejska M, Radimersky T, Klimes J, Friedman M, Aarestrup F M, Hasman H and Cizek A 2009 Antimicrobial-resistant fecal *Escherichia coli* in wild mammals in central Europe: multiresistant *Escherichia coli* producing extended-spectrum beta-lactamases in wild boars. *J. Appl. Microbiol.* **108** 1702

[14] Kohlmann R, Hoffmann A, Geis G and Getermann S 2015. MALDI-TOF mass spectrometry following short incubation on a solid medium is a valuable tool for rapid pathogen identification from positive blood cultures. *Int. J. Med. Microbiol.* **305** 469

[15] Dallenne C, Da Costa A, Decré D, Favier C and Arlet G 2010 Development of a set of multiplex PCR assays for the detection of genes encoding important β-lactamases in *Enterobacteriaceae*. *J. Antimicrob. Chempther.* **65** 490

[16] Woodford N 2010 Rapid characterization of beta-lactamases by multiplex PCR. *Methods Mol. Biol.* (Clifton, N.J.) **642** 181

[17] Farmer J J, Davis B R, Hickman-Brenner F W, McWhorter A, Huntley-Carter G P, Asbury M A, Riddle C, Wathen-Grady H G, Elias C and Fanning G R 1985 Biochemical identification of new species and biogroups of *Enterobacteriaceae* isolated from clinical specimens. *J. Clin. Microbiol.* **21** 46

[18] Bollet C, Gainnier M, Sainty J M, Orhesser P and De Micco P 1991 *Serratia fonticola* isolated from a leg abscess. *J. Clin. Microbiol.* **29** 834

[19] Agudelo Higuita N I, Huycke M M 2014 Enterococcal Disease, Epidemiology, and Implications for Treatment. In: Gilmore MS, Clewell DB, Ike Y, et al., editors. Enterococci: From Commensals to Leading Causes of Drug Resistant Infection [Internet]. Boston: Massachusetts Eye and Ear Infirmary; 2014. Available from: https://www.ncbi.nlm.nih.gov/books/NBK190429/

[20] Gilliver M A, Bennett M, Begon M, Hazel S M and Hart C A 1999 Antibiotic resistance found in wild rodents. *Nature* **401** 233

[21] Hernandez J, Bonnedahl J, Eliasson I, Wallensten A, Comstedt P, Johanson A, Granholm S, Melhus A, Olsen B and Drobni M 2010 Globally disseminated human pathogenic *Escherichia coli* of O25b-ST131 clone, harbouring blaCTX-M15, found in glaucous-winged gull at remote Commander Islands, Russia. *Environ. Microbiol. Rep.* **2** 329

IOP Conf. Series: Earth and Environmental Science **85** (2017) 012084 doi:10.1088/1755-1315/85/1/012084

Effect of vacuum and modified atmosphere packaging on microbiological properties of cold-smoked trout

J Đorđević[1], N Pavlićević[2], M Bošković[1], J Janjić[1], M Glišić[1], M Starčević[3] and M Ž Baltić[1]

[1] Faculty of Veterinary Medicine, University of Belgrade, Bulevar oslobodjenja 18, 11000 Belgrade, Serbia
[2] Veterinary Specialty Institute, Segedinski put 88, 24000 Subotica, Serbia
[3] Serbian Armed Forces

Email: jasnalonce@gmail.com

Abstract. Because of the importance of different packaging methods for the extension of fish shelf life, as a highly perishable food, the aim of the present study was to examine the effect of vacuum and modified atmosphere packaging on the total *Enterobacteriaceae* and lactic acid bacteria counts of cold-smoked Salmon trout (*Oncorhynchus mykiss*) stored at 3°C during six weeks. Trout fillets were vacuumed packaged (VP) or packaged in one of two different modified atmospheres, with gas ratio of 50%CO_2/50%N_2 (MAP1) and 90%CO_2/10%N_2 (MAP2) and analysed on days 0, 7, 14, 21, 28, 35 and 42. Both the total *Enterobacteriaceae* and total lactic acid bacteria counts increased in the trout fillets in all packaging types during storage. A significantly lower total *Enterobacteriaceae* count was determined in the MAP fish compared to the VP fish, with the weakest growth rate and lowest numbers attained in MAP2 fillets. The lactic acid bacteria count was higher in trout packaged in MAP compared to VP, with the highest number in the MAP with 90% CO_2 (MAP2).

1. Introduction

Fish are gradually becoming the favoured food in many countries because of their favourable content of proteins, minerals, vitamins and essential fatty acids, as well as their positive ratio of n-6/n-3 fatty acids [1]. However, fish is highly perishable. Quality loss and subsequent spoilage of fish and fish products are mainly the result of microbial spoilage [2]. Fish preservation, including smoking as one of the oldest preservation methods, dates from prehistory, and is also used in today's production process [3]. Because the modern consumer is looking for safe, high-quality food, with retained sensory characteristics and nutritional value of the raw material, as well as longer shelf life, various new packaging systems to delay the spoilage and extend the shelf life of fish and meat and their products were developed in the last decade [4].

In vacuum packaging (VP), air removal ensures anaerobic/microaerophilic conditions, increases the CO_2 and reduces pH of the product. O_2 in the packaging is converted into CO_2 due to respiration of meat tissue and bacterial activity. This prevents the growth of aerobic bacteria and allows the growth of facultative anaerobes. Regular vacuuming extends shelf life up to three weeks, but the food can become dry [5]. Therefore, food packaging in a gas mixture, i.e. modified atmosphere packaging (MAP), became the leading packaging technology in the 21st century. MAP basically acts similarly to VP, with the difference that in VP, the gas composition that inhibits microorganisms develops in the package during storage, while in MAP, the gas mixture in the package is modified in order to create the same conditions. MAP can contain a mixture of different gases (O_2, CO_2, N_2, CO, etc.), is very popular for fresh meats and is frequently used because it suppresses bacterial spoilage. Depending on the type of meat or meat product, the gases in the packaging can be used individually or can be combined in different proportions in order to achieve the desirable effect. The most frequently used gas in MAP is CO_2, which possesses strong antimicrobial activity, especially on Gram negative bacteria, resulting in extended bacterial lag phase and generation time. Its inhibitory effect on the growth of microorganisms is concentration-dependent, and the higher the concentration of CO_2 in the system, the higher is the inhibitory effect observed [6]. For fish and fishery products, most gas mixtures do not include O_2 because high-O_2 systems, in most cases, provide limited benefit to shelf life extension of fishery products. The reduction of O_2 slows down lipid oxidation and the development of rancidity [2]. The role of N_2, also used in MAP systems, is to prevent fat oxidation and package collapse [7,8].

Because of the importance of the different packaging methods for extension of fish shelf life as well different amounts and types of gases in MAP, the aim of the present study was to investigate the effect of VP and MAP with two different gases on microbiological properties (total *Enterobacteriaceae* and lactic acid bacteria counts) of cold-smoked trout.

2. Materials and methods

2.1. Cold-smoked trout preparation

Salmon trout (*Oncorhynchus mykiss*) used in the present study originated from the Bočac fish farm (Tropic, Banja Luka, Republika Srpska, Bosnia and Herzegovina), where all fish were grown and kept under the same conditions. Live fish were transported from the fish farm to the processing plant in specialised vehicles and placed into the open pools at the processing plant yard, with necessary conditions for proper accommodation of the fish (constant water flow, controlled oxygen and temperature), without feed. Trout were primarily processed, stunned, bleed and eviscerated in the manner usual for industrial plant. Fish weighed about 1 kg. After primarily processing, the spine was removed by cutting parallel to it, whereby the left and right halves were obtained. Trout halves were rinsed with water and immersed in barrels with 9% salt brine (wet salting) and rosemary (*Quantum satis*) during 24 hours at 4°C. After brining, fish were drained for one hour at 20°C and then smoked for eight hours in an automatic smoked chamber at 28°C. Smoke was produced by combustion of beech sawdust in a generator separated from the smoking chamber. After smoking, fish were cooled at 2°C for ten hours, then residual parts of the ribs were removed and fish halves were sliced from the medial side, into thin fillets, up to 0.5cm thick, each weighing about 75g.

2.2 Cold-smoked trout fillet packaging

Cold-smoked fillets were split into three groups. The first group was vacuum packaged (VP), the second was packaged in modified atmosphere with 50% CO_2 and 50% N_2 (MAP1) and the third group was packaged in modified atmosphere with 90% CO_2 and 10% N_2 (MAP2). MAP was conducted using a Multivac machine (Multivac C350, D-87787 Wolfertschwenden, Germany). The packaging foil was OPA/EVOH/PE (oriented polyamide/ethylene vinyl alcohol/polyethylene, UPM-Kymmene, Walki Films, Finland) with low gas permeability (degree of permeability to oxygen $5cm^3\ m^{-2}\ day^{-1}$ at 23°C, nitrogen $1cm^3\ m^{-2}\ day^{-1}$ at 23°C and water vapour $15g\ m^{-2}\ day^{-1}$ at 38°C). Packages were filled with commercial gas mixture (Messer Tehnogas, Serbia). The ratio of gas:fish in the packages was 2:1. After packaging, all groups of the fish fillets were stored at 3°C for six weeks.

2.3. Microbiological analyses

Cold-smoked trout packaged in VP or MAP was analysed for total *Enterobacteriaceae* count and lactic acid bacteria (LAB) count on day 0 and on days 7, 14, 21, 28, 35 and 42 of storage. For bacterial enumeration, 10g of fish meat was weighed out aseptically after package opening, transferred into sterile Stomacher bags and 90 ml of Maximum Recovery Diluent (MRD; Merck, Germany) was added to each sample. Samples were homogenized in a Stomacher blender (Stomacher 400 Circulator, Seward, UK) for 2 min. Serial decimal dilutions were prepared and 0.1 ml of appropriately diluted suspension was inoculated directly on the surface of appropriate media. *Enterobacteriaceae* were enumerated on Violet Red Bile Glucose Agar (VRBGA, Merck, Germany) after incubation at 37°C for 24h according to ISO 21528-2:2004 [9], while LAB were enumerated on MRS (Merck, Germany) following incubation at 30°C for 72h according to ISO 15214:1998 [10]. After an appropriate period of incubation for each type of bacteria, plates were examined visually for typical colony types and morphological characteristics associated with each growth medium, number of colonies was counted, and results were recorded as colony forming units per g (CFU/g).

2.4. Statistical analyses

The experiment was conducted in a completely randomized design, six repetitions were carried out for each treatment and the treatments were arranged in a 3 x 7 factorial design (3 treatments, 7 storage periods). Numbers of microorganism were transformed into logarithms (log). Statistical analyses of the results were conducted using the software GraphPad Prism version 6.00 (GraphPad Software, San Diego, California USA, www.graphpad.com). The results were expressed as the mean±standard deviation. The effects of different treatments during storage period were appraised by one-factor analysis of variance-ANOVA with Tukey's multiple comparison test at 95% confidence level (difference considered significant if $p<0.05$). Linear regression was used to establish the statistical relationship between the number of microorganisms in different packaging and storage period.

3. Results and Discussion

The total *Enterobacteriaceae* and LAB counts and significant differences between all groups of cold-smoked trout fillets during six weeks of storage are shown in Table 1.

Table 1. Total *Enterobacteriaceae* count and lactic acid bacteria (LAB) count (log CFU/g) in smoked trout fillets during six weeks of storage (mean±SD)

Day of	Total *Enterobacteriaceae* count			LAB count		
Storage	VP	MAP1	MAP2	VP	MAP1	MAP2
0	1.65^{A}±0.20	2.51^{A}±0.24	2.03±0.29	1.03^{A}±0.25	1.38^{A}±0.29	1.28±0.21
7	2.99^{AB}±0.82	2.29^{A}±0.41	2.23^{B}±0.49	2.17^{A} ±0.32	1.77±0.47	1.57^{A}±0.39
14	4.67^{AB}±0.19	2.87^{A}±0.80	2.33^{B}±0.15	2.47^{AB}±0.42	1.64^{A}±0.46	1.50^{B}±0.40
21	5.70^{AB}±0.49	3.46^{A}±0.78	3.15^{B}±0.50	3.58^{AB}±0.22	4.40^{AC}±0.14	4.59^{BC}±0.10
28	6.08^{AB}±0.15	1.91^{AC}±0.30	3.51^{BC}±0.56	5.03^{A}±0.26	4.88^{B}±0.10	6.01^{AB}±0.56
35	6.36^{A}±0.31	5.93^{B}±0.37	2.86^{AB}±0.20	5.09^{AB}±0.45	7.02^{A}±0.19	7.11^{B}±0.16
42	7.01^{A}±0.53	6.60^{B}±0.47	4.75^{AB}±0.17	6.20^{AB}±0.30	6.88^{AC}±0.38	7.32^{BC}±0.19

$^{A-C}$ same uppercase letter within rows within the same bacterial group indicates significant difference $p<0.05$ between different packaging types

In all packaging types, significant increases of the total *Enterobacteriaceae* count were noted, which can also be seen based on the regression equation of each packaging type (Figure 1). The positive values of coefficient b in the regression equations for all packaging types indicate the trend of constantly increasing total *Enterobacteriaceae* counts. Coefficient b had the highest value in the regression equation for the VP fillets, and the lowest for the MAP2 fillets. This indicated the weakest growth rate of *Enterobacteriaceae* occurred in MAP2 fillets, so this type of packaging provided the most pronounced antimicrobial effect of CO_2.

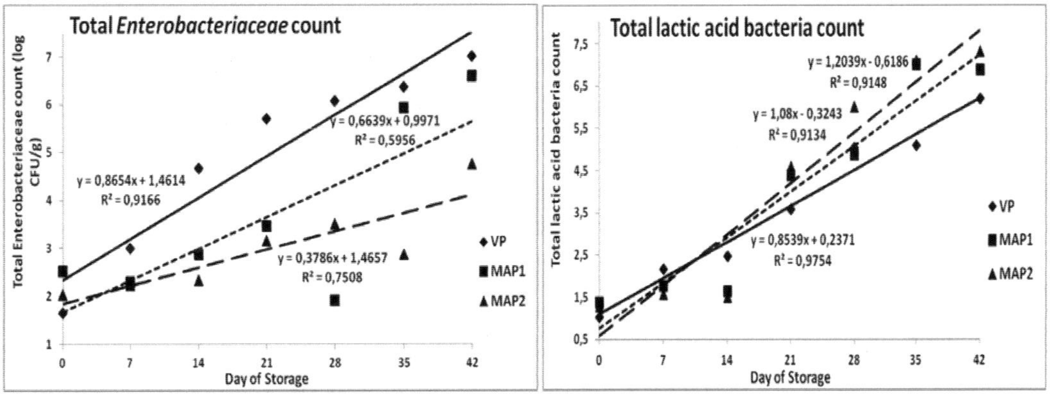

Figure 1. Regression equations for total *Enterobacteriaceae* and lactic acid bacteria counts (log CFU/g) in smoked trout fillets during six weeks of storage

On the one hand, some authors claim that the spoilage of fish meat is correlated with the *Enterobacteriaceae* count and that the *Enterobacteriaceae* are the main cause of spoilage of fish meat and cold-smoked fish meat products, because of the ability of this family of microorganisms to grow at low temperatures [11,12]. On the other hand, some other authors consider that *Enterobacteriaceae* count cannot be the only indicator of spoilage, and that it can only be used as an indicator of the quality of fish and cold-smoked fish products [3]. The initial number of *Enterobacteriaceae* which will be present in meat is mostly affected by hygiene during the production process, but also the types and forms of meat. In spoilage, these organisms degrade amino acids to sulphur compounds and diamines (volatile, malodorous components), which affect sensory properties of the meat, on the basis of which the final estimate of the spoilage is made [12,13]. The results of the present study showed total *Enterobacteriaceae* counts were between 6 and 7 log CFU/g in the fish from VP and MAP1, while in the fish from MAP2, numbers were lower (below 5 log CFU/g) at the moment when the trout fillets were considered unacceptable on the basis of sensory analysis (data not shown; [14]). Although the total *Enterobacteriaceae* count found in the present study was far more than 10^3 log CFU/g, which according to the Dutch standard was considered as limit value for the total number of *Enterobacteriaceae* in fresh fish [15,16], results are in accordance with the fact that CO_2, primarily in high concentrations, acts inhibitively on Gram negative bacteria, including *Enterobacteriaceae*. In accordance with the present study are results of others [17,18,19,20,21], in which the influence of MAP on fresh fish was examined, and which came to the conclusion that high levels of CO_2 in the package (>50%) have a protective effect on the product and consequently, reduce both the total *Enterobacteriaceae* count as well as the total number of bacteria.

On the basis of the positive values of the coefficient b in the regression equations, it can be seen that during the storage, the total number of LAB showed a tendency to increase in fish fillets in all packaging types (Figure 1). These results are in accordance with research of Arkoudelos *et al.* [19], Dondero *et al.* [22] and Masniyom *et al.* [23].

The dominant microbiota of fish cold-smoked products packaged in a mixture of gases undoubtedly depends on the gas composition. The dominant microbiota in cold-smoked fish products packaged in a modified atmosphere with CO_2 is CO_2-resistant microbiota [6]. Gram positive bacteria such as LAB, primarily *Lactobacillus* spp., *Leuconostoc* spp. and *Brochothrix thermosphacta*, are not sensitive to CO_2, which is why they become dominant in MAP [24-27]. Also, the fact that LAB adapt very quickly to the environmental conditions (salt content, pH value, cold storage regime) enable these bacteria to become the dominant microbiota in cold-smoked fish products during storage [3]. Although LAB are not typical causative agents of spoilage of cold-smoked fish products, when they are in greater numbers they can affect the sensory properties by creating volatile components, and consequently create typical odour defects. Even so, Patsias *et al.* [28] consider that the spoilage that occurs due to an increase in LAB numbers is less offensive than in the case of spoilage caused by the growth of aerobic bacteria. The results of the present study showed higher LAB counts in fish packaged in MAP compared to VP, with the highest number in the trout packed in MAP with 90% CO_2. The explanation may be the fact that CO_2 inhibited the activity of some groups of potentially competitive microorganisms, which led to uninterrupted growth of LAB. Competition for substrates and microbial antagonism are considered to be the most important factors in the survival of specific microorganisms/microbiota in a particular ecological niche [29]. The lack of competition among the various bacterial species, due to the environment conditions, likely promoted undisturbed growth of LAB on fillets packaged in MAP, which is the case in the present study [3].

Acknowledgment

This paper was supported by the Ministry of Education, Science and Technological Development, Republic of Serbia, through the funding of the Project – Selected biological hazards to the safety/quality of food of animal origin and the control measures from farm to consumer (No 31011).

References

[1] Baltic ZM, Kilibarda N, Teodorovic V, Dimitrijevic M., Karabasil N and Djuric J 2010 Značaj ribe u ishrani ljudi. *XXX seminar inovacije znanja veterinara, Fakultet veterinarske medicine, Beograd, Zbornik predavanja* pp. 39-58
[2] DeWitt CAM and Oliveira A 2016 Modified atmosphere systems and shelf life extension of fish and fishery products *Foods* **5**(3) 48
[3] Gómez-Guillén MC, Gómez-Estaca J, Giménez B and Montero P 2009 Alternative fish species for cold-smoking process *Int. J. Food Sci. Technol.* **44**(8) 1525-1535
[4] Bøknæs N, Jensen KN, Guldager HS, Østerberg C, Nielsen J and Dalgaard P 2002 Thawed chilled Barents Sea cod fillets in modified atmosphere packaging-application of multivariate data analysis to select key parameters in good manufacturing practice. *LWT-Food Sci. Technol.* **35**(5) 436-443
[5] Radetić P, Milijašević M, Jovanović J and Velebit B 2007 Pakovanje svežeg mesa u modifikovanoj atmosferi-trend koji traje *Tehnologija mesa* **48**(1-2) 99-108
[6] Sivertsvik M, Jeksrud WK and Rosnes JT 2002 A review of modified atmosphere packaging of fish and fishery products–significance of microbial growth, activities and safety. *Int. J. Food Sci. & Technol.* **37**(2) 107-127
[7] Skandamis PN and Nychas GJE 2002 Preservation of fresh meat with active and modified atmosphere packaging conditions. *Int. J. Food Microbiol.* **79**(1) 35-45
[8] Singh P, Wani AA, Saengerlaub S and Langowski HC 2011 Understanding critical factors for the quality and shelf-life of MAP fresh meat: a review *Crit. Rev. Food Sci. Nutr.* **51**(2) 146-177
[9] ISO 21528-2:2004 – Microbiology of food and animal feeding stuffs – Horizontal methods for the detection and enumeration of Enterobacteriaceae – Part 2: Colony count method.
[10] ISO 15214:1998 – Microbiology of food and animal feeding stuffs – Horizontal method for the enumeration of mesophilic lactic acid bacteria – Colony-count technique at 30 degrees C.
[11] Ólafsdóttir G, Chanie E, Westad F, Jonsdottir R, Thalmann CR, Bazzo S, Labreche S, Marcq P, Lundby F and Haugen JE 2005 Prediction of microbial and sensory quality of cold smoked Atlantic salmon (*Salmo salar*) by electric nose *J. Food Sci.* **70**(9) 563-574
[12] Säde, E, Murros, A and Björkroth J 2013 Predominant enterobacteria on modified-atmosphere packaged meat and poultry *Food Microbiol.* **34**(2) 252-258
[13] El-Gendy NM, Ibrahim HA, Al-Shabasy NA and Samaha IA 2014 *Enterobacteriaceae* in beef products (Luncheon, Pasterma, Frankfurter and Minced meat) from Alexandria retail outlets *Alexandria J. Vet. Sci.* **41**(1) 80-86

[14] Djordjevic J, Pavlićevic N, Boškovic M, Baltic T, Ivanovic J, Glamoclija N and Baltic ZM 2016 Sensory evaluation of cold-smoked trout packaged in vacuum and modified atmosphere. *Meat Technol.* **57**(2) 141-149
[15] Lyhs U, Lahtinen, J and Schelvis-Smit R 2007 Microbiological quality of maatjes herring stored in air and under modified atmosphere at 4 and 10 C *Food Microbiol.* **24**(5) 508-516
[16] Vaz-Pires P and Barbosa A 2004 Sensory, microbiological, physical and nutritional properties of iced whole common octopus (*Octopus vulgaris*) *LWT - Food Sci. Technol.* **37** (1) 105-114
[17] Torrieri E, Cavella S, Villani F and Masi P 2006 Influence of modified atmosphere packaging on the chilled shelf life of gutted farmed bass (*Dicentrarchus labrax*) *J. Food Eng.* **77**(4) 1078-1086
[18] Ježek F and Buchtová H 2007 Physical and chemical changes in fresh chilled muscle tissue of common carp (*Cyprinus carpio L.*) packed in a modified atmosphere *Acta Vet. Brno* **76**(8) 83-92
[19] Arkoudelos J, Stamatis N and Samaras F 2007 Quality attributes of farmed eel (*Anguilla anguilla*) stored under air, vacuum and modified atmosphere packaging at 0 degrees C *Food Microbiol.* **24**(7) 728-735
[20] Pantazi D, Papavergou A, Pournis N, Kontominas MG and Savvaidis IN 2008 Shelf-life of chilled fresh Mediterranean swordfish (*Xiphias gladius*) stored under various packaging conditions: Microbiological, biochemical and sensory attributes *Food Microbiol* **25**(1) 136-143
[21] Stamatis N and Arkoudelos J 2007 Quality assessment of *Scomber colias japonicus* under modified atmosphere and vacuum packaging *Food Control* **18**(4) 292-300
[22] Dondero M, Cisternas F, Carvajal L, Simpson R 2004 Changes in quality of vacuum-packed cold-smoked salmon (*Salmo salar*) as a function of storage temperature *Food Chem.* **87** 543-550
[23] Masniyom P, Benjakul S and Visessanguan W 2002 Shelf-life extension of refrigerated seabass slices under modified atmosphere packaging *J. Sci. Food Agric.* **82** 873-880
[24] Limbo S, Torri L, Sinelli N, Franzetti L and Casiraghi E 2010 Evaluation and predictive modelling of shelf life of minced beef stored in high-oxygen modified atmosphere packaging at different temperatures *Meat Sci.* **84** 129-136
[25] McMillin KW 2008 Where is MAP going? A review and future potential of modified atmosphere packaging for meat *Meat Sci.* **80** 43-65
[26] Sanjeev K and Ramesh MN 2006 Low oxygen and inert gas processing of foods *Crit. Rev. Food Sci. Nutr.* **46**(5) 423-451
[27] Sivertsvik M, Rosnes TJ and Kleiberg GH 2003 Effect of Modified Atmosphere Packaging and Superchilled Storage on the Microbial and Sensory Quality of Atlantic Salmon (Salmo salar) Fillet *Food Microbiol. Safety* **68**(4) 1467-1472
[28] Patsias A,Chouliara I, Badeka A, Savvaidis IN and Kontominas MG 2006 Shelf-life of a chilled precooked chicken product stored in air and under modified atmospheres: microbiological, chemical, sensory attributes *Food Microbiol.* **23** 423–429
[29] Gram L and Huss HH 1996 Microbiological spoilage of fish and fish products *Int. J. Food Microbiol.* **33**(1) 121-137

IOP Conf. Series: Earth and Environmental Science **85** (2017) 012085 doi:10.1088/1755-1315/85/1/012085

Somatic cell counts in bulk milk and their importance for milk processing

N R Savić[1], D P Mikulec[2], R S Radovanović[3]

[1] Eko-lab Ltd, Padinska Skela, Serbia

[2] Scientific Veterinary Institute of Serbia Beograd, Department for food hygiene, Autoput 3, New Belgrade, Serbia

[3] Department of Food Hygiene and Technology, Faculty of Veterinary Medicine, University of Belgrade, Bulevar Oslobođenja 18, 11000 Belgrade, Serbia

E-mail: mimica@vet.bg.ac.rs

Abstract. Bulk tank milk somatic cell counts are the indicator of the mammary gland health in the dairy herds and may be regarded as an indirect measure of milk quality. Elevated somatic cell counts are correlated with changes in milk composition The aim of this study was to assess the somatic cell counts that significantly affect the quality of milk and dairy products. We examined the somatic cell counts in bulk tank milk samples from 38 farms during the period of 6 months, from December to the May of the next year. The flow cytometry, Fossomatic was used for determination of somatic cell counts. In the same samples content of total proteins and lactose was determined by Milcoscan. Our results showed that average values for bulk tank milk samples were 273,605/ml from morning milking and 292,895/ml from evening milking. The average values for total proteins content from morning and evening milking are 3,31 and 3,34%, respectively. The average values for lactose content from morning and evening milking are 4,56 and 4,63%, respectively. The highest somatic cell count (516,000/ml) was detected in bulk tank milk sample from evening milk in the Winter and the lowest content of lactose was 4,46%. Our results showed that obtained values for bulk tank milk somatic cell counts did not significantly affected the content of total proteins and lactose.

1.Introduction

Bovine mastitis is the most prevalent and costly disease, affecting dairy farms worldwide. Economic losses associated with mastitis derive mainly from a decrease in milk production and to a lesser extent, from the culling of chronically infected cows, cost of veterinary treatment, and penalties on milk quality [1]. Bovine mastitis is characterized by inflammation of the mammary gland. The inflammation severity depends on the causative agent and the host response. Resident and recruited cells together play an essential role in immediate defense against local infection [2]. Somatic cells are therefore a reflection of the inflammatory response to an intramammary infection. The somatic cell counts (SCC) for an uninfected quarter is approximately 70, 000 cells. There is variation around this mean, so its value can increase with age, decreasing milk production and days in milk period [3]. Bulk tank SCC is a general indicator of the udder health in a herd and it is also regarded as an indirect measure of milk quality [4]. Elevated SCC, are correlated with changes in milk composition, casein and more serum-derived whey proteins, as well as increased proteolytic and lipolytic activities [5]. Researches has defined the "gold standard" somatic cell counts up to 100,000 cells / ml, and values greater than 100,000 cells / ml are considered to result in the reduction of production and quality of milk during processing [6]. The increase in the number of somatic cells in the milk is followed by change in the chemical composition of milk by lowering casein, lactose, calcium, and an increase of sodium, chloride and serum proteins. In dairy herds, where the somatic cell counts are greater than 300,000/ml milk, are proposed to be present a mastitis problem in herd, and such milk is less suitable for the processing [7] Due to the reduced secretory capacity of the mammary gland cells in mastitis, the percentage of lactose in milk is decreased. In addition, milk with high somatic cell counts can contain microorganisms whose presence in milk can pose a risk to human health especially to vulnerable population [8] The aim of this study was to assess the number of somatic cells that significantly affect the quality of milk.

2. Materials and methods

The material represented 38 bulk milk samples of morning and evening milking from 38 farms during the period of 6 months, from December to the May of the next year. The samples were taken from the tank, not preserved and the analyses started in the laboratory in 6 hours after sampling. The content of total proteins, lactose and somatic cell counts were determined in the samples. Somatic cell counts were determined by flow-cytometry by Fossomatic 5,000. The total content of proteins and lactose in bulk tank milk samples were determined by Milcoscan.

3. Results and discussion

The results of determination of total proteins, lactose content and SCC counts from morning milking and evening milking are presented in the Table 1 and 2.

Table 1 Results of determination of total proteins, lactose content and somatic cell counts from morning milking

Stat. param.	Total proteins (%)	Lactose (%)	SCC/ml
Average	3,31	4,56	273,605
Standard dev.	0,11	0,05	80,357
Min	3,02	4,49	75,000
Max	3,47	4,67	446,000
Varition (%)	3,21	1,03	29,370

Stat. param.	**Total proteins (%)**	**Lactose (%)**	**SCC/ml**
Average	3,31	4,56	273,605
Standard dev.	0,11	0,05	80,357
Min	3,02	4,49	75,000
Max	3,47	4,67	446,000
Varition (%)	3,21	1,03	29,370

The obtained results showed that the average value for SCC in the winter from morning milking was 273,605 cells per ml, and the content of total proteins and lactose was 3.31% and 4.56% respectively. The somatic cell counts ranged from 75 to 446,000/ml. Nineteen bulk tank milk samples had lower somatic cell count from 250,000 cells /ml and only 1 sample had a value of less than 100,000 and 3 samples of over 400,000 (Table 3).

Table 2 Results of determination of total proteins, lactose content and somatic cell counts from evening milking

Stat. param.	**Total proteins (%)**	**Lactose (%)**	**SCC/ml**
Average	3,34	4,63	292,895
Standard dev.	0,19	0,23	72,743
Min	3,06	4,47	168,000
Max	4,41	5,56	516,000
Varition (%)	5,89	4,97	24,836

Stat. param.	Total proteins (%)	Lactose (%)	SCC/ml
Average	3,34	4,63	292,895
Standard dev.	0,19	0,23	72,743
Min	3,06	4,47	168,000
Max	4,41	5,56	516,000
Varition (%)	5,89	4,97	24,836

The obtained results showed that the average value for SCC in the winter from evening milking was 292.895 cells per ml, and the content of total proteins and lactose was 3.34% and 4.63% respectively. The somatic cell counts ranged from 168 to 516 cells × 10^3 / ml. Twelve samples of raw milk had lower somatic cell count of 250,000 cells /ml. in no one saple the number was lower than 100,000., and in four samples the number was higher than 400,000/ml. Total protein content ranged from 3.06 to 4.41, and lactose from 4.47 to 5.56.

The content of total proteins and lactose is slightly higher in bulk tank milk samples from evening milking as result in season variation [3].[9] reported a significant decrease in casein content when either Holstein or Guernsey milk exceeded SCC of 500,000; depression was greater above one million SCC. SCC above 500,000 has been associated with poor quality cheese because of increased rennet to cutting time and lower curd firmness. The mastitis or elevated SCC is associated with a decrease in lactose, α-lactalbumin, and fat in milk because of reduced synthetic activity of mammary tissue [10]. With higher SCC, the concentrations of serum albumins and immunoglobulins increase which reduces heat stability of mastitis milk, causing coagulation, or flocculation during pasteurization. [11] investigated the relationship between SCC and composition (total solids, fat, protein and lactose content) of milk from individual Holstein cows and indicated that SCC of individual cow's milk significantly correlated with a decrease in milk constituents only under conditions of average SCC in bulk milk above 1,000,000 /ml. Negative correlation between lactose and chloride content was detected in cows diagnosed with subclinically mastitis with high SCC [12].

In Table 3 are presented the results of somatic cell counts in bulk tank milk samples from the morning and evening during six months-

Table 3 The results of somatic cell counts in bulk tank milk samples from the morning and evening during six months

Stat. param.	December m	January e	February m	March m	April m	May e
Average	273,605	292,895	270,026	254,974	256,316	295,895
Standard dev.	80,357	72,743	87,765	93,736	93,310	74,539
Min	75,000	168,000	101,000	111,000	58,000	142,000
Max	446,000	516,000	499,000	483,000	463,000	496,000
<100	1	0	0	0	1	0
<250	19	12	18	21	16	10
>250	15	22	16	13	18	24
>400	3	4	4	4	3	4
Varition	29,370	24,836	32,502	36,763	36,404	25,191

Legend: m-morning milking; e-evening milking

The highest value for SCC (516,000/ml) was in bulk tank milk samples from evening milking in the Winter (the January) and the lowest (58,000/ml) in the April.

The process of using bulk tank milk analysis in useful tool in improving milk quality and herd mammary health. It should be kept in mind that although individual cow samples for milk culture and SCC are more definitive for diagnosis and monitoring of udder health, Bulk tank milk analysis is less expensive, more convenient, and faster than testing milk samples from individual animals or groups of cows. When the SCC of bulk tank milk is interpreted within the context of the farm's management practices, this information provides a basis for evaluating current and potential milk quality and mastitis problems in a herd.

4. Conclusions

The process of using bulk tank milk analysis in useful tool in improving milk quality and herd mammary health. Our results showed that average values for bulk tank milk samples was 273,605/ml from morning milking and 292,895/ml from evening milking. The average values for total proteins content from morning and evening milking are 3,31 and 3,34%. The average values for lactose content from morning and evening milking are 4,56 and 4,63%. The highest SCC (516,000/ml) was detected in bulk tank milk sample from evening milk in Winter. The highest SCC (516,000/ml) was detected in bulk tank milk sample from evening milk and the lowest content of lactose was 4,46%. The obtained values of bulk tank milk somatic cell counts did not significantly affect the content of total proteins and lactose.

Proper milking techniques, improved sanitation, effective use of teat, dipping and dry period therapy and improvement in management may reduce SCC by reducing the spread of new infections.

References

[1] Seegers H, Fourichon C and Beaudeau F 2011 Production effects related to mastitis and mastitis economics in dairy cattle herds. *Vet. Res.* **34** 475–491

[2] Rainard P, Riollet C. 2006 Innate immunity of the bovine mammary gland. *Vet. Res.* **37** 369–400

[3] Schepers AJ, Lam, TJGM, Schukken YH, Wilmink JBM & Hanekamp WJA 1997 Estimation of various components for somatic cell counts to determine the threshold for uninfected quarters. *Journal of Dairy Science* 80 1833–1840

[4] Schukken, YH, Wilson, DJ, Welcome, F, Garrison-Tikofsky L and Gonzalez RN 2003 Monitoring udder health and milk quality using somatic cell counts. *Veterinary Research.* **34** 579-596

[5] Auldis MJ and Hubble IB 1998 Effects of mastitis on raw milk and dairy products. *Aust. J. Dairy Tech* **53** 28-36.

[6] Hamann J 2002 Relationships between Somatic cell counts and milk composition. *Bulletin of the IDF* **372** 56-59.

[7] Pitkala A, Haveri M, Pyorala S, Myllys V and Honkanen-Buzalski T. 2004 Bovine Mastitis in Finland 2001—Prevalence, Distribution of Bacteria, and Antimicrobial Resistance. *J. Dairy Sci.* **87** 2433–2441 American Dairy Science Association

[8] NMC Board of directors report 2005 Human health risks associated with high somatic cell count milk. Symposium summary *website: www.nmconline.org.*

[9] Haenlein G FW, Schultz LH and Hansen, LR 1973 Composition of proteins in milk with varying leukocytes counts. *Journal of Dairy Science* **56** 1071-1024

[10] Harmon R J 1994 Physiology of mastitis and factors affecting somatic cell counts. *Journal of Dairy Science* **77** 2103-2112

[11] Fernandes AM, Oliveira CAF, Tavolaro P 2004 Relationship between somatic cell counts and composition of milk from individual Holstein cows. *Arq. Inst. Biol. Sao Paulo* **71** 163-166.

[12] Ahmad J, Hussain I., Mahmood N. and Munir R. 1988 Lactose determination as an aid to subclinical mastitis diagnosis. *Pakistan Vet. J.* **8** 25-28

[13] Sharif A.and Muhammad G 2008 Somatic cell count status as an indicator of udder health status under modern dairy production: a Review.*Pakistan Vet. J.* **28** (4) 194-200

[14] Jayarao BM, Wolfgang DR. 2003 Bulk-tank milk analysis. A useful tool for improving milk quality and herd udder health. *Vet Clin North Am Food Anim Pract.* **19** (1):75-92.

IOP Conf. Series: Earth and Environmental Science 85 (2017) 012086 doi:10.1088/1755-1315/85/1/012086

Factors affecting elimination of polycyclic aromatic hydrocarbons from traditional smoked common carp meat

J Babić[1], S Vidaković [1], S Škaljac[2], B Kartalović[1], D Ljubojević[1], M Ćirković[1], V Teodorović[3]

[1] Scientific Veterinary Institute "Novi Sad", Novi Sad, Serbia
[2] Faculty of Technology, University of Novi Sad, Bulevar Cara Lazara 1, Novi Sad, Serbia
[3] Faculty of Veterinary Medicine, University of Belgrade, Bulevar oslobođenja 18, Belgrade, Serbia

E-mail: jelenababic@niv.ns.ac.rs

Abstract. Smoking techniques have been progressively improved and different procedures have been developed in different regions for treating fish. In these times, the technology is mainly used for enrichment of fish with specific taste and odour, to extend the shelf-life of these perishable products and appearance required widely on the market. A lot of chemical contaminants such as polycyclic aromatic hydrocarbons (PAHs) are formed during the combustion of fuel in the smoking process. PAHs are a group of compounds that have been the subject of great concern in the recent years due to their toxic, mutagenic and/or carcinogenic potentials to humans. These fact can have a significant impact on the acceptance of these products by consumers. In this review article, the objective is to describe factors affecting elimination of polycyclic aromatic hydrocarbons from traditional smoked common carp meat.

1. Smoked common carp meat

Food smoking is one of the oldest food technologies which mankind has used for at least 10 000 years [1,2]. Production of smoked meat is very popular and smoked meat presents a significant part of the human diet in Serbia and our region, which makes smoked products very popular and consumed quite often and traditional uncontrolled smoke kilns are still widely being used [3].

The fish processing industry is not well developed in our country because for many years the amount of fish was just not enough and also because of the consumers' habits [4]. The consumption of smoked fish in our country is lower compared to other countries in the EU, but it shows the tendency for the significant increase [5]. Within this type of the fish products, the smoked carp, silver carp and trout meat is available in our country and the most famous smoked fish meat is smoked common carp meat [6].

Consumers prefer smoked fish due to several reasons such as taste, amount of essential fatty acids, and readiness to eat [7]. In these times, the technology is mainly used for enrichment of fish with specific taste and odor, to extend the shelf-life of these perishable products and appearance required widely on the market. About 15% of the total quantity of fish for human consumption in Europe is offered on the market in the form of either cold- or hot-smoked products [8]. Traditional smoking involves treating of pre-salted, whole, eviscerated or filleted fish with wood smoke. The most often,

IOP Conf. Series: Earth and Environmental Science **85** (2017) 012086 doi:10.1088/1755-1315/85/1/012086

smoke is produced by smouldering wood and shavings or sawdust in the oven, directly below the hanging fish or fillets, laid out on mesh trays.

Fish contains n-3 polyunsaturated fatty acids (PUFAs) that appear to play several useful roles for human health [9]. Conversely, potential health hazards could be associated with smoked foods may be caused by carcinogenic components of wood smoke – mainly polycyclic aromatic hydrocarbons (PAHs), derivatives of PAHs, such as nitro-PAHs or oxygenated PAHs, and to a lesser extent also N-nitroso compounds and heterocyclic aromatic amines [10]. These fact can have a significant impact on the acceptance of these products by consumers.

2. Polycylic aromatic hydrocarbons - PAHs

A lot of chemical contaminants such as polycyclic aromatic hydrocarbons (PAHs), dioxins, formaldehyde, nitrogen and sulphur oxides are formed during the combustion of fuel in the smoking process [11]. PAHs are a group of compounds that have been the subject of great concern in the recent years due to their toxic, mutagenic and/or carcinogenic potentials to humans. PAHs comprise the largest class of chemical compounds known to be cancer causing agents. Some, while not carcinogenic, may act as synergists. The main route of exposure to PAHs for non-smokers and non-occupationally exposed individuals is through food consumption [12, 13].

PAHs have been detected in food, both raw/non-processed and processed foods. The presence of PAHs in raw foods is associated with environmental pollution [14]. PAHs are found ubiquitously, such as water, air, soil, and, therefore also in food [7]. Due to their lipophilicity, persistence and high toxicity, a lot of studies have shown that in the aquatic habitat, many organisms, such as fish and shellfish, readily accumulate PAHs from the environment and store them in their tissues reaching levels higher than those in the ambient medium [15,16,17,18].

In fact, the International Agency for Research on Cancer classifies some PAHs as known, possibly, or probably carcinogenic to humans (Group 1, 2A or 2B). Among these are benzo[a]pyrene (Group 1), naphthalene, chrysene, benz[a]anthracene, benzo[k]fluoranthene and benzo[b]fluoranthene (Group 2B) [19]. Some PAHs are well known as carcinogens, mutagens, and teratogens and therefore pose a serious threat to the health and the well-being of humans.

3. Factors affecting occurrence and concentrations of PAHs in smoked products

Smoke production in modern smoking ovens is closely controlled and the removal of PAHs and other undesirable compounds is facilitated by the smoke generators being separated from the smoking chamber. In contrast, in traditional smoking conditions, very high combustion temperatures are reached and the foodstuff is in direct contact with all components of the smoke generated. Direct exposition of meat products to smoke brings about higher concentrations of PAHs as compared to indirect methods, when PAHs are partially eliminated by condensation in tars [2,20]. Reports that PAH levels in traditional smoked foods can reach high levels have in recent years prompted considerable interest in their quantification and control.

The composition and amount of PAHs depend upon numerous factors, such as the composition and type of wood as well as moisture content, oxygen accessibility, the temperature of smoke generation, smoking duration etc. [2, 21, 18]. Also, deposition and penetration of smoke components into smoked fish depends on natural content of PAHs in raw fish, water activity of the food, fat content, heat source, distance of heating, design of the food device, drainage of fat etc. [21, 7]. Even, there is evidence showing that female fish exhibited significantly lower mean ΣPAH concentrations than male in all examined the species, except for *Liza abu* [22].

Wood composition, especially lignin content, also influences the levels of PAHs produced [20]. Also, the use of hardwoods instead of softwoods has been recommended to reduce the presence of PAHs in smoke and in smoked foods too [21].

Temperature of smoke generally plays a very important role, because the amount of PAHs in smoke formed during pyrolysis increases linearly with the smoking temperature within the interval

400–1000°C [20]. In addition, PAHs can also be formed at lower temperatures [18]. Also, concentration of PAHs in the smoke increased when the materials were burnt with flames [23, 24].

Simko et al. (1991) have shown the decrease in benzo[a]pyrene content caused by dehydration of product that confirmed the effect of moisture content on PAHs concentration in smoked food [25].

There is a positive relationship between lipid content and PAH residual levels. PAHs are lipophilic in nature and usually accumulate in the fatty tissues of organisms. This provides further evidence that the lipid content of the tissues is the determining factor in the bioaccumulation of PAHs by fish. The formation of PAHs is known to occur through pyrolysis of fat at temperatures above 200 °C and it is highly stimulated at temperatures over 700 °C [7]. Pyrolysis of other organic matter such as proteins and carbohydrates might be involved, but the greatest concentrations of PAHs have been shown to arise from fat pyrolysis. This provides further evidence that the lipid content of the tissues is the determining factor in the bioaccumulation of PAHs by fish [26, 27, 28].

Group of authors Babic et al. have proved that application of charcoal filter [4], zeolite filter [3] and gravel filter [29] in production of smoked common carp meat decrease the PAHs content.

The package of smoked food into appropriate packaging material could also remarkably decrease the PAH content [2]. The highest concentration of PAHs in smoked food usually occur at the food surface and maximum concentration of PAHs is immediately after finishing the smoking, then it decreases due to light decomposition and interaction with present compounds [28].

4. Conclusions

Studies focused on the effects of smoking on the potential changes, increases or decreases, of chemical contaminants such as PAHs in smoked common carp meat are very important because this may be a contributing factor to the recent increases in prevalence rates in cases of cancer in our country.

There is therefore a need to educate manufacturers about safe smoking practices, and also most importantly to adopt their procedure that would reduce PAHs content in smoked common carp meat with traditional kilns in order to ensure the health safety of consumers. Levels of contamination can be significantly reduced under controlled conditions accepting good manufacturing practice principles using current knowledge and appropriate technological equipment.

References

[1] Simko P 2002. Determination of polycyclic aromatic hydrocarbons in smoked meat products and smoke flavouring food additives. *J Chromatogr B Analyt Technol Biomed Life Sci.* **770** 3-18

[2] Simko P 2005. Factors affecting elimination of polycyclic aromatic hydrocarbons from smoked meat foods and liquid smoke flavorings. *Mol. Nutr. Food Res.* **49** 637-647

[3] Babić J, Kartalović B, Petrović J, Okanović Đ, Novakov N, Ljubojević D, Ćirković M, Teodorović V 2017. The application of zeolit filter in production of smoked common carp meat. *Proceedings of V International congress „ Engineering, enviroment and materials in processing industry"*, Jahorina 15th-17th March 2017, Republic of Srpska, Bosnia and Herzegovina, 189-196

[4] Babić J, Kartalović B, Petrović J, Okanović Đ, Novakov N, Ljubojević D, Ćirković M, Teodorović V 2017. The application of charcoal filter in production of smoked common carp meat. *Proceedings of V International congress „ Engineering, enviroment and materials in processing industry"*, Jahorina 15th-17th March 2017, Republic of Srpska, Bosnia and Herzegovina, 181-189

[5] Ćirković M, Ljubojević D, Okanović Đ, Đorđević V, Novakov N 2014. Effect of hot smoking process on processing yield and microbiological safety of Common Carp, *II International Congress "Food Technology, Quality and Safety" (FoodTech2014), Proceedengs*, 100-104

[6] Pavlićević N, Đorđević V, Dimitrijević M, Karabasil N, Baltić M, Bošković M, Petrović J 2013. Effect of pre-processing of trout by freezing on the characteristics of smoked trout fillets, *Tehnologija mesa* **54** 57–68

[7] Bansal V, Kim K-H 2015. Review of PAH contamination in food products and their health hazards. *Environ Int.* **84** 26–38
[8] Stolywo A, Sikorski Z 2004. Polycyclic aromatic hydrocarbons in smoked fish – a critical review, *Food Chemistry* **91** 303-311
[9] Liu J, Ma D W L 2014. The Role of n-3 Polyunsaturated Fatty Acids in the Prevention and Treatment of Breast Cancer. *Nutrients* **6** 5184–5223
[10] Shen G, Tao S, Wei S, Zhang Y, Wang R, Wang B, Li W, Shen H, Huang Y, Chen Y, Chen H, Yang Y, Wang W, Wang X, Liu W, Simonich SLM 2012. Emissions of Parent, Nitro, and Oxygenated Polycyclic Aromatic Hydrocarbons from Residential Wood Combustion in Rural China. *Environ. Sci. Technol.* **46** 8123–8130
[11] CAC/RCP 68-2009 Code of practice for the reduction of contamination of food with polycyclic aromatic hydrocarbons (pah) from smoking and direct drying processes. World Health Organization & Food and Agriculture Organization of the United Nations
[12] Ma Y, Harrad S 2015. Spatiotemporal analysis and human exposure assessment on polycyclic aromatic hydrocarbons in indoor air, settled house dust, and diet: A review. *Environ Int.* **84** 7-16
[13] Domingo J, Nadal M 2015. Human dietary exposure to polycyclic aromatic hydrocarbons: A review of the scientific literature. *Food and Chemical Toxicology* **86** 144-153
[14] Guillen MD, Sopelana, P, Partearroyo MA 2000. PAH in liquid smoke flavorings obtained from different types of wood. Effect of storage in polyethylene flasks on their concentrations. *Journal of Agricultural and Food Chemistry* **48** 5083–5087
[15] Shadi A, Mazandarani MK, Nikpour Y 2012. Concentrations of polycyclic aromatic hydrocarbons (PAHS) in sediments of Khowre-Musa System (Persian Gulf). *World* **4** 83–86
[16] Johnson LL, Arkoosh MR, Bravo CF, Collier TK, Krahn MM, Meador JP, Myers MS, Reichert WL, Stein JE 2008. The Effects of Polycyclic Aromatic Hydrocarbons in Fish from Puget Sound, Washington, ed Di Giulio RG and Hinton DE (Taylor & Francis Group, LLC) 873-919
[17] Dhananjayan V, Muralidharan S 2012. Polycyclic Aromatic Hydrocarbons in Various Species of Fishes from Mumbai Harbour, India, and Their Dietary Intake Concentration to Human. *International Journal of Oceanography 2012*, 6 pages
[18] Abdel-Shafy H I, Mansour M S M 2016. A review on polycyclic aromatic hydrocarbons: Source, environmental impact, effect on human health and remediation. *Egyptian Journal od Petroleum* **25** 107-123
[19] IARC (International Agency for Research on Cancer) 2010. Some non-heterocyclic polycyclic aromatic hydrocarbons and some related exposures *Monogr Eval Carcinog Risks Hum* **92** 765-771
[20] Nollet LML 2007. Handbook of Meat, Poultry and Seafood Quality. Blackwell Publishing
[21] García-Falcón M S, Simal-Gándara J 2005. Polycyclic aromatic hydrocarbons in smoke from different woods and their transfer during traditional smoking into chorizo sausages with collagen and tripe casings. *Food Additives & Contaminants* **22** 1-8
[22] Rahmanpour S, Ghorghani NF, Masoumeh S, Ashtiyani L 2014. Polycyclic aromatic hydrocarbon (PAH) in four fish species from different trophic levels in the Persian Gulf. *Environmental Monitoring and Assessment* **186** 7047–7053
[23] Lijinsky W 1991. The formation and occurrence of PAH associated with food. *Mutation Research* **259** 251–261
[24] Rey Salgueiro L, Garcı´a Falco´n MS, Soto Gonza´lez B, Simal Gandara J 2004. Procedure to measure the level of PAH in wood ashes used as fertilizer in agroforestry soils and their transfer from ashes to water. *Journal of Agricultural and Food Chemistry* **52** 3900–3904
[25] Simko P, Karovicová J, Kubincová M 1991. Changes in benzo[a]pyrene content in fermented salami. *Z. Lebensm. Unters. Forsch.* **193** 538-540
[26] Basak S, Sengor G F, Karakoc F T 2010. The Detection of Potential Carcinogenic PAH Using

HPLC Procedure in Two Different Smoked Fish, Case Study: Istanbul/Turkey. *Turkish Journal od Fisheries and Aquatic Science* **10** 351-355

[27] Xu FL, Wu WJ, Wang JJ, Qin N, Wang Y, He QS, He W, Tao S 2011. Residual levels and health risk of polycyclic aromatic hydrocarbons in freshwater fishes from Lake Small Bai-Yang-Dian, Northern China. *Ecol Model* **222** 275–286

[28] Essumang DK, Dodoo DK, Adjei JK 2014. Effective reduction of PAH contamination in smoke cured fish products using charcoal filters in a modified traditional kiln, *Food Control* **35** 85-93

[29] Babić J, Kartalović B, Petrović J, Okanović Đ, Božić B, Pelić M, Ćirković M 2015. Ecological safe production of smoked common carp meat. *„One Health – New Challenges“, Novi Sad, Serbia. First International Symposium of Veterinary Medicine (ISVM2015), Proceedings* 142 – 147

Attitudes of Serbian food technology students towards surgical and immunocastration of boars and their sensitivity to androstenone and skatole

I Tomasevic[1], I Đekić[1], M Font-i-Furnols[2], M Aluwé[3], M Čandek-Potokar[4], M Bonneau[5] and U Weiler[6]

[1] Faculty of Agriculture, University of Belgrade, Nemanjina 6, 11080 Belgrade, Serbia
[2] IRTA-Product Quality, Finca Camps i Armet, 17121, Monells, Girona, Catalonia,Spain
[3] Institute for Agricultural and Fisheries Research (ILVO), Animal Sciences Unit, Scheldeweg 68, 9090 Melle, Belgium
[4] Agricultural Institute of Slovenia, Hacquetova ulica 17, 1000 Ljubljana, Slovenia
[5] IFIP, The French Pork and Pig Institute, Le Rheu, France
[6] Universitat Hohenheim, PO Box 700562, 70593 Stuttgart, Germany

E-mail: tbigor@agrif.bg.ac.rs

Abstract. Various European Union pork chain actors and stakeholders agreed in 2010 on a road map to voluntarily abandon piglet castration by 1 January 2018. Because currently in Serbia, male piglets are surgically castrated and consumers are not used to the boar taint odour and flavour, the introduction of boar meat may modify the acceptability of pork. The objective of the study was to investigate the attitudes, awareness and opinions of future Serbian food technologist towards surgical castration of boars and its alternatives, and to test their sensitivity to androstenone and skatole. We found that they were concerned about the animal welfare issues and that they were willing to pay a little more for meat from animals treated with dignity. This was more so if they were females and less so if they had had a rural upbringing. They strongly believed that surgical castration is painful for the animals, but at the same time agreed that meat from castrated pigs is of better quality. Their ambiguous attitudes regarding efficacy and quality of alternatives to surgical castration clearly indicated the knowledge gap that must be filled by appropriate modifications of the curriculum. Students demonstrated average sensitivity to both androstenone and skatole. Females exhibited higher intensities of difference in both cases.

1. Introduction

Boar taint is an offensive odour and flavour present in meat from some non-castrated male pigs. This defect has principally been ascribed to androstenone (sex steroid produced by testes) and skatole (product of bacterial degradation of tryptophan in the gut). Surgical castration of male piglets is a standard practice in pork production to prevent this boar taint in meat. Due to the different odours of androstenone (urine or sweat like) and skatole (faecal), it is likely that both substances contribute separately in an additive manner to off-odour, but they differ in their origin, chemical properties and the likelihood of consumer reactions. Their relative contribution is determined by both the sensitivity of consumers and the frequency of carcasses with high concentrations of the respective substances [1]. In contrast to androstenone, skatole is perceived by 99% of consumers, but the percentage of boar carcasses high in skatole concentrations is much lower than the percentage of carcasses with high androstenone concentrations [2]. Previous study also showed that only those consumers who perceive the odour of pure androstenone as unpleasant clearly differentiated tainted from non-tainted entire male pork [3].

In Europe, the proportion of male pigs that are left entire (uncastrated) has been high for many years in the British Isles and Iberian Peninsula, and has recently increased in The Netherlands and to a lesser extent in Germany, Belgium and France. Various European Union pork chain actors and stakeholders agreed in 2010 on a road map to voluntarily abandon piglet castration by 1 January 2018 [4]. With regard to meat quality, the main challenges in pork production with entire males concern firstly, the reduction of boar taint and secondly, altered meat technological properties [5-7]. As no rapid online detection method for boar taint is yet available, complete non-castration poses risks because consumers could reduce their consumption of pig meat due to the presence of boar-tainted meat [8].

In Serbia, surgical castration is most frequently performed without anaesthesia. The procedure, as traditionally performed, negatively affects performances and is painful [9], the latter reason being why surgical castration is performed with pain treatment or has been abandoned in several countries where animal welfare has been a growing concern [10]. When abandoning castration, it is necessary to find alternative solutions to prevent this taint defect in meat. One of the promising alternatives to surgical castration effectively preventing boar taint is immunocastration [11]. Today, immunocastration is an alternative to the production of entire male pigs. The main problem for the wider use of immunocastration seems to be the fear related to consumer acceptance [5].

Because today in Serbia, male piglets are surgically castrated and as the consumers are not used to the boar taint odour and flavour, the introduction of boar meat could modify consumer acceptability of pork. The data collected by the Serbian Chamber of Commerce shows that, in 2015 alone, Serbia imported 20,100 tons of pork (€40.8 million) mostly from Spain, Hungary and Germany. The same year, Serbia exported 12,000 tons of pork valued at €24.3 million [12]. The lack of knowledge of future Serbian food technologist about the production of entire males (EM) or immunocastrates (IC) would lead to their inability to cope with the challenges in management of product quality (detecting and reducing boar taint, coping with extreme leanness), further compromising the competitiveness of the Serbian pork sector in the future.

The objective of the present study was, therefore, to investigate the attitudes, awareness and opinions of future Serbian food technologist towards surgical castration of boars and its alternatives, and to test their sensitivity to androstenone and skatole. We believe that insights in their attitudes towards this important and contemporary animal welfare and meat technology issue and the identification of probable knowledge gaps will be useful to better coordinate and target their upcoming training on this topic.

2. Materials and methods

2.1. Survey

The survey of 100 randomly selected food technology students (undergraduate and postgraduate studies) at the University of Belgrade – Faculty of Agriculture – on the perception of castration of male pigs was conducted during 2017. A brief 15 minute introduction about surgical and immunocastration of piglets was presented to the students prior to the survey.

A structured questionnaire was developed and consisted of three sections. The first section included general demographic information about the respondents. The second section explored 13 statements related to the castration of male pigs from the welfare (6 items) and economic (2 items) points of view and meat preferences and beliefs (5 items) giving the respondents the opportunity to rate their degree of agreement on a seven-point Likert scale: 1 "Disagree very strongly", 2 "Disagree strongly", 3 "Disagree", 4 "Neither agree nor disagree", 5 "Agree" 6 "Agree strongly" to 7 "Agree very strongly" (Table 1). The third section gave the respondents the opportunity to state their attitudes towards surgical castration and immunocastration using a 7-point scale (1 – "Harmful / Easy / Bad" 4 – "Neither harmful nor beneficial / Neither easy nor difficult / Neither bad nor good", 7 – "Beneficial / Difficult / Good").

2.2. Triangle test

Triangle tests were performed to evaluate whether assessors (students) could differentiate paper strips with spiked solution of androstenone (A) and skatole (S) from blank paper strips. Sixteen assessors (8 male and 8 female postgraduate students) participated in the triangle test, and replicated their evaluations twice. This made a total of 32 evaluations for each compound. Half of the assessors (4 male and 4 females) participated in first test where the odd sample (paper strip with compound "B"; blank paper strip "A") had the following combination for the two replications: (AAB, ABA, BAA). The second test was presented to the other half of the assessors with the following combination of strips for the two replications (AAB, ABA, BAA, BBA, BAB, ABB) [13, 14].

Instructions to the participants included (i) a presentation of the task; (ii) the obligation to evaluate the samples in the imposed order; (iii) the obligation to give a response; (iv) the possibility of giving their opinion on the degree of difference between the sample they chose and the others by circling one of the following descriptors which most closely described the intensity of difference (0 = none; 1 – very slight; 2 – slight; 3 – moderate; 4 – large; 5 – extreme).

2.3. Statistical processing

The 13 statements considered in the survey were subjected to a principal component analysis (PCA) to gain a better understanding of the overall correlations in the data set. A Varimax orthogonal rotation was employed to aid interpretability. The Mann-Whitney U test was carried out to determine if statistically significant differences exist between genders or place where students were born and raised (rural or urban).

For the triangle test, the null hypothesis is H0: nc = 1/3 n, with nc = observed number of correct responses and n = total number of responses. The alternative hypothesis is H1: nc > 1/3 n. The critical number of correct responses to reject H0 at $\alpha = 0.05$ in favour of H1 was obtained from the binomial distribution [13, 14].

All data were analysed using SPSS Statistics 17.0 (Chicago, Illinois, USA) data analysis software.

3. Results and discussion

3.1. Attitudes of students towards surgical and immunocastration of boars

Out of 100 students who participated in the study, three questionnaires were excluded because of failure to answer 50% or more of the questions, and 97 questionnaires were further processed. Demographic profiling showed that female students prevailed (63.9%). Approximately half of respondents grew up in rural surroundings (52.6%) while 47.4% of them grew up in urban areas.

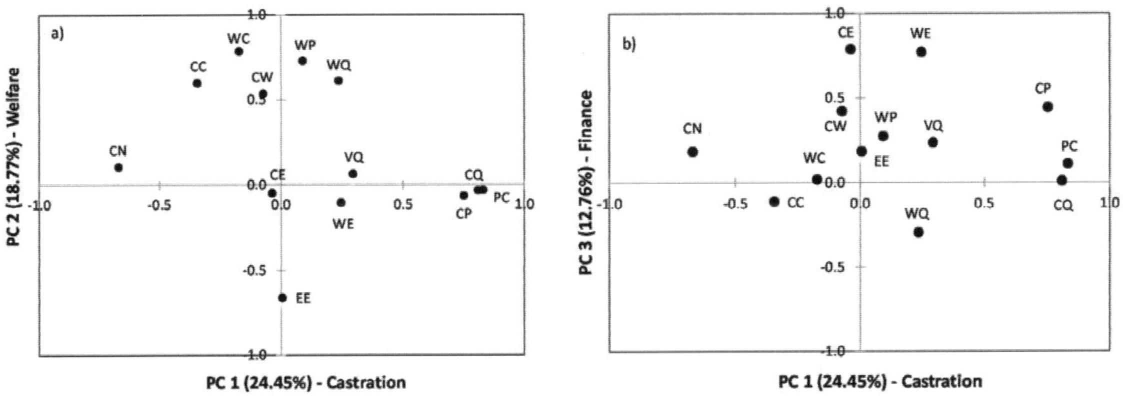

Figure 1. Principal component analysis loadings plot.
(Explanation of the abbreviations given in Table 1)

A PCA was run on the 13 statements about castration of male pigs. The suitability of PCA was assessed prior to analysis. The overall Kaiser-Meyer-Olkin (KMO) measure was 0.688. Bartlett's test of sphericity was statistically significant ($p < 0.0005$), indicating that the data was likely factorizable. PCA revealed three components that had eigenvalues greater than one. Visual inspection of the scree plot indicated that these components should be retained [15]. In addition, a three-component solution met the interpretability criterion. As such, three components were retained and this solution explained 56% of the total variance. The rotated solution exhibited 'simple structure' [16]. The interpretation of the data was consistent with the statements the questionnaire was designed to measure, with strong loadings of "Castration" items on Component 1, "Welfare" items on Component 2, and "Finance" items on Component 3 (Figure 1). Only the item 'pig castration with vaccines improves pork quality' is not well explained by any of the three principal components.

On average, Serbian food technology students strongly agreed that surgical castration produces pain to the animals (6.1) and believe that taking care of animal welfare produces meat of higher quality (5.9) (Table 1). They were willing to pay a little more for meat from animals treated with dignity (5.3) and they were worried about welfare of animals for human consumption (5.2). They consider that meat from castrated male pigs is of better quality (5.1) while believing that surgical castration of animals is cruel (5.1). Female students were more concerned about animal welfare issues and were significantly more convinced that taking care of animal welfare produces meat of higher quality.

On average, students had no preference about eating meat of castrated pigs (4.2), or willingness to pay a little more for meat from castrated pigs (4.1). They neither agreed nor disagreed whether meat from castrated pigs is more expensive (4.1), whether castration of animals for human consumption is necessary (3.9), or if pig castration with vaccines improves pork quality (3.7) and whether ensuring animal welfare means to eat meat that is more expensive (3.7). Finally, they strongly disagreed with the statement that it does not matter if we mistreat the animals because at the end we eat them. Students who were raised in rural environments were less convinced that surgical castration is cruel, while the statement that ensuring animal welfare means to eat meat that is more expensive was more agreeable to students raised in urban environments (data not shown).

Table 1. Statements about the castration of boars given by the male and female food technology students

Statements	**Male**	**Female**	**Total**
N	35	62	97
Surgical castration produces pain to the animal **(CW)**	6.0 ± 1.1	6.1 ± 1.2	6.1 ± 1.2
Taking care of animal welfare produces meat of higher quality **(WQ)**	5.5 ± 1.5^a	6.1 ± 1.1^b	5.9 ± 1.3
Meat from castrated pigs is of better quality **(CQ)**	5.3 ± 1.5	5.0 ± 1.3	5.1 ± 1.4
The meat from castrated pigs is more expensive **(CE)**	4.2 ± 1.4	4.1 ± 0.9	4.1 ± 1.1
I am willing to pay a little more for meat from castrated pigs **(CP)**	4.3 ± 1.6	4.0 ± 1.4	4.1 ± 1.5
I am worried about welfare of animals for human consumption **(WC)**	4.9 ± 1.4	5.4 ± 1.1	5.2 ± 1.2
The castration is not necessary **(CN)**	3.9 ± 1.6	4.0 ± 1.8	3.9 ± 1.7
I am willing to pay a little more for meat from animals treated with dignity **(WP)**	4.9 ± 1.4^a	5.6 ± 1.3^b	5.3 ± 1.4
The surgical castration is cruel **(CC)**	4.6 ± 1.9^a	5.4 ± 1.2^b	5.1 ± 1.6
Does not matter if we mistreat the animals because at the end we eat them **(EE)**	2.1 ± 1.0	1.9 ± 1.5	2.0 ± 1.3
I prefer to eat meat of castrated pigs **(PC)**	4.5 ± 1.4	4.1 ± 1.6	4.2 ± 1.5
To ensure animal welfare means to eat meat that is more expensive **(WE)**	3.7 ± 1.7	3.8 ± 1.6	3.7 ± 1.6
Pig castration with vaccines improves pork quality **(VQ)**	3.6 ± 1.7	3.8 ± 1.4	3.7 ± 1.5

Note: Items denoted with different letters within a row are significantly different at the level of 5%.
(1) "Disagree very strongly", (2) "Disagree strongly", (3) "Disagree", (4) "Neither agree nor disagree", (5) "Agree", (6) "Agree strongly", (7) "Agree very strongly".

Students also believed that surgical castration is neither harmful nor beneficial (4.3 ± 2.2; 1 "Harmful", 4 "Neither harmful nor beneficial", 7 "Beneficial"), that surgical castration is neither easy nor difficult (3.7 ± 2.2; 1 "Easy", 4 "Neither easy nor difficult", 7 "Difficult") and that pig castration with vaccines is neither good nor bad (4.3 ± 2.1; 1 "Bad", 4 "Neither good nor bad", 7 "Good").

3.2. Triangle test

The sensory analysis showed that the paper strips loaded with androstenone can be distinguished as different (n=32, nc=23) at the level of α=0.05. With 95% confidence level, at least 38% of the consumers could distinguish the two samples. It seems that the percentage of consumers sensitive to androstenone assessed in our study was a little lower compared to the general average of 45% [17]. The average degree of difference for correct answers was 1.9 ± 1.6 with no significant difference between correct and incorrect answers ($p \geq 0.05$). Female assessors ranked the degree of difference higher compared to male assessors (2.00 opposed to 1.75). Our result is in agreement with a previous general conclusion that sensitivity of consumers to androstenone is higher in women than in men [17].

The sensory analysis showed that the paper strips loaded with skatole can be distinguished as different (n=32, nc=24) at the level of α=0.05. With 95% confidence level, at least 44% of the consumers could distinguish the two samples. The average degree of difference for correct answers was 4.0 ± 1.0 with no significant difference between correct and incorrect answers ($p \geq 0.05$). Female students who noticed the differences responded that the degree of difference was higher (4.50) compared to male students (3.42).

4. Conclusions

It is obvious that future Serbian food technologist are concerned about the animal welfare issues and that they are willing to pay a little more for meat from animals treated with dignity. This was more so if they were females and less so, if they had had a rural upbringing. They strongly believed that surgical castration is painful for the animals, but at the same time, agreed that meat from castrated pigs is of better quality. Their ambiguous attitudes regarding efficacy and quality of alternatives to surgical castration clearly indicate a knowledge gap that must be filled by appropriate modifications of the curriculum. Students demonstrated average sensitivity to both androstenone and skatole. Females exhibited higher intensities of difference in both cases.

Acknowledgements
The authors would like to acknowledge networking support by the COST Action CA 15215 "Innovative approaches in pork production with entire males".

References

[1] Weiler U, Font i Furnols M, Fischer K, Kemmer H, Oliver M A, Gispert M. 2000 Influence of differences in sensitivity of Spanish and German consumers to perceive androstenone on the acceptance of boar meat differing in skatole and androstenone concentrations. *Meat Science* **54**(3) pp 297-304
[2] Weiler U, Dehnhard M, Herbert E, Claus R. 1995 Einfluβ von Geschlecht, Genotyp und Mastendgewicht auf die Androstenon- und Skatolkonzentrationen im Fett von Mastschweinen. Die Ebermast, Angewandte Wissenschaft. Landwirtschaftsverlag Münster pp 14-33
[3] Bonneau M, Chevillon P. 2012 Acceptability of entire male pork with various levels of androstenone and skatole by consumers according to their sensitivity to androstenone. *Meat Science* **90**(2) pp 330-337
[4] Bee G, Chevillon P, Bonneau M. 2015 Entire male pig production in Europe. *Animal Production Science* **55**(12) pp 1347-59
[5] Candek-Potokar M, Skrlep M, Lukac N B. 2015 Raising Entire Males or Immunocastrates – Outlook on Meat Quality. *Procedia Food Science* 5 pp 30-3
[6] Škrlep M, Čandek-Potokar M, Lukač N B, Povše M P, Pugliese C, Labussière E. 2016 Comparison of entire male and immunocastrated pigs for dry-cured ham production under two salting regimes. *Meat Science* **111** pp 27-37
[7] Aluwe M, Langendries K C, Bekaert K M, Tuyttens F A, De Brabander D L, De Smet S. 2013 Effect of surgical castration, immunocastration and chicory-diet on the meat quality and palatability of boars. *Meat Science* **94**(3) pp 402-7
[8] Font-i-Furnols M, Aaslyng M D, Backus G B C, Han J, Kuznetsova T G, Panella-Riera N. 2016 Russian and Chinese consumers' acceptability of boar meat patties depending on their sensitivity to androstenone and skatole. *Meat Science* **121** pp 96-103
[9] Aluwe M, Tuyttens F A, Millet S. 2015 Field experience with surgical castration with anaesthesia, analgesia, immunocastration and production of entire male pigs: performance, carcass traits and boar taint prevalence. *Animal : An International Journal of Animal Bioscience* **9**(3) pp 500-8
[10] Aleksić J. 2012 The investigation into the efficacy of immunocastration aimed at the prevention of sex odour in boar's meat. PhD Thesis, Faculty of Veterinary Medicine, University of Belgrade.
[11] Aluwe M, Degezelle I, Depuydt L, Fremaut D, Van den Broeke A, Millet S. 2016 Immunocastrated male pigs: effect of 4 v. 6 weeks time post second injection on performance, carcass quality and meat quality. *Animal : An International Journal of Animal Bioscience* **10**(9) pp 1466-73
[12] Bjelotomic S. 2016 Serbia exports shockingly little meat. *Vecernje Novosti* 09.11.2016;Sect. Economy.

[13] ISO 2004 ISO 4120:2004 Sensory analysis — Methodology — Triangle test. Geneva, Switzerland: International Organization for Standardization.
[14] Sauvageot F, Herbreteau V, Berger M, Dacremont C. 2012 A comparison between nine laboratories performing triangle tests. *Food Quality and Preference* **24**(1) pp 1-7
[15] Cattell R. 1966 The scree test for the number of factors. *Multivariate Behavioural Research* **1** pp 245-76
[16] Thurstone LL. 1947 Multiple factor analysis. Chicago, USA: University of Chicago Press.
[17] Font-i-Furnols M. 2012 Consumer studies on sensory acceptability of boar taint: a review. *Meat Science* **92**(4) pp 319-29.

Institute of Physics
Dirac House, Temple Back
Bristol BS1 6BE UK

ISSN: 1755-1307
ISBN 978-1-5108-4891-7
9 781510 848917